BUILDINTECH BIT 2020. Innovations and Technologies in Construction

IOP Conference Series: Materials Science and Engineering
Volume 945

Belgorod, Russia
8 – 9 October 2020

Editors:

Klyuev Sergey Vasil'yevich Lesovik
Valeriy Stanislavovich Vatin
Nikolay Ivanovich

ISBN: 978-1-7138-4778-6
ISSN: 1757-8981

Printed from e-media with permission by:

Curran Associates, Inc.
57 Morehouse Lane
Red Hook, NY 12571

CURRAN ASSOCIATES INC. proceedings.com

Some format issues inherent in the e-media version may also appear in this print version.

Institute of Physics
Dirac House, Temple Back
Bristol BS1 6BE UK

Phone: 44 1 17 929 7481
Fax: 44 1 17 920 0979

techtracking@iop.org

Additional copies of this publication are available from:

Curran Associates, Inc.
57 Morehouse Lane
Red Hook, NY 12571 USA
Phone: 845-758-0400
Fax: 845-758-2633
Email: curran@proceedings.com
Web: www.proceedings.com

TABLE OF CONTENTS

IOP Conf. Series: Materials Science and Engineering **945** (2020) 011001 doi:10.1088/1757-899X/945/1/011001

PREFACE

BUILDINTECH BIT 2020. INNOVATIONS AND TECHNOLOGIES IN CONSTRUCTION

IOP Conf. Series: Materials Science and Engineering **945** (2020) 011001 doi:10.1088/1757-899X/945/1/011001

PREFACE

The International Scientific Conference "BUILDINTECH BIT 2020. INNOVATIONS AND TECHNOLOGIES IN CONSTRUCTION" is held from 08 to 09 October 2020 on the basis of Federal State Budgetary Educational Institution of Higher Education Belgorod State Technological University named after V.G. Shukhov. BUILDINTECH BIT is an annual conference.

The work of the conference is aimed at providing an opportunity for scientists, graduate students and representatives of the construction industry to generalise the recent achievements results in the field of industrial and civil construction (design, organisation of all types of work, construction materials, etc.), discussing promising directions for the construction industry development and establishing useful partner relationships for future interaction.

The conference is held on numerous topics:

Session 1. Building Materials

Session 2. Building Constructions, Industrial and Civil Construction

Session 3. Structural Mechanics and Theory of Structures

Session 4. Computational Methods, Mathematical Models and Information Technologies in Construction

Session 5. Organization and Technology of Construction Production.

Session 6. Innovations in utilities: heating, gas supply, ventilation and air conditioning, water supply and sanitation.

Session 7. Operation and reliability of transport and technological machines and complexes.

Session 8. Economics in construction.

420 participants from 30 regions of the Russian Federation and 19

participants from 7 countries of the near and far abroad (China, Iraq, Kazakhstan, Tajikistan, the Ukraine, Uzbekistan, Vietnam) were registered for the conference in 2020.

The important problems of Building Constructions, Building Materials, Industrial and Civil Construction and Structural Mechanics and Theory of Structures (from ideas and projects to immediate implementation) are raised and the ways of their solution and optimization are suggested in the reports presented in frameworks of these topics. Many reports were prepared as a benefit of the state support of science in forms of grants, federal target programs, resolutions of the Government of the Russian Federation and other programs.

All articles were peer-reviewed by specialists with a degree. 85 best scientific manuscripts corresponding to the profile of the conference and reflecting the results of theoretical and experimental studies of the authors are recommended for publication.

Annual holding of such conferences is planned to be organized on the main topics of the construction industry.

Editors:

Associate Professor **Klyuev Sergey Vasil'yevich**	Belgorod State Technological University, 308036 Belgorod, Russia
Dr. **Lesovik Valeriy Stanislavovich**	Belgorod State Technological University, 308036 Belgorod, Russia
Dr. **Vatin Nikolay Ivanovich**	Peter the Great St. Petersburg Polytechn, 195251 St. Petersburg, Russia

Organizing Conference Committee:

Evtushenko E.I. – Doctor of Engineering Sciences (Advanced Doctor), Professor, Belgorod State Technological University named after V.G. Shukhov,
Uvarov V.A. – Doctor of Engineering Sciences (Advanced Doctor), Professor, Belgorod State Technological University named after V.G. Shukhov,

IOP Conf. Series: Materials Science and Engineering **945** (2020) 011001 doi:10.1088/1757-899X/945/1/011001

Scientific Conference Committee:

IOP Conf. Series: Materials Science and Engineering **945** (2020) 011001 doi:10.1088/1757-899X/945/1/011001

All conference participants express deep gratitude to the Science team.

IOP Conf. Series: Materials Science and Engineering **945** (2020) 011002 doi:10.1088/1757-899X/945/1/011002

Peer review declaration

BUILDINTECH BIT 2020. INNOVATIONS AND TECHNOLOGIES IN CONSTRUCTION

All papers published in this volume of IOP Conference Series: Materials Science and Engineering have been peer reviewed through processes administered by the Editors. Reviews were conducted by expert referees to the professional and scientific standards expected of a proceedings journal published by IOP Publishing.

- **Type of peer review:** Double-blind. Both the reviewer and author identities are concealed from the reviewers, and vice versa, throughout the review process.

- **Conference submission management system:** Papers are submitted through the site. Every participant in the conference served by this site, "in any capacity," must maintain a login and password. The same login must be used for all conferences. Follow the link to register a login, to check and update your personal information, or to retrieve your password. To comply with the General Data Protection Regulation, the first time when you login to a new account, you will be directed to the required opt-in consent form before you can use your account in any capacity for products and services offered through this site. All papers submitted to the conferences supported by this site need to be prepared as compliant PDF files and must be formatted according to the template.
- **Number of submissions received: 205**

- **Number of submissions sent for review: 205**

- **Number of submissions accepted: 85**

- **Acceptance Rate (Number of Submissions Accepted / Number of Submissions Received X 100): 41**

- **Average number of reviews per paper: 2**

- **Total number of reviewers involved: 80**

- **Any additional info on review process:**

All papers submitted to the «Modern Scientist» journal pass the procedure of reviewing according to the order established by editorial board. Members of the editorial board and leading Russian and international experts in corresponding areas of life sciences, invited as independent readers, perform peer reviews. Conference Editors choose readers for peer review. In the case of the manuscript relevance to the conference profile, established rules and requirements, it is accepted by the editorial board and is sent for the reviewing. Otherwise the article is rejected without further reviewing. Reviewers prepare reviews of articles on a voluntary basis. Reviewers should submit

IOP Conf. Series: Materials Science and Engineering **945** (2020) 011002 doi:10.1088/1757-899X/945/1/011002

their reviews in a scanned form. The reviewer makes a conclusion about the possibility of publishing an article:

- "recommended for publication";
- "recommended taking into account the correction of the comments made by the reviewer";
- "article must be sent for revision";
- "not recommended for publication".

The articles are reviewed and accepted only after at least two positive reviews of three reviewers. After making a decision by the editorial board of the conference to admit the article for publication, the author is informed about it.

- **Contact person for queries: Klyuyev Sergey <klyuyev@yandex.ru>**

IOP Conf. Series: Materials Science and Engineering **945** (2020) 012001 doi:10.1088/1757-899X/945/1/012001

Modern methods of intensification of heat exchange processes in plate apparatuses

N Yu Savvin[1*], L A Kushchev[1] and A I Alifanova[1]

[1]Department of heat and gas supply and ventilation, Belgorod State Technological University named after V.G. Shukhov, Kostyukov St., 46, Belgorod, 308012, Russia

E-mail: n-savvin@mail.ru

Abstract. The heat supply system of the Russian Federation is characterized by significant wear and tear of equipment: heating networks and boilers. At the same time, an important element of heat supply systems equipment is heat exchangers of various designs, which are used in boilers, IHP, thermal power plants, and nuclear power plants. Various heat exchange equipment used in modern heat power engineering is considered. It is established that the use of plate heat exchangers is a more appropriate engineering solution than shell-and-tube heat exchangers, due to economic factors and operational advantages. The method of intensification of heat transfer of the heated liquid by means of technological depressions of spherical shape located according to the linear law is considered.

1. Introduction

In the Russian Federation, centralized heat supply is the most widespread, as this type of heat supply reduces fuel consumption and operating costs, makes it possible to use low-grade fuel, reduces the degree of pollution of the air basin and improves the sanitary condition of settlements.

However, in the second half of the XX century and at the beginning of the XXI century, decentralized, autonomous and individual heat supply systems are becoming more widespread. This is primarily due to intensive suburban house construction both in the Russian Federation and in developed countries.

In the Russian Federation, housing and communal services are one of the most important branches of the national economy, which is confirmed by the state policy. To form a strong and sustainable growth of the economy, the modernization of heating networks is carried out in accordance with Federal Law No. 261 of 23.11.2009 “On energy saving and improving energy efficiency” [1], as well as the Order of the Government of the Russian Federation of 13.11.2009 N 1715-R “On the Energy strategy of Russia for the period up to 2030”[2].

For 2018, according to the Ministry of Energy, the total sales of heat in the country is 2060 million. Gcal/ year, including the housing sector and the budget sector consume 1086 million Gcal, industry and other consumers 974 million Gcal. More than 400 million t.c.f./year is spent on heat supply [3].

IOP Conf. Series: Materials Science and Engineering **945** (2020) 012001 doi:10.1088/1757-899X/945/1/012001

2. Materials and methods

2.1. Materials

The main type of heat exchange equipment in the production of heat energy are plate and shell-and-tube heat exchangers (figure 1). This equipment is used in boiler houses, individual heat points (IHP), as well as in thermal power plants and nuclear power plants.

a) b)

Figure 1. a – shell and tube heat exchanger; b – plate heat exchanger.

Housing and communal services in the Russian Federation traditionally use shell-and-tube heat exchangers. This is primarily due to their low cost and ease of operation, but they are characterized by a relatively low coefficient of heat transfer. When comparing two heat exchangers of the same thermal power Q, W, the heat transfer coefficient of the plate apparatus is at least 2.5 times higher [4].

To increase the heat transfer coefficient of the shell and tube heat exchanger an original design of the heat exchanger with increased turbulence of the heated liquid is proposed [5, 6].

A new solution in the heat supply systems of the Russian Federation is the use of plate heat exchangers, which are characterized by higher values of the heat transfer coefficient with significantly smaller dimensions. These devices can also work with various types of heat carrier (water, gas, oil, etc.), which expands the scope of application of these devices [7]

2.2. Methods

In Russia and developed industrial countries, active theoretical and experimental research is being conducted on the intensification of heat exchange in plate heat exchangers for various industries. The intensification of heat transfer is considered in the work of A.Yu. Maskinskaya [8]. A method for increasing the efficiency of a plate heat exchanger by using a surface with holes is proposed. It was possible to achieve an increase in the heat transfer coefficient by 25% compared to a smooth surface. However, when using a surface with holes, the hydraulic resistance increased.

In [9], the use of cationic and magnetic filters for cleaning the working fluid in plate heat exchangers is proposed. Due to these measures, there was no reduction in efficiency and, as a result, it allowed increasing it. However, the technology of installing a cationic filter requires a large amount of reagents, so it is better to use a magnetic filter.

As A.A. Gukhman noted [10] - the main task of the intensification of convective heat transfer is such an impact on the boundary layer that would make it thinner or partially destroy it. It is confirmed that increasing the speed of the incoming flow reduces the thickness of the boundary layer, but, in turn, there is a rapid increase in hydraulic resistance. In addition, the use of this method is limited by the growth of energy costs.

Studies on the influence of a heat exchange surface with a geometrically altered shape, namely rough surfaces, on the flow in the wall area, and, consequently, on the total hydrodynamic characteristics of streamlined bodies were first conducted by William Froude, the work was performed in 1869 in Great Britain [11].

Samara State Technical University conducted research on the flow of heat exchange surfaces with welded separation spikes [12]. This modernization allowed not only preventing deformation of the plates under the influence of pressure, but also turbulating further the flow of the coolant. It is important to note that the height of the spikes determines the height of the channel, and the frequency of their location (longitudinal and transverse step) – the allowable pressure in the device. The disadvantages of such a design solution are the complexity of manufacturing, cleaning and repair.

In many developed countries - the United States, Canada, the European Union (Germany, France), as well as China and India, intensive theoretical and experimental research is being conducted to improve the efficiency of plate devices [13]. This work is particularly active in Denmark (firms “Ridan”, “Alfa Laval”, “Danfoss”). The devices of these manufacturers are widely used in the heat power industry of the Russian Federation. Largely, the research is aimed at increasing the turbulence of the liquid flow both at the heated surface and at the cooled surface because of original technical solutions.

E.A. Mikulich defined ring stresses on the border of holes and radial stresses in plate elements. The analysis of numerical calculations showed that the action of the pulse load when the holes come together increases the ring stresses at their borders and reduces the radial stresses at the internal points of the plate elements [14].

Qian H. studied the effect of pulsations on the heat transfer characteristics of plate heat exchangers. Thus, with constant mass flow of hot water, at a ripple frequency of 1.78 Hz, it was possible to achieve the maximum value of the heat transfer coefficient of 4415.73 $W/(m^2 \cdot K)$. It was found that the average increase in the heat transfer coefficient at the transition to the pulsating mode was 20% [15].

According to A.A. Zhukauskas’s fundamental research, the amount of heat energy transferred from a heated solid to a lower-temperature liquid is directly dependent on the nature of the fluid flow around this body [16]. Therefore, for Reynolds numbers $2 \cdot 10^3 - 10^4$ on flat surfaces, the increase in heat transfer is more than 10%.

L.M. Kovalenko and A.F. Glushkov showed the basics of calculation and design of heat exchangers with a complex shape of the heat exchange surface are given. The possibilities of reducing the metal and energy consumption of industrial heat exchange equipment are shown, and the problem of rational, careful attitude to fuel is considered [17].

It is important to improve the designs of existing plate heat exchangers. Convective heat exchange with surfaces of complex shape made of a thin sheet is of particular interest. The main task of intensifying convective heat transfer is to influence the boundary layer in a way that would make it thinner or partially destroy it. This is shown in the works of L.G. Loytsyansky [18] and G. Shlichting [19].

Thus, to intensify heat exchange in a plate heat exchanger, it is necessary to increase the turbulence of the coolant flowing around the heat exchange surface.

3. Results

3.1. Facility

The Heat and Gas Supply and Ventilation Department of the Belgorod State Technological University named after V.G. Shukhov conducts research on the intensification of the plate heat exchanger. The semi-industrial facility “Independent heating system of a residential building”, shown in figure 2,was developed.

The main elements of which are 1 - supply pipeline from the heat supply source; 2 - pump; 3-regulating device; 4 - flow meter; 5 - temperature sensor; 6 - supply pipeline from the heat exchanger; 7 - heating system consumers; 8 - return pipeline (from the heating system to the heat exchanger); 9 - plate heat exchanger; 10 - return pipeline to the heat supply source; 11 - heat supply source; 12 - ball valve; 13 - pressure gauge; 14 - heat calculator

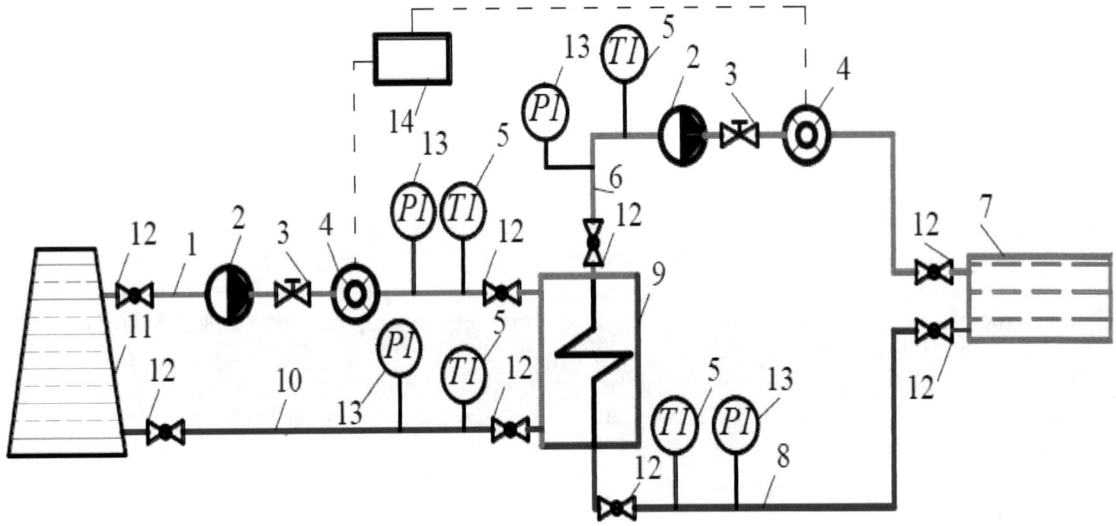

Figure 2. Scheme of semi-industrial facility "Independent heating system of a residential building".

3.2. The heat transfer coefficient

The research results are shown in figure 3. All the studies discussed above are aimed at turbulization of the coolant flow at the heat exchange surface in the heating and heated circuits. As a result, the heat transfer coefficient K, W/(m^2 K), plate heat exchanger increases, which leads to the reduction of metal consumption (reducing the size), increasing efficiency of heat-exchange equipment.

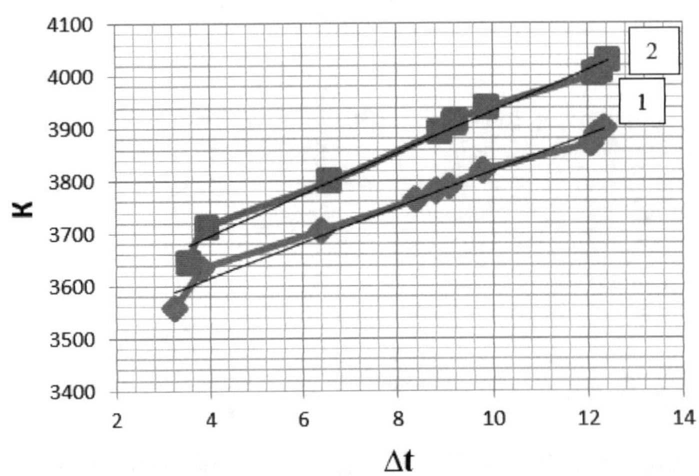

Figure 3. Graph of the dependence of the heat transfer coefficient on the temperature pressure: 1- serial heat exchanger; 2 - intensified heat exchanger.

The graph shows that the heat transfer coefficient K, W/(m^2K) increases almost linearly with an increase in the average temperature pressure Δt, and on average up to 5% more than that of a serial heat exchanger.

Conclusions. As a result of the conducted research, it is confirmed that the use of an intensified heat exchanger leads to an intensification of the heat exchange process. The increase in the heat transfer coefficient with a seasonal increase in the temperature pressure is more intense (pay attention to the trend line) in the device under study than in the serial one. A high value of K will eventually reduce the metal consumption of the device.

Operation of an intensified heat exchanger, in our opinion, will increase the advantages of the plate device, reduce the cost of current and planned repairs, and will increase the operational reliability of heat supply systems.

4. Discussion

All the studies discussed above are aimed at turbulization of the coolant flow at the heat exchange surface in the heating and heated circuits. As a result, the heat transfer coefficient K, W/(m^2K), plate heat exchanger increases, which leads to the reduction of metal consumption (reducing the size), increasing efficiency of heat-exchange equipment.

In our opinion, studies aimed at improving the usability, extending the service life and intensifying thermal processes associated with changes in the geometry of the heat exchange surface of the device are the most promising, which will result in increased turbulence of the liquid flow at the heat exchange surface.

5. Summary

The increase in the heat transfer coefficient with an increase in the temperature pressure occurs more intensively in the device with modified plates than in the serial one. A high value of K will reduce the metal consumption of the device.

6. References

[1] 2009 Law of the Russian Federation "On energy saving and energy efficiency improvement and on amendments to certain legislative acts of the Russian Federation" *Collection of legislation of the Russian Federation* **261** 312

[2] 2009 Order of the Government of the Russian Federation "On the Energy strategy of Russia for the period up to 2030" **1715** 612

[3] Delyagin G N, Lebedev V I, Permyakov B A 2010 Heat generating plants *Textbook for universities* 623

[4] Kushchev L A, Nikulin N Yu, Feoktistov A Yu 2019 Highly efficient shell and tube heat exchangers for housing and communal services *Scientific journal of construction and architecture* **2(54)** 59-67

[5] Nikulin N Yu, Kushchev L A, Feoktistov A Yu 2015 Shell and tube heat exchanger *Pat. 149737 Russian Federation IPC F28D 7/00* **2014134083/06** 2-3

[6] Kushchev L A, Nikulin N Yu 2015 High-speed heat exchanger *Pat. 158070 Russian Federation, МПК F28F 1/14* **2015110302/06** 35-39

[7] Kushchev L A, Nikulin N Yu, Alifanova A I 2017 Modern methods of heat exchange intensification in housing and tube heat exchangers *Bulletin of BSTU named after V.G. Shukhov* **9** 73-79

[8] Maskinskaya A Yu 2004 Increasing the efficiency of heat exchange devices due to the intensification of heat exchange on the surface with wells *Industrial heat power* 140

[9] Gubina K A, Pashnin S V, Mayorov M A 2019 Analysis of reducing the heat transfer coefficient of a plate heat exchanger *Heat Supply News* **32** 41-47

[10] Gukhman A A 1977 Intensification of convective heat exchange and the problem of comparative evaluation of heat exchange surfaces *Heat Power Engineering* **4** 5-8

[11] Froude W 1869 On some difficulties in the received view of fluid friction *British Assotiation*

[12] Gorshenin A S, Kudinov V A 2009 Methods of heat transfer intensification *Textbook* 82

[13] Martynenko O G, Mikhalevich A A, Shikov V K 1995 *Handbook for heat exchangers* 528

[14] Mikulich O A 2019 Study of the interaction of defects in plate elements under non-stationary load *Scientific Notes* **65** 177-182
[15] Qian H, Kudashev S V, Plotnikov V A 2019 Experimental study on heat transfer of pulsating flow enhanced the plate heat exchanger *Ogarev Mordovia State University* 81-92
[16] Zhukauskas A A 1982 Convective transfer in heat exchangers *Science* 472
[17] Kovalenko L M, Glushkov A F 1986 Heat exchangers with heat transfer intensification *Energoatomizdat* 240
[18] Loytsyansky L G 1973 Fluid mechanics and gas *Science*
[19] Schlichting G 1974 Boundary layer theory *Science* 712

Acknowledgments
This work was realized under the support of the President Scholarship; in the framework of the Program of flagship university development on the base of the Belgorod State Technological University named after V. G. Shukhov, using equipment of High Technology Center at BSTU named after V.G. Shukhov.

IOP Conf. Series: Materials Science and Engineering **945** (2020) 012002 doi:10.1088/1757-899X/945/1/012002

Features of mathematical formulation and solution of the problem of optimal division of funds in the construction business

A P Chernyaev[1*], A Yu Meerson[2], I V Sukhorukova[2] and G P Fomin[2]

[1]Moscow Institute of Physics and Technology (State University), Russia, Dolgoprudny, Moscow Region, Institutsky per. 9
[2]Plekhanov Russian University of Economics, Stremyanny lane, 36, Moscow, 117997, Russia

Email: chernyaev49@yandex.ru

Abstract: The paper considers a microeconomic model of the construction business economy, assuming that it is a stable object in a crisis. The paper presents the distinctive features of the microeconomic formulation of the problem of finding a method for optimal separation of construction materials into consumed and accumulated parts. This area is constantly paid attention to and described in the media, sometimes with much unexpected outcome for investors. It is quite natural that consumption is optimized, just as in macroeconomics, but not in its pure form. The integral discounted utility of consumption is maximized. The communication equation in this paper differs from the similar macroeconomic equation, as the construction business exists and survives even in crisis conditions not in the same way as it does in social environments.

1. Introduction

The construction business is one of the key factors of sustainable economic development. In the conditions of economic crises, construction business demonstrates stable survival rate, which can be envied by numerous enterprises. The efficiency of the construction industry is determined primarily by the prompt solution of all negative processes that arise. The specifics of the construction sector of the Russian Federation are that there are certain difficulties in obtaining government contracts when conducting electronic auctions. The public procurement website is a priority when carrying out orders in the construction sector, as it allows saving budget funds significantly, ensures the competitiveness of contractors, and minimizes the bureaucratic official component. Auctions are held automatically, which reduces the risks of unfair preliminary collusion between construction companies. However, there is a very significant disadvantage of this procedure. Preference for a construction company is always given only based on the criterion - the price of the contract. This circumstance contributes to numerous abuses. Low-skilled workers, often without experience in the performance of complex construction works, are recruited for the execution. Of course, the customer needs to reduce operating costs to ensure the profitability of the business. Unaffordable requirements are often set for the performer to exclude the majority of tenderers. Numerous one-day fake companies are created. When performing construction work, outdated equipment is applied and manual labor is widely used. Many construction companies are forced to pay salaries using gray schemes to hide their real income.

Therefore, when choosing a contractor, it is necessary to be guided not only by the cost of the contract, but also by competencies and economic quality priorities. It should be borne in mind that the full set of financial statements of a construction company includes: balance sheet; statement of financial results; statement of changes in equity; statement of cash flows; accounting policy, and is included in the main balance sheet equation: Assets = Liabilities + Equity + Income-Expenses + Investments-Withdrawals. In this form, balance sheet equation clearly demonstrates not only the process of increasing capital as a result of the construction company's own activities, but also the opportunities for changing it, brought from outside. It provides a general view of the funds, operations and results of the construction organization and their reflection in the financial statements.

2. Materials and methods

Similarly to the problem of optimal division of material resources into consumed and accumulated parts [1], it is possible to maximize consumption, but not in its pure form. As in works [2-4], we will maximize the integral discounted utility of consumption. However, in the considered mathematical model of the construction business economy, the ordinary differential balance equation, which plays the role of the connection equation in the optimal control problem, has expressed microeconomic specificity. The economic characteristics encountered in the course of research, both known and unknown, will be considered continuous functions of time. The variation method is used in this work.

3. Results and discussion

1. Connection equation of the problem of optimal division of material resources into consumed and accumulated parts in the construction business. In contrast to [2-4], we will proceed from the basic balance equation [1]

$$Y = I + C, \tag{1}$$

where Y – income, I – investment and C – consumption. Let us suppose that now

$$Y = P + \rho x, I = x'_t, \tag{2}$$

where P – proceeds from investors, ρ – bank rate, x – accumulated capital of the construction business, and t – time. Thus, the banking system directs the accumulated capital to the construction business. Substituting (2) in (1) we get

$$P + \rho x = x'_t + C,$$

or

$$x'_t = P + \rho x - C. \tag{3}$$

Equation (3) can also be understood in a different balance-sheet sense than (1), namely: the rate of increment of accumulated capital is equal to the instantaneous difference between income and expenditure [1, 3]. It is (3) that acts as the connection equation in the optimal control problem.

IOP Conf. Series: Materials Science and Engineering **945** (2020) 012002 doi:10.1088/1757-899X/945/1/012002

For equation (3), it is natural to set the Cauchy problem for the initial moment of time t_0 :

$$x(t_0) = x_0 > 0 \quad (4)$$

2. Maximization of the penalty function. We take the integral discounted utility of consumption as a penalty function

$$J(C) = \int_{t_0}^{t_1} u(C(t))\exp(-\delta t)\mathrm{d}t \quad (5)$$

Here t_1 – finite time point, u – utility function δ – discount factor, and (5) – the functional that we need to maximize. Following [4], we consider that the construction business assesses the utility of consumption by function $u(C)$ that describes a constant risk aversion according to Arrow-Pratt:

$$a = -\frac{u''(C)C}{u'(C)} \geq 0 \quad (6)$$

Following [11-13] from (6) we get

$$u'(C) = \frac{\gamma}{C^a} = \gamma C^{-a}, \quad (7)$$

$$u(C) = \begin{cases} \dfrac{\gamma C^{1-a}}{1-a} + \chi, a \neq 1; \\ \gamma \ln C + \chi, a = 1; \end{cases} \gamma = \text{const} > 0, \chi = \text{const}. \quad (8)$$

Further, the integration constants in (8) will be considered as given ones.

Considering the increment of the functional (5): $J(C(t)+h(t)) - J(C(t))$, adding a condition at the right end

$$x(t_1) = x_1 > 0, \quad (9)$$

and following [3], we obtain the existence of the maximum of the functional (5), and the validity of the Euler equation

$$u'(C(t))\rho(t)e^{-\delta t} + \frac{d}{dt}\left[u'(C(t))e^{-\delta t}\right] = 0. \quad (10)$$

Replacement

$$w = u'(C(t))e^{-\delta t} \tag{11}$$

simplifies Euler equation (10)

$$\frac{dw}{dt} + \rho(t)w = 0.$$

The latter is simply integrated:

$$w = De^{-\int_{t_0}^{t}\rho(s)ds}, D = \text{const}. \tag{12}$$

From (11) and (12), we have

$$u'(C(t))e^{-\delta t} = De^{-\int_{t_0}^{t}\rho(s)ds}.$$

Expressing the derivative of the utility function from the last equation and comparing it with equation (7), we get

$$u'(C(t)) = De^{\delta t - \int_{t_0}^{t}\rho(s)ds} = \frac{\gamma}{[C(t)]^a}. \tag{13}$$

Expressing the consumption from (13) we will have

$$[C(t)]^a = \frac{\gamma}{D}e^{\delta t - \int_{t_0}^{t}\rho(s)ds}, C(t) = \left(\frac{\gamma}{D}\right)^{\frac{1}{a}} e^{\frac{\delta}{a}t - \frac{1}{a}\int_{t_0}^{t}\rho(s)ds}. \tag{14}$$

From the left formula (14), you can get the right one if the Arrow–Pratt risk aversion (6) is strictly positive.

For the final solution of the problem it remains to find a constant for integration D, which first occurs in equation (12).

3. Use of the condition at the right end. Equation (3) is linear and following [1] we can write

$$x(t) = \int_{t_0}^{t} e^{\int_{\tau}^{t}\rho(s)ds}\left[P(\tau) - C(\tau)\right]d\tau + x_0 e^{\int_{t_0}^{t}\rho(s)ds}. \tag{15}$$

Formula (15) gives a solution to the Cauchy problem (3), (4) and is checked directly. The fact that (15) satisfies (4) is obvious, and that (15) satisfies equation (3) follows from the Leibniz formula [3]. Substituting in (15) $t = t_1$ and using (9) it can be written

$$x_1 = \int_{t_0}^{t_1} e^{\int_{\tau}^{t_1}\rho(s)ds}\left[P(\tau) - C(\tau)\right]d\tau + x_0 e^{\int_{t_0}^{t_1}\rho(s)ds}. \tag{16}$$

Now let us transform (16) as follows

$$\int_{t_0}^{t_1} e^{\int_{\tau}^{t_1}\rho(s)ds} C(\tau)d\tau = \int_{t_0}^{t_1} e^{\int_{\tau}^{t_1}\rho(s)ds} P(\tau)d\tau + x_0 e^{\int_{t_0}^{t_1}\rho(s)ds} - x_1 . \tag{17}$$

Now let us substitute the expression we have obtained for consumption (14) in (17)

$$\left(\frac{\gamma}{D}\right)^{\frac{1}{a}} \int_{t_0}^{t_1} e^{\int_{\tau}^{t_1}\rho(s)ds} e^{\frac{1}{a}\int_{t_0}^{\tau}\rho(s)ds - \frac{\delta}{a}\tau} d\tau = \int_{t_0}^{t_1} e^{\int_{\tau}^{t_1}\rho(s)ds} P(\tau)d\tau + x_0 e^{\int_{t_0}^{t_1}\rho(s)ds} - x_1 .$$

Raising the latter to $a > 0$ degree we will have

$$\frac{\gamma}{D}\left(\int_{t_0}^{t_1} e^{\int_{\tau}^{t_1}\rho(s)ds} e^{\frac{1}{a}\int_{t_0}^{\tau}\rho(s)ds - \frac{\delta}{a}\tau} d\tau \right)^{a} = \left(\int_{t_0}^{t_1} e^{\int_{\tau}^{t_1}\rho(s)ds} P(\tau)d\tau + x_0 e^{\int_{t_0}^{t_1}\rho(s)ds} - x_1 \right)^{a}$$

So, from the last equation, we can express the desired constant

$$D = \frac{\gamma \left(\int_{t_0}^{t_1} e^{\int_{\tau}^{t_1}\rho(s)ds} e^{\frac{1}{a}\int_{t_0}^{\tau}\rho(s)ds - \frac{\delta}{a}\tau} d\tau \right)^{a}}{\left(\int_{t_0}^{t_1} e^{\int_{\tau}^{t_1}\rho(s)ds} P(\tau)d\tau + x_0 e^{\int_{t_0}^{t_1}\rho(s)ds} - x_1 \right)^{a}}, \tag{18}$$

of course, if the denominator of the fraction (18) is different from zero.

4. Summary

As in the conditions of economic chaos and instability, the construction business demonstrates a stable survival rate, which can be envied by many enterprises; the study of various aspects of the economy of the construction business is very relevant. In the very important problem of optimal division of material resources into consumed and accumulated parts, both in microeconomics and in macroeconomics, there must be and are similarities and differences. As in macroeconomics [1] it is necessary to maximize consumption, but not in its pure form. As in [5-9], we maximized the integral discounted utility of consumption. In the considered mathematical model of the construction business economy, the ordinary differential balance equation, which plays the role of a communication equation in the optimal control problem, has expressed microeconomic specificity due to the specific nature of the business. However, this work is aimed at identifying and studying general economic trends, both in micro and macro cases. As in work [1], the interesting variation method was used here. Thus, it is proposed to integrate the successes and advantages of various methods of modeling the economic processes of construction business development for monitoring, control and formation of adequate management decisions in advance.

5. References

[1] Chernyaev A P, Meerson A Yu, Sukhorukova I V, Fomin G P 2020 Methods for optimal Separation of Income in Consumable and Accumulated Parts *International Journal of Recept Technology and Engineering* **8** 797–801

[2] Sukhorukova I V, Fomin G P, Ekareva I L, Mushrub V A 2019 Hybrid Method for Multi-

Criteria Risk Minimization *Espacios* **40** 14-25

[3] Guriev S M, Pospelov I G 1994 General equilibrium model of the transition economy *Mathematical modeling* **6** 3–21

[4] Meerson A Yu, Chernyaev A P 2018 Options for setting optimization problems for generalized utility of consumption in the Solow model with various restrictions *Fundamental research* **7** 121–125.

[5] Romer P 1987 Growth Based on Increasing Returns Due to Specialization *American Economic Review* **77** 56-62

[6] Sukhorukova I V Chistyakova N A 2018 Optimization of the Formation of the Capital Structure of the Insurance Company, Taking into Account the National Specifics of Insurance *Journal of Reviews on Global Economics* **7** 146-151

[7] Nordhaus W 2017 Integrated assessment models of climate change *NBER Reporter* **3** 16-21

[8] Wang Y, Luo Y 2009 On Rank Reversal in Decision Analysis *Mathematical and Computer Modelling* **49** 1221-1229

[9] Popov V A 2018 Inflation and consumer basket *Journal of Reviews on Global Economics* **7** 453-456

IOP Conf. Series: Materials Science and Engineering **945** (2020) 012003 doi:10.1088/1757-899X/945/1/012003

Comparative analysis of the parameters of generators with circular and asymmetric oscillations

M D Gerasimov[1*], N S Lubimyi[1] and E A Yakovlev[1]

[1]Department of Hoisting and Road Machines, Belgorod State Technological University named after V.G. Shukhov, Kostyukov St., 46, Belgorod, 308012, Russia
[2]Department of Roads and Railways, Belgorod State Technological University named after V.G. Shukhov, Kostyukov St., 46, Belgorod, 308012, Russia

E-mail: nslubim@bk.ru

Abstract. In the paper the calculation and comparative analysis of the technical characteristics and main parameters of a vibrator with circular oscillations with a driving force equal to 5 kN and a vibrating device with asymmetric oscillations with a similar driving force for two, three and four stages of vibration blocks with directional vibrations are performed based on the design methodology of vibration devices with asymmetric oscillations. The aim of this paper is to identify the positive and negative effects in the transfer of the vibrator with circular oscillations on the vibrating device with asymmetric oscillations. The paper uses a method for selecting rational ratios of static moments of debalances, which provides the highest coefficient of azimuth of the driving force of both individual stages of vibration blocks, and the coefficient of the total driving force of the entire device as a whole. This paper will be useful for manufacturers of vibration machines for construction and road vehicles.

1. Introduction

Modern rates of development of methods of design and creation of construction and road equipment and technology dictate the need to search for and implement ideas that significantly increase the speed of their work. The speed of work performed by machines, with high quality, becomes a priority factor when choosing machines and their manufacturers.

In the production of construction and road-building materials, construction and road construction works the use of vibration equipment, vibration devices, vibrators and vibration blocks plays a significant role in enhancing the speed of production.

According to the classification [1,2], vibrators and vibrating devices with circular and directional oscillations are used in the industry.

A significant way to increase the efficiency of vibration machines, and thus increase the speed of work performed, is their ability to work near the resonance [3,4]. However, as noted in a number of publications [5,6,7], creating conditions close to resonance is a much easier task than maintaining these conditions with stable output parameters over a long period of machine operation. Thus, the use of the resonance phenomenon in construction and road vehicles, or close to it, is currently problematic.

However, there is a method for generating mechanical oscillations that simulates conditions close to resonance to a high degree. This is a method for generating asymmetric oscillations by vibrating devices consisting of vibration blocks with directed oscillations along a single line of action, Fig. 1, [8,9].

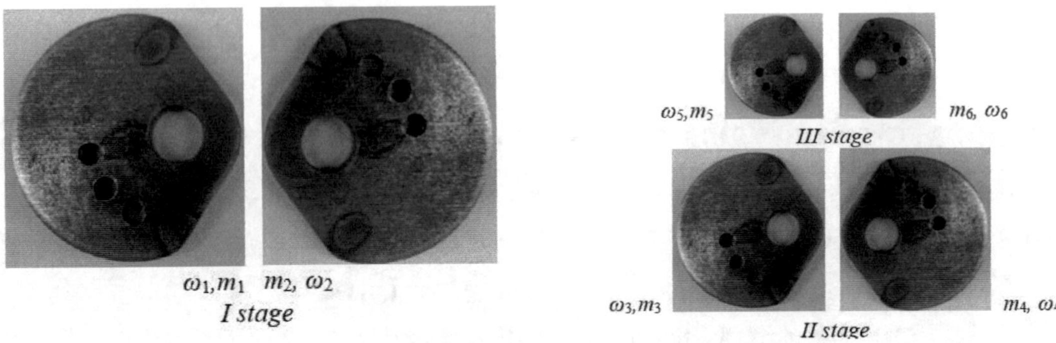

Figure 1. Scheme of debalance vibration blocks consisting of three stages of two-shaft vibration blocks with debalance shafts rotating in opposite directions.

The method of obtaining the total driving force for asymmetric oscillations is described in several sources [5, 6, 10, 11, 12]. Thus, the total effect of the asymmetry of the driving force estimated by the quantitative index – coefficient of asymmetry of the driving force (k_a), or, equivalently the coefficient of dynamic system with an asymmetric oscillations (k_d). Numerically, the coefficient of asymmetry of the driving force is the ratio of the amount of the driving force acting in one direction, usually in the direction of performing useful work ($F_{u.w.}$) to the amount of the driving force acting in the opposite direction, i.e., in the direction of idle speed ($F_{id.sp.}$):

$$k_a = k_d = \frac{F_{u.w.}}{F_{id.sp.}} \quad (1)$$

It is found out [13] that the numerical value of k_a for a vibrating device with asymmetric oscillations consisting of two stages of vibration blocks with directional oscillations can be: $k_a \leq 2.0$. The numerical value of k_a for a vibrating device with asymmetric oscillations consisting of n stages of vibration blocks with directional oscillations can be: $k_{an} < n$. This means that with the correct and rational choice of the initial parameters of the vibration block of each stage, we can get an asymmetry coefficient of 2.0. If we deviate from the rational parameters, the value of the asymmetry coefficient of a two-stage vibrating device will be within the range of the numerical value k_a for a vibrating device with asymmetric oscillations consisting of two stages of vibration blocks with directional vibrations can be: $k_a < 2.0$. Similarly, for a multi-stage one: $k_{an} < n$.

The aim of this work is to calculate the rational parameters of a vibrating device with asymmetric oscillations on the example of using the technical parameters of an industrial vibrator with circular oscillations consisting of two, three and four stages of vibration blocks with directional oscillations and to identify trends in quantitative and qualitative parameters in comparison with the basic sample.

2. Materials and methods

The paper uses classical universal and special research methods. The first stage includes the creation and processing of databases of technical solutions in the field of generating asymmetric oscillations for use in construction and road equipment. Based on the obtained databases, we selected patents and printed works that contain elements of the design methodology for vibration devices with asymmetric oscillations. Design methods are classified according to the directions that reflect their significant differences. At the second stage, the design and creation of a calculation block for processing numerical values and parameters of harmonic and asymmetric oscillations are completed by the specialized program Delpfy, which is designed to determine the optimal parameters of the vibrating device with asymmetric oscillations with an arbitrary number of vibration stages with directional oscillations. At the third stage, the method [10-13] was used to solve numerically and obtain rational parameters of the stages of vibration blocks included in a vibrating device with asymmetric oscillations in order to obtain the highest value of the asymmetry coefficient of the total driving force. Numerical calculations are performed using Excel spreadsheets included in the instrumental calculation apparatus.

3. Results

For comparative analysis, we take a vibrator with circular oscillations, widely used in industry IV-05-50 [14].

From the technological characteristics for further calculations and analysis, we take: the maximum driving force – F = 5.0 kN; oscillation frequency – f = 50 Hz (n_0= 3000 rpm); static moment of debalances – $M_{st.0}$= 5.1 kg·cm, 0.051 kg·m; engine power – N = 0.5 kW; weight – m_v= 15 kg.
The paper does not imply any goals in the part – what vibrator is better. Each design is made to perform its own tasks and occupies its own market sector.
We check the calculated value of the driving force:

$$F = m_0 \cdot r_0 \cdot \omega_0^2 = M_{\text{CT.0}} \cdot \omega_0^2 = 0.05 \cdot 314^2 = 5018\ N \qquad (2)$$

where $\omega_0 = \frac{\pi \cdot n}{30} = \frac{3.14 \cdot 3000}{30} = 314\ s^{-1}$ - angular speed of rotation of unbalanced shafts, m_0 - the total mass of debalances, kg.

We take 5.0 kN according to the technical characteristics. The projection of the driving force on the vertical axis is defined by the expression:

$$F = m_0 \cdot r \cdot \omega_0^2 \cdot \cos(\omega_0^2 \cdot t + \varphi_0) \qquad (3)$$

where t – current time, s; φ_0 – initial phase, deg.

The graph of changes in the value F within the period (T_0 = 1/50=0.02 s) is shown in Fig. 2a.
The result is obtained at the sum of the mass of the debalances: m_0= 2.54 kg, at the eccentricity: r = 0.02 m (2 cm) and at the speed of the debalance shaft n = 300 rpm. If it is necessary to calculate the parameters of a vibration block with directional oscillations consisting of two rotating counterbalance shafts at the same initial parameters, we should take the mass of the debalances on each shaft (Fig. 1) equal to m_1=m_0/2=1.27 kg. The coefficient of asymmetry of the driving force is: $k_a = \frac{F^+}{|F^-|} = \frac{5^+}{|5^-|} =$ 1.0, as the value of the driving force on the positive and negative sides of the graph is equal in magnitude to each other.

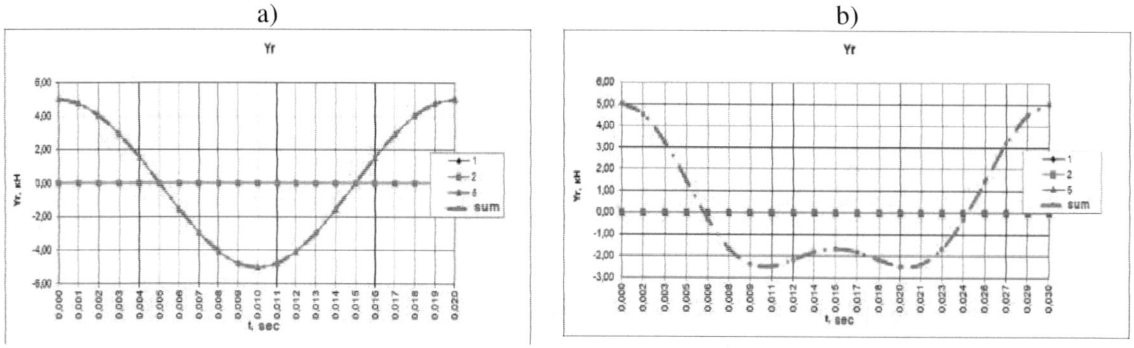

Figure 2. a - Graph of changes in the magnitude of the projection of the driving force F=5.0 kN on the vertical axis (Yr, kN) of the vibrator with circular oscillations within the period (T_0 = 1/50=0.02 s); b - Graph of the change in the projection of the total driving force F=5.0 kN on the vertical axis (Yr, kN) of a two-stage vibration device with asymmetric oscillations within the period (T_0 = 0.03 s).

In accordance with the method for a vibrating device with asymmetric oscillations consisting of two stages of vibration blocks with directed oscillations [13], we divide the total driving force F=5.0 kN into two components: F_1=3.33 kN and F_2=1.67 kN. The rotation speeds of the unbalanced shafts are accepted: n_1=2000 rpm and n_2=4000 rpm. The static moments of debalances are determined from ratios:

$$M_{st1} = \frac{F_1}{\omega_1^2} \; and \; M_{st2} = \frac{F_2}{\omega_2^2} \tag{4}$$

The initial parameters for calculating a two-stage vibrating device with asymmetric oscillations are summarized in table 1.

Table 1. Initial parameters for the calculation of the two-stage vibrating device with asymmetric oscillations.

Constants		Parameter	№ of vibrator	
			1	2
π	3.1416	Weight (kg)	1.9	0.475
		Radius (cm)	2	1
		Init. phase (deg.)	0	0
T	0.03	Speed (rpm)	2000	4000
		Ri (m)	0.02	0.01
dt	0.0015	φi (rad)	0.00	0.00
		ωi (1/s)	209.4	418.88

Here, T = 0.03 s, the period of debalance oscillation at n = 2000 rpm; dt = 0.0015 s - the time, duration, length of time, by dividing the duration of the period on 20 segments; φi = the initial phase of the i-th vibration unit composed of the vibration device with asymmetric oscillations; Ri is the eccentricity of debalances of the i-th vibration unit composed of the vibration device with the asymmetric oscillations; ωi – the angular velocity of rotation of the debalance shafts.
The calculation results are shown in table. 2.

Table 2. Results of calculating parameters of a two-stage device with asymmetric oscillations.

№	t	F1	F2	Sum
0	0.000	3.33	1.67	5.00
1	0.002	3.17	1.35	4.52
2	0.003	2.70	0.52	3.21
3	0.005	1.96	-0.52	1.44
4	0.006	1.03	-1.35	-0.32
5	0.008	0.00	-1.67	-1.67
6	0.009	-1.03	-1.35	-2.38
7	0.011	-1.96	-0.52	-2.47
8	0.012	-2.70	0.52	-2.18
9	0.014	-3.17	1.35	-1.82
10	0.015	-3.33	1.67	-1.67
11	0.017	-3.17	1.35	-1.82
12	0.018	-2.70	0.52	-2.18
13	0.020	-1.96	-0.52	-2.47
14	0.021	-1.03	-1.35	-2.38
15	0.023	0.00	-1.67	-1.67

16	0.024	1.03	-1.35	-0.32
17	0.026	1.96	-0.52	1.44
18	0.027	2.70	0.52	3.21
19	0.029	3.17	1.35	4.52
20	0.030	3.33	1.67	5.00
	Max	3.33	1.67	5.00
	Min	-3.33	-1.67	-2.47

The asymmetry coefficient of the total driving force is:
$k_a = \frac{5,0}{|-2,47|}$ = 2,02 kN. Total weight of required debalances is 2,375 kg.

The graph of the change in the projection of the total driving force on the vertical axis is shown in Fig. 2 b.

The asymmetry coefficient of the total driving force for the calculation of a three-stage vibrating device with asymmetric oscillations is: k_a =3.01 kN. The total weight of required debalances is 3.56 kg. The values of the components of the total driving force have the following values: $P_1 + P_2 + P_3$=2.5+1.67+0.83=5.0 kN.

A graph of the change in the value of the projection of the total driving force on the vertical axis is shown in Fig. 4 a.

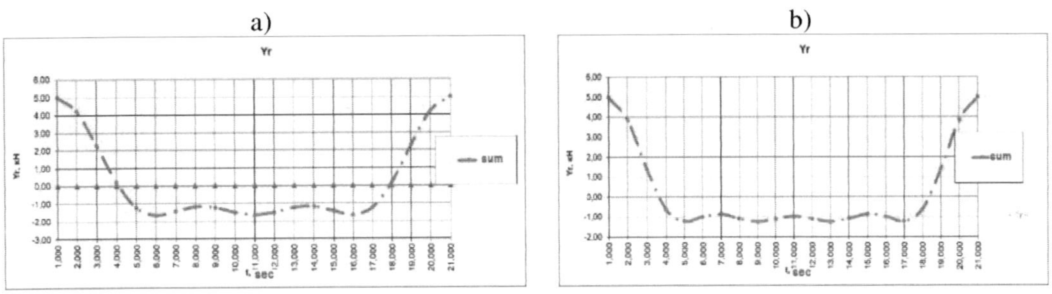

Figure 3. a - Graph of changes in the projection of the total driving force F=5.0 кН on the vertical axis (Yr, кН) of a three-stage vibrating device with asymmetric oscillations within a period (T_0 = 0.04 s); b - Graph of the change in the projection of the total driving force F=5.0 kN on the vertical axis (Yr, kN) of a four-stage vibrating device with asymmetric oscillations within the period (T_0 = 0.06 s).

The asymmetry coefficient of the total driving force for the calculation of a four-stage vibrating device with asymmetric oscillations is: k_a =kN. The total weight of required debalances is 4.0 kg. The values that make up the total driving force have the following values: $P_1 + P_2 + P_3 + P_4 = 2.0 + 1.5 + 1.0 + 0.5 = 5.0$ кН.

The graph of the change in the value of the projection of the total driving force on the vertical axis is shown in Fig. 3 b.

IOP Conf. Series: Materials Science and Engineering **945** (2020) 012003 doi:10.1088/1757-899X/945/1/012003

4. Summary

Based on the results of calculations, we can present a table of comparative parameters of vibration devices for obtaining a driving force of 5 kN, acting in the direction of performing useful work *F*u.w., table 3.

Table 3. Results of the comparative parameters of the vibration devices.

№	Parameter	Basic sample	Number of stages of the vibration device 2	3	4
1	Component of the driving force, $F_{u.w.}$ kN	5.0	5.0	5.0	5.0
2	Component of the driving force acting in the direction of idle speed, $F_{idl.sp.}$ kN	5.0	2.47	1.66	1.25
3	Asymmetry coefficient of the driving force, k_a	1.0	2.02	3.01	4.0
4	Total weight of debalances, $\sum m_0$	2.54	2.375	3.56	4.002
5	Oscillation period, T, s	0.2	0.3	0.4	0.6
6	Time of action of the force $F_{u,w.}$, s	0.01	0.011	0.012	0.016
7	Time of action of the force $F_{idl.sp.}$, s	0.01	0.019	0.028	0.044

The value of the driving force component in the direction of idle speed, at the accepted parameters of individual stages of vibration blocks with directional oscillations, is reduced from 5.0 kN to 1.25 kN. The asymmetry coefficient of the driving force, which increases from 1.0 in the basic sample to 4.0 in a four-stage vibration device, with a small degree of conditionality, can be considered a coefficient that evaluates the effect of the system resonance. The time of action of the force *F*u.w. does not change significantly, about 9 ... 10% when changing the number of stages. It is noteworthy that with the increase of number of stages increases the total weight of debalances. However, the method allows, within a static moment of debalances to reduce the total weight of them by increasing the radii of the displacement of the centers of gravity of respective debalances.

The comparative calculations show that in necessary and justified conditions [15, 16], in which it is necessary to have high values of the driving force in the direction of performing useful work and low values of the driving force in the direction of performing idle speed, it is justified to switch to the design, creation and use of machines with vibration devices that generate asymmetric oscillations. The method presented in the paper allows selecting the main technological and design parameters of vibration blocks of a multi-stage vibration device that provide the highest coefficient of asymmetry of the total driving force.

5. References

[1] Chelomei V.N. 1981 Vibration processes and machines *M Mechanical Engineering* **(6)** 509
[2] Bauman V.A. 1970 Vibration machines in the construction and production of building materials *M. Mechanical Engineering* 632
[3] Astashev V.K. 2011 About new directions of using the resonance phenomenon in machines *Bulletin of scientific and technological development. National Technology Group* **8(48)** 10 – 15
[4] Blekhman I.I. 2013 The theory of vibration processes and devices *Vibration mechanics and vibration technology. - SPb., Publishing House "Ore and Metals"* 640
[5] Nejshtadt A.I. 1975 Passing through resonance in a two-frequency problem *Reports of the USSR Academy of Sciences* **221** 301-304

[6] Pechenev A.V. 1986 On the motion of an oscillatory system with limited excitation near resonance *Reports of the USSR Academy of Sciences* **290(1)** 12-15
[7] Balendra T., Swaddiwudhipong, S. & Quek, S. T. 1984 Free vibration of asymmetric shear wall-frame buildings *Earthquake Engineering and Structural Dynamics* **12(5)** 629-650
[8] Ermolenko V.N., Nasonov I.V., Nesterenko P.N. Toothed inertial self-balancing mechanism. Ptent RF, no. 2350806, 2009
[9] Kleibl A, Heichel Ch. Vibration Generator. Patent US, no. 7804211, 2009
[10] Gerasimov M.D. 2016 Addition of vibrations in vibration absorbers *Bulletin of the Belgorod State Technological University named after. V G Shukhov* **3** 116-121
[11] M.D. Gerasimov, Oleg V. Mkrtychev, Sergey N. Glagolev, Dmitriy M. Gerasimov, Sergey S. Latyishev 2016 Method of determination of vibrating screens oscilation's amplitude in a characteristic point for plane motion *ARPN Journal of Engineering and Applied Sciences* **11(20)** 12295-12301
[12] M.D. Gerasimov, Glagolev S.N., Gerasimov D. M., Mkrtychev O. V. 2015 Determination of the driving force's asymmetry factor and the vibrostand's work's analysis *International Journal of Applied Engineering Research* **10(24)** 45392-45398
[13] M. D. Gerasimov, M A Romanovich, N D Vorobiev, Ebrahim Amini 2018 Results of research to improve efficiency of vibrating machines. International Conference "Complex equipment of quality control laboratories" *IOP Conf. Series: Journal of Physics: Conf. Series* 12015-12019
[14] Industrial vibrators and vibration equipment 2018 *Product Catalog. Yaroslavl factory «KRASNYJ MAYAK»* 41
[15] Evlakhova E.Yu., Ivanov A.V., Zheleznyakov V.A. 2019 Analysis of the results of monitoring of dynamic impacts during experimental immersion of driving piles at the construction site of the campus of sfu *Construction Materials and Products* **2(6)** 12-19
[16] Odenbakh I.A., Taurit E.B. 2020 Some new technologies in road construction *Construction Materials and Products* **3(1)** 62-69

Acknowledgments
The work was realized as part of performing service task in the framework of the Program of flagship university development on the base of Belgorod State Technological University named after V.G. Shukhov.

IOP Conf. Series: Materials Science and Engineering **945** (2020) 012004 doi:10.1088/1757-899X/945/1/012004

Determination of thermal parameters of a shell and tube heat exchanger with increased turbulization of the working fluid

N Yu Nikulin[1*], L A Kushchev[1] and A Yu Feoktistov[1]

[1]Department of Heat and Gas Supply and Ventilation, Belgorod State Technological University named after V.G. Shukhov, Kostyukov St., 46, Belgorod, 308012, Russia

E-mail: Nikulin@mail.ru

Abstract. Shell and tube heat exchangers are still widely used. This equipment is widely used in heat supply systems, food, chemical and petroleum industries. Therefore, conducting research to improve their performance is an important area. By changing the geometry of the heat exchange surface, the flow of the working fluid is further turbulated and the heat exchange area increases. The heat exchanger with a developed heat exchange surface is proposed, which differs from the serial device (with smooth tubes). Thus, a plate equipped with a cylindrical edge is installed on the heat exchange tube. The working fluid in the inter-tube space, flowing around the edges, is turbulated. This reduces the thickness of the laminar sublayer on the plate and edge. Accordingly, the heat transfer from the heat exchange surface to the streamlined fluid increases. When new elements appear in the design of the heat exchange surface, it is necessary to develop the calculation of temperature parameters for these surfaces. Formulas for calculating the average surface temperature of the edge and plate under conditions of turbulent flow around these surfaces are obtained analytically. The temperatures of the plate and edge will determine the Prandtl criterion Pr. In the future, this is necessary to find the main parameter of the heat exchanger – the heat transfer coefficient.

1. Introduction

Improvement of heat supply systems is an urgent task at the present time. Therefore, the design of boilers, heat networks, heat points, etc. is made using modern high-performance equipment and materials [1]. Such equipment also includes heat exchangers used for heating, municipal and technological hot water supply.

Shell and tube heat exchangers are still popular equipment in heat supply (figure 1). These devices are widely used in various industries. This spread of shell and tube heat exchangers is explained by their resistance to pressure drops, the ability to work in environments with high pressure values (up to 14 MPa) and temperatures (up to 600 °C), a slight decrease in the heat transfer coefficient with moderate overgrowth of the heat exchange surface [2]. Improving the technical and operational characteristics of these devices requires additional research and justification.

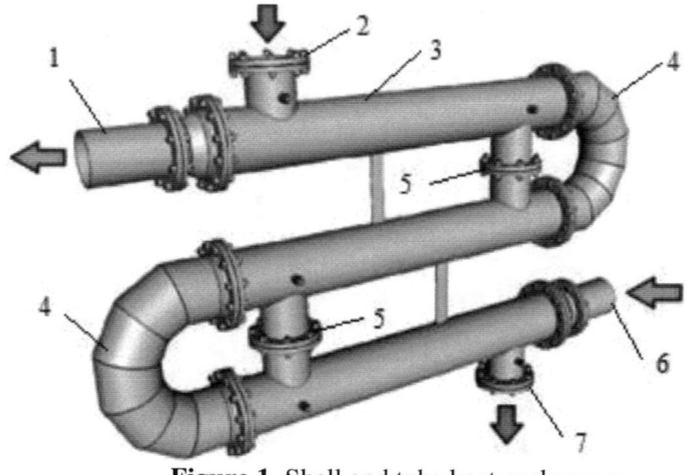

Figure 1. Shell and tube heat exchanger:
1 – cooling liquid outlet pipe (pipe space); 2 - pipe for the supply of heated fluid; 3 – body; 4 – connecting roll; 5 - pipes connecting the inter-pipe space; 6 - pipe for hot (heating) fluid supply;7 – outlet pipe of the heated fluid.

2. Materials and methods

In Russia and abroad, to improve the thermal characteristics, work is being actively carried out to improve the designs of heat exchangers [3]. According to the method of modernization of the structure for the intensification of heat exchange, we proposed the following classification:

1) constructive solutions:
- arrangement of movement of the heat carrier in the body;
- changing the geometry of the heat exchange surface;
2) the use of external physical fields;
3) adding chemicals to the heat carrier.

It should be noted that of all the ways to intensify heat transfer, "changing the geometry of the heat transfer surface" is the most common. This can be represented as a change in the geometry of the surface of the heat exchange tube: the use of spherical holes on the inner surface [4], biconcave spherical recesses [5]; semicircular hollows [6]. It is also possible to use tubes of spiral surface [7, 8], one smooth tube in the form of a spiral [9, 10]; in the form of a conical spiral [11].

We proposed an original design of a shell and tube heat exchanger [12]. The main difference from the serial [13] is that the plate (mounted on the tube) is equipped with a cylindrical edge (figure 2).

This allows increasing the heat exchange surface. Also, the working fluid is turbulated when it flows around the edge. This creates additional turbulization of the heated liquid on the plate and edges (in the inter-tube space). In turn, turbulization will contribute to the destruction of the wall-mounted laminar layer on the plate and edges, which means that heat transfer from the heat exchange surface to the heated fluid will increase [14]. In the end, this will lead to the increase in the performance of the heat exchanger for the heat carrier: hot or superheated water.

It is important to note that when new elements of the heat exchange surface – edges and plates – occur, it is necessary to develop a method for calculating the average temperature of these surfaces.

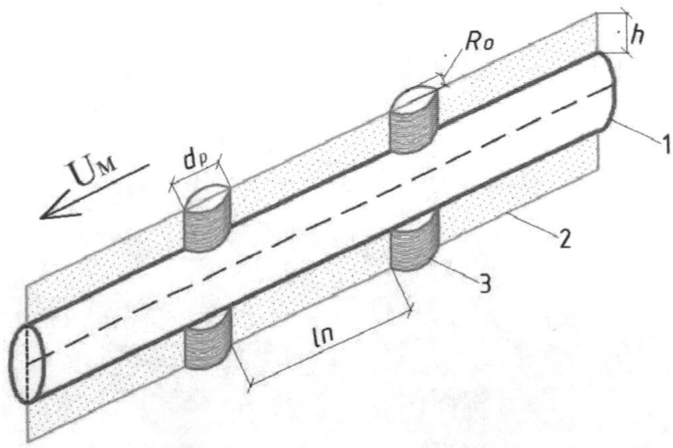

Figure 2. Heat exchange surface of the shell and tube heat exchanger: 1 – tube; 2 – plate; 3 – edges of cylindrical shape; $U_м$ – velocity of the fluid in the inter-tube space, m/s; R_0 – edge radius, d_p – edge diameter, l_n – length of the plate between the edges, h – height of the plate.

To find the temperature change inside a cylindrical edge with radius R_0, m, and length h, m, with a forced heat source q_0, W/m^3, represented as a thin plate located along the axis of the cylinder (figure 3), we use the stationary Fourier thermal conductivity equation in cylindrical coordinates [15]:

$$\frac{\partial^2 T}{\partial r^2}+\frac{1}{r}\frac{\partial T}{\partial r}+\frac{\partial^2 T}{\partial z^2}+\frac{q_0}{\lambda}=0, \tag{1}$$

where λ – coefficient of thermal conductivity of a cylindrical body, W/(m•K).

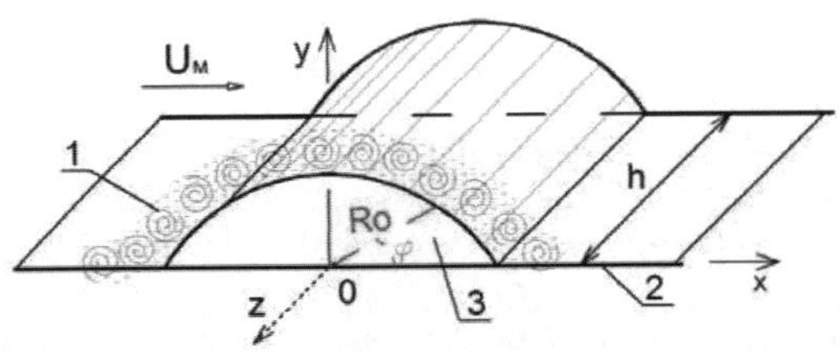

Figure 3. Scheme of the cylindrical edge for determining the temperature inside the cylindrical body with the heat source in the form of a plate: 1 – twirl, 2 – plate, 3 – edge.

It should be noted that due to the axial symmetry, the temperature distribution function inside the cylinder does not depend on the coordinate φ.

In the state of thermodynamic equilibrium, the amount of heat allocated per unit of time in a unit of volume of a cylindrical body is determined by the ratio:

$$q_0=\frac{q_{пл}}{V}, \tag{2}$$

where V – edge volume, m^3, $q_{пл}$- the amount of heat generated by the plate, W.

It is important to note that based on [15], the amount of heat generated by the plate is determined by the formula:

$$q_{пл}=-\lambda_0 S\frac{\partial T}{\partial z}, \tag{3}$$

where λ_0- coefficient of thermal conductivity of the plate, W/(m K), S – surface area of the plate, m^2, $\partial T / \partial z$ - temperature gradient along the z axis, K/m.

The volume of a cylindrical edge is determined by:

$$V = \pi R_0^2 h . \tag{4}$$

The surface area of the plate is equal to:

$$S = 2R_0 h . \tag{5}$$

Then equation (1) taking into account formulas (2) – (5) takes the form:

$$\frac{\partial^2 T}{\partial r^2} + \frac{1}{r}\frac{\partial T}{\partial r} + \frac{\partial^2 T}{\partial z^2} - \frac{2\beta}{\pi R_0}\frac{\partial T}{\partial z} = 0 , \tag{6}$$

where the following mark is introduced:

$$\beta = \frac{\lambda_0}{\lambda} . \tag{7}$$

Using the Fourier method [16], we look for the solution of equation (6) in the form:

$$T(r, z) = T_r(r) \cdot T_z(z) . \tag{8}$$

After the transformations, we present the final equation of temperature change T(R_0, z), K, over the surface of a cylindrical edge [17]:

$$T(R_0, z) = T_0 I_0 \left(\beta \sqrt{\left(1 + \frac{R_0 \ln\left(\frac{T_0}{T_1}\right)}{\beta h}\right)^2 - 1} \right) \cdot \left(\frac{T_0}{T_1}\right)^{-\frac{z}{h}} . \tag{9}$$

where T_0 – temperature at the point of contact of the plate and tube with the coordinate z=0, T_1 – temperature at the end of the plate with the coordinate z=h.

The obtained dependence (9) allows determining the average temperature on the edge surface using the formula:

$$T_{cp}^{p} = I_0 \beta \sqrt{\left(1 + \frac{R_0 \ln\left(\frac{T_0}{T_1}\right)}{\beta h}\right)^2 - 1} \cdot \frac{T_0 - T_1}{\ln\left(\frac{T_0}{T_1}\right)} . \tag{10}$$

where I_0 – the Bessel function of the zero order of the first row.

To determine the average surface temperature of the plate with height h and thickness 2δ, m, we place the plate as shown in figure 4.

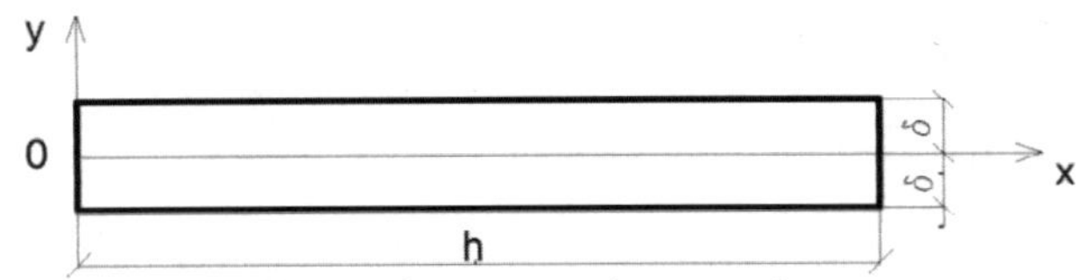

Figure 4. Heat exchanger plate.

The temperature distribution over the height of the plate, which is flowed by the turbulent flow of fluid, is determined by the following dependence [18]:

$$T_{пл} = T_0 \sum_{n=0}^{\infty} a_n \left(e^{-\overline{\nu_n} \frac{x}{\delta}} + \varepsilon_n e^{\overline{\nu_n} \frac{x}{\delta}} \right) \cos\left(\overline{\nu_n} \frac{y}{\delta} \right), \tag{11}$$

where value a_n is equal to:

$$a_n = \frac{2 \sin c(\overline{\nu_n})}{(1 + \sin c(2\overline{\nu_n}))(1 + \varepsilon_n)}. \tag{12}$$

Parameter $\overline{\nu_n}$ is a dimensionless value and is found as:

$$\overline{\nu_n} = \pi n + \alpha_n. \tag{13}$$

Value α_n is determined from the tables offered in [18], and depend on dimensionless value Λ:

$$\Lambda = \frac{\lambda_{ж} \delta}{\delta_c \lambda}. \tag{14}$$

where $\lambda_{ж}$ - coefficient of thermal conductivity of the fluid, W/(m K); δ_c – thickness of the boundary layer, m.

Value ε_n is determined as:

$$\varepsilon_n = \frac{\overline{\nu_n} - \Lambda}{\overline{\nu_n} + \Lambda} e^{-2\overline{\nu_n} \frac{h}{\delta}}. \tag{15}$$

Then the average value of the plate surface temperature T^n_{cp}, K, can be calculated as follows:

$$T^n_{cp} = \frac{T_0}{h} \frac{1}{2\delta} \int_0^h \left(\int_{-\delta}^{\delta} \sum_0^4 a_n \left(e^{\frac{-\overline{\nu_n} x}{\delta}} + \varepsilon_n e^{\frac{\overline{\nu_n} x}{\delta}} \right) \cos\left(\frac{\overline{\nu_n} y}{\delta} \right) dy \right) dx. \tag{16}$$

As a result of converting expression (16), we obtain a formula for determining the average temperature of the surface of the plate that is flowed by a turbulent flow:

$$T^n_{cp} = \frac{T_0}{h} \sum_0^4 \frac{a_n \sin \overline{\nu_n}}{\overline{\nu_n}^2} \left(e^{\frac{-\overline{\nu_n} h}{\delta}} + \varepsilon_n e^{\frac{\overline{\nu_n} h}{\delta}} - 1 - \varepsilon_n \right). \tag{17}$$

IOP Conf. Series: Materials Science and Engineering **945** (2020) 012004 doi:10.1088/1757-899X/945/1/012004

3. Results

Thus, important thermal parameters of the new elements of the heat exchange surface of the shell and tube apparatus with increased turbulization of the working fluid in the inter-tube space were determined: the average surface temperatures of the cylindrical edge T^p_{cp} (formula 10) and the plate T^n_{cp} (formula 17). The values of these temperatures are necessary to find the Prandtl numbers Pr for the edge and plate. Knowing The Pr value, it is possible to determine the Nusselt number Nu for new elements of the heat exchange surface of a shell and tube apparatus with increased turbulization of the working fluid.

It is known that the coefficient of heat transfer from the heat exchange surface is calculated based on the geometric dimensions (for the plate – length, m, for the edge – diameter, m), the coefficient of thermal conductivity of the fluid (W/(m K)) and the number Nu [15]. Hence, knowing the coefficient of heat transfer from the edge α_p, W/(m^2 K) and a plate of $\alpha_{пл}$,, and the coefficient of heat transfer from the outer surface of the tube α_m [19], it is possible to calculate the average heat transfer coefficient (according to M.A. Mikheev's research) from the outer, heating surfaces of heat transfer α_2 with the changed geometry:

$$\alpha_2 = \frac{\alpha_m F_m + \alpha_{пл} F_{пл} + \alpha_p F_p}{F_m + F_{пл} + F_p}. \quad (18)$$

where F_m, $F_{пл}$, F_p - the area of the outer surface, pipe, plate, edge respectively, m^2.

It is important to note that by defining the parameter α_2, it is possible to calculate the heat transfer coefficient of a shell and tube heat exchanger with increased turbulization using the well-known formula [19].

4. Summary

Thus, the proposed procedure for calculating the temperature parameters of the heat exchange surface is necessary for further development of the method for selecting heat exchangers with increased turbulization of the working fluid. The development of a software package for selecting the standard size and calculating the design and technological parameters of such heat exchangers is also of interest.

5. References

[1] Anikanova T V and Pogoromsky A S 2018 The use of semi-aquatic calcium sulfate to intensify the processes of hardening of structural heat-insulating foamed concrete *Construction Materials and Products* **1(3)** 25 – 32

[2] Lavygin V M, Sedlov A S and Tsanev S V 2009 Thermal power stations (Moscow: Publishing house of MEI) p 466

[3] Kuchev L A, Nikulin N Yu, Feoktistov A Yu 2019 Hydraulic heat tests of the shell and tube heat exchanger using heated liquid of higher turbulization *IOP Conf. Series: Materials Science and Engineering* **560** 012079.

[4] Kuntysh V B 2014 Thermal efficiency of whirl intensification of gas flow heat transfer in longitudinal and transverse flow around round-tube surfaces *Proceedings of higher educational institutions of the CIS* **2** 68–75

[5] Ilyinkov A V, Takmovtsev V V, Shchukin A V, Zaripov I Sh and Rybakov A S 2018 Thermohydraulic characteristics of biconcave recesses *Bulletin of Kazan State Technical University named after A.A. Tupolev* **4** 61–64

[6] Wagh P 2015 Optimization of a Shell and Tube Condenser using Numerical Method *Int. Journal of Engineering Research and Applications* **7** 9–15

[7] Itaya Y, Yamada M, Marumo K and Kobayashi N 2017 Heat and mass transfer through spiral tubes in absorber of absorption heat pump system for waste heat recovery *Propulsion and Power Research* **2** 140–146
[8] Karimov R M, Zaplatin A V and Tashbulatov R R 2019 Coiled Heat Exchanger with Small Radius Bent Tubes for Controlled Heat Treatment of High Viscosity Waxy Oil *IOP Conf. Series: Earth Environ. Science* **272** 022193
[9] Kruglov G A, Bakunin V V and Andreeva M V 2015 Theoretical studies of the relationship between flow turbulization and heat transfer coefficient *Bulletin of KrasSAU* **6** 67–73
[10] Alimoradi A, Lele M M and Gupta R K 2018 Investigation of exergy efficiency in shell and helically coiled tube heat exchangers *Case Studies in Thermal Engineering* **10** 1–8
[11] Purandare P S 2016 Experimental investigation on heat transfer and pressure drop of conical coil heat exchanger *Thermal Science* **6** 2087–2099
[12] Kushchev L A, Nikulin N Yu, Alifanova A I, Yakovlev E A and Shchekin I I 2017 Modern methods of heat exchange intensification in shell and tube heat exchangers of housing and communal services *Bulletin of BSTU named after V.G. Shukhov* **9** 73–79
[13] Sokolov E Ya 2001 Central heating and heat pipelines (Moscow: Publishing house of MEI) P 472
[14] Anisin A A 2008 Intensification of heat exchange in * profiled channels of plate heat exchangers (Bryansk: BSTU publishing house) 152
[15] Nashchokin V V 2013 Technical thermodynamics and heat transfer (M.: Book on demand) 496
[16] Grebennikova I V 2016 Equations of mathematical physics (Yekaterinburg: UFU) 164
[17] Nikulin N Yu 2019 Study of heat transfer in an intensified shell and tube apparatus *Bulletin of BSTU named after V.G. Shukhov* **4** 77–82
[18] Alkhasova D A 2009 Research and hydrodynamic calculations of downhole heat exchangers with longitudinal edges: dissertation (Makhachkala) 143
[19] Nikolaeva A A 2017 Designer's Guide. Design of heat networks (Moscow: Book on demand) 359

Acknowledgements
This work was realized in the framework of the Program of flagship university development on the base of the Belgorod State Technological University named after V G Shukhov, using equipment of High Technology Center at BSTU named after V G Shukhov.

Analysis of profitability of production in the construction industry of the Irkutsk region

E A Filatov[1]

[1]Irkutsk Scientific Center of the Siberian Branch of the Russian Academy of Sciences, 134, Lermontov str., 664033, Irkutsk, Russian Federation

E-mail: johnru3000@rambler.ru

Abstract. In economic practice, many specific indicators of profitability are used. They may differ in the complexity of the calculation, the availability of necessary information, and the usefulness of the analytical conclusions process. To measure the size of profit and the amount of funds used to achieve it in the industry economy, the indicator profitability of production is used. Profitability of production is an important parameter for determining the efficiency of the economy. Profitability of production is the most general, qualitative indicator of economic efficiency of production, efficiency of functioning of companies in the industry. This indicator is very important for making current and strategic decisions. The article presents the author's model for analyzing the profitability of production. The article reveals the influence of factors affecting the change in profitability and are more methodical approaches to its calculation (methods of factor analysis developed by Filatov E. A.). The article presents the author's analytic, systematic statistical material for the analysis of key indicators revealing the impact on the change of profitability of the construction industry in Irkutsk region of the Russian Federation.

1. Introduction

In economic analysis, the performance of various economic entities can be assessed by indicators such as revenue or profit (loss). However, the values of these indicators are not enough to form an opinion about the effectiveness of economic entities. This is due to the fact that these indicators are absolute characteristics of the activities of economic entities, and their correct interpretation for evaluating performance can only be carried out in conjunction with other indicators that reflect the funds invested in the company. Therefore, in economic analysis, profitability indicators are calculated to characterize the overall performance and profitability of various activities (economic, financial, and entrepreneurial).

The term profitability originates from «rent», which literally means income. Thus, profitability in the broad sense of the word means profitability, profitability [1].

Profitability indicators are widely used for evaluating financial and economic activities of enterprises in all industries. For example, profitability indicators are used for comparative evaluation of the performance of individual enterprises and industries that produce different volumes and types of products. These indicators characterize the profit received in relation to the production resources spent. The most commonly used indicators are profitability of sales and profitability of production [2-7].

IOP Conf. Series: Materials Science and Engineering **945** (2020) 012005 doi:10.1088/1757-899X/945/1/012005

Profitability of production is the most generalized, qualitative indicator of economic efficiency of production, efficiency of functioning of economic entities. The profitability of production just measures the amount of profit received with the size of the means by which it is obtained.

2. Research questions

The essence of the problem of increasing the profitability of production is to increase the economic results for each unit of costs in the process of using available resources.

Therefore, the task of economic science is to increase attention to the analysis of profitability as an economic phenomenon, to the forms of its manifestation, and methods of accounting in practice. The relevance of the topic is explained by the fact that the market economy makes it necessary to increase production efficiency.

The problems of increasing the efficiency of production and increasing the efficiency of individual economic entities in particular have been studied in the works of foreign and domestic scientists. Among them in the Russian economic literature, first of all, we can distinguish the works of: A.I. Anchishkin, A.M. Burman, P.G. Bunich, K.M. Velikanov, L.A. Gatovsky, B.C. Kabakov, A.E. Karlik, B.C. Nemchinov, A.V. Panskov, V.A. Lantsov, A.I. Muraviev, V.V. Novozhilov, B.P. Plyshevsky, S.G. Strumilin, K.G. Tatevosov, N.N. Ukhov, G.S. Khachaturov, N.A. Tsagolov and others, as well as research by foreign scientists I. Ansoff, B. Karloff, M. Porter, V. Hoyer, and others.

In Russia, the problems of integrated assessment of the performance of companies are presented in publications at the level of familiarity with a special direction in economic science. There are few monographs on the problems of integrated performance analysis, and none where these problems are generalized at all. The most noteworthy conceptual approaches to integrated analysis of the performance of A.D. Sheremet, A.N. Romanov, Yu.A. Lvov, A.N. Bogatko, J.V. Bogatin, E.Yu. Duhonina, D.V. Isayev, G.G. Fetisov.

Thus, the lack of development of issues related to the comparative evaluation of various methods for evaluating the performance of companies, and the urgent need to resolve them, indicate the relevance and practical significance of the chosen topic of the article.

3. Materials and methods

Five indicators were used for factor analysis of production profitability: revenue, cost, net profit, asset value, and cost of equity.

Further, based on the methods of deterministic (functional) factor analysis developed by the author [8], we will assess the degree of influence of four factors on the change in the profitability of the construction industry in the Irkutsk region of the Russian Federation.

Source data for alternative factor analysis of profitability in the construction industry (all legal entities) Irkutsk region of the Russian Federation are presented in table 1 based on data from 4 statistical bulletins of Irkutskstat (1. Financial results of organizations. Volume 2-large and medium-sized organizations. 2017; 2. Financial results of organizations. Volume 3-small businesses. 2017; 3. Financial results of organizations. Volume 2-large and medium-sized organizations. 2018; 4. Financial results of organizations. Volume 3-small businesses. 2018).

Table 1. Baseline data for the factor analysis.

No.	Indicators	№ fact or's	2017 (0) *	2018 (I) **	Deviation (Δ) ***
1	*V* – Net revenue, thousand rubles		119165673	110973284	-8192389
	including:				
	- large and medium-sized organizations, thousand rubles		49976459	45595417	-4381042
	- small businesses, thousand		69189214	65377867	-3811347

	rubles				
2	***SS*** – Cost of work, thousand rubles		114180693	105115881	-9064812
	including:				
	- large and medium-sized organizations, thousand rubles		48483420	43859651	-4623769
	- small businesses, thousand rubles		65697273	61256230	-4441043
3	***P*** – Net profit, thousand rubles		1667614	3008741	1341127
	including:				
	- large and medium-sized organizations, thousand rubles		160323	302092	141769
	- small businesses, thousand rubles		1507291	2706649	1199358
4	***A*** – Asset value (average annual rate), thousand rubles		116062474	108937693	-7124781
	including:				
	- large and medium-sized organizations, thousand rubles		43115408	35393264	-7722144
	- small businesses, thousand rubles		72947066	73544429	597363
5	***SK*** – Cost of equity (average annual rate), thousand rubles		13230536	14441173	1210637
	including:				
	- large and medium-sized organizations, thousand rubles		6090313	5503905	-586408
	- small businesses, thousand rubles		7140223	8937268	1797045
6	**POP** – Profitability of production (3/2)		0.014605	0.028623	**0.014018**
7	**ROE** – Return on equity (3/5)	F_1	0.126043	0.208345	0.082302
8	**FIR** – Financial independence ratio (5/4)	F_2	0.113995	0.132564	0.018569
9	**RI** – Resource intensity (4/1)	F_3	0.973959	0.981657	0.007698
10	**YOP** – Yield of production (1/2)	F_4	1.043659	1.055723	0.012065

where: * **0** – past (basic) period (year), taken as reference base; ** **I** – reported (current) period (year); *** **Δ** – change for the period, calculated as the difference between the fact and the plan (**I** – **0**).

The original formula derived by the author for factor analysis of production profitability will have the following form (formula 1):

$$POP = \frac{P}{SK} * \frac{SK}{A} * \frac{A}{V} * \frac{V}{SS} = F_1 * F_2 * F_3 * F_4 = \prod_{n=1}^{4} F_n \quad (1)$$

The author's model of production profitability consists of 4 factors:

F_1 – ROE – return on equity, calculated as the ratio of net profit (***P***) to equity (***SK***). This is one of the key indicators of financial profitability. This indicator is important for the owners of the company, since it characterizes the profit that the owner will receive from the unit of money invested in the organization.

IOP Conf. Series: Materials Science and Engineering **945** (2020) 012005 doi:10.1088/1757-899X/945/1/012005

F_2 – FIR – financial independence coefficient or autonomy coefficient, which describes the ratio of equity (***SK***) to the total amount of capital (assets) (***A***) of the organization. The autonomy coefficient shows how independent an organization is from creditors.

F_3 – RI – resource intensity the reverse indicator of resource return or capital return. If the resource return (capital return) indicates the efficiency with which the company uses its resources to produce products and has an economic interpretation, showing how much revenue from sales (***V***) accounts for one monetary unit of investment in the assets (***A***) of the organization. The resource intensity characterizes the ratio of assets (***A***) to one monetary unit of sales revenue (***V***). Resource intensity – the amount of resources used to produce a unit of final product, i.e. the ratio between resources consumed and products produced (in physical form or in the form of services). It includes both components that are part of the final product and those consumed during its production. Resource intensity is expressed in the number of units of resources spent per unit of production. This indicator is used for analytical comparisons for various purposes.

F_4 – YOP – yield of production, close to the indicator of profitability of production (***POP***). Profitability of production is calculated as the ratio of net profit to cost. Profitability of production shows how much a company has net profit (***P***) per monetary unit of cost (***SS***). The author introduced the indicator of yield of production (***YOP***), which is calculated as the ratio of sales revenue (revenue) (***V***) per unit cost (***SS***). This indicator shows the ratio of revenue received from sales (***V***) to the amount of funds spent on its receipt (cost – ***SS***).

The author's indicator of yield of production (***YOP***) and the well-known indicator of profitability of production (***POP***) characterize the effectiveness of the company's production and commercial activities.

Based on the data in table 1, it can be seen that:

- the return on equity of the construction industry of the Irkutsk region in 2017 was 12.60%, while in 2018 it was 20.83%, an increase of 8.23%;
- the coefficient of financial independence of the construction industry of the Irkutsk region in 2017 was 11.40%, while in 2018 it was 13.26%, an increase of 1.86%;
- the resource intensity of the construction industry in the Irkutsk region in 2017 was 97.40%, while in 2018 it was 98.17%, an increase of 0.77%;
- the yield of production of the construction industry in the Irkutsk region in 2017 was 104.37%, while in 2018 it was 105.57%, an increase of 1.21%.

The resulting indicator of the author's model of profitability of production in the construction industry of the Irkutsk region in 2017 was 1.46%, while in 2018 it was 2.86%, an increase of 1.40%.

The total deviation for the resulting indicator (Δ***POP***) is determined by the formula 2:

$$\Delta POP = \sum_{n=1}^{4} \Delta POP\,(F_n) = \Delta POP\,(F_1) + \Delta POP\,(F_2) + \Delta POP\,(F_3) + \Delta POP\,(F_4) \qquad (2)$$

Auxiliary data on comparative coefficients for factor analysis are presented in table 2.

Table 2. Multiple comparative coefficients by one factor.

Comparison of factors	Designation of comparative coefficients	Value	The product of coefficients (value)
F_{11} / F_{10}	A_1	1.652967	1.0
F_{10} / F_{11}	A_2	0.604973	
F_{21} / F_{20}	A_3	1.162890	1.0
F_{20} / F_{21}	A_4	0.859926	
F_{31} / F_{30}	A_5	1.007904	1.0
F_{30} / F_{31}	A_6	0.992158	
F_{41} / F_{40}	A_7	1.011560	1.0
F_{40} / F_{41}	A_8	0.988572	

The author's (alternative) methods of factor analysis are presented in the table 3.

Method No. 1.1 (formulas 1.1–1.4 in the table 3) is based on the difference between the effective planned indicators, which are adjusted on the comparative coefficients (A_1, A_3, A_5).

Method No. 1.2 (formulas 2.1–1.4 in the table 3) is based on the difference between the effective actual indicators, which is adjusted on the comparative coefficients (A_4, A_6, A_8).

Method No. 2.1 (formulas 3.1–3.4 in the table 3) is based on the ratio of the departure of the original factor to the original plan factor multiplied by the planned performance indicator, which is adjusted on the comparative coefficient (A_1, A_3, A_5).

Method No. 2.2 (formulas 4.1–4.4, in the table 3) is based on the ratio of the departure of the original factor to the original actual factor multiplied by the actual performance indicator, which is adjusted on the comparative coefficient (A_4, A_6, A_8).

Method No. 3.1 (formulas 5.1–5.4 in the table 3) is based on the difference between the effective actual and planned indicators, which is adjusted on the comparative coefficients (A_1, A_3, A_5).

Method No. 3.2 (formulas 6.1–6.4 in the table 3) is based on the difference between the effective actual and planned indicators, which is adjusted on the comparative coefficients (A_4, A_6, A_8).

Method No. 4.1 (formulas 7.1–7.4 in the table 3) is based on the ratio of deviation of the effective factor to the difference between the effective actual and planned factors, which is adjusted on the comparative coefficient (A_1, A_3, A_5).

Method No. 4.2 (formulas 8.1–8.4 in the table 3) is based on the ratio of deviation of the effective factor to the difference between the effective actual and planned factors, which is adjusted on the comparative coefficients (A_4, A_6, A_8).

Method No. 5.1 (formulas 9.1–9.4 in the table 3) is based on the ratio of deviation of the effective factor to the difference between the actual performance factors, which is adjusted on the comparative coefficients (A_1, A_3, A_5).

Method No. 5.2 (formulas 10.1–10.4 in the table 3) is based on the ratio of deviation of the effective factor to the difference between the planned performance factors, which is adjusted on the comparative coefficients (A_4, A_6, A_8).

Table 3. Methods of alternative factor analysis using comparative coefficients.

№ formulae	formulas / calculations	
	the main part of the formula	**adjustment factors**
1.1	$\Delta POP\ (F_1) = POP_0{*}(A_1) - POP_0$	–
1.2	$\Delta POP\ (F_2) = (POP_0{*}(A_3) - POP_0){*}$	A_1
1.3	$\Delta POP\ (F_3) = (POP_0{*}(A_5) - POP_0){*}$	$A_1{*}A_3$
1.4	$\Delta POP\ (F_4) = (POP_0{*}(A_7) - POP_0){*}$	$A_1{*}A_3{*}A_5$
2.1	$\Delta POP\ (F_1) = (POP_I - POP_I * (A_2)){*}$	$A_8{*}A_6{*}A_4$
2.2	$\Delta POP\ (F_2) = (POP_I - POP_I * (A_4)){*}$	$A_8{*}A_6$
2.3	$\Delta POP\ (F_3) = (POP_I - POP_I * (A_6)){*}$	A_8
2.4	$\Delta POP\ (F_4) = POP_I - POP_I * (A_8)$	–
3.1	$\Delta POP\ (F_1) = (\Delta F_1/F_{1\,0}) * POP_0$	–
3.2	$\Delta POP\ (F_2) = (\Delta F_2/F_{2\,0}) * POP_0{*}$	A_1
3.3	$\Delta POP\ (F_3) = (\Delta F_3/F_{3\,0}) * POP_0{*}$	$A_1{*}A_3$
3.4	$\Delta POP\ (F_4) = (\Delta F_4/F_{4\,0}) * POP_0$	$A_1{*}A_3{*}A_5$
4.1	$\Delta POP\ (F_1) = (\Delta F_1/F_{1\,I}) * POP_I{*}$	$A_8{*}A_6{*}A_4$
4.2	$\Delta POP\ (F_2) = (\Delta F_2/F_{2\,I}) * POP_I{*}$	$A_8{*}A_6$
4.3	$\Delta POP\ (F_3) = (\Delta F_3/\ F_{3\,I}) * POP_I{*}$	A_8
4.4	$\Delta POP\ (F_4) = (\Delta F_4/\ F_{4\,I}) * POP_I$	–
5.1	$\Delta POP\ (F_1) = POP_I{*}(B_8) - POP_0$	–
5.2	$\Delta POP\ (F_2) = (POP_I{*}(B_6) - POP_0){*}$	A_1

5.3	$\Delta POP\ (F_3) = (POP_I*(B_4) - POP_0)*$	A_1*A_3
5.4	$\Delta POP\ (F_4) = (POP_I*(B_2) - POP_0)*$	$A_1*A_3*A_5$
6.1	$\Delta POP\ (F_1) = (POP_I - POP_0*(B_7))*$	$A_8*A_6*A_4$
6.2	$\Delta POP\ (F_2) = (POP_I - POP_0*(B_5))*$	A_8*A_6
6.3	$\Delta POP\ (F_3) = (POP_I - POP_0*(B_3))*$	A_8
6.4	$\Delta POP\ (F_4) = POP_I - POP_0*(B_1)$	–
7.1	$\Delta POP\ (F_1) = \Delta POP - (POP_I - (POP_0*A_1)$	–
7.2	$\Delta POP\ (F_2) = \Delta POP - (POP_I - (POP_0*A_3))*$	A_1
7.3	$\Delta POP\ (F_3) = \Delta POP - (POP_I - (POP_0*A_5))*$	A_1*A_3
7.4	$\Delta POP\ (F_4) = \Delta POP - (POP_I - (POP_0*A_7))*$	$A_1*A_3*A_5$
8.1	$\Delta POP\ (F_1) = \Delta POP - ((POP_I *A_2) - POP_0)*$	$A_8*A_6*A_4$
8.2	$\Delta POP\ (F_2) = \Delta POP - ((POP_I *A_4) - POP_0)*$	A_8*A_6
8.3	$\Delta POP\ (F_3) = \Delta POP - ((POP_I *A_6) - POP_0)*$	A_8
8.4	$\Delta POP\ (F_4) = \Delta POP - ((POP_I *A_8) - POP_0)*$	–
9.1	$\Delta POP\ (F_1) = \Delta POP - (POP_I - (POP_I*B_8))$	–
9.2	$\Delta POP\ (F_2) = \Delta POP - (POP_I - (POP_I*B_6))*$	A_1
9.3	$\Delta POP\ (F_3) = \Delta POP - (POP_I - (POP_I*B_4))*$	A_1*A_3
9.4	$\Delta POP\ (F_4) = \Delta POP - (POP_I - (POP_I*B_2))*$	$A_1*A_3*A_5$
10.1	$\Delta POP\ (F_1) = \Delta POP - ((POP_0*B_7) - POP_0)*$	$A_8*A_6*A_4$
10.2	$\Delta POP\ (F_2) = \Delta POP - ((POP_0*B_5) - POP_0)*$	A_8*A_6
10.3	$\Delta POP\ (F_3) = \Delta POP - ((POP_0*B_3) - POP_0)*$	A_8
10.4	$\Delta POP\ (F_4) = \Delta POP - ((POP_0*B_1) - POP_0)$	–

4. Results and discussions

The result of methods 1.1, 2.1, 3.1, 4.1, 5.1 is presented in table 4, the result of methods 1.2, 2.2, 3.2, 4.2, 5.2 is presented in table 5.

Table 4. The result on methods 1.1, 2.1, 3.1, 4.1, 5.1.

No.	the main part of the formula	adjustment factors		result
1	$\boldsymbol{\Delta POP}$ (F_1) = 0.009537	–		0.009537
2	$\boldsymbol{\Delta POP}$ (F_2) = 0.002379	1.652967	A_1	0.003932
3	$\boldsymbol{\Delta POP}$ (F_3) = 0.000115	1.922219	A_1*A_3	0.000222
4	$\boldsymbol{\Delta POP}$ (F_4) = 0.000169	1.937412	$A_1*A_3*A_5$	0.000327
	0.012200			**0.014018**

Table 5. The result on methods1.2, 2.2, 3.2, 4.2, 5.2.

No.	the main part of the formula	adjustment factors		result
1	$\boldsymbol{\Delta POP}$ (F_1) = 0.011307	0.843433	$A_8*A_6*A_4$	0.009537
2	$\boldsymbol{\Delta POP}$ (F_2) = 0.004009	0.980820	A_8*A_6	0.003932
3	$\boldsymbol{\Delta POP}$ (F_3) = 0.000224	0.988572	A_8	0.000222
4	$\boldsymbol{\Delta POP}$ (F_4) = 0.000327	–		0.000327
	0.015868			**0.014018**

Factor analysis allows you to obtain a quantitative assessment of the impact of factor deviations on the deviation of the value of the indicator under study. As can be seen from the final result of tables 1, 4, 5, the purpose of the analysis is achieved-the determination of the influence of factors is disclosed without deviations.

IOP Conf. Series: Materials Science and Engineering **945** (2020) 012005 doi:10.1088/1757-899X/945/1/012005

According to the results of the analysis, the following factors influenced the change in the profitability of production (ΔPOP) in the construction industry of the Irkutsk region in the amount of 1.40%:

- the increase in the return on equity of the construction industry of the Irkutsk region (F_1) by 8.23% increased the studied indicator by 0.95%;
- an increase in the financial independence coefficient of the construction industry of the Irkutsk region (F_2) by 1.86% increased the studied indicator by 0.40%;
- an increase in the resource intensity of the construction industry in the Irkutsk region (F_3) by 0.77% increased the studied indicator by 0.02%;
- the increase in yield of production of the construction industry in the Irkutsk region (F_4) by 1.21% increased the studied indicator by 0.03%.

Table 1 shows that revenue for the construction industry in the Irkutsk region has decreased by about 8.2 billion rubles over the last period under study. At the same time, the revenue of small enterprises decreased by 3.8 billion rubles, and large and medium-sized organizations by 4.4 billion rubles, or about 15% more than that of small enterprises.

The cost of the construction industry in the Irkutsk region for the last study period decreased by about 9 billion rubles or more than revenue by 0.8 billion rubles or 10%. That is, despite the decline in the construction industry in the Irkutsk region, enterprises have become more efficient at managing costs. At the same time, the revenue of small enterprises decreased by 4.4 billion rubles, and large and medium-sized organizations by 4.6 billion rubles, or about 5% more than that of small enterprises.

Net profit in the construction industry of the Irkutsk region increased by 1.34 billion rubles over the last period under study. At the same time, the net profit of small businesses increased by 1.2 billion rubles, while large and medium-sized organizations increased by 0.14 billion rubles. Small businesses added 8.6 times more to the net profit of the construction industry in the Irkutsk region than large and medium-sized organizations.

The cost of the total capital of construction industry enterprises in the Irkutsk region has decreased by about 7.1 billion rubles over the last study period. At the same time, the cost of total capital for small enterprises increased by 0.6 billion rubles, while for large and medium-sized organizations it decreased by 7.7 billion rubles.

The cost of equity capital of construction companies in the Irkutsk region increased by 1.21 billion rubles over the last study period. At the same time, the cost of equity for small enterprises increased by 1.8 billion rubles, while for large and medium-sized organizations it decreased by 0.59 billion rubles. In other words, large and medium-sized organizations have become more dependent on creditors.

Table 6. Structure of absolute indicators for the production profitability model, %.

No.	Indicators	2017	2018	Δ
1	***V*** – Net revenue	100.00	100.00	0.00
	including:			
	- large and medium-sized organizations	41.94	41.09	-0.85
	- small businesses	58.06	58.91	0.85
2	***SS*** – Cost of work	100.00	100.00	0.00
	including:			
	- large and medium-sized organizations	42.46	41.73	-0.74
	- small businesses	57.54	58.27	0.74
3	***P*** – Net profit	100.00	100.00	0.00
	including:			
	- large and medium-sized organizations	9.61	10.04	0.43
	- small businesses	90.39	89.96	-0.43

4	***A*** – Asset value (average annual rate)	100.00	100.00	0.00
	including:			
	- large and medium-sized organizations	37.15	32.49	-4.66
	- small businesses	62.85	67.51	4.66
5	***SK*** – Cost of equity (average annual rate)	100.00	100.00	0.00
	including:			
	- large and medium-sized organizations	46.03	38.11	-7.92
	- small businesses	53.97	61.89	7.92

Table 6 shows that in the structure of income and costs of the construction industry in the Irkutsk region, small enterprises occupy about 60%, and large and medium-sized organizations about 40%. The deviation during the last study period was less than 1%.

In the structure of the final financial result (net profit) for the construction industry of the Irkutsk region, small enterprises account for about 90%, and large and medium-sized organizations for about 10%. The deviation during the last study period was less than 0.5%.

In the structure of the total assets of the construction industry of the Irkutsk region, small enterprises on average make up 2/3, and large and medium-sized organizations about 1/3. At the same time, the value of total assets for small enterprises increased by an average of 5% during the last study period, while for large and medium-sized organizations, respectively, it decreased by 5%.

In the structure of the equity capital of the construction industry of the Irkutsk region, small enterprises on average make up 60%, and large and medium-sized organizations about 40%. At the same time, the cost of equity for small enterprises increased by an average of 8% during the last study period, while for large and medium-sized organizations, respectively, it decreased by 8%.

In contrast to absolute indicators such as revenue, cost or profit, profitability is a relative indicator, which makes it possible to compare businesses of different business lines and different sizes. The cost-effectiveness of production (or cost-effectiveness) can be compared to the economic efficiency coefficient.

Based on the data in table 1, we calculate the profitability of the construction industry in the Irkutsk region by organizations (table 7).

Table 7. Profitability of production in the construction industry of the Irkutsk region, %.

No.	Indicators	2017	2018	Δ
1	all organization	1.46	2.86	**1.40**
2	- large and medium-sized organizations	0.33	0.69	**0.36**
3	- small businesses	2.29	4.42	**2.12**

Table 7 clearly shows that small businesses are more effective in the construction industry of the Irkutsk region than large and medium-sized organizations.

Integrated assessment of the company's economic condition is important primarily for comparative analysis of the efficiency of enterprises.

5. Summary

Business performance has always been the focus of attention for managers, economists, and researchers. This is an actual problem, it is the basic subject of study of economic science and practice.

The global financial and economic crisis of 2020, which the author previously wrote about [9], the bursting of the financial bubble on stock markets, and the bankruptcy of many companies have put the problem of evaluating the company's performance once again in the center of attention of the world community of economists.

The indicator of profitability of production is particularly important in modern economic conditions, when it is necessary to constantly make a number of extraordinary decisions to ensure profitability, and, consequently, financial stability [10-16].

Factors that affect the profitability of production are numerous and diverse. Some of them depend on the activities of specific teams, others are related to the technology and organization of production, the efficiency of the use of production resources, the introduction of scientific and technological progress.

6. References

[1] Abryutina M S 2010 Economics of the enterprise. Moscow, Publishing house «Business and Service», 585

[2] Avdeenko I A, Kaev Yu A 2016 Factors of increasing profit and profitability *New Science: From Idea to Result* **5-1 (84)** 112-114

[3] Lbova N O 2017 Management of business profitability *Actual Issues of Economic Sciences* **16-1** 105-109

[4] Lupil'ceva M A 2016 Bases and methods of the factorial analysis of profitability *Right. Economy. Security* **3 (9)** 89-92

[5] Nikonenko S Yu 2016 System of forming indicators of profitability of a commercial organization and factors of its growth *Actual Directions of Scientific Research of the XXI Century: Theory and Practice* **4 (24)** 139-143

[6] Berdnikova T B 2018 Analysis and diagnostics of financial and economic activity of the enterprise. Moscow, UNITY, 391

[7] Korotkova E M 2018 Anti-crisis management. Moscow, INFRA-M, 432

[8] Filatov E A 2020 Factor analysis of the knowledge intensity of the invested capital of Russia by Filatov's methods *Advances in Social Science, Education and Humanities Research* **128** 964-972 DOI: https://doi.org/10.2991/aebmr.k.200312.132

[9] Filatov E A 2019 Forecast of the beginning of the fundamental global financial and economic crisis of the third Millennium *Management of Economic Systems* **10**

[10] Certo S C 2003 Modern management. Prentice Hall, 595

[11] Farnham P G 2005 Economics for managers. Prentice Hall, 558

[12] Hill C W L, Jones G R 2004 Strategic management. An Integrated approach. Houghton Mifflin Company, 1182

[13] Naylor J 2004 Management. Prentice Hall, 668

[14] Sydsæter K, Hammond P 2002 Essential mathematics for economic analysis. Prentice Hall, 684

[15] Thompson A A, Fulmer W E, Strickland III A J 1990 Readings in strategic management. Richard D. IRWIN, Inc., 511

[16] Zikmund W G 2003 Business research methods. THOMSON, 748

Acknowledgments
The paper is produced in the framework of the scientific project of the Siberian branch of the Russian Academy of Sciences No. XI.174.1.4 «Activation of the internal development potential of regions of the resource specialization (on the example of the Baikal region)».

IOP Conf. Series: Materials Science and Engineering **945** (2020) 012006 doi:10.1088/1757-899X/945/1/012006

Energy efficiency improvement of aerated concrete block wall fences

Lyudmila Suleymanova[1]*, Inna Pogorelova[1] and Karim Suleymanov[1]

[1]Belgorod State Technological University named after V.G. Shukhov, Belgorod, Russian Federation

E-mail: ludmilasuleimanova@yandex.ru

Abstract. The development of effective protecting structures is currently one of the most popular areas in the construction industry. In order to ensure energy efficiency and environmental safety in the construction of civil buildings, masonry made of aerated concrete blocks with high thermal protection properties is used as protecting structures. As glue and cement-sand solutions in masonry have low thermal conductivity and are temperature bridges, the issue of filling through seams of aerated concrete masonry is very acute. The authors have developed a double row of concrete masonry, having a minimum number of cross-cutting horizontal seams, which not only improves thermal insulation properties, but also reduces material consumption and, consequently, the complexity of the arrangement of protecting structures of civil buildings. The influence of through horizontal seams made of cement-sand grout and block binding on the resistance to heat transfer of masonry made of aerated concrete blocks is considered.

1. Introduction

Reducing the material intensity and, consequently, the labor intensity of construction while maintaining the operational characteristics of buildings and structures has become the main direction of construction development. In order not to increase material consumption at increasing thermal protection properties new construction materials, products and structures are developed and introduced, the emergence of which allows combining the main trends of joint construction: reducing material consumption as the main economic task, and increasing the thermal protection properties of protecting structures as a component of the overall task of improving energy efficiency of buildings [1-4].

Cellular concrete is one of the main effective building materials used in the construction of external walls of buildings [5-9]. Construction systems made of cellular concrete blocks provide the required thermal insulation with a minimum thickness of the structure. However, the factor that reduces the energy efficiency of protecting structures made in the form of masonry of cellular concrete blocks is the filling of masonry seams with heat-conducting materials [10].

At arranging masonry from cellular concrete blocks, three main variants of masonry seams are used: on polyurethane glue, cement-sand grout and cement glue [11, 12].

The use of polyurethane glue as a joint filler for cellular concrete masonry is the most cost-effective due to the higher performance of work during installation and a lower cost of construction in the ratio of the cost of filling masonry seams. This type of masonry allows for high thermal uniformity due to the small thickness of the horizontal seam (less than 1 mm) and high thermal conductivity of the seam material. However, due to the low shear stiffness and high deformability, polyurethane adhesive can only be used when installing curtain walls of buildings.

Masonry on a cement-sand grout is the most material-intensive due to the arrangement of seams with a thickness of 8...10 mm, while this material is the least heat-conducting, which reduces the heat-protective properties of the entire enclosing structure.

Cement glue as a filler for masonry seams is the most optimal material due to its high thermal uniformity with a seam thickness of 1...2 mm, while having a low deformability and high strength.

The presence of heat-conducting inclusions, which are temperature bridges, in the protecting structures leads to additional losses of heat energy during the operation of buildings during the heating period and, as a result, to additional operating costs. However, failure to take these factors into account may lead to an unjustified underestimation of the design capacity of heating devices.

Taking into account the fact that the main heat losses occur through the through seams of wall structures, the development of cellular concrete masonry walls with a minimum number of through seams is one of the main tasks of ensuring the efficiency of buildings.

As the protecting structure of one-dimensional cellular concrete blocks, the most common masonry with cross-binding of blocks through a row is used (Fig. 1, *a*). However, this masonry is not energy efficient enough due to the presence of a large number of through horizontal seams that reduce the thermal uniformity of the protecting structure.

The authors developed an energy-efficient double row aerated concrete masonry walls (Fig. 1, *b*), which is characterized by high thermal uniformity due to reducing the number of through horizontal seams [13].

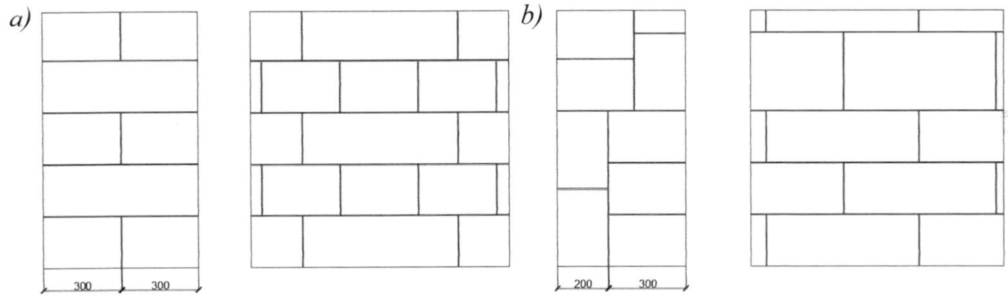

Figure 1. Cross section and general view of cellular concrete block masonry *a*) masonry with cross-binding of blocks in a row *b*) double row cellular concrete masonry.

Energy-efficient double row aerated concrete masonry consists of two vertical rows of one-size blocks and is represented by alternating tiers made in the form of two sections with a die binding between them. The lower section is represented by an internal vertical row of blocks laid down with a stretcher, and an external row of blocks laid down with a bedside. The upper section is represented by an inner row of blocks laid with the bedside down, and an outer row laid with the stretcher down.

This technological solution allows reducing the number of through horizontal seams by 5.6 times, thereby increasing the thermal uniformity, and reducing the material consumption and, as a result, the labor intensity of the masonry device by 20 % due to the lower thickness of the protecting structure.

2. Methods

The study of the thermal properties of aerated concrete masonry, taking into account heat-conducting inclusions in the form of through seams, is quite important for the effective use of masonry, often as the only thermal insulation layer of a wall structure. Therefore, increasing the thermal uniformity of masonry from aerated concrete blocks while ensuring the required performance characteristics determines the energy and economic efficiency of the arrangement of double row cellular concrete masonry developed by the authors.

As part of a comprehensive study of the types of masonry and their performance characteristics the authors conducted computational and experimental studies of the effect of through seams on thermal performance [14-20].

For the calculation and experimental evaluation of the thermal performance of the wall structure, the authors considered fragments of aerated concrete masonry, represented as a section of through seam and a section of block binding. By mechanical processing, models of masonry (Fig. 2) with dimensions of 250×250×50 mm are made, consisting of two rows of fragments of aerated concrete blocks connected with a cement-sand grout. Models *a* and *c* (Fig. 2) are represented by sections of binding of fragments of blocks, models *b* and *d* (Fig. 2) are represented as one and two through seams, respectively. All seams are 5 mm thick.

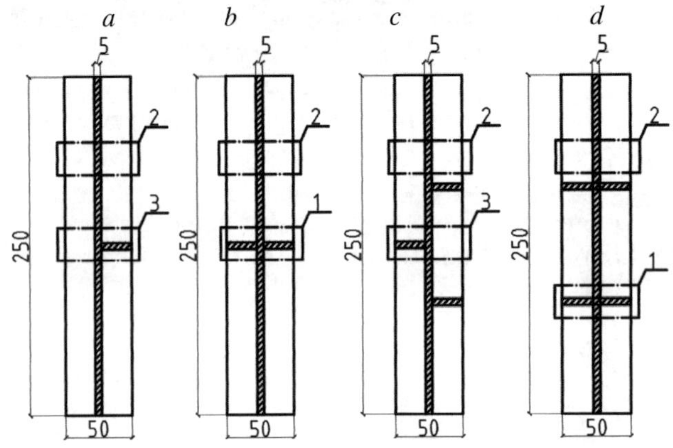

Figure 2. Cross-sectional view of masonry models: *a*) in the form of binding of fragments of blocks with a single non-through seam; *b*) in the form of a single through seam; *c*) in the form of binding of fragments of blocks with three non-through seams; *d*) in the form of two through seams; *1* – the site of blocks binding; *2* – site of layered construction; *3* – site in the form of a through seam.

The calculation of the given heat transfer resistance of masonry models (Fig. 2) is made using the formula:

$$R_{ro} = \Sigma A_i / (\Sigma A_i/R_{o,i} + \Sigma L_j\psi_j + \Sigma N_k K_k) \qquad (1)$$

where A_i – the area of the structure of the i-th type in the fragment under consideration, m^2;
L_j – the length of all j-type joints in the fragment under consideration, m;
N_k – the number of point thermal inhomogeneities of the k-th type in the considered fragment, PCs;
$R_{o,i}$ – heat transfer resistance of a homogeneous part of the *i*-th type structure, (m^2·°C)/W;
ψ_j – additional specific linear heat loss through the junction of the *j*-th type, W/(m·°C);
K_k – additional specific heat losses through point heat engineering inhomogeneity of the *k*-th type, W/°C.

The fragments under consideration are represented by three types of homogeneous part of the structure:

– masonry section in the form of a through seam (Fig. 2, pos. *1*);

– masonry section in the form of a layered structure consisting of two vertical rows of fragments of aerated concrete blocks connected by a layer of cement-sand grout (Fig. 2, pos. *2*);

– section of binding of seams of masonry fragments (Fig. 2, pos. *3*).

3. Results and discussion

Experimental studies were conducted on the thermal conductivity meter ITP MG-4. The temperature difference between the refrigerator and the heater of the device during the experiment was 30 °C. The results of the computational and experimental evaluation, including the determination of the reduced resistance to heat transfer, linear losses through the joint and the coefficient of thermal conductivity, are shown in Fig. 3.

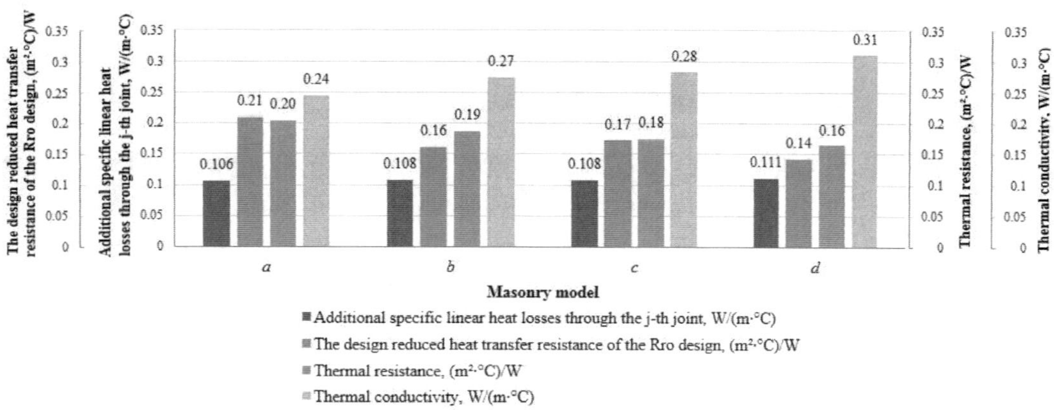

Figure 3. Average values of the main heat engineering indicators.

From the data presented in Fig. 3, it can be concluded that the best thermal performance of the masonry model *a* that is due to the absence of through seams in this model, which are heat-conducting inclusions, by 10...30% compared to the masonry model *b*, which has a single through seam. It should be noted that the thermal performance of the masonry model *c*, which has three sections of block binding, is comparable to the performance of the masonry model *b* with a single through seam. Also, according to the research, the least energy-efficient model is the masonry *d*, whose thermal characteristics are inferior to the model of masonry *c* with non-through seams by 10…25 %.

4. Summary

Energy-efficient double row aerated concrete masonry for the construction of protecting wall structures with a minimum number of through seams has been developed and designed. To determine the effect of through seams on the thermal performance of masonry, construction models were erected on the through seam section and on the block binding section. After conducting computational and experimental studies and processing the results, it was found that reducing the number of through horizontal seams increases the energy efficiency of wall structures by 10...30 %.

The use of the energy-efficient double row aerated concrete masonry developed by the authors allows to increase the thermal protection properties of the wall structure by reducing the heat-conducting inclusions in the form of through seams, as well as to reduce the material consumption and, consequently, the labor intensity of the construction of the protecting structure by reducing the thickness of the masonry while ensuring the required performance characteristics.

5. References

[1] Grinfeld G I 2013 Perfomance Characteristics of Autoclaved Aerated Concrete with Density 400 kg/cub.m *Construction of Unique Buildings and Structures* **5(10)** 28-57

[2] Gushchin S V, Semenenko A S and Shen Chkhao 2020 World trends in the development of energy-saving technologies *Bulletin of BSTU named after V.G. Shukhov* **5** 31-43

[3] Karpov D F 2019 Application of active and passive thermal control in the flaw detection of building materials and products, protecting structures of buildings and structures *Construction Materials and Products* **4** 39-44

[4] Stritih U 2003 Heat Transfer Enhancement in Latent Heat Thermal Storage System for Buildings *Energy and Buildings* **35(11)** 1097-1104

[5] Suleimanova L A 2012 Non-Autoclaved Aeroted Concrete at Composite Binding *Internationale Baustofftagung* 2-0830-2-0835

[6] Suleymanova L A, Pogorelova I A and Marushko M V 2014 Theoretical Basis of Formation Highly Organized Porous Structure of Aerated Concrete *Materials Science Forum* **945** 309-317

[7] Suleymanova L A, Lesovik V S, Kara K A, Malyukova M V andSuleymanov K A 2014 Energy-Efficient Concretesfor Green Construction *Research Journal of Applied Sciences* **12** 1087-1090

[8] Ahmed A 2017 Sustainable Construction Using Autoclaved Aerated Concrete (Aircrete) Blocks *Research & Development in Material Science* **1(4)**

[9] Yang F, Sun L Z and Xie Z L 2012 Theoretical Study on Optimal Design of Thermal Performance of Aerated Concrete-Based Composite Thermal Insulation Wall *Advanced Materials Research* **450-451** 663-666

[10] Gorshkov A S, Rymkevich P P and Vatin N I 2014 Simulation of Non-Stationary Heat Transfer Processes in Autoclaved Aerated Concrete-Walls *Magazine of Civil Engineering* **8** 38-48

[11] Gorshkov A S and Gladkikh A A 2010 Influence of grout joints of masonry on the parameters of thermal uniformity of walls made of aerated concrete *Civil Engineering Journal* **3(13)** 39-42

[12] Gorshkov A S, Rymkevich P P and Vatin N I 2013 Properties of the Wall Structures Made of Autoclaved Cellular Concrete Product on the Polyurethane Foam Adhesive *MagazineofCivilEngineering* **5** 5-19

[13] Suleymanova L A, Ryabchevsky I S, Kolomatsky A S, Pogorelova I A, Marushko M V and Bazhenova O O 2019 Energy-efficient masonry walls made of cellular concrete blocks *Patent of the Russian Federation № 2019141142,. Bul 7*

[14] Karpov D 2019 The Active Method of Control the Thermal Conductivity of Building Materials and Products *Bulletin of BSTU named after V G Shukhov* **4(7)** 57-62

[15] Girault M and Petit D 2005 Identification methods in nonlinear heat conduction. Part II: inverse problem using a reduced model *Int. Journal of Heat and Mass* **48(1)** 119-133

[16] Siu W M 2001 Discrete Formulation Using Thermal Resistance for Conduction Heat Transfer Analysis of Sphere Packings *The Hong Kong University of Science and Technology Library*

[17] Korniyenko S V, Vatin N I andGorshkov A S 2016 Thermophysical Field Testing of Residential Buildings Made of Autoclaved Aerated Concrete Blocks *Magazine of Civil Engineering* **64(04)** 10-25

[18] Peng P F, Qin X M and Wu Y S 2014 Performance Study on Masonries of Different Aerated Concrete Blocks *Key Engineering Materials* **633** 299-302

[19] Tsilingiris P T 2006 Parametric Space Distribution Effects of Wall Heat Capacity and Thermal Resistance on the Dynamic Thermal Behavior of Walls and Structures *Energy and Buildings* **38(10)** 1200-1211

[20] Pilipenko N V and Gladskikh D A 2014 Determination of the Heat Losses of Buildings and Structures by Solving Inverse Heat Conduction Problems *Measurement Techniques* **57(2)** 181-186

Acknowledgements
This work was realized in the framework of the Program of flagship university development on the base of the Belgorod State Technological University named after V G Shukhov, using equipment of High Technology Center at BSTU named after V G Shukhov.

IOP Conf. Series: Materials Science and Engineering **945** (2020) 012007 doi:10.1088/1757-899X/945/1/012007

Compounding features of composite gypsum binders for porous composites in construction printing technologies

N V Chernysheva[1] and S V Shatalova[1*]

[1]Department of building materials science, products and structures, Belgorod State Technological University named after V.G. Shukhov, Kostyukov St., 46, Belgorod, 308012, Russia

E-mail: shatalova.sv@yandex.ru

Abstract. At this stage of development of construction technologies and widespread digitalization, one of the most promising and interesting areas of the Russian and international scientific community is construction printing. The main idea of construction printing is to create a finished product or structure with a single installation. The most important task now is to master printing with porous concrete to give structures the required thermal characteristics, without which additive construction cannot fully compete with conventional technologies in the construction of wall structures. One of the promising materials in this case is foam concrete, in which the rapid fixation of the porous structure and the given shape is provided by the use of a composite gypsum binder. The paper considers the possibility of moving away from traditional composition that include active mineral additives by using Portland cements with various mineral additives. The influence of the used foaming agent on the hardening process of the gypsum-Portland cement system with various mineral additives was studied. This solution is designed to simplify the technology, make it more attractive for small and medium-sized businesses, increase the compatibility of the binder with foaming agents, as well as its activity to ensure high mechanical performance of foam concrete.

1. Introduction

One of the most advanced trends in the development of construction technologies is construction printing. According to a large number of publications, its advantage is an increase in the construction rate, the ability to create structures of rational sections of the most diverse configuration [1,2], etc.

Existing industrial models of 3D printers mainly use concrete mix as a building material, which is extruded horizontally on the layer previously laid out by the printer. The structure erected in this way must meet the required regulatory requirements, the most important of which are strength and thermal. If the strength indicators of the structures being built are sufficient for the construction of load-bearing walls of low-rise buildings, the required thermophysical indicators are achievable only with the use of insulation. However, such solutions go against the main ideas of 3D printing - creating a finished product with a single installation [3-6].

In this regard, the development of printing with porous materials that have indicators comparable to those traditionally used is very promising. This will make the technology attractive for small and medium-sized businesses on the one hand, and on the other hand, it will open up new facets of the use of additive technologies.

From this point of view, such materials as expanded clay concrete, foam concrete, non-autoclaved aerated concrete are of interest. According to the analysis of publications in the public domain, this direction is in its infancy. The studies on the possibility of digital printing with expanded clay concrete, foam concrete [7] and non-autoclaved aerated concrete are known [8].

In our opinion, for these purposes, as an alternative to non-autoclaved aerated concrete and other specified materials, it is advisable to use foam concrete, in view of the simplicity of technology, cost-effectiveness, the ability to build load-bearing walls and structural elements with the same properties as on the basis of traditional materials, namely, high thermal insulation, sound absorption and fire resistance.

The foam concrete technology itself has been sufficiently studied [9-11] and others. The main problems for its adaptation to construction printing, in our opinion, are the need for rapid fixation of the structure and the acquisition of the load-bearing capacity of the structure, so that the laid layers are not afraid of external influences. The ability of materials to harden under adverse site conditions, such as rapid dewatering, is also a serious problem that has received little attention.

The use of widely used, traditional hydraulic binders is not rational due to their slow hardening, even when using accelerators. The use of air binders, namely gypsum, is not optimal due to its low water resistance. In this regard, in our opinion, the introduction of fast-hardening composite gypsum binders (CGB) of increased water resistance with active mineral additives of different genesis can be very promising for these purposes [12,13].

However, mineral additives with high activity are expensive, and those that are waste products often have a non-permanent composition and therefore do not have stable properties. In addition, the multi-component composition complicates the production technology and increases the cost of additive construction, which is not profitable for small businesses. Based on this, it is very tempting to obtain CGB without a mineral additive, by mixing only two components, which will increase the content of active components and potentially the strength of foam concrete.

It is known that an active mineral additive in the composition of CGB is necessary for the binding of $Ca(OH)_2$, thus it eliminates the conditions for the formation of ettringite and high-base calcium hydroaluminates [13]. Therefore, in order to exclude these phenomena, it was hypothesized that in highly porous systems, when using Portland cements containing mineral additives, these processes will be minimal.

The second problem in obtaining foam concrete is to which extent the used foaming agent affects the hardening process of the binder.

In this regard, the main task of the study was to study the influence of the type of Portland cement and foaming agent on the strength and stability of CGB.

2. Materials and methods

2.1. Materials

In this work, as the binding base of CGB we used: α-modification gypsum binder GVVS-16 (G-16) of JSC "Samara gypsum plant", Portland cement CEM I 42.5 N JSC "Novoroscement" (CEM I) GOST 31108-2016, Portland cement CEM II/A-P 42.5 SS JSC "Novoroscement" (CEM II(SS)) GOST 22266-2013, Portland cement CEM II/B-Sh 42.5 N LLC "South Ural GPC", GOST 31108-2016 (CEM II(Sh)). As a pore-forming additive, the foaming agent "Penostrom" of LLC "SHIELD" of Shebekino, Belgorod region, corresponding to TU 0250-001-22299560-97 was used.

2.2. Methods

To model the behavior of the system under study, a two-factor three-level experiment was performed for each composition of CGB with different types of Portland cement. The conditions for planning an experiment are shown in table. 1. The amount of the closing water was assumed to be constant W/T=0.45.

Table 1. Conditions for the experiment planning.

Factor		Level of variation			Variation range
Natural form	Encoded form	-1	0	1	
Percentage of gypsum in the mix, %	x_1	25	50	75	25
Dosage of foaming agent, %	x_2	0	0.3	0.6	0.3

The output parameters were the strength of the hardened composite at the age of 7, 14 and 63 days when stored in air-wet conditions. The choice of these time intervals is determined by the fact that the system indicators at the age of 7 days allow assessing the contribution of gypsum to the formation of composite strength. In addition, the age of 7 days, as practice shows, is the maximum before critical dehydration of the material in unfavorably dry conditions. The increase in the strength of the composite in the interval from 7 to 14 days is mainly due to cement, while the development of destructive phenomena is still minimal. The time interval of up to 63 days provides both the maximum possible hydration of cement in specific conditions, and the development of the main volume of destructive processes. Tests performed at this age are necessary to assess the stability of the resulting system.

3. Results

On the basis of regression equations, the dependences of the dynamics of the strength set of a stone based on CGB with various types of Portland cement were obtained (Fig. 1). The amount of foaming agent (factor x_2= -1) is zero. Factor x_1 (the percentage of gypsum in the mix) varied throughout the considered range. The Figure 2 shows the dependence of the strength of the stone based on CGB with a foaming agent in the amount of 0.3% of the weight of the binder (factor x_2=0). The study was conducted on dense, rather than porous systems in order to avoid the influence of fluctuations in the value of the achieved porosity on the strength indicators.

It is known that when gypsum interacts with the main components of Portland cement clinker, along with hydration products that are favorable for structure formation, compounds are formed in the form of hydrosulfoaluminates and calcium hydrosulfoferrites, which, depending on the conditions of their occurrence, either contribute to strengthening the structure of the hardening binder, or weaken or even destroy it [12]. Therefore, one of the main factors is the content of C_3A and C_4AF. In the cements used in the work, their number decreases in the series CEM I - CEM II (Sh) - CEM II (SS).

It is also known that in the "Portland cement – gypsum – active silica contained in mineral supplements – water" instead of ettringite hydrosilicates and calcium hydroaluminate with reduced basicity, hydrogenate, hydroxyltoluene, hydrochloridehemical can be formed. These new formations are characterized by a lower content of hydrated water in their molecules compared to its content in the ettringite molecule [12].

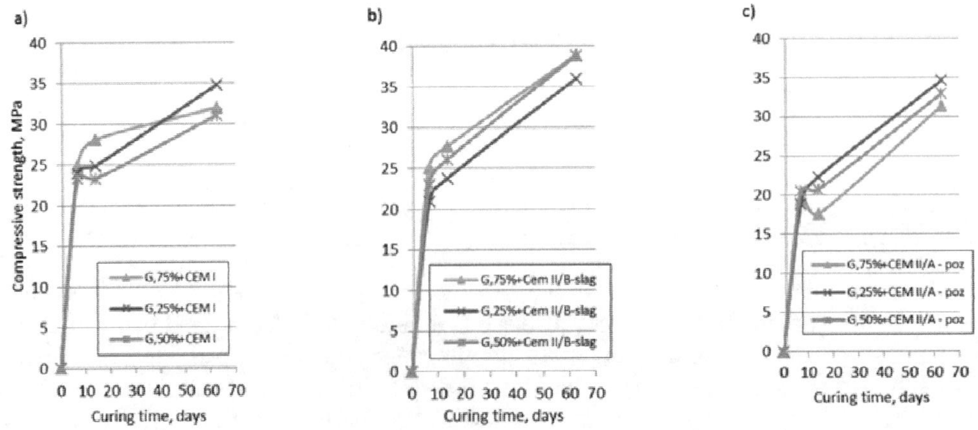

Figure 1. Dynamics of gypsum cement stone hardening on various types of Portland cement: a) CEM I; b) CEM II (Sh), c) CEM II (SS).

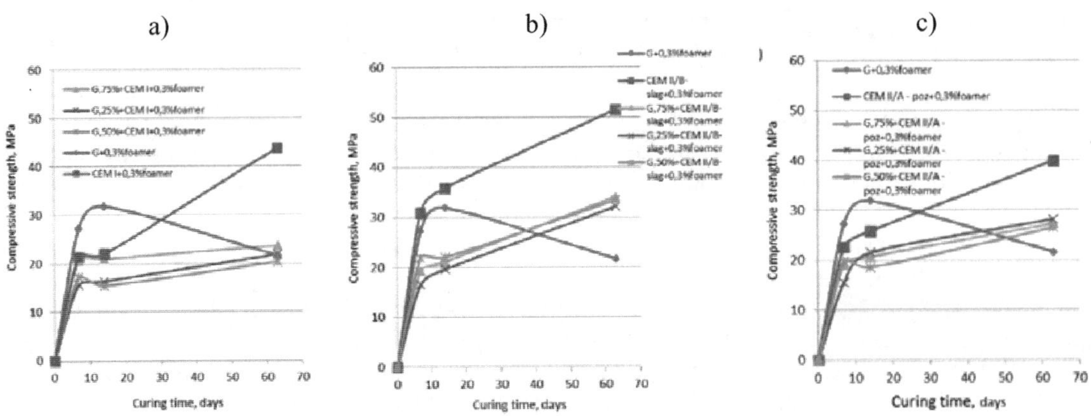

Figure 2. Dynamics of gypsum cement stone hardening on various types of Portland cement at the concentration of foaming agent – 0.3%: a) CEM I; b) CEM II (Sh), c) CEM II (SS).

Comparing the obtained compositions in the ratio "gypsum : clinker : mineral additive" based on CEM II (Sh) (50:20:30) and CEM II (SS) (50:30:20) with the compositions of CGB, where heavy concrete (HC) and quartzite sandstone (QS) were used as mineral additives in the ratio "gypsum : Portland cement : mineral additive" (60:20:20), we have comparable strength indicators of the hardened stone at the age of 28 days (Fig. 3 (a)) and softening coefficients (Fig.3 (b)).

The second factor affecting the hardening processes of the system under study is the action of the applied foaming agent. According to [14], anionic foaming agent molecules, having a negatively charged group – SO_3^-, are adsorbed on calcium hydroaluminate particles having a positive surface charge. Therefore, as the C_3A content in the cement increases, the content of non-adsorbed foaming agent in the liquid phase decreases, which worsens the properties of the mixture.

a) b)

IOP Conf. Series: Materials Science and Engineering **945** (2020) 012007 doi:10.1088/1757-899X/945/1/012007

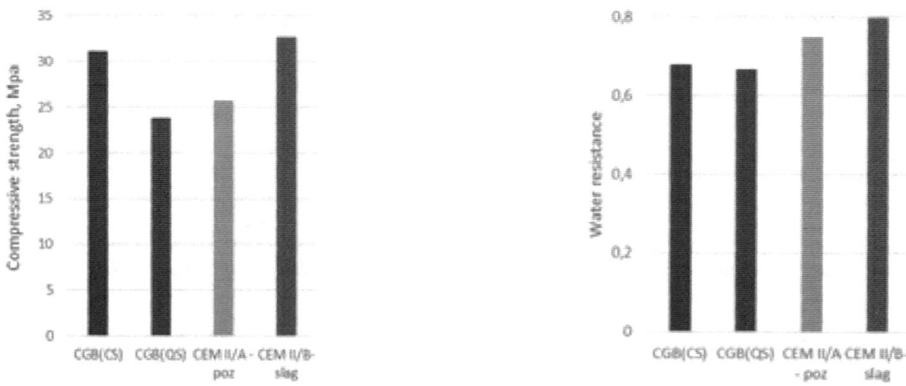

Figure 3. Strength indicators of hardened stone at the age of 28 days (a); values of softening coefficients (b).

4. Discussion

At the age of 63 days for all cements, there is a tendency to increase the strength at the maximum dosage of the foaming agent. Perhaps this is due to the fact that a large amount of foaming agent, slowing down the hydration of cement, stretches the development of destructive processes over time, and reduces their negative impact on strength. This factor manifests itself the more strongly, the worse the cement relates to joint hardening with gypsum in the series CEM II (Sh) - CEM II (SS) - CEM I. At lower dosages, the prevailing factor is the inhibition of cement hardening, at higher dosages - the suppression of destructive processes. Accordingly, it is acceptable and appropriate to use increased dosages of foaming agent in porous systems on CGB, as this simplifies the manufacturing process and suppresses negative phenomena. The inevitable loss of strength in this case for CEM II (Sh) and CEM II (SS) will be 10...15%, for CEM I up to a maximum of 40...50%.

5. Summary

Based on the conducted research, it was concluded that the use of CEM II (Sh) composition "gypsum – clinker – mineral additive" (50:20:30) is suitable from the used cements for foam concrete on CGB. This allows not implementing a separate mineral additive, which simplifies the production technology. At the same time, there is a minimum of destructive phenomena and a neutral attitude to any dosage of the foaming agent is observed. The main disadvantages of CEM II (Sh), such as slow gripping, low initial speed of hardening are fully compensated by a gypsum binder.

6. References

[1] Luneva D. A., Kozhevnikova E. O., Kaloshina S. V. 2017 Application of 3d printing in construction and prospects for its development *Bulletin of the Perm national research Polytechnic University Construction and architecture* **8(1)** 90-101

[2] Yi Wei Daniel Tay, Biranchi Panda, Suvash Chandra Paul, NisarAhamed Noor Mohamed, Ming Jen Tan&Kah Fai Leong. 2017 3D printing trends in building and construction industry: a review *Virtual and Physical Prototyping* **12(3)** 261-276

[3] Elistratkin M.Y., Lesovik V.S., Alfimova N.I., Shurakov I.M. 2019 On the question of mix composition selection for construction 3d printing. Materials Science Forum 945, 218-223

[4] Klyuev S.V., Klyuev A.V., Shorstova E.S. 2019 The micro silicon additive effects on the fine-grassed concrete properties for 3-d additive technologies *Materials Science Forum* **974** 131 – 135

[5] Lesovik V. S., Absemetov M. V., Elistratkin M. Yu., Pospelova M. A., Shatalova S. V. 2019 On the issue of studying the features of structure formation of composite carriers for non-autoclaved aerated concrete *Construction Materials and Products* **2(3)** 41-47

[6] Klyuev S.V., Klyuev A.V., Shorstova E.S. 2019 Fiber concrete for 3-d additive technologies *Materials Science Forum* **974** 367 – 372

[7] Viacheslav Markin, Venkatesh Naidu Nerella, Christof Schröfl, Gyunay Guseynova Viktor Mechtcherine 2019 Material design and performance evaluation of foam concrete for digital fabrication *Materials* **2433** 12-32

[8] Glagolev E. S., Ermolaeva A. E., Elistratkin M. Yu. 2018 The use of cellular concrete in the 3d additive industry *Russian online conference dedicated to the day of science* 75-82

[9] Shakhova L. D., Lesovik V. S. 2008 Features of obtaining heat-insulating foam concrete on synthetic foaming agents *News of higher educational institutions. Construction* **3 (591)** 51-56

[10] Alfimova N. I., Kireeva S. Yu., Gudkov D. V., Shurakov I. M., Korbut E. E. 2018 Optimization of recipe and technological parameters for manufacturing cellular concrete mix *Construction Materials and Products* **1(2)** 30-36

[11] Chernysheva N.V., Shatalova S.V., Drebezgova M.Y., Lesnichenko E.N. 2020 Thermal insulating and constructive foamed concrete on a composite gypsum binder *Materials Science Forum* **974** 125-131

[12] Drebezgova M. Yu., Chernysheva N. V., Shatalova S. V. 2017 Composite gypsum binder with multicomponent mineral additives of different Genesis *Bulletin of BSTU named after V.G. Shukhov* **10** 27-34

[13] Chernysheva N. V., Shatalova S. V., Evsyukova A. S., Fischer Hans-Bertram 2018 Features of selecting a rational composition of a composite gypsum binder *Construction Materials and Products* **1(2)** 45-52

[14] Rakhimbayev sh. M., Tarasenko V. N., Degtev I. A., Tverdokhlebov D. V. 2005 Influence of the solid phase on the properties of foam concrete. Materials of the international scientific and practical conference "Porobeton 2005» 80-84

Acknowledgments

The work was carried out as part of the implementation of the Development Program of the reference university on the basis of BSTU named after V.G. Shukhov until 2021. Contract No. A-82/20. Research topic: Aerated concrete based on water-resistant gypsum binders for heat-insulating and heat-insulating structural layers of building envelopes constructed using the construction printing method.

IOP Conf. Series: Materials Science and Engineering **945** (2020) 012008 doi:10.1088/1757-899X/945/1/012008

Simulation of accelerated concrete destruction by inoculation of an association of bacteria of the genus *Thiobacillus*

E V Fomina[1*], S V Sverguzova[1], E N Goncharova[1] and I Flores-Vivián[2]

[1]Belgorod State Technological University named after V.G. Shoukhov, 46, Kostukova Str., Belgorod, 308012, Russia
[2] The Autonomous University of Nuevo León, Pedro de Alba s/n, San Nicolás de Los Garza, Nuevo León, Mexico

E-mail: fomina.katerina@mail.ru

Abstract. The mechanisms of biogenic destruction of concrete under the influence of microbiological corrosion are described. An accelerated method of inoculation of the association of sulfur-oxidizing thionic bacteria *Thiobacillus thioparus* and *Thiobacillus thiooxidans* was used. Beyerink solution was used as a nutrient medium. The impact of two types of microorganisms in model systems imitates microbiocorrosion conditions; this allows reducing the test time and predicting the mechanisms of concrete destruction during operation. The active development of *Thiobacilus thioparus* promotes the leaching of concrete, further bacteria of the genus *Thiobacilus thiooxidans* appear with a decrease in pH, and both species actively synthesize sulfuric acid. The kinetics of the bio corrosion process was monitored for 60 days. The bacterial adaptation period is 16–18 days, after which the pH of the aqueous medium decreases. The rate of formation of a biological acid in an aqueous medium determines the leaching of Mg^{2+}, Ca^{2+}, and SO_4^{2-}ions from concrete. Concrete when exposed to bio acids for two months loses 30% of its strength. The microstructure analysis of bio corrosion products on the concrete surface was carried out with the help of a scanning electron microscope.

1. Introduction

Building materials are exposed to the environment effect during operation that causes their destruction. Corrosion of building materials, among of which a person spends up to 80% of his life is one of the main problems in the context of globalization of technological development. Among the most aggressive factors causing the degradation of materials the impact of microorganisms, amplifying with simultaneous fluctuations in temperature and humidity is included [1, 2].

Porosity and microzonal colonization of concrete surface by bacteria leads to a change in pH and destruction of the surface layer, calcium carbonate in particular. Destructive microbes are extremely diverse. The most important ones, according to many researchers are nitrifying and sulfur-oxidizing bacteria, as well as microscopic fungi [3], the growth rate of which, the specific composition of metabolites, the concentration of aggressive substances, and the duration of exposure determine the intensity and degree of damage to objects. Aerobic thionic bacteria are considered [4] especially aggressive, exothermically oxidizing the reduced forms of sulfur with the access of free oxygen:

$$(HS, H_2S) \xrightarrow{O_2} S° + H_2O \quad (1)$$

$$(S_2O_3^{2-}, S°) \xrightarrow{O_2+H_2O} H_2SO_4 \quad (2)$$

The action of these bacteria can significantly reduce the service life of concrete structures, for example, from the expected 100 years to 30-50 years [5]. Among the standards for testing materials for microbiological stability, there are no regulatory documents for such tests of building materials [6]. Meanwhile, lightweight concretes and concretes using techno genic raw materials and secondary raw materials are exposed to microbiological corrosion [7–11]. Under natural operating conditions, these processes have a long period, and long-term testing is not suitable for standards. In this regard, rapid methods of microbiological corrosion of concrete are needed.

The purpose of this work is to study the kinetics of the bio corrosion process, since with a wide variety of publications. There is no single view on its mechanism in the literature. The effects of microbiological destruction of concrete were studied by the inoculation method of the association of thionic bacteria *Thiobacillus thioparus* and *Acidithiobacillus thiooxidans. Thiobacillus thioparus* bacteria predominate in the environment and on the surface of building materials [12]. *Thiobacillus thiooxidans* predominate in the destroyed facilities and sewage disposal facilities, i.e. bacterial species that prefer lower pH values (Table 1).

Table 1. Physiological characteristics of the bacteria [5].

Type of bacteria	pH for development	Sulfate substrate	Metabolism products
Thiobacillus thioparus	6-10	H_2S, S^0, $S_2O_3^{2-}$	Polythionic acids, S^0
Acidithiobacillus thiooxidans	< 3	H_2S, S^0	S^0, $S_2O_3^{2-}$

After lowering the pH to 10, *Th. thioparus* sulphate bacteria can colonize and react with concrete surfaces, lowering the pH. Then, as the pH decreases, the bacteria *Ac. Thiooxidans* are included. It should be noted that *Ac. thiooxidans* remains viable at a pH close to 1.0–1.5, which corresponds to a 1.0 N solution of sulfuric acid. The pH limits of the development of various types of bacteria are not fully established and are the subject of research.

2. Materials and methods

For the preparation of samples, Portland cement CEMI 42.5H manufactured by PLC Belgorodsky Cement and standard monofractional quartz sand of the Volsky deposit were used. The ratio of quartz sand to cement was taken 3:1 in accordance with Russian Standard GOST 30744-2001.

To study concrete, beam samples 4 × 4 × 16 cm in size were formed. The samples were placed in closed desiccators. Beyerink solution was used as a nutrient medium (Table 2).

Table 2. The composition of the Beyerink solution, g / l.

$Na_2S_2O_3 \cdot 5H_2O$	NH_4Cl	$NaHCO_3$	$Na_2HPO_4 \cdot 12H_2O$	$MgCl_2 \cdot 6H_2O$	$FeSO_4 \cdot 7H_2O$	pH
5	0.1	1	0.2	0.1	traces	8

In order to intensify microbiological corrosion of the samples, the associations of *Thiobacillus thioparus* and *Acidithiobacillus thiooxidans* cultures were introduced into the desiccator; the number of bacteria per unit volume was 10^6 cl/ml. As control compositions were samples that did not contain bacteria. In a blank experiment, the behavior of bacteria in the absence of concrete samples was evaluated.

Incubation was carried out in a thermostat at a temperature of 30 °C. The content of calcium, magnesium, sulfates in solution and samples was determined according to Russian Standard GOST 5382-91. pH was measured using a pH-150M ionomer. The microstructure of concrete was studied using a high-resolution scanning electron microscope TESCANMIRA 3 LMU.

3. Results and discussion

Evaluation of the leaching rate of mineral constituent and hydrated phases of concrete is important to establish timeframe when the strength loss occurs during exploitation. Number of microorganisms can live on surface of without revealing any vital activity, but capable of rapid growth and destroying building materials under appropriate conditions. Therefore, the understanding of vital activity of the microorganisms is crucial to address durability issues of building materials.

In Figure 1, thionic bacteria in a blank experiment cause acidification of the medium, starting after 8–10 days and reaching pH = 2–3 after 20 days of the experiment. A stable pH value of 8–9 for concrete was observed in the control experiment. In the presence of microorganisms, after a period of their adaptation (16–18 days), the pH of the medium drops sharply and on the 24th day, pH = 3.

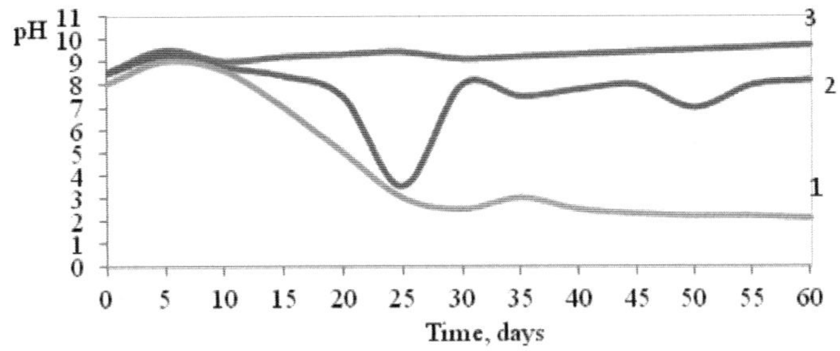

Figure 1. Measurement of pH in an aqueous medium containing: 1 - thionic bacteria without samples (blank experiment); 2 - concrete with bacteria; 3 - sample of concrete without bacteria.

The influence of an alkaline environment formed in the presence of concrete promotes the active development of *Thiobacilus thioparus*, further bacteria of the genus *Acidithiobacillus thiooxidans* appear with acidification of the medium, and both species actively synthesize sulfuric acid.

Sulfuric acid of bacterial origin interacts with the components of concrete - unreacted oxide of calcium, magnesium, aluminum with the formation of medium and acid sulfates; calcium carbonate, sulfate, and orthosilicate to form hydrocarbonates, hydrosulfates, and hydroorthosilicates.

These reactions cause alkalization and there are jumps in increasing the pH to 8.4. In the future, the general direction of the process remains, although some emissions on the kinetic curve and the appearance of sulfur flakes in the solution on the 30th day and their subsequent disappearance on the 48th day are associated with the oxidation of thiosulfate ions $S_2O_3^{2-}$ and the subsequent process of the formation of sulfuric acid by the reaction Eq. (3) [13]:

$$S + H_2O + 3/2O_2 \xrightarrow{bacteria} H_2SO_4 \quad (3)$$

Leaching of calcium and magnesium ions from the samples in the absence of bacteria does not occur (Fig. 2).

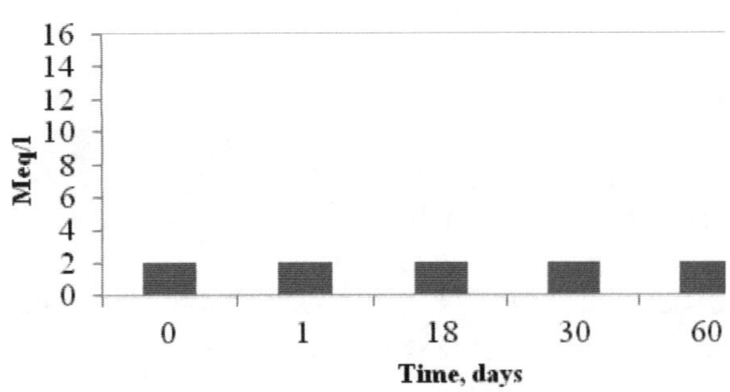

Figure 2. Leaching of ions in an aqueous medium with concrete.

IOP Conf. Series: Materials Science and Engineering **945** (2020) 012008 doi:10.1088/1757-899X/945/1/012008

However, in the presence of microorganisms with the completion of their adaptation (18 days), the release of Mg^{2+} ions into the solution begins, reaching up to 14 meq / l to 60-th day (Fig. 3).

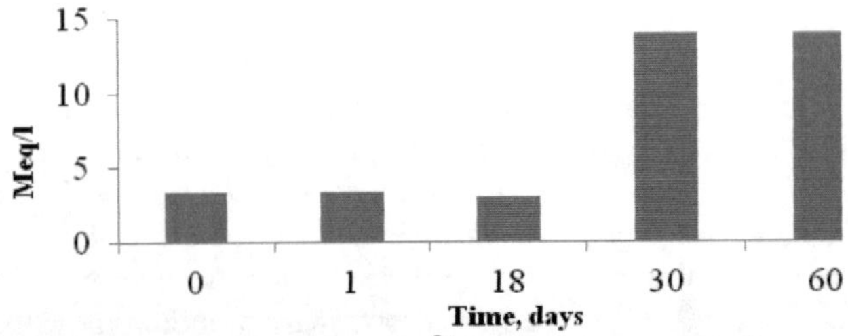

Figure 3. The content of magnesium ions Mg^{2+} in an aqueous medium containing thionic bacteria and concrete.

In the future, hydrolysis probably occurs with the formation of magnesium salts in the aquatic environment. These reactions in concrete in the presence of alkalis are undesirable, since they lead to deformation of concrete [14].

Leaching of calcium salts as a result of bacterial activity is significantly less pronounced. The concentration of Ca^{2+} ions up to 13 meq / l is noted to 60-th day.

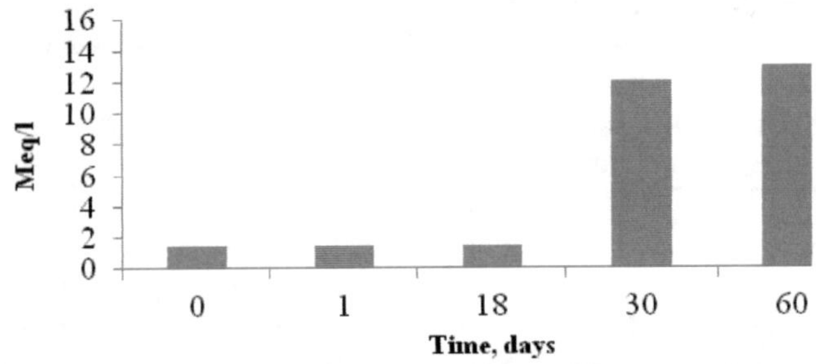

Figure 4. The content of calcium ions in an aqueous medium containing thionic bacteria and concrete.

The increment in the concentration of sulfate ions in the initial solution in the presence of bacteria and concrete from the very beginning of the experiment was caused by the period of bacterial adaptation. With the further formation of calcium, magnesium, and aluminum sulfates and hydrosulfates, the sulfate ions consumed for these reactions from the surrounding solution decrease, which is noted in the figure when the concentration decreases from 3.3 to 2.4 meq/l (Fig. 5).

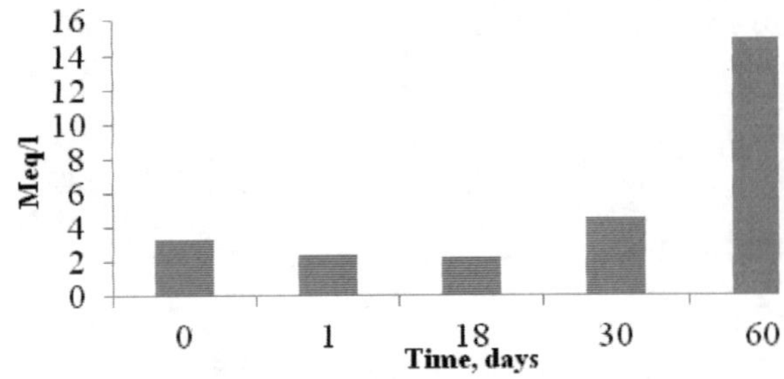

Figure 5. The content of sulfate ions in an aqueous medium containing thionic bacteria and concrete.

In a further experiment, starting from 18-th day, there is a dynamics of an increase in sulfate concentration, which is associated with the development of bacteria and the synthesis of sulfuric acid.

After 60 days of conducting the experiment, a visual analysis of the surface layers of concrete showed a white coating up to 1 mm thick after exposure to bio corrosion. Ingrown crystals and white salt deposits were visible on the surface (Fig. 6 a).

a

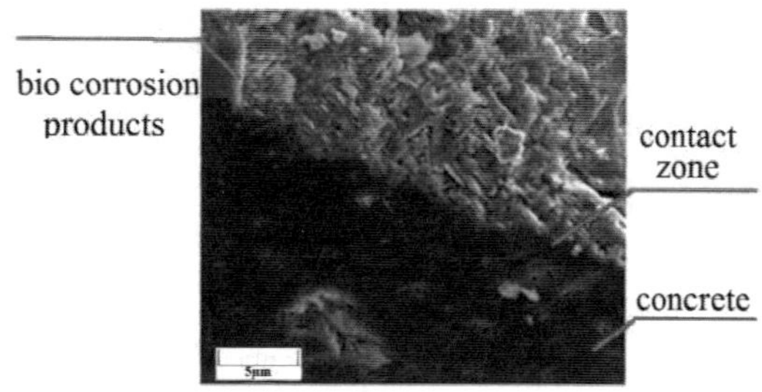

b

Figure 6. Concrete samples: a – white coating - bio corrosion products; b – SEM analysis of concrete surface after bio corrosion.

SEM analysis made it possible to establish the morphology of crystals on the surface of the sample with a predominance of tabular, columnar, prismatic neoplasms characteristic of sulfates (Fig. 6 b). The bacteria's production of sulfuric acid leads to the dissolution of the cement matrix and the formation of sulfate compounds in the form of gypsum and ettringite. The growth of these crystals causes cracking in concrete and its destruction. In addition, one can observe crystals of calcium carbonate, which have a pronounced lamellar morphology.

Corrosion resistance was evaluated by the coefficient of resistance, which is the ratio of the strength of the test sample to the control one, which was not influenced by bacteria during the experiment. The corrosion resistance after two months of the experiment was decreased as indicated by 30 % strength reduction.

4. Summary

The simultaneous effect of two types of microorganisms in model systems simulating microbiocorosion conditions allows reducing the test time and evaluating the expected course of destruction and leaching of the main components of concrete. Effect of the bacterial acid *Thiobacillus thioparus* and *Acidithiobacillus thiooxidans* increases the solubility of the concrete surface, making it vulnerable to mass transfer and structural degradation. The data obtained can be used to model the processes of corrosion of concrete caused by the action of sulfuric acid of biological origin.

5. References

[1] Stanaszek-Tomal E, Fiertak M 2016 Biological Corrosion in The Sewage System and The Sewage Treatment Plant *Procedia Engineering* **161** 116–120

[2] Goncharova E N, Chaplin Yu V, Krut V V 2004 Protection of building materials from bio-corrosion *Bulletin of BSTU named after V.G. Shukhov* **8** 48

[3] Tokach Y E, Goncharova E N 2016 Creating Bioresistant Technogenic Waste Basted Coatings for Construction Materials *Procedia Engineering* **150** 1547–1552

[4] Munyao O M,Thiong'o J K, Muthengia J W, Mutitu D K et al 2020 Study on the effect of *Thiobacillus* intermedius bacteria on the physico-mechanical properties of mortars of ordinary portland cement *Heliyon* **6(1)** e03232

[5] Wu M, Wang T, Wu K, Kan L 2020 Microbiologically induced corrosion of concrete in sewer structures: A review of the mechanisms and phenomena *Construction and Building Materials* 239

[6] Bertron A 2014 Understanding interactions between cementitious materials and microorganisms: a key to sustainable and safe concrete structures in various contexts *Materials and Structures* **47** 1787–1806
[7] Kozhukhova N I, Lebedev M S, Vasilenko M I, Goncharova E N 2016 Ecology-toxicology study of low-calcium solid wastes from power plants *International Journal of Pharmacy and Technology* **8(3)** 15349–15360
[8] Kozhukhova N I, Lebedev M S, Vasilenko M I, Goncharova E N 2018 Fly Ash Impact from Thermal Power Stations on the Environment *Journal of Physics: Conference Series* **1066(1)** 012010
[9] Kozhukhova N I, Lebedev M S, Vasilenko M I, Goncharova E N 2019 Toxic effect of fly ash on biological environment *IOP Conference Series: Earth and Environmental Science* **272** 022065
[10] Fomina E V, Kozhukhova N I 2014 Phyto-testing of Korkino coal field based composite binders *Bulletin of BSTU named after V.G. Shukhov* **3** 54–57
[11] Sverguzova S V, Shaykhiev I G, Tom Otiti, Sapronova Zh A 2018 Increasing the strength and frost resistance of ceramic products at using melasses bards as plasticizer *Construction Materials and Products* **1(2)** 19–29
[12] Vasilenko M I, Goncharova E N 2013 Microbiological features of process of damage of concrete surfaces *Basic researches* **4(4)** 886–91
[13] Jensen H S, Lens P N L, Nielsen J L 2011 Growth kinetics of hydrogen sulfide oxidizing bacteria in corroded concrete from sewers *Journal of hazardous materials* **189(3)** 685–91
[14] Yousefi A, Allahverdi A, Hejazi P 2014 Accelerated biodegradation of cured cement paste by *Thiobacillus* species under simulation condition *International Biodeterioration & Biodegradation* **86** 317–326

Acknowledgments
The reported study was funded by RFBR according to the research project № 18-03-00352. The work is realized in the framework of the Program of flagship university development on the base of the Belgorod State Technological University named after V.G. Shukhov, using equipment of High Technology Center at BSTU named after V.G. Shukhov.

IOP Conf. Series: Materials Science and Engineering **945** (2020) 012009 doi:10.1088/1757-899X/945/1/012009

Accounting policy of construction companies in accordance with International Financial Reporting Standards

G G Derzayeva[1]

[1]Kazan (Volga) Federal University, 18, Kremlevskaya str., KAZAN, 420008, Republic of Tatarstan, Russian Federation

E-mail: guzelchan@mail.ru

Abstract. This paper examines the practical experience of construction companies in terms of their disclosure of information about significant elements of accounting policy in the notes to their financial statements prepared in accordance with International Financial Reporting Standards. The aim of the study is to identify the main problematic issues that arise in construction companies when preparing accounting policies in accordance with International Financial Reporting Standards and find ways to solve them. Using such methods as analysis and synthesis, comparison, logical and systematic approaches, the author identifies the main problems that occur in construction companies when drafting accounting policy, as well as offers solutions to them. As a result of the research, the paper compares the practice of forming accounting policy by leading construction companies, identifies the main problematic aspects and suggests ways to solve them. The article concludes on the uniformity of approaches to disclosure of information about significant elements of accounting policy, proves the identity of most of the problematic aspects that arise in the formation of accounting policy in accordance with International Financial Reporting Standards by construction companies, and substantiates the analogy of ways to solve them.

1. Introduction

Issues related to the formation of accounting policy for the preparation and presentation of financial statements by construction companies are regulated at the international level in International Financial Reporting Standards (hereinafter referred to as IFRS), at the Russian level - by the Ministry of Finance of Russia, which develops Federal Accounting and Reporting Standards for organizations based on IFRS. Currently, the procedure for standardization of accounting (financial) statements of organizations in Russia is expressed in the implementation of a number of practical steps:

- the Russian Ministry of Finance has published a translation of the document "Conceptual framework for the presentation of financial statements", which means its implementation in Russia;
- the Ministry of Finance of the Russian Federation has adopted regulations that introduce IFRS and IAS in Russia;
- the Ministry of Finance of the Russian Federation has adopted regulations that give effect to IFRS clarifications in Russia - IFRIC Clarifications and SIC Clarifications;
- Official Letters of the Ministry of Finance of the Russian Federation have been published that answer practical questions that arise when organizations prepare and submit IFRS reports;
- Federal Accounting Standards have been adopted;

- a number of draft Federal Accounting Standards are under public discussion;
- research has already been conducted on practical and theoretical aspects of the formation of financial statements by construction companies under IFRS and the results of these studies done by such authors as E.S. Druzhilovskaya [1], L.I. Kulikova [2], A.Yu. Sokolov [3], A.V. Ivanovskaya [4], M.R. Safiullin [5], G.G. Derzaeva [6, 7], O.V. Dedova [8], E.N. Zaporozhtseva [9], T.N. Shilova [10] and others have been published [11, 12].

According to IFRS, a reporting company is a company that is responsible for (or voluntarily decides to) submit general-purpose financial statements, which in turn include regulated general-purpose financial statements that disclose the assets, liabilities, equity, income and expenses of that company. In addition, a general purpose financial report is a report that provides financial data about the reporting company's economic capabilities, the company's obligations, and changes in economic capabilities and obligations that users will need to make decisions about providing resources to that company.

2. Materials and methods

The subject area of this research is the largest Russian construction companies. Let us consider the general purpose financial statements published in accordance with International Financial Reporting Standards as of December 31, 2019, by six major Russian construction companies:

Construction company PJSC "LSR Group" - hereinafter LSR [13].
Construction company PJSC "PIK Group" - hereinafter PIK[14].
Construction company JSC "Soyuzdorstroy" - hereinafter SDS [15].
Construction company PJSC "IDCC of the North Caucasus" - hereinafter IDCC [16].
Construction company PJSC "Setl Group of Companies" - hereinafter Setl [17].
Construction company JSC "RBI Group" - hereinafter RBI [18].

When preparing financial statements in accordance with IFRS, significant attention should be paid to the disclosure of information in the notes to the financial statements. Information that is reflected in the notes to the financial statements should include the following elements: the conditions of the construction company, significant accounting policies, financial risks that may have a significant impact on the company's activities, and so on. The most important information that is disclosed in the notes to the financial statements of construction companies is information about the significant accounting policies of the organization, as it is necessary for users to evaluate the company's work and understand the main indicators of reporting.

Russian construction companies use IFRS (IAS) 8 "Accounting policies, changes in accounting estimates and errors" to form accounting policies for the purpose of preparing and presenting financial statements in accordance with IFRS. In IFRS 8, accounting policies are defined as the specific methods, assumptions, bases, methods and rules used by a company to prepare and present financial statements.

In table 1, we will consider the content of the accounting policy sections reflected by the above-mentioned construction companies in the financial statements, and how each of the Russian construction companies under consideration reflects information about accounting objects in the notes to the statements prepared under IFRS in terms of disclosure of information about accounting policies.

Table 1. Information about the disclosure of accounting policies in the notes to the financial statements made according to IFRS by Russian construction companies.

Content of accounting policy sections	LSR	PIK	Setl	IDCC	SDS	RBI
1. Principles of consolidation	+	+	+	+	+	+
2. Foreign currency	+	+	+	+	+	+
3. Financial tools	+	+	+	+	+	+
4. Authorized capital	+	+		+	+	+
5. Fixed assets	+	+	+	+	+	+
6. Intangible assets	+	+	+	+	+	+

7. Advance payments given		+		+		
8. Value added tax				+	+	
9. Leased assets	+	+	+			
10. Investment property	+	+		+	+	+
11. Inventory	+	+	+	+	+	+
12. Assets and liabilities under contracts	+	+				+
13. Impairment of assets	+	+	+	+	+	+
14. Non-current assets held for sale	+					+
15. Employee benefits	+	+	+	+	+	+
16. Reserves	+	+	+	+	+	+
17. Revenue	+	+	+	+	+	+
18. Government subsidies		+		+		
19. Social payments		+		+		
20. Other expenses	+	+	+			
21. Finance income and expenses	+	+	+	+	+	+
22. Income tax expense	+	+	+	+	+	+
23. Profit per share	+			+		
24. Segment reporting	+	+	+			
25. New standards and interpretations not adopted yet	+		+		+	+

The study of disclosure in the notes to the financial statements on significant accounting policies made in table 1 allows making a conclusion about the objects that are included in the Russian construction companies in the accounting policies based on IFRS, as well as significant identity of the majority of approaches to formation of accounting policy according to IFRS. The study also shows that the analyzed Russian construction companies do not apply the new accounting standards that have already been approved by the IASB, but their use for the preparation of financial statements has not been required yet.

The research conducted in table 1 on the disclosure of information in the notes to the financial statements about significant elements of accounting policies suggests that Russian construction companies do not seek to include additional totals in the reporting, for example, such as EBITDA. Although IFRS does not require the inclusion of such indicators, it does not prohibit them. It seems that their reflection in the reporting would increase the usefulness of reporting information for users in making management decisions.

3. Results

The study in table 1 of the disclosure of information in the notes to the financial statements about significant elements of accounting policies suggests that not all of the information disclosed in the financial statements of construction companies relates to the data that should be disclosed in the accounting policy. A significant part of the data provided is redundant. For example, this "extra" data includes the inclusion of inventory valuation rules in the accounting policy. A description of the inventory valuation rules is neither required nor necessary, as all of these rules are not accounting options developed or selected by the company itself, but regulations of the corresponding IAS 2 "Inventory". For the same reason, according to the study conducted in table 1, it is "unnecessary" to disclose in the notes to the financial statements presented under IFRS, in terms of significant accounting policies, such information as:

- characteristics of the equity method that accounts for investments in associates and the definition of associates (as such data duplicates word for word the provisions of IAS 28 "Investments in associates and joint ventures");

IOP Conf. Series: Materials Science and Engineering **945** (2020) 012009 doi:10.1088/1757-899X/945/1/012009

- definition of related parties (as this definition duplicates the rules of IAS 24 "Disclosure of information about related parties" and IFRS 12 "Disclosure of information about participation in other organizations");
- determining the functional currency (as this information copies the provisions of IAS 21 "Effect of changes in exchange rates");
- the composition and general rules for the assessment and offsetting of financial instruments (as they duplicate the requirements of IAS 39 "Financial instruments: recognition and assessment" and IFRS 9 "Financial tools");
- defining financial and operating leases and listing the terms of their recognition (as this information duplicates the provisions of IFRS 16 "Leases");
- determination of investment property, terms of initial recognition (as such information copies the provisions of IAS 40 "Investment property");
- description of the accounting rules for impairment and the order of de-recognition of fixed assets and intangible assets (as these rules word-for-word repeat of the requirements in IAS 36 "Impairment of assets", and also the aforementioned IAS 16 "Fixed assets" and IAS 38 "Intangible assets");
- definition of non-current assets and discontinued operations, and the conditions for their recognition (as this information repeats the provisions of IFRS 5 "Non-current assets held for sale and discontinued operations");
- description of defined contribution and defined benefit programs, the difference between long-term and short-term employee benefits (as this information repeats the provisions of IAS 19 "Employee Benefits" and IAS 26 "Accounting and reporting on pension programs");
- description of the conditions for creating each of the reserves (as this information is similar to the provisions of IAS 37 "Estimated liabilities, contingent liabilities and contingent assets");
- general rules for accounting the revenue (as this information repeats the relevant provisions of IFRS 15 "Revenue from contracts with customers" and IAS 18 "Revenue");
- principles and conditions for the recognition of government grants (as this information duplicates IAS 20 "Accounting for government grants and disclosure of government assistance");
- description of the calculation of current and deferred income tax, deferred tax assets and liabilities (as this information copies the provisions of IAS 12 "Income Taxes");
- definitions of common and preferred shares, formulas for calculating diluted and basic earnings per share (as this information repeats the provisions of IFRS 2 "Share-based Payments" and IAS 33 "Profit per share");
- definitions of the operating segment, conditions for recognizing the segment as an operating one (as this information is similar to the provisions of IFRS 8 "Operating segments").

Let us summarize the above results of the study of information disclosure in the notes to the financial statements about significant elements of accounting policies by construction companies in table 2.

Table 2. Summary of the results of the analysis of accounting policy data of construction companies that are not subject to disclosure in financial statements prepared in accordance with IFRS.

Accounting policies	Duplicated standards
Definition and description of the evaluation rules	IAS 2 "Inventory"
Characteristics of the equity method, definition of associates	IAS 28 "Investments in associates and joint ventures"
Definition of a related party	IFRS 12 "Disclosure of information about participation in other organizations"
Defining the functional currency	IAS 21 "Effect of changes in exchange rates"

Composition and general rules for evaluating and offsetting financial instruments	IFRS 9 “Financial tools”
Definition of financial and operating leases, listing the conditions for their recognition	IFRS 16 “Leases”
Definition of investment property, conditions for initial recognition	IAS 40 “Investment property”
Description of the rules for accounting for impairment and the procedure for de-recognition of property, plant and equipment and intangible assets	IAS 36 “Impairment of assets” IAS 16 “Fixed assets” IAS 38 “Intangible assets”
Definition of non-current assets and discontinued operations, conditions for their recognition	IFRS 5 “Non-current assets held for sale and discontinued operations”
Definition of defined contribution and defined benefit programs, distinction between long-term and short-term employee benefits	IAS 19 “Employee benefits” IAS 26 “Accounting and reporting on pension programs”
Description of conditions for creating reserves	IAS 37 “Estimated liabilities, contingent liabilities and contingent assets”
Description of revenue accounting rules	IFRS 15 “Revenue from contracts with customers” IAS 18 “Revenue”
Description of the principles and conditions for recognizing government grants	IAS 20 “Accounting for government grants and disclosure of government assistance”
Description of the calculation of current and deferred income tax, deferred tax assets and liabilities	IAS 12 “Income Taxes”
Definitions of common and preferred shares, formulas for calculating diluted and basic earnings per share	IFRS 2 “Share-based Payments” IAS 33 “Profit per share”
Definition of an operating segment, conditions for recognizing a segment as an operating one	IFRS 8 “Operating segments”

4. Summary

It seems that the inclusion of data duplicating the regulations of the IFRS standards leads to “cumbersome” accounting policies in the accounting policy, that is, their overload, redundancy and inconvenience for use. However, provided that this data is not information about the accounting options selected by the organization from the accounting options allowed by the IFRS standards or the accounting methods developed by the organization independently (if they are not included in these standards). Disclosure of unnecessary information in the accounting policy makes it very difficult to find really significant and necessary information about the accounting methods selected and independently developed by the organization.

So, construction companies do not need to disclose extra data in their accounting policies that repeat the requirements of accounting standards.

Thus, the conducted study allowed comparing the significant accounting policies disclosed in the notes to the financial statements prepared under IFRS by the six largest Russian companies, to identify the identity of many approaches to formation of accounting policy according to IFRS for Russian construction companies, to identify the main problematic aspects and to suggest possible solutions to improve the quality of the information disclosed in the notes to the financial statements prepared under IFRS.

IOP Conf. Series: Materials Science and Engineering **945** (2020) 012009 doi:10.1088/1757-899X/945/1/012009

5. References

[1] Druzhilovskaya E S 2016 Practice of forming accounting policies according to IFRS OS by Russian organizations (correlation with accounting policies compiled according to other standards by Russian and foreign organizations) *Accounting in budget and non-profit organizations* **20** 2–8

[2] Kulikova L I, Gubaidullina A R 2016 Information disclosure influence about financial statement risk on strategic managerial decision *Academy of Strategic Management Journal* **15** 131-138

[3] Gafurov I, Kulikova L, Sokolov A, Shaykhutdinova K, Negreeva V 2019 Depreciation in the aspect of matching revenues and expenses of the company *E3S Web of Conferences* **110** 02046-02050

[4] Kulikova L I, Akhmedzyanova F N, Ivanovskaya A V 2016 Ways of assets value misstatement that companies use when making financial statements *International Business Management* **10** 5705-5709

[5] Safiullin M R, Derzayeva G G, Elshin L A 2013 About assessment of budgetary policy efficiency of municipalities *World Applied Sciences Journal* **27** 299-304

[6] Derzayeva G G, Akhmadieva G G 2014 Budgetary policy efficiency of municipalities in the field of education *ERPA INTERNATIONAL CONGRESS ON EDUCATION (ERPA CONGRESS 2014)* **152** 1148–1153

[7] Derzayeva G G 2019 Pricing at the cost of production for budget institutions *Eurasian legal journal* **11** 391–394

[8] Dedova O V 2019 Tax reporting of the organization *Russian Economic Bulletin* **2 (3)** 61-66

[9] Zaporozhtseva E N, Stitsenko N N 2020 Simplified accounting system and preparation of accounting reporting for small and medium enterprises *Russian Economic Bulletin* **3 (2)** 234-238

[10] Shilova T N, Ostapchuk T V 2019 Peculiarities of accounting of accounts receivable debt in the implementation calculations with deferred payment *Modern Economy Success* **4** 104-108

[11] Klyuev S V, Klyuev A V, Shorstova E S 2019 The micro silicon additive effects on the fine-grassed concrete properties for 3-d additive technologies *Materials Science Forum* **974** 131–135 doi:10.4028/www.scientific.net/MSF.974.131

[12] Klyuev S V, Klyuev A V, Shorstova E S 2019 Fiber concrete for 3-d additive technologies *Materials Science Forum* **974** 367–372 DOI: 10.4028/www.scientific.net/MSF.974.367

[13] Financial statements prepared in accordance with International Financial Reporting Standards of PJSC LSR Group as at 31 December 2019 URL: https://www.lsrgroup.ru/investors-and-shareholders/disclosure-of-information

[14] Financial statements prepared in accordance with International Financial Reporting Standards of PJSC PIK Group PIK as at 31 December 2019: https://www.pik.ru/about/news-and-reports/financial-results

[15] Financial statements prepared in accordance with International Financial Reporting Standards of JSC Soyuzdorstroy as at 31 December 2019 URL: http://www.avtoban.ru/investors/disclosure/

[16] Financial statements prepared in accordance with International Financial Reporting Standards of PJSC IDCC of the North Caucasus as at 31 December 2019 URL:https://www.rossetisk.ru/shareholders_and_ investors /raskrytie informatsii/mezhdunarodnaya_otchetnost/

IOP Conf. Series: Materials Science and Engineering **945** (2020) 012009 doi:10.1088/1757-899X/945/1/012009

[17] Financial statements prepared in accordance with International Financial Reporting Standards of PJSC Setl Group of Companies as at 31 December 2019 URL: https://www.setlcity.ru/investors
[18] Financial statements prepared in accordance with International Financial Reporting Standards of JSC RBI Group as at 31 December 2019 URL: https://group.rbi.ru/about/financial-statements/

IOP Conf. Series: Materials Science and Engineering **945** (2020) 012010 doi:10.1088/1757-899X/945/1/012010

Soundproofing materials in construction using polymer composites

L V Denisova[1*], N V Klyuchnikova[1] and S V Emelyanov[2]

[1]Department of heat and gas supply and ventilation, Belgorod State Technological University named after V.G. Shukhov, Kostyukov St., 46, Belgorod, 308012, Russia
[2]Department of Physics and Chemistry of Materials named after Dogadkin B.A., MIREA - Russian Technological University

E-mail: loveden13@mail.ru

Abstract. Today, obtaining products for recycling of polymer materials is of great interest, including for the construction industry. At the same time, not only new materials are created with the specified performance characteristics, but also the environmental problem is solved, nature is preserved and natural resources are saved. In this paper, it is proposed to use a fabric cord of used tires as a component of soundproofing material. Various types of latex were used as a binder: acrylic, natural, and styrene-butadiene. It was found that all the obtained composites have soundproofing properties. The best soundproofing properties are found in a material containing styrene-butadiene latex. Thus, it is shown that materials based on latex of various nature and textile products of tire processing can be used for soundproofing materials in the construction industry.

1. Introduction

Currently, polymer materials are increasingly being introduced into various industries. One of the main consumers of polymer composite materials is the construction industry [1]. Such an active introduction of polymer materials into the construction industry is due to the fact that by introducing various fillers and modifiers into the polymer matrix, it is possible to give composites the required performance properties, varying each time in accordance with narrowly directed requirements [3-4].

Today, obtaining products for recycling of polymer materials is of great interest, including for the construction industry. At the same time, we not only create new materials with the specified performance characteristics, but also solve the environmental problem. Today, the issue of recycling car tires is particularly acute [5]. For example, every year an incredibly huge amount of used tires is sent to landfills. In Europe, about 9 million tons of tires are sent for write-off every year, and in the United States for 2019, the number of worn tires exceeded 3.9 million tons. And if rubber tires are processed, for example, into rubber crumb, then the fabric cord, which is carefully and not without problems separated from rubber and metal, does not have a large-scale application, which to a greater extent worsens the economic situation, in terms of profitability. It is necessary to search for possibilities of using fabric cord [6].

That is why the recycling of tires for use in the construction industry is of interest, in particular, for the production of soundproofing screens [7].

2. Materials and methods

2.1. Materials

For making compositions, components such as binders, fillers, and plasticizers are necessary. Natural, styrene-butadiene and acrylic latex, bituminous emulsion, shungite filler, C3 plasticizers and neonol were used to create a soundproofing material.

Natural latex is obtained by coagulation from the milky juice of Brazilian hevea with the addition of acetic or formic acids. The resulting gel is washed with water and sheets are rolled on rollers, then they are dried [8].

Acrylic latexes include pure acrylic and acrylic-styrene copolymers. The final characteristics of latex can be adjusted during production due to the excellent copolymerization of acrylic and methacrylic esters.

The range of applications of such latexes is very wide, ranging from soft and pressure-sensitive paints or other coatings, ending with the reinforcement of non-woven materials.

Crosslinking is possible under various conditions: at high temperatures, using UV rays, or other options. Products using acrylic dispersions can withstand harsh weather conditions, temperature changes, etc. Dispersions can have different types: paste, foam, spray solution [9].

Styrene-butadiene latexes are general-purpose latexes that replace natural latex in the tire industry, for example, for impregnating the cord [10].

Styrene-butadiene latexes are resistant to storage, dilution, and mechanical stress, this is due to the high content of emulsifier and stabilizer and the small particle size of such latexes. High viscosity is determined by the small size of the particles and their large number.

Crystals of crushed, finely ground shungite have a high level of adhesion, as they have well-defined bipolar properties and are soluble in almost all organic and inorganic substances. In addition, shungite has a wide range of bactericidal properties. Its properties are determined by the nanostructure and composition of its constituent elements: shungite carbon is evenly distributed in a silicate framework of fine quartz crystals, which is confirmed by studies of ultrathin shungite sections by transmission electron microscopy and scanning electron microscopy in absorbed and backscattered electrons [11]

Fast-, medium - and slow-disintegrating emulsions are obtained by adjusting the concentration of the emulsifier. In order to obtain a superstable emulsion, a large amount of emulsifier must be added. Bituminous emulsions, in contrast to bitumen, have a lower viscosity at indoor temperature, which promotes penetration into smaller pores, and they have a high adhesive ability to surfaces of different structures [12].

Plasticizers must be well compatible with the polymer, have low volatility, be resistant to extraction from the polymer by liquid media, be chemically inert, and have no smell [13]. Plasticizer C3 has high plasticizing properties and an effective water-reducing effect, also increases the strength, water resistance, reduces the time and intensity of vibration, and reduces the temperature of isothermal heating [14]. Neonol is a white oily liquid. It has a melting point of 1°C. It is soluble in water, methanol, xylene, and insoluble in kerosene. The specific viscosity is 240 units at a temperature of 25°C. The specific percentage of water content is 0.5%. The product flash point is 94°C.

2.2. Methods

As the developed composite material will be used in the construction industry as a soundproofing material, it is necessary that this composite has good soundproofing properties. In addition, when creating a composite, a high-quality impregnation of fabric cord and shungite with elastomer emulsions plays an important role. The wetting edge angle was measured using the LC-1 goniometer, which measures the edge angle using the lying drop method. The angle was determined based on the description of the shape of the drop contour using the Laplace equation and the tangent method. Today, there are technologies that allow recording the image of a drop and, with the help of programs, get all the necessary data. This is the modern goniometer used for measuring the edge angle [15]. The soundproofing properties of the material are equally important. To measure the noise level by an objective method, a noise meter device was used, in which noise is perceived using a broadband microphone that converts sound vibrations into electrical ones.

Conducting all the above tests of the resulting composite material is a prerequisite for analyzing the performance properties of PCM, which will be used in the construction industry as a finishing soundproofing material.

3. Results

Three samples were prepared for research.

Sample №1

To prepare the first sample acrylic latex, cord fiber, shungite and neonol were taken.

At first, acrylic latex was mixed with neonol and shungite. Then, cord fiber was impregnated with the resulting mixture. The resulting material was dried.

Sample №2

For the preparation of the second sample natural latex, cord fiber, crumb rubber, shungite and bitumen emulsion were taken.

Initially, bituminous emulsion, shungite and natural latex were mixed. After that cord fiber was impregnated with the mixture. Then the impregnated fiber was mixed with a rubber crumb on the rollers. The resulting material was dried.

Sample №3

For the preparation of the third sample styrene-butadiene latex, cord fiber, shungite, bitumen emulsion and plasticizer-C3 were taken.

Initially, styrene-butadiene rubber, bituminous emulsion, shungite and plasticizer were mixed. Then cord fiber was impregnated with the mixture. The resulting material was dried.

In the course of research, it was necessary to find out the value of the edge angle of wetting latex mixed with shungite. Shungite is fine; the degree of grinding is 0.05 microns. To determine the required amount of shungite, a calculation was performed, which showed that it is necessary to enter no more than 25 mass parts of shungite into latex in relation to the content of rubber in this latex. For the transition from volume to mass, we used the density of latex, which depends on its concentration. The density of the rubber phase is 914 kg/m^3, the dispersion medium – sulfur – 1020 kg/m^3. Based on this, the density of 1000 kg/m was conditionally accepted. The volume of latex is the same – 10 ml.

After calculating the required amount of shungite, mixtures were prepared and the edge angle was measured (table 1).

Table 1. The wetting angle of viscose fabric with a mixture of latex and shungite.

Experience number	Type of mixture		
	Natural latex + shungite	Styrene-butadiene + shungite	Acrylic latex + shungite
1	128.702	124.018	161.050
2	112.034	115.062	154.621
3	127.213	122.765	152.325
4	115.342	121.622	149.023
5	126.669	111.284	157.902
6	124.645	111.940	149.303
7	126.948	129.822	148.337
8	122.341	111.093	150.740
9	126.732	128.782	151.049
10	122.264	126.806	140.210
Average value	121.289	120.319	151.456

The results of table 1 show that the edge angle is smaller in a mixture of styrene-butadiene latex with shungite. This is explained by the fact that shungite particles have a diphilic nature and have a negative charge, which allows good distribution in the volume of latex, but the negative charge prevents good wettability with viscose. We also see that the values of the edge angle of pure latex are lower than the values of the edge angle of mixtures. The decrease in wettability is explained by an increase in viscosity.

As the wetting angle is the lowest for a mixture of styrene-butadiene latex with shungite, further studies were conducted with this mixture.

Next, it was necessary to determine the optimal amount of shungite, in mass fractions, which should be added to improve the properties of the material, but, at the same time, avoid a large decrease in wettability and excessive viscosity (figure 1).

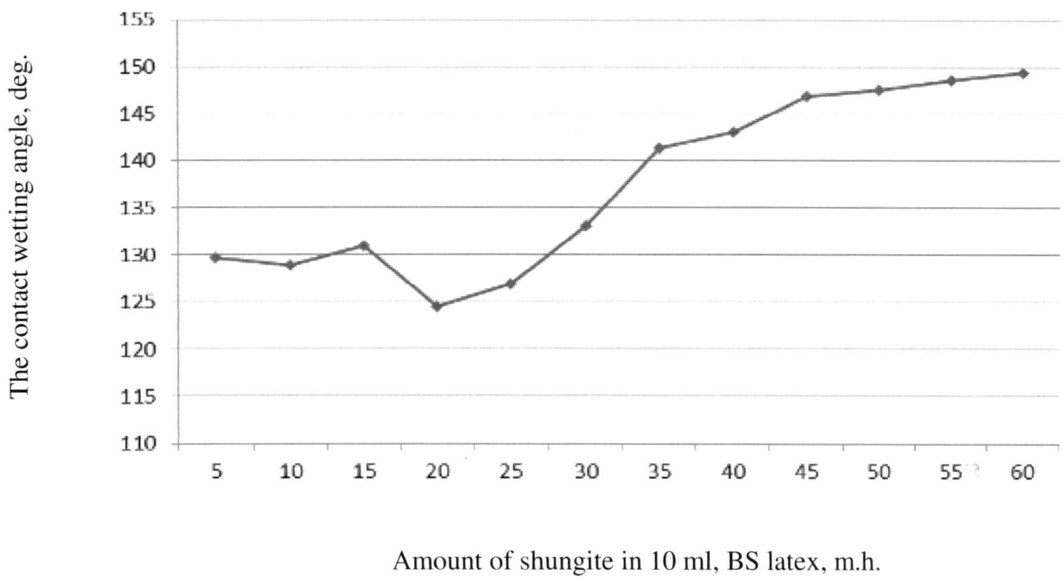

Figure 1. The dependence of the wetting angle on the amount of schungite introduced into the mixture.

From these data, we see that with increasing number of injected shungite the value of the contact wettability angle grows, which is associated with an increase in the difference of charges when the number of shungite grows and an increase in viscosity of the mixture.

However, the obtained values show that the wettability of the latex and shungite mixture is too low. As in tire factories automobile fabric cord is necessarily impregnated with a compound that increases the adhesion to the rubber, cord regenerate contains this composition. Therefore, we impregnated the viscose fabric with styrene-butadiene latex. Butadiene-styrene latex, is a part of the impregnation composition, and can be used in our experiment for a closer assessment of the wetting of the cord regenerate. After impregnation, the edge angle was measured (table 2).

Table 2. The wetting angle of the impregnated viscose fabric with latex.

Experience number	Type of mixture	
	BS latex	BS latex + shungite
1	60.731	69.209
2	47,747	51.236
3	50.234	60.527
4	50.865	56.197

Experience number	Type of mixture	
	BS latex	BS latex + shungite
5	46.781	69.818
6	54.121	58.890
7	57.977	57.332
8	47.131	61.130
9	48.029	49.674
10	47.105	50.870
Average value	51.072	58.488

Based on the obtained results, we can see that the values of the edge angle decreased almost twice, compared to the values when measuring non-impregnated fabric, and for pure latex, and for the mixture, which is associated with an increase in affinity.

The use of an objective noise meter allowed determining approximate values of noise volume levels, which is due to the limited frequency response characteristics. When measuring constant or stationary noise, noise levels were measured for 10 minutes. During this time, several readings of the device arrow were taken. Then the maximum and minimum values were selected from the found values, which were used to calculate the average noise level.

In the course of conducting a research of the acoustic properties of the manufactured material, the following results were obtained, shown in figure 2.

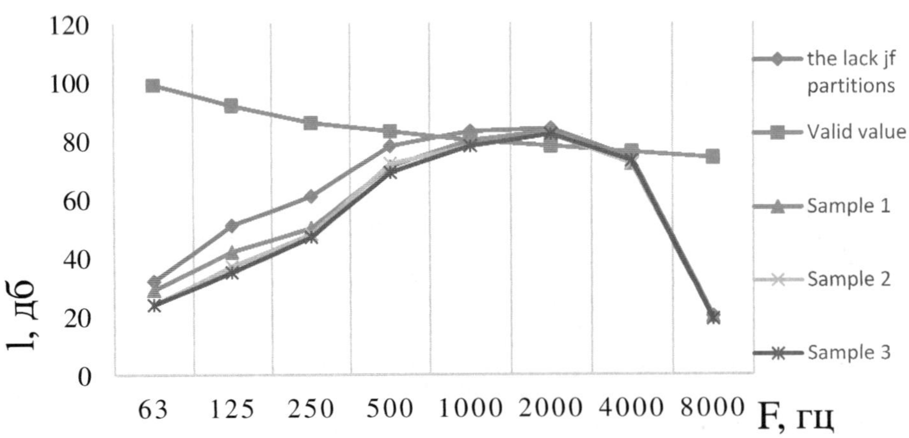

Figure 2. Sound pressure level versus frequency.

As we can see, all samples have soundproofing properties, and sample №3 has the best soundproofing characteristics.

4. Discussion

Research showed that in order to give the necessary properties without creating significant difficulties during processing, 25 mass parts of shungite must be added to the composite material. It was found that the mixture of shungite and styrene-butadiene latex has better wettability compared to mixtures based on acrylic and natural latex.

The best soundproofing properties are found in a material containing styrene-butadiene latex.

5. Summary

Thus, it is shown that materials based on latex of various nature and textile products of tire processing can be used for soundproofing materials in the construction industry.

6. References

[1] Kalinin Yu.K., Kalinin A.I., Skorobogatov G.A. 2008 Shungites of Karelia – for new building materials in chemical synthesis, gas treatment, water treatment and medicine *Chemistry* **6** 69

[2] Anisimov, V.A. 2013 Polymer composite materials *Chemistry* **103**

[1] Utyuzh A., Volchkova I., Tikhonova Y., Klyuchnikova N. 2019 Abrasion-resistant elastic coatings technology for dental implants *Journal of Global Pharma Technology* **6** 200-205

[2] Bessmertny V.S., Kochurin D.V., Bondarenko D.O., Bragina L.L., Varfolomeeva S.V. 2019 Thermal insulation block materials with protective and decorative coatings *Construction Materials and Products* **2(1)** 4-10

[3] Wolfson S.I., Fafurina E.A., Fafurin A.V. 2011 Methods of utilization of tires and rubber products *Bulletin of Kazan Technological University* **1** 74-79

[4] Drozdovsky V.F., Razgon D.R. 1995 Processing and use of worn tires (directions, economy, ecology) *Caoutchouc and rubber* **2** 5

[5] Okuneva G.A., Radoutsky V.Yu., Shaptala V.G. 2008 Study of soundproofing properties of building materials and constructions based on foam glass *Bulletin of BSTU named after V.G. Shukhov* **4** 45-48

[6] Cheremisina O.A., Sychev M.M., Sirotinkin N.V., Korsakov V.G. 2009 Polymer compositions based on latex and dispersed materials *News of St. Petersburg State Technical Institute* **5** 62-64

[7] Martin D.M., Smith U.K. 2006 Production and application of rubber products: equipment, mixing, processing technologies, finishing *Profession* 477

[8] Kornev A.E., Bukanov A.M., Sheverdyaev O.N. 2005 Technology of elastomeric materials *NPPA Istek* 508

[9] Mosin O.V. 2013 Shungite and its properties *Journal of physical chemistry* **2** 123-127

[10] Sorokina O.V., Potapov E.E., Reznichenko S.V., Bobrov A.P., Smal V.A., Yadykina V.V., Tikunova I.V. 2018 Investigation of properties of high-filled composites based on bitumen and shungite (carelite) *Caoutchouc and rubber* **2(77)** 92-96

[11] Hong C. K., Isayev A. I. 2002 Ultrasonic Devulcanization of Unfilled SBR Under Static and Continuous Conditions *Rubber Chemistry & Technology* **1(75)** 133-142

[12] Jent A.N., Mark D. 2011 Caoutchouc and rubber. Science and technology monograph *Intelligence* 767

[13] Kiselev M.G, Savich V.V., Pavich T.P. 2006 Determination of the edge angle of wetting on flat surfaces *Science and technology* **1** 38-41

Acknowledgements

This work was realized in the framework of the Program of flagship university development on the base of the Belgorod State Technological University named after V G Shukhov, using equipment of High Technology Center at BSTU named after V G Shukhov.

IOP Conf. Series: Materials Science and Engineering **945** (2020) 012011 doi:10.1088/1757-899X/945/1/012011

Finishing materials in construction using polymer composites

N V Klyuchnikova[1*], M A Klepikova[1] and L V Denisova[1]

[1]Department of Theoretical and Applied Chemistry, Belgorod State Technological University named after V.G. Shukhov, Kostyukov St., 46, Belgorod, 308012, Russia

E-mail: 4494.55@mail.ru

Abstract. Polymer composite materials are increasingly replacing traditional composites such as wood, metal, and natural fibers from the world market of building materials. Polymer composite materials are widely used in the construction industry. Currently, the development of advanced polymer-based composites is a very relevant area. The promising method for modification of epoxy polymer composite with mineral filler schungite is considered. It was found that the introduction of mineral filler increases the composite resistance to impact forces. It is proved that schungite-filled epoxy composites have increased strength and performance properties, which is very important when they are used in the construction industry. After carrying out a number of physical, mechanical and operational studies, it can be concluded that polymer composite materials based on ED-20 resin with the content of mineral filler shungite have the best performance, compared to non-filled composite materials. Thus, the resulting PCM can be used in the construction industry as a finishing material, as it has the necessary complex of strength and performance properties.

1. Introduction

To date, there is not a single branch of human life where materials and products based on polymers are not used. The main consumer of polymer composites, over the past decade, is the construction industry [1]. This is due to the fact that polymer composite materials can be modified in various ways, which makes it possible to obtain construction materials with a wide range of necessary technological and operational properties, which vary simply [2, 3]. Improving previously used recipes allows creating materials for the construction industry with completely new and unique properties.

This interest in polymer-based materials is fully justified, as polymer composites have increased performance and strength characteristics. Such materials are able to maintain their properties in a wide temperature range, in various chemically aggressive mediums, as well as in various operating conditions: multiple loads and deformations [4]. That is why the construction industry needs to develop and improve finishing materials based on polymers, thereby making polymer composites very popular and promising materials [5].

2. Materials and methods

2.1. Materials

Polymer composite materials based on epoxy resin are widely used in the construction industry. For the production of a polymer composite, which will be used as a building finishing material, an epoxy

IOP Conf. Series: Materials Science and Engineering **945** (2020) 012011 doi:10.1088/1757-899X/945/1/012011

resin of the ED-20 brand was selected. This is due to the fact that this epoxy resin has a number of advantages over other brands [6]. It is characterized by low color (i.e. it is transparent), has no impurities and water inclusions, has high strength, excellent adhesive and water-proof properties, resistance to most chemically aggressive environments, a small mass of finished products, which is very important in the production of PCM for the construction industry [7].

It is worth noting that an important component in epoxy compositions is a hardener, the choice of which largely determines the future properties of the polymer composite [8]. Polyethylene polyamine (PEPA) was chosen as the hardener, as the introduction of this component gives the resulting composition a set of necessary properties.

In addition to the hardener, the polymer composite material includes a plasticizer, dibutyl phthalate as which was chosen. The introduction of this plasticizing additive allows reducing the viscosity of the composition, increasing its plasticity, thereby facilitating the processing of the polymer composition [9]. In addition, the introduction of plasticizer allows increasing the frost resistance of the product, which is very important when using materials in the construction industry.

However, despite the importance of the hardener and plasticizer in PCM, the most significant component in the epoxy composite is the filler. It is the introduction of this component that makes it possible to improve the strength, performance, and physical and mechanical properties of composite materials [10].

The introduction of filler into the polymer matrix is the most common way to change the properties of polymers purposefully [11]. Fillers are structurally active additives with a developed specific surface area, which can be used quite simply to vary the properties of the composition.

The most common fillers for PCM are mineral additives such as chalk, kaolin, and feldspar [12]. The introduction of dispersed fillers in a polymer composite material allows solving several problems simultaneously: reducing the cost of the composite; increasing its strength properties; changing the temperature range in which the product can be operated; increasing the performance properties [13].

In addition to these traditional mineral fillers, the natural mineral shungite has recently become very popular [14]. Due to its unusual and unique structure, shungite gives increased performance and technological properties to compositions where it was introduced as filler.

Shungite is an amorphous rock with an unusual structure. The structure of the mineral consists in the fact that allotropic carbon is found as a modification of fullerenes and fullerene-like forms [15]. This mineral is formed from organic deposits and has a gray-black color. The matrix of a shungite rock is a molecular crystal that is a transitional configuration between organic and inorganic matter (Fig. 1).

A characteristic feature of shungite crystals is the presence of fullerene-like structures, which increases the resistance to oxidation, heat and heat resistance of polymer composites where this filler was introduced [16]. The natural mineral shungite consists entirely of bipolar particles, due to this fact, this filler is quite simply combined with both polar and non-polar polymer matrices, which makes it easier to work with this filler [17].

Figure 1. Matrix of shungite rock.

Therefore, to obtain a polymer composite material, which will be used as a finishing material in the construction industry, the choice of raw materials was stopped on: epoxy resin of the ED-20 brand, the hardener polyethylene polyamine, the plasticizer dibutyl phthalate and the natural mineral filler shungite.

2.2. Methods
As the resulting composite material will be used in the construction industry as a finishing material, it is necessary to this composite to have the necessary complex of strength and performance properties. That is why it is necessary to conduct a number of physical and mechanical tests of the resulting material to make sure that it is suitable for use in the construction industry [18].

The resulting composite material based on epoxy resin and the natural mineral shungite was subjected to a number of physical and mechanical tests, which were carried out according to GOST methods.

One of the important physical and mechanical indicators is the determination of the bending strength of the material. This test allows determining how well the material resists bending, that is, what is the rigidity of the material.

In addition to the rigidity of the material, its ability to restore its shape is also a very important indicator [19]. This ability of the material can be determined by examining the elastic modulus of the polymer composite. Determination of the elastic modulus in tension is carried out in accordance with GOST on a universal testing machine.

As PCM based on epoxy resin and shungite will be used as a building finishing material, it is highly likely that this composite will encounter chemically aggressive medium. That is why it is necessary to make sure that the resulting composite is resistant to alkaline and acidic aggressive medium. This study was conducted in accordance with GOST, holding samples in H_2SO_4 and KOH medium. The resistance of the material to chemically aggressive medium is determined by the mass that the sample loses during holding it in a solution of alkali or acid. The smaller the mass lost during the test, the more resistant the PCM is to a particular environment. This test helps determine whether it is possible to operate this polymer composite material in conditions where either alkaline or acidic components will be present.

It is also necessary to take into account the fact that if PCM is used as a building material, it will be in contact with moisture. That is why building finishing materials based on polymer composites must have water resistance. The ability of the material to be used in high humidity conditions is determined according to GOST.This method makes it possible to determine the water absorption of samples based on epoxy resin with the content of mineral filler shungite, as well as without the use of filler. Based on the data obtained, it is possible to conclude about the stability of PCM in environments with high humidity.

Carrying out all the above tests of the resulting composite material is a prerequisite for the analysis and study of strength, physical, mechanical and operational properties of PCM, which will be used in the construction industry as a finishing material.

3. Results
When conducting a study of the dependence of the bending strength of the material on the content of the filler, it was found that this indicator decreases with an increase in the content of the natural mineral shungite (Figure 2).

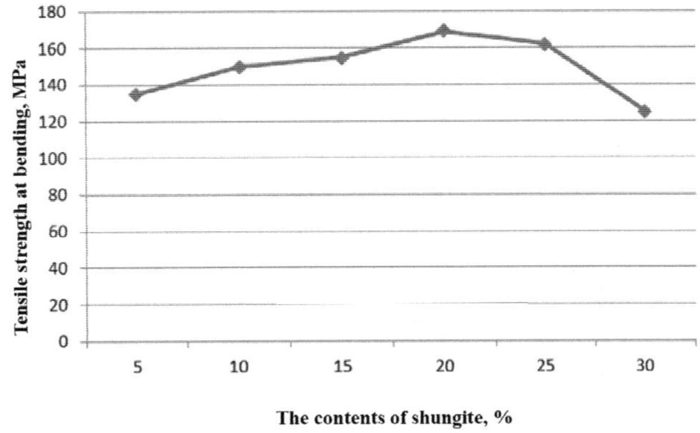

Figure 2. Dependence of the bending strength on the content of the mineral filler shungite.

It is worth noting that the maximum value of the bending strength is achieved when the content of 5% mineral filler in the polymer composite, that is, with this content of filler, the material has the greatest rigidity.

When determining the dependence of the elastic modulus of tension on the content of the filler, it can be concluded that with an increase in the content of the filler in the composite, this indicator increases (Figure 3).

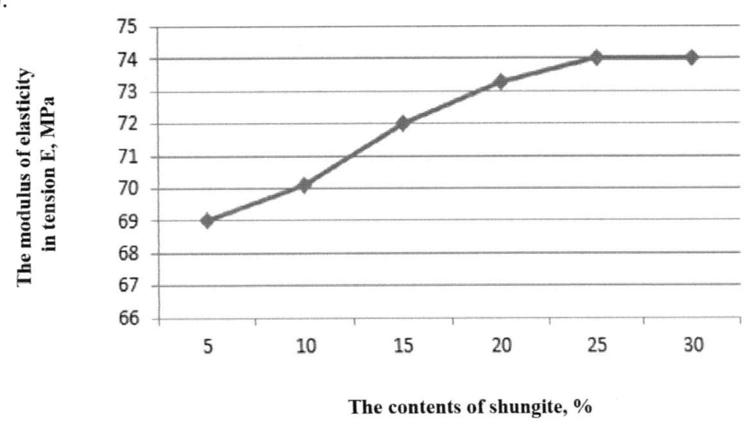

Figure 3. Dependence of the elastic modulus under tension on the content of shungite in a polymer composite material.

Based on the data obtained (Figure 3), it can be concluded that with an increase in the content of shungite, the ability of the epoxy polymer composite to restore its original shape increases.

When testing PCM for stability in aggressive environments, it was concluded that the introduction of a mineral filler in the epoxy composite provides its strengthening up to 25%, as well as an increase in the microhardness of the material up to 30% (Figure 4).

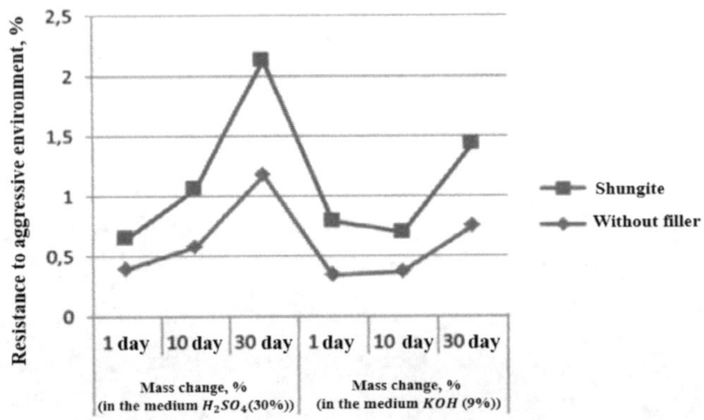

Figure 4. Influence of aggressive environment on composites with and without mineral filler.

Also, the introduction of shungite allows increasing the compressive strength without compromising the elasticity of the material, which provides increased adhesion strength of the material to the treated surfaces.

When determining the water absorption of the material, it was found that epoxy PCM filled with shungite have less water absorption, compared to samples without filler. This indicates that filled polymer composites have the highest resistance in conditions with high humidity.

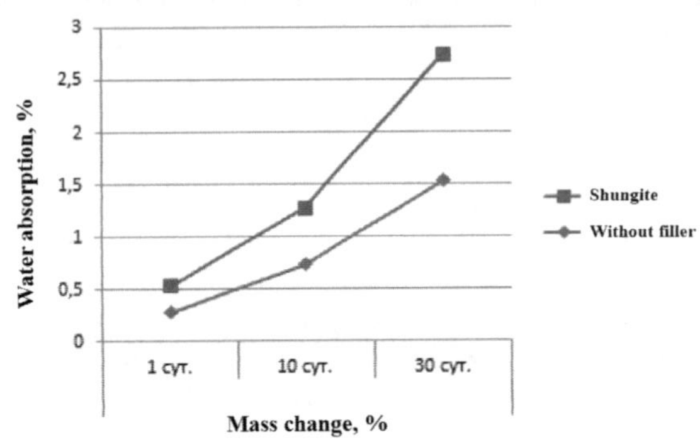

Figure 5. Change in water absorption in epoxy PCM with and without filler

4. Discussion

After carrying out a number of physical, mechanical and operational studies, it can be concluded that polymer composite materials based on ED-20 resin with the content of mineral filler shungite have the best performance, compared to non-filled composite materials.

Thus, the resulting PCM can be used in the construction industry as a finishing material, as it has the necessary complex of strength and performance properties.

5. Summary

The introduction of natural mineral filler shungite in the epoxy polymer matrix allows obtaining a polymer composite material that has all the properties that are necessary for finishing materials used in the construction industry.

6. References

[1] Kalinin Yu K, Kalinin A I, Skorobogatov G A 2008 Shungites of Karelia – for new building materials in chemical synthesis, gas treatment, water treatment and medicine *Chemistry* **6** 69
[2] Anisimov V A 2013 Polymer composite materials *Chemistry* 103
[3] Utyuzh A, Volchkova I, Tikhonova Y, Klyuchnikova N 2019 Abrasion-resistant elastic coatings

technology for dental implants *Journal of Global Pharma Technology* **6** 200-205

[4] Ksantes M, Kulezneva V N 2010 Functional fillers for plastics *Scientific bases and technologies* **5(17)** 462

[5] Osipov O N 2008 Problems of recycling and processing of polymers *Pakkermash* 53

[6] Bondarenko N I, Bondarenko D O, Burlakov N M, Bragina L L 2018 Study of the effect of plasma chemical modification on the macro- and microstructure of the surface layer of autoclave wall materials *Construction Materials and Products* **1 (2)** 4-10

[7] Muhin A M, Galimov E R 2013 Functional based on modified polyvinylchloride *Journal of international Scientific Publications: Materials, Method and Technologies* **7(3)** 130-136 (Bulgaria)

[8] Khachaturov A A, Potapov E E, Fionov A S, Kolesov V V, Prut E V 2018 Functional elastomeric materials based on styrene-butadiene rubbers and magnetite *Materials of the reports of the VII all-Russian conference "Caoutchouc and rubber 2018: traditions and innovations"* Moscow, Expocenter, April 25-26 **8** 64

[9] Klyuchnikova N V, Gordeev S A, Gordienko M D 2017 Polymer composite material based on thermoplastic polyimide. *Bulletin of BSTU named after V G Shukhov* **12** 126-129

[10] Solomatov V I 2001 Epoxy-carbon composite for the protecting building constructions *Protecting composite material and third millenium technologies: II International Scientific and Practical Conference "Composi 2001"* **135** 18-20

[11] Bobrov A P, Kablov W F, Smail V A 2015 Study of the abrasive activity of natural carbon-containing mineral Compound ingredients of Polymer composites *Interm. Polym. Sci. and Technol* **42(9)** 68-71

[12] Potapov E E, Sorokina O V, Shelukhina A A 2018 Study of the influence of shungite (Carelite) on the properties of polymer-bituminous composites. *Conference "Caoutchouc and rubber - 2018: traditions and innovations"* **178** 86

[13] Mosin O V 2013 Shungite and its properties Journal of physical chemistry **(2)**123-127

[14] Klyuchnikova N V, Sokolenko I V, Evtushenko E 2014 Influence of Metal Component on Caking of Metal-Ceramic Composites Research *Journal of Pharmaceutical, Biological and Chemical Sciences* **5(5)** 1637

[15] Kovalevsky V V 2001 Shungite rocks-prospects and problems of use in composite materials. *Theory and practice of production technologies for products made of composite materials and new metal alloys* **8(44)** 303

[16] Potapov N N, Komova E V, Prut V I, Solodilov A N, Kovaleva 2017 Express method for evaluating the activity of shungite filler in elastomeric composite materials *Caoutchouc and rubber* **2** 92-96

[17] Khachaturov A A, Potapov E E, Kolesov V V, Fionov A S, Bobrov A P, Smal V A, Prut E V, Shevchenko V G, Tikunova I V 2018 Study of electrophysical and acoustic properties of polymer composite materials based on SKEPT and shungite *Caoutchouc and rubber* **2(77)** 96-101

[18] Potapov E E 2010 Shungit-a promising ingredient of rubber mixtures for the tire industry and RTI industry Conference *"Caoutchouc and rubber - 2018: traditions and innovations"* **7** 3-5

[19] Sorokina O V, Potapov E E, Reznichenko S V, Bobrov A P, Smal V A, Yadykina V V, Tikunova I V 2018 Investigation of properties of high-filled composites based on bitumen and shungite (carelite) *Caoutchouc and rubber* **2(77)** 92-96

Acknowledgements
This work was realized in the framework of the Program of flagship university development on the base of the Belgorod State Technological University named after V G Shukhov, using equipment of High Technology Center at BSTU named after V G Shukhov.

IOP Conf. Series: Materials Science and Engineering **945** (2020) 012012 doi:10.1088/1757-899X/945/1/012012

Lean construction – an effective management system in the construction industry

E N Parfenova[1*], Zh N Avilova[2] and A N Ganzha[3]

[1]Department of Economics and Management, Belgorod State University, Pobedy St., 85, Belgorod, 308015, Russia
[2]Department of Economics and management, Belgorod State Technological University named after V.G. Shukhov, Kostyukov St., 46, Belgorod, 308012, Russia
[3]Department of Heat Engineering and Energy Efficient Technologies, National Technical University «Kharkiv Polytechnic Institute», Kyrpychova St., 2, Kharkiv, 61002, Ukraine

E-mail: parfenova@bsu.edu.ru

Abstract. The paper discusses the possibilities and prospects of using methods and tools of lean construction to improve the arragement of the construction business in the Belgorod region. Lean construction is considered as a system of innovative methods of process management in construction. The use of this system is an effective tool for optimizing the results of work: the time and cost of work is reduced by constantly improving the quality of the final product. Exploring the methodology and tools of lean construction, we can identify four main areas of use: employee engagement, cost reduction, optimization of construction processes, and loss prevention. The paper presents the results of using the concept of lean construction in foreign countries, as well as methods and tools aimed at involving employees in lean production, reducing costs and preventing losses, optimization, automation and informatization of the production process at the level of strategic and operational management. The specificity of the choice of tools and methods of work performed at all levels of management for the construction industry is determined. The application of the concept of lean construction is justified, which allows organizations to increase their competitiveness and business efficiency.

1. Introduction

In modern economic conditions, the basis for the success of a construction company is a high-level organization. The main quality is to know the responsibility of each employee in the value chain of the final product. The concept of lean construction provides these opportunities. It depends on each construction company how to use them, and currently the organization activities in the process of modern management will be effective if the management processes are modernized to improve the construction process and overall economic development in order to increase the scale and technical capabilities of the organization.

Currently, the country's construction companies face the task of optimizing organizational methods of construction management in accordance with the best world practices [1, 10, 11]. The stages of development for improving the organization differ due to significant gaps between them, so there is an

IOP Conf. Series: Materials Science and Engineering **945** (2020) 012012 doi:10.1088/1757-899X/945/1/012012

increasing need to implement modern management methods using the latest tools. One of these concepts is lean construction as a system of innovative methods for managing processes in construction. The use of this system is an effective tool for optimizing the results of work: the time and cost of work is reduced by constantly improving quality of the final product.

2.Materials and methods

The methodological basis of the research was a systematic approach to the research object, the principle of specificity, principle of development and the principle of objectivity. The study used methods of situational, structural and functional, economic and statistical, comparative analysis, tabular and graphical interpretation of empirical and factual information. The subject of the research is the concept, methods, tools and practice of using lean production in the construction industry.

The study of the concept of lean production and justification of the possibility and feasibility of its application in the construction industry was carried out on the basis of the Russian standard GOST R 56020-2014 "Lean production. Basic provisions and dictionary".

The application of lean production philosophy, values, principles and tools is based on presenting the business as a stream of value creation for the consumer, flexibility, identifying and reducing losses, and continuous improvement of all activities at all levels of the organization, involvement and development of personnel in order to increase customer satisfaction and other stakeholders.

3. Results and discussions

The main goal of lean construction is continuous improvement. The concept of lean construction is borrowed from the theory of lean production. At the same time, the construction industry has adapted a number of lean methods to reduce losses and increase profits. The lean manufacturing system itself is not just a technology; it is a whole management philosophy that focuses all employees on development and achieving maximum results [2].

The practice of lean production includes many methods and tools for managing manufacturing. At the same time, 27% of the total number of employees involved in this process. This fact indicates that the system of lean production pays considerable attention to the formation of a common corporate goal, the creation of a lean culture at the enterprise (table 1).

Table 1. Methods and tools for involving employees in lean production.

Tools and methods	Function
5S	A tool that allows arranging the workplace of an employee efficiently, both linear and administrative
TQM	A method designed to improve the quality of organizational and managerial processes
Gemba	A tool aimed at involving administration in the production process
Hoshin Kanri	A tool that allows establishing communication between managers and ordinary employees quickly
Kaizen	A tool designed to combine employees' efforts to minimize costs

In addition, loss prevention is an important area in lean production. This toolkit account for a large share of tools (33%), as the main task of lean production is to optimize manufacturing processes in order to reduce costs and prevent losses (table 2).

IOP Conf. Series: Materials Science and Engineering **945** (2020) 012012 doi:10.1088/1757-899X/945/1/012012

Using lean production tools and methods, you can imagine the process as a value stream, identifying inefficient operations that should be excluded from the process, as the customer does not want to pay for them [3, 9].

To do this, they map the value stream and use loss analysis to determine the processes that need to be removed from the stream. Future enhanced flows should be taken into account when considering the question of the elimination or reduction of losses at the map of future flows.

Table 2. Methods and tools aimed at reducing costs and preventing losses.

Tools and methods	Function
«Just-in-time»	A method that allows fulfilling orders "just in time", using raw material stocks and financial flows rationally
PDCA	A method that allows solving problems and making changes using a simple algorithm of actions
OEE	A method for detecting losses from improper use of machines and equipment
SMART	A tool that allows setting goals correctly, preventing production losses
Heijunka	A tool that allows planning the inventory, bringing the frozen assets to the minimum
Continuous flow	A tool that allows building the production process optimally, preventing downtime in the movement of material flows in it
Poka-Yoke	A tool aimed at protecting against errors leads to minimizing costs
Bottleneck analysis	A tool for identifying production bottlenecks
Muda	A tool that allows determining losses that lead to an improvement in the quality of employees' work

The remaining tools and methods of lean production (40%) are aimed at optimizing the production process, increasing productivity, automation and informatization in the company (table 3).

Table 3. Methods and tools of lean production for optimization, automatization and informatization of the production process.

Tools and methods	Function
Visual management	A tool for collecting information about all operations and processes
Standardize work	A tool designed to standardize all production operations and processes
Jidoka	A method of partial automation of machines and equipment that allows to identify problems and start solving them at the detection stage
SMED	Method for quick changeovers of machines and equipment, moving from their stop to the problem solving of equipment in operation
ISO	Global quality standards
TPM	A method that allows extending the life of machines and equipment by informing all employees of the production process about the rules of equipment maintenance
Kanban	A tool that allows arranging production on the "just in time" principle
KPI	A productivity indicator tool that allows assessing risks and achieving strategic goals in a timely manner
VSM	A tool that allows to identify processes that create and do not create value

	visually
TFM	A flow management method that relies on planning and managing operations and supply chains
Takt time	A tool that allows planning specific performance for different departments depending on orders

For the construction industry, to improve the quality of work, the choice of tools and methods is determined by the specifics of the work performed, which facilitate its implementation at all levels

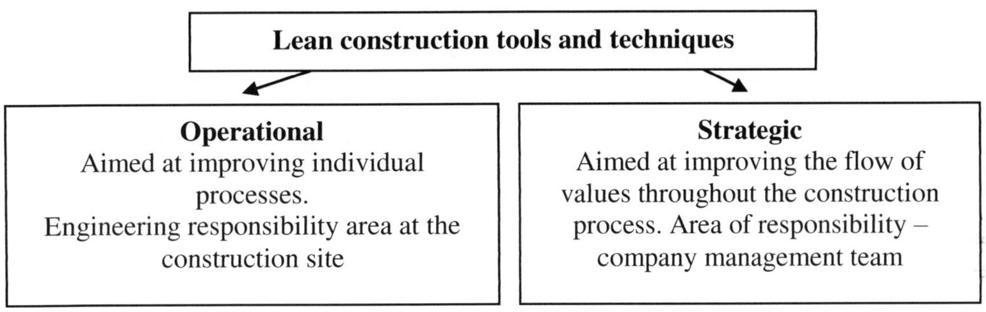

Figure 1. Tools and methods of lean construction.

At the level of operational management, effective methods and tools for lean construction will be: 5S, standardization, mapping, submission of optimization proposals, and "pull" planning. At the level of strategic management, the most effective methods and tools are: information modeling, just-in-time deliveries, TQM, and a histogram of the value stream. The main activities in lean construction are: establishing the value of each specific product; identifying the value stream; support for the functioning of the value stream; "pulling" the necessary product by the consumer; continuous improvement. The effectiveness of this practice has been repeatedly proven in a number of countries of the world's leading economies as a real tool for modernization, both for the construction process and for the company's management as a whole (table 4).

Table 4. Results of using the concept of lean construction in foreign countries.

Country	Main directions of using the concept of lean production	Methods and tools of lean construction	The result of using the concept of lean production
Israel Austria Belgium	cost reduction, employee engagement	Just-in-time, Kaizen, Kanban, 5S system, visual management, standardize work, TQM, TPM	A systematic and integrated approach to reducing costs and developing lean production, working with staff to create a "lean" culture
USA	process optimization cost reduction loss prevention	Kaizen, Pull system, visual management2, standardize work and safety, LEED	Eliminating waste throughout the value chain

China	process optimization cost reduction employee engagement	Kaizen, Kanban11, Just-in-time, 5S system, visual management, standardize work, TPM, TQM, SMED	A comprehensive approach using proprietary methods to reduce the time of equipment changeover.
Great Britain SAR	process optimization cost reduction loss prevention	Kaizen, 5S system, Jidoka, SMED	supply chain management, creating long-term relationships based on increasing value while reducing core costs
Italy	process optimization	Just-in-time, TPM, ISO, Kaizen, standardize work and safety	Focus on product quality, process standardization, and environmental care
Finland	cost reduction process optimization	Kaizen, Just–in–time, TQM, ISO	Emphasis on the choice of suppliers and the quality of raw materials, rhythmic work, speed of customer service

To date, some Russian construction companies have already applied the concept of lean construction [5]. However, many companies were not able to build the right work on the implementation of methods and tools of the concept of lean construction, so in the end they had to freeze their implementation. An analysis of the practice of Russian enterprises that initiated the introduction of the concept of lean construction, but did not implement it, shows that many of them make the same mistakes [8].

Firstly, thoughtless implementation of lean construction tools and methods leads to the fact that huge efforts are spent on eliminating all problems. Often, some of the tools can be implemented at a later stage, or they may not be needed at all.

Secondly, employees refuse to learn the principles of lean construction. An untrained person who works for a long time in one company often does not see any problems. However, this does not mean that there are no problems, so the process of implementing the concept of lean construction is delayed.

Thirdly, the implementation of the lean construction process is delegated to a group of employees. At the same time, the rest of the team does not participate in changes and actively opposes the introduction of innovations. It is necessary to involve all staff, actively demonstrating the benefits of changes – reducing labor costs and eliminating useless work.

Fourthly, management often neglects material incentives for participating in the introduction of innovations in the enterprise. Today, the reality is that without an understanding of personal gain, staff will not actively participate in changes. Employees should clearly know how they will be able to earn extra money, and this incentive is still the strongest.

IOP Conf. Series: Materials Science and Engineering **945** (2020) 012012 doi:10.1088/1757-899X/945/1/012012

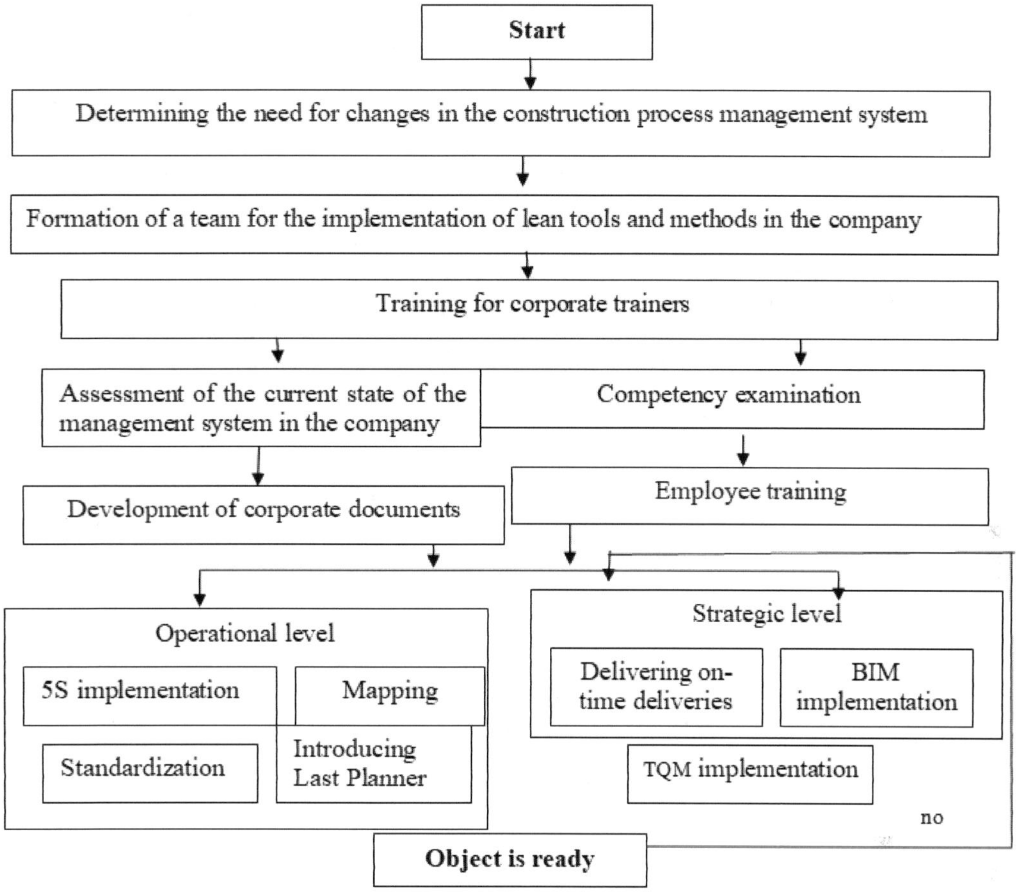

Figure 2. Algorithm for implementing the concept of lean construction in construction companies.

Fifthly, management expects quick results from implementing the concept of lean construction. It is patience and consistency that is the secret to the success of change.

Sixthly, it is often not correct to involve consultants in the implementation of the concept of lean construction. It is necessary to avoid theorists, and to engage practitioners.

Seventhly, they shift the responsibility to the consultants. The most professional third-party expert will not do their job for the management and employees. It is necessary to have a personal interest and desire of all personnel to accept and then implement new technologies in construction.

Eighthly, there is often the optimization of operations and processes, which should not be. This is one of the most frequent and dangerous mistakes. Instead of wasting time and effort on optimizing useless processes, they should be eliminated.

Ninthly, employees are often afraid of optimizing the number of employees. Increasing efficiency and reducing costs will logically lead to a review of the staff number, but often enterprises do not go to the reduction, so all efforts to implement the concept of lean construction are slowed down [6].

We must admit that there are some companies that have introduced the concept of lean construction and improved their skills in organizing construction. The result of this work became the annual increase in productivity, reduced production cycle, increase the turnover of funds, and increase the level of customer satisfaction, development of personnel motivation system [4]. To ensure success in the implementation of the improvement program, an algorithm for implementing the concept of lean construction in modern construction enterprises is proposed (figure 2).

It should be noted that all of the above stages are designed for a continuous process, as training personnel in the principles of lean construction is the basis for identifying and solving problems to optimize construction processes [7].

4. Summary

Implementing the concept of lean construction takes a long time, and many problems arise. At the same time, the use of this concept has significant material advantages and benefits over competitors. To eliminate the identified losses, as well as reduce costs, they need to use the created value stream algorithm. Thus, the proposed algorithm is effective, as the integrated use of the optimal set of methods and tools of the concept of lean construction, taking into account their logical and temporal ordering, will lead to a gradual improvement of management and construction processes and will become a key way for the company to survive in modern market conditions [7, 10, 12].

5. References

[1] Chernykh E A 2009 Operational planning and construction quality: domestic and foreign experience. *Quality management* **4** 270–279

[2] Gorelik P I 2014 Lean construction as an innovative method of construction management *Construction of unique buildings and structures* **12(27)** 40–48

[3] Chetaeva E A 2018 Foreign and domestic practice of using the lean production system *Vector of economy* **1 (19)** 55-60

[4] Isakov I V 2019 Current state and features of production application in Russia and abroad. In the collection: Collection of selected articles on the materials of scientific conferences of the SRI "National Development" 256-264

[5] Maysky R A, Ismagilova V S, Ismagilova A R 2016 Modern problems of development of the world experience of "lean production" in Russia *Bulletin of Economics and Management* **3(4)** 36-43

[6] Popova E S 2016 Lean construction methodology as an innovative approach to managing the construction industry *Master's Journal* **1** 141-146

[7] Parfenova E N Avilova Zh N 2018 Need for portfolio project management in construction *Bulletin of BSTU named after V.G. Shukhov* **1** 106-110

[8] Doroshenko Yu A Klimashevskaya A A 2017 Analysis of the scientific and technical potential of enterprises in the construction materials industry in the context of assessing the need for technological modernization in the industry *Bulletin of BSTU named after V.G. Shukhov* **1** 214-218

[9] Doroshenko Y A Malykhina I O Somina I V 2018 Problems and prospects of scaling of high-tech-business as an innovativeinvestment driver of economic systems development *Espacios* **39(47)** 23

[10] Doroshenko Y A Minaeva L A Avilova Z N Somina I V Leonova O V 2015 Socio-economic aspects of development of small innovative business in Russia *Journal of Applied Engineering Science* **13(4)** 293-298

[11] Subbotin O S 2019 Building materials and technologies of modern housing: architectural and ecological aspects *Construction Materials and Products* **2(4)** 84 – 88

[12] Manzhilevskaya S E 2019 Organizational and economic problems of ecological safety in construction *Construction Materials and Products* **2(4)** 73 –78

Acknowledgements

This work was realized in the framework of the Program of flagship university development on the base of the Belgorod State Technological University named after V G Shukhov, using equipment of High Technology Center at BSTU named after V G Shukhov.

IOP Conf. Series: Materials Science and Engineering **945** (2020) 012013 doi:10.1088/1757-899X/945/1/012013

Methods of risk minimization in investment and construction projects

I V Sukhorukova[1]*, D A Maksimov[1] and G P Fomin[1]

[1]Plekhanov Russian University of Economics, Stremyanny lane, 36, Moscow, 117997, Russia

Email: suhorukovaira@yandex.ru

Abstract. The paper puts forward a hypothesis about the possible integration of risk assessment methods in order to minimize them in investment and construction projects, for that the problems of risk minimization in the management and implementation of construction projects are considered. A critical review of existing methods of risk analysis is carried out and a method of risk minimization is proposed based on the creation of a "Construction risk Informatoria" and methods for optimizing the reallocation of resources according to the minimum time criterion, which determines the risk of lost profits. The example of the use of risk minimization methods in the optimization procedure for network models of low-rise building construction is given, and the composition of the application of mathematical methods and models that contribute to increasing the economic effect in this area is described. As a result, effective recommendations were developed in the form of tools for monitoring the dynamics of construction work to prevent risk situations.

1. Introduction

Creating an effective model for managing the implementation of construction projects is one of the main basic problems facing the modern manager. In these conditions, in order to overcome crisis situations and ensure the continuous functioning of the business, the necessary criterion is to develop mechanisms and search for options for the most optimal use of the existing material and production resource base, scientific, technical and organizational and economic capacities. Managing business in a crisis situation is quite a difficult task, as it is necessary to make most management decisions in the shortest possible time. It is necessary to analyze a significant amount of business information when developing optimal behavior, so the price of an error made by the manager increases critically. It is very important for a modern manager to learn not only to solve emerging problems in short time, but it is much more important to warn and prevent their occurrence timely. This can only be done using modern and effective management tools. The most important problem, which is primarily solved by the manager, is the analysis of the effectiveness of the labor activity of the team in carrying out construction work. Implementation of any investment project in the construction sector is associated with various types of risks due to uncertainty conditions. Investment and construction projects are characterized by the uncertainty of the project implementation conditions, which is due to the possibility of negative situations and consequences, the probability of adverse situations and consequences, and the adverse situations and

IOP Conf. Series: Materials Science and Engineering **945** (2020) 012013 doi:10.1088/1757-899X/945/1/012013

consequences themselves, which are directly related to damages, economic, social or environmental losses.

It should be noted that modern projects in the investment and construction sector are a very complex system that includes a large number of participants: the customer, general contractor and subcontractor construction and installation organizations; commercial banks and financial bodies and organizations; design and research institutes; suppliers of construction materials, structures, parts and semi-finished products, technological equipment; organizations and bodies that carry out various types of control and supervision of construction; divisions that operate construction equipment, machinery, and vehicles [1-5]. Existing publications illustrate the use of various methods to minimize risks in investment and construction projects, but unilateral decisions have not created convincing outlook for development prospects yet, so it is necessary to continue research in this difficult area.

2. Materials and methods

The modern scientific method of management in construction is a method of modeling future development in investment and construction projects, which allows integrating the advantages of existing methods and models in a successful combination and simultaneously eliminate the disadvantages.

The reasons for the risks are incomplete knowledge of events that may or should occur in the future. In this regard, the types and causes of risks of production and economic activities of a construction company in the market of low-rise housing construction were studied and a continuously replenished accumulator was created – “Construction Risk Information Center”. Various types of classification of construction risks, hierarchical and facet ones, are presented as the main platform for collective discussion with indication of their possible manifestations in construction. All this is graphically linked to the main tasks of organization, planning and management of low-rise construction and construction and installation works; preparation of construction sites; construction of load-bearing structures of buildings; construction of enclosing structures of buildings; arrangement of external engineering networks and equipment; protection of structures and equipment; finishing works; doing ground works; construction of roads, their bases and surfaces; construction and repair of buildings and structures. Network models in the field of modeling and scheduling of discrete technological processes can describe the relationship between work and a certain class of organizational and technological schemes of construction processes.

3. Results

It should be noted that the existing PERT-COST technology assumes the possibility of reducing the time required to complete a set of works by increasing costs. In addition, the PERT-COST technology is a method that allows the manager to minimize the total cost of a project if it is completed within a specified time frame, as well as to minimize the duration of the project in conditions of a limited budget. A typical mistake in the presentation of PERT technology is to assess the risk of project failure solely by the duration of the critical path. So, for example, a set of works must be completed in no more than X weeks, and the variance of the critical path duration is known and the mean square deviation of the critical path duration is equal to σ, then we find the probability that the critical path length will be no more than X weeks, and the normal distribution function with mathematical expectation a and variance σ2 is used for risk assessment:

$$F(x,a,\sigma)=\frac{1}{\sqrt{2\pi\sigma}}\int_{-\infty}^{x}\exp\left(-\frac{(t-a)^2}{2\sigma^2}\right)dt$$

Calculations in the study show that the probability of failure of complex works within the period is 34.5%, which is of course a lot, but it happens in practice, and it is not determined by the duration of the critical path, and duration of the path is close to critical. Thus, it is shown that when calculating the risk of failure to complete an investment project on time, it is necessary to take into account the paths that are close to critical. This led to the need to perform a complete optimization of the entire complex of construction works.

Any organizational and technological model should describe the list of construction and installation works, the order of their implementation, the nature of the relations between the works, reflecting the specifics of construction technology, building regulations, and the need for rational use of resources. The challenge is to analyze the network model in order to identify reserves and "weak points", and then perform optimization. At the same time, it is important to use the advantages and eliminate the disadvantages of various modeling methods and integrate them [6-10]. It should be remembered that the discovered reserves allow managing the complex of work more flexibly by reassigning them from one work to another, more stressful, not arbitrarily, but reasonably according to the chosen optimization methods [11-12].

Thus, we build and analyze the network model in order to identify reserves and "weak points". At the same time, the tabular method for calculating parameters allows solving this problem, but the graphical method of analysis still provides greater visibility and credibility, which makes it easier to proceed to the optimization process, and the combination of various modeling methods allows combining their advantages. Labor intensity shows the time spent to perform a single operation and is inversely proportional to labor productivity. To justify the business plan, as well as to analyze how effectively the labor force is used and how much labor intensity is affected by a variety of different reasons: the level of qualification of personnel, the degree of technical equipment of production, the complexity of manufacturing, the degree of automation and working conditions, we suggest using existing optimization methods in an integrated version as part of network methods and models on the example of low-rise construction, for which in table1 a structural and time table of the list of works and resources for the construction of a three-story residential building using the "Construction Risk Information Center" is presented.

Table 1. Structural and time table of residential building construction works.

Work content	Denotation a_i	Basic a_i	Cost b_i	Cost rubles	Man-hour	Hours t_i
Ground works	a_1	-	b_1	9680	390	27.62
Construction of foundation						
-formwork installation	a_2	a_1	b_2	802150	360	2479.53
-filling the foundation	a_3	a_2	b_3	450800	440	1140.11
Basement utility room	a_4	a_1	b_4	535600	620	961.32
Wall construction						
- exterior walls	a_5	a_3a_4	b_5	2557000	2310	1231.79
- interior walls	a_6	a_5	b_6	1231900	2530	541.84
Installation of partition walls	a_7	a_3a_4	b_7	3092000	4380	785.57
Installation of stairs	a_8	a_6a_7	b_8	100500	280	399.42
Mounting of platforms	a_9	a_8	b_9	98500	220	498.23
Mounting of balcony slabs	a_{10}	a_9	b_{10}	890500	325	3049.07
Mounting of floors	a_{11}	a_6a_7	b_{11}	1905000	1100	1927.17
Mounting of thermal insulation	a_{12}	$a_{10}a_{11}$	b_{12}	7510650	8000	1044.73
Installation of canopies	a_{13}	$a_{10}a_{11}$	b_{13}	850000	100	9458.80
Mounting of openings	a_{14}	a_{13}	b_{14}	1708030	1550	1226.26
Mounting of blind area	a_{15}	a_{14}	b_{15}	450000	550	910.47
Mounting of porches	a_{16}	a_{15}	b_{16}	730000	400	2030.86
Installation of the roof	a_{17}	a_{16}	b_{17}	1834700	4500	453.70

Finishing works	a_{18}	$a_{12}a_{17}$	b_{18}	3341500	1025 0	362.77
Mounting of water pipes						0.00
- installation of standpipes	a_{19}	$a_{12}a_{17}$	b_{19}	320500	280	1273.76
- the layout in the flats	a_{20}	a_{19}	b_{20}	10500	470	24.86
Mounting of the sewage system						
- installation of standpipes of the sewage system	a_{21}	a_{20}	b_{21}	560500	370	1685.74
- the layout in the flats	a_{22}	a_{21}	b_{22}	250500	750	371.68
Mounting of heating						
-pipe layout	a_{23}	a_{22}	b_{23}	950000	810	1305.14
- installation of tubes and radiators	a_{24}	a_{23}	b_{24}	250500	1530	182.19
Mounting of ventilation						
-installation of lattices	a_{25}	a_{24}	b_{25}	260000	820	352.84
- installation of air ducts	a_{26}	a_{25}	b_{26}	570500	680	933.61
Electrical installation	a_{27}	a_{26}	b_{27}	2236200	3400	731.90
Lightning protection						
-installation of tower	a_{28}	$a_{18}a_{27}$	b_{28}	850500	100	9464.36
- installation of connectors	a_{29}	a_{28}	b_{29}	79500	130	680.52
Grounding						
- excavation	a_{30}	$a_{18}a_{27}$	b_{30}	60500	200	336.62
- backfill	a_{31}	a_{30}	b_{31}	70500	90	871.69
Mounting of floors	a_{32}	$a_{31}a_{29}$	b_{32}	1761000	2630	745.11
Mounting of doors	a_{33}	a_{32}	b_{33}	1156200	1820	706.93

We perform optimization based on the time criterion, in accordance with the target function of the following type:

$$T_{\text{KP}} = \sum_{i=1}^{n} t_{\text{KP}}^{i} \to min.$$

For the purposes of objectivity, we will analyze the network model, starting with determining the minimum amount of working time for performing a complex of construction works. Therefore, let us trace possible paths from the initial event to the final one. There are thirty-two such paths, of which we will select the longest - the critical path, then we will determine the time reserves for each path relative to it:

$$R(L_i) = T_{\text{KP}} - T_i.$$

Then we build a network graph on a time scale, starting with the work of the critical path to establish the locations of reserves. On this basis, due to the reserves found on non-critical works, the duration of the critical path T_{KP} can be reduced. By transferring some of the resources to critical work, they can reduce the time required to complete the work and thus get new deadlines for the entire complex. The optimal plan for performing construction work will be such a plan, when T_{KP}^{0} will be the smallest of all possible. But the mechanism for redistributing funds involves reducing the part of funds for non-critical work (i, j) by a certain amount $x_{ij} < b_{ij}$, defined by special restrictions, which naturally leads to an increase in the time of its execution, then it is fair to write:

$$t'_{ij} = f(x_{ij}) > t_{ij}.$$

IOP Conf. Series: Materials Science and Engineering **945** (2020) 012013 doi:10.1088/1757-899X/945/1/012013

At the same time, part of the transferred resources $x_{ij} = x_{hk}$, invested in critical work (h, k), lead to a decrease in its execution time, which can be described by the following expression:

$$t'_{рл} = f(x_{hk}) > t_{hk}.$$

The duration of the work depends on the amount of resources allocated, obviously, the "time – money" model works, and does not depend on how these resources were invested. Of course, this dependence is nonlinear and is described by the following expression:

$$t = \lim_{n\to\infty} t_0 \left(1 \pm c\frac{x}{n}\right)^n = t_0 \exp(cx) = t_0 \exp\left(\frac{x}{b}\right),$$

which indicates on an exponential form of connection, but in practice, when performing calculations, this function is upgraded by an approximate linear expression:

$$t'_{ij} = t_{ij}(1 \pm x_{ij}) = t_{ij} \pm t_{ij}\frac{x_{ij}}{b_{ij}};$$

Due to the fact that the allocated resources are limited, the condition for their preservation must be met, therefore, the amount of funds withdrawn from the work (i, j) must be equal to the amount of funds transferred to the work (h, k):

$$\sum_{СН}^{M} x_{ij} = \sum_{ПЕР}^{N} x_{hk},$$

where M – the number of jobs that funds were withdrawn from;
N – the number of jobs that funds were transferred to.

In the process of reallocation of funds, they must comply with the restriction on the amount of transferred funds x_{ij} from work (i, j), which is determined by the availability of a free reserve $r_{ij}^{СВ}$ in this work according to the formula:

$$x_{ij} \leq \frac{r_{ij}^{СВ}}{t_{ij}c_{ij}}.$$

In accordance with the above models, we optimize and build the optimal network plan for the complex of works for the construction of a low-rise building, then get the location of all works on critical paths that do not have reserves. This provision made it possible to realize the risk of lost profits - savings of an absolute value of R=1726 hours, which is very significant. Thus, the described network methods and optimization models allow solving problems of optimal distribution and redistribution of limited resources, aimed at minimizing risks for resources of different nature, including the time of completion of the entire complex of construction works.

4. Summary

Thus, modern optimization methods are integrated into a set for solving problems of distribution and redistribution of limited resources, aimed at minimizing the time of completion of the entire complex of works and thereby realizing the risk of lost profits. As a result, we developed effective recommendations for creating a tool called "Construction Risk Information Center", which provides continuous accumulation of risks to ensure continuous monitoring, as well as visual support of construction operations. In addition, it is very important that all participants in the construction of a residential building are equally informed to monitor the dynamics of the development of construction operations to minimize risks in investment and construction projects.

5. References

[1] Camilleri D H 2006 Tsunami construction risks in the Mediterranean –Ounlning Maltas scenario *Disaster prevention and management* **15** 146-162

[2] Sukhorukova I V Chistyakova N A 2018 Optimization of the Formation of the Capital Structure of the Insurance Company, Taking into Account the National Specifics of Insurance *Journal of Reviews on Global Economics* **7** 146-151

[3] El-Adaway I H, Kandil A A 2010 Construction risks single versus portfolio insurance *Journal of management in engineering* **26** 2-8

[4] Sukhorukova I V, Chistyakova N A 2019 Insurance of the termination risk of projects with

IOP Conf. Series: Materials Science and Engineering **945** (2020) 012013 doi:10.1088/1757-899X/945/1/012013

joint companion activity *Journal of Reviews on Global Economics* **8** 269-274
[5] Popov V A 2018 Inflation and consumer basket *Journal of Reviews on Global Economics* **7** 453-456
[6] Dikmen I, Birgonul M T, Sapci N E, Ozorhon B. 2010 Using analytic network process to assess business failure risks of construction firms *Engineering Construction & Architectural Management* **17** 369-386
[7] Ginzburg A V, Ryzhkova A I 2018 Assessment of construction project “pure” risks Materials Science Forum **931** 1245-1248
[8] Logvinova I L, Rubin Y B, Sherstiuk A E 2018 Mutual insurance of transport infrastructure construction risks as an inherent part of competitive environment *European Research Studies Journal* **21** 460-473
[9] Borkovskaya V G, Passmore D 2020 Behavioral engineering model to identify risks of losses in the construction industry *Smart innovation ,systems and technologies* **138** 243-250
[10] Vegas-Fernández F, Rodríguez-López F. 2019 Methodology for determining the most severe risks of a construction project and identification of risky projects *Revista de la construccion* **17** 423-435
[11] Chernyaev A P, Meerson A Yu, Sukhorukova I V, Fomin G P 2020 Methods for optimal Separation of Income in Consumable and Accumulated Parts *International Journal of Recept Technology and Engineering* **8** 797–801
[12] Sukhorukova I V, Fomin G P ,Ekareva I L, MushrubVA 2019 Hybrid Method for Multi-Criteria Risk Minimization *Espacios* **40** 14-25

IOP Conf. Series: Materials Science and Engineering **945** (2020) 012014 doi:10.1088/1757-899X/945/1/012014

One of the approaches to the concrete strength mathematical model

V M Nikonorov[1*] and V V Nikonorov[2]

[1]Higher School of Management and Business, Peter the Great St.Petersburg Polytechnic University, Polytechnicheskaya St., 29, St. Petersburg, 195251, Russia
[2] Department of Roads, Bridges and Tunnels, St. Petersburg State University of Architecture and Construction, 2 nd Krasnoarmeyskaya St., 4, St. Petersburg, 190005, Russia

E-mail: nikanorv@mail.ru

Abstract. The importance of concrete as a structural material is undeniable. The key indicator of concrete as a construction material is strength. Creating and solving a concrete strength mathematical model will allow optimizing raw material costs, increasing the financial result of concrete production, and managing the concrete strength at the stage of its production. The authors reviewed existing concrete strength mathematical models. In the considered models, the concrete strength was influenced by various factors – water-cement ratio, C3S hardness, etc. The authors used the kinetic theory of solid destruction as the basis for their mathematical model of concrete strength. The process of destruction of the sample is considered without deformation. Applying the concept of strained and broken bonds, the authors constructed and solved a concrete strength mathematical model. The main parameters of the model are the speed formation coefficient of the stressed bonds and the speed formation coefficient of destroyed bonds. The model allows estimating the time required for sample destruction at a given load. This is a non-trivial conclusion.

1. Introduction

Relevance. Concrete is a key material in construction. Reasons: significant reserves of raw materials, relatively low production costs; the ability to give the desired shape to the finished product; coupling with metal. The significance of concrete (reinforced concrete) for construction is not in doubt. The most important indicator of a structural material is strength. Accordingly, the concrete strength mathematical model will allow highlighting the factors that affect the strength and the degree of influence of each factor separately. This, in turn, will lead to more precise management of the required concrete strength and possible savings in the cost of concrete production.

The research object is concrete.

The research subject is the concrete strength.

The research aim is to construct and solve a concrete strength mathematical model.

Studies of concrete strength are presented in [1,2,3,4,5].

"Concrete: an artificial stone-like building material obtained by forming and hardening a rationally selected and compacted concrete mixture" [6].

The formula for the compressive strength of concrete [7].

$$S = \alpha \frac{F}{A} K_w \tag{1}$$

S – axial compressive strength of concrete, MPa;
F – breaking load, N;
A – sample area, mm^2;
α – scale coefficient;
K_w – correction coefficient (for cellular concrete).

The axial compressive strength of concrete is measured after 28 days of hardening under normal conditions.

There are concrete strength mathematical models in the form of a multiple nonlinear regression equation (a 2nd degree polynomial).

$$S = \alpha + \sum_{i=1}^{n} \alpha_i X_i + \sum_{i=1}^{n} \sum_{j=1}^{n} \alpha_{ij} X_i X_j + \sum_{i=1}^{n} \alpha_{ii} X_i^2 \tag{2}$$

In these models, strength is a function of the volume fraction of the cement stone, the true water-cement ratio (total water-cement ratio), and so on.

The classical Abrams model is well known [8].

$$S = \frac{k}{A^x}; x = \frac{V_в}{V_ц} \tag{3}$$

k – strength coefficient (cement activity);
A – constant;
Vв – volume of the shutting water, m^3;
Vц – volume of cement (ρ=1500kg/m^3), m^3;
x – water-cement ratio.

Later, after a series of transformations under the hands of researchers, this formula was transformed into the M. Bolomey's formula. [9]

$$S = K(\frac{1}{x} - \frac{1}{2}) \tag{4}$$

K – constant.

In [10] the concrete strength is a polynomial of the 2nd degree of x (water-cement ratio)

$$S = 50.525 - 2.915x - 0.975x^2. \tag{5}$$

The model of the dependence of concrete strength on porosity P is known [11].

$$S = S_0^{-4.84P}. \tag{6}$$

In [12], the time dependence is proposed for predicting the concrete strength without waiting for 28 days

$$S = \alpha \ln t + \beta \tag{7}$$

α, β – constants.

The origins of formula (7) can be seen in Sh. Popovich's work [13].

$$S = c_1 \log t + c_2 \tag{8}$$

c1, c2 – constants that depend on the type of cement, concrete curing, testing, etc.

Also Sh. Popovich gives the formula for the hardness of alite C3S (the main component of concrete) (9). Alite is a tricalcium silicate 3CaO x SiO2. In the first 28 days of hardening, alit makes the main contribution to the concrete strength.

$$\frac{d^2 S_1}{dt^2} = a_1^2 (s_0 - s_1) \tag{9}$$

S1 – hardness C3S in cement paste;
S0 – hardness C3S at the initial time;

a_1^2 constant, depending on the fineness of the grinding and the cement composition, etc.

[14] provides an interpretation of formula (9), which is extended to the concrete strength as such:

$$\frac{dS}{dt} = K(S_{\lim} + S_0 - S) \tag{10}$$

$$K = a_1 \exp(\frac{-a_2}{t}) \tag{11}$$

S – concrete strength, MPa;
$S_{\lim}$ – limit value S;
S_0 – initial value S;
K – depends on the product temperature.

2. Methods

The main hypothesis of the research: to build a concrete strength mathematical model, the kinetic theory of solid destruction is applied. We will consider the process of destruction without taking into account the deformation. We start from the strained and broken bonds in the concrete sample. Strained bonds occur in a concrete sample when a destructive load F is applied. Because of the concrete heterogeneity (filler, cement stone), local overvoltage occur, bonds are torn. We assume that in the process of destruction, strained connections turn into broken ones. Let us apply mathematical modeling methods: we create a concrete strength mathematical model.

Model assumptions.

$$\frac{dN_t}{dt} = \gamma N_t \tag{12}$$

$$\frac{dN_b}{dt} = \varphi(N_t - N_b) \tag{13}$$

$$N_t = \chi F \tag{14}$$

$$N_b(t=0) = 0; N_b(t=T) = \lambda V \tag{15}$$

where Nt – the number of strained bonds;
Nb – number of broken bonds;
γ – speed formation coefficient of strained bonds;
φ – speed formation coefficient of broken bonds;
χ – the proportionality coefficient between the applied load F and the number of strained bonds;
N – the total number of bonds in the concrete sample. It is assumed that at application F all bonds become strained, after that in the places of local overvoltage these bonds become broken;
λ – the constant that defines the number of bonds per volume unit.
V – sample volume.

(12) – the rate of increase in the number of strained bonds is proportional to the number of strained bonds. Their number increases in time with the action F.

(13) – the rate of increase in the number of broken bonds is proportional to the difference between the strained and broken bonds.

(14) – the number of strained bonds is proportional to F

(15) – in the initial moment of time the number of broken bonds is equal to zero. At finite time T, the number of broken bonds is proportional to the sample volume, and all the bonds become broken. It is assumed that there was a complete destruction of the sample.

IOP Conf. Series: Materials Science and Engineering **945** (2020) 012014 doi:10.1088/1757-899X/945/1/012014

(12)-(15) – a concrete strength mathematical model on the basis of kinetic theory.

The type of mathematical model indicates that (12) and (13) can be reduced to a linear homogeneous 2nd order differential equation with constant coefficients (hereinafter – HDE). The method of solving such an equation becomes clear – calculating the roots of the characteristic equation.

3. Results

We introduce the symbols:

$$N_t = x; N_b = y \tag{16}$$

Then (12) and (13) we write in the form

$$\frac{dx}{dt} = \gamma x \tag{12'}$$

$$\frac{dy}{dt} = \varphi(x - y) \tag{13'}$$

After transformations, the system (11)-(12) is reduced to the form

$$y'' + y(\varphi - \gamma) - \gamma\varphi y = 0 \tag{17}$$

We obtained a linear homogeneous differential equation with constant coefficients. As the initial conditions are set, we have the Cauchy problem. Accordingly, the type (17) and the presence of initial conditions indicate that the solution is unique. Roots of the characteristic equation

$$A = \frac{\gamma - \varphi}{2}(1 + \sqrt{1 + \frac{4\gamma\varphi}{(\varphi - \gamma)^2}}; \quad B = \frac{\gamma - \varphi}{2}(1 - \sqrt{1 + \frac{4\gamma\varphi}{(\varphi - \gamma)^2}}$$

We apply the symbols

$$\alpha = \frac{\gamma - \varphi}{2}; \beta = \sqrt{1 + \frac{4\gamma\varphi}{(\varphi - \gamma)^2}}$$

Solution of the HDE we write in the form

$$N_b = C_1 e^{\alpha(1+\beta)t} + C_2 e^{\alpha(1-\beta)t} \tag{18}$$

From (15)

$$N_b(t=0) = C_1 + C_2 = 0$$

$$N_b(t=T) = C_1 e^{\alpha(1+\beta)T} + C_2 e^{\alpha(1-\beta)T} = \lambda V$$

$$C_1 + C_2 = 0; C_1 = -C_2 = C \tag{19}$$

$$C = \frac{\lambda V}{e^{\alpha(1+\beta)T} - e^{\alpha(1-\beta)T}} \tag{20}$$

$$N_b = \frac{\lambda V(e^{\alpha(1+\beta)t} + e^{\alpha(1-\beta)t})}{e^{\alpha(1+\beta)T} - e^{\alpha(1-\beta)T}} \tag{21}$$

According to the authors, (21) is a solution to the concrete strength mathematical model (12)-(15) based on kinetic theory. It follows from (21) that if γ and φ (α=0) are equal, Nb tends to infinity, which contradicts (15). To avoid this contradiction, it should be assumed that destruction occurs when a part of the strained bonds becomes broken, and not all the strained bonds become broken. In this case, destruction means that the sample broke into fragments, the original shape is lost. But inside the fragments, there are unbroken bonds. This assumption also removes the "idem per idem" problem in (21).

4. Summary

(21) allows estimating the time required for sample destruction. As the number of broken bonds Nb in a concrete sample can be calculated using its volume. The model can be complicated by assuming that it is enough to break part of all the bonds to destroy it. If we compare the moment of destruction of the sample with the percentage of broken bonds from the number of all bonds (the number of all bonds depends on V), we can estimate at what time point the destruction of the sample will occur.

5. References

[1] Sprince A, Pakrastins L, Gailitis R Long-Term Parameters of New Cement Composites – [Electronic resource] URL: https://www.scopus.com/inward/record.uri?eid=2-s2.0-85075857384&doi=10.1007%2f978-3-030-33342 (access date 11.05.2020)

[2] Klyuev S V, Klyuev A V, Shorstova E S (2019) The micro silicon additive effects on the fine-grassed concrete properties for 3-d additive technologies *Materials Science Forum* **974** 131 – 135

[3] Klyuev S V, Klyuev A V, Shorstova E S (2019) Fiber concrete for 3-d additive technologies. Materials Science Forum **974** 367 – 372

[4] Zharikov I S, Laketich A, Laketich N (2018) Impact of concrete quality works on concrete strength of monolithic constructions *Construction Materials and Products* **1 (1)** 51 – 58

[5] Elistratkin M Yu, Kozhukhova M I (2018) Analysis of the factors of increasing the strength of the non-autoclave aerated concrete *Construction Materials and Products* **1 (1)** 59 – 68

[6] Heavy and fine-grained concretes. Technical conditions. GOST 26633-2015. – [Electronic resource] URL: http://docs.cntd.ru/document/1200133282 (access date 11.05.2020)

[7] Concretes. Methods for determining the strength of control samples. GOST 10180-2012. – [Electronic resource] URL: http://docs.cntd.ru/document/1200100908 (access date11.05.2020)

[8] Abrams D A (1918) Design of concrete mixtures. Bulleten 1: Structural Materials Research Laboratory. Chicago: Lewis Instityte

[9] Bolomey J (1942) Deformation elastigues, plastigues et de retrait de guelgues betons // Bulleten technique de la Suisse Romande **(15)**

[10] Erofeev V T, Boldina I V, Rodin A I (2014) Determination of design characteristics of concrete produced by LLC "KSM" *Fundamental research* **(9)** 1935-1941

[11] Nesvetaev G V, Kardumyan G S (2015) Influence of own deformations on porosity and properties of cement stone *Construction materials* **9** 38-42

[12] A Allahverdi, M Mahinroosta (2013) A model for prediction of compressive strength of chemically activated high phosphorous slag content cement *International journal of civil engineering* **12 (4)** 481-487

[13] S. Popovics. History of a Mathematical Model for Strength Development of Portland Cement Concrete [Electronic resource]. URL: https://www.concrete.org/publications/internationalconcreteabstractsportal/m/details/id/401 (access date 11.05.2020)

[14] Gromov Yu Yu, Ivanova O G, Lagutin A V, Lutkhon T (2001) Mathematical model of concrete hardening in the conditions of heat treatment of reinforced concrete plants *Bulletin of TSU* **6 (3)** 344-345

IOP Conf. Series: Materials Science and Engineering **945** (2020) 012015 doi:10.1088/1757-899X/945/1/012015

Development trend of the construction market in Russian regions

E V Lapteva[1]*, E P Ogorodnikova[1] and V F Remizova[1]

[1]Plekhanov Russian University of Economics in Orenburg (Orenburg, Russia)

E-mail: lapa1984@inbox.ru

Abstract. Construction in Russia is a dynamically developing type of economic activity. This is facilitated by various programs that support construction at both the municipal and federal levels. In the paper the method of economic-statistical analysis application of the construction market of the Orenburg region is formed, which will allow identifying patterns and factors influencing the development of the construction market of the Orenburg region, studying trends in the volume and structure of the construction market in order to identify promising directions of the construction market development in the region. As a result of applying methods of economic and statistical analysis (structure, dynamics and forecasting and factor analysis), structural and dynamic changes in the regional construction market are identified, trends in the dynamics of construction volume and factors affecting the growth rate (decline) of the construction market in the region are determined.

1. Introduction

Construction in Russia is a dynamically developing type of economic activity. Throughout the history of Russia, uneven construction volumes have been observed over time. In the 20th century, the maximum input of habitation is considered to be the period of 60-80s, when the country's leadership introduced the so-called standard building. The 1990s can be described as destructive in all areas of the economy, including construction. The pace of residential real estate construction declined sharply during this period. With the beginning of the new century, a new era in construction began, and new tools for implementing construction projects appeared, such as the Federal law on mortgages and shared construction. In the Orenburg region over the past two decades, the construction market has undergone significant changes, which has increased the volume of construction work, commissioning of housing. This is facilitated by various programs that support construction at both the municipal and federal levels. The development of construction improves the quality of life of the population, as the need for housing is the basis of their life for most people.

These points determine the relevance of the research topic, as well as determine its aim and objectives.

The aim of the research is to provide a comprehensive economic and statistical analysis of changes in the construction market in the Orenburg region.

Within the framework of the selected research, the following tasks were set and solved in connection with the aim:

- the theoretical foundations of construction research were studied;
- the analysis of the current state of construction in the Orenburg region was carried out;

– the analysis of housing commissioning in the Orenburg region was carried out;
– the analysis of the dynamics of housing prices in the primary market of the Orenburg region was done;
– forecast of housing input volumes using the trend equation, multiple regression equation, regression equation, and trend equation was done;
– the factors that have the greatest impact on the development of the construction market were identified.

The subject of the research is the construction market of the Orenburg region.

The object of research is the Orenburg region.

The choice of the research period (2001-2018) is justified by the presence of crisis phenomena in the economy during this period, including those related to the introduction of US and European sanctions.

2. Materials and methods

In the process of studying the current state of the construction market of the Orenburg region, the authors used such methods of economic and statistical research as tabular, graphical, analysis of structural changes, analysis of dynamics series, methods of forecasting using trend models and factor analysis.

The authors suggest the following scheme of economic and statistical research of the state of the external debt of the Russian Federation (Fig. 1).

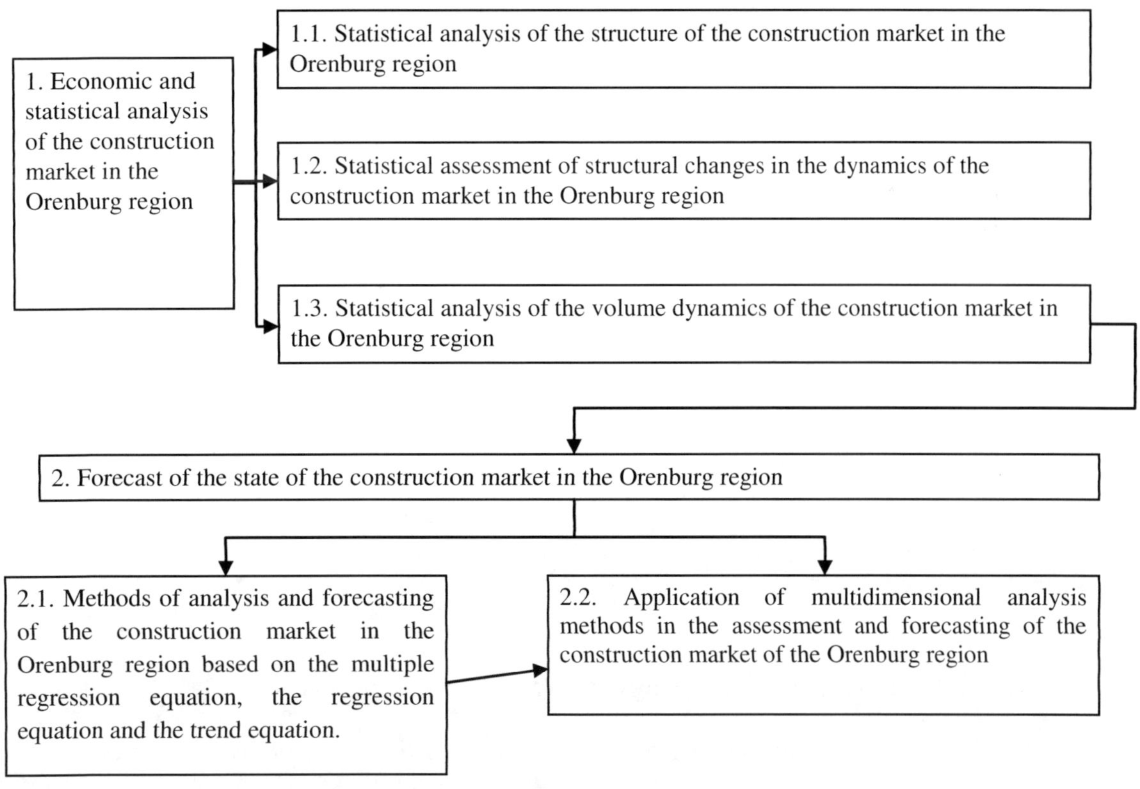

Figure 1. Scheme of economic and statistical research of the state and prospective development of the construction market of the Orenburg region [5].

As a result of using the proposed scheme of economic and statistical research of the state and prospective development of the construction market, it is possible to obtain comprehensive assessment of the dynamics and structure of construction in the region, forecast indicators and evaluate the prospects for development.

3. Results

To analyze the state of the construction market in the Orenburg region, it is important to consider its development dynamics (Fig. 2). The growth rate in 2007 compared to 2006 was 20.5%, this was largely due to the development of mortgage lending. However, under the influence of the global financial crisis in 2008, the Russian economy entered a recession, which could not but affect construction. [3] The volume of construction work in 2009 compared to 2008 reduced by almost a third. The recovery of construction rates occurred only by 2011, when it was possible to reach the level of 2008. In the future, there is an increase in the volume of construction work. Even the negative economic events of 2015-2016 did not have a critical impact on construction. In 2016, the volume of construction work amounted to 49750.8 million rubles, and in 2018 it reached more than 66 billion rubles. Let us analyze the construction trends for 2005-2018. [6]

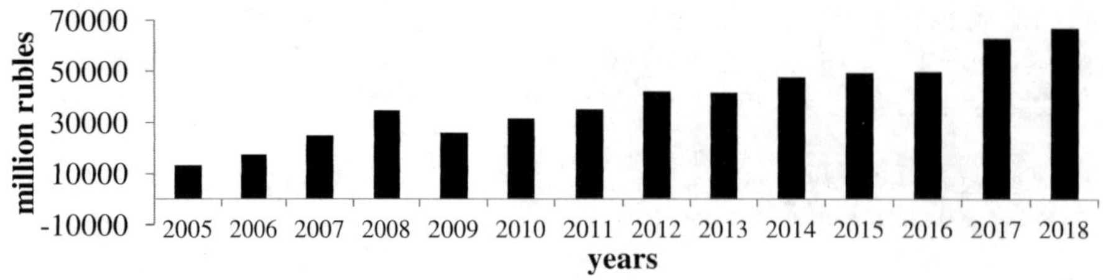

Figure 2. The volume of work performed by the type of economic activity "Construction" in the Orenburg region, million rubles.

The steady economic development of Russia since the beginning of the 2000s has been reflected primarily in housing construction. Over the past fourteen years, every year there has been an increase in the area of residential buildings in the region (Fig. 3). The fundamental factor in this situation is mortgage lending. The long-term nature of mortgage loans has had a positive impact on the effective demand of the population. [7] But since 2008 in Russia, as in the whole world, there was a financial and economic crisis, which negatively affected the commissioning of residential buildings and housing construction in general, as well as the solvency of the population.

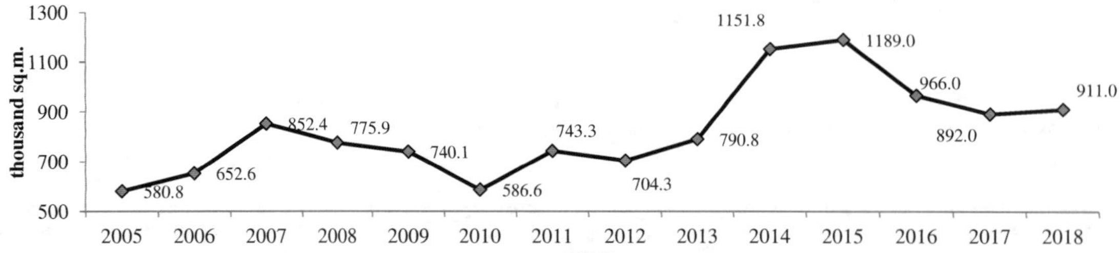

Figure 3. Dynamics of commissioning of residential buildings in the Orenburg region in 2005-2018.

In 2009, housing commissioning decreased in comparison with 2008 by 35.8 thousand m^2, and in 2010 by 189.3 thousand m^2. But in 2011, due to state support, there was an increase in the commissioning of housing by 156.7 thousand m^2. State programs to provide housing for the population began to be developed. These include, in particular, programs to relocate residents of the region from dilapidated and emergency housing, providing housing for socially vulnerable categories of citizens, such as orphans and the disabled, eliminating queues for housing among military personnel, etc. And, starting from 2014, there was a sharp jump in the commissioning of residential buildings in the Orenburg region compared to 2013 by 361 thousand m^2, and in 2015 it increased by another 37.2 thousand m^2. In 2015, the maximum value of housing commissioning was reached, which was 1190.1 thousand sq. m.

However, at the end of 2015, the country was again subjected to crisis phenomena, this time related to economic sanctions imposed against the Russian Federation, as well as the destabilization of the national currency and oil exchange rates. This negatively affected both the population's ability to pay and the region's ability to implement social programs. Thus, the introduction of housing in 2017 compared to 2015 decreased by almost 300 thousand sq. m. And only in 2018, there was a new trend to increase the volume of housing construction. [8]

One of the factors that influenced the increase in demand for housing was the development of mortgage lending. As a result, there was an increase in the effective demand of the population. The main part of commissioning of residential buildings is due to own and borrowed funds in urban areas.

The important indicator of the economic state of the region is the market value of housing. Housing affordability is one of the key needs of the population, and has a great impact on the social security of citizens. Here there are data on the cost of housing in the primary and secondary real estate market for 2007-2018.

During the study period, fluctuations in the housing market prices are similar to changes in the pace of construction. If the average difference in the cost of sq. m. is about 8%, in some years (2014-2015), the gap reached 17-18%. In the future, there is a gradual smoothing of the difference and in 2017 the cost of housing in the primary and secondary markets is almost the same. In 2018, the housing market has again seen an increase in the cost of housing. At the same time, prices in the primary and secondary housing markets are comparable. So, the cost of 1 sq. m. in the new building on average was 39138 rubles, in the secondary market 39875 rubles.

Due to the introduction of Federal law 214-FZ into legislative proceedings, the situation is changing significantly. In 2018, the main part of construction projects in the Orenburg region was built on the basis of the groundwork made in the period preceding the entry into force of the new law. [17]

In the period from 2007 to 2018, there is an increase in the cost of primary housing in the Orenburg region. In comparison with 2007, in all subsequent years, the cost of housing increased from 18.4% to 43.58%. In 2018 compared to 2008 the price of primary residential real estate in the Orenburg region increased by 11,142.4 rubles or 39.8%. The maximum increase in the price of 1 sq. m. falls on 2014 and is 43.58% in relation to 2007. [18] (Fig. 4).

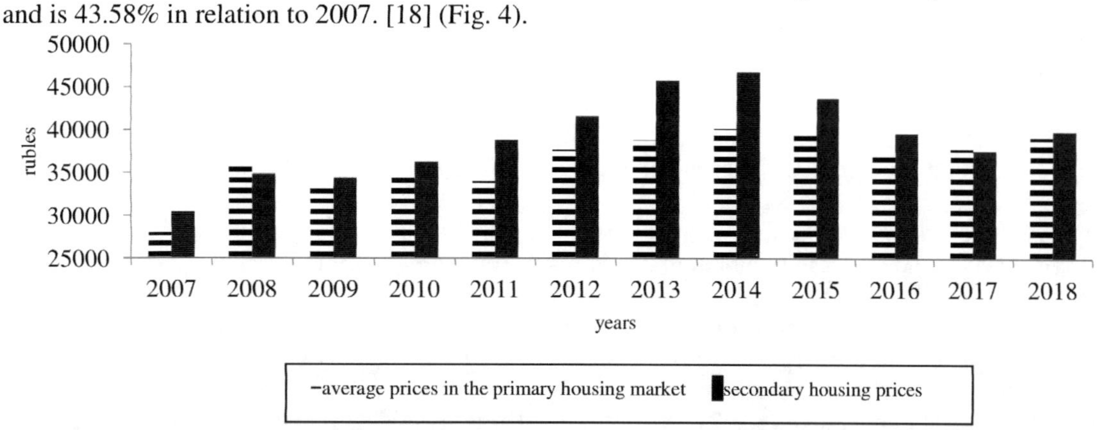

Figure 4. Average cost in the primary and secondary housing market in the Orenburg region, rubles/sq. m.

The largest increase in the cost of housing occurred in 2008 compared to 2007 by 28.06%. Also in 2012, the cost of primary residential real estate increased by 10.66% compared to 2011. In other periods, the changes are less expressed. The cost of housing has been declining since 2015, in 2017 it increased slightly, and only in 2018 there was a trend to increase the cost of real estate in the housing market. The increase in 2018 was 3.58.

To determine the effective trend equation, we need to pay attention to the approximation coefficient. Among the given trends, the power and polynomial functions have a higher indicator than the others. To make decisions and forecast, we will use the power function trend.

Let us check the selected model for adequacy by calculating the average relative error using formula 1.The result of calculating the average relative error showed that the power function has an error of 4.04%. So we can say that the function has high accuracy.

Let us check the selected model for adequacy using the Durbin-Watson criterion according to the formula:

$$d = \frac{\sum_{t=2}^{n}(e_t - e_{t-1})^2}{\sum_{t=1}^{n} e_t^2} \approx 2(1-r), \quad (1)$$

where r_t – first-order autocorrelation.

The Durbin-Watson criterion d=1.99. We will determine the values of the Durbin–Watson statistics from the table at the 5% level of significance of the Durbin-Watson criterion, the upper and lower confidence limits. According to the Durbin-Watson table for n=12 and k=1, we find: d1 = 0.97 and d2 = 1.33. Therefore, there is no autocorrelation and the model is adequate. The resulting model can be used for forecasting. [11]

Thus, we can conclude that in the Orenburg region, construction plays an important role in the development of the socio-economic aspect of the population's life. Crisis processes in the country's economy as a whole have a negative impact on the financial sector of the region. However, despite the negative phenomena, the region managed to maintain a high volume of construction work in 2015-2017, and significantly increase it in 2018. [12]

One of the indicators of the economic activity "Construction" is the level of commissioning of residential buildings. In 2001-2018, there was a significant change in the dynamic range of the above indicator.

Commissioning of housing in the analyzed period was generally increasing. The first maximum peak occurred in 2007, when 852/4 thousand square meters of housing were introduced. Further, under the influence of the global financial crisis in 2008, commissioning of housing was decreasing and in 2010 this figure was only 586.6 thousand square meters.

The second maximum peak of housing commissioning was observed in 2015, and amounted to 1189 thousand square meters. This is 602 thousand square meters more than in 2010, i.e. more than twice. This phenomenon indicates a favorable financial situation of the population and an increase in the number of transactions for the purchase of housing. [21]

Since the beginning of 2016, the country's economy has entered a period of recession, which has affected the number of built and commissioned meters of residential real estate. In 2017, the volume of commissioned housing was 892 thousand square meters, which is almost 300 thousand square meters less than in 2015.

The increase in commissioning of residential buildings in the Orenburg region is an indicator of the stable financial situation of the population, which resulted in an increase in the number of transactions for the purchase of housing.

During the period under review, there are periods of significant decline in the dynamics of housing stock commissioning; this is due to depressive phenomena in the economy. [10]

If the above trend continues, the volume of housing commissioning in 2019 may be 958.10 thousand sq.m. and in the range (883.31;1032.89), in 2020 it may be 973.30 thousand sq. m. and in the range (897.50;1049.10), in 2021 it may be 987.99 thousand sq. m. and in the range (911.18;1064.80), in 2022 it may be 1002.20 thousand sq. m. and in the range (924.40;1080.01), and in 2023 it may amount to 1015.97 thousand sq. m. and be in the range (937.18;1094.76).

The forecast of housing commissioning for 2019-2023 was carried out using the trend equation and the regression equation. [20]

The trend of housing commissioning is best described by a power trend. According to the forecast, the volume of housing commissioning will increase, and in 2019 it may reach 958.10 thousand sq. m. and be in the range (883.31;1032.89), and in 2023 it may reach 1015.97 thousand sq.m. and be in the range (937.18;1094.76) (table 1).

Table 1. Forecast values of residential real estate input in the Orenburg region (point and interval forecast), thousand sq. m.

Years	Lower confidence limit	Forecast	Upper confidence limit
2019	883.31	958.10	1032.89
2020	897.50	973.30	1049.10
2021	911.18	987.99	1064.80
2022	924.40	1002.20	1080.01
2023	937.18	1015.97	1094.76

Source: compiled by the authors

The correlation and regression analysis revealed a strong dependence of housing commissioning on the level of per capita income of the population and the consumer price index. Using the trend equation, the values of the above parameters are predicted for 2019-2023. Based on this, the forecast of housing commissioning is made using the regression equation. According to the forecast, housing commissioning in the Orenburg region will increase in the next five years. In 2019, housing commissioning may amount to 1017.18 thousand sq. m. and be in the range (947.36; 1086.99), and in 2023 the level of housing commissioning can reach 1129.8 thousand sq. m. and may be in the range (1059.99; 1199.61).

The market value of housing is also an important indicator of the socio-economic development of the region, as it is the availability of housing that has a great impact on the social security of citizens.

4. Summary

Thus, we can conclude that the forecast values of the volume of housing commissioning in the Orenburg region are increasing. According to calculations based on the trend equation housing commissioning in the region in 2023 may reach 1015.97 thousand square meters.

The forecast of changes in the cost of housing in the primary market in the period 2019-2023 was made using the trend equation. According to the forecast, the cost of housing in the primary market will increase, and in 2019 it may be 40150.96 rubles/sq. m. and be in the range (37681.39;42620.53), and in 2023 it may reach 41420.05 rubles/sq.m. and be in the range (37711.26;45128.84).

Correlation and regression analysis was used to determine the relationship between performance and factor characteristics. Based on the regression equation, a forecast of changes in the value of primary real estate was made. According to it, the growth rate of the cost of primary housing in the Orenburg region with realistic forecast may be 102.3 % and be in the range (101.7;102.9), with optimistic forecast may reach 104.3% and be in the range (103.7;104.9), with pessimistic forecast may be 95.4% and may be in the range (94.8; 96.0) [22].

In the Orenburg region, construction activity is one of the most progressive, effective and large-scale types of activity, but there are problems such as:

- lack of qualified workers who leave for other, more developed regions on the background of the economic decline;
- the cost of building materials has increased, which leads to higher prices for residential buildings;
- irrational territorial placement of production facilities leads to high transport costs, which fall on the cost of construction projects.

To solve these problems, the Orenburg region needs to continue implementing new technologies and its own production of construction materials, which will simplify the process of building houses and make housing more affordable for the population. It is also necessary to create competitive jobs for high-quality professionals in order to build high-quality, comfortable and aesthetic housing.

5. References

[1] Akulova A Sh, Golchenko E I 2016 Analysis of the current state and main trends in the development of the construction market and its segments (on the example of the Orenburg region) (Scientific periodical) **11** 33-35

[2] Alexandrova L A, Lapteva E V, Ogorodnikova E P 2019 Digital transformation of the Russian economy, features and ways of development *Materials of the international research competition: YEAR EXPERT 2019* 4-9

[3] Evseeva E I 2015 Modern problems of housing construction development in Russia *Volga Region Scientific Bulletin* **6-2 (46)** 16 – 18

[4] Kitaev M A 2015 Advantages and problems of low-rise construction in Russia *Volga Region Scientific Bulletin* **7** 82 – 87

[5] Lapteva E V, Portnova L V Practical training in statistics: theory of statistics and economic statistics. Orenburg: IPK "University" 227

[6] Kogan Yu V 2019 Topical issues of interconnected development of housing stock and social infrastructure *Housing construction* **9** 48-52

[7] Kozhukhar A O 2016 Analysis of problems and main trends in the development of housing construction in Russia *Problems of Economics and Management* **12 (64)** 46 – 49

[8] Buzyrev V V, Plakhiy A Yu 2016 Elimination of the housing problem based on the revival of industrial housing construction in the regions of Russia *Problems of the modern economy* **1** 128 – 130.

[9] Ogorodnikova E P, Petrina A V 2017 Assessment of tax potential as a component of investment attractiveness of the Orenburg region *Competitiveness in the global world: economy, science, technology* **12 (59)** 970-976.

[10] Korotin S A 2013 Features and development trends of the regional market of low-rise housing construction *Regional economy* **1 (22)** 93 – 96

[11] Saenko I A 2013 Analysis of the dynamics and structure of housing commissioning in the Russian Federation *Economic analysis: theory and practice* **37(292)** 60- 65

[12] Svenik T V 2015 Adjustment of housing construction strategies in the context of the crisis *Proceedings of Irkutsk State Economic Academy* **6** 941 – 946

[13] Soldatova L A, Soldatov S A, Davydova N Yu 2015 On the analysis of the housing construction market in the Orenburg region *Housing strategies* **2 (2)** 101-108

[14] Starkova N O, Mirzoyan A R 2014 Current state and features of housing construction development in the Russian Federation *Science and modernity* **33** 222-226

[15] Shubin I L, Umnyakova N P, Matveeva I V, Andrianov K A 2019 The quality of the building shell – the basis for creating environmentally safe living environment *Housing construction* **6** 10-15

[16] Kannan M R, Santhi M H 2013 Constructability Assessment of Climbing Formwork Systems Using Building Information Modeling *Procedia Engineering* **64** 1129-1138

[17] Lyalin D O, Mashtaler S N, Dmitrenko E A 2017 Application of the Autodesk Revit software complex in project activities *Proceeding of the Donbas National Academy of Civil Engineering and Architecture* **3** 23-27

[18] Martin D D 2017 Evans External balances, trade and financial conditions *The Journal of International Economics* **107** 165-184

[19 Roos G, Pike S, Fernström L 2010 Intellectual capital: management approach *Saint Petersburg: Higher School of Management* 436

[20] Sadkova E 2013 Out of 100,000 migrants having a work permit, only 352 persons are qualified specialists URL: https://ok-inform.ru/obshchestvo/3702-v-peterburg-chashche-vsego-priezzhayut-uzbeki.html (access date: 15.04.2019)

[21] Combes P P, Mayer T, Thisse J F 2008 Economic geography. The integration of regions and nations. Princeton University Press 399

[22] Tsyganov A 2018 *Control over Foreign Investments in the Economic Entities of Strategic Importance* [Electronic resource] URL: www.usrbc.org/pics/File/tsyganov/FZ-57%20USRBC%202012%20%5 BCompatibility%20 Mode%5D.pdf (access date: 15.10.2019).

IOP Conf. Series: Materials Science and Engineering **945** (2020) 012016 doi:10.1088/1757-899X/945/1/012016

Obtaining sorption material from sorghum for aqueous solutions purification from heavy metals

E V Fomina[1*], S V Sverguzova[1], ZH A Sapronova[1] and M I Kozhukhova[2]

[1]Belgorod State Technological University named after V.G. Shoukhov, 46, Kostukova Str., Belgorod, 308012, Russia
[2]Department of Civil&Environmental Engineering, University of Wisconsin Milwaukee, P.O. Box 413, Milwaukee, WI 53201, USA

E-mail: fomina.katerina@mail.ru

Abstract. The paper presents the results of using straw (stalks) of sorghum as a sorbent for removing heavy metal ions of Cu (II) from aqueous solutions. The physical and chemical analysis of sorghum has been carried out. The modification of plant materials by exposure to a temperature of 130-150 ° C is proposed. Thermal activation allows obtaining a highly developed surface of the sorghum microstructure and increasing the total pore volume and their average diameter. The regularities of the effect of the concentration of the modified sorbent and the differences of its dispersed particles on the efficiency of water purification are investigated. The purification of an aqueous solution containing 50 mg/dm^3 of Cu^{2+} ions was 95% under using a thermo-treated sorghum of 0.63–0.1 mm in an amount of 2.5 g per 100 ml of an aqueous solution. Thermal modification of sorghum allows you to get an effective sorbent and immediately solve two environmental problems, such as wastewater treatment and utilization of plant waste.

1. Introduction

At the present stage of the development of human civilization, techno genic pollution has affected almost all vital resources. Atmosphere, soil, surface and unprotected groundwater were especially polluted. Interest in water use is excessively high, as industry develops and megalopolises grow. The accumulation of excess heavy metal ions, one of which is copper, is of great concern due to toxicity. Heavy metals cause damage by accumulating in humans and animals. This can lead to serious illnesses such as mental confusion, allergies, memory loss, irritability, and balance in digestive problems. Effective removal of heavy metal ions from aqueous systems is very important for the protection of the environment and public health [1, 2].

Adsorption is an advantageous option for water and wastewater treatment due to its convenience, simplicity and ease of design. In addressing this issue, plant waste can be viewed as a sorbent obtained with minimal effort. The production of organic sorbents from renewable plant materials corresponds to current research trends [3]. Agricultural waste is varied in structure and properties, therefore, there are differences in sorption capacity, pore volume, efficiency of extraction of various substances from aquatic environments, therefore, research on the features and modification methods for their use in water purification does not lose its relevance [4-9] . The advantages of using plant waste for wastewater treatment are as following: relatively simple processing technology, good adsorption potential, selectivity for heavy metal ions, low cost, availability, ease of regeneration [10].

Sorghum has a variety of forms and varietal crops, both annual and perennial. In general, sorghum belongs to the grass plants of the family of cereals. Sorghum grains are processed into cereals, flour and starch. Wickerwork, paper, brooms are made of straw. In addition, sorghum straw is the main source of natural fibers as an alternative raw material for bio composites. For example, it is considered as a biopolymer for the production of microfiber [11]. Sorghum straw is a source of getting cellulose production [12]. Sorghum straw is a renewable, cheap and widespread resource. These materials, due to their high porosity, are of interest as feedstock for the production of sorbents.

2. Materials and methods

Thermal Analysis (DTA) was carried out with the help of derivatograph of model 431Q -1500 and device of synchronous thermal analysis STA 449 F1

The analysis of peculiarities of the chemical composition and structure of the CCP samples has been carried out by using the scanning electron microscope of high resolution «TESCAN MIRA 3 LMU».

To determine the bulk density the method GOST 19440-94 was used. The moisture of material, water absorption and ignition losses were determined using standard methods [13].

For roasting sorghum, a Liop LF-7/13-G2 muffle furnace was used.

To study the specific surface area and porosity of the sorbent at the micro level, we used the method of gas adsorption using the SoftSorbi – II ver.1.0 device.

The purification efficiency of model solutions containing Cu^{2+} ions was carried out in a static mode as follows: the sorghum sample was placed in 100 ml of an aqueous solution with a copper ion concentration of 50 mg/dm^3 and it was mixed for 20 minutes in a conical flask. Then the contents of the flask were filtered through a paper filter. The concentration of Cu^{2+} ions in solutions before and after purification was determined by a colorimetric method according to the method GOST R 4388-72 with sodium diethyldithiocarbamate indicator at a wavelength of 440 nm.

After that the contents of the flask were left for sedimentation for 15 minutes. Clarified water was examined for the petroleum oil concentration, using concentration meter КН-3 (Russia)

The purification efficiency (E) was calculated according to the formula:

$$E = (C_1 - C_2)/C_1 \cdot 100\%$$

where C_1 and C_2 - the concentrations of substances before and after of water purification, respectively.

The granulometric compositions of materials were determined using the method of sieve analysis. The principle of sieve analysis is to find the amount of material retained by a sieve with holes of a certain size. The sieves are arranged one above the other sequentially as the size of the holes in the grid decreases. 2.5-5.0; 2.0-2.5; 1.4-2.0; 1.4-1.25; 0.63-1.25; 0.63-1mm.

The grinding of sorghum samples was carried out in a porcelain mortar.

3. Results and discussion

The physical and chemical characteristics of sorghum were investigated in the work (Table 1). Sorghum water retention is 35%. Ignition losses are 98%, which indicates the predominant content of organic substances.

Table 1. Physical and chemical properties of Sorghum.

pH aqueous extract	7,0
Loss on ignition, %	98
Moisture content, %	35
Bulk density, kg / m^3	180
Humidity, %	8,5

IOP Conf. Series: Materials Science and Engineering **945** (2020) 012016 doi:10.1088/1757-899X/945/1/012016

A significant content of organic substances in the form of an amorphous phase is confirmed by the results of x-ray structural analysis (Fig. 1). The roentgenogram of the sample has a characteristic appearance. It is a broad line (gallo) with an angular width of $2\theta = 17 - 28$ °. The amorphous fraction in sorghum is represented by lignin and hemicellulose, which covers the semi-crystalline fraction of cellulose [14].

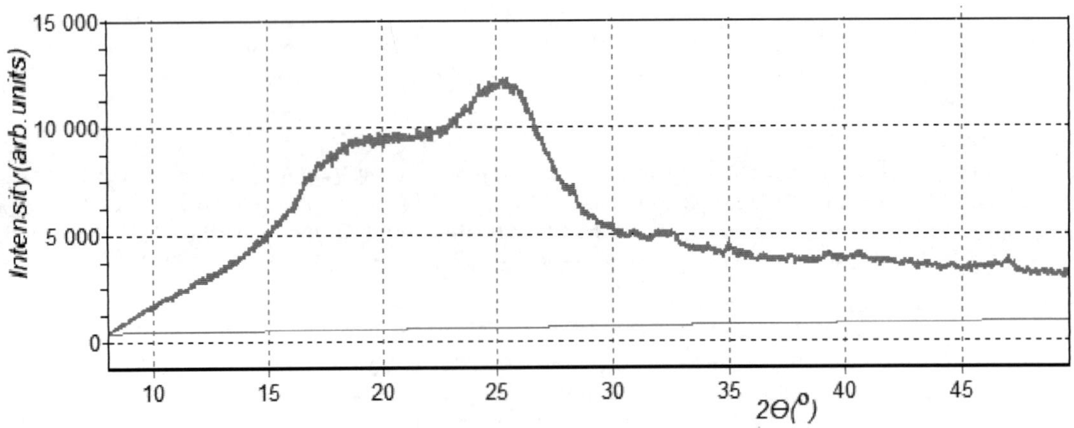

Figure 1. Radiograph of stalks of sorghum.

The thermal characteristics of sorghum has been studied using thermo gravimetric analysis (TGA), which was determined in the temperature range of 0 – 700 °C (Fig. 2).

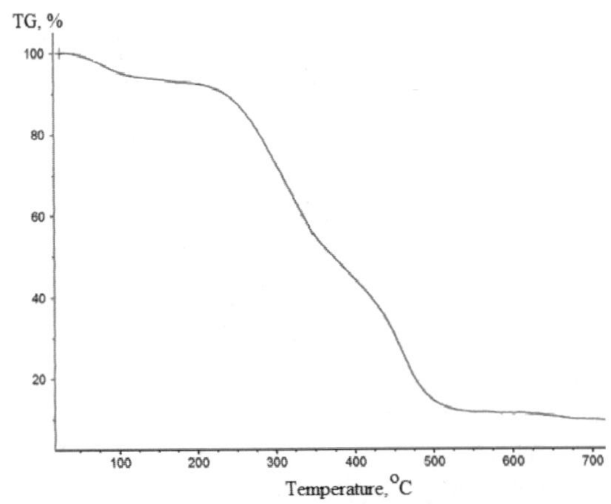

Figure 2. Thermogravimetric analysis of sorghum (TGA).

Fig. 2 shows the variety of the fiber mass due to temperature. As the temperature rises, the weight of the fiber decreases. A significant jump in weight reduction is observed at temperature above 250 °C. This temperature was chosen as the maximum for further thermal modification of sorghum.

The modification was carried out as follows: stalks of sorghum of 5 mm length, dried for 3 days at a temperature of 20 – 23 °C, air humidity of about 30%, were burned in a muffle furnace at various temperatures for 20 minutes. It was experimentally confirmed that when conducting thermal activation with a temperature above 150 °C, the dried stalks are charred and burned. The optimal activation temperature was 130 – 150 °C.

Images of the sorghum particles surface were obtained by the scanning electron microscopy method.

IOP Conf. Series: Materials Science and Engineering **945** (2020) 012016 doi:10.1088/1757-899X/945/1/012016

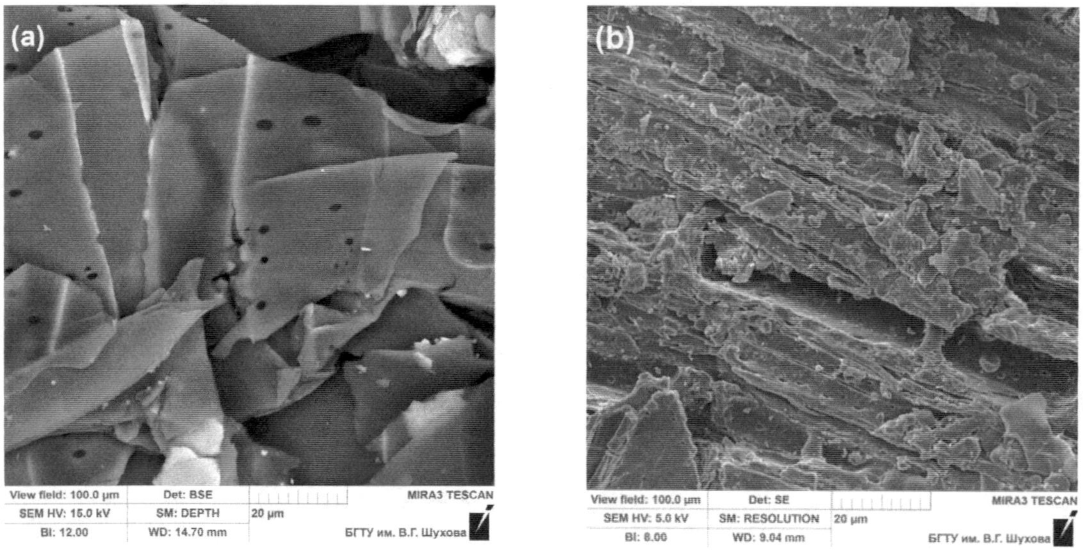

Figure 3. Microstructure of sorghum surface: a - unmodified sorghum; b - modified sorghum.

Based on the microstructure, the crushed sorghum stalks are a smooth surface with pores (Fig. 3 *a*). During heat treatment, dirt and wax are removed from the surface of the sorghum stalk, thus creating a rough, highly developed surface (Fig. 3 *b*). According to the results obtained as a result of adsorption / desorption of nitrogen, it was found that the specific surface according to the BET method (SBET) of the initial sorghum was 0.9 m^2/g (Tab. 2). After heat treatment, the specific surface of the sorption material increased by 42%.

Table 2. The change in the physical properties of the material.

Sample	S_{BET} (m^2/g)	Square of pores surface, S_{por} (m^2/g)	The total pore volume with a diameter of from 1.7 to 300 nm, $V_{por} \times 10^{-3}$ (cm^3/g)
Sorghum	0.9	0.19	0.53
Modified sorghum	0.14	0.27	0.58

According to the analysis of the pores distribution at the micro level, an increase in the total porosity of the modified material is observed with an increase in the number of pores on the nanometric range from 1.7 to 300 nm (Tab. 2). The increase in total porosity and their specific surface is an important factor in adsorption processes.

The results of the study of the efficiency of purification of an aqueous solution containing Cu^{2+} ions with sorbents with different particle diameters are shown in Figures 4 and 5.

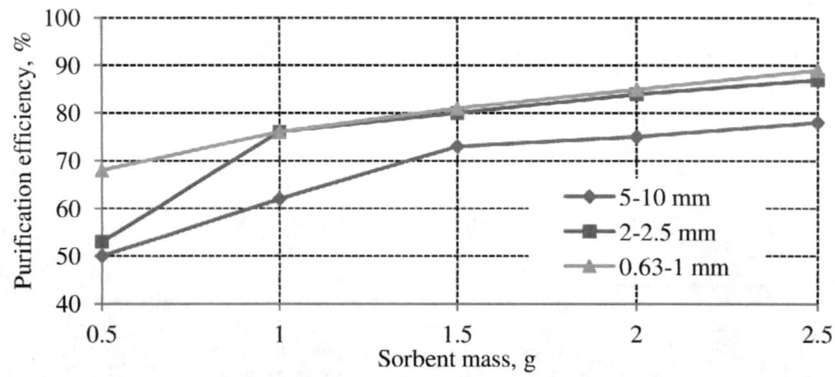

Figure 4. Dependence of the purification efficiency of an aqueous solution containing 50 mg / dm^3 of Cu^{2+} ions on the mass and dispersity of sorghum particles without modification.

From the presented illustrations, it is obvious that with an increase in the specific surface of sorghum particles due to its preliminary grinding, the degree of purification increases. Sorghum particles with a size of 5–10 mm have the lowest degree of water purification of 78%. The sorghum fraction with a particle range of 0.63–1 mm increases the degree of water purification from heavy metals to 89%.

Significant changes in water purification are observed when modified sorghum is used (Fig. 5).

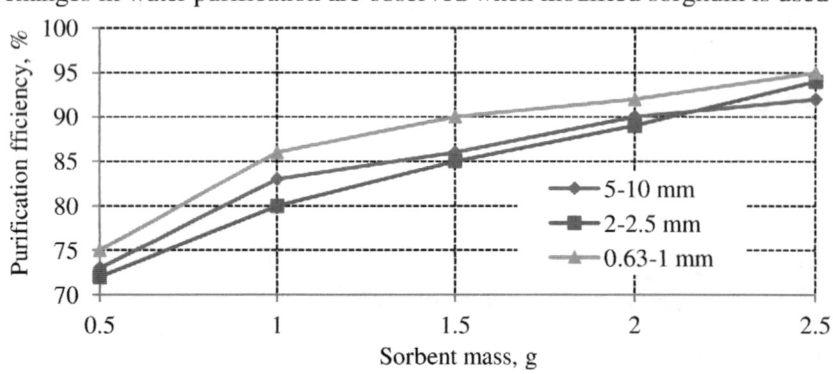

Figure 5. Dependence of the purification efficiency of an aqueous solution containing 50 mg/dm^3 of Cu^{2+} ions on the mass and dispersity of particles of thermal modified sorghum.

The use of modified sorghum increases the efficiency of water purification with all the differences in the dispersity of particles.

The highest rates of purification of an aqueous solution containing 50 mg/dm^3 of Cu^{2+} ions were 95% using thermal-treated sorghum of 0.63–1 mm in the amount of 2.5 g per 100 ml of an aqueous solution.

4. Summary

As a result of the experiments, it was proved that it is possible to use sorghum plant waste to purify wastewater from heavy metal ions Cu^{2+} Processing plant raw materials with a temperature of 130 – 150 °C for 20 minutes allows you to increase the activity of the sorbent by obtaining a highly developed surface, increasing the total porosity. The high purification efficiency of a 95 % solution containing Cu^{2+} ions is possible due to the use of a modified sorghum of fraction 0.63–1 mm.

IOP Conf. Series: Materials Science and Engineering **945** (2020) 012016 doi:10.1088/1757-899X/945/1/012016

5. References

[1] Baken K A, Sjerps R M A, Schriks M, Wezel A P 2018 Toxicological risk assessment and prioritization of drinking water relevant contaminants of emerging concern *Environment International* **118** 293–303

[2] Huang G, Yang C, Zhang K, Shi J 2009 Adsorptive removal of copper ions from aqueous solution using cross-linked magnetic chitosan beads *Chinese Journal of Chemical Engineering* **17(6)** 960–966

[3] Yingjie Dai, Qiya Sun, Wensi Wang, Lu Lu 2018 Utilizations of agricultural waste as adsorbent for the removal of contaminants: A review *Chemosphere* **211** 235–253.

[4] Tang C, Potter I D, Xu J, Bandara T 2020 Mechanisms for the removal of Cd(II) and Cu(II) from aqueous solution and mine water by biochars derived from agricultural wastes *Chemosphere* **254** 126745

[5] Abdus-Salam N, Adekola S K 2018 Adsorption studies of zinc(II) on magnetite, baobab (*Adansonia digitata*) and magnetite–baobab composite *Applied Water Science* **8** 222

[6] Sverguzova S V, Sapronova Z A, Svyatchenko A V, Otiti T 2018 Adsorption of spun oil with native and thermal modified of chestnut tree waste *Construction materials and products* **1(1)** 4–11

[7] Sapronova Z A, Sverguzova S V, Fomina E V 2017 Nanocomposite carbon-bearing sorption material *Advances in Engineering Research* **133** 728–733

[8] Sverguzova S V, Shaykhiev I G, Grechina A S, Shaykhieva K I 2018 The use of waste from the processing of oat biomass as a sorption material for the removal of pollutants from aquatic environments (literature review) *Construction and environmental management* **2(67)** 51–60

[9] Elnikov D A, Sverguzova ZH A and Sverguzova S V 2011 The influence of the temperature treatment of the defect on the purification efficiency of model solutions from dyes *Bulletin of BSTU named after V.G. Shukhov* **2** 144–147

[10] Singh R 2020 Recycling of Agricultural Waste for Wastewater Treatment *Encyclopedia of Renewable and Sustainable Materials* **2** 514–519

[11] Handayani S, Husnil Y A, Handayani A S, Ismojo and Chalid M 2019 Application of waste sorghum stem (sorghum bicolour) as a raw material for microfibre cellulose *IOP Conf. Series: Materials Science and Engineering* **509** 012015

[12] Alves J A A, Santos M D L, Morais C C et al 2019 Sorghum straw: Pulping and bleaching process optimization and synthesis of cellulose acetate *International Journal of Biological Macromolecules* **135** 877–886

[13] Butt Y M, Timashev V V 1973 Practicum on chemical technology of binder materials *High School* 503

[14] Dhyani V, Kumar J, Bhaska T 2017 Thermal decomposition kinetics of sorghum straw via thermogravimetric analysis *Bioresource Technology* **245(A)** 1122–1129

Acknowledgments

This work was supported by a grant from the President of the Russian Federation for state support of young Russian scientists – candidates of sciences and doctors of sciences and leading scientific schools of the Russian Federation, application number МД-1249.2020.5. The work is realized in the framework of the Program of flagship university development on the base of the Belgorod State Technological University named after V.G. Shukhov, using equipment of High Technology Center at BSTU named after V.G. Shukhov.

IOP Conf. Series: Materials Science and Engineering **945** (2020) 012017 doi:10.1088/1757-899X/945/1/012017

Sustainability of functioning of enterprises of the building materials industry of the Belgorod region in 2018-2019

I A Slabinskaya[1] and O B Benderskaya[2*]

[1] Department of Accounting and Auditing, Belgorod State Technological University named after V.G. Shukhov, Kostyukov St., 46, Belgorod, 308012, Russia
[2] Department of Accounting and Auditing, Belgorod State Technological University named after V.G. Shukhov, Kostyukov St., 46, Belgorod, 308012, Russia

E-mail: obenderskaya@gmail.com

Abstract. Ensuring the sustainability of functioning under all changes in the economic situation in Russia and in the world continues to be one of the most important goals of managing an enterprise in any sector of the real sector of the economy. This article investigates the stability of the functioning of a sample of the most famous enterprises in the construction materials industry of the Belgorod region: OAO Belgorodasbestocement, ZAO Belgorodskij cement, OAO Belgorodstrojdetal', OAO Zavod ZHBK-1 and AO Strojmaterialy, using the author's methods for assessing the stability of functioning. The study period is 2018-2019. The informational basis of the study is the public financial statements of enterprises for 2019. The assessment methodology is based on a system of 12 indicators-indicators of the stability of functioning. For a comparative intercompany assessment of enterprises, the author's method of calculating a comprehensive assessment was used – a modified method of scoring. Based on the obtained values of comprehensive assessments, stability ratings for the functioning of enterprises for 2018 and for 2019 are compiled.

1. Introduction

Ensuring the sustainability of functioning continues to be one of the most important management goals and one of the main problems for enterprises in the real sector of the economy of the Russian Federation. Various aspects of this issue are highlighted in the works of such Russian scientists as S M Bukhonova, Yu A Doroshenko, E N Chizhova [3-6, 8, 9]. The subject of this article is the intercompany comparative comprehensive assessment (CA) of the sustainability of enterprises. Objects of research – five enterprises of the building materials industry of the Belgorod region. This is one of the leading industries in the region. This is due to the fact that in the Belgorod region there is a serious scientific school in this direction [1, 7, 13, 15]. The objectives of the study are to assess the level of sustainability of enterprises in the period 2018-2019 and compiling sustainability ratings of enterprises over the years.

2. Methods and techniques

The main method of the study is a modified method of scoring, one of the methods for calculating a comprehensive assessment [10, 12]. The study is also based on a systems approach, methods of analysis and synthesis (generalization), it uses spatial multidimensional comparisons, the coefficient method, and statistical methods (calculation of average values and growth rates).

IOP Conf. Series: Materials Science and Engineering **945** (2020) 012017 doi:10.1088/1757-899X/945/1/012017

The methodology for assessing the sustainability of the functioning of enterprises is based on a system of twelve indicators-indicators of the stability of functioning [11]. Based on them, using the modified method of scoring, a comprehensive assessment is calculated that provides a quantitative assessment of the level of stability of the enterprise for a given period of time. CA focus – the more the better; the range of possible values is from 0 to 10 points. Values 0 and 10 are determined by the best and worst sample values of the indicators. The sustainability rating of enterprises is based on calculated CA values.

3. Results

As can be seen from table 1, in 2018, the share of stable sources of capital varied from 20 % to 94 % in the sample. The range of values of provision with stable sources of stocks of enterprises was also large – from complete insecurity to four times coverage. The degree of coverage of current liabilities with current assets ranged from 40 % to 1051 %. Four out of five enterprises were solvent and financially stable. The asset turnover rate ranged from 0.55 revolutions to 1.42 revolutions per year. All enterprises grew sales revenue; three of them showed growth in retained earnings and equity. The main activity was profitable at three out of five enterprises. OAO Belgorodstrojdetal' was unprofitable in all profit indicators, and AO Strojmaterialy only in profit from sales [14].

Note that the best values of indicators in the tables are highlighted, and the worst are underlined.

Table 1. Indicators, comparative CA and stability rating of the functioning of a sample of Belgorod enterprises of the building materials industry in 2018.

Indicators, units	Indicator values				
	OAO Belgorod-asbestocement	ZAO Belgorodskij cement	OAO Belgorod-strojdetal'	OAO Zavod ZHBK-1	AO Strojmaterialy
The share of stable sources of capital (average annual value)	0.75	0.88	0.90	**0.94**	<u>0.20</u>
Stable sources share in stock financing (average annual value)	0.88	**4.03**	1.98	2.53	<u>-3.04</u>
Coverage of current liabilities with current assets (average annual value)	3.08	6.13	6.76	**10.51**	<u>0.40</u>
The number of turns of total assets for the year	**1.42**	<u>0.55</u>	1.0	0.64	1.01
The cost of 1 rub. revenue, rub.	0.95	**0.82**	1.02	0.94	<u>1.13</u>
Revenue growth rate, %	<u>108.76</u>	112.84	**127.40**	115.35	108.88
Retained earnings growth rate, %	99.51	105.69	<u>96.01</u>	103.45	**173.79**
The growth rate of equity, %	99.54	**105.15**	<u>96.37</u>	103.25	103.76
Profitability of sales on profit from sales,%	4.63	**17.87**	-2.20	5.83	<u>-12.87</u>
Return on sales in net profit, %	2.95	4.34	<u>-3.03</u>	**4.77**	0.70
Return on total assets, %	**5.54**	1.58	<u>-3.04</u>	3.89	0.28
Return on equity, %	**5.74**	5.02	<u>-3.49</u>	3.29	3.69
Comparative CA	5.72	6.86	3.74	6.57	3.35
Sustainability rating of enterprises	3	1	4	2	5

The most stable in 2018 was the ZAO Belgorodskij cement, as evidenced by the highest value of its CA. The plant met the standards [11] with the values of all indicators, and in four indicators it was ahead of other enterprises. It can be concluded that its stability of functioning was absolute.

In second place in the rating is OAO Zavod ZHBK-1. He also had all the evaluation indicators in line with the standards, and the company was the leader in three indicators. The value of its CA was only 0.29 points lower than that of ZAO Belgorodskij cement. All this allows us to make a conclusion about the absolute stability of the functioning of OAO Zavod ZHBK-1.

OAO Belgorodasbestotsement took the third place in the rating. It showed the best sample values of three indicators. But due to the fact that the plant did not meet the standards for two evaluation indicators, its CA was 0.85 points lower than that of OAO Zavod ZHBK-1. The above indicates that the stability of its functioning is normal.

OAO Belgorodstrojdetal' and AO Strojmaterialy were at the bottom of the rating. They significantly lagged behind the rest of the enterprises in terms of CA. OAO Belgorodstrojdetal' did not meet the standards for the values of seven evaluation indicators, and AO Strojmaterialy did not meet the standards for the values of five indicators. OAO Belgorodstrojdetal' had a normal financial condition – it was solvent and even had signs of financial stability. But the enterprise was unprofitable. AO Strojmaterialy had an unsatisfactory financial condition and loss on core activities. Based on all of the above, we can conclude that both AO Strojmaterialy and OAO Belgorodstrojdetal' functioned unstable.

To assess the dynamics of the stability of the functioning of enterprises, we calculated the corresponding indicators in table 2. Their values show how many of the 12 indicators of the stability of functioning for the estimated period have better values than for the other of the two studied periods. It is easy to see that the situation improved in 2019 compared to 2018 in most indicators (ten out of twelve) only at OAO Zavod ZHBK-1. Accordingly, only it increased the stability of functioning [2]. About OAO Belgorodstrojdetal', we can say that its level of stability has not changed in dynamics (six out of twelve indicators have improved, and the other six have become worse). The stability level of other enterprises decreased in the dynamics.

Table 2. Indicators of the dynamics of the sustainability of enterprises in 2019 compared to 2018.

Enterprises	OAO Belgorod-asbestocement		ZAO Belgorodskij cement		OAO Belgorod-strojdetal'		OAO Zavod ZHBK-1		AO Strojmaterialy	
Periods	2018	2019	2018	2019	2018	2019	2018	2019	2018	2019
Indicators	10	2	8	4	6	6	2	10	7	5

In 2019, the level and range of values of most assessment indicators did not change significantly (table 3). Four out of five enterprises remained solvent and financially stable. For two enterprises, revenue growth was replaced by its decrease; only two enterprises maintained the growth of retained earnings and equity. The main activity has become profitable for all enterprises. OAO Belgorodstrojdetal' and AO Strojmaterialy lost losses, and ZAO Belgorodskij cement became unprofitable.

Table 3. Indicators, comparative CA and stability rating of the functioning of a sample of Belgorod enterprises of the building materials industry in 2019.

Indicators, units	Indicator values				
	OAO Belgorod-asbestocement	ZAO Belgorodskij cement	OAO Belgorod-strojdetal'	OAO Zavod ZHBK-1	AO Strojmaterialy
The share of stable sources of capital (average annual value)	0.73	0.91	0.87	**0.94**	0.19
Stable sources share in stock financing (average annual value)	0.83	4.40	1.64	**2.71**	-3.44
Coverage of current liabilities with current assets (average annual value)	2.69	7.54	5.88	**11.09**	0.40
The number of turns of total assets for the year	**1.51**	0.49	0.96	0.64	1.08
The cost of 1 rub. revenue, rub.	0.96	**0.86**	0.99	0.91	0.99
Revenue growth rate, %	106.46	93.22	93.44	105.11	**112.11**
Retained earnings growth rate, %	97.18	98.74	98.35	**105.90**	101.04
The growth rate of equity, %	97.43	98.86	98.51	**105.57**	100.09
Profitability of sales on profit from sales, %	3.58	**14.09**	1.41	8.64	1.27
Return on sales in net profit, %	2.31	-1.09	0.16	**7.95**	0.02
Return on total assets, %	4.64	-0.78	0.60	**6.60**	0.29
Return on equity, %	4.86	-1.15	0.18	**5.51**	0.09
Comparative CA	4.63	4.15	2.80	8.10	2.33
Sustainability rating of enterprises	2	3	4	1	5

In 2019, the range of CA values of enterprises increased. This suggests that the difference in the level of sustainability of their functioning has become more significant.

In 2019, the leader in the sustainability of operation was OAO Zavod ZHBK-1. As noted above, this happened as a result of an improvement in most of its indicators. Accordingly, the stability of the operation of the plant is absolute, as in 2018.

OAO Belgorodasbestocement moved from third place in the ranking to second, despite the fact that its stability in dynamics has decreased. This happened due to the weakening of stability of ZAO Belgorodskij cement. OAO Belgorodasbestocement, as in 2018, does not meet the standards for two assessment indicators, and its CA value is almost two times lower than that of the rating leader (and below the average level). So the stability of its functioning can only be considered normal with a tendency to weaken.

ZAO Belgorodskij cement rolled back to third place in the ranking due to the net loss in 2019. At the same time, the enterprise has the lowest level of production costs in the sample and, accordingly, the highest level of its profitability, as well as normal all indicators of financial condition. So we can say that the level of stability of functioning of ZAO Belgorodskij cement is quite satisfactory.

OAO Belgorodstrojdetal' and AO Strojmaterialy remained in the fourth and fifth places of the rating, respectively. In 2019, each of them does not comply with the standards for the values of three assessment indicators. This is better than in 2018. But the level of production costs of enterprises is critically high, and the level of profitability indicators is close to zero, which led to a very low level of

IOP Conf. Series: Materials Science and Engineering **945** (2020) 012017 doi:10.1088/1757-899X/945/1/012017

their CA. So, the stability of functioning at AO Strojmaterialy and OAO Belgorodstrojdetal' in 2019 is still absent.

4. Summary

According to the results of a study of a sample of enterprises of the building materials industry of the Belgorod region for the period 2018-2019 revealed that most of them are functioning stably. At the same time, some enterprises lack stability, which puts them at risk in the event of a deterioration in the situation in the building materials market or in the event of a general deterioration in the economic situation that the world faced in 2020 due to the COVID-19 pandemic.

It should be noted that 2019 was less successful for the building materials industry than 2018: for the vast majority of enterprises in the studied sample, in 2019 there was a weakening in the functioning stability, that is, they came up to 2020 crisis with a "reduced immunity".

If the sample studied were representative, based on the results obtained in this work, generalizations could be made about the state and dynamics of the development of the building materials industry in the Belgorod region as a whole. However, unfortunately, the lack of open access to the financial statements of many business entities does not allow constructing representative samples and conducting more accurate studies.

5. References

[1] Ahmed A A, Fedyuk R S, Liseytsev Yu L, Timokhin R A, Murali G 2020 Use of Iraq concrete scrap as filler and aggregate of heavyweight and lightweight concrete *Construction Materials and Products* **3(3)** 28-39

[2] Bryantseva T A, Shevchenko M V 2019 Innovative leadership of the region: assessment issues and strengthening factors (on the example of the Belgorod region) *Problems of the modern economy* **4** 128-130

[3] Bukhonova S M, Doroshenko Yu A, Trunova E V 2004 Application of sustainability performance indicators in company management *Economic analysis: theory and practice* **9** 10-21

[4] Bukhonova S M, Sergeeva S A 2017 Problems of reindustrialization of Russian industry on the new technological principles *Bulletin of BSTU named after V.G. Shukhov* **10** 194-203

[5] Chizhova E N, Shevchenko M V 2011 Integral efficiency of management system of the industrial enterprise *Bulletin of BSTU named after V.G. Shukhov* **1** 95-99

[6] Chizhova E N, Sorokina V Yu, Vesnina O O, Osychenko E V 2018 The concept of monitoring the management of innovative industrial enterprise *Bulletin of BSTU named after V.G. Shukhov* **11** 156-163

[7] Dmitrieva T V, Markova I Yu, Strokova V V, Bezrodnykh A A, Kutsyna N P 2020 Effeciency of stabilizers of various compositions for strengthening soils with mineral binders *Construction Materials and Products* **3(1)** 30-38

[8] Doroshenko Yu A, Golubotskikh M 2017 Factors of competitive sustainability and assessment of their influence on enterprises within building materials industry *Bulletin of BSTU named after V.G. Shukhov* **1** 26-31

[9] Doroshenko Yu A, Klimashevskaya A 2017 The analysis of scientific and technical potential of the enterprises of construction materials industry in the context of assessment of need of carrying out technological modernization for the industry *Bulletin of BSTU named after V.G. Shukhov* **1** 214-218

[10] Slabinskaya I A, Benderskaya O B 2017 Sustainability of Belgorod region enterprises of building materials and building structures industry under sanctions *Advances in Engineering Research: Proceedings of the International Conference AIME 2017* **133** 102-107

[11] Slabinskaya I A, Benderskaya O B 2019 Quality of functional stability management of Belgorod food industry enterprises *Herald of the Belgorod University of Cooperation, Economics and Law* **6** 307-321

IOP Conf. Series: Materials Science and Engineering **945** (2020) 012017 doi:10.1088/1757-899X/945/1/012017

[12] Slabinskaya I A, Benderskaya O B, Mitrokhin A A, Truhin A S 2015 Methods of Company Financial Stability Monitoring *International Business Management* **9(6)** 1091-1096
[13] Tolstoy A D 2020 Fine-grained high-strength concrete *Construction Materials and Products* **3(1)** 39-43
[14] Usatova L V, Arskaya E V, Budnichenko O 2014 The importance of profit and mechanism of its distribution in modern conditions *Belgorod Economic Bulletin* **3(75)** 175-186
[15] Volodchenko A A, 2020 Influence of artificial calcium hydrosilicates on the hardening processes and properties of non-autoclave silicate materials based on unconventional aluminosilicate raw materials *Construction Materials and Products* **3(2)** 19-28

Acknowledgments
This work was realized under the support of the President Scholarship; in the framework of the Program of flagship university development on the base of the Belgorod State Technological University named after V G Shukhov.

IOP Conf. Series: Materials Science and Engineering **945** (2020) 012018 doi:10.1088/1757-899X/945/1/012018

Investigation of stresses and deformations of ball mill support axles taking into account the temperature field by numerical method

J A Bondarenko[1], T A Duyun[2] and O V Bestuzheva[3*]

[1]Professor of Department of mechanical engineering, BSTU named after V.G. Shukhov, Belgorod, Russia
[2]Head of Department of mechanical engineering, BSTU named after V.G. Shukhov, Belgorod, Russia
[3]Docent of the Department of applied informatics and information technologies, Belgorod National Research University, Belgorod, Russia

E-mail: bestuzheva@bsu.edu.ru

Abstract. In the article, the stresses and deformations of the ball mill support axes are studied numerically, taking into account the temperature field. Expressions are given that determine the values of the stress-strain state of the external surface of the ball mill trunnion, depending on the temperature of the internal and external surfaces. The analysis of the behavior of stresses and deformations when changing the temperature parameters of the trunnion is based on the numerical method. Constant values and parameters of the trunnion are set, and the temperature ranges of the external and internal surfaces of the loading and unloading trunnions are set. A numerical simulation of deformations and stresses is presented, described in the form of dependency graphs with varying parameters-the temperature of the internal and external surface of the trunnion within specified limits. The dependences of stresses and deformations of the external surface on temperature changes are analyzed. It is shown that as the temperature difference increases, the stresses in both the circumferential and axial directions increase, and the deformations in the circumferential and axial directions monotonously increase. The unloading axle is subject to a more complex stress-strain state under the influence of a temperature field, and the external surface of the axle has the highest values of stresses and deformations.

1. Introduction

When operating a ball mill, the dangerous cross-section of the bottoms is the transition point of the cylindrical part of the trunnion to the conical one [1, 2]. Under the influence of constant loads of the housing with grinding bodies and material, the simultaneous action of gravity and rotation due to the moment of external forces, a bend occurs [3].

The trunnion has the form of a hollow cylinder, which is pinched at the end and rotates around the longitudinal axis [4]. The stress-strain state of the ball mill pin is estimated on the basis of a mathematical model that includes a complete system of equilibrium equations that determine the

IOP Conf. Series: Materials Science and Engineering **945** (2020) 012018 doi:10.1088/1757-899X/945/1/012018

relations of elastic-plastic deformation, taking into account the effects of cyclic loading of the material, with appropriate initial and boundary conditions.

According to [5, 6, 7], in the first approximation, the solution of this problem can be obtained using engineering approaches based on the simplest relations of the theory of elasticity, resistance of materials and mechanics of materials [8, 9, 10]. The bending of a rotating axle is caused by the simultaneous action of gravity and rotation due to the moment of external forces, rotation when bending a cylindrical part leads to the formation of a complex stress-strain state [8]. Modeling such a process requires the use of the Euler-langrangian approach, tracing the history of deformation of material particles, including the loading process and elastic-plastic loading processes with the appearance of secondary plastic deformations and changes in the yield strength of the material, taking into account the features of alternating deformation.

The stresses of the external surface in the axial direction are calculated based on [9] with the summation of stresses arising under the influence of gravity and rotation:

$$\sigma_z = \frac{\alpha E\left(T_{R_2}-T_{R_1}\right)}{1-\nu}\left[\frac{R_1^2}{R_2^2-R_1^2}-\frac{1}{2\ln\left({}^{R_2}/_{R_1}\right)}\right]+\sigma_{z_0}. \tag{1}$$

According to [9], the stresses on the outer surface of the ball mill shaft in the circumferential direction are determined:

$$\sigma_\varphi = \frac{\alpha E\left(T_{R_2}-T_{R_1}\right)}{1-\nu}\left[\frac{R_1^2}{R_2^2-R_1^2}-\frac{1}{2\ln\left({}^{R_2}/_{R_1}\right)}\right]. \tag{2}$$

Deformations of the external surface of the axle in the axial, circumferential and radial directions are found according to [9]:

$$\varepsilon_z = \frac{\alpha\left[\left(T_{R_2}-T_{R_1}\right)\left(R_2^2-R_1^2\right)+2\left(R_1^2T_{R_1}-R_2^2T_{R_2}\right)\ln\left({}^{R_2}/_{R_1}\right)\right]}{2\left(R_2^2-R_1^2\right)\ln\left({}^{R_2}/_{R_1}\right)}+\varepsilon_{z_0}, \tag{3}$$

$$\varepsilon_\varphi = \frac{\alpha}{2(1-\nu)}\left[\frac{(1-3\nu)\left(R_2^2T_{R_2}-R_1^2T_{R_1}\right)-(1+\nu)\left(R_1^2\left(T_{R_1}-T_{R_2}\right)+T_{R_2}\left(R_2^2-R_1^2\right)\right)}{R_2^2-R_1^2}-\frac{(1-\nu)\left(T_{R_2}-T_{R_1}\right)}{\ln\left({}^{R_2}/_{R_1}\right)}\right], \tag{4}$$

$$\varepsilon_r = \frac{\alpha}{2(1-\nu)}\left[\frac{(1-3\nu)\left(R_2^2T_{R_2}-R_1^2T_{R_1}\right)+(1+\nu)T_{R_2}\left(R_2^2-R_1^2\right)}{R_2^2-R_1^2}+\frac{2\nu\left(T_{R_2}-T_{R_1}\right)}{\ln\left({}^{R_2}/_{R_1}\right)}\right]. \tag{5}$$

These expressions (1) – (5) determine the values of the stress-strain state of the external surface of the ball mill pin, depending on the temperature of the internal and external surfaces.

2. Methods and materials

The analysis of the behavior of stresses and deformations when changing the temperature parameters of the trunnion on the basis of (1) – (5) is carried out by the numerical method [10]. When conducting the study, the values of the constant and parameters of the trunnion are set numerically.

A mill with a standard size with diameter of 3.2×15 m with the following parameters is considered:

- the size of the axle:

$$d = 1.4 \text{ m}, \; l = 0.92 \text{ m}. \tag{6}$$

- the radii of the inner and outer surfaces are equal:

$$R_1 = 0.525 \text{ m}, \; R_2 = 0.7 \text{ m}. \tag{7}$$

Constant values for the common axle material – 35L steel have the following values, according to [11]:

IOP Conf. Series: Materials Science and Engineering **945** (2020) 012018 doi:10.1088/1757-899X/945/1/012018

- Poisson's ratio:

$$\nu = 0.25; \quad (8)$$

- elastic modulus:

$$E = 201\ 000 \text{ MPa}. \quad (9)$$

Coefficient of linear expansion of steel 35L for temperatures from 20°C to 200°C – α= 12 deg^{-1}.

To determine the stresses and strains, you need to set the temperatures of the internal and external surfaces of the trunnion. The temperature of the inner surface depends on the temperature of the feed material:

- for the loading pin – the clinker temperature varies within:

$$T_{R_1} = 70...100°C;$$

- for discharge trunnion is the temperature of the cement varies between:

$$T_{R_1} = 100...150°C.$$

To measure the temperature of the external surface of the ball mill trunnion, thermocouples are used.

The temperature of the external surface of the loading and unloading trunnion varies within:

- loading axle: $T_{R_2} = 35...45°C$;
- unloading of the axle: $T_{R_2} = 40...65°C$.

In the framework of the mathematical package Maple, equations (1) – (5) are solved when the parameters – the temperature of the internal and external surface of the trunnion are varied within the specified limits:

$$T_{R_1} = 70...150°C, \quad (10)$$

$$T_{R_2} = 30...70°C. \quad (11)$$

The solution of numerical modeling of strains and stresses, according to the equations (1) to (5) described in the form of graphs according to the temperature of the inner and outer surfaces in figures 1 – 2.

The graphs of stresses in the outer surface of the axle in the axial and circumferential direction is shown in figure 1. Voltage σ_z slightly exceed the voltage σ_φ, this is because in the axial direction is added to the load of gravity σ_{z_0} (figure 2), accounting for 1% of σ_z.

Analysis of the dependence of external surface stresses on temperature changes indicates that as the temperature difference between the internal and external surfaces of the trunnion increases, the stresses in both the circumferential and axial directions increase. The minimum voltage values are taken when the temperature of the inner surface approaches the temperature of the outer surface – when the temperature is equal to $T_{R_1} = T_{R_2}$, the voltage in both directions is zero – $\sigma_\varphi = \sigma_z = 0$. The maximum value of the external surface stress under the influence of the temperature field in the circumferential and axial directions at a given temperature range is $\sigma_\varphi = \sigma_z = 172$ MPa, which is achieved at a temperature difference of $\Delta T_R = 120$ °C.

The change in the deformations of the external surface of the trunnion in the axial, circumferential and radial directions when the temperatures of the external and internal surfaces change is shown in figure 2.

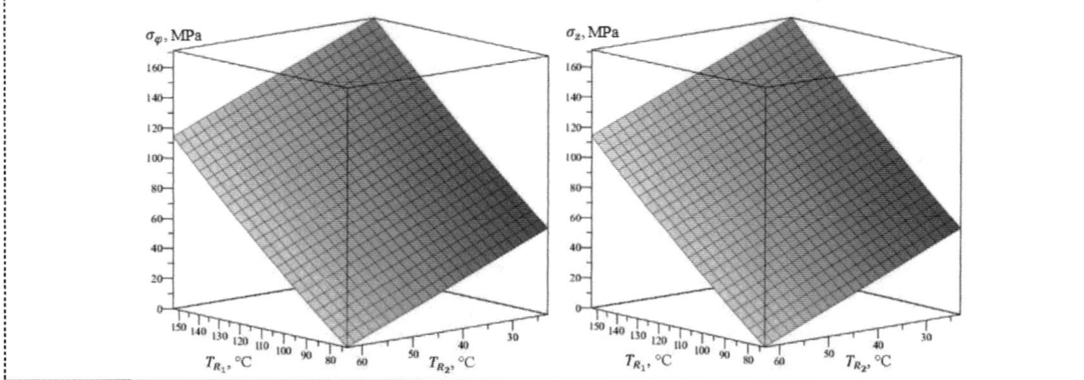

a b

Figure 1. Changes in the stress of the external surface of the trunnion with changes in the temperature of the external and internal surface: a) in the circumferential direction; b) in the axial direction.

The dependence of deformations in the radial direction under the influence of the temperature field is significantly inhomogeneous: a decrease in the temperature of the inner surface leads to a decrease in deformations, while an increase in the temperature of the outer surface leads to an increase in deformations. Thus, the deformations of the external surface in the radial direction increase from the internal cylindrical surface of the trunnion to the external: the maximum values in the specified temperature range are $\varepsilon_r = 4.102$ mm at the temperature of the internal surface $T_{R_1} = 67$°C and the external surface $T_{R_2} = 68$°C.

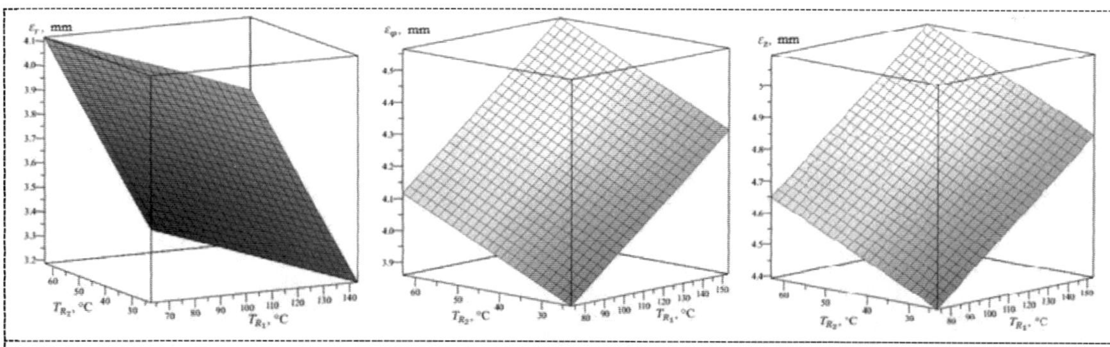

a b c

Figure 2. The change in deformation of the outer surface of the axle with changes in temperature of internal and external surfaces: a) in the radial direction; b) circumferential direction; c) in the axial direction.

Graphs of the dependence of deformations in the circumferential and axial directions with varying temperatures indicate that the deformations monotonously increase with increasing temperatures of the internal and external surfaces of the trunnion. The minimum values of deformations at a given temperature range – $\varepsilon_\varphi = 3.86$ mm, $\varepsilon_z = 4.39$ mm are assumed at the internal surface temperature $T_{R_1} = 73$°C and external surface $T_{R_2} = 23$°C, maximum values – $\varepsilon_\varphi = 4.6$ mm, $\varepsilon_z = 5.8$ mm at $T_{R_1} = 165$°C and $T_{R_2} = 55$°C.

The temperature of the external surface in production conditions is assumed to be constant on the entire external surface along the axis,as the cooling liquid is poured. Let's consider the change in the stress-strain state of the loading and unloading trunnion at a constant temperature of the external

IOP Conf. Series: Materials Science and Engineering **945** (2020) 012018 doi:10.1088/1757-899X/945/1/012018

surface, averaged according to the readings of the sensors of the ball mill No. 13 with the standard size with diameter of 3.2×15 m at JSC "Belgorod cement".

The average external surface temperature values determined from the sensor readings have the following values:

- loading axle:

$$T_{R_2} = 42°\text{C}; \tag{12}$$

- unloading of the axle:

$$T_{R_2} = 58°\text{C}. \tag{13}$$

The temperature of the inner surface of the trunnion axis is changed: when loading increases along the advancing material from the heat to the body, the value of T_{R_1} varies according to (10), while unloading decreases along the progress of the material from the housing to the discharge pipe, while T_{R_1} varies according to (11).

3. Results and discussions

According to (1) – (5), graphs of stress and strain dependences of the loading and unloading trunnion of a ball mill with a standard size with diameter of 3.2×15 m at an assumed constant temperature of the external surface (12) – (13) and variation of the internal surface temperature in a given limit (10) are constructed (figures 3 – 4).

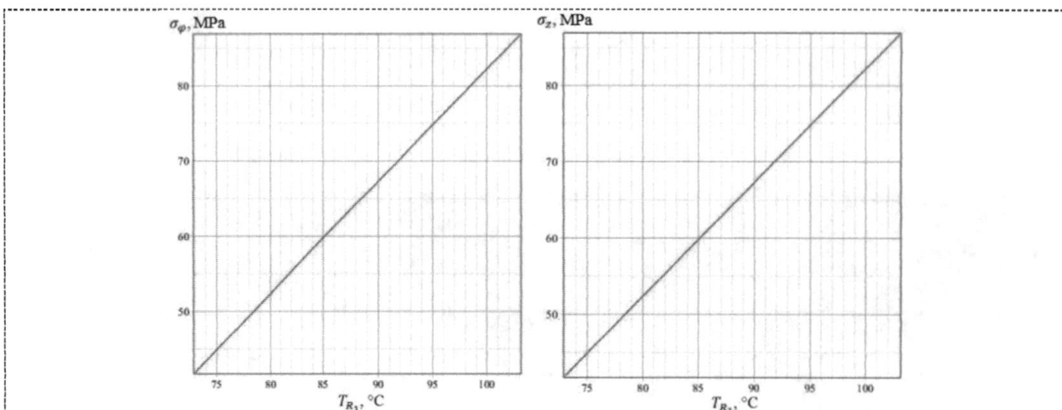

a b

Figure 3. Graphs of the dependence of the external surface stresses of the loading pin on the internal surface temperature at a constant temperature of the external surface T_{R_2} = 42°C: a) in the circumferential direction; b) in the axial direction.

The graphs of the dependence of the stresses on the external surface of the loading pin in the circumferential and axial directions when the temperature of the internal surface varies, shown in figure 3, indicate a linear character: with an increase in the temperature T_{R_1}, the stresses in these directions σ_φ and σ_z grow. The maximum value of the voltage $\sigma_\varphi = \sigma_z =$ 87 MPa is assumed at T_{R_1} = 103°C.

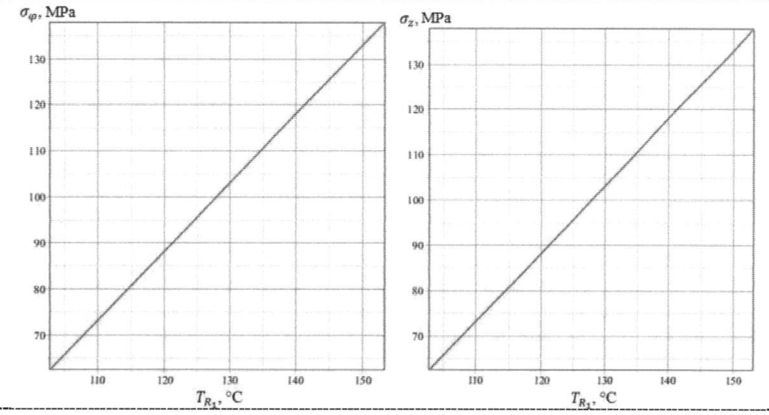

a b

Figure 4. Graphs of the stress dependence of the external surface of the discharge pin on the internal surface temperature at a constant temperature of the external surface $T_{R_2} = 58°C$: a) in the circumferential direction; b) in the axial direction.

The graphs of the dependence of the stresses on the outer surface of the unloading pin in the circumferential and axial directions when the temperature of the inner surface varies, shown in figure 4, also indicate a linear character: as the temperature of the inner surface increases T_{R_1}, the stresses in these directions σ_φ and σ_z increase. The maximum value of the voltage $\sigma_\varphi = \sigma_z = 138$ MPa is assumed at $T_{R_1} = 153°C$.

Graphs of the dependence of deformations of the external surface of the loading and unloading trunnions when the temperature of the internal surface varies in the radial, circumferential and axial directions are shown in figures 5 – 6.

Deformations in the radial direction of the loading and unloading trunnions have an inverse relationship: when the temperature of the inner surface increases, the deformations monotonously decrease. Maximum strain values are reached at the internal surface temperature T_{R_1}= 73°C for the loading axle and T_{R_1}= 102°C for the unloading axle (figure 5).

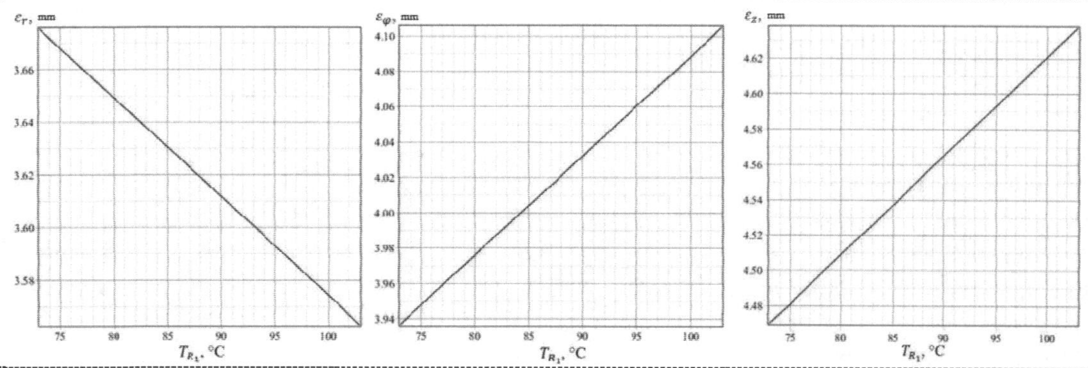

a b c

Figure 5. Graphs of the dependence of deformations of the external surface of the loading pin when the temperature of the internal surface varies: a) in the radial direction; b) in the circumferential direction; c) in the axial direction.

The analysis of dependences of deformations of the external surface in the circumferential and axial directions is monotonously increasing: as the temperature of the internal surface increases, the deformations increase (figure 6). The maximum values of deformations for a given range of internal surface temperature have the following values:

- in the circumferential direction: $\varepsilon_\varphi = 4.12$ mm at $T_{R_1} = 104$°C – for the loading axle, $\varepsilon_\varphi = 4.49$ mm at $T_{R_1} = 154$°C – for the discharge axle;
- in the axial direction: $\varepsilon_z = 4.63$ mm at $T_{R_1} = 104$°C – for the loading axle, $\varepsilon_z = 5.25$ mm at $T_{R_1} = 154$°C – for the discharge axle.

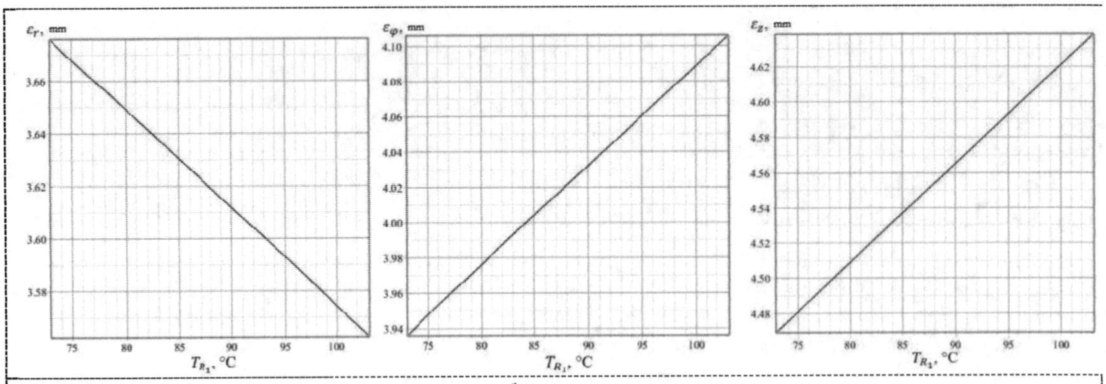

a b c

Figure 6. Graphs of the dependence of deformations of the external surface of the unloading pin when the temperature of the internal surface varies: a) in the radial direction; b) in the circumferential direction; c) in the axial direction.

The graphs of dependencies of stresses and deformations of the external surface presented in figures 3-6 prove that the discharge axle is more susceptible to a complex stress-strain state under the influence of a temperature field: the largest values of stresses and deformations are on the external surface of the discharge axle.

4. Summary

Using a numerical method for a mill size with diameter of 3.2x15 m, the analysis of the dependence of stresses and deformations occurring on the external surface of the trunnion on the temperature difference of its internal and external surfaces was carried out. It was found that with an increase in the temperature difference in $\Delta T_R = 120$°C ($T_{R_1} = 157$°C, $T_{R_2} = 37$°C), the stresses in both the circumferential and axial directions grow and amount to $\sigma_\varphi = \sigma_z = 172$ MPa, the deformations in the circumferential and axial directions monotonously increase and reach $\varepsilon_\varphi = 4.6$ mm, $\varepsilon_z = 5.8$ mm. The maximum strain value in the radial direction is $\varepsilon_r = 4.102$ mm. The unloading axle is subject to a more complex stress-strain state under the influence of a temperature field, and the external surface of the axle has the highest values of stresses and deformations.

5. References

[1] Markova O, Aulov V, Lavrenchuk A and Fedorenko M 2012 Analysis of methods for calculating wear of contact surfaces of friction *International scientific and practical conference of students, postgraduates and young scientists "Youth and scientific and technical progress" (Gubkin)* vol 1 (Gubkin: Publishing house of BSTU) 65-69

[2] Fedorenko M and Markova O 2014 Oscillatory processes in grinding mills of the cement industry *International scientific and practical conference "Technique and technology of modern production"* (Penza) 113-116

[3] Bogdanov V, Ilyin A and Semikopenko I 2008 *Basic processes in the production of construction materials* (Belgorod: Publishing house BSTU named after V. G. Shukhov) 550

[4] Gologorsky E, Dotsenko A and Ilyin A 2006 *Operation and repair of equipment of construction industry enterprises* (Moscow: Architecture) 503

[5] Boyarshinov M 2013 Estimation of the stress-strain state of a rotating long cylinder *Bulletin of PNRPU Mechanics* **1** 25-38

[6] Birger I, Shorr B and Iosilevich G *1979 Calculation of the strength of machine parts* (Moscow: Mashinostroenie) 702

[7] Erdedi A and Erdedi N 2012 *Theoretical mechanics. Resistance of materials* (Moscow: Akademiya) 320

[8] Svetlitsky V 2009 *Construction mechanics of machines. Mechanics of rods* (Moscow: Fizmatlit) 408

[9] Bestuzheva O, Bondarenko J and Khanin S 2019 Mathematical description of the stress-strainstate of the ball mill axle under the action of gravity and rotation *Bulletin of the Belgorod state technological University named after V. G. Shukhov* **3** 128-133

[10] Bakhvalov N, Zhidkov N and Kobelkov G 2003 *Numerical methods* (Moscow: Nauka) 630

[11] Argatov I and Dmitriev N 2003 *Fundamentals of the theory of elastic discrete contact* (Moscow: Politechnika) 240

IOP Conf. Series: Materials Science and Engineering **945** (2020) 012019 doi:10.1088/1757-899X/945/1/012019

Research on parameters of the bladed pneumatic screw mechanism of the digging and transport machine

A A Kovalev[1]*, V V Kutuzov[1] and I I Elkina[1]

[1]V.I. Vernadsky Crimean Federal University, ProspektVernadskogo 4, Simferopol, 295007, Republic of Crimea, Russian Federation

E-mail:kovaland59@mail.ru

Abstract. The main intensifiers that increase the efficiency of the bulldozer are considered. The concept of the digging and transport machine for bulk material is proposed, the design of which combines the positive qualities of pneumatic and bulldozer equipment. The machine is mainly intended for use in reclamation construction for filling holes, ravines, deep wells, filling the water-resistant layer of the bottom of a pond or reservoir, and also for supplying finely divided stone mining waste (sawdust, coal powder, limestone flour, carbonate sand, etc.) in the back of a dump truck or to the blade. The machine can also be used as auxiliary equipment and to fight a fire in the field by supplying a bulk fire-extinguishing agent directly to the edge of the flame front

1. Introduction

The bulldozer is one of the main machines used in industrial, civil, road, reclamation construction, as its design is simple, versatile and has a low cost of work.

Improving the design of the working body of the bulldozer, according to the works [1,2, 3, 7, 10], it is one of the main ways to improve its performance. It is known that the main advantage of pneumatic mechanical equipment is the ability to simultaneously perform operations of grinding and moving bulk material (for example, sand) in the screw-pressure mechanism, as well as mixing it with compressed air in the mixing air chamber and further supplying the aeromaterial mixture to the warehouse.

In the patent RU 2 414273 "Fire extinguishing composition, method of its preparation and method of fire extinguishing" [11] it is noted that crushed limestone (calcium carbonate) in the form of limestone flour is a well-known safety tool in coal mining. Explosive fire can be suppressed by scattering limestone flour in the air. Under strong heating, limestone decomposes chemically: $CaCO_3 \Rightarrow CaO+CO_2$. This decomposition reaction requires energy and therefore produces a cooling effect. Therefore, such a digging and transport machine (DTM) can be used as an auxiliary equipment and to extinguish a fire in the field by supplying bulk fire extinguishing agent directly to the edge of the flame front.

The working name of the machine is pneumatic-screw bulldozer. A special feature of the design is the presence of a loading rectangular hole in the plane of the bulldozer blade and the location of a pneumatic-mechanical installation behind it.

The purpose of the study was to assess the possibility of implementing bladed – pneumatic and screw device in the construction of digging and transport machine (DTM) and to determine the theoretical relationship of main parameters of pneumatic and screw and bulldozer equipment of DTM.

IOP Conf. Series: Materials Science and Engineering **945** (2020) 012019 doi:10.1088/1757-899X/945/1/012019

2. Materials and methods

In work [2] it is noted that "one of the topical issues is the creation of universal construction equipment that works on new physical and technical principles of interaction with the environment and has high environmental qualities".

Performance of the bulldozer blade (see [8]) when digging is

$$P = B \cdot h \cdot \vartheta_\sigma \cdot C_l \qquad (1)$$

where B - width of the bulldozer blade, m; h - depth of ground digging, m; ϑ_σ - speed of the base machine, m/s; C_l - the coefficient of the soil loosening.

The capacity of screw auger intensifier (see [8]) is:

$$P_{SC} = 3600 \cdot \frac{\pi \cdot (D_{ST}^2 - d_S^2)}{4} \cdot S \cdot n_{SC} \cdot \psi \qquad (2)$$

где D_{ST} - outer diameter of screw turns, m; d_S - the shaft diameter of the screw, m; S- the step of the screw winding, m; n_{SC} - the frequency of the screw rotation, s^{-1}; ψ - the coefficient of filling space between turns in the screw.

Formula for determining the relation between the rotation speed of the screw auger intensifier from its geometric parameters and the speed of the base machine:

$$n_{SC} = \frac{B_{BL} \cdot h \cdot \vartheta_\sigma \cdot C_l}{900 \cdot \pi \cdot (D_{ST}^2 - d_S^2) \cdot S \cdot \psi}. \qquad (3)$$

The performance of the pneumatic installation in the pneumatic lift mode is noted in [6], determined by the last pressure turns of the screw according to the formula:

$$\Pi_{SC} = 0.785 \cdot (D_{SC}^2 - d_S^2) \cdot S_P' \cdot C_{FR} \cdot \rho_o \cdot n\,/60 \qquad (4)$$

where D_{SC} – screw diameter, m; d_S - the shaft diameter of the screw, m; $S_{\text{Н}}'$ - step of the last pressure turn without the turn thickness; C_{FR} - the friction coefficient of the screw; ρ_o - volume mass of the material,t/m^3; n - the frequency of the screw rotation, min^{-1}.

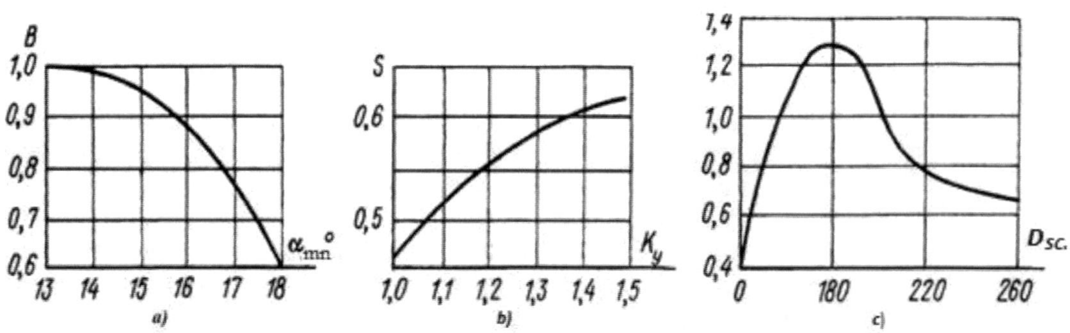

Figure 1. Dependence of the components of the sliding coefficient:***a)B*** on lift angle $\boldsymbol{\alpha_{mn}}$ of the screw line of the last pressure turn of the screw; ***b) S*** on the compression ratio of the screw $\boldsymbol{K_y}$;***c)*** ***ξ***on the screw diameter $\boldsymbol{D_{SC}}$.

Power consumption (kW) per screw drive

$$N_{SC} = 430 \cdot p_C \cdot D_{SC}^2 \cdot n \qquad (5)$$

where 430 cm^2·m^{-1}·s^{-1} - coefficient; p_C - overpressure in the mixing chamber of the feeder, MPa.

3. Results

The previously considered intensifiers of the working process of the bulldozer, in general, accelerated the process of forming the drawing prism when used on various soils.

A distinctive feature of the proposed DTM intensifier (see Fig. 2) consists, firstly, in the fact that the loading window 2 in the design of the blade 1 plays the role of a flow forming element, and secondly, it is used for finely ground and loosened stone crushing waste (hereinafter referred to as bulk material). Pneumatic mechanical installation (PMI) 3 is located behind the bulldozer blade 1. The plane of the opening of the receiving chamber 4 of PMI 3 is located horizontally (in its normal working position). The loading window 2 and the receiving chamber 4 are connected by hopper 8.

a) b)

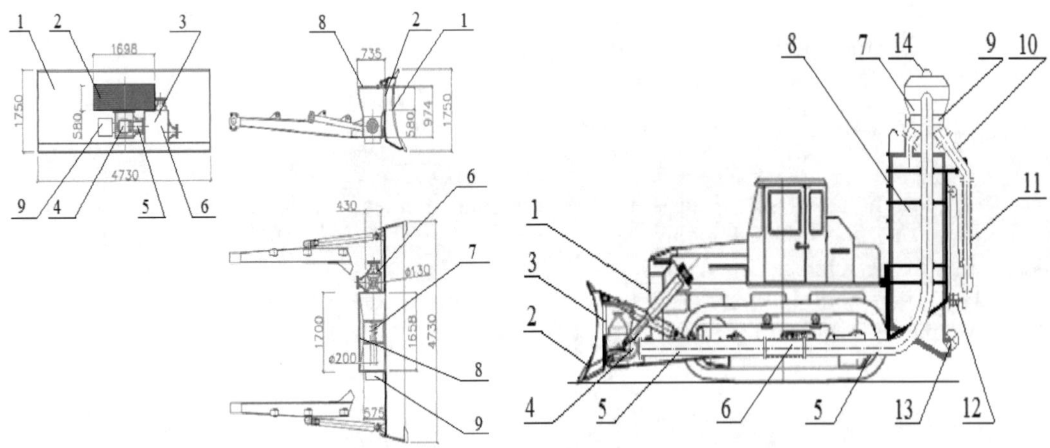

Figure 2. a) scheme of the bladed pneumatic screw mechanism of the digging and transport machine: 1 - bulldozer blade; 2 – loading window of the bulldozer blade; 3 - pneumatic screw equipment (pneumatic screw installation); 4 – receiving chamber; 5 - cylindrical body (cylinder); 6 – universal mixing air chamber; 7 – screw; 8 – hopper; 9 – drive;
b) general scheme of the mechanism of a bladed pneumatic screw digging and transport machine: 1- basic machine; 2 – blade; 3 – loading window; 4 – pneumatic and mechanical installation with a unified mixing chamber; 5 – discharge line; 6 – flexible pipeline section; 7 – hopper-extinguisher;8 - main hopper; 9 – switcher of the direction of the material flow; 10 – nozzle for material flow outlet; 11 – flexible pipeline; 12 - device for lifting flexible pipeline; 13 – side pneumatic unloader; 14 – nozzle for air outlet to the filter.

The PMI contains in its design a cylindrical body 5 (hereinafter referred to as the cylinder) with a screw 7 and a universal mixing air chamber 6.

The layout scheme of the bladed pneumatic screw mechanism of the DTM is shown in Fig. 2.

To describe the working process of the machine, we use Fig. 2 b).

At the first stage of the working process, a layer of pre-loosened bulk material moves along the surface of the blade 2. Then the bulk material moves to the loading window 3 and enters the hopper through it. In the lower part of the hopper is the PMI receiving chamber.

At the second stage, the loosened material is sent from the receiving chamber by the screw along the cylindrical body to the universal mixing air chamber, where it is mixed with air. In the process of moving along the cylindrical body, the screw grinds and crushes the lumps of material being moved. Then the crushed material is sent to the mixing air chamber.

At the third stage in the mixing air chamber, the flow of particles of crushed material is mixed with compressed air and enters the discharge line 5.

Then the aeromaterial mixture under the influence of overpressure of compressed air moves through the hopper-extinguisher 7 to the main hopper 8. The flow switcher 9 directs the flow to the nozzle 10 and further to the flexible pipeline 11 in the case of outlet of the material into the car.

A side pneumatic unloader 13 is also used to unload the hopper 8. Lifting and lowering of the flexible pipeline 11 is performed using a system of blocks and polispasts 12. Air is diverted through the nozzle 14 for cleaning.

1. Assumptions underlying the calculation model:

IOP Conf. Series: Materials Science and Engineering **945** (2020) 012019 doi:10.1088/1757-899X/945/1/012019

1.1. The design of the bulldozer blade contains a hole through which the material cut off by the blade during the course of the bulldozer enters in a steady mode;
1.2. Further transportation of the material is carried out by a single-or double-support screw (having two opposite (or counter) directed screw surfaces);
1.3. The material is collected in a hopper located behind the bulldozer.

2. The main balance ratio.

From the assumption (1.1) about the steady mode of movement of material to the screw mechanism, it follows that Q_0=Q_{SC}, where Q_0 is the volume of cut material, Q_{SC} is the volume of material transported by the screw mechanism, per unit of time (in calculations– for 1 second).

3. Geometric parameters of elements of the bladed mechanism.

$h -$ depth of the cut layer(m); $v -$ speed of the bulldozer ($m \cdot s^{-1}$);
$L -$ the width of the blade (m)

$$\text{Hence: } Q_b = Lvh \tag{6}$$

Geometric and kinematic parameters of the screw mechanism: $D -$ screw diameter(m); $d_{SC} -$ the shaft diameter of the screw (m); $s -$ screw step (m); n– the rotation speed of the shaft of the screw, rpm, $\omega -$ speed rotation of the screw shaft, rps

Volume capacity of one screw (with one cylindrical body)

$$Q^1_{SC} = 0.785 \cdot (D^2 - d^2_{SC}) \cdot n \cdot S'_P \cdot C_{FR}/60 \tag{7}$$

with the two screws (with the two cylindrical bodies)

$$Q^2_{SC} = 1.57 \cdot (D^2 - d^2_{SC}) \cdot n \cdot S'_P \cdot C_{FR}/60 \tag{8}$$

Preliminary discussion of possible technical solutions.

Parameters h, L, v, m (number of cylindrical bodies (cylinders) - 1 or 2), D, d, s, n, S'_p , C_{FR} cannot be set arbitrarily. The ratio (1) that coordinates their interaction sharply narrows the range of possible technical solutions. Moreover, depending on the application conditions of the device, various combinations of these parameters are possible.

In this regard, we note the fact that the listed parameters differ significantly in their role in the balance dependence: h, v have a continuous nature of change and well-defined ranges of change: $h =$ 0.15 ... 0.2 (m), v =0.25 … 0,75 (m/s).

At the same time, the parameters L, m, D, d, s, n take discrete series of values, this is due to the presence of specific devices with known geometric and kinematic values.

To ensure compatibility, we equate expressions (6) and (7).

The depth of the cut-off soil layer h of a bulldozer with a bladed pneumatic screw pressure mechanism is determined depending on D, d_{sc}, S'_p, n, L, v, according to the formula:

$$h = \left[0.785(D^2 - d^2_{sc}) \cdot n/60 \cdot S'_p \cdot C_{FR}\right]/(L \times v) \tag{9}$$

Let us substitute dependency (5) in expression (14) and get

$$h = \left[0.785(D^2 - d^2_{SC}) \cdot 16.67 \cdot S'_P \cdot B \cdot S \cdot (10 \cdot P_C)^{\xi}\right]/(L \times v) \tag{10}$$

where P_C - excess pressure in the mixing chamber of PMI, MPa.

We fix the parameter values D =0.2m, d_{SC}=0.1m, S'_P = 0.12m,ω = 16.67rps,v = 0.25 m/s, L = 4.25m.

$$\text{Then we can write: } h = 0.044 \cdot B \cdot S \cdot (10 \cdot P_C)^{\xi} \tag{11}$$

Next, we take the value of the screw compaction coefficient K_y= 1.1 and according to graph a) in Fig.1, we get S =0.52.

Then, we take the lifting angle of the helical line of the turn by the average radius α_{cp}=18° and according to the graph b) in Fig. 1, we get the value *B*=0.6.

Then for screw D = 0.2 m we get $h = 0.014 \cdot (10 \cdot P_C)^{\xi}$, where ξ= 1 (see the graph c) Fig. 1). This means that $h = 0.014 \cdot (10 \cdot P_C)$.

Variation range ***h***= 0.02 m; 0.25 m; 0.3 m; 0.35; 0.4 m. P_C= 0.11 MPa; 0.12 MPa; 0.13 MPa; 0.14 MPa; 0.15 MPa.

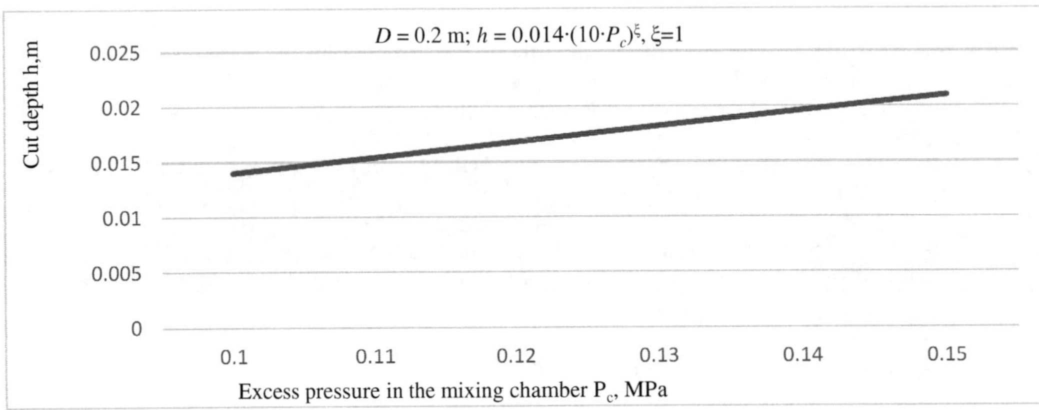

Figure 3. The dependence of the cut depth ***h*** on the excess pressure P_c in the mixing chamber of the PMI of DTM.

Let us fix the value of the parameters *D* = 0.18m, d_{SC}= 0.1m, S'_P = 0.096m, ω = 16.67rps, v = 0.25 m/s, *L* = 4.25m.

Next, we take the value of the screw compaction coefficient K_y = 1.136 according to graph a) in Fig.1, we get S =0.47.

Then, we take the lifting angle of the helix of the turn at the average radiuses α_{cp} =18° and according to the graph b) in Fig. 1 get the value *B*= 0.6.

Then for screw D = 0.18 m we get $h = 0.007 \cdot (10 \cdot P_C)^{\xi}$, where ξ= 1.28 (see the graph c) Fig. 1).

Therefore $h = 0.007 \cdot (10 \cdot P_C)^{1.28}$.

Let us fix the value of the parameters *D* = 0.25 m, d_{SC}= 0.1 m, S'_P = 0.12m, ω = 16.67 rps, v = 0.25 m/s, L = 4.25 m.

Next, we take the value of the screw compaction coefficient *Ky* = 1.136 according to graph a) in Fig.1, we get S = 0.47.

Then, we take the lifting angle of the helix of the turn along the average radius α_{cp}=18° and according to graph b) in Fig. 1, we get the value B = 0.6.

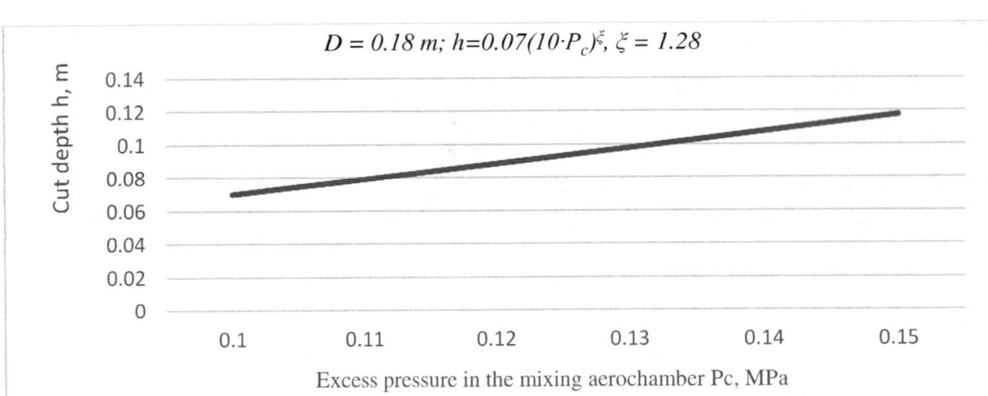

Figure 4. The dependence of the cut depth ***h*** on the excess pressure ***Pc*** in the mixing chamber

of the PMI of DTM.

Then for screw D = 0.25 m we get $h = 0.06 \cdot (10 \cdot P_C)^{\xi}$, where ξ= 0.67 (see the graph c) Fig. 1). Hence $h = 0.06 \cdot (10 \cdot P_C)^{0.67}$.

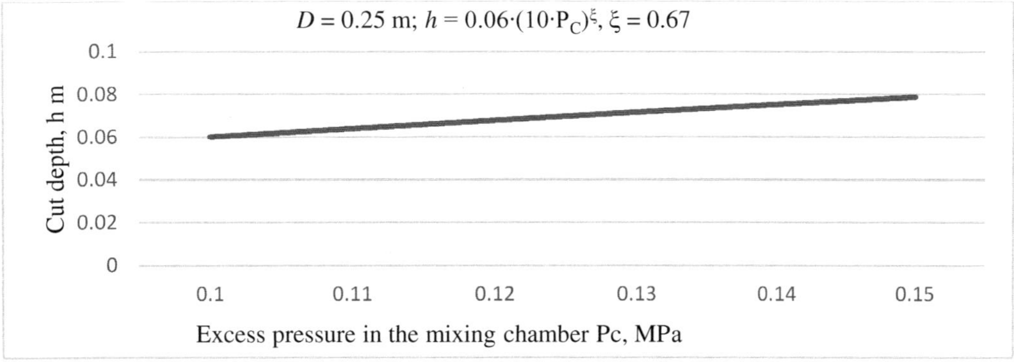

Figure 5. The dependence of the cut depth ***h*** on the excess pressure ***Pc*** in the mixing chamber of the PMI of DTM.

4. Summary

1. Based on the review of scientific literature [1, 2, 3, 5, 7, 9] and patent research [4] it was found that one of the directions of development of digging and transport machines is the use of various modulators of the workflow, the designs of which are based not only on physical-mechanical soil properties, but also the nature of the physical effects underlying the technical solutions, for example, from the use of compressed air;
2. Features of the concept of the bladed pneumatic screw mechanism of the proposed DTM is that a loading window is made in the center of the bulldozer blade, and the receiving chamber of the pneumatic and mechanical installation is located under the transition hopper.
3. In this study, the power dependence of the depth of the section of loosened material on the pressure of compressed air in the mixing air chamber of the bladed pneumatic screw digging and transport machine is established.

Let us note that a special role in the degree of non-linearity of the dependence of the depth of the cut of bulk material on the pressure of compressed air in the mixing air chamber has a parameter such as the diameter of the screw of the bladed pneumatic DTM. The dividing line is the size of the screw diameter D = 0.2 m.

4. Introduction of bladed pneumatic screw devices in the structures of bulldozer blades allows not only to increase the productivity of digging and transport machines, but also to create environmentally safe equipment.

5. References

[1] Balovnev V I 2001 Road-building machines and systems: textbook / under the general editorship of V I. Balovnev 2nd ed., add. and rev. SibADI Publishing House 525

[2] Balovnev V I, Khmara L A 1993 Intensification of soil development in road construction M.: Transport 383

[3] Bochkareva T M 2015 Technology of planning and digging works: educ.-method. manual Perm: Publishing house of Perm National Research Polytech 132

[4] Goncharova O V, Vakhrushev S I Patent research on improving the design of bulldozer blades *Online journal of Perm National Research Polytechnic University "Construction and architecture. Experience and modern technologies: based on the materials of the VIII All-Russian youth conference of postgraduates, young scientists and students"* Access mode:http://sbornikstf.pstu.ru/council/?n=6&s=302

[5] Evtyukov S A Shatunov M M 2005 Handbook of pneumatic complexes and pneumatic transport

IOP Conf. Series: Materials Science and Engineering **945** (2020) 012019 doi:10.1088/1757-899X/945/1/012019

equipment Under of general editorship M.M. Shapunov. SPb.: “Publishing house of DNA” 456

[6] Kalinushkin M P, Koppel M A, Seryakov V S, Shapunov M M 1986 Pneumotransport equipment: DirectoryInternational Journal of Engineering; under the general ed. of M.P. Kalinushkin L.: mechanical engineering, Leningrad 286

[7] Pavlov V P, Minin V V, Baykalov V A, Artemyev M I 2011 Machines for construction and maintenance of roads and airfields: Research, calculation, construction: textbook. manual Krasnoyarsk: Siberian Federal University 196

[8] Khmara L A, Krol R N 2005 Theoretical study of the operating modes of the screw intensifier *Interstroymekh - 2005: proceedings of the international scientific and technical conference Part 1. Tyumen* 262 – 266

[9] Shestopalov K K 2017 Lifting and transport, construction and road machinery and M.: Academia 414

[10] Yanson R A, Agapov A B, Demin A A, Koshkarev E V, Petrenko V F 2012 Machines for earthwork and construction and installation works. Educational edition Moscow: ASV Publishing House 358

[11] Fire extinguishing agent, its preparation method and fire extinguishing method: pat.RU 2 414273 Rus.Federation BorkYan. 2008110691/05; declared 22.09.06; published 10.11 2009 Byul **31** **(2)**

Modeling of organizational and functional components of investment and construction controlling in the reproduction of eco-residential real estate

A M Krygina[1*], I P Avilova[1], M I Oberemok[1] and A G Grebenik[2]

[1] Department of Construction Management and Real Estate, Construction Engineering Institute, Belgorod State Technological University named after V.G. Shukhov, Kostyukov St., 46, Belgorod, 308012, Russia

[2] Department of Power engineering and automatics, Institute of Power engineering, Information technologies and Management systems, Belgorod State Technological University named after V.G. Shukhov, Kostyukov St., 46, Belgorod, 308012, Russia

E-mail: kriginaam@mail.ru

Abstract. This article is devoted to the development of a structural and functional model for the formulation of high-performance regional business systems in the context of implementing eco-projects at various levels in a situation where supply exceeds demand. This approach generally corresponds to the current problems of macro and micro levels in the domestic and global economy and takes into account its negative impact on the construction industry. The theory is based on situational modeling with the identification of two key situations according to the principle of transformational development of local internal competition. The paper proposes zoning of a standardized life cycle of planning and controlling of the regional environmental programs. A structural and functional model of system formation of consortium-type organizational structures has been developed to regulate local and cluster competitiveness in construction on the meso-level. It suggests a principal situational economic model of functioning of mortgage and investment projects and programs for the development of competitiveness in territorial reproduction systems. It also provides the conceptual model of the life cycle of managing the cost of mortgage and investment programs and observes two periods of the life cycle of managing investment projects.

1. Introduction

Today's complex micro and macroeconomic situation, further negative forecast scenario of development of the world and national economy significantly complicate prospects of Russian construction industry [1]. The domestic construction complex is replete with not only constraining problems [2], but also reducing the competitiveness of production and economic activities in comparison with foreign ones [3]. To ensure competitiveness, a construction company needs a certain innovative potential [4]. The main goal is an optimal combination of all organizational, technical and managerial aspects [5], which are the target criteria for long-term planning of construction production and its innovative development. Competition determines the key direction of development of construction production and acts as a mechanism for regulating industry proportions [6]. As a market institution, competition

IOP Conf. Series: Materials Science and Engineering **945** (2020) 012020 doi:10.1088/1757-899X/945/1/012020

helps economic entities of the investment and construction complex to respond objectively to market signals [7], including information about demand for finished construction products [8]. Among other market institutions, competition is characterized by the absence of self-organization properties [9] and stability [10].

Formation and support of competition in the market is one of the key current tasks of regional authorities in the field of economic development. In this regard, there is a problem of providing the regional economy with institutional, market, information and other types of infrastructure that form a competitive environment. Its solution in the regions is carried out with the help of various measures and tools that improve the business environment within the framework of current federal legislation.[11]. It was tasked to develop the structural-functional model on ecological levels in the situation of supply exceeding demand. In this regard, the study is based on the research of domestic and foreign experience in the development of the competitiveness of long-term mortgage and investment projects and sectoral programs in construction [12,13]. The implementation of such programs should be performed as a capital budget planning system with limited financial resources and monitoring indicators in the form of investment controlling for essential components of regulation.

2. Materials and methods

As a theoretical basis for structural and functional modelling, two principal situations are identified:

the first situation (W_1) is characterized by the formation and development of local internal competitiveness in territorial reproduction systems $KTVS_{I-J}$ based on competitive interactions of investment and construction enterprises in the production and technological regional integration of territorial construction complexes (TCC);

the second situation (W_2) is the formation of high-performance and highly effective $KTVS_{I-J}$ reproduction systems W_1 through the transformational development of local internal competitiveness W_1 into external highly active and highly competitive cluster business systems in eco-building. In this case, competitive advantages appear due to more active and aggressive competitive strategies due $KTVS_{I-J}$ involvement into interregional, national and international territorial reproduction systems such as construction consortium.

The development of this model is based on the division into two stages:

I. Start-up short-term mortgage and investment eco-projects and programs of the consortium type $MICP_{I-J}$(c) [14] for the formation and development of competitive advantages of $KTVS_{I-J}$ in local territorial markets of the municipal and regional levels - W_c [I];

II. Long-term mortgage and investment sustainable complexes (ecopark development) and consortium-type programs of mortgage and investment projects of the consortium type $MICP_{I-J}$ (dk) for the transformational development of local internal competitiveness of W_c[I] into external highly active and highly competitive cluster business systems in ECOstructure - W_{dk} [II].

According to the accepted research paradigm, the following analytical structural blocks are accepted:

1. The program implementation areas:

A. Territorial-industrial reproduction business systems $KTVPS_{I-J}$,[M];

B. Socio-infrastructure urban planning system [SGS];

C. Engineering and transport urban infrastructure system [ITI];

D. Territorial eco-housing and urban planning systems [THS].

2. Control indicators:

R_1 - Managing $KTVS_{I-J}$ productivity competitiveness;

R_2 - Cost management of $KTVS_{I-J}$;

R_3 - Reliability management of $KTVS_{I-J}$.

3. Structural and functional modelling of content processes of programs:

3.1. Formation of local competitiveness of $KTVS_{I-j}$;

3.2. Regional competitiveness of $KTVS_{I-j}$;

3.3. Interregional high-performance cluster-type $KTVS_{I-j}$ business systems;

3.4. Federal high-performance cluster-type $KTVS_{I-j}$ business systems;

3.5. International $KTVS_{I-j}$.

4. Organizational structures for program implementation:

4.1. Formation of separate local organizational structures of the consortium type: LLC, CJSC, Funds, Agencies, etc.;

4.2. Formation of regional structures for investment projects and programs management;

4.3. Formation of interregional management organizations, consortiums, associations, etc.;

4.4. Formation of Federal structures, cluster business systems, and public self-regulatory bodies;

4.5. Formation of international organization structures of the consortium type.

5. Objects of reproduction in the system of $W_c\,[I]$ and $W_{dk}\,[II]$ which are proposed to be considered as territorial-portfolio land and property complexes of real estate of various industries and infrastructure as key elements of property and national wealth.

$$TPLPC_{I-J} = \sum(\mathrm{A})KTVPS_{\mathrm{i}} + \sum(\mathrm{B})SGS_{\mathrm{i}} + \sum(\mathrm{C})\mathrm{ITI}_{\mathrm{i}} + \sum(\mathrm{D})\mathrm{THS}_{\mathrm{i}} \quad (1)$$

6. Subjects of reproduction in the $W_c\,[I]$ and $W_{dk}\,[II]$ systems – separate enterprises, business systems of territorial construction complexes (TCC) and partners of the consortium for the implementation of $MICP_{I-J}$ programs (banks, developers, state and municipal authorities, etc.).

7. Life cycles of investment projects, programs and competitive advantages in the system $W_c\,[I]$ and $W_{dk}\,[II]$.

As a result of the analysis, taking into account the accepted basic elements of program modelling in the form of analytical structural blocks, a generalized structural and functional model of the evolutionary development of mortgage and investment housing programs in the formation of the eco-housing market in construction was developed (figure 1) The resulting model allows to predict the system of evolutionary development of programs to build competitiveness at the meso-level of construction and identify promising areas for the formation of highly competitive and high-performance territorial production systems for managing the market of affordable eco-housing.

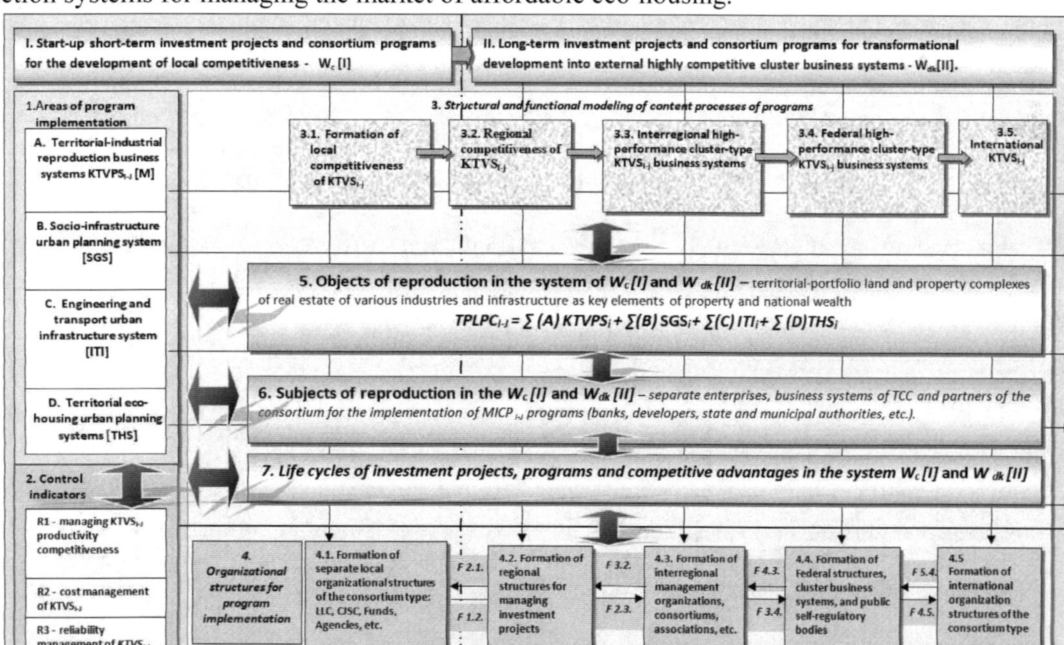

Figure 1. Structural and functional model of system formation of consortium-type organizational structures for regulating local and cluster competitiveness in construction at the meso-level.

The catalyst for the acceleration of reproduction processes in $TPLPC_{I-J}$ is individual investment projects within the framework of programs for the development of industrial competitiveness of

IOP Conf. Series: Materials Science and Engineering **945** (2020) 012020 doi:10.1088/1757-899X/945/1/012020

$KTVS_{I-J}$, as well as the completion of the processes of stabilization of regional industrial reproduction systems as an external, more highly integrated reproductive organizational and production system that generates the flow of goods, services and income.

A reliable basis for the management effect is the methodological foundations of the theory of life cycles, which must be combined with modern theories of cost management. In this regard, it is proposed to form a single vertically integrated structure of competitiveness management when managing the cost of the software systems under study, in order to achieve maximum efficiency in the use of both budget funds and the maximum effectiveness of the target management functions of all partners in the consortium.

Investment programs must necessarily pass an examination to ensure compliance with the conditions for entering self-financing, using refinancing systems, calculating the payback period of investment funds, and other economic parameters for managing the cost of programs and investment projects [15,16].

The analysis showed that for a standardized life cycle of planning and controlling of environmental programs, it is advisable to allocate 6 typical zones:

1 zone - the "starting" zone of the eco-program, where it may be a significant predominance of budget funds and credit funds of temporary partners over non-budgetary ones. At this stage, it is proposed to determine the planned indicators for the development of programs and standards for the participation of federal, regional and extra-budgetary financing sources;

2 zone - a zone of stable inflow of extra-budgetary funds of temporary and permanent partners in order to ensure the commercial success of the eco-program. The level of return on participation in the project at the initial stage of the development program should be higher than the average regional return on investment due to planned instruments of state regulation. This ensures that budget funds from the commercial success of the program's projects are refinanced both back into this program and will ensure their repayment. The level of repayment of budget funds in the presence of non-commercial components (interest-free loans or social grants) will be less than the initially invested amount of budget funding. At this stage, a certain level of return on funds from temporary partners, especially state-owned ones, may already be planned for their accumulation and subsequent "launch" of other problematic socio-economic investment projects based on innovative competitive advantages [17,18].

3 zone - zone of balance of funds of temporary partners of the consortium and extra-budgetary sources of business systems of the developer type with increasing trends of withdrawal of budget and credit funds from project financing [19];

4 zone – zone of transition to the point of zero participation of funds of temporary partners, completion of the process of refinancing funds from the project and exit of temporary partners from the project;

5 zone – zone of the expiry of the refund time partners, including budget and credit transfer programs on extrabudgetary self-financing and self-development;

6 zone - zone of withdrawal of the state, municipality and other temporary partners from support of this program in view of its self-development on the basis of self-financing and economic self-regulation with the performance of the residual function of state monitoring [20].

The final result of the research is a basic situational economic model of the functioning of mortgage and investment projects and programs $MICP_{I-J}$ for the development of competitiveness in $KTVS_{I-J}$ within the framework of the consortium system of investment controlling, which is shown in Fig. 2.

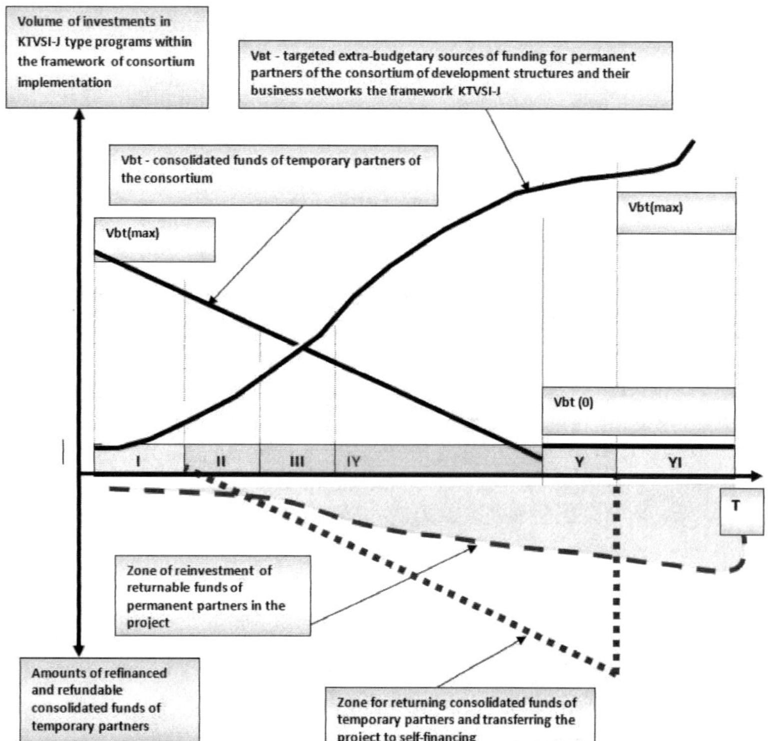

Figure 2. Principal situational economic model of functioning of mortgage and investment projects and programs $MICP_{I\text{-}J}$ for the development of competition in $KTVS_{I\text{-}J}$ in the framework of the consortium system of investment controlling.

3. Results

It is proposed to define the market value of the programs of the $MICP_{I\text{-}J}$ (MVP) type as a multi-component structure of all participants of the investment consortium with the allocation of the total cost of the MVP (T) consisting of:

I. The consolidated value of temporary participants in the consortium (CT_p):

1.2 state budget (SB),

1.2 municipal budget (MB),

1.3 cost of banking resources (BR);

1.4 cost of competitive innovation (CI);

II. Consolidated market value of participation of development structures and their business networks as permanent members of the consortium (CD).

$$\text{MVP (0)} = \text{MVP (CTp)} \{\text{SB; MB; BR; CI}\} + \text{MVP (CD)}, \tag{2}$$

$$\text{MVP (CTp)} = \textstyle\sum_0^t V_{bt} \cdot {}^{1}/_{(1+i_{cb})^{t}}, \tag{3}$$

V_{bt} is the amount of eco-program funding from the consolidated budgets of temporary participants of the investment consortium SB; MB; BR; CI as their sum for years from *0* to *t*;

$1/(1+i_{cb})^t$ – weighted average discount rate of consolidated budgets of temporary participants of the investment consortium at the planned differentiated cost of budget sources of i_{cb};

$$\text{MVP (CD)} = \textstyle\sum_0^t \text{V}_{\text{bt}} \cdot {}^{1}/_{(1+i_{cd})^{t}}, \tag{4}$$

V_{knt} - the amount of funding for the program from the consolidated budget of development structures and their business networks as permanent members of the consortium;

$1/(1+i_{cd})^t$ - discount rate of funds from development structures and their business networks as permanent members of the consortium at planned average weighted cost i_{cd}..

4. Summary

Thus, the research results allow us to form a conceptual picture of increasing competitive advantages in terms of productivity and efficiency in territorial production business systems of the developer type through investment consortium in the life cycles of investment projects and programs.

The analysis of models shows that state and municipal regulation, as well as innovative, mortgage and investment partnership development of competitive advantages, are characterized by cycles of activity with different parameters of cost management.

5. References

[1] Malysheva T V, Shinkevich A I, Zelenkina E V, Denisov I S, Vorotyntseva T M, Dmitrieva O A, Kurdyumov V I 2017 Development and concentration efficiency study of enterprises innovation activity in real sector of economy *Eurasian Journal of Analytical Chemistry* **12(7)** 1347-1356

[2] Krygina A M 2013 Conceptual framework of development and transformation of organizational competitiveness of the territorial-reproducing systems in construction *Basic research* **10-5** 996-1000

[3] Belyaeva S, Voronov D, Erypalov S 2017 Methodical principles of evaluation of competitive ability of construction industry and real estate development company *MATEC Web of Conferences* 08033

[4] Maslyukova E, Anoshina Y, Khakimov A, Aleksandrovskiy M 2018 Methodology for the development of innovation projects on the basis of competitiveness indexes *MATEC Web of Conferences* 08012

[5] Levitsky T Yu 2016 Competitive strategy of construction production in the conditions of innovative development of the industry *Economics, management, law realities and perspectives* 191-195

[6] Gumba Kh M, Belyaeva S V, Voronov D S, Erypalov S E 2017 Assessment of the competitiveness of the construction industry and enterprises: methodology and practice *Economy and entrepreneurship* **3-1(80)** 894-900

[7] Tsatkhlanova T T, Ubushayeva B G, Erdnieva E V 2018 Formation of a competitive environment in the economy of the region *Municipal Academy* **3** 93-99

[8] Krygina A M 2014 Formation of organizational and economic decisions in innovative housing construction *Creative economy* **7(91)** 86-99

[9] Kudratova G M, Makarova L V 2018 Comprehensive analysis of the competitiveness of products and construction industry enterprises *Science alley* **3-5(21)** 382-387

[10] Uchinina T V, Molokova T V 2019 Formation of a competitive strategy of a development company in the construction of multi-storey residential real estate *Housing strategies* **6-2** 175-198

[11] Yusufova A M, Cetin T E 2018 The practice of formation of economic conditions for stimulation of construction production development in Russia *Regional problems of economic transformation* **12(98)** 90-98

[12] Gusakova E A, Krygina A M 2012 Applying an innovative approach to managing changes in the implementation of investment and construction projects *Real estate: Economics, management* **2**, 84-86

[13] Krygina A M 2013 Modeling of program-target organization and management of competitiveness of territorial reproduction systems in construction *Industrial and civil construction* **10** 59-62

[14] Krygina A M, Krygina N M, Pozdnyakova E V 2016 Public-private partnership as an effective tool for the development of low-rise housing eco-construction *Economics and entrepreneurship* **2-2 (67-2)** 412-416

[15] Naumov A E, Koshlich Yu A, Oberemok M I, Belousov A 2019 Comparative analyzes for increasing the energy efficiency of civil constructions *19th International Multidisciplinary*

IOP Conf. Series: Materials Science and Engineering **945** (2020) 012020 doi:10.1088/1757-899X/945/1/012020

Scientific GeoConference SGEM 2019 Conference Proceedings 277-284

[16] Oberemok M I, Naumov A E Shchenyatskaya M A 2019 Qualitative analysis of view characteristics of residential property *Bulletin of BSTU named after V.G. Shukhov* **3 (4)** 44-51

[17] Oberemok M I, Naumov A E 2019 Improving the implementation of space syntax concept in the design of commercial property *Real estate: economics, management* **4** 58-62

[18] Avilova I P, Naumov A E, Krutilova M O 2019 Methodology of GHG emissions assessment caused at the construction of energy facilities. case study: hydropower *IOP Conference Series: Materials Science and Engineering* **552** 012018

[19] Ananyeva E S, Korshunova N N 2020 Smart house as a new type of housing *Construction Materials and Products* **3 (1)** 83 – 88

[20] Otsokov K A 2020 Innovative technologies in construction and their use in organizational and technological events *Construction Materials and Products* **3 (1)** 7-13

Acknowledgments

The work is realized in the framework of the implementation of a comprehensive project to create high-tech production “Development of new methods and tools for management of property in the budget sector and their implementation in the software package of the information-analytical system for centralized management of property owned by the constituent entities of the Russian Federation, municipalities, as well as state property companies” (agreement No. 074-11-2018-026 of 07/11/2018).

IOP Conf. Series: Materials Science and Engineering **945** (2020) 012021 doi:10.1088/1757-899X/945/1/012021

Management mechanisms for resolving organizational conflicts in the context of implementing the economic opportunities of organizations in the construction industry

T V Tselyutina[1*], Zh N Avilova[2] and O A Timokhina[1]

[1]Department of Economics and Management, Belgorod State University, Pobedy St., 85, Belgorod, 308015, Russia
[2]Department of Economics and management, Belgorod State Technological University named after V.G. Shukhov, Kostyukov St., 46, Belgorod, 308012, Russia

E-mail: tselyutina@bsu.edu.ru

Abstract. The paper presents a critical analysis of the functional representation of the mechanism for managing social and labor conflicts in the process of organizational changes. Network mechanisms and their functionality in the management system of organizational conflicts are updated. The relationship between the level of staff involvement in work and the level of conflict in the organization is determined. To ensure effective management of organizational conflicts, it is necessary to take into account the level of involvement of the personnel of the organization in processes at different levels, as well as to create conditions and economic methods of motivation that ensure the necessary level of involvement. "Portfolio" of empirical tools for comprehensive research of issues and a mechanism for managing organizational conflicts are developed, taking into account the specifics of the activities of construction organizations.

1. Introduction

The problem of studying mobile mechanisms for managing organizational conflicts in organizations in the construction industry is primarily determined by the very rapid pace of transition to new management technologies in the process of social and labor activity. As a result, there is the staff's resistance to organizational changes, "hidden" conflicts, and the problem of reducing the staff. Therefore, it is important that organizations have effective management mechanisms for conflict prevention and management, based on a program-oriented approach and involvement of staff in organizational changes. The need to find an innovative, more adaptive approach to managing organizational conflicts is dictated by the variability of the environment and the need to ensure the effectiveness of strategic development of more effective solution to the contradiction between available resources and the required innovative proposals of external environment.

2. Materials and methods

The methodological basis of the research was a systematic approach to the research object, the principle of specificity, principle of development and the principle of objectivity. Research methods: organizational-comparative; group of empirical methods – archival (primarily, document analysis),

IOP Conf. Series: Materials Science and Engineering **945** (2020) 012021 doi:10.1088/1757-899X/945/1/012021

survey methods (questionnaires, interviews), observation; processing methods – qualitative and quantitative analysis; interpretation methods – structural. The subject of the research is socio-economic and managerial relations that arise in the process of forming mechanisms for resolving social and labor conflicts in construction organizations.

3. Results and discussions

The manifestation of VUCA economy trends in the construction industry because construction, as the Russian economy as a whole is facing deep and systemic challenges reflecting the uncertainty and instability of the global trends, so the complexity and contradictions of internal barriers to development of Russian industry. Among the most significant challenges, along with strengthening global competition, innovation waves of technological change (digitalization, bio - and nanotechnology), reorientation of financing of housing construction, it is necessary to highlight the increasing role of human capital as a key factor of economic development of the country. These challenges have determined the need and expediency of developing management mechanisms for resolving organizational conflicts in the context of implementing the capabilities of the VUCA economy in construction [1].

In modern conditions, the role of innovation is increasing in all spheres of public life. This is a reflection of the growing understanding of society that the socio-economic development of the country and certain spheres of life is impossible without innovations in various spheres of public life [2]. The toughening competitive situation for many Russian enterprises has put them in front of the problem of effective operational management aimed at reducing production costs, as well as ensuring optimal use of human resources, which in turn implies the search for new, innovative approaches to managing organizational conflicts [3, 4]. The analysis shows that in economic practice, construction organizations implement various innovations by type of novelty, depth of changes made, scale and scope of application [5].

Management of organizational conflicts occupies one of the central places in the system of personnel management of modern construction organizations. Moreover, taking into account the importance of labor resources, modern companies are aimed not only at getting benefits from the labor potential of the company, but also at acquiring experience, new knowledge, and the latest technologies by employees. In addition, modern managers are interested in the growth of competencies and personal motives for the work of their employees. The future competitiveness of the company and further development depend on effective management of personnel formation. In the environment where it is clear that the development of the personnel of the company affects its market stability, it is necessary to recognize the need for a measured strategy for developing human resources [6].

Within the framework of the issue we define the essence of social-labor conflict as the polar collision of certain subjects of social-labor relations, which resulted in the observed open limit aggravation of contradictions in social and labor spheres of the organization, and involving local management or strategic action (decisions) that are necessarily directed to the achievement of the goals and objectives by each of the actors, the gap and protect of their interests or the interests of their group (department, team, etc.).

3.1. Recommended "portfolio" of empirical tools for complex research issues

The first direction of the study is to assess the satisfaction with the personnel's work of all categories.

The second direction of the research is the assessment of conflict in the construction industry organizations, based on the identification of an integral indicator of social tension. This indicator is detected using factor analysis. A special place in the management of social and labor conflicts is occupied, first of all, by their prediction and prevention based on the control of the level of social tension. The higher the social tension, the more negative, adverse conditions and factors, at any stage of origin and course, can lead to a conflict situation [7].

The main problem – it is necessary to provide effective procedures for forecasting and preventing possible social and labor conflicts, taking into account the specifics of a particular construction

IOP Conf. Series: Materials Science and Engineering **945** (2020) 012021 doi:10.1088/1757-899X/945/1/012021

organization. It is necessary to carry out systematic monitoring of the level of social tension, which will be one of the tools of the social and labor conflict management system, namely, prevention.

The third direction of research is expert interviewing of managers and specialists as the main initiators of supporting measures for managing social and labor conflicts in order to study existing conflict management procedures. The task is to get an expert assessment of the main signs of social tension (absence at work; increase in interpersonal conflicts; increase in dismissals on their own; dissemination of false information, rumors; collective failure to comply with orders of senior management, etc.). The aim is expert assessment (including consulting resources) of current measures and procedures for conflict prevention; the degree of involvement of personnel (staff) in the system of resolving social and labor conflicts in the construction industry [8].

The fourth direction of research is the analysis of the socio-psychological climate as an indicator and resource for preventing the development of social and labor conflicts. We recommend using the methodology for studying the socio-psychological climate in the labor team of O. Mikhalyuk and A. Shalyto; the methodology for diagnosing the types of behavior of personnel in a conflict situation – the questionnaire of K. Thomas "Determining ways to regulate conflicts"; the methodology for interpersonal diagnostics of relations of T. Leary.

The general set of research on the impact of the level of staff involvement on the number of organizational conflicts was made up of 9 construction organizations in Belgorod (figure 1).

a *b*

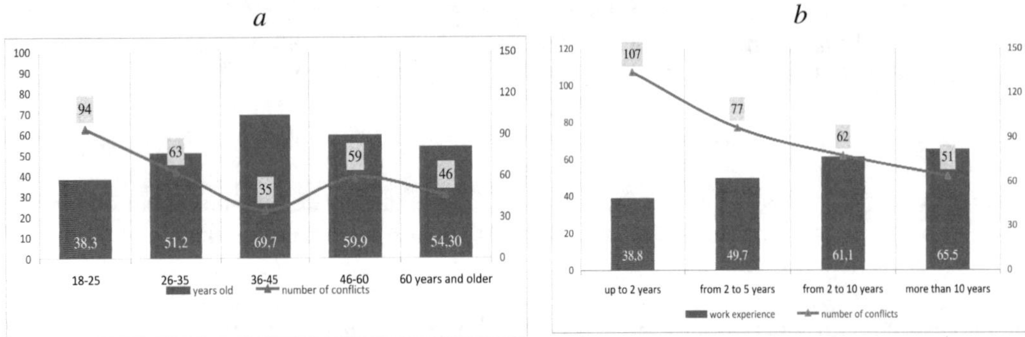

Figure 1. Employee engagement index a) by age and number of conflicts and b) by length of service and number of conflicts.

Research results: a significant part of organizational conflicts is related to interpersonal conflicts (70%); the largest number of conflicts occurs in the group of 36-45 years old, while this group also has a significant decrease in the level of involvement; with a long period of work of employees, their level of involvement decreases significantly, while the number of interpersonal conflicts between them increases (this is a very poor indicator).

Table 1 shows the distribution of correlations between factors that influence conflicts and the degree of conflict in a particular construction organization. The higher the value of the correlation indicator, the stronger the influence of the factor.

IOP Conf. Series: Materials Science and Engineering **945** (2020) 012021 doi:10.1088/1757-899X/945/1/012021

Table 1. Distribution of correlations between factors that influence conflicts and the conflict degree in the organization.

Factors	Conflict level			
	High	Above average	Average	Low
Relations with colleagues and management	0.539	0.418	0.420	0.401
Content and nature of the work	0.225	0.128	0.118	0.001
Career and development	0.347	0.301	0.223	0.0008
Economic motivation	0.478	0.489	0.289	0.167
Changes and innovations	0.151	0.114	0.112	0.018
Loyalty to the company	0.012	0.125	0.005	-0.109
Culture of the construction organization	0.189	0.111	0.187	0.008

Conclusion – the factor "relations with colleagues and management" has the maximum impact on the number of conflicts. Regardless of how respondents determined the degree of conflict in a particular construction organization, there is a clear correlation between this factor and the occurrence of conflicts. It is also worth noting such factors as "economic motivation", primarily "salary" and "career and development". In addition, a negative (inverse) correlation dependence between the level of conflict and the degree of staff involvement was found (-0.282), which reflects the fact that when one variable grows, the second one decreases. The greater the number of conflicts in a construction organization, the lower the level of staff involvement, and the more dissatisfaction with economic methods of motivation.

At forming a comprehensive mechanism for managing social and labor conflicts for organizations in the construction industry, we propose to apply a program-target approach that can combine the object and subject of management, a system for implementing management decisions in the conditions of socio-economic transformation of the VUCA environment.

The model of managing social and labor conflicts in the process of organizational changes is shown in figure 2.

The set of reasons underlying the social and labor conflict determines its general nature, configuration, features of its course, as well as the basic conditions for managing it:

– ignoring, incompetence, ignorance of the norms of current labor legislation in construction by managers;

– reduction of social and labor guarantees for all categories of personnel;

– low wages and their late payment (long delays in payments), unfair rewards and bonuses;

– corrupt, bureaucratic attitude of the top management to the social and labor interests of employees; irrational social policy and social infrastructure of the construction organization;

– the lack of policies to ensure decent working conditions;

– "veiled" attempts by the administration to dismiss employees illegally;

– devaluation of corporate and labor culture values as a social and managerial resource.

The mechanism is based on the following sequence of actions:

– combining the concept of the model and the theory of the stakeholder network explain the formation of the stakeholder system, as well as the roles that stakeholders play in the process of organizational change;

– the importance of resistance factors (both organizational and individual) as the cause of conflict is emphasized;

– using network mechanisms and strategies to intervene in the conflict system for management and resolution;

– providing a methodological basis for the process of identification and elimination in the context of organizational changes at various levels of complexity of the organization with the participation of individuals, divisions, departments.

IOP Conf. Series: Materials Science and Engineering **945** (2020) 012021 doi:10.1088/1757-899X/945/1/012021

<table>
<tr><td>Participants
- Parties (opponents)
- Instigators
- Abettors
- Organizers (they plan but do not always participate in the clash)</td><td>Subject
- difference of views of opponents on the object (real or imaginary)</td><td>Object
- a scarce resource; limitations, values, etc.</td></tr>
<tr><td colspan="3">Conflict situation (origin of the conflict, latent conflict)</td></tr>
<tr><td colspan="3">Incident (an event or circumstance that caused the clash)</td></tr>
<tr><td colspan="2">Intentional interference management in the implementation of the interests of the opponent</td><td>Escalation of conflict</td></tr>
<tr><td colspan="3">Crisis (constructive or destructive break in the relationship of opponents)</td></tr>
<tr><td colspan="2">Confrontation</td><td>Negotiations, truces, compromise</td></tr>
<tr><td colspan="3">The end of the conflict (resolution, truce, “impasse”)</td></tr>
<tr><td colspan="2">Destructive conflict
- Separation of participants
- Eliminating the initiator-leader of the conflicting group
- Elimination of the subject
- Truce</td><td>Constructive conflict
- Self-destruct of the object of conflict or the circumstances of its use</td></tr>
</table>

Figure 2. Mechanism for managing social and labor conflicts in the process of organizational changes (compiled by the authors).

On the one hand, resistance is a phenomenon that can both undermine organizational change by delaying or slowing its start, which will make it difficult to implement and increase costs, and serve as a source of additional information for developing a more successful change process [9, 4].

The factors that cause resistance must be identified and evaluated in order to manage them effectively as part of an overall conflict management strategy. Factors that influence the resistance process can be classified into individual and organizational ones. Individual resistance factors include: selective perception and retention, self-interest, frustration, fear of the unknown, low motivation, feelings of failure, self-doubt, conservatism, and loss of control. Organizational factors of resistance include: compliance with norms and values, culture, past experience, and so on.

The proposed model of conflict management in the process of organizational changes involves the allocation of a “network of small niches” for the purpose of more detailed consideration and identification of participants in the conflict (active or passive). It is important to answer questions related to individual or organizational sustainability, as well as to get a complete picture of the system of their interaction. The formation of networks of this type can affect both the perception and opinions, and the interpretation of emerging events and phenomena, which in turn can be used in conflict management by providing a point influence on the perception of conflict participants in the process of organizational changes (table 2).

Table 2. Network mechanisms and their functionality in the conflict management system.

Network mechanism	The functionality of the mechanism
Communication proximity	This mechanism considers an organization as a communication network in which stakeholders repeatedly interact (directly or indirectly) to process resources and information.
Positional proximity	This applies to a network of structurally independent individuals who may not be connected to each other, but they have similar attributes – roles, obligations, states, and expectations.
Spatial proximity	Spatial proximity is based on the probability of interaction and dissemination of social information between parties in close proximity, which subsequently affects the formation of relationships. In contrast to direct interaction, this can affect the processing of social information through the impact or unavailability of individuals to the organizational subclimate, targeted materials, and activities.

Implementation of network mechanisms should be carried out in the process of making changes through "niches". In the context of the change process, the intervention should have an impact on the specific individual directly involved or expected to participate in the changes. Network mechanisms, together with appropriate intervention strategies, will have a direct impact on the attitudes and behaviors of conflict participants. The network mechanism will ensure the flow of information, while the nature of the intervention and the roles played by public opinion leaders during these events will ensure the implementation of the process of change [10].

The proposed conflict management mechanism takes into account the constantly increasing complexity of information systems and the growing number of contacts, which in turn requires the staff of modern organizations to have appropriate knowledge in the field of modern communications [11].

It is worth noting that one of the advantages of the proposed mechanism for managing organizational conflicts is the creation of more favorable conditions for the development of intellectual capital, which, according to I.V. Somina, in the last decade is considered as one of the main factors for ensuring the competitiveness of our country, regions and individual companies [12, 13, 14, 15, 16].

4. Summary

The main methodological approach to the solution and prevention of organizational conflicts, which we define as the most effective in obtaining a social effect is the method of choosing strategic means of resolving conflict situations, taking into account the specifics of the activities of a particular construction organization. The proposed mechanism of conflict management – this information management system, comprising a set of target programs and flexible socio-managerial mechanisms for timely resolution, first of all, social and labor conflicts, the task of which is to ensure that on the basis of diagnostic and prognostic analysis to trace the emergence and deployment of conflict processes in social and labor sphere of specific construction company, and depending on their nature put forward reasonable proposals, specific procedures, methods for prevention, localization and resolution of organizational conflicts.

5. References

[1] Strategy for the development of the construction industry of the Russian Federation until 2030: approved by Order of the Government of the Russian Federation of May 10, 2016 No. 868-r https://www.srosp.ru/upload/files/documents/strategy_text_13112019.pdf.

[2] Podborodnikov I S 2019 Impact of the level of innovation activity in the region on the degree of development of small and medium-sized businesses, education, health care in the region *Russian Economic Bulletin* **2(3)** 78–90

[3] Minaeva L A, Glagolev S N, Doroshenko Yu A 2019 Practice of applying innovative solutions in operational management. Actual problems of economic development: Collection of reports of the X International correspondence scientific and practical conference. Publishing house: Belgorod State Technological University named after V.G. Shukhov 269-272 https://elibrary.ru/item.asp?id=41326232

[4] Doroshenko Y A, Minaeva L A, Avilova Z N, Somina I V, Leonova O V 2015 Socio-economic aspects of development of small innovative business in Russia *Journal of Applied Engineering Science* **13(4)** 293-298

[5] Bukhonova S M, Kamyshinsky K O, Kurlov D A 2006 Development of a mechanism for providing investment resources for innovative activities of industrial enterprises. Innovations. Publishing house: LLC “Transfer-Innovations” (Saint Petersburg) **7(105)** 117-120 https://elibrary.ru/item.asp?id=12937468

[6] Cherepanova T G, Arbenina T I, Makhinova N V 2019 Methodological approaches to the formation of enterprise personnel *Modern Economy Success* **1** 120-127

[7] Hachaturova M 2010 The role of personality traits in choosing of coping strategies in an organizational conflict Department of Psychology, University - Higher School of Economics, Moscow. World academy of science, engineering and technology **3778** 735-736 https://elibrary.ru/item.asp?id=18087061

[8] Avilova Zh N, Tselyutina T V 2017 Consulting resources as a component of the innovative potential of the construction sector in the region *Bulletin of BSTU named after V.G. Shukhov* **10** 209-216

[9] Tselyutina T V, Vlasova T A, Timokhina O A, Maslova Y V 2019 Convergence of institutes of functional representation of group interests under transformation of social processes, quality of life, and human capita *Humanities and Social Sciences Reviews* **7(6)** 111-117 http://giapjournals.com/index.php/hssr

[10] Tselyutina T V, Timokhina O A, Vlasova T A, Maslova Y V 2019 Development of the Personnel Risks Assessment and Supply Chain Strategy as a Basis of the Risk Management System of Modern Organizations *International Journal of Supply Chain Management* **8(5)** 1030-1038 https://ojs.excelingtech.co.uk/index.php/IJSCM/article/view/3912

[11] Doroshenko Yu A, Logachev K I, Antipov E A 2011 Improvement of management of strategic development of industrial holding *Belgorod Economic Bulletin* **4(64)** 3-8 https://elibrary.ru/item.asp?id=20770079

[12] Somina I V, Chernyshova D I 2019 Role of intellectual capital in the development of innovation science-intensive technologies and innovations (XXIII scientific readings) International scientific and practical conference dedicated to the 65th anniversary of BSTU named after V.G. Shukhov. Publishing house: Belgorod State Technological University named after V.G. Shukhov 121-124 https://elibrary.ru/item.asp?id=39254861

[13] Otsokov K A 2020 Innovative technologies in construction and their use in organizational and technological measures *Construction materials and products* **3(1)** 7-13 https://doi.org/10.34031/2618-7183-2020-3-1-7-13

[14] Ismailov R I 2020 Making production decisions using the expert evaluation procedure *Construction Materials and Products* **3(1)** 49-56 https://doi.org/10.34031/2618-7183-2020-3-1-49-56

[15] Subbotin O S 2019 Building materials and technologies of modern housing: architectural and ecological aspects *Construction Materials and Products* **2(4)** 84-88

[16] Manzhilevskaya S E 2019 Organizational and economic problems of ecological safety in construction *Construction Materials and Products* **2(4)** 73-78

Acknowledgements
This work was realized in the framework of the Program of flagship university development on the base of the Belgorod State Technological University named after V G Shukhov, using equipment of High Technology Center at BSTU named after V G Shukhov.

IOP Conf. Series: Materials Science and Engineering **945** (2020) 012022 doi:10.1088/1757-899X/945/1/012022

Using of automated risk assessment systems to ensure the safety of personnel at construction sites

A Yu Semeykin[1*], E V Klimova[2], E A Nosatova[3] and Yu V Khomchenko[4]

[1, 2, 3] Department of Life Safety, Belgorod State Technological University named after V.G. Shukhov, 46, Kostyukov Str., Belgorod, 308012, Russia

[4] D.O.O. Artika, Herceg Novi, Montenegro

E-mail: semeykin.ay@bstu.ru

Abstract. The development of automated risk assessment systems is a promising area for scientific research in the field of OSH management. The paper analyses the main problems of the functioning of the OSH management system and identifies the need for its improvement in order to reduce occupational injuries. As a way to improve it, we propose the development and implementation of an automated system for forecasting and managing professional risks (AS FMPR), which implements an algorithm for multivariate analysis of the causes of occupational injuries. The stages of risk assessment are considered. The aim of this work is to assess the applicability of automated risk assessment systems to ensure the safety of personnel at construction sites. The results of the distribution of the probability of injury to workers depending on the factors studied are presented. This system will bring the procedure for assessing occupational risks in construction to a whole new level.

1. Introduction

At present, in Russia, as in many other developing countries, the construction industry is one of the most traumatic industries. According to the International Labor Organization, annually the number of victims of accidents in the construction industry is more than 450,000 people a year [1]. The types of incidents that result in personal injury to workers include the following [2-7, 24]:

- fall of victims from a height;
- falling objects on the victims;
- electric shock;
- the impact of rotating, moving parts of machines and mechanisms;
- jamming between a moving and a stationary object;
- transport accidents;
- collapse of the soil during earthwork, etc.

Table 1. The dynamics of industrial injuries by type of economic activity "Construction" in the Russian Federation for 2017-2018.

The average number of employees (thousand people)		The number of victims of industrial accidents with loss of ability to work for 1 working day or more and fatal				The number of victims of fatal accidents at work			
		number of victims		per 1000 employees		number of victims		per 1000 employees	
2017	2018	2017	2018	2017	2018	2017	2018	2017	2018
1192.2	1150.3	2146	1930	1.8	1.7	214	190	0.179	0.165

IOP Conf. Series: Materials Science and Engineering **945** (2020) 012022 doi:10.1088/1757-899X/945/1/012022

According to the data of table 1, in Russia there is a tendency to reduce the level of injuries in construction, but so far this industry remains the leader in the number of injured. The main type of incident is the fall of the victim from a height [8-9].

The data of the international labor organization and Eurostat on the level of injuries in the EU countries indicate that the share of accidents pertaining to various types of work on construction sites is 61.7% (for cases without fatal outcome) and 52.9% (cases with fatal outcome) [1, 10, 11] (Table 2).

Table 2. Industrial injuries in Europe.

NACE (Section)	Non-fatal accidents at work								
	Total	Industrial site	Tertiary (office, amusement area, miscellaneous)	Public area	Construction site, opencast quarry or mine	Health establishment	Farming, fish farming, forest zone	In the home	Other or no information
NUMBER	Non-fatal accidents at work (thousands)								
Total (all activities)	3344.5	995.5	615.1	333.0	307.5	227.3	106.5	80.6	678.9
Construction (F)	376.4	51.4	6.9	22.9	232.4	0.9	1.7	8.3	51.9
SHARE	Non-fatal accidents at work (%)								
Total (all activities)	100.0	29.8	18.4	10.0	9.2	6.8	3.2	2.4	20.3
Construction (F)	100.0	13.7	1.8	6.1	61.7	0.2	0.5	2.2	13.8
NUMBER	Fatal accidents at work (number)								
Total (all activities)	3 552	637	174	1 017	518	29	287	58	832
Construction (F)	733	61	8	135	388	2	10	10	119
SHARE	Fatal accidents at work (%)								
Total (all activities)	100.0	17.9	4.9	28.6	14.6	0.8	8.1	1.6	23.4
Construction (F)	100.0	8.3	1.1	18.4	52.9	0.3	1.4	1.4	16.2

The main reasons for injuring people at construction sites are non-compliance with safety rules, non-use of personal protective equipment and others. But the main reason is the erroneous actions of workers.

Erroneous actions or omissions of an employee are associated with non-compliance with safety rules, including with getting used to dangerous conditions, with disruption of the normal flow of the technological process, with errors of the management, lack of application or lack of personal and collective protective equipment, etc. Therefore, the selection of modern tools and technologies is required to establish the conformity of the psychological characteristics of the person with the activities performed, assessment criteria his psycho-emotional state, their accounting and analysis, together with factors affecting the creation of a risky situation [8, 9, 14, 15].

2. Materials and methods

Modern OSH management systems (OSH) are based on the principle of hazard prevention. Risk category - is one of the hazard characteristics currently in use [9].

The most promising and effective approach to the implementation of "hazard prevention" in practice is the process of identification, assessment and management of professional risks. In foreign countries, this tool allows you to achieve exceptionally high results and the desired goal - jobs without injuries and occupational diseases [10]. According to the ILO, the most dangerous are the manufacturing and construction sectors, followed by transport, agriculture and mining.

At present, in Russia, the transition to a professional risk management system is only advisory in nature. At the same time, the definition of occupational risk management given in the Labor Code of the Russian Federation indicates that the introduction of this system entails the need to change the OSH management system as a whole.

IOP Conf. Series: Materials Science and Engineering **945** (2020) 012022 doi:10.1088/1757-899X/945/1/012022

Assessment of occupational risks, ensuring health and labor safety are an important part of the entire enterprise management, the development and implementation of which can be ensured on the basis of the international standard OHSAS 18001: 2007 "Occupational health and safety management systems" and a number of domestic laws, taking into account economic aspects related to the compensation of costs for improving the conditions and labor protection of employees [12, 13].

For effective risk management, a more adequate and comprehensive risk assessment is necessary, taking into account the professional and personal qualities of the personnel, their psychophysiological state, and factors of working conditions in real time. At the same time, the enterprise management should optimally assess the production risk for the safety and health of employees, taking into account the nature of their work in the enterprise and all types of impacts in the performance of their duties, and for this an ongoing risk management process should be introduced.

At the same time, it is necessary to take into account a significant degree of uncertainty of factors that determine the level of professional risk, and the limited capabilities of specialists of enterprises responsible for labor safety. In this regard, the use of modern information technologies and automation systems for assessing, forecasting and managing professional risks (AS FMPR) will significantly accelerate this process [12, 13].

Modern tools for predicting incidents and unsafe employee behaviour are algorithms created on the basis of various technologies that combine the capabilities of a computer and the Internet. At the same time, they use intelligent systems based on neural network models, Markov networks, Bayes networks, and other models.

The work of Russian scientists is mainly aimed at assessing and managing occupational health risks for workers, including the use of bio information technologies [12]. The theory of functional systems of P.K. Anokhin was chosen as the methodological basis for managing occupational risks to the health of a human operator. Based on intelligent neural systems, cognitive technologies have been developed to assess the impact of harmful production factors [13].

Abroad, there are attempts to create models of unsafe behaviour of workers based on the analysis of indicators of their behaviour and the working environment, such as the level of management control, working conditions, safety management system, level of employee participation, level of knowledge of safety requirements, attitude to safety, motivation, allocation of resources and work stress.

The foreign models of analysis of employee behaviour are based on Bayesian network theories [16–20, 24]. The following technologies are used at the heart of foreign expert systems:

- Fault Tree Analysis (FTA) analysis methods;
- Analysis of errors and their consequences (Failure Mode and Effects Analysis (FMEA));
- Analysis of processes based on fuzzy approaches (Fuzzy Analytic Network Process);
- Forecasting events based on Bayesian Networks;
- Analysis of events and processes using GERT networks (Graphical Evaluation and Review Technique);
- Forecasting events using the technology of neural networks (Neural Networks).

An analysis of these works showed that they are based on predictive modelling of incidents based on known statistics for certain types of jobs and professions, but they do not take into account the constantly changing factors of the work environment, the labor process, and changes in the psychoemotional and physiological state of the employee. It should be noted that the results of well-known domestic and foreign studies have not been brought to practical implementation in the form of expert labor safety management systems implemented in real enterprises.

3. Results and discussions

The analysis of accidents made it possible to single out the following causes of work-related injuries: technical, sanitary-hygienic, and organizational (including ergonomic), geographical, social, psychophysiological, biographical, and economic.

IOP Conf. Series: Materials Science and Engineering **945** (2020) 012022 doi:10.1088/1757-899X/945/1/012022

The developed risk assessment system includes the maximum possible number of assessment criteria to reduce errors in the assessment.

The effectiveness of the proposed risk assessment system depends on the choice of modelling method, forecasting and calculation of the probability of incidents with employees of the enterprises and the magnitude of professional risk.

To determine the most effective incident modelling method, it is necessary to calculate the probability of an incident taking into account the influence of the factors studied.

For this, it is proposed to use the following methods of modelling and forecasting events and incidents:

- logical and probabilistic modelling by the fault tree method;
- analysis of processes based on fuzzy approaches (Fuzzy Analytic Network Process);
- a method for constructing and calculating probability using Bayesian networks.

An example of event modelling (industrial accident) can be the construction of a fault tree (Fig. 1).

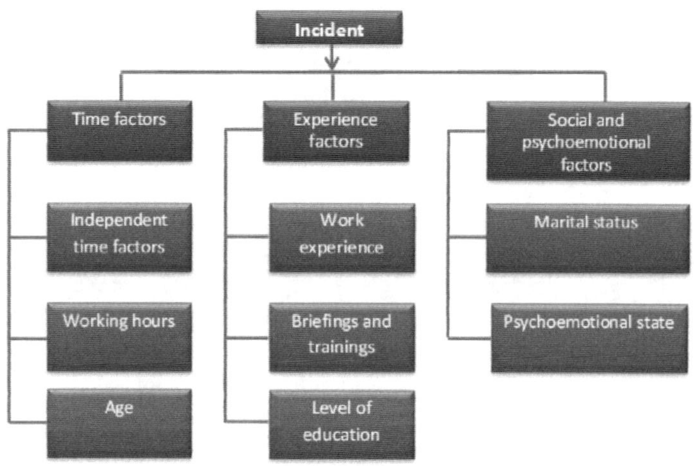

Figure 1. The scheme for calculating the probability of injury to an employee, taking into account time, social factors, the level of competence and the psychoemotional state of a worker.

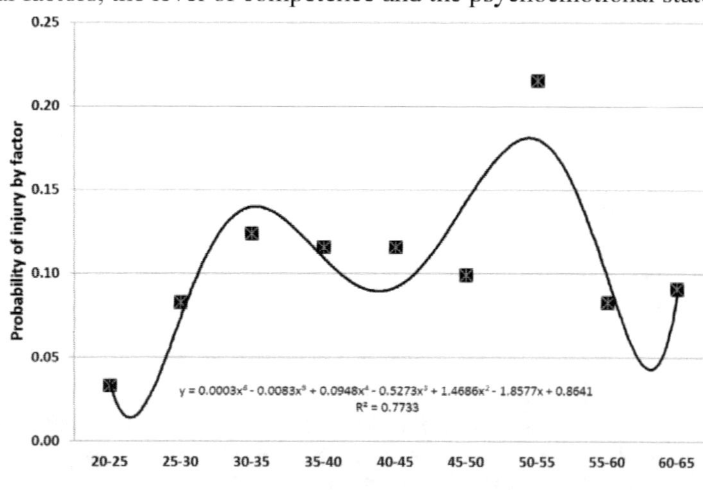

Figure 2. The distribution of the probability of injury to workers by age factor.

IOP Conf. Series: Materials Science and Engineering **945** (2020) 012022 doi:10.1088/1757-899X/945/1/012022

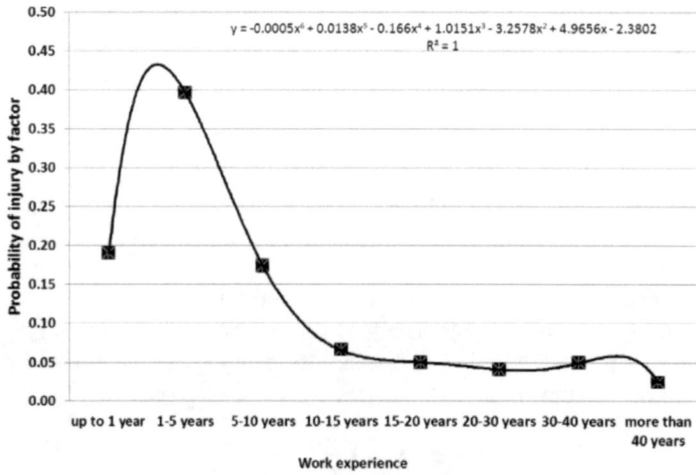

Figure 3. The distribution of the probability of injury to workers by the factor of work experience by profession.

Taking into account the data on the probability of injury to workers for each factor studied, it is possible to determine the probability of an incident (injury or death of an employee), taking into account their joint influence on the final probability of the event. At the first stage of the process of incident probability determination, a retrospective analysis of the statistical indicators of injuries is carried out by means of a quantitative statistical assessment of the factors causing occupational injuries. At the second stage, a predictive analysis of the magnitude of the indicators of injuries and occupational risk is carried out on the basis of fuzzy logic methods. The third stage involves the quantification of traumatic factors based on an expert method. The fourth stage is the identification of problem jobs. The final stage is responsible for choosing a managerial decision to manage the level of professional risk (a database of decision options for managers and specialists of enterprises responsible for ensuring labor safety) [21-23, 25].

For example, for an employee aged 45 who has 5 years of work experience, the level of education is the average professional for the current date and time (June 18, 2020, 14:00 00), the probability of an incident will be 0.046. For comparison, for an employee aged 55 years, the work experience in the profession is 5 years, the level of education is secondary, and the probability of an incident will be 0.16.

This model can serve as a base for its further refinement and expansion, taking into accounts the influence of factors such as gender, elapsed time after checking the knowledge of an employee on labor protection, the class of working conditions, type of production, profession, qualification, the presence of harmful factors, etc.

4. Summary

Prediction and management of occupational risks will positively affect the overall safety and the level of occupational injuries of the enterprise, and will allow employers to significantly increase the effectiveness of the processes for assessing, forecasting and managing occupational risks in enterprises or specific workplaces.

To improve the safety of workers performing work at heights, it is proposed to apply new training technologies in the field of labor protection, in particular, pre-shift testing, as well as encouraging the use of safe working practices and observing safety rules; an assessment of the psychoemotional state of the employee to reduce the risk of injury is recommended to be introduced in real time, i.e. introduce an individual integrated diagnosis of the body, an assessment of the level of fatigue, stress, the influence of emotions; also in real time to keep records and analysis of parameters characterizing working conditions.

IOP Conf. Series: Materials Science and Engineering **945** (2020) 012022 doi:10.1088/1757-899X/945/1/012022

Automated risk assessment system, which has been developed at BSTU named after V.G. Shukhov allows us to reliably calculate the risk and select the best measures to reduce it.

5. References

[1] Eurostat / Non-fatal and fatal accidents at work, by working process and economic activity, EU-28, 2017 Available at: https://ec.europa.eu/eurostat/statistics-explained/index.php?title=File:Non-fatal_and_fatal_accidents_at_work,_by_working_process_and_economic_activity,_EU-28,_2017.png

[2] Chi C F, Chang T C and Ting H I 2005 Accident patterns and prevention measures for fatal occupational falls in the construction industry *Applied Ergonomics* **36(4)** 391–400

[3] Chi S, Han S and Kim D Y 2013 Relationship between unsafe working conditions and workers' behavior and impact of working conditions on injury severity in U.S. construction industry. *Journal of Construction Engineering and Management* **139(7)** 826–838

[4] Cheng C W, Leu S S, Lin C C and Fan C 2010 Characteristic analysis of occupational accidents at small construction enterprises *Safety Science* **48(6)** 698–707

[5] Huang X, Hinze J 2003 Analysis of construction worker fall accidents *Journal of Construction Engineering and Management* **124(1)** 262–271

[6] Gibb A, Lingard H, Behm M and Cooke T 2014 Construction accident causality: Learning from different countries and differing consequences *Construction Management and Economics* **32(5)** 446-459

[7] Mitropoulos P, Cupido G and Namboodiri M 2009 Identifying root causes of construction injuries *Journal of Construction Engineering and Management* **135(9)** 881–889

[8] Garrett J W, Teizer J 2009 Human factors analysis classification system relating to human error awareness taxonomy in construction safety *Journal of Construction Engineering and Management* **135(8)** 754-763

[9] Zhang M, Shi R and Yang Z 2020 A critical review of vision-based occupational health and safety monitoring of construction site workers *Safety Science* **126** 104658

[10] Fang W, Ding L, Luo H and Love P E D 2018 Falls from heights: A computer vision-based approach for safety harness detection *Automation in Construction* **91** 53-61

[11] Fang W, Ding L, Zhong B, Love P E D and Luo H 2018 Automated detection of workers and heavy equipment on construction sites: A convolutional neural network approach *Advanced Engineering Informatics* **37** 139-149

[12] Klimova E V, Semeykin A Yu and Nosatova E A 2019 Information System for Forecasting and Management of Occupational Risks *IOP Conf. Ser.: Earth Environ. Sci.* **272** 022198

[13] Behm M 2005 Linking construction fatalities to the design for construction safety concept *Safety Science* **43(8)** 589-611

[14] Nikulin A N and Nikulina A Y 2017 Assessment of occupational health and safety effectiveness at a mining company *Ecology, Environment and Conservation* **23(1)** 351–355

[15] Nikulin A N, Ikonnikov D and Dolzhikov I 2019 Smart Personal Protective Equipment in the Coal Mining Industry *International Journal of Civil Engineering and Technology* **10(4)** 852–863

[16] Kinn S, Khuder S A, Bisesi M S and Woolley S 2000 Evaluation of safety orientation and training programs for reducing injuries in the plumbing and pipefitting industry *Journal of Occupational and Environmental Medicine* **42(12)** 1142-1147

[17] Cheng T, Teizer J 2013 Real-time resource location data collection and visualization technology for construction safety and activity monitoring applications *Automation in Construction* **34** 3-15

[18] Tang N, Hu H, Xu F and Zhu F 2019 Personalized safety instruction system for construction site based on internet technology *Safety Science* **116** 161-169

[19] Amiri M, Ardeshir A, Hossein M, Zarandi F 2017 Fuzzy probabilistic expert system for

occupational hazard assessment in construction *Safety Science* **93** 16–28

[20] Ismailov R I 2020 Making production decisions using the expert assessment procedure. *Construction Materials and Products* **3(1)** 49-56

[21] Semeykin A Yu, Klimova E V, Nosatova E A, Kochetkova I A 2019 Expert information-analytical decision support system for professional risk management based on the database of real cases of industrial injuries *IOP Conf. Ser.: Earth Environ. Sci.* **224** 012011

[22] Semeykin A Yu, Klimova E V, Kochetkova I A 2020 Forecasting and managing professional risks using information analytical systems based on fuzzy logic methods *IOP Conf. Ser.: Earth Environ. Sci.* **459** 052083

[23] Semeykin A Yu, Kochetkova I A, Klimova E V, Nosatova E A and Drozdova A O 2019 Decision-making process modeling for occupational safety management system based on a fuzzy analysis of production incidents *J. Phys.: Conf. Ser.* **1333** 072021

[24] Lapina A P, Ponomarenko A V, Shentsova K V and Kotesova A A 2019 Analysis of the causes of accidents at different stages of life cycle of a construction object *Construction Materials and Products* **2(2)** 17-22

[25] Manzhilevskaya S E 2019 Environmental monitoring of ecological safety in areas of construction, reconstruction and operation of objects *Construction Materials and Products* **2(3)** 78-84

Acknowledgments

The work is realized in the framework of the Program of flagship university development on the base of the Belgorod State Technological University named after V.G. Shukhov, using equipment of High Technology Center at BSTU named after V.G. Shukhov.

IOP Conf. Series: Materials Science and Engineering **945** (2020) 012023 doi:10.1088/1757-899X/945/1/012023

Portland cement foam concrete with a zeolite-containing mineral additive

V N Rozhin[1]

[1]North-Eastern Federal University named after M.K. Ammosov, Belinsky street, 58, Yakutsk, 677000, Russia

E-mail: barcelon@mail.ru

Abstract. The possibility of obtaining high-quality cement by an environmentally friendly (without CO_2 emission into the atmosphere) method of joint grinding of imported Portland cement clinker and local natural mineral raw materials in hard-to-reach areas of the North-East of the Russian Federation is substantiated. The results of an experimental study of the compositions and technological parameters of the manufacture of cement compositions based on portland cement clinker and active mineral additives from zeolite-containing rocks of Khonguruu, allowing non-autoclaved foam concrete with a density of 600-800 kg / m^3 and compressive strength of 1,8-2,8 MPa with stable quality parameters are presented (monodisperse pore distribution and hardened inter-pore skeleton, providing stable values of density, strength and thermal conductivity).

1. Introduction

The development strategy of the Arctic and the North (the Far North according to Soviet terminology) provides for the development of the building materials industry with the maximum use of local raw materials [1, 2]. The main material for the purpose and scope of use for the construction of buildings and structures is structural and structural-heat-insulating (including non-autoclave foam concrete) cement concrete. High strength and durability of concrete, as well as reliability in the operation of structures and buildings in general, depends on the quality of Portland cement used. The most significant disadvantage of foam concrete on cement is the low degree of hydration of the binder, in contrast to autoclave production. A low degree of cement hydration, which is a consequence of insufficient crystallization of hydration products and can cause instability of density, structure and strength, shrinkage of non-autoclaved foam concrete during molding and hardening. Therefore, to eliminate these shortcomings, hydration and hardening accelerators, active mineral additives, special additives that reduce the water-hard ratio, and others are introduced into the initial foam concrete mixture.

Ensuring high and stable quality of Portland cement delivered to regions of the Arctic and the North remote from industrial centers is a difficult task due to the interaction of cement with air moisture (hydration) and loss of activity during transportation and long-term storage. The activity of Portland cement when stored in packaging bags in a marine climate decreases to 32% within 4 months, and after 62 months to 62% [3, 4].

In connection with this circumstance, it seems that in hard-to-reach regions of the Arctic and the North it is advisable to produce high-quality cement using environmentally friendly (without CO_2

emission into the atmosphere) by co-grinding imported Portland cement clinker and local natural mineral raw materials containing amorphous silica, glassy silica or silica in compounds (silicates, aluminosilicates, hydrosilicates and etc.) [5].

Based on the analysis of scientific and technical literature [6-12], the main working hypothesis is put forward on the possibility of ensuring the stable quality of foam concrete using Portland cement with active mineral additives from local raw materials due to their modifying effect on the basic properties of cement and concrete, related to the peculiarity of their chemical and mineralogical compositions.

Object of study - non-autoclave foam concrete on Portland cement with natural active mineral additives.

The subject of the study is natural raw materials with a high content of amorphous modifications of silica and zeolite mineral, the formation processes of the phase composition, structure and properties of foam concrete on Portland cement with natural mineral additives.

2. Materials and methods

As part of the development program for the Far Eastern region of the Russian Federation, it is planned to set up production of Portland cement on the territory of the ALROSA diamond mining company near natural bulk tonnage - zeolite-containing rocks of the Khonguruu deposit.

Therefore, we chose a zeolite-containing rock (hereinafter, zeolite) of this field as a mineral additive to cement. The chemical composition of the studied zeolite is characterized by a high content of silicon oxide. According to the mineralogical composition and X-ray diffraction pattern, it was found that the raw material is clean because it contains a small amount of clay impurities, mica and sand, belongs to the clinoptilolite (60.4% wt.) Type of mineral raw material and has pozzolanic activity [3]. The accelerating effect of zeolite on the hydration of C_3S and the early formation of C-S-H in cement stone was established in [7, 8].

Portland cement clinker JSC PA "Yakutcement". The mineralogical composition of clinker, % by weight: C_3S – 62–65; C_2S – 10–13; C_3A – 6–7 и C_4AF – 10–13. The chemical composition of clinker, % by weight: CaO – 66,0; SiO_2 –19,0; Fe_2O3 – 3,9; AI_2O_3 – 3,7; MgO – 2,6; K_2O – 1,4; Na_2O – 1,2; SO_3 – 0,7.

As a regulator of setting time in Portland cement, gypsum stone of the Olekminsky deposit was used. The content of $CaSO_4 \times 2H_2O$ is 80,80 %.

For comparative analysis, control tests of samples of Portland cement JSC AO "Yakutcement" with a shelf life of 6 months or more in the accredited testing laboratory "Yakutsk-Expert" were carried out. It was found that the loss of activity of the brand CEM I 32.5 R is 31.7%, brand CEM I 42.5 R - 40.3%.

To establish the possibility of using zeolite as an active mineral additive in cement production, standard tests were carried out in accordance with GOST 25094-2015 (Russian standard certification). According to a statistical evaluation of the significance of differences in compressive strength of samples with the addition of zeolite and samples with standard multi fraction sand, it was found that mineral additives are active and can be used for the production of cements.

The effect of the content of mineral additives and the specific surface area of cement based on Portland cement clinker, gypsum stone, additives from zeolite on the properties of cement paste was investigated. During research, the content of mineral additives varied from 5 to 15%, and gypsum - 3% by weight of cement. For comparison, we also studied the properties of control samples based on non-additive Portland cement.

The following cement compositions were used, % by weight

1– clinker - 92, gypsum – 3, zeolite - 5;

2 – clinker - 87, gypsum – 3, zeolite - 10;

3 – clinker - 82, gypsum – 3, zeolite - 15;

The control composition is clinker 97, gypsum 3.

The reduction in the setting time of the cement paste and the acceleration of the initial structure formation of the stone are presumably due to the interaction of calcium hydroxide formed during alite hydrolysis with active additives from zeolite with the formation of low-basic hydrosilicates and calcium hydroaluminates. In accordance with the classification according to GOST 30515-2013 (Russian standard certification), the studied Portland cement with mineral additives in an amount of 5-15% is normally set.

The dependences of the compressive strength of a cement-sand mortar on the content of mineral additives are determined for various specific cement surfaces. It was found that the maximum strength of cement samples (45.6 MPa) is achieved for cement with 15% zeolite with a specific surface of 415 m^2 / kg. At the same time, Portland-pozzolana cement of the CEM II/A-P 42.5 N grade was obtained (GOST 31108-2016) (Russian standard certification) [3].

3. Results

Cement-based foam concrete with the addition of zeolite have compressive strength at 28 days of age 15-20% higher than foam concrete on factory cement CEM I 32.5 R (Fig. 1 a), and 11-15% higher than cement CEM I 42.5 R (Fig. 1 b). With an increase in the specific surface of cements from 315-330 m^2 / kg to 390-415 m^2 / kg, the strength characteristics of foam concrete do not change significantly, the increase in compressive strength is only 2.5-5.8%.

a b

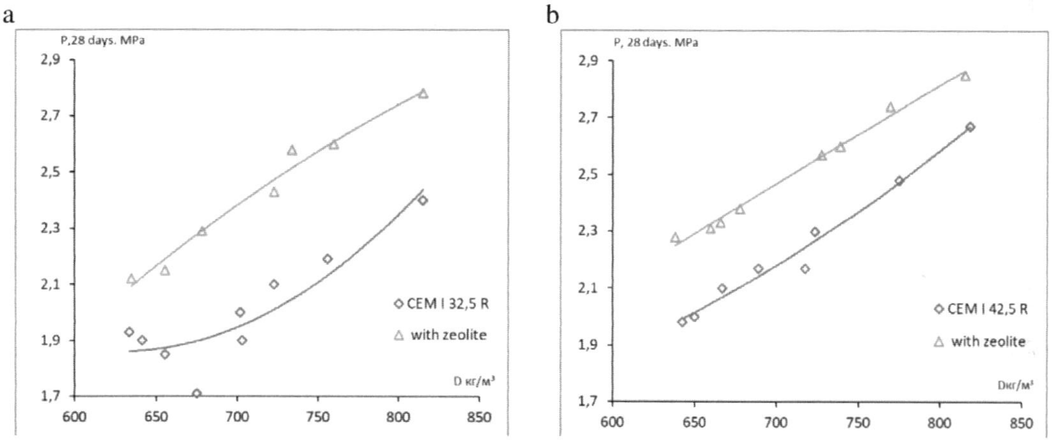

Figure 1. Change in compressive strength and average density of foam concrete at different finenesses of grinding Portland cement with the addition of zeolite a - 315-330 m^2 / kg; b - 390-415 m^2 / kg.

The curing strength of foam concrete samples at the age of 7 days is practically the same. Strength at 28 days of age for foam concrete on cement with a zeolite content is slightly higher. (Fig. 2 a)
The results of experimental data indicate a uniform increase in shrinkage deformations during drying over the entire period. Shrinkage deformations of foam concrete (Fig. 2 b) occur intensively during the first 30 days, and then slow down. Shrinkage deformations of foam concrete on cements with mineral additives more than 30 days old are 0.2-0.35 mm / m less than foam concrete on cement CEM I 32.5 R. The developed non-autoclave foam concrete based on cement with a zeolite mineral additive has the following quality parameters: general shrinkage 2.2–2.6 mm / m, sorption moisture less than 7%, vapor permeability coefficient 0.14–0.15 mg/(m·h·Pa), frost resistance F25-F35, which meets the requirements of GOST 25485-89 (Russian standard certification).

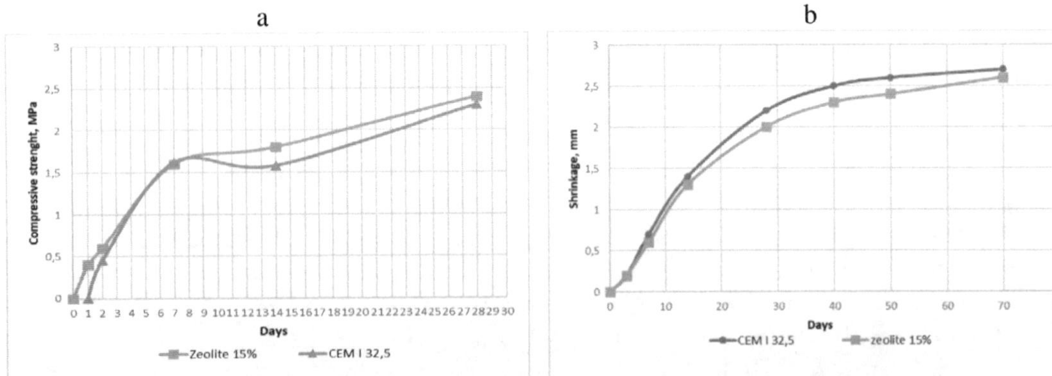

Figure 2. Kinetics of the set of strength (a) and shrinkage deformation (b) of samples of foam concrete D600 on cement with zeolite in an amount of 15% wt.

The main parameters for the stability of the quality of foam concrete are the uniformity of the compressive strength index (GOST 18105-2010). According to the results of processing statistical data, it was found that in foam concrete with a medium density grade D600 on cement with zeolite addition, the coefficient of variation in compressive strength decreases from 15.6 to 5.6% (compared to foam concrete on CEM I 32.5 R more than 6 months of storage), and with a density of D800 - from 12.1 to 4.3%, which made it possible to obtain concrete with a compressive strength class of B1.5 and B2.5, respectively.

The increase in the strength characteristics of foam concrete can be justified on the basis of the theory of hydration of cement systems and experimental data obtained on the basis of x-ray phase analysis [3] and computer analysis of microphotographs (Fig. 3, d, e, f). As can be seen from them, in a cement stone having a coagulation microstructure, a decrease in the amount of ettringite is observed due to the binding of calcium hydroxide during the initial structure formation of foam concrete into sparingly soluble low basic hydrosilicates due to the presence of amorphous silica in quartz-feldspar sand.

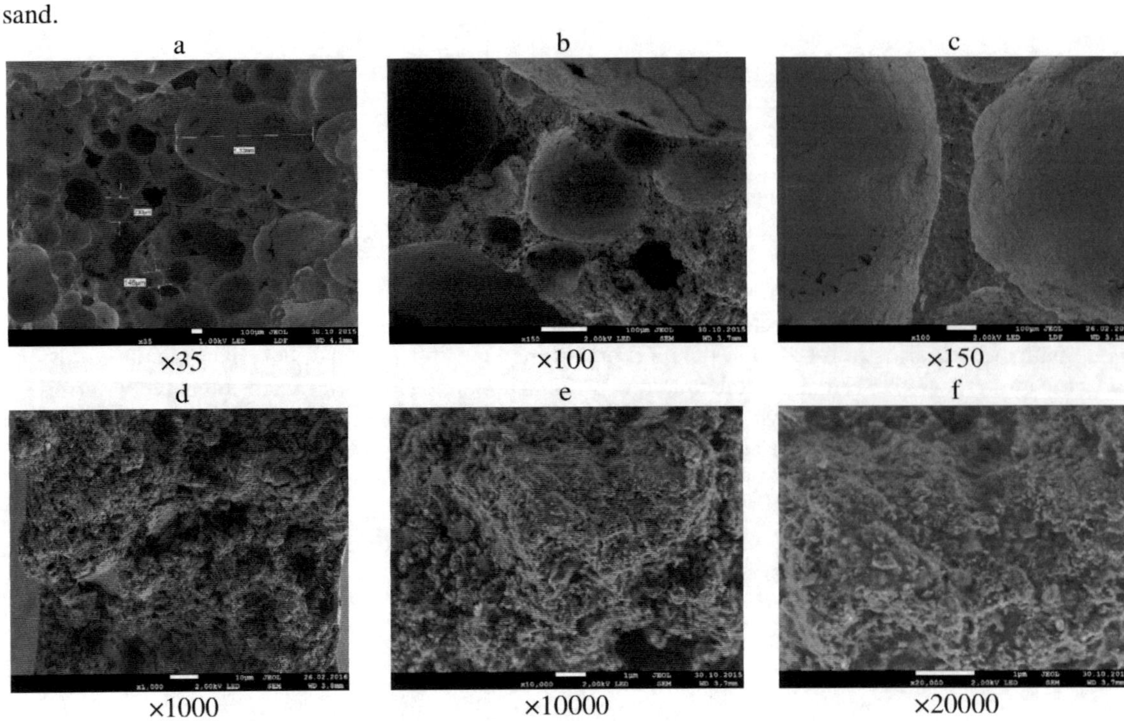

Figure 3. Microstructure of pores and inter-pore septa.

The interaction of the cement composition with the surface of fine-grained sand introduced into the foam concrete mixture as a filler is clearly seen in the image of the microstructure of the foam concrete with a particle of sand (Fig. 4 a, b, c).

a b c

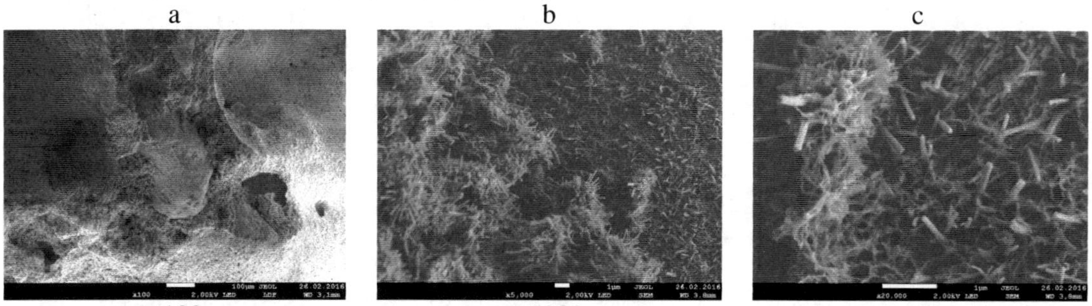

×100 ×5000 ×20000

Figure 4. Microstructure of neoplasms at the phase boundary "sand aggregate-cement composition".

The developed foam concrete based on cement from imported Portland cement clinker and mineral additives from zeolite is recommended for use in the production of structural and heat-insulating wall materials for the construction of buildings in areas of the Arctic and the North that are remote from industrial centers.

4. Summary

The results of experimental studies substantiate the working hypothesis about the modifying effect of an active mineral additive from a zeolite-containing rock of the Khonguruu deposit with a high content of a zeolite mineral - montmorillonite on the structure and properties of Portland cement and foam concrete based on them due to its pozzolanic activity.

Compositions and technological methods for preparing a foam concrete mixture based on a cement composition produced by joint grinding of Portland cement clinker (82-92% wt.), Gypsum stone (3% wt.) active mineral additives from zeolite-containing rocks of the Khonguruu field (5-15% wt.), with a fineness of grinding 315-415 m^2 / kg, fine aggregate (30% wt. over) in the form of fine sand, an aqueous solution of Etalon foaming agent (1:50), allowing to obtain non-autoclaved foam concrete with a density of 600-800 kg / m^3 depending on W/T with a strength of 1.8-2.8 MPa with stable quality parameters (monodisperse pore distribution and hardened inter-pore skeleton, providing stable values of density, strength and thermal conductivity).

5. References

[1] Sarvut T O 2018 The principles of habitat formation in the Arctic region *Vestnik MGSU* **13 (2)** 130-140

[2] Smirnova O O, Lipina S A, Kudryasheva E V 2016 Formation of support zones in the Arctic: methodology and practice *Arctic and North* **25** 148-157

[3] Mestnikov A E, Kudyakov A I, Rozhin V N 2019 Portland cement with natural active mineral additives *Bulletin of Tomsk State University of Architecture and Civil Engineering* **21 (2)** 192–201 DOI: 10.31675/1607-1859-2019-21-2-192-201

[4] Berdov G I, Ilyina L V 2010 Activation of cements by the action of mineral additives *International Journal of Applied and Basic Research* **9** 55-58

[5] Leonovich S N, Sviridov D V, G.L. Schukin 2015 Composition of dry mix for non-autoclave foam concrete of natural hardening *Building Materials* **5** 70-73

[6] T Perraki, E Kontori, S Tsivilis, G Kakali 2010 The effect of zeolite on the properties and hydration of blended cements *Cement and Concrete Composites* **32** 128–133

[7] Martin Bohá, Dana Kubátová, Radovan Neasa, Anežka Zezulová, Alexandra Masárová, Radoslav Novotnýb 2016 Properties of cement pastes with zeolite during early stage of

hydration *International Conference on Ecology and new Building materials and products, ICEBMP 2016 Procedia Engineering* **151** 2-9

[8] Klyuev S V, Klyuev A V, Shorstova E S 2019 Fiber concrete for 3-d additive technologies *Materials Science Forum* **974** 367–372 DOI: 10.4028/www.scientific.net/MSF.974.367

[9] Mestnikov A, Semenov C, Strokova V, Nelubova V 2016 Autoclave foam concrete: Structure and properties Citation. AIP Conference Proceedings 2. Сер. "Advanced Materials in Technology and Construction, AMTC 2015: Proceedings of the II All-Russian Scientific Conference of Young Scientists *Advanced Materials in Technology and Construction* 070010.

[10] Klyuev S V, Klyuev A V, Shorstova E S 2019 The micro silicon additive effects on the fine-grassed concrete properties for 3-d additive technologies *Materials Science Forum* **974** 131–135 doi:10.4028/www.scientific.net/MSF.974.131

[11] Bazhenov Y M, Zagorodnjuk L H, Lesovik V S, Yerofeyeva I V, Chernysheva N V, Sumskoy D A 2016 Concerning the role of mineral additives in composite binder content *International Journal of Pharmacy and Technology* **8(4)** 22649-22661

[12] Volodchenko A A, Lesovik V S, Zagorodnjuk L H, Volodchenko AN, Aleksandrovna K A 2016 The control of building composite structure formation through the use of multifunctional modifiers *Research Journal of Applied Sciences* **10(12)** 931-936

IOP Conf. Series: Materials Science and Engineering **945** (2020) 012024 doi:10.1088/1757-899X/945/1/012024

The dynamics of the stability of the functioning of ZAO Belgorodskij cement in 2016-2019

I A Slabinskaya[1] and O B Benderskaya[2*]

[1] Department of Accounting and Auditing, Belgorod State Technological University named after V G Shukhov, Kostyukov St., 46, Belgorod, 308012, Russia

[2] Department of Accounting and Auditing, Belgorod State Technological University named after V G Shukhov, Kostyukov St., 46, Belgorod, 308012, Russia

E-mail: obenderskaya@gmail.com

Annotation. Stability of functioning is one of the main goals of managing economic entities, whether they are independent business units or divisions of business associations. The complexity of this economic category determines the need to improve scientifically grounded instruments for measuring, assessing and monitoring the level of sustainable functioning. One such tool is the comparative comprehensive assessment (CA) of the stability of functioning, which is described in this article. The assessment author's methodology is based on a system of 12 indicators of the stability of functioning. For a dynamic comparative assessment of enterprise, the author's method of calculating a comprehensive assessment is used - a modified method of scoring. The object of the research is ZAO Belgorodskij cement, one of the largest enterprises in the building materials industry in the Belgorod region. The subject of the research is the dynamics of the stability of the functioning of ZAO Belgorodskij cement in the period 2016-2019. During this period, there was a constant struggle between the local authorities and the management of the enterprise for the closure of the plant, which the plant managers oppose. The objectives of the study are to improve the methodology for a comprehensive assessment of the stability of functioning and assess the impact of uncertainty in the position of the plant on the stability of its functioning.

1. Introduction

ZAO Belgorodsky Cement, part of the Eurocement Group holding, is one of the most famous enterprises in the building materials industry in the Belgorod region [8, 11]. The cement plant was built immediately after the Great Patriotic War and played an important role in the restoration of Belgorod, destroyed by the war, in its development at all subsequent stages of history and in the development of scientific schools of technologies for the production of cement and building materials and products from it [6, 7, 10, 14]. However, a dispute has been going on between the leadership of the Belgorod region and the holding company to close the enterprise over the past five years. Local authorities believe that the plant, with its outdated equipment and spent quarries, is causing environmental damage to the city [1]. The holding's management does not want to part with a profitable asset. While the plant continues to operate, changing managers, stopping (according to the holding's management) most of the furnaces and reclaiming the open pits. In this article, we find out

IOP Conf. Series: Materials Science and Engineering **945** (2020) 012024 doi:10.1088/1757-899X/945/1/012024

how this affects the level of stability of its functioning. The research period is 2016-2019. Information sources of the research - annual accounting financial statements of the enterprise from open sources.

The article is also based on the works of such domestic scientists as SM Bukhonova, E N Chizhova and Yu A Doroshenko [2-5, 9].

2. Methods and techniques

The main method of the study is the author's modified method of scoring, one of the methods for calculating a comprehensive assessment [13]. The type of CA used is dynamic comparative CA. It allows you to quantitatively measure the level of sustainability of the enterprise in each year from the chronological series, as well as assess the change in the level of sustainability over time.

The study is also based on a systems approach, methods of analysis and synthesis (generalization), it uses dynamic multidimensional comparisons, the coefficient method, and statistical methods (calculation of average values and growth rates).

The methodology for assessing the sustainability of the functioning of enterprise is based on a system of twelve indicators of the stability of functioning [12]. These indicators can be seen in table 1. The following values are considered normal:

- the share of stable sources of capital – above 0.5;
- stable sources share in stock financing – at least 0.6;
- coverage of current liabilities with current assets – at least 2;
- cost per one ruble of revenue – less than 1;
- growth rates – at least 100 %;
- profitability indicators – above 0.

It should be noted that a specific standard cannot be defined for the number of turns of total assets for the year. Its focus – the more the better.

According to the modified scoring method, the comprehensive assessment is calculated using the formula

$$\mathrm{CA}_i = \frac{\sum_{j=1}^{m} b_{ij}}{m}, \qquad i = \overline{1,n}, \tag{1}$$

where

$$b_{ij} = b_{\min} + \frac{a_{ij} - a_{\mathrm{worst}j}}{a_{\mathrm{best}j} - a_{\mathrm{worst}j}} (b_{\max} - b_{\min}) \cdot \tag{2}$$

A dynamic complex assessment is calculated for an individual enterprise over a number of periods. In this case, n is the number of compared periods (in our case, $n = 4$); m is the number of indicators (in our case, $m = 12$); a_{ij} – value of the j-th indicator for the i-th period; b_{ij} – score of the j-th indicator for the i-th period; $a_{\mathrm{worst}j}$ and $a_{\mathrm{best}j}$ – respectively, the worst and the best among the values of the j-th indicator; $b_{\min}$ and $b_{\max}$ are the minimum and maximum values of the score, respectively, which correspond to the worst and best values of the indicators. We have set $b_{\min} = 0$; $b_{\max} = 10$. In this case, the value of the complex assessment (1) can vary from 0 to 10. CA = 0, if in the estimated period the values of all indicators of enterprise stability are worse than in other periods; CA = 10, if in the estimated period the values of all indicators of enterprise sustainability are better than in other periods. Thus, CA (1) determines the level of stability of functioning on a 10-point scale relative to the best and worst values of the assessment indicators a_{ij} for the entire studied period.

3. Results

Table 1 shows the values of indicators of stability of the functioning of ZAO Belgorodskij cement in the period under study. The best values of indicators in the table are highlighted, and the worst are underlined.

Table 1. Indicators of the stability of functioning of ZAO Belgorodskij cement in 2016-2019.

Indicators, units	Indicator values			
	2016	2017	2018	2019
1) the share of stable sources of capital (average annual value)	0.59	0.84	0.88	**0.91**
2) stable sources share in stock financing (average annual value)	1.52	3.37	4.03	**4.40**
3) coverage of current liabilities with current assets (average annual value)	2.54	4.08	6.13	**7.54**
4) the number of turns of total assets for the year	**0.70**	0.51	0.55	0.49
5) cost per one ruble of revenue, rub.	0.85	0.84	**0.82**	0.86
6) revenue growth rate, %	91.48	69.70	**112.84**	93.22
7) retained earnings growth rate, %	**132.84**	118.66	105.69	98.74
8) the growth rate of equity, %	**128.09**	116.56	105.15	98.86
9) profitability of sales on profit from sales,%	14.56	16.08	**17.87**	14.09
10) return on sales in net profit, %	12.48	**13.52**	4.34	-1.09
11) return on total assets, %	**12.50**	8.60	1.58	-0.78
12) return on equity, %	**24.63**	15.30	5.02	-1.15

The first four indicators characterize the financial condition of the enterprise. During 2016-2019, their values were in line with the standards. Thus, the share of stable sources of financing was at least 0.59, grew in dynamics and in 2019 reached 0.91. This indicates a high degree of financial independence and leads to a low risk of insolvency. The share of long-term sources in financing production and other reserves has also consistently increased. During 2016-2019 the reserves were fully funded by stable sources. This indicates a low risk of production interruptions and also contributes to solvency. The enterprise was solvent for current liabilities, as the degree of their coverage with current assets was sufficient and grew in dynamics. So, in 2019, current assets were 7.54 times higher than the value of current liabilities. Deterioration in dynamics was observed only in terms of the indicator of the business activity of the enterprise – in terms of the rate of turnover of assets. In general, we can conclude that the financial condition of ZAO Belgorodskij cement in 2016-2019 was normal and improved over time.

The level of the cost of production of the plant in 2016-2019 was also normal and decreased until 2018 inclusive. Some growth in the indicator in 2019 was insignificant. Thus, the main activity of the plant was quite efficient.

Revenues from the sale of ZAO Belgorodskij cement products in the period under study decreased, except for 2018, when their growth was 12.84%.

According to the change in the level of production costs, the return on sales, calculated by profit from sales, in 2017 and 2018 grew, and in 2019 decreased, but remained at a good level – 14.09 %. The rest of the profitability indicators were positive during 2016-2018, but their level by 2018 decreased significantly, and in 2019 the final financial result was a loss, which led to negative values of these indicators.

The growth of retained earnings and equity continued throughout 2016-2018, although a slowdown was observed. In 2019, due to a loss, retained earnings and equity decreased [15].

Thus, in 2016-2019, against the background of strengthening financial condition and rather high efficiency of production at ZAO Belgorodskij cement, financial results have consistently deteriorated. According to the values of the dynamic complex assessment of the stability of functioning in the last row of table 2, it is clear that this led to a significant decrease in the level of stability of ZAO

Belgorodskij cement in 2019. If in 2016-2018 CA values changed, respectively, by -0.21 and by 0.46 points, then in 2019 there was a decrease immediately by 2.87 points.

Table 2. Scores of indicators and dynamic comparative CA of ZAO Belgorodskij cement in 2016-2019.

Indicator numbers	Scores of indicators			
	2016	2017	2018	2019
1)	0	7.96	9.04	10
2)	0	6.42	8.70	10
3)	0	3.08	7.17	10
4)	10	1.02	2.80	0
5)	1.25	5.27	10	0
6)	5.05	0	10	5.45
7)	10	5.84	2.04	0
8)	10	6.06	2.15	0
9)	1.25	5.27	10	0
10)	9.28	10	3.71	0
11)	10	7.06	1.78	0
12)	10	6.38	2.39	0
Dynamic comparative CA	5.57	5.36	5.82	2.95

The relatively low CA values are explained by the fact that for the period 2016-2019 there has never been a year in which the level of stability indicators of ZAO Belgorodskij cement would have been much better than in other years.

As noted above, CA defines a relative, not an absolute level of resilience. In order to correctly assess its values, it is necessary to analyze the compliance of the assessment indicators with the standards.

ZAO Belgorodskij cement had the highest level of operational stability in 2018. At that time, all stability conditions were met. The values of the three assessment indicators were the best over the entire study period. This allows us to assess the level of stability of ZAO Belgorodskij cement in 2018 as absolute. In another article, we compared the stability of the functioning of five enterprises of the building materials industry in the Belgorod region in 2018 and found that ZAO Belgorodskij cement had the highest stability, which confirms the conclusion about absolute stability.

In 2016, only one condition for the sustainability of functioning was violated – revenue declined. The values of the five assessment indicators were the best over the entire study period. The CA value was only 0.25 below the best value. This level of sustainability can be considered excellent.

In 2017, one condition for the stability of functioning was also violated. Only one assessment indicator had the best value. The CA value was 0.46 below the best value. Let us estimate this level of stability of functioning as normal.

In 2019, six assessment indicators did not meet the standards. However, their values were quite a bit worse than the normative level. At the same time, the values of the other three indicators were the best for the entire studied period. Therefore, it would be wrong to say that ZAO Belgorodskij cement was unstable in 2019. This level can be described as quite satisfactory.

The reasons for the significant decrease in the stability of the plant's operation in 2019 deserve a more detailed consideration. As mentioned above, they are in receiving a net loss at the end of 2019. The immediate reason for the loss was the increase in the amount of interest paid compared to 2018. It can be assumed that this is interest on long-term borrowed funds, which, since 2016, constitute more than 40 % of all sources of financing for ZAO Belgorodskij cement. Knowing the procedure for financing production units adopted in holdings, it can be assumed that the plant pays interest on long-term loans from the parent company, and the amount of interest paid is regulated at the discretion of the holding's management. This can be a convenient lever for regulating the profitability of an enterprise. One can only assume what goals this regulation pursues: minimizing tax payments to the local budget or accumulating arguments for closing production.

4. Summary

Based on the results of the study, it can be concluded that ZAO Belgorodskij cement is a successful company: during 2016-2019 it functions steadily, its financial condition is strengthening, the volume of assets is growing. It can be concluded that the uncertainty of the plant's position does not affect its economy. True, there are periodic fluctuations in the level of stability of the plant, the amplitude of which is growing in dynamics. If the identified trend continues, in 2021 the plant should be expected to lose its ability to function sustainably. However, it seems to us that the management of ZAO Belgorodskij cement fully controls the economic situation at the plant. Analysis of the data from the financial statements of the enterprise suggests that the nature of fluctuations in the level of its stability is controlled by the holding's management. Therefore, further dynamics will be determined not by previous trends, but by political motives and the impact of the COVID-19 pandemic on the building materials industry.

5. References

[1] Bryantseva T A, Shevchenko M V 2019 Innovative leadership of the region: assessment issues and strengthening factors (on the example of the Belgorod region) *Problems of the modern economy* **4** 128-130

[2] Bukhonova S M, Doroshenko Yu A, Trunova E V 2004 Application of sustainability performance indicators in company management *Economic analysis: theory and practice* **9** 10-21

[3] Bukhonova S M, Sergeeva S A 2017 Problems of reindustrialization of Russian industry on the new technological principles *Bulletin of BSTU named after V G Shukhov* **10** 194-203

[4] Chizhova E N, Shevchenko M V 2011 Integral efficiency of management system of the industrial enterprise *Bulletin of BSTU named after V G Shukhov* **1** 95-99

[5] Chizhova E N, Sorokina V Yu, Vesnina O O, Osychenko E V 2018 The concept of monitoring the management of innovative industrial enterprise *Bulletin of BSTU named after V G Shukhov* **11** 156-163

[6] Klyuev S V, Klyuev A V, Shorstova E S 2019 The micro silicon additive effects on the fine-grassed concrete properties for 3-d additive technologies *Materials Science Forum* **974** 131–135 doi:10.4028/www.scientific.net/MSF.974.131

[7] Dmitrieva T V, Markova I Yu, Strokova V V, Bezrodnykh A A, Kutsyna N P 2020 Effeciency of stabilizers of various compositions for strengthening soils with mineral binders *Construction Materials and Products* **3(1)** 30-38

[8] Doroshenko Yu A, Golubotskikh M 2017 Factors of competitive sustainability and assessment of their influence on enterprises within building materials industry *Bulletin of BSTU named after V G Shukhov* **1** 26-31

[9] Doroshenko Yu A, Klimashevskaya A 2017 The analysis of scientific and technical potential of the enterprises of construction materials industry in the context of assessment of need of carrying out technological modernization for the industry *Bulletin of BSTU named after V G Shukhov* **1** 214-218

[10] Lesovik V S, Absimetov M V, Elistratkin M Yu, Pospelova M A, Shatalova S V 2019 For the study of peculiarities of structure formation of composite binders for non-autoclaved aerated concrete *Construction Materials and Products* **2(3)** 41-47

[11] Slabinskaya I A, Benderskaya O B 2017 Sustainability of Belgorod region enterprises of building materials and building structures industry under sanctions *Advances in Engineering Research: Proceedings of the International Conference AIME 2017* **133** 102-107

[12] Slabinskaya I A, Benderskaya O B 2019 Quality of functional stability management of Belgorod food industry enterprises *Herald of the Belgorod University of Cooperation, Economics and Law* **6** 307-321

[13] Slabinskaya I A, Benderskaya O B, Mitrokhin A A, Truhin A S 2015 Methods of Company Financial Stability Monitoring *International Business Management* **9(6)** 1091-1096

[14] Tolstoy A D 2020 Fine-grained high-strength concrete *Construction Materials and Products* **3(1)** 39-43

[15] Usatova L V, Arskaya E V, Budnichenko O 2014 The importance of profit and mechanism of its distribution in modern conditions *Belgorod Economic Bulletin* **3(75)** 175-186

Acknowledgments
This work was realized under the support of the President Scholarship; in the framework of the Program of flagship university development on the base of the Belgorod State Technological University named after V G Shukhov.

IOP Conf. Series: Materials Science and Engineering **945** (2020) 012025 doi:10.1088/1757-899X/945/1/012025

Efficiency of using shungite filler for modifying organic binder

V V Yadykina[1*], K S Vyrodova[1] and E E Potapov[2]

[1] Belgorod State Technological University named after V.G. Shukhov, Kostyukov St., 46, Belgorod, 308012, Russia

[2] Lomonosov Moscow State University of Fine Chemical Technologies, Vernadsky prospect, 86, Moscow, 119571, Russia

E-mail: vvya@intbel.ru

Abstract. Modification of bitumen with various additives, including polymer ones, is one of the most effective ways to regulate the performance properties of asphalt concrete. Due to the high cost of polymers, a promising area of research is to reduce the polymer content, for example, by introducing mineral filler into its composition. A wide range of properties and unique structure of fullerene-containing natural shungite allow it to be used in various branches of science, industry and technology to create a variety of materials and determine the search for new areas of its application. One of the ideas that are currently under development is the use of shungite in conjunction with an organic binder used for the production of asphalt concrete. The work is devoted to the study of the influence of fine shungite on the characteristics of polymer-bituminous binder (PBB) and asphalt concrete. It is shown that the introduction of shungite into the PBB increases its viscosity, adhesive and cohesive properties, and the homogeneity of the mixture. This is due to the unique composition, structure and properties of shungite. The positive influence of the use of polymer shungite binder on the quality of asphalt concrete was established. With the introduction of 5% shungite in the PBB composition, the strength at 20 and 50°C, the strength of water-saturated samples, and the shear stability of polymer asphalt concrete significantly increased. This made it possible to reduce the polymer content of PBB by 10%.

1. Introduction

One of the most effective ways to regulate the physical, mechanical and operational properties of asphalt concrete is the modification of bitumen with various additives. Currently, the most common method of improving the quality of bituminous binders is the introduction of polymers in them. The use of polymer modifiers reduces the sensitivity of bitumen to temperature changes, extends the temperature range of its operation, and increases the cohesive strength and elasticity. However, this technology is quite expensive [1-3].

One of the promising ways to reduce the cost of polymer-bituminous binders (PBB) is to reduce the polymer content, for example, by introducing mineral filler. The use of calcium carbonate, clays, metal oxides, and carbon nanotubes is known as polymer matrix filler [4-5]. The use of natural shungite, which has a unique structure and properties: high adsorption activity, bactericidal, catalytic, and reducing properties is of significant interest in this direction [6-8].

IOP Conf. Series: Materials Science and Engineering **945** (2020) 012025 doi:10.1088/1757-899X/945/1/012025

The positive effect of shungite when used in bituminous binders and asphalt concrete is known from the literature [9-11].

The purpose of this work is to obtain a polymer-bituminous binder with a reduced amount of polymer while maintaining or improving the quality of PBB and asphalt concrete with its use.

2. Methods and materials

This task was solved by introducing a polymer-bituminous binder heated to a temperature of 160°C, a fine dispersed filler – shungite. BND 90/130 bitumen was used as the initial binder, polymer-styrene-butadiene thermoelastolayer of Korea Kumho Petrochemical Co., plasticizer – Uniplast produced by Selena LLC (Shebekino, Belgorod region). The filler was fine dispersed shungite (Karelite), provided by the Karelian investment company "RBK".

PBB and asphalt concrete characteristics were determined using standard methods. Additionally, the bitumen cohesion according to ASTM D3910-98 and the PBB microstructure were studied using the AXIO SCOPE A1 (Carl Zeiss) laboratory research microscope in transmitted light.

3. Results and discussions

In order to find out the rational amount of shungite that allows replacing an expensive polymer in the composition of PBB, experiments were conducted on the selection of PBB compositions with different amounts of polymer and shungite. The amount of SBS was from 3.5 to 2.97 %, shungite - from 3 to 10 % relative to the mass of BND 90/130 bitumen.

From the data obtained, it follows that when shungite is introduced into the polymer-bituminous binder, its softening temperature remains at the level of the initial PBB. The depth of penetration of the needle, which determines the plasticity of bitumen at 25 and 0°C, decreases with the introduction of shungite, This trend is observed for all compositions with different polymer content (3.5; 3.15 and 2.97 %), which indicates a significant structuring effect of the filler. For the same reason, which is quite logical, there is a decrease in extensibility at 25 and 0 °C after the introduction of shungite. This is especially expressed when the content of shungite in the amount of 10 % of the mass of PBB. The elasticity of the PBB remains at the level of elasticity of the original composition. At the same time, due to the introduction of shungite, complete homogeneity of the modified binder is achieved (in comparison with the binder without an additive), which was estimated from micrographs (Fig. 1).

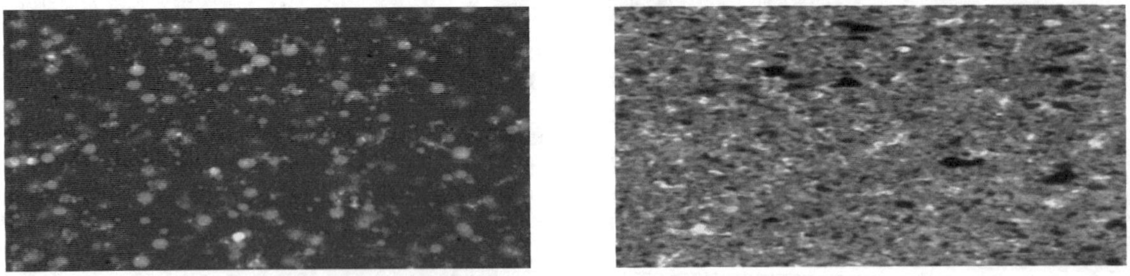

Figure 1. Micrographs of polymer-bituminous binder a) without the introduction of shungite; b) with the introduction of shungite.

It also follows from the results that increasing the amount of shungite by more than 5% in the PBB is impractical. Although the introduction of 10 % shungite allows getting a binder that meets the regulatory requirements, however, the depth of penetration of the needle does not correspond to the PBB 60 brand. When using shungite in the amount of 3%, this indicator is at the lower limit of the requirements of GOST R 52056-2003, so reducing the amount of filler is also impractical.

From the results obtained, it follows that the compositions containing 3.15% SBS and 3 and 5% shungite are rational. This allows for a 10% reduction in the concentration of the polymer compared to the control compositions. All indicators of these compositions meet the requirements of GOST R

52056-2003. Thus, it is shown that it is possible to reduce the amount of polymer by 10 % in the PBB due to the introduction of shungite.

Research was conducted on the effect of shungite on the cohesive strength properties of PBB. The definition of cohesion is based on measuring the shear strength of a thin layer of bitumen (10 mc) at a constant rate of load increase. The binder compositions are shown in table 1.

Table 1. Compound of polymer-bitumen binder.

Type of additive	№ of compound	The amount of SBS from the mass of the initial bitumen, %	Shungite content, %
BND 90/130, without additive	1	0	0
SBS 3.5 %	2	3.5	0
	3	3.5	5
SBS 3.15 %	4	3.15	0
	5	3.15	5

Test results are presented on Fig. 2.

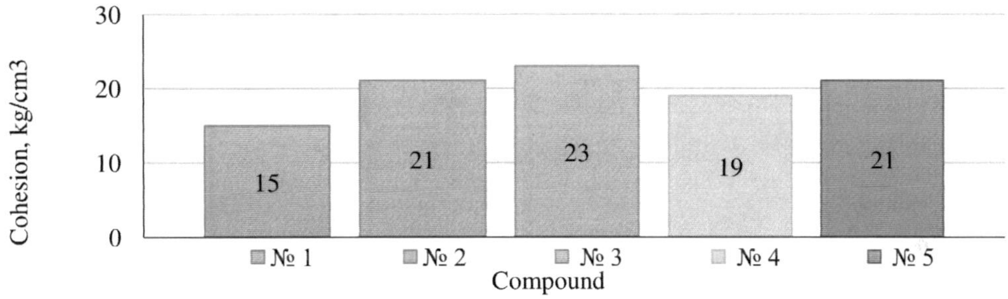

Figure 2. Cohesive strength properties of the binder.

Cohesion is similar in nature to viscosity and is determined by intermolecular bonding forces and the structure of bitumen. The results show that shungite has a significant structuring effect. The cohesion of composition 5 with 5 % shungite and with lower polymer content (3.15%) is equal to this indicator for composition 2, which contains a high concentration of SBS (3.5%). Accordingly, it can be assumed that the obtained PBB compositions filled with shungite will have higher strength characteristics than the original ones.

The adhesion of bitumen-polymer-shungite binder to the surface of crushed stone was also evaluated.

The quality of adhesion was assessed visually by the degree of preservation of the film of bituminous binder on crushed stone grains, prepared beforehand in accordance with GOST 12801-98, after boiling it in distilled water for 30 minutes. The bonding quality of the binder was determined for the same compositions as the cohesion.

Test results are given on Fig. 3.

IOP Conf. Series: Materials Science and Engineering **945** (2020) 012025 doi:10.1088/1757-899X/945/1/012025

Figure 3. Adhesion of test samples to granite crushed stone.

The results indicate that PBB using shungite has better adhesion compared to the control composition, which is due to the positive effect of the filler on the properties of bitumen: increased viscosity and cohesive strength. In addition, shungite contained in bitumen gives the binder increased activity due to the high concentration of paramagnetic centers. This should have a positive impact on the properties of the road surface.

The influence of PBB filled with shungite on the properties of asphalt concrete coverage was studied using the example of polymer asphalt concrete. Samples of polymer asphalt concrete were formed on PBB with a concentration of SBS polymer of 3.5 and 3.15 % and on PBB with the same SBS concentrations, but additionally containing 5 % shungite. The compositions are shown in table 1.

Fig. 4 shows the results of changing the strength of samples of polymer asphalt concrete prepared on modified bitumen compared to samples on the original BND 90/130 bitumen.

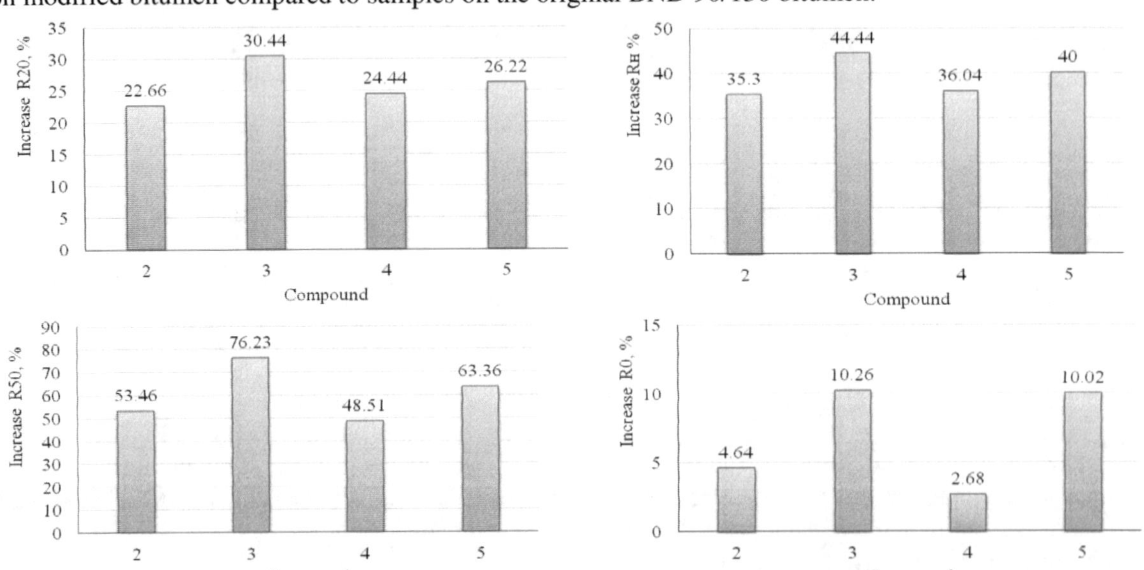

Figure 4. Changes in strength characteristics of polymer-asphalt concrete samples.

The analysis of the obtained data shows that there was an increase in the main strength characteristics of the polymer-asphalt concrete mixture for all compositions using the studied additives.

A significant increase in the strength at 20 and 50˚C, as well as the strength of water-saturated samples, was found. Thus, the compressive strength at 20˚C when using SBS thermoelastolayer as an additive increases by 22.66 and 24.44 % (compounds 2 and 4), and when introducing shungite in the amount of 5 % (compounds 3 and 5) – by 30.44 and 26.22%, respectively. The introduction of SBS, and especially shungite, had a much greater impact on the strength at 50˚C. The compressive strength at 50˚C when used as an additive of SBS thermoelastolayer increases by 53.46 and 48.51% (compounds 2 and 4), and when introducing shungite in the amount of 5 % (compounds 3 and 5) – by 76.23 and 63.36%, respectively. This is due to an increase in the viscosity, softening temperature, cohesion and adhesion of the binder when using the studied additives and should have a positive effect on the stability of the asphalt concrete coverage to rutting.

As expected, the introduction of shungite had a significant impact on the adhesion at shifting (Fig. 5). This indicator increased with the introduction of SBS in bitumen by 53.3 and 56.6 %, and with the use of shungite by 80 %.

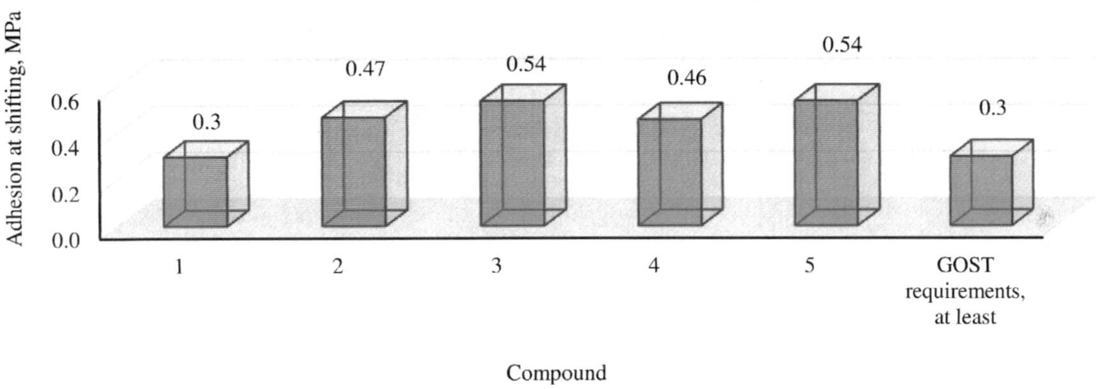

Figure 5. Adhesion at the shifting of asphalt concrete on various compounds of the binder.

Thus, the introduction of shungite in the PBB containing 3.5 and 3.15 % SBS leads to an increase in the strength of polymer asphalt concrete at 20 and 50 °C, water resistance, and shear stability.

4. Summary

It is established that the use of shungite as a part of polymer-bituminous binder leads to an increase in its viscosity, cohesive and adhesive characteristics, and the uniformity of PBB. This is due to the high adsorption capacity of shungite, which is associated with the presence of natural fullerenes, surface structure, and the presence of active adsorption centers. The introduction of 5 % shungite with a reduction in the amount of SBS from 3.5 to 3.15 % allows getting asphalt concrete, the characteristics of which are not inferior to those for a compound containing 3.5 % SBS, but without shungite.

Thus, when using shungite, it is possible to reduce the amount of SBS in the mixture by 10 % without losing the quality of asphalt concrete.

5. References

[1] Yarmolinsky V A, Tsupikova L S 2018 Polymer-bituminous binders in road construction. *In the collection: the Far East. Highways and traffic safety international collection of scientific papers* 200-206

[2] Yadykina V V, Gridchin A M, Trautvain A I, Verbkin V I 2015 Study of the influence of various polymers and plasticizers on the properties of BND 60/90 bitumen and asphalt concrete based on it *Bulletin of BSTU named after V G Shukhov* **6** 40-45

[3] Odenbach I A, Taurit E B 2020 Some new technologies in road construction *Construction Materials and Products* **3(1)** 62-69

[4] Nechesova Yu M 2015 Production of elastomeric compositions filled with modified calcium carbonate at the latex stage *Author. dis. Cand. of Eng. Sciences,* Voronezh 20

[5] Shekhovtsova S Yu, Vysotskaya M A 2017 Influence of single-wall carbon nanotubes on the temperature range of performance of polymer-bituminous binders and asphalt concretes

based on them. *KSUAE proceedings* **4(42)** 335-342

[6] Korneev Yu V, Yanovsky Yu G, Boyko O V, Chirkunova S V 2013 Nanodisperse mineral shungite – a new reinforcing filler of elastomeric materials. *III Russian conference "Caoutchouc and rubber - 2013: traditions and innovations"* 70-71

[7] Solovyova A B., Rozhkova N N., Glagolev N N 1999 Organic matter of shungite as a factor determining the physical and chemical activity of shungite filler in polymer composites. *Journal of physical chemistry* **73 (2)** 299-306

[8] Potapov E E., Valia A, Bobrov A P., Prekop Sh, Shechter V E, Reznichenko S V, Morozov Yu L 2010 Shungit-a promising ingredient of rubber mixtures for the tire industry and the RTI industry. *Theses of reports of April 19 -22 conference* 289-295

[9] Vysotskaya M A, Rusina S Yu, Belyaev D, Kiselev O 2015 Shungite - as a component of a bituminous composition for the road industry. *Collection of papers and reports of the annual scientific session of the Association of asphalt concrete researchers* 18-26

[10] Sorokina O V, Potapov E E, Reznichenko S V, Bobrov A P, Smal V A, Yadykina V V, Tikunova I V 2018 Investigation of properties of high-filled composites based on bitumen and shungite (Karelite) *Caoutchouc and rubber* **77 (2)** 92-94

[11] Podolsky V P, Chernousov D I 2008 Influence of shungite mineral powder on changes in the structural and mechanical properties of asphalt concrete over time *Science and innovation in construction* 394-399

Acknowledgments

This work was realized in the framework of the Program of flagship university development on the base of the Belgorod State Technological University named after V G Shukhov, using equipment of High Technology Center at BSTU named after V G Shukhov.

IOP Conf. Series: Materials Science and Engineering **945** (2020) 012026 doi:10.1088/1757-899X/945/1/012026

The modern information technologies in construction for improving occupational safety

E V Klimova[1*], A Yu Semeykin[2], D A Sinebok[3] and Yu V Khomchenko[4]

[1, 2, 3]Department of Life Safety, Belgorod State Technological University named after V.G. Shukhov, 46, Kostyukov Str., Belgorod, 308012, Russia
[4]D.O.O. Artika, Herceg Novi, Montenegro

E-mail: klimova.ev@bstu.ru

Abstract. Today, construction takes first place in the number of industrial accidents, including severe and fatal ones. Inconsistency of jobs, work at height, work in extreme conditions, adverse weather factors, all this and much more is the cause of traumatic factors. The article discusses the implementation aspects of modern information technologies that allow you to assess the psycho-emotional state of the employee in real time and make timely decisions on the organization of work safety. Based on the results of the analysis, a "Psychophysiological Monitoring System" is proposed. Existing analogues of such systems are not used in production, more often they are designed to assess the physiological state of athletes. The technology of analysis and assessment of the state of workers is presented, which a complex of interrelated parts is such as: the main sensor module, local receiving and transmitting device, stationary receiving and processing device, sensors for evaluating external conditions, a database and a server for data transmitting and processing. Based on the developed hardware-software complex, measures are proposed for the prevention of injuries in construction during construction and installation works on height.

1. Introduction

Today, occupational risk assessment is an integral part of labor protection at the enterprise. The significance of this innovation is determined by the transition of the domestic economy to a risk-based approach.

The popularization of professional risk assessment is associated with industrialization, during the active use of technologies, equipment, biochemical substances, various energy sources, all areas of activity are at risk.

A statistical study of the causes of injuries shows that the introduction of improved equipment and mechanisms is not a key factor in reducing accidents and injuries. Only part of the technical imperfection is associated with a significant number of accidents, the main cause of technological accidents is a person (80-90%), namely, the adoption by him of incorrect or inappropriate actions in this situation [1]. Erroneous actions are usually unintentional, that is, a person is making deliberately erroneous actions, regards them as correct and suitable for a given situation.

The reasons for this behavior can be both insufficient qualifications and lack of sufficient information to make a decision. This is especially clearly seen in conditions of time pressure, the need for quick decision-making in an emergency.

IOP Conf. Series: Materials Science and Engineering **945** (2020) 012026 doi:10.1088/1757-899X/945/1/012026

To prevent erroneous actions, timely identification of potentially dangerous conditions of people is necessary, namely when a person is confused or excited and may make a mistake. At the moment, labor protection specialists do not have enough information to identify these conditions. Effective forecasting requires the development of information support for timely decision-making and the prevention of erroneous actions that result in an accident or injury.

2. Methods

Modern tools for predicting incidents and unsafe employee behavior are algorithms created on the basis of various technologies that combine the capabilities of a computer and the Internet. At the same time, intelligent systems based on neural network models, Markov networks, Bayesian networks and other models are used. There are attempts to create models of unsafe behavior of workers based on an analysis of indicators of their behavior and working environment, such as the level of management control, working conditions, the safety management system, the level of employee participation, knowledge of safety requirements, safety attitudes, motivation, resource allocation and production stress.

The following technologies are used in the basis of foreign expert systems [2-13]:

– Fault Tree Analysis (FTA);
– Failure Mode and Effects Analysis (FMEA);
– Fuzzy Analytic Network Process;
– Bayesian Networks;
– Graphical Evaluation and Review Technique;
– Neural Networks.

All of the above listed methods of analyzing and forecasting incidents, erroneous actions of personnel, accidents and emergency situations make it possible to create integrated software products that are designed to facilitate occupational safety control and management processes in enterprises, hazardous production facilities, etc.

3. Results and discussions

The relevance of occupational injuries affects all countries of the world. According to the ILO, annually about 2.3 million people die from industrial accidents, which is about 3.4% of the total mortality per year [1].

In terms of occupational injuries, Russia is 2.5–3 times ahead of the EU countries. 6 deaths per year per 100 thousand people, while for for EU countries the data is from 0–3, for the UK – 0.8, Ukraine – 3.8, Kyrgyzstan – 4.1, Lithuania – 4.2 Kazakhstan – 5, Moldova – 5.2. Despite the improving dynamics, our country remains the leader among post-Soviet countries in terms of mortality in the workplace [1].

According to the ILO, the most dangerous are the manufacturing and construction sectors, followed by transport, agriculture and mining.

One of the main factors holding back the technological and economic process in Russia is the percentage of occupational injuries.

The International Labor Organization provides the following statistics, according to the number of victims, the dynamics are presented in Figure 1.

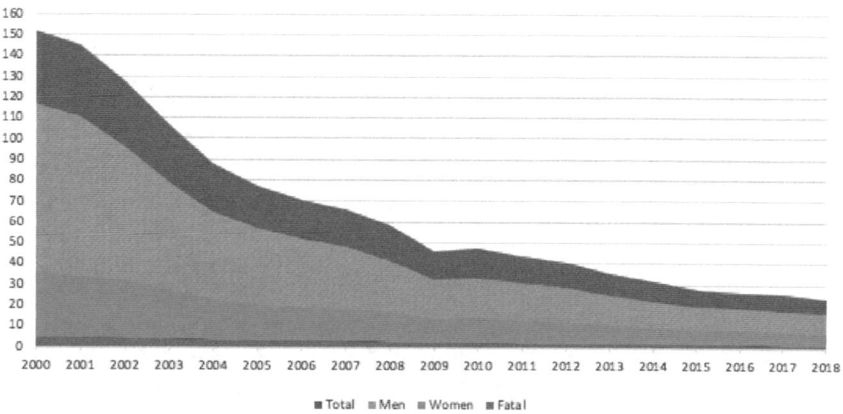

Figure 1. The dynamics of industrial injuries in 2000–2018 years.

And despite the increase in cash funding for this period spent on labor protection measures per worker, the problem of injuries remains relevant.

This can be explained by the fact that the main cause of an industrial accident, including fatal cases, is the human factor, which includes a lack of qualified personnel, a lack of knowledge in the field of safe work, personal reasons, psychoemotional causes of the individual, lack of motivation and much other [2-5, 24].

The distribution of accidents by cause is provided in Figure 2.

Accordingly, the main cause of all accidents is the person himself and to reduce and measures to reduce injuries should be aimed at the employee [6, 7].

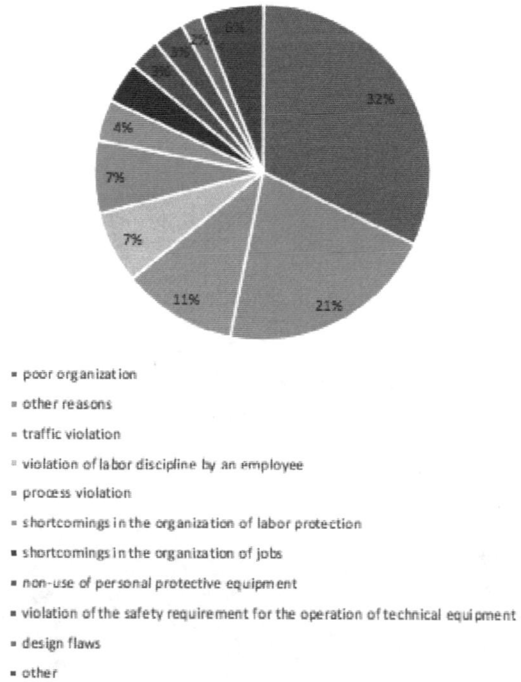

Figure 2. Distribution of causes of severe accidents.

IOP Conf. Series: Materials Science and Engineering **945** (2020) 012026 doi:10.1088/1757-899X/945/1/012026

The proposed project relates to medical equipment, and more specifically to the technology of industrial medicine, and specifically to a system for monitoring physiological parameters in order to determine the psychophysiological state of workers in the construction industry [8-25].

The feasibility of this development is as follows:

1) Cost reduction (transportation of sick employees, payment of her disability, compensation, the search for new personnel, and so on)

2) Increasing the efficiency and productivity of production (the need to timely carry out preventive measures to support the health conditions of employees for their effective work).

The closest analogues in terms of functionality, but not in scope are the ZephyrTM PSM Training system and BioHarness®.

These systems are aimed at tracking the performance of athletes, to create effective training and the correct redistribution of the body's forces. Accordingly, these systems include the necessary sensors of physiological indicators that are in demand for tracking the progress of training, but they cannot be adapted to control production workers, since their functionality is not adapted for this. Therefore, the development of the "Psychophysiological Monitoring System" was based on the results of existing developments, but it is a unique system adapted for monitoring psychophysiological indicators, which takes into account environmental parameters and statistical data. The list of patents is RU 2677061, US 20130023739, US 20140340219, US 20140343448, US20130144130, US 20120165645.

The psychophysiological monitoring and warning system may include one or more wireless sensors that perceive and process information from sensors about the physiological state of a person. The processing performed by the sensors includes a function for comparing values, after which a signal is generated on the device of the controlling person. To reduce the amount of information, a signal can be received by the controlling person only when the indicators of the state of human health deviate from standard values. The controlling person may be the person responsible for labor protection or a specially authorized employee who regulates work efficiency and monitors the health of the worker. The transmission of contextual signals about the deviation of the psychophysiological state of the employee will allow the controlling person to take preventive measures before the onset of injury.

An example of a remote system for monitoring physiological parameters is shown in Figure 3.

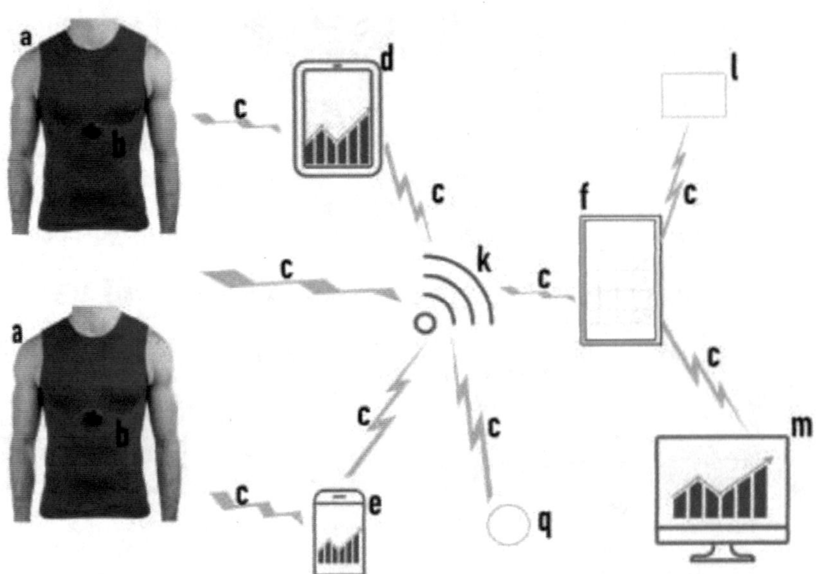

Figure 3. The general principle of the "Psychophysiological monitoring system".

The system includes people "a", each of whom wears a sensor "b". The "b" sensors transmit signals through the "c" wireless network. The transmitted signals can be transmitted to the local computing devices "e", "d". As these devices smartphones, tablets and any other convenient computing devices connected to the server "f" through the network "k" can be presented. These devices are equipped with specialized software that processes and visualizes the received data (for example, ProRAE Guardian or Welltory programs). Additional "q" sensors that measure the microclimate of the working medium (air temperature, relative humidity, resulting room temperature, air velocity and local asymmetry of the resulting temperature) can also interact with server "f" through the network "k". The server "k" stores the received data and performs a detailed analysis of the monitoring data of physiological parameters, and then provides them to the local devices "e", "d" at the request of the controlling person. Further, the server can carry out further communication with a remote computer, which will allow remote monitoring to an interested person, such as the head of the enterprise. Also, the server "f" can communicate with various databases in which the collected data can be stored.

Each sensor unit "b" is capable of simultaneously sensing several physiological parameters. Since the device is intended for people employed in work with special conditions, the sensor should include all the necessary sensors of physiological indicators sufficient for quick response, and the device should not be bulky and uncomfortable when conducting work. These indicators are:

- Heart rate
- Breathing rate
- Body temperature
- Heart rate variability
- Skin-galvanic reaction
- Body position in space

The sensor module is equipped with radio communications for wireless data transmission and command reception, which can be done using the personal Bluetooth frequencies (for example, the RN4020 model module), as well as these units will be equipped with an individual notification system of the one directly working by sending an audio signal when registering hazardous to humans health conditions. The sensor module is also equipped with electronic circuits, including microcontrollers, memory, and sensors for measuring physiological parameters, a power source, sound and light indicators and power buttons.

4. Summary

As a result, a "Psychophysiological Monitoring System" was developed, the purpose of which is to continuously record the physiological indicators of the worker associated with his psycho-emotional state, transfer, analyze and display the results on local or stationary devices to employees responsible for safety. If a deterioration in the state of health is detected, depending on the indicator, the strength of its change and individual indicators, measures should be taken to provide medical or psychological assistance or other organizational measures should be taken (break, transfer to another type of work, exemption from work for a certain time, etc.).

5. References

[1] International Labor Organization / World statistics Available at:: https://www.ilo.org/moscow/areas-of-work/occupational-safety-and-health/WCMS_249276/lang--en/index.htm

[2] Chi C F, Chang T C and Ting H I 2005 Accident patterns and prevention measures for fatal occupational falls in the construction industry *Applied Ergonomics* **36(4)** 391–400

[3] Chi S, Han S and Kim D Y 2013 Relationship between unsafe working conditions and workers' behavior and impact of working conditions on injury severity in U.S. construction industry. *Journal of Construction Engineering and Management* **139(7)** 826–38

[4] Cheng C W, Leu S S, Lin C C and Fan C 2010 Characteristic analysis of occupational accidents at small construction enterprises *Safety Science* **48(6)** 698–707

[5] Huang X, Hinze J 2003 Analysis of construction worker fall accidents *Journal of Construction Engineering and Management* **124(1)** 262–71
[6] Gibb A, Lingard H, Behm M and Cooke T 2014 Construction accident causality: Learning from different countries and differing consequences *Construction Management and Economics* **32(5)** 446-59
[7] Mitropoulos P, Cupido G and Namboodiri M 2009 Identifying root causes of construction injuries *Journal of Construction Engineering and Management* **135(9)** 881–9
[8] Klimova E V, Semeykin A Yu and Nosatova E A 2019 Information System for Forecasting and Management of Occupational Risks *IOP Conf. Ser.: Earth Environ. Sci.* **272** 022198
[9] Zhang M, Shi R and Yang Z 2020 A critical review of vision-based occupational health and safety monitoring of construction site workers *Safety Science* **126** 104658
[10] Fang W, Ding L, Luo H and Love P E D 2018 Falls from heights: A computer vision-based approach for safety harness detection *Automation in Construction* **91** 53-61
[11] Fang W, Ding L, Zhong B, Love P E D and Luo H 2018 Automated detection of workers and heavy equipment on construction sites: A convolutional neural network approach *Advanced Engineering Informatics* **37** 139-49
[12] Semeykin A Yu, Kochetkova I A, Klimova E V, Nosatova E A and Drozdova A O 2019 Decision-making process modeling for occupational safety management system based on a fuzzy analysis of production incidents *J. Phys.: Conf. Ser.* **1333** 072021
[13] Semeykin A Yu, Klimova E V, Kochetkova I A 2020 Forecasting and managing professional risks using information analytical systems based on fuzzy logic methods *IOP Conf. Ser.: Earth Environ. Sci.* **459** 052083
[14] Garrett J W, Teizer J 2009 Human factors analysis classification system relating to human error awareness taxonomy in construction safety *Journal of Construction Engineering and Management* **135(8)** 754-63
[15] Behm M 2005 Linking construction fatalities to the design for construction safety concept *Safety Science* **43(8)** 589-611
[16] Nikulin A N and Nikulina A Y 2017 Assessment of occupational health and safety effectiveness at a mining company *Ecology, Environment and Conservation* **23(1)** 351–5
[17] Nikulin A N, Ikonnikov D and Dolzhikov I 2019 Smart Personal Protective Equipment in the Coal Mining Industry *International Journal of Civil Engineering and Technology* **10(4)** 852–63
[18] Kinn S, Khuder S A, Bisesi M S and Woolley S 2000 Evaluation of safety orientation and training programs for reducing injuries in the plumbing and pipefitting industry *Journal of Occupational and Environmental Medicine* **42(12)** 1142-47
[19] Cheng T, Teizer J 2013 Real-time resource location data collection and visualization technology for construction safety and activity monitoring applications *Automation in Construction* **34** 3-15
[20] Tang N, Hu H, Xu F and Zhu F 2019 Personalized safety instruction system for construction site based on internet technology *Safety Science* **116** 161-9
[21] Amiri M, Ardeshir A, Hossein M, Zarandi F 2017 Fuzzy probabilistic expert system for occupational hazard assessment in construction *Safety Science* **93** 16–28
[22] Klyuev S V, Klyuev A V, Shorstova E S 2019 Fiber concrete for 3-d additive technologies *Materials Science Forum* **974** 367–372 DOI: 10.4028/www.scientific.net/MSF.974.367
[23] Ismailov R I 2020 Making production decisions using the expert assessment procedure. *Construction Materials and Products* **3(1)** 49-56
[24] Lapina A P, Ponomarenko A V, Shentsova K V and Kotesova A A 2019 Analysis of the causes of accidents at different stages of life cycle of a construction object *Construction Materials and Products* **2(2)** 17-22

[25] Manzhilevskaya S E 2019 Environmental monitoring of ecological safety in areas of construction, reconstruction and operation of objects *Construction Materials and Products* **2(3)** 78 – 84

Acknowledgments
The work is realized in the framework of the Program of flagship university development on the base of the Belgorod State Technological University named after V.G. Shukhov, using equipment of High Technology Center at BSTU named after V.G. Shukhov.

IOP Conf. Series: Materials Science and Engineering **945** (2020) 012027 doi:10.1088/1757-899X/945/1/012027

Dynamic test method for full body harnesses exploited in cold climates

Prostakishin Dmitry[1*] and Nam Thanh Pham[1]

[1]National Research Moscow State University of Civil Engineering, Moscow, Russia

E-mail: ppe_mgsu@mail.ru

Abstract. In this paper, a method of testing the dynamic strength of full body harnesses is proposed. The possibility of conducting the method in laboratory conditions is determined by testing the method while maintaining the specified conditions of low temperatures. The main attention is paid to the effect of the set temperature on the full body harnesses, when dynamic shock loads occur in user operation mode. The advantage of the developed method of dynamic testing of full body harnesses is the simplicity in bringing the test sample to the climatic parameters of operation by pre-conditioning. This makes it possible to implement dynamic testing under specified climatic conditions within mass laboratories in the chain of conformity confirmation. Pre-conditioning allows you to conduct research on the influence of climate factors on various characteristics of safety harnesses. The interaction of the binding elements with each other, previously reduced to the specified temperature parameters, is also studied. It is necessary to pay special attention to the resulting impact loads during dynamic tests, and their relationship when conditioning products at different temperature.

1. Introduction

According to the statistics of industrial injuries, falling from a height takes the leading place among all dangerous industrial factors in the number of accidents and emergencies. They account for approximately 39.5%, if we add to these numbers accidents that occurred when objects fell from a height (an approximate figure of 10.5%), we will find that half of all accidents at work are related to factors that occur when the force of attraction, that results to the user falling, or the object on the user [1, 2].

An important aspect of ensuring the safety of employees at work is to provide them with personal protective equipment (PPE) and collective protective equipment (CPE) against falling from a height, as well as safety helmets. The widespread introduction and monitoring of mandatory use of protective equipment has significantly reduced the number of incidents related to the influence of hazardous production factors. The introduction of new rules on labor protection in the production of work at height has also saved many lives. Regulatory activities by the enforcement of the law allowed to structure and designate areas of responsibility to officials, and the necessary mandatory training in safe methods and techniques for performing work at height contributed to increasing the user competence [3, 4], despite the fact that there is a huge number of organizations that issue certificates without appropriate training. Another important reason for the increase in the level of safety and reduction in the number of accidents at work was the introduction of mandatory PPE certification, which includes an assessment of the state of production with on-site inspection control by an expert at the

IOP Conf. Series: Materials Science and Engineering **945** (2020) 012027 doi:10.1088/1757-899X/945/1/012027

manufacturer's enterprise, and certification tests [5] by an independent testing laboratory in accordance with the test methods included in the list of technical regulations. This type of activity is strictly controlled by the structures of the Federal accreditation service, which serves as filters for unscrupulous organizations, that periodically pass the barriers of regulatory authorities, violating the law for the sake of financial profit, reducing the overall level of safety and quality of products being turned up on the market.

The usage of CPE from falling from a height at the moment does not provide the necessary level of user safety, due to the lack of a clear regulatory framework and quality control of products [6]. Manufactured CPE are not subjected to mandatory certification, which in turn cannot have a positive impact on their quality.

The PPE used is in combination to create a security system in the form of system components. GOST R 58208-2018/EN 363:2008 "System of standards for labor safety (occupational safety). Personal protective equipment against falling from a height. Personal protection systems against falling from a height. General technical requirements" defines the following types of safety systems for work at height: holding system, workplace positioning system, rope access system, safety system, rescue system [7-10]. These designs include a device to support the body and a connecting system that connects to a secure anchor point. A body support device may be, for example, a full body harness for fall arest, work positioning and suspension, harness for rescue operations, evacuation triangles. The most widely used are safety tethers, because they are mandatory for all situations in which the user may fall from a height. These tethers ensure that the user is correctly stopped when falling, and then held in the head-up position. The use of such tethers in the fall arrest system significantly reduces the risk of injury to the user during the overload they experience. To date, test methods represent the minimum safety requirements that can provide PPE in case of their correct operation. These methods do not confirm the quality parameters of products and the ability of products to resist harmful effects on them [11,12]. If there are no mandatory requirements for testing products in their operating conditions, the possibility of their safe use is questioned [13]. An important aspect of the use of PPE is the climatic conditions of operation, which differ sharply from the laboratory conditions (23±5°C) under which certification tests of PPE are carried out, which in turn puts the life and health of users at risk. PPE against falling from a height must protect the user from the risks associated with falling at all operating temperature ranges specified by the manufacturer and regulatory standards. The impact of temperatures on PPE against falling from a height is currently studied [14, 15] and it is necessary to develop such experience starting with research and control tests.

The purpose of this work is to ensure the safety of people who meet the factor of "height" and the dangers associated with it in their production activities. By developing new PPE testing methods that are close to real operational parameters, it will be possible to achieve a higher level of safety of the products themselves, as well as the preservation of human health and life. Information about the results and methods of testing will allow competent specialists in the field to build a more competent security system.

To achieve the above goals, the following tasks were completed:

- development of a method for testing products at low temperatures, based on current regulatory documents;
- testing of the method;
- analysis of results, comparative assessment of existing possible models for conducting such tests;
- the formation of findings and conclusions which will provide the necessary recommendations.

2. Methods and materials

Before performing dynamic tests, samples of full body harnesses were pre-conditioned in a climate chamber at a temperature of -60°C for 4 hours (figure 1). The time of conditioning in the climate chamber was chosen according to one four-hour work shift.

Figure 1. Conditioning of full body harness sample in the climate chamber.

Tests of full body harnesses were carried out on a stand for dynamic tests of PPE against falling from a height [16]. Dynamic tests of full body harnesses were performed after conditioning for 90 seconds from the moment the sample was removed from the climate chamber. The safety leash was tightly put on a 100 kg dummy. As a sling, a dynamic rope with a diameter of 11 mm was used, meeting the requirements of EN 892-2016 "Mountaineering equipment - Dynamic mountaineering ropes - Safety requirements and test methods". The total length of the sling from the place of attachment to the tether to the place of attachment to the anchor point of the test stand was 2 m. The dummy was lifted by the upper ring of the fastening bolt (feet down) (figure 2) to such a height that the free fall of the dummy was approximately 4 m before the sling begins to restrain the fall of the test dummy. By means of the release device, the dummy was allowed to go into free fall without initial velocity. After repeated conditioning of the same sample in the climate chamber, the dynamic test was repeated with the dummy, but suspended by the lower ring (head down).

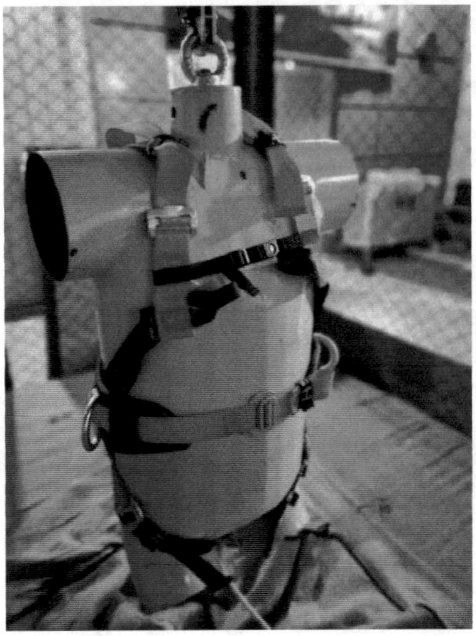

Figure 2. Position of the dummy in the starting position.

IOP Conf. Series: Materials Science and Engineering **945** (2020) 012027 doi:10.1088/1757-899X/945/1/012027

3. Results and discussions

According to the requirements of GOST R EN 361-2008 "Occupational safety standards system. Personal protective equipment against falls from a height. Full body harnesses. General technical requirements. Test methods", for dynamic tests with a 100 kg dummy, the safety harness must withstand two consecutive drop tests with a free fall distance of 4 m without releasing the dummy (one drop test with the dummy's feet down and one drop test with the dummy's head down). After each fall, the dummy must stop in the head-up position and the angle between the longitudinal axis of the plane of the mannequin's back and the vertical is no more than 50°.

As part of the study, tests were conducted on 3 samples. During tests, samples of safety full body harnesses withstood falls, without releasing the dummy. After the fall, the dummy stopped in the head-up position and the angle between the longitudinal axis of the plane of the dummy's back and the vertical was within the allowed range (figure 3).

Fig. 3. Position of the dummy after the fall.

The advantage of the developed method of dynamic testing of full body harnesses is the simplicity in bringing the test sample to the climatic parameters of operation by pre-conditioning. This makes it possible to carry out dynamic tests under specified climatic conditions in the framework of mass laboratories in the chain of conformity confirmation, and the implementation of the method in the mass testing sector is an integral part of quality control and safety of the necessary technical parameters of the above-mentioned products.

The following disadvantages of this dynamic test method of the full body harnesses are distinguished:

- different changes in the temperature of samples from the set temperature during conditioning from the moment of sample extraction from the climate chamber, depending on the manufacturing materials and the preparation process;
- too short period of time from the moment the sample is removed from the climate chamber until the dummy is dropped;

- the full body harness is put directly on the dymmy, which does not take into account the presence of protective warm clothing on the user under real conditions of a cold climate.

4. Summary

Pre-conditioning is allowed to conduct research on the influence of climate factors on various characteristics of full body harnesses. When testing for dynamic strength, the interaction of the binding elements with each other, previously reduced to the specified temperature parameters, is also studied. It is necessary to pay special attention to the resulting impact loads during dynamic tests, and their relationship when conditioning products at different temperature parameters.

In the future, it is planned to conduct research on the influence of climate factors on other PPE against falling from a height, both with the help of pre-conditioning, and with the direct exposure of climate factors during static strength tests. Analysis of methods of direct conditioning and pre-conditioning of products with subsequent tests for impact and static loads, is potentially able to show the degree of simulation and reproducibility of such effects, as well as to correlate them with actual situations when the user falls into operating conditions at extremely low/high temperature parameters. It is also worth noting the importance of fixing the peak dynamic load on the product and anchor point when stopping the fall and its display in the test reports to obtain a more complete picture of the force load on the product and the user, since this is a very important and basic characteristic in systems for stopping the fall [17-19].

5. References

[1] Khamidullina E A , Tolstikhina Yu A, Povetkina P N 2018 System approach to high-rise operations safety *XXI century. Technosphere Safety* **3**(3) 24–35

[2] Rahitov G S, Abdyukova G M 2015 Analysis of occupational injuries at construction sites. *Collection of Scientific Articles of Participants of the All-Russian Scientific-Practical Conference with International Participation* **2** 286

[3] Vasilenko V, Korolchenko D and Pham N T 2018 Definition of the inspection criteria for personal protective equipment (for work at heights) on example of full body harnesses *MATEC Web of Conferences* **251** 02042

[4] Stupakov A A and Lelikov G D 2014 Calculation of risks from the use of personal protective equipment against falls from a height *Mechanization of construction* **12**(846) 50-54

[5] Pham N T, Vasilenko V and Korolchenko D 2018 Test and certification procedures of pulleys as a part of personal fall arrest system *IOP Conference Series: Materials Science and Engineering* **365**(4) 042057

[6] Pham N T, Vasilenko V V and Korolchenko D A 2017 Actualization of the dynamic test methods to be systemized for temporary edge protection systems *Pozharovzryvobezopasnost/Fire and Explosion Safety* **26**(12) 35-44

[7] Stupakov A A, Lelikov G D, Semenov P A and Vasilenko V V 2015 Inspection and restoration of high-rise facilities by the method of industrial high rise works *Mechanization of construction* **2**(848) 48-52

[8] Korolchenko D A and Pham N T 2019 Loads in rope access system when working at heights *IOP Conference Series: Materials Science and Engineering* **603**(5) 052010

[9] Pham N T, Lelikov G and Korolchenko D 2018 Improvement of the safety systems for working at heights on transmission towers *IOP Conference Series: Materials Science and Engineering* **365**(4) 042054

[10] Senchenko V A, Lelikov G D and Glumov E A 2018 Attachment to the ladder, Patent No. RU 184480 U1

[11] Vasilenko V V, Lelikov G D, Ovchinnikova T A and Korolchenko D A 2019 Determination of criteria for assessing the effect of inorganic acids on synthetic ropes in order to improve the safety of high-altitude works *Pozharovzryvobezopasnost/Fire and Explosion Safety* **28**(6) 35-51

[12] Vasilenko V V, Lelikov G D and Zherdev K V 2020 Effect of acid solutions on the residual strength of safety and rescue ropes *Occupational Safety in Industry* **2** 38-44

[13] Baszczynski K, Jachowicz M 2009 The Effect of the Use of Full Body Harnesses on Their Protective Properties *International journal of occupational safety and ergonomics* **15**(4) 435-446

[14] Baszczynski K 2004 Influence of weather conditions on the performance of energy absorbers and guided type fall arresters on a flexible anchorage line during fall arresting *Safety science* **42**(6) 519-536

[15] Senchenko V A, Pushenko S L and Staseva E V Concept of technical safety measures when performing high-rise works on the roofs of buildings in the Arctic North *Bulletin of the Volgograd state University of architecture and construction. Series: Construction and architecture* **51**(70) 178-185

[16] Stupakov A A, Kapyrin P D, Lelikov G D, Semenov P A and Vasilenko V V 2015 Stands for the study of personal protective equipment against falling from a height *Bulletin of MSUCE* **8** 130-39

[17] Lelikov G D and Vasilenko V V 2017 Analysis of the use of safety synthetic rope slings as PPE from falling from a height *Construction — the formation of the living environment. Electronic Resource: Collection of works of the XX International Interuniversity Scientific and Practical Conference of Students, Undergraduates, Postgraduates and Young Scientists* 475-77

[18] Vasilenko V V 2017 Actualization of methodology for dynamic testing of shock absorbers as personal protective equipment against falling from a height *Construction — the formation of the living environment. Electronic Resource: Collection of works of the XX International Interuniversity Scientific and Practical Conference of Students, Undergraduates, Postgraduates and Young Scientists* 439-41

[19] Korolchenko D, Vasilenko V and Lelikov G 2018 Problems of the dynamic test method for individual protection equipment (shock absorbers) *MATEC Web of Conferences* **193** 05034

IOP Conf. Series: Materials Science and Engineering **945** (2020) 012028 doi:10.1088/1757-899X/945/1/012028

Mixing process simulation of the initial building materials components using the DEM Solution EDEM system

P S Samsonova[1*], S Y Lozovaya[1], N E Bogdanov[1] and N M Lozovoi[2]

[1]Department of Mechanical Equipment, Belgorod State Technological University named after V.G. Shukhov, Kostyukov St., 46, Belgorod, 308012, Russia
[2]Department of City Cadastre and Engineering Surveys, Belgorod State Technological University named after V.G. Shukhov, Kostyukov St., 46, Belgorod, 308012, Russia

E-mail: samsonova.polinka@gmail.com

Abstract. Quality of building materials, products and constructions depends on the organization of technological process of their production, and it, as a rule, concerns preparatory operations, such as crushing and in a greater degree mixing of initial compound components of the future products and constructions.Production of high-quality building mixtures is one of the most actual tasks of modern building technologies. Today, the mixing of bulk materials has evolved into a special branch of technological knowledge, which is based on mechanical processes, whose goal is to ensure the highest possible degree of alignment of individual components in the mixture. Due to the development of digital technology, modern industry is able to abandon expensive and long field experiments and tests in favor of digital experiments. The results of digital experiments have a high degree of convergence with real experiments, but are carried out many times faster. To model the influence of the material loading method on the heterogeneity factor of the mixture, digital sampling was carried out using EDEM software in a screw type mixer.

1. Introduction

Mixing is one of the most important technological processes, since the quality of finished products and structures in the future depends on the homogeneity of the mixture from the original components. There are several factors that affect the quality of the mixture in the mixing process, such as the condition and concentration of the components of the mixture, the construction of the mixer (the form of the working body, the form of the mixing chamber, options for loading the mixed components), the technological mode of the mixing process.

In any volume of ideally homogeneous mixture from the phenomenological point of view, there should be particles of all components in quantities determined by their given ratio. However, this ideal arrangement of the particles in the mixture volume is not observed in reality. In randomly selected volumes of the mixture there can be a large number of combinations of particles of components, i.e. their distribution in the mixture by chance, which leads to the use of a static approach [1] to estimate the quality of the mixture.

The most common and frequently used criteria for assessing the quality of a mixture [2] is the inhomogeneity factor:

$$V_c = \frac{100}{\bar{c}}\sqrt{\frac{\sum_{i=1}^{n}(c_i-\bar{c})^2}{n-1}}, \quad (1)$$

where $\bar{c}$ – the arithmetic mean of the key component concentration in the samples;
c_i – the concentration value of the key component in the i sample;
n – the number of samples to be analyzed.

EDEM is the leading software product on the market to simulate the behaviour of bulk material particles (dry and wet) during various processes. Developed using advanced discrete element modeling (DEM) technology, EDEM quickly and accurately simulates and analyses the behaviour of bulk materials such as rocks, dry powders, coal, soil, pellets, building material components, etc.

2. Materials and methods

2.1. Materials

Screw type mixer (Fig. 1) consists of a conical casing 1, cover 4, ledge 7, drive screw 3, drive ledge 2, screw 5, locking mechanism 6. Screw 5, rotating around its own axis driven by an engine 3 through a pair of conical gears, makes a rotation on a planetary trajectory around the axis of the casing mixer from the drive 2, a pair of gears and ledge 7. The upper end of the screw 5 is connected to the driver 7 by a coupling, freely suspended in the lower section of the mixer housing. Screw and driver drives are mounted on the cover 4 of the mixer casing. [1].

Material is loaded through the spigots in the top cover, unloaded through the discharge box. One of the most important components of the mixer is the swivel support of the screw, as it is placed directly in the mass of the bulk material, so its design should provide operation without lubrication and prevent abrasion of the shaft neck by particles of the mixing material [3].

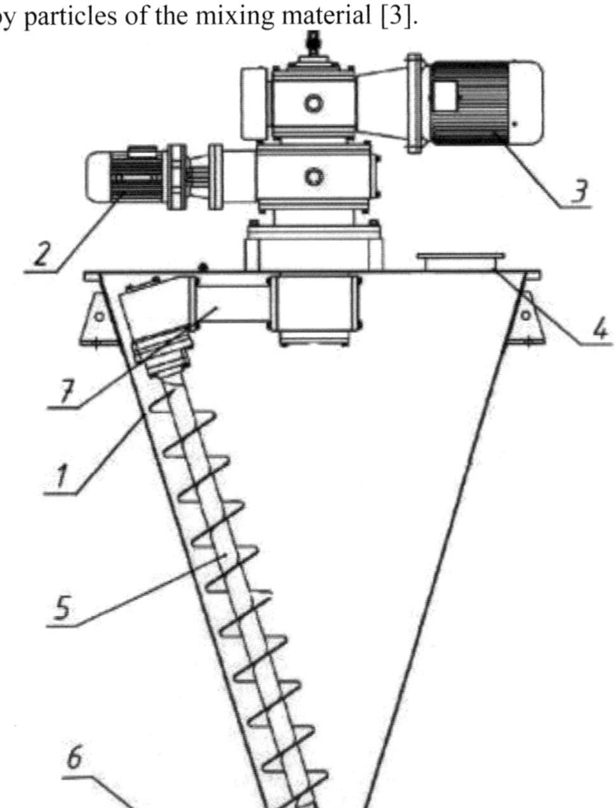

Figure. 1. Screw type mixer.

IOP Conf. Series: Materials Science and Engineering **945** (2020) 012028 doi:10.1088/1757-899X/945/1/012028

As the screw rotates around the axis of the body and its own axis, particles of the material are lifted up by its windings, also some particles are carried away along the walls of the body, and the rest of the particles, located closer to the center and opposite side of the cone from the screw, are lowered down. In this way, three material trajectories are formed in the mixing chamber: upwards along the screw turns, along the wall of the mixing chamber and downwards in the space free from the screw. This results in a quality mixing of the components of the mixture.[4].

2.2. Methods

For researching the parameters of heterogeneity of the mixture, can be using EDEM software, which allows to choose a wide range of physical properties of mixed mate-rials, such as the simulation of any type and shape of material: large lumps or fine powders, dry or wet materials. All this makes it possible to simulate processes and with this help to make conclusions about how different bulk and wet materials will interact with each other as well as with elements of the designed equipment under different technological conditions [5].

The discrete element method (DEM) is a numerical method used to simulate the mo-tion of particles interacting with each other through a collision. In fact, EDEM simu-lates the dynamics of each particle individually and numerically integrates their ac-celerations, which depend on the sum of all forces, including contact and gravitation-al forces. The position of each particle is captured and their interactions are evaluat-ed at each time step. Then all forces acting on each particle are calculated and New-ton's second law is used to determine accelerations to find the particle's speed and position at each moment of time, this process is repeated until the end of the simula-tion [6-8]. Therefore, the discrete element method (DEM) is a reliable and high-performance tool to simulate various processes with bulk and wet materials, in par-ticular, this method is widely used to simulate the processes occurring in different types of mixers [7,8] to obtain quality building materials, products and structures.

3. Results

3.1. Coefficient of inhomogeneity

With the help of EDEM software in the Analyst application, sampling for digital estimation of the inhomogeneity factor in the area of the mixing chamber centre (Figures 2, 3) and radial sampling for different loading options for the initial components of the mixtures (Figures 5, 6) - horizontal (Figures 2, 5) and vertical (Figures 3, 6) - for further theoretical calculations of determining the particle behavior of the mixed material has been simulated.

In the Analyst module, using the "Setup Selection" section, four "Total Mass Sensor" material groups have been created, which show the weight of the entire load that is part of this group, as well as the weight of each component of the mixture that is in the group volume. The groups are located in the centre of the mixing chamber circle and rise up the centre axis of the chamber. Based on the results of modeling with the help of Maple analytic package, in time steps 5, 9, 13 and 17 sec the digital coefficient of heterogeneity according to formula (1) was calculated for the received samples (Table 1).

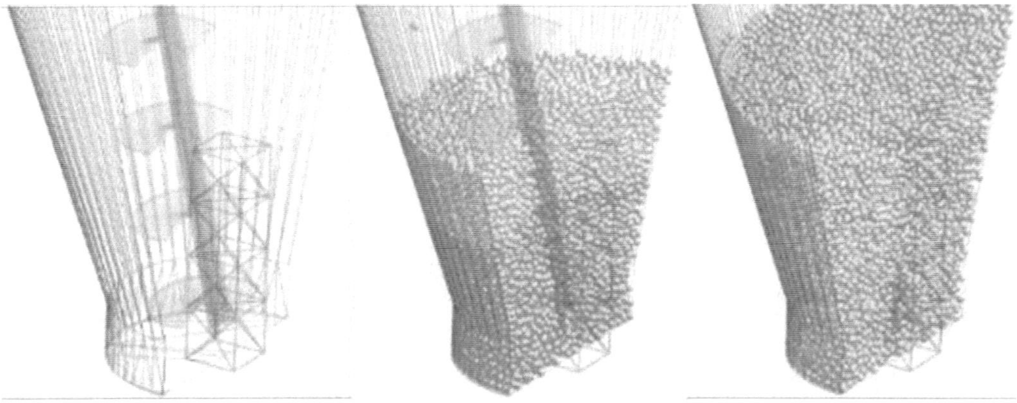

Figure 2: Horizontal loading of material.

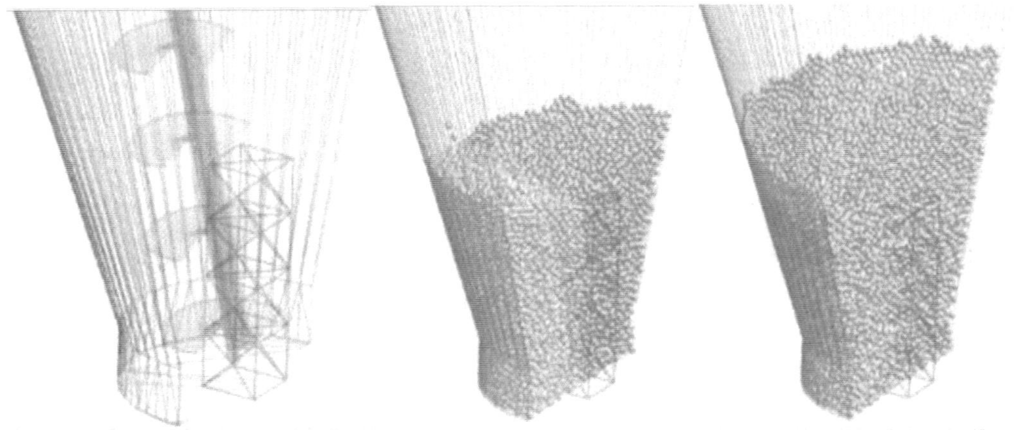

Figure. 3: Vertical material loading.

Table 1. Coefficient of inhomogeneity in the central sample.

Sec.	Horizontal loading, V_C,%	Vertical Loading, V_C,%
5	10	23
9	12	14
13	16	16
17	16	9

Based on the results of the calculations, a graph of the change in the numerical heterogeneity coefficient (Fig. 4) from the mixing time is plotted. All lines characterizing the change of the coefficient values have a decreasing sinusoidal character. This is explained by the fact that when moving the shaft around the axis of the cone body and its axis, at a certain period of rotation, the screw, capturing the material, raises it up, then under the surfaces of the coils formed voids, which are filled with the material as from the upper layer, as well as related sector.

IOP Conf. Series: Materials Science and Engineering **945** (2020) 012028 doi:10.1088/1757-899X/945/1/012028

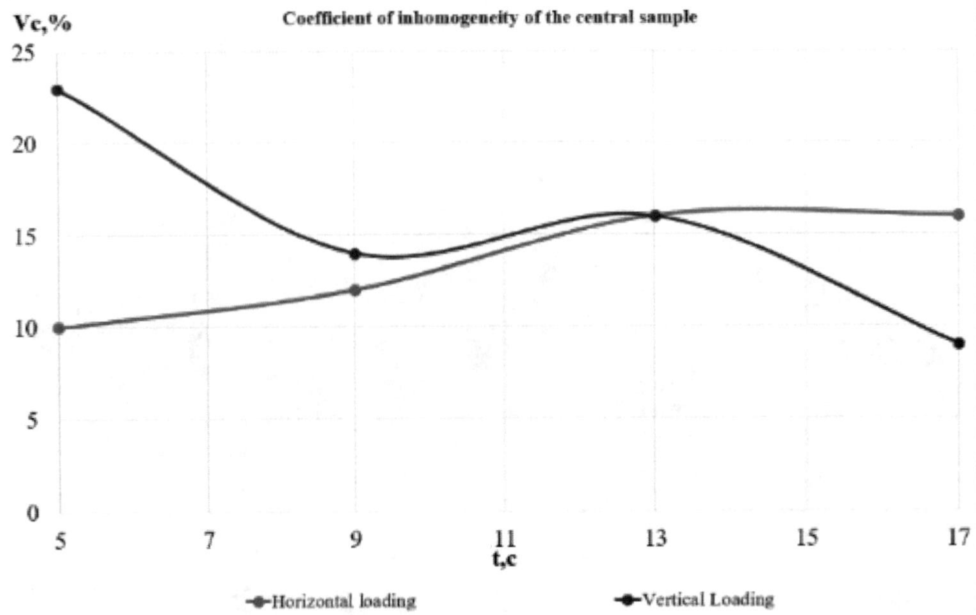

Figure. 4: Graph of the digital heterogeneity factor of the central sample:
1 - horizontal loading of material, 2 - vertical loading of material.

Modeling of obtaining digital radial samples by known methods at mixing with horizontal (Fig. 5) and vertical loading (Fig. 6) camera material is made. Then the numerical heterogeneity coefficients are calculated (Table 2), and a graph is drawn (Fig. 7).

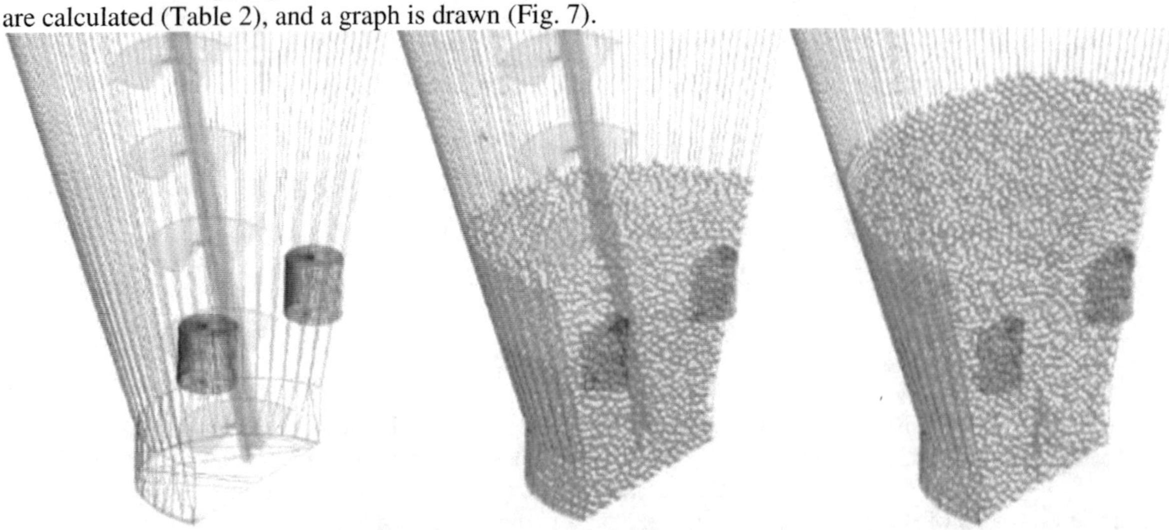

Figure. 5: Horizontal loading of material.

IOP Conf. Series: Materials Science and Engineering **945** (2020) 012028 doi:10.1088/1757-899X/945/1/012028

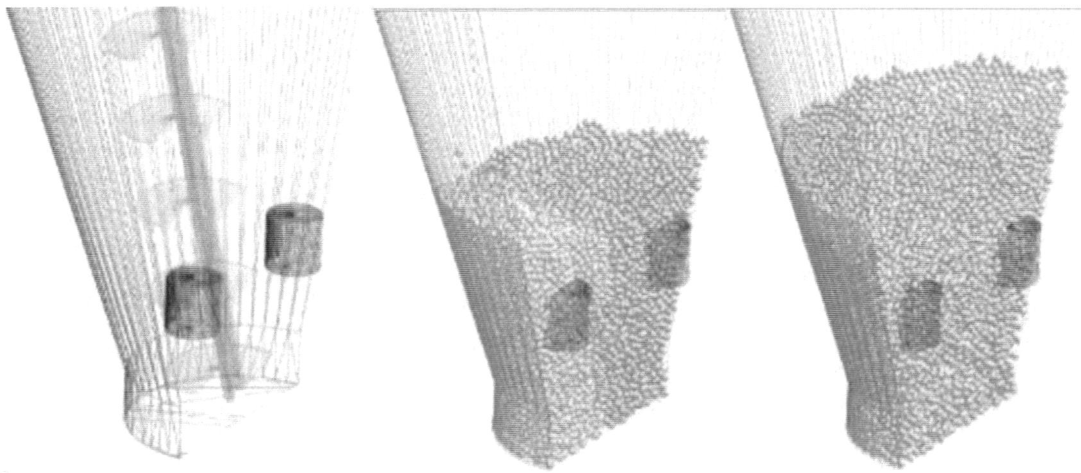

Figure. 6: Vertical loading of material.

Table 2. Homogeneity factor of radial sample.

Sec.	Horizontal loading, V_C,%	Vertical Loading, V_C,%
5	23	4
9	23	23
13	7	16
17	12	7

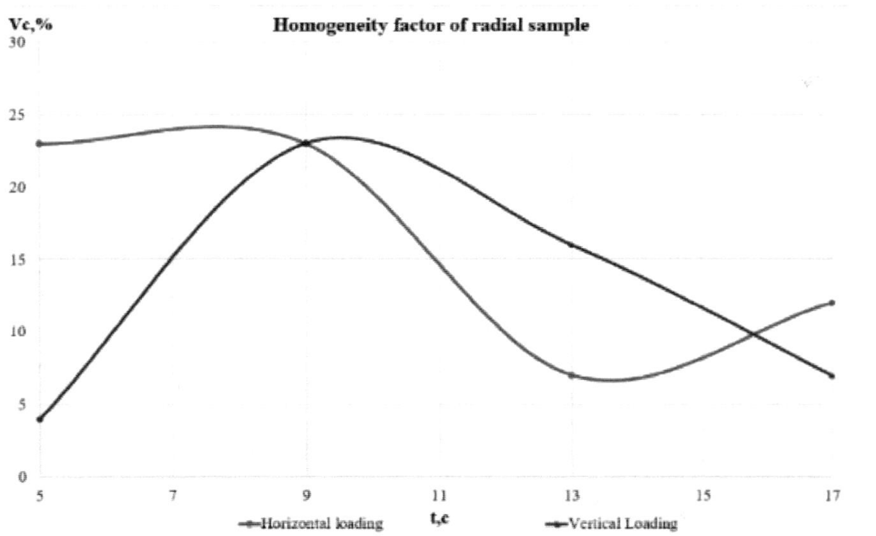

Figure. 7: Diagram of the digital heterogeneity factor of the radial sample:
1 - horizontal material loading, 2 - vertical material loading.

4. Discussion

The analysis of the graph showed that, as with the central sample, all lines are sinusoidal, which is associated with the nature of movement due to particles of conicity of the chamber and the impact of the working body on the material. At the same time the lines characterizing mixing at horizontal loading have the most intensively decreasing character, than at vertical loading. Thus, the particles on layers move the most intensively at horizontal loading.

5. Summary

The combined analysis of the results of modeling of mixing process and further selection of central and radial electronic samples, calculation of heterogeneity coefficients for each loading variant at several time steps showed that the most intensive mixing process and the lowest values of heterogeneity coefficient of the finished product are achieved exactly at horizontal location of raw materials in the mixing chamber of vertical screw mixer. This is due to the fact that the main part of the mixing cycle is the transportation of material up the screw, due to which the material from the bottom layer is introduced into the upper, and the upper layer particles fill the voids from the raised material. In this process, the particles of the mixture are moved radially by the screw movement around the axis of the housing, which intensifies the movement of the particles in the load cross section.

Thus, using the proposed tools, it is possible to simulate the processes of mixing the original components in the mixing equipment with horizontal chambers for further production of building materials of products and constructions.

6. References

[1] Generalov M B 2004 Machines and devices of chemical and petrochemical production, 832

[2] Lozovoi N M, Radinskaya S Y, Lozovaya S Y 2020 Experimental studies of the dependence of the heterogeneity coefficient of the mixture on the technological parameters in the mixer with a variable working chamber *International Scientific Conference Interstroymeh IOP Conf. Series: Materials Science and Engineering* **786** 012049

[3] Makarov Y I 1973 Machines for bulk materials mixing, 216

[4] Samsonova P S, Lozovaya S Y, Lozovoi N M 2020 Particle kinetics of the two-component mixtures in a mixer with a screw vertical working body *Materials, equipment and resource-saving technologies: materials of the international scientific and technical conference* 62–64

[5] EDEM Software. 08.06.2020: https://www.edemsimulation.com

[6] Krenzer K, Mechtcherine V, Palzer U 2019 Simulating mixing processes of fresh concrete using the discrete elementmethod (DEM) under consideration of water addition and changes inmoisture distribution *Cement and Concrete Research* **115** 274–282

[7] Golshan S, Zarghami R, Norouzi N, Mostoufi N 2017 Granular mixing in Nauta blenders *Powder Technology* **305** 279–288

[8] Rydunskaya L V, Lozovaya S Y 2013 Experimental studies on the management of energy characteristics of grinding devices *Bulletin of BSTU named after V G Shukhov* **2** 177-180

[9] Lozovaya S Y, Sablin V S, Tkacheva O V 2014 Experimental studies of homogenization processes in a static mixer *Bulletin of BSTU named after V G Shukhov* **2** 84-88

[10] Lozovoi N M, 2012 Study of the mixing process in a mixing and grinding device with a changeable working chamber using modern computer technologies *Bulletin of BSTU named after V G Shukhov* **1** 79-81

[11] Otsokov K.A. 2020 Innovative technologies in construction and their use in organizational and technological events *Construction Materials and Products* **3 (1)** 7 – 13

[12] Odenbakh I.A., Taurit E.B. 2020 Some new technologies in road construction *Construction Materials and Products* **3 (1)** 62 – 69

Acknowledgments

This work was realized under the support of the President Scholarship; in the framework of the Program of flagship university development on the base of the Belgorod State Technological University named after V G Shukhov, using equipment of High Technology Center at BSTU named after V G Shukhov.

IOP Conf. Series: Materials Science and Engineering **945** (2020) 012029 doi:10.1088/1757-899X/945/1/012029

Experimental studies of obtaining biogas from waste of meat processing enterprises

D Y Suslov[1*] and P S Sedyh[1]

[1]Department of heat and gas supply and ventilation, Belgorod State Technological University named after V.G. Shukhov, Kostyukov St., 46, Belgorod, 308012, Russia

E-mail: Suslov1687@mail.ru

Abstract. A promising direction for the development of gas supply systems in rural areas is the production and use of biogas. In the developed countries of the world, biogas production and use systems are widely used. The amount of biogas produced depends on the type and chemical composition of the feedstock. The aim of this work is to study the dependence of the specific yield and chemical composition of biogas on the type of substrate used and the temperature of biomass in the digester. Waste from meat processing enterprises was used as an initial substrate for research: technical fat; intestines; bird blood; pulp. Two temperature settings were chosen for the study: 38 ° C and 42 ° C. As a result of the studies, the values of the optimal duration of fermentation of the studied types of substrate were determined: bird blood - 21 days, pulp - 25 days, intestines - 25 days, technical fat - 40 days. It has been established that the maximum value of the specific yield of biogas is observed during the fermentation of technical fat - 1509.8 l / kg, and the smallest of bird blood - 833.4 l / kg. At the same time, the highest methane content is observed during fermentation of technical fat - 66%, and the smallest from bird blood - 53.5%. It was also found that the biogas yield at a fermentation temperature of 42 ° C is 7-8% higher than the biogas yield obtained at a temperature of 38 ° C.

1. Introduction

At present, the level of gasification in the Russian Federation is 68.5%, including 72% in cities and urban-type settlements, and 59.2% in rural areas. At the same time, in all regions of the country, there is a lower level of gasification in rural areas compared to cities [1].

A promising direction for the development of gas supply systems in rural areas is the production and use of biogas in centralized gas supply systems. In the developed countries of the world, biogas production and use systems are widely used. The leaders in terms of the volume of biogas produced are China, Germany, Finland and Great Britain [2-4].

In operating biogas plants, waste from livestock, poultry and plant growing enterprises is often used as an initial substrate [5, 6]. The volume and composition of the produced biogas depend on the chemical composition of the type of substrate, namely on the content of proteins, fats and carbohydrates. It is known that the greatest amount of biogas is released during anaerobic fermentation of fats [7, 8]. A promising direction of use as initial substrates is the use of wastes from meat processing enterprises, which contain a large amount of fat [9-11]. At the same time, the waste of meat processing enterprises is often used as part of cosubstrates with other organic waste (manure, grass, fruits and vegetables). This allows you to get more methane than the fermentation of one type of substrate [12-14].

Also an important factor influencing the efficiency of the biogas production process is the temperature of the biomass in the digester. It is known [15, 16] that the process of anaerobic fermentation of wastes from meat processing enterprises is more stable, and the biogas yield is higher in the mesophilic temperature regime, as compared to the thermophilic regime.

The aim of this work is to study the dependence of the specific yield of biogas and its chemical composition on the type of substrate used and the temperature of the biomass in the digester.

2. Materials and methods

Waste from meat processing enterprises of the Belgorod region was taken as the initial substrate for research:
- Technical fat;
- Intestines;
- Bird blood;
- Pulp.

Two temperatures were chosen for the study: 38 ° C and 42 ° C.

At the first stage of the experiment, the dry matter content in the original substrates was determined (Table 1).

Table 1. Determination of the proportion of dry matter in the test substrates.

Name of substrate	The mass of ceramic crucibles, kg			Moisture W,%	Dry matter DM,%
	Empty, m_3	With initial substrate, m_1	With dried substrate, m_2		
Technical fat	52.65	91.15	90.92	0.60	99.40
Intestines	52.65	92.47	62.73	74.69	25.31
Bird blood	52.65	98.45	60.82	82.17	17.83
Pulp	52.65	94.95	63.04	75.43	24.57

The study of the biogas production process was carried out on an experimental plant with a digester with a volume of 150 liters. The experimental setup is shown in Figure 1. The installation consists of a digester, a gasholder, devices for heating and mixing biomass, and instrumentation. The digester contains a mixing and heating system, a pipe for supplying the initial substrate, a pipe for draining biomass, and a pipe for removing the resulting biogas [17].

The main indicator of the biogas production process is the specific biogas yield, measured in liters per kilogram of dry matter (DM). When assessing the chemical composition of biogas, the main indicator is the percentage of methane and carbon dioxide.

Specific biogas yield is determined by the formula V (l/kg):

$$V = \frac{V_B}{m \cdot DM} \tag{1}$$

where V_B – the volume of biogas produced, l; m – the mass of the substrate, kg; DM - dry matter content,%.

To determine the volume of biogas produced, a VK-G1.6T gas meter is used. The FP-34 gas analyzer is used to determine the chemical composition of biogas. This type of equipment is a portable measuring device for continuous monitoring of CH4 and CO2 concentration.

The moisture content of the substrate is determined by the formula:

$$W = \frac{m_1 - m_2}{m_1 - m_3}, \tag{2}$$

where m_1 – is the mass of crucibles with the initial substrate, kg; m_2 – is the mass of crucibles with dried substrate, kg; m_3 – is the mass of crucibles without substrate, kg.

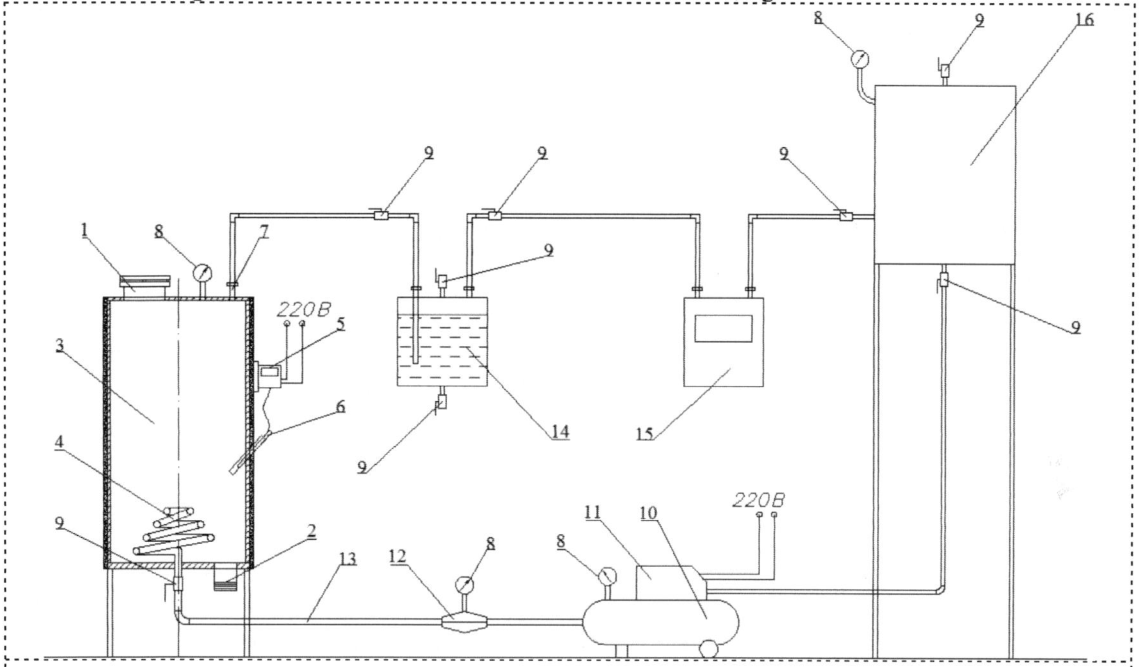

Figure 1. The experimental setup: 1 - branch pipe for loading the initial substrate; 2 - biomass discharge branch pipe; 3 - anaerobic digester; 4 - bubbling pipeline; 5 - temperature regulator; 6 - thermometer; 7 - biogas outlet branch pipe; 8 - manometer; 9 - ball valve; 10 - compressor; 11 - electric motor; 12 - pressure regulator; 13 - gas supply pipeline for mixing biomass; 14 - biogas purification filter; 15 - gas meter; 16 - gas holder.

To determine the moisture, empty dried ceramic crucibles were taken and weighed. Then, the crucibles filled with the substrate were placed in a drying cabinet at a temperature of 106 ° C. for 4 hours. Next, the crucibles with the dried substrate were re-weighed.

The dry matter content was determined from the moisture:

$$DM = 100 - W \quad (3)$$

To determine the mass and moisture of the raw materials, laboratory electronic scales NVT6401 / 2 and drying cabinet ШС-80-01 were used.

3. Results and discussion

3.1. Technical fat

The results of studying the specific yield and composition of biogas depending on the temperature and duration of fermentation of technical fat are presented in Figure 2 and Figure 3.

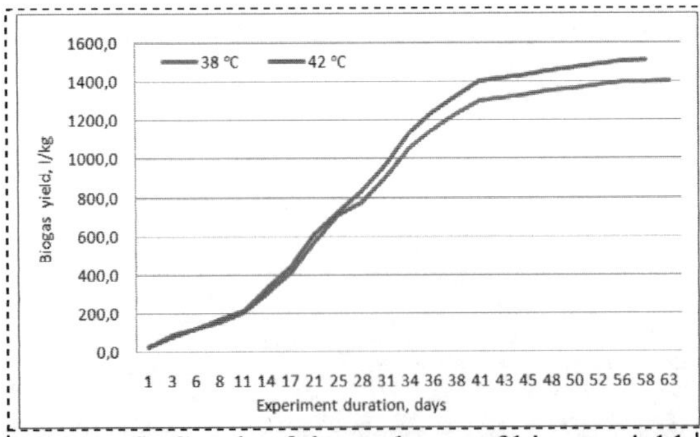

Figure 2. Graph of dependence of biogas yield on the duration of fermentation of technical fat.

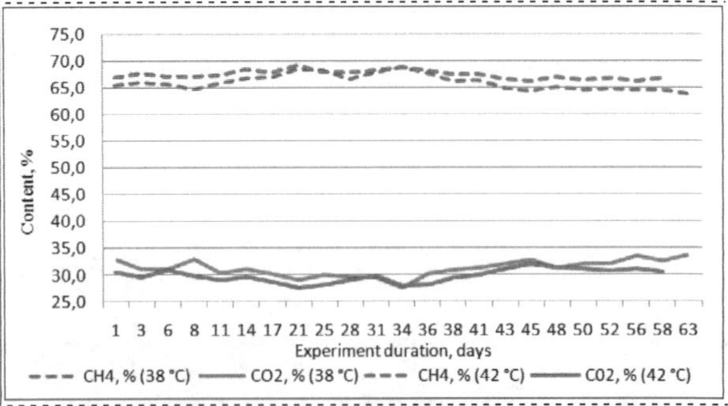

Figure 3. Graph of dependence of biogas composition on the duration of fermentation of technical fat.

From Figure 2, the maximum daily values of biogas yield are observed from 1 to 41 days of the experiment, which is 93% of the total biogas yield. At the same time, the total biogas yield obtained at a temperature of 42 ° C was 1509.8 l / kg, which is 7.77% more than at a temperature of 38 ° C (1400.9 l / kg).

Analysis of the chemical composition of biogas shows (Figure 3) that the methane content has maximum values in the period from 1 to 41 days of the experiment, then a drop in the methane content is observed. At the same time, the maximum methane values were recorded on day 34 at 38 ° C and on 21 days at 42 ° C, and are 68.8% and 69.2%, respectively. The average methane content at 42 ° C was 67.4%, and at 38 ° C - 66%. Thus, we can conclude that the optimal duration of fermentation of technical fat is a period of 40 days.

3.2. Intestines

The results of the study of the specific yield and composition of biogas depending on the temperature and duration of the experiment of anaerobic fermentation of the intestines are presented in Figure 4 and Figure 5.

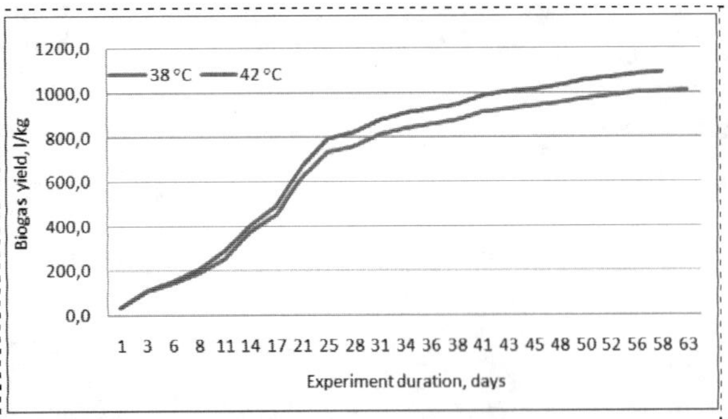

Figure 4. Graph of the dependence of biogas yield on the duration of intestinal fermentation.

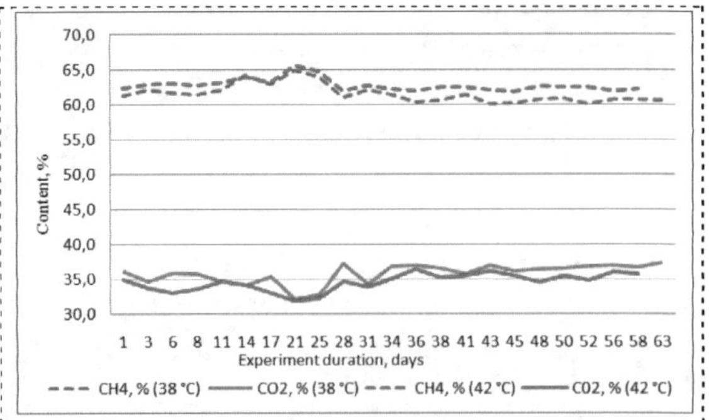

Figure 5. Graph of the dependence of the biogas composition on the duration of intestinal fermentation.

Figure 4 shows that the maximum daily readings of biogas yield are observed from 1 to 25 days of the experiment, which is 73% of the total biogas yield. In this case, the total yield of biogas obtained at a temperature of 42 ° C was 1091.7 l / kg, which is 8.22% more than at a temperature of 38 ° C (1008.8 l / kg).

Analysis of the chemical composition of biogas shows (Figure 5) that the methane content has stable values from 1 to 25 days, then a drop in the methane content is observed. At the same time, the maximum value of methane at temperatures of 38 ° C and 42 ° C for 21 days is 64.9% and 65.5%, respectively. The average methane content at 42 ° C was 62.7%, and at 38 ° C - 61.5%. Thus, it can be concluded that the optimal duration of gut fermentation is 25 days.

3.3. Bird blood

The results of the study of the specific yield and composition of biogas depending on the temperature and duration of the experiment of anaerobic fermentation of bird blood are presented in Figure 6 and Figure 7.

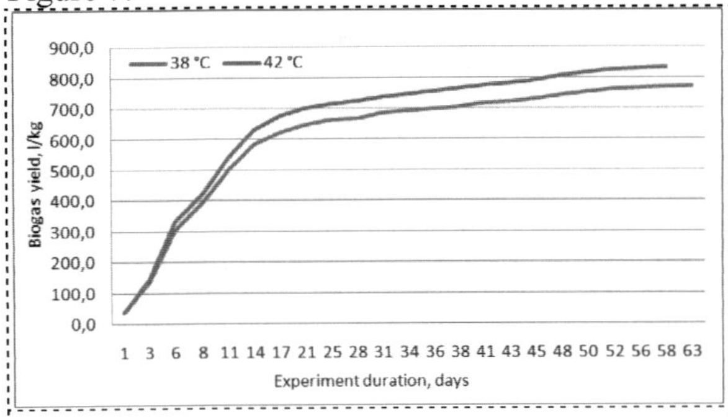

Figure 6. Graph of the dependence of biogas yield on the duration of fermentation of bird blood.

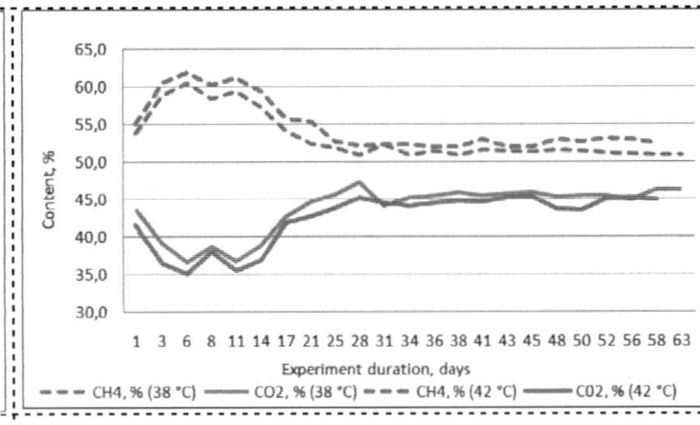

Figure 7. Graph of the dependence of the biogas composition on the duration of fermentation of bird blood.

Figure 6 shows that the maximum daily readings of biogas yield are observed from 1 to 21 days of the experiment, which is 84% of the total biogas yield. At the same time, the total biogas yield obtained at a temperature of 42 ° C was 833.4 l / kg, which is 7.94% more than at a temperature of 38 ° C (772.1 l / kg).

Analysis of the chemical composition of biogas shows (Figure 7) that the methane content readings have stable values from 1 to 21 days, then a drop in the methane content is observed. At the same time, the maximum value of methane was recorded at temperatures of 38 ° C and 42 ° C on the 6th day, and is 60.6% and 62.0%, respectively. The average methane content at 42 ° C was 54.8%, and at 38 ° C - 53.3%. The absolute difference in methane yield at different temperature conditions was 1.5%. Thus, it can be concluded that the optimal fermentation period for bird blood is 21 days.

3.4. Pulp

The results of the study of the specific yield and composition of biogas, depending on the temperature and duration of the experiment of anaerobic fermentation of pulp are presented in Figure 8 and Figure 9.

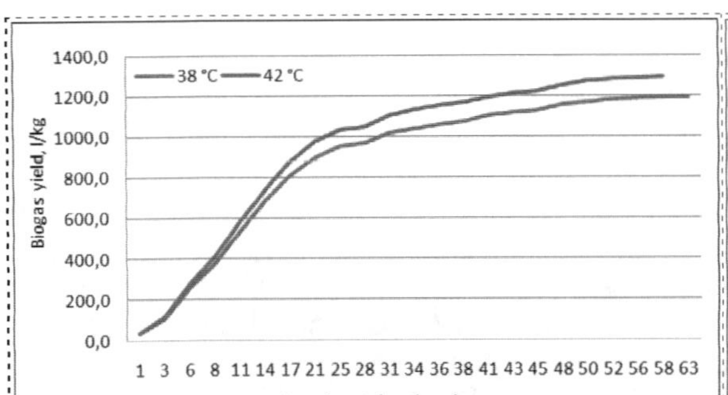

Figure 8. Graph of the dependence of biogas yield on the duration of pulp fermentation.

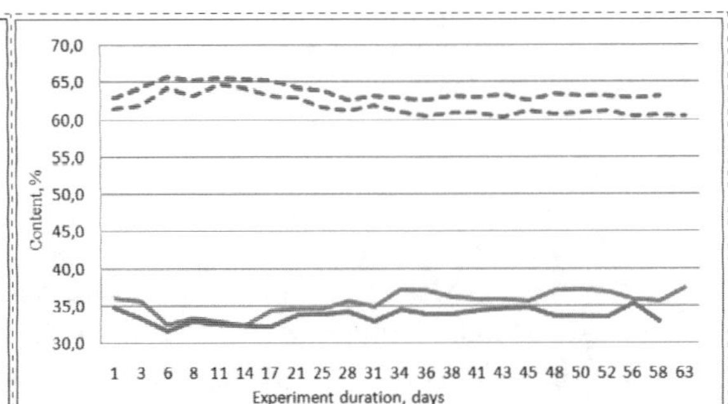

Figure 9. Graph of the dependence of the biogas composition on the duration of pulp fermentation.

Figure 8 shows that the maximum daily readings of biogas yield are observed from 1 to 25 days of the experiment, which is 80% of the total biogas yield. At the same time, the total biogas yield obtained at a temperature of 42 ° C was 1292.2 l / kg, which is 8.21% more than at a temperature of 38 ° C (1194.1 l / kg).

The analysis of the chemical composition of biogas shows (Figure 9) that the methane content is stable from 1 to 25 days, then a drop in the methane content is observed. At the same time, the maximum methane value was recorded at temperatures of 38 ° C and 42 ° C on the 6th day, and is 64.7% and 65.7%, respectively. The average methane content at 42 ° C was 63.7%, and at 38 ° C - 61.7%. The absolute difference in methane yield at different temperature conditions was 2.0%.
Thus, it can be concluded that the optimal duration of pulp fermentation is 25 days.

4. Summary

Experimental studies of the biogas production process were carried out on an experimental installation with a digester volume of 150 liters. The nature of the influence of the process temperature on the specific yield and chemical composition of biogas during anaerobic fermentation of wastes from meat processing enterprises in a mesophilic temperature regime has been established. Specific biogas yield at a fermentation temperature of 42 ° C is 7 ... 8% higher than the biogas yield obtained at a temperature of 38 ° C. It should be noted that the process of biogas evolution at 42 ° C stops 5 days earlier.

Graphical dependences of the specific yield of biogas and methane content on the duration and temperature of the fermentation process were obtained. It was found that the highest value of the specific yield of biogas is observed during the fermentation of technical fat - 1509.8 l / kg, and the lowest from bird blood - 833.4 l / kg. At the same time, the highest methane content is observed during fermentation of technical fat - 66%, and the smallest from bird blood - 53.5%.

The values of the optimal duration of fermentation of the studied types of substrate have been determined. The shortest duration is characterized by bird blood - 21 days, and the greatest duration is technical fat - 40 days. Duration of fermentation of pulp and intestines is 25 days.

The results obtained on the biogas yield and methane content showed that the wastes of meat processing enterprises are good sources of biogas with a high methane content. These data allow us to conclude that it is advisable to widely use waste from meat processing enterprises to obtain an alternative source of energy - biogas.

5. References

[1] Suslov D and Temnikov D 2020 [The optimum location of the biomethane plant and the station for feeding biomethane to the gas supply systems]. *Vestnik Volgogradskogo gosudarstvennogo arhitekturno-stroitel'nogo universiteta. Seriya: Stroitel'stvo i arhitektura [Bulletin of Volgograd State University of Architecture and Civil Engineering. Series: Civil Engineering and Architecture]* **1** 196-203

[2] Scarlat N, Dallemand J-F and Fahl F 2018 Biogas: Developments and perspectives in Europe *Renewable Energy* **129(A)** 457-72

[3] Linke B 2015 *Country Report Germany IEA Bioenergy* **T 37** (Berlin, Germany)

[4] Lambert M 2017 Biogas: a significant contribution to decarbonising gas markets? Oxford Energy Insight 15

[5] Karim K., Klasson K, Hoffmann R, Drescher S, DePaoli D and Al-Dahhan M 2005 Anaerobic digestion of animal waste: Effect of mixing *Bioresource Technology* **96(14)** 1607–1612

[6] Torrijos M 2016 State of Development of Biogas Production in Europe. *Procedia Environmental Sciences* **35** 881-889

[7] Eder B and Schulz H 2007 Biogas-Praxis: Grundlagen, Planung, Anlagenbau, Wirtschaftlichkeit, Beispiele [Biogas practice: basics, planning, plant construction, economy, examples] (Staufen im Breisgau: Ökobuch)

[8] Paterson M and Kuhn W 2010 *Guide to Biogas - From production to use* (Gülzow: FNR)

IOP Conf. Series: Materials Science and Engineering **945** (2020) 012029 doi:10.1088/1757-899X/945/1/012029

[9] Hamawand I 2015 Anaerobic digestion process and bio-energy in meat industry: A review and a potential *Renewable and Sustainable Energy Reviews*, 37-51

[10] Marcos, Al-Kassir A, Mohamad A, Cuadros F and López-Rodríguez F 2010 Combustible gas production (methane) and biodegradation of solid and liquid mixtures of meat industry wastes *Applied Energy* **87** (**5**) 1729-1735

[11] Ware A and Power N 2016 Biogas from cattle slaughterhouse waste: Energy recovery towards an energy self-sufficient industry in Ireland, *Renewable Energy* **97** 541-549

[12] Buendía I, Fernández F, Villaseñor J and Rodríguez L 2009 Feasibility of anaerobic co-digestion as a treatment option of meat industry wastes *Bioresource Technology* **100** (**6**) 1903-1909

[13] Agabo-García C, Solera R and Pérez M 2020 First approaches to valorizate fat, oil and grease (FOG) as anaerobic co-substrate with slaughterhouse wastewater: Biomethane potential, settling capacity and microbial dynamics *Chemosphere* **259** 127474

[14] Salehiyoun A, Di Maria F, Sharifi M, Norouzi O, Zilouei H and Aghbashlo M 2020 Anaerobic co-digestion of sewage sludge and slaughterhouse waste in existing wastewater digesters, *Renewable Energy* **145** 2503-2509

[15] Bayr S, Rantanen M, Kaparaju P and Rintala J 2012 Mesophilic and thermophilic anaerobic co-digestion of rendering plant and slaughterhouse wastes *Bioresource Technology* **104** 28-36

[16] Hejnfelt A and Angelidaki I 2009 Anaerobic digestion of slaughterhouse by-products *Biomass and Bioenergy* **33** (**8**) 1046-1054

[17] Suslov D 2020 Investigation of the effect of mixing and biomass temperature in the anaerobic digester on the volume of biomethane *IOP Conf. Ser.: Mater. Sci. Eng.* **791** 012059

Acknowledgments

The study was carried out in the framework of the scholarship of the President of the Russian Federation SP-180.2019.1.

IOP Conf. Series: Materials Science and Engineering **945** (2020) 012030 doi:10.1088/1757-899X/945/1/012030

Lightweight materials based on wood waste

A S Yespayeva[1*], Z N Altayeva[1], B K Sarsenbayev[2], G O Karshyga[3] and G R Sauganova[2]

[1] T. Basenov Institute of Architecture and Construction, Satbayev University, 22a Satpayev Str., Almaty, 050013, Kazakhstan
[2] M. Auezov South Kazakhstan State University, 5 Tauke Khan Avenue, Shymkent, 160012, Kazakhstan,
[3]Korkyt Ata Kyzylorda University, 29a Ayteke bi Str., Kyzylorda, 12001, Kazakhstan

E-mail: stroitelstvo_ukgu@mail.ru

Abstract. This paper discusses ways of recycling woodworking waste into light aggregate, optimal compositions of modified binders, as well as the mechanism of cement hydration in the presence of dispersion polymer powders. The methods of obtaining light aggregate based on woodworking waste by granulating in a disc granulator, which is a plate with a tilt angle of 45^0 and a number of turnovers of 17-19 rpm, are considered. Methods for obtaining granulated aggregate using sawdust of various fractions by the roasting free method are presented, optimal granulation modes are determined, and the main characteristics of the resulting aggregate and concrete based on them are studied. The optimal composition of granulated aggregate on modified cement provides 40 ... 60 % of wood processing waste, 60 ... 40 % of cement, 25 ... 32 % of water and granulation time - 7 ... 9 minutes. With this composition, the bulk density of the resulting aggregate corresponds to 850...750 kg/m^3, the maximum strength of the granules when compressed in the cylinder is 3.5...4.1 MPa. The optimal content of the polymer additive in the binder is in the range of: mowilith - 0.4...0.7 % and tylose - 0.21...0.13 %. Granulated aggregate on modified aggregate has a frost resistance in the range of F35... F50. The choice of granulated aggregate on modified cement showed the effectiveness of the aggregate for light concretes. The strength characteristics of modified concrete of 25-75 brands were studied. The mechanical properties of modified concrete depending on the average density were studied. Concrete with an average density of 1000 kg/m^3 has M25 brand, at 1200 kg/m^3 - M35, at 1300 kg/m^3 - M50, at 1500 kg/m^3 - M75.

1. Introduction

The problem of low waste production of building materials in many aspects coincides with the problem of rational use of mineral resources [1-5].

Among the main directions for implementing the principles of waste-free technology in the production of building materials is the processing of waste from various industries in order to save natural resources.

The technologies for producing light aggregates, mainly by firing are known. Due to the increase in the cost of energy resources, there is an urgent need to develop a technology for obtaining light aggregates using the roasting free method [6-9].

IOP Conf. Series: Materials Science and Engineering **945** (2020) 012030 doi:10.1088/1757-899X/945/1/012030

A simple technology for obtaining filler – roasting free one is known [10-12]. Production of aggregate using this technology is based on granulation of a mixture consisting of ash and its activators, which can be used as lime, cement, cement dust, gypsum. Granule hardening occurs in natural conditions and during hydrothermal treatment, which reduces fuel and electricity consumption, and does not require large investments in comparison with traditional technology.

The roasting free light aggregate based on ash, lime and slag-Portland cement was obtained [13-15]. Pulverized ash mixed with clinker binder or lime is crushed on a vibrating mill and granulated with water on a disc granulator, and then the resulting granules are steamed at 85^0C in the mode 4+10+2H. The strength of the granules is 50-70 kgf/cm^2 with a bulk density of 700-730 kg/m^3, water absorption in the range of 30-34%.

Kazakhstan has accumulated a huge amount of various wastes, including woodworking waste, which is increasing every year. However, the use of wood waste in the production of building materials is very low, which requires the development of cost-effective technological methods for their processing.

However, the use of woodworking waste to produce light concrete and its use in the construction industry is not widespread.

Recycling of woodworking waste for the production of construction materials contributes to the creation of low- and waste-free technologies, reducing the cost of production, reducing the area of land for dumps [16-18].

Thus, the solution of the problem of using woodworking waste for the production of light materials, the analysis of the main physical and mechanical properties and the nature of the hardening process by developing new effective technologies for processing these wastes is relevant.

2. Materials and methods

2.1. Materials

As raw materials for obtaining light aggregates we used: Portland cement (table 1), dispersion polymer powders mowilith DM 2072 and tylose MV 5009 P2, wood waste – sawdust fraction of 1.25 mm or less, sand with a size modulus of 2.05.

Table 1. Physical and mechanical properties of cement.

Cement properties	Measure unit	Values
normal density	%	25.1
gripping time: beginning ending	hour-min	 2-45 3-15
specific surface	cm^2/g	3100
strength after steaming at: compression bending	 MPa MPa	 33.9 6.0
strength at 28 day age of hardening at: compression bending	 MPa MPa	 40.85 6.5
uniformity of volume change		resist
brand		M 400

Mineralogical composition of cement, mass. %: C_3S – 50; C_2S – 20…30; C_3A=8, C_4AF – 10…13, the rest is glass phase.

Mowilith DM 2072 is a water-redispersible synthetic polymer based on homo -, co - and triple polymers of vinyl acetate ($CH_3COOCH{=}CH_2$). The composition of mowilith – "know-how", has a high adhesive ability. Firm "Clariant" (Germany) produces mowilith of different brands.

IOP Conf. Series: Materials Science and Engineering **945** (2020) 012030 doi:10.1088/1757-899X/945/1/012030

MV 5009 P2 tylose is also produced by the firm "Clariant" (Germany). In the composition of tylose, presumably the main component is ethylcellulose $[C_6H_7O_2(OH)_{3-v}(OC_2H_5)_v]_n$. Physical and chemical studies (ICS) revealed the presence of absorption bands of C-H, O-H and C-O – bonds. But unlike the latter, in the IR spectrum of tylose, the frequency of the absorption band of deformation vibrations of CH_2 groups corresponds to the influence of a double bond ($=CH_2$), and the splitting of the absorption band of the methyl group is characteristic of the isopropyl group $(CH_3)_2$ CH [19]. Changes in the structure, as well as the "know – how" of the company characterize tylose as a water-retaining additive. Figure 1 shows the structure of the tylose.

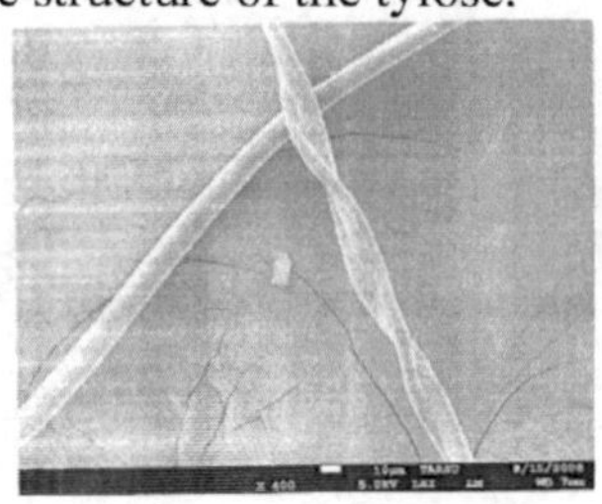

Figure 1. Structure of the dispersion polymer additive– tylose at the nanoscale.

The additive was introduced into the cement paste and concrete mix in a dry state.

To obtain a granular light aggregate, we used woodworking waste - sawdust of a fraction of 1.25 mm or less.

Sawdust of 1.25 mm fraction was previously crushed into smaller particles to obtain granulated aggregate. Waste fractions of 0.63 mm were also used for granulation.

Sand with a size modulus of 2.05 is used as filler.

2.2. Methods

The chemical and mineralogical composition of raw materials, as well as the influence of additives and aggregates on the processes of hydration and hardening of cement were studied using physical and chemical analysis methods.

The morphology, phase composition, contact zones and phase relationships, and the distribution of elements over the area were studied using analytical scanning electron microscopy (ASEM).

We used the DRON-3M x-ray unit, the MOM-1500 derivative graphic unit, the Specord m-80 infrared spectrophotometer, and the MIN-8 microscope.

Light roasting free granulated aggregate based on woodworking waste is obtained by granulation in a disc granulator, which is a plate with a tilt angle of 45^0 and a number of turnovers of 17-19 rpm (figure 2).

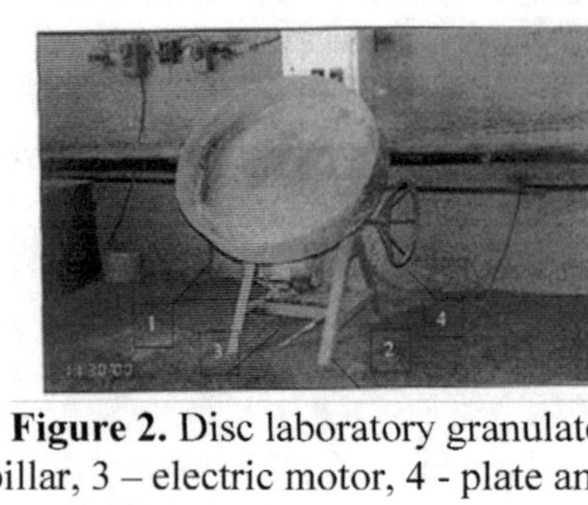

Figure 2. Disc laboratory granulator
1 – plate, 2- pillar, 3 – electric motor, 4 - plate angle setting stop.

Methods for obtaining granulated aggregate by granulation in a disc granulator.

First method. The crushed waste of the 1.25 fraction was pre-soaked in water to reduce extractives for 0.5-1 hour. After the pre-soaking time, the waste was dried at a temperature of 100^0C to a humidity of 8 %. Then in the mixer, the waste was thoroughly mixed with 2...3 parts of cement,

IOP Conf. Series: Materials Science and Engineering **945** (2020) 012030 doi:10.1088/1757-899X/945/1/012030

gradually adding water to the mixture humidity of 25 ... 35 %. The prepared mixture in the amount of 4 ... 5 kg was given to a disc laboratory granulator. During the pelletizing process, the resulting granules were powdered with the remains of cement.

Second method. To obtain granulated aggregate, waste fractions of 0.63 mm were used. The mixer was loaded with woodworking waste and 3 parts of cement, thoroughly mixed with the subsequent addition of water to the mixture humidity of 35 ... 40 %. The resulting mixture was given in an amount of 4 ... 5 kg to a disc laboratory granulator followed by powdering the pellets with the remains of cement.

Third method. In this way the mixer was charged with 50 % of crushed waste fraction of 1.25 mm and 50 % fractions 0.63 mm, after thorough mixing they add 2...3 parts of cement, 0.5 % polymer additive of the quantity of dry mix and re-produce thorough mixing. Water is added to the mixture with constant stirring until the humidity reaches 30 ... 35 %. The prepared raw material mixture in the amount of 4 ... 5 kg is given to a disc laboratory granulator to produce pellets, followed by powdering them with cement. Mowilith and tylose were used as polymer additives.

Granulation depends on the following factors: the humidity of the mixture, the tilt angle of the granulator, the degree of filling of the plate and the granulation time. Therefore, along with the change in the granulometry of the mixture, the granulation time varied within 7 ... 9 min.

The compositions of mixtures and their densities are shown in table 2.

Table 2. Content of components in the raw mix.

№ of composition	Amount of components, mass. %						
	cement	water	polymer additives		sawdust fractions, mm		
			mowilith	tylose	1.25	0.63	crushed
1*	65	29	-	-	55	-	-
2**	55	26	0.4	-	-	-	45
3*	50	30	-	-	-	55	-
4***	40	35	0,9	0.3	60	-	-
5***	45	30	0.75	0.27	-	20	35
6**	50	35	-	0.24	50	-	-
7**	55	28	0.6	-	25	20	-
8***	60	27	0.4	0.21	-	-	40
9***	50	32	0.5	0.18	25	-	25
10***	40	25	0.7	0.13	20	20	20

Note: * - according to I method; ** - according to II method; *** - according to III method.

3. Results

At containing a small amount of cement, the strength of the granules in all cases was low, so when selecting the optimal composition, the cement consumption varied within 40 ... 70 %. The amount of wood processing and water waste, respectively, varied between 60 ... 30 % and 35...25 %. The content of polymer additives for the third composition varied in quantity: mowilith - 1.5...3 % of the dry weight; tylose - 0.5...1.0 % of the dry weight.

When 50% of the cement was added to the mix, the raw granule density was 650 g/gran. With the introduction of 70% cement, the density of granules doubled. The optimal composition corresponds to the consumption of materials in the following amounts: 55 % of wood waste, 45 % of cement and 30 % of water. The presence of a polymer additive in the mixture increases the strength of the granules, which explains the increase in the strength of the granulated aggregate. Wood waste particles are wetted with an aqueous solution when wet, forming films. Water films under the influence of surface tension are connected, bringing together the particles of waste. Because of repeated impacts, the particles are compacted, and excess water appears on their surface, which attracts new particles and, at the same time, cement grains. When the crushed waste is constantly rolled in the granulator with cement grains, the granules increase in size and are compacted.

At 35% humidity of the mixture, the raw density of the granules was 450 g/cm^3, and at 30 % humidity it was 580 g/cm^3, which is quite acceptable. When 65...45% of cement is added to the mixture, the same pattern is observed. Increasing the amount of cement to a certain limit (70 %) improves the granularity of the mixture by increasing the dispersion of the mass and the interaction of cement with the components of crushed waste.

As the granulation time increases, the granules decrease in size and their bulk density increases. High strength of granules is obtained at a granulation time of 7 ... 9 min, and this time is accepted as optimal for all compositions. An increase in the time above the optimal one leads to the emergence of cracks.

Table 3 shows the dependence of the raw density of granules and the strength of the finished product on the granulation time. When granulating the cement-wood mixture for 3...5 minutes, the granules are obtained larger and less durable.

Table 3. Effect of the granulation time of the mixture on the properties of gravel.

Granulation time, min	Raw density, g/cm^3	Bulk density, kg/m^3	Tensile strength at compressing in the cylinder, MPa
3	500	800	1.1
5	450	630	1.5
7	700	880	2.9
9	650	800	2.5
12	750	810	2.1

Compliance with the optimal technological parameters allowed obtaining granulated aggregate on modified cement of the following grain composition (table 4). The test results showed that the aggregate is mainly dominated by granules of the 5-10mm fraction.

Table 4. Grain composition of granulated aggregate.

№ compound	Residues on the sieves, mass. %, fraction, mm								Pass through a sieve of 0.14,%	
	10		5.0		2.5		1.25 and less			
	part	full	part	full	part	full	part	full	part	full
8	19.81	19.81	47.79	67.9	26.81	94.41	4.19	98.6	1.4	100
9	17.23	17.23	45.82	63.05	28.63	91.68	7.42	99.1	0.9	100
10	18.51	18.51	46.19	64.70	28.01	92.71	6.09	98.8	1.2	100

Table 5 shows the main physical and technical characteristics of granulated aggregate compositions 8 ... 10. Granulated aggregate of composition 7 containing mowilith in an amount of 2.25 %, based on crushed waste from fractions 1.25 and 0.63 mm, has a higher bulk density of 1000 kg/m^3 than composition 1 obtained on a cement binder. In this regard, the brand of granulated filler on modified cement (900...1100), respectively, higher than that of the cement (500,600), the strength of the granules from the first ranges from 4.5...5.2 MPa, and the second two and a half times lower (1.7...2.0 MPa), the coefficient of the first resistance is 1.2...1.7. Low water resistance of the granulated filler compositions 8…10 due to the content in the binder dispersion of the polymer, which forms a kind of film on the surface of the grains of cement and waste wood.

Table 5. Physical and technical properties of the filler.

Main indicators	Granulated fraction filler, mm, compositions (№)					
	5		6		7	
	5-10	10-20	5-10	10-20	5-10	10-20
Bulk density, kg/m^3	950	900	1000	950	1100	1000
Compressive strength of granules (MPa) in the state of:						
dry	4.8	4,5	5.1	4.6	5.2	4.9
saturated with water	3.9	3,6	4.5	3.9	4.3	4.0
Water resistance coefficient	1.2	1,3	1.6	1.7	1.4	1.5
Water absorption, %	18.0	19,5	13.7	14.9	16.1	16.0
Intergranular voidness, %	33.3	41,2	36.2	39.8	37.4	38.2
Frost resistance, cycles	40	35	45	50	50	45
Thermal conductivity, W/m*deg	0.11	0.10	0.09	0.09	0.091	0.092

Granulated aggregate on modified aggregate has frost resistance in the range of F35... F50. All this generally determines the choice of granulated aggregate on modified cement as the most effective aggregate for light concrete.

The strength characteristics of modified concrete 25-75 brands were studied.

The mechanical properties of modified concrete depend mainly on the average density. Concrete with an average density of 1000 kg/m^3 has M25 brand, at 1200 kg/m^3 - M35, at 1300 kg/m^3 - M50, at 1500 kg/m^3 - M75. The strength characteristics of the modified concrete meet the requirements of the standards for light concrete. The prismatic strength and bending tensile strength for the above brands fluctuates between 2.7 – 7.0 MPa and 0.79 - 1.8 MPa, respectively, and grows with increasing average density of modified concrete (table 6). However, the strength coefficients decrease with increasing compressive strength of concrete.

Table 6. Physical and mechanical properties of concrete containing mowilith.

Brand	Average density, kg/m^3	Tensile strength, MPa, at:		
		compression, R_{com}	prismatic, R_{prism}	bending, R_{bend}
M25	1000	3.5	2.7	0.79
M35	1200	5.0	3.8	1.15
M50	1300	7.5	5.5	1.65
M75	1500	10.0	7.0	1.80

Table 7 shows experimental data on determining the degree of hydration of Portland cement of the Karaganda cement plant for a test of normal density with a polymer additive of tylose in comparison with non-additive cement. The samples were solidified in water at a temperature of 20^{0}C.

Table 7. The degree of hydration of the modified cement.

№ of composition	Type and quantity of additives, tylose, %	W/C	Chemically bound water, %, at the age, days			The degree of hydration of cement, %, at the age, days		
			7	28	90	7	28	90
1	-	0.26	12.8	15.5	18.7	0.53	0.61	0.72
2	0.5	0.2	11.0	12.5	15.3	0.45	0.51	0.62

It should be noted that there is a slight decrease in the amount of chemically bound water and the degree of hydration of cement. The discrete distribution of dispersions of the polymer additive - tylose is additively realized with the active centers of the clinker phases of cement grains.

4. Discussion

The high physical and technical properties of a granular aggregate based on woodworking waste obtained on a binder containing dispersed polymer powders can be explained in such a way that the polymer phase in the cement stone forms an organic structure. In this case, the hydrate phases form a crystallization-coagulation structure, which is strengthened in defective places, such as pores, cracks, and the polymer component, which causes the formation of a more durable and elastic structure. With a higher ratio of polymer to cement (P/C), a continuous polymer mesh is formed.

The hydration kinetics of clinker minerals changes in the presence of polymer additives. The process of C_3S hydration in aqueous solutions of most organic additives slows down somehow[19]. In this case, the phase composition of cement stone is represented by fibrous hydrosilicates and calcium hydroxide. Organic additives have almost no effect on the hydration of C_2S, as this mineral is characterized by low activity in the initial period, and the effect of the polymer component is reduced in the later period.

Polymer additives have the greatest influence on the interaction of C_3A with water, and not only the speed of the process changes, but also the phase composition of the cement stone. As a result of C_3A and C_3S hydration, part of the water binds to hydroaluminates and calcium hydrosilicates, and partial water evaporation occurs. Increasing the concentration of the solution leads to coagulation of the polymer phase and the formation of membranes between the hydrated and initial particles of cement and aggregate. In the subsequent period, two phases (inorganic and polymer) germinate: the polymer component fills the pore space and the resulting defective places, compacting and connecting additional elements of the cement stone structure.

The use of dispersion polymer powders increases the adhesion of cement stone to other materials. Water-soluble polymer additives, introduced in an amount of up to 3%, plasticize the concrete mixture, increase its uniformity and workability, and reduce water separation.

In some polymer systems, the coagulation is preceded by the formation of a framework of cement stone. In this case, the hardening process slows down, but there is a greater mutual penetration of the two phases, which provides high performance properties of the composition. The polymer phase in cement stone has a microheterogenic structure, as it contains inorganic inclusions of hydrated phases, non-hydrated cement particles or thin fractions of aggregates. Reinforcement of the polymer component improves the properties of the component itself and the system as a whole.

Cement concrete of the third millennium is a modified concrete. Features in the mechanism of hydration and hardening of cement are enhanced by the action of additives in concrete - a water-reducing effect, i.e. the effect of reducing the water demand of the concrete mixture. Modifiers of cement concretes – natural or artificial chemicals introduced into the composition of concrete during manufacture, significantly improve the technological properties of the concrete mix, physical and technical indicators of concrete, reducing its cost and increasing its durability [20, 21].

The effect of the modifying additive in the cement system is manifested by the following features: the ability to absorb at the interface of phases and participate in the formation of spatial coagulation structures, both in the volume and in the surface layers [22].

Exchange processes during cement hydration occur by the donor-acceptor mechanism. The strength and crack resistance of concrete increases significantly. The effect of the short-range selective action of the modifier of the cement matrix structure with the maximum technical and technological action is observed. This process is associated with the dissolution of the tylose additives with the mixing water, with the formation of a liquid phase with a low surface tension and the subsequent manifestation of Deryagin- Rehbinder disjoining effect at defective surface sites of cement grains. The adsorption-energy "capture" of the liquid phase should be attributed to the mechanism of cement activation during hydration [23- 25].

5. Summary

The optimal composition of granulated aggregate on modified cement provides 40 ... 60 % of wood processing waste, 60 ... 40 % of cement, 25 ... 32 % of water and granulation time - 7 ... 9 minutes. With this composition, the bulk density of the resulting aggregate corresponds to 850...750 kg/m^3, the maximum strength of the granules when compressed in the cylinder is 3.5...4.1 MPa. The optimal content of the polymer additive in the binder is within the range of: mowilith - 0.4...0.7 % and tylose - 0.21...0.13 %.

The mechanism for forming a modified active layer on light aggregate granules from wood waste is as follows:

- the result of the mechanism provides the creation of a protective hydrophobic polymer film based on a complex additive of mowilith and tylose, each of which has a double effect, regulating the water-holding capacity and creating the surface section of phases;
- the formation of cement hydration products occurs on film surfaces in the form of nanoscale defect-free nanocrystals of hydrosilicates that provide the strength of the surface film;
- it was found that the water-holding capacity of the dispersion polymer additive tylose is regulated by the microstructure of the cellulose fiber that is part of it.

When forming the nanostructure of modified concrete, the effective role of the polymer additive should be noted. In order to distinguish elements of unity in the mechanism of active formation of such diverse structural initial phases of cement stone and concrete, a sharp change in the rheological properties of cement paste and concrete mix in our experiments was achieved by a significant change in the amount of distribution of dispersed particles in the volume of the dispersion medium.

6. References

[1] Shevchuk M O 2017 Problems and achievements of processing of vegetable raw materials *Bulletin of Polotsk State University. Series B. Industry. Applied science. Chemical technology. Labour protection* **11** 95–102

[2] Grinshpan D, Savitskaya T, Tsygankova N, Makarevich S, Kimlenka I, Ivashkevich O 2017 Good real world example of wood-based sustaina-ble chemistry *Sustainable Chemistry and Pharmacy* **5** 1 – 13 doi.org/10.1016/j.scp.2016.11.001

[3] Dubovoy E V, Sviridov E B, Shcherbak N V, Dubovoy V K 2017 Energy-saving, environmentally safe technology of air cooling by evaporative type devices Saint-Petersburg: Publishing house of Polytechnic University 286

[4] Litvyak V V, Ospankulova G H, Shaymerdenova D A, Yurkshtovich N K, Butrim S M, Roslyakov Yu F 2016 Atlas: morphology of polysaccharides Astana: LLP "EDIGE" 335

[5] Medvedeva G A, Lifantyeva A F 2020 The research of multilayer outer fencing including materials using ash and slag waste of thermal power plants *Construction Materials and Products* **3 (2)** 29 – 35 DOI: 10.34031/2618-7183-2020-3-2-29-35

[6] Barkhatov V I, Dobrovolsky I P, Kapkaev Yu Sh 2015 Rational use of natural resources of the Chelyabinsk region: monograph.I. Barkhatov, Chelyabinsk: Publishing House of Chelyabinsk State University 624

[7] Lesovik V S, Absimetov M V, Elistratkin M Yu, Pospelova M A, Shatalova S V 2019 For the study of peculiarities of structure formation of composite binders for non-autoclaved

aerated concrete *Construction Materials and Products* 2 (3) 41 – 47

[8] Klyuev S V, Khezhev T A, Pukharenko Yu V, Klyuev A V 2018 To the question of fiber reinforcement of concrete *Materials Science Forum* **945** 25 – 29

[9] Orentlicher L P, Soboleva G N 2000 Roasting free composite porous aggregate from wet asbestos-cement waste and light concrete based on it *Construction materials* **7** 18–20

[10] Bondarenko N I, Bondarenko D O, Kochurin D V, Bragina L L, Varfolomeeva S V 2018 Technology of plasma metallization of the wood and fibrous board *Construction Materials and Products* **1(3)** 4 – 10

[11] Bessmertny V S, Kochurin D V, Bondarenko D O, Bragina L L, Yalovenko T A 2018 Vitreous protective and decorative coverings on wood particle board *Construction Materials and Products* 1 (4) 4 – 12

[12] Klyuev S V, Khezhev T A, Pukharenko Yu V, Klyuev A V 2018 Experimental study of fiber-reinforced concrete structures *Materials Science Forum* **945** 115 – 119

[13] Tolstoy A D 2020 Fine-grained high-strength concrete *Construction Materials and Products* **3 (1)** 39 – 43 DOI: 10.34031/2618-7183-2020-3-1-39-43

[14] Ivanova V N 2017Processing of high-yield fibrous semi-finished products *IVUZ Forest journal* **6** 15-17

[15] Ivanova V N, Makhotina L G 2018 Prospects for the use of commercial types of cellulose for the production of products with high added value *Chemical fibers* **4** 25-26

[16] Khviyuzov S S, Bogolitsyn K G, Gusakova M A , Zubov I N 2015 Estimation of lignin content in wood by IR spectroscopy *Fundamental research* **9** 87-90

[17] Shapovalova I, Vurasko A, Petrov L, Kraus E, Leicht H, Heilig M, Stoyanov O 2018 Hybrid composites based on technical cellulose from rice husk *Journal of Applied Polymer Science* **135 (5)** 45796

[18] Smolin A S, Komarov V I 2017 Role of lignin in the technology of materials for corrugated cardboard *Problems of mechanics of pulp and paper materials: Proceedings of the IV International scientific and technical conference. - Arkhangelsk: Northern (Arctic) Federal University named after M.V. Lomonosov* 29-34

[19] Klyuev S V, Klyuev A V, Shorstova E S 2019 The micro silicon additive effects on the fine-grassed concrete properties for 3-d additive technologies *Materials Science Forum* **974** 131–135 doi:10.4028/www.scientific.net/MSF.974.131

[20] Aymenov A Zh, Khudyakova T M, Sarsenbayev B K 2018 Studying the mineral additives effect on a composition and properties of a composite binding agent *Oriental journal of chemistry* **34(4)** 1945-1955

[21] Taimassov B T, Sarsenbayev B K, Khudyakova T M, Kolesnikov A S, Zhanikulov NN Development and testing of low-energy-intensive technology of portlandcement»//«Eurasian chemico-technological journal» -2017y. Volume 19 –Number 4. Pg.347-355.

[22] Aymenov A.Zh., Sarsenbayev N.B., Khudyakova T.M., Sarsenbayev B.K., Kopzhassarov B.T. 2016 Effect of additive of polymetallic ores' tailings on properties of composite cements *Eurasian chemico-technological journal* 153-160

[23] Klyuev S V, Klyuev A V, Shorstova E S 2019 Fiber concrete for 3-d additive technologies *Materials Science Forum* **974** 367–372 DOI: 10.4028/www.scientific.net/MSF.974.367

[24] Klyuev S V, Khezhev T A, Pukharenko Yu V, Klyuev A V 2018 Fiber concrete for industrial and civil construction *Materials Science Forum* **945** 120 – 124

[25] Elistratkin M Yu, Minakov S V, Shatalova S V 2019 Composite binding mineral additive influence on the plasticizer efficiency *Construction Materials and Products* **2 (2)** 10 – 16

IOP Conf. Series: Materials Science and Engineering **945** (2020) 012031 doi:10.1088/1757-899X/945/1/012031

Influence of the composition of cement systems on the activity of modifying additives

N N Onoprienko[1*] and O N Salnickowa[1]

[1]Department of city cadastre and engineering survey, Belgorod State Technological University named after V.G. Shukhov, Kostyukov St., 46, Belgorod, 308012, Russia

E-mail: dstt_80@mail.ru

Abstract. Modifiers of cement systems with different compositions of functional groups of polymers were studied: carboxymethylcellulose, methylcellulose, polyvinyl alcohol, polyacrylamide. The results of studies of the influence of active ions on the viscosity of aqueous solutions of water-soluble polymers using viscometry methods are presented. The influence of the type and composition of functional groups of polymer additives on the resistance to calcium, sodium and potassium ions is analyzed. It is shown that in the presence of calcium, sodium or potassium ions in the liquid phase of cement solutions, it is possible to reduce the effectiveness of polymers as stabilizers of binding systems. The most resistant to electrolytes are polymers with dominated hydroxyl groups. The composition of functional groups of water-soluble polymers and the possibility of their interaction with various ions contained in cement, aggregate, filler, in the closing water, in the composition of soils have a decisive influence on the compatibility of the polymer and the binder, the properties of cement-polymer compositions. The principles of regulating the rheological and technological properties of cement-polymer compositions, deformations of dispersed systems by selecting the rational composition of the dispersed mixture, mineral and chemical additives are given. The regularities of compatibility of components of cement systems, mineral and chemical additives in order to increase the adhesion and strength characteristics of materials based on them in the development of compositions of masonry solutions, injection and other compositions for chemical fixing of soils, in individual housing construction, in order to carry out repair and restoration work to strengthen soil bases and stone structures are presented.

1. Introduction

Conducting high-tech research to establish patterns of compatibility of components of cement systems in order to regulate their rheological and technological parameters in the direction of improving the adhesion and strength characteristics of materials based on them is an urgent task in various sectors of construction production [1-10].

Compositions based on cement systems are widely used in construction works to fix chemically the soils. To improve the technological characteristics of mortar mixtures, physical and mechanical properties of solutions and structures based on them, various additives are added to cement dispersions: water-soluble polymers, electrolytes, which increase the level and quality of construction work, reduce shrinkage deformations and stresses, and increase the load-bearing capacity of soils in the conditions of existing building [3, 4, 11-13]. The search for effective chemical additives and rational compositions of dispersed systems and products based on them is constantly conducted by national and foreign researchers and scientists [14-16].

Polymer modifying additives are added to binding systems that are used in thin-layer technologies of masonry and finishing works, in modern finishing compositions, recipes for chemical fixing of soils, dry building mixes, compositions for construction printing, self-leveling floors, etc. [3-6, 8, 13, 17, 18]. In this regard, research is needed to ensure compatibility and synergy of components of such building mixes to ensure their high technological and construction-technical characteristics. The issue is relevant both from the point of view of commercial interest, and in order to ensure the safety of structures under construction, including the exclusion of various cases of accidents and destruction of both natural and technogenic nature [3, 7]. To ensure the normal operation of objects, it is important to take into account all possible deformations of the ground bases, which can cause unacceptable deformations in the structures of buildings and structures [3, 7]. A number of works highlighted the special role of engineering and geological surveys on the construction site, including soil consolidation. Chemical methods of fixing soils by existing injection methods, has well-defined limits of application and in some cases are practically impractical. In this regard, further research is being conducted on ways to fix the soil, for example, the drilling method, with the creation of cement.

Issues of preserving the activity of modifying additives in the composition of multicomponent dispersions play an important role in such processes and phenomena as adhesion, shrinkage, dilatancy [3-6, 9, 12]. In [3] it is shown that calcium ions are the most active components of the cement system. However, in addition to Ca^{2+}, the composition of cement systems may contain other reactive cations or anions, currently the effect of which on the properties of cement-polymer compositions is insufficiently studied.

The aim of this work is to study the resistance of modifying additives of water-soluble polymers to the activity of certain chemical elements contained in cement systems. This will allow developing ways to control the properties of dispersed systems by selecting a rational composition of the binder, mineral components, polymer additives and other modifiers of cement systems. The obtained regularities will allow making scientific and technical choice of components of cement-polymer systems in order to strengthen the soil bases, strength of stone structures and their adhesive properties, regulate the rheological and technological properties of compositions, deformations of dispersed systems by selecting a rational composition of the dispersed mixture, mineral and chemical additives.

2. Materials and methods

2.1. Materials

For the research, modifying additives were selected - water-soluble polymers of national production: methylcellulose (MC), carboxymethylcellulose (CMC), polyacrylamide (PAA), polyvinyl alcohol (PVA). Drinking water was used, which meets the requirements of GOST 23732-2011 "Water for concrete and mortar. Technical conditions". Cement CEM I 42.5 N according to GOST 31108-2003 of JSC "Belgorod cement".

The studied polymer additives had different composition of functional groups: methylcellulose (MC) - methoxyl group ($-OCH_3$); carboxymethylcellulose (CMC) - carboxylate ($-COO^-$); polyacrylamide (PAA) - amide group (- $CONH_2$); polyvinyl alcohol (PVA) - hydroxyl group (-OH).

Chlorides were used to assess the resistance of water-soluble polymers to various active chemical elements. Water solutions of $CaCl_2$, NaCl, and KCl were used as a source of calcium, sodium, and potassium ions. The dosage of chlorides was 10, 20 and 30% of the weight of the polymer additive in terms of dry matter. As protective reagents for carboxylate additives, soda ash (Na_2CO_3) of the ChDA brand was used.

IOP Conf. Series: Materials Science and Engineering **945** (2020) 012031 doi:10.1088/1757-899X/945/1/012031

2.2. Methods

The viscosity of aqueous solutions of polymers of various concentrations was measured using capillary viscometers VPJ-2 (GOST 10028-81) according to the instructions specified in the passport of the device. The capillary diameters of viscometers were 1.31 and 0.56 mm.

The resistance of aqueous polymer solutions to the action of Ca^{2+}, Na^{+} and K^{+} ions was estimated by the relative viscosity ($\eta_{rel} = \eta_1/\eta_2$). This value is equal to the ratio of the viscosity of the polymer solution when adding an electrolyte (η_1, MPa·s), to the viscosity of an aqueous polymer solution of the same concentration (η_2, MPa·s).

It is proposed to predict the behavior of polymer additives with different composition of functional groups in the liquid phase of binders by the relative viscosity of the aqueous polymer solution.

3. Results

Some water-soluble salts, including sodium, potassium, calcium, and magnesium chlorides, are aggressive to building materials. Mg^{2+}, Ca^{2+}, Na^{2+}, K^{+} cations are small in size and can be contained as part of the cement system, and enter it from the external environment. For example, K^{+} and Na^{+} cations are present in cement as sulfates in the silicate and aluminate phases. Active elements can enter cement systems from mineral and chemical additives or aggregates [18-22]. The technical closing water may also contain the above-mentioned ions in various dosages.

The results of studies of the influence of calcium, sodium and potassium cations on the viscosity of water solutions of water-soluble polymers are presented below (table 1).

The composition of their functional groups plays an important role in the resistance of water-soluble polymers to the components of cement or other matrix. In this regard, it is necessary to pay special attention to the study of the properties of additives-modifiers, taking into account the chemical nature of the polymer.

Table 1. Dependence of relative viscosity of water solutions of polymers on the content of potassium, sodium, calcium and sodium carbonate chlorides.

Relative viscosity (η_{rel}) of aqueous polymer solutions	Electrolyte content, % by weight of the polymer additive			
	0	10	20	30
		$CaCl_2$		
Carboxymethylcellulose (CMC)	1.000	0.888	0.769	0.715
Methylcellulose (MC)	1.000	0.989	0.989	0.989
Polyvinyl alcohol (PVA)	1.000	1.080	1.080	1.000
Polyacrylamide (PAA)	1.000	0.502	0.443	0.438
		NaCl		
Carboxymethylcellulose (CMC)	1.000	1.000	0.962	0.938
Methylcellulose (MC)	1.000	1.018	0.978	0.945
Polyvinyl alcohol (PVA)	1.000	1.026	1.022	1.011
Polyacrylamide (PAA)	1.000	0.538	0.453	0.369
		KCl		
Carboxymethylcellulose (CMC)	1.000	1.050	0.969	0.945
Methylcellulose (MC)	1.000	0.957	0.929	0.961
Polyvinyl alcohol (PVA)	1.000	1.000	1.000	1.000
Polyacrylamide (PAA)	1.000	0.473	0.394	0.385
		Na_2CO_3 + 20%$CaCl_2$		
Carboxymethylcellulose (CMC)	1.000	1.389	1.146	0.911

IOP Conf. Series: Materials Science and Engineering **945** (2020) 012031 doi:10.1088/1757-899X/945/1/012031

Table 1 shows that the least resistant to the action of calcium ions are water-soluble polymers carboxymethylcellulose (CMC) and polyacrylamide (PAA). Methylcellulose (MC) and polyvinyl alcohol (PVA) practically do not lose their activity in this medium. The viscosity of almost all non-ionic cellulose esters under study increases with growing KCl concentration. The viscosity of aqueous solutions of carboxymethylcellulose (CMC), which is a non-ionic polymer, decreases when interacting with potassium ions largely than with sodium ions. Aqueous solutions of polyacrylamide (PAA) sharply lose their viscosity under the action of sodium and potassium cations. The most resistant to electrolytes are polyvinyl alcohol (PVA), which is dominated by hydroxyl groups.

Studies on the stabilization of polymers containing carboxyl group (COO^-) with additional protective additives are shown on the example of increasing the stability of carboxymethylcellulose CMC to calcium ions with the use of soda ash (Na_2CO_3) ChDA brand (table 1, 2). Data of viscometric studies (table 1) are consistent with the physical and mechanical properties of modified cement-sand mortar (table 2), confirming the stabilization effect of soda polymer additives.

Table 2. Physical and mechanical characteristics of solutions with additives of carboxymethylcellulose (CMC) and soda (Na_2CO_3).

Content of CMC, %	Content of Na_2CO_3, %	Water-cement ratio (W/C)	Coagulation of the polymer	Adhesion strength, MPa, at the age of 28 days
-	-	0.48	-	0.09
1	-	0.64	yes	0.12
1	1	0.66	no	0.26
1	2	0.64	no	0.33

Stabilization of solutions with the addition of 1% CMC with Na_2CO_3 soda (table 2) significantly increases the adhesion strength of cement-sand solutions (the ratio of cement: sand 1:2.5 by weight) with silicate bricks in comparison with solutions without the addition of a protective reagent.

4. Discussion

Thus, in the presence of calcium, sodium and potassium ions in the liquid phase of cement solutions, it is possible to reduce the effectiveness of polymers as stabilizers of binding systems.

All water-soluble polymers, the main functional group of which is carboxylate (-COO-), are deactivated in the cement medium mainly due to Ca^{2+} ions, while they practically cease to function, while non-ionic water-soluble polymers remain active in the cement medium.

Polymers containing carboxylate groups are poorly compatible with Portland cement, and also require the introduction of additional protective additives.

Solutions of methylcellulose and polyacrylamide are highly sensitive to monovalent potassium and sodium ions. However, studies showed that potassium was more aggressive than sodium.

Polyacrylamide (PAA) is insoluble to calcium, potassium and sodium ions, so it is not recommended to use it in cement-polymer compositions.

There is a special sensitivity of methylcellulose to sodium and potassium ions.

It is shown that the most resistant to electrolytes are low-viscosity polymers - polyvinyl alcohol, where hydroxyl groups dominate. PVA reacts weakly to sodium and potassium cations and is the most stable among the studied polymers.

According to experimental studies, additives of methylcellulose and polyvinyl alcohol are the most resistant to the action of aggressive ions among other additives of water-soluble polymers studied in the work, which determines the prospects for their use in construction solutions.

5. Summary

It is established that the composition of functional groups of water-soluble polymers and the possibility of their interaction with various ions contained in the cement, aggregate, filler, in the closing water, in the composition of soils have a decisive influence on the compatibility of the polymer and the binder, the properties of cement-polymer compositions.

The conducted research allows giving some principles for selecting rational compositions of cement-polymer dispersed systems, taking into account the established regularities of the influence of the composition of functional groups, the dosage of water-soluble polymers and electrolytes on the properties of cement-polymer dispersions. Thus, water-soluble polymers containing carboxylate groups are incompatible with Portland cement and require the introduction of additional protective reagents. It follows from the research that the properties of cement-polymer compositions with the addition of carboxylate polymers and protective reagents made on various binders may differ significantly, which should be taken into account when developing formulas of cement-polymer dispersions. Additives of non-ionic polymers are the most resistant to the action of aggressive ions and, therefore, are the most promising for use in construction solutions for various purposes.

When selecting the rational composition of a dispersed mixture, mineral and chemical additives, taking into account the laws of compatibility of components, it is possible to regulate in the right direction the rheological, technological, adhesive, physical and mechanical properties of cement-polymer compositions, deformations of dispersed systems.

The research results are expected to be used in the development of masonry mortar compositions, injection and other compositions for chemical soil fixing, in individual housing construction, in order to carry out repair and restoration work to strengthen the soil bases and stone structures.

6. References

[1] Elistratkin M Yu, Minakova A V, Jamil A N, Kukovitsky V V, Issa Jamal Issa Eleyan 2018 Composite binders for finishing compositions *Construction Materials and Products* **1(2)** 37-44

[2] Alfimova N I, Pirieva S Yu, Gudov D V, Shurakov I M, Korbut E E 2018 Optimization of receptural-technological parameters of manufacture of cellular concrete mixture *Construction Materials and Products* **1(2)** 30-36

[3] Onoprienko N N, Rakhimbaev Sh M 2016 Building solutions and dry mixes with additives of water-soluble polymers of domestic production: monograph *Belgorod: BSTU Publishing House* 155

[4] Onoprienko N N, Salnikova O N Improving the efficiency of mortars of domestic production 2019 *Bulletin of BSTU named after V.G. Shukhov* **8** 22-29

[5] Onoprienko N N, Salnikova O N 2019 Interconnection of Intrinsic Deformations and Adhesive Phenomena in Modified Disperse Systems *Materials Science Forum* **974** 440-445

[6] Onoprienko N N 2019 To the Question of the Cement-Polymer Composite Materials' Rheological Properties *Materials Science Forum* **974** 156-161

[7] Kalachuk T G, Kara K A 2018 General calculation principles of structurally-unstable soils deformation *Advances in Engineering Research* 306-310

[8] Elistratkin M Yu, Lesovik V S, Alfimova N I, Glagolev E S 2018 About developing of the building press technologies *Bulletin of BSTU named after V.G. Shukhov* **5** 11-19

[9] Strokova V V, Nelyubova V V, Rykunova M D 2019 Resistance of cement stone in solutions of sanitation facilities *Engineering and construction journal* **6 (90)** 72-84

[10] Bespalov V V, Yucher D, Salmanov I D, Kurbanov I N, Kupavykh S V 2018 Deformational compatibility of masonry with composite materials *Engineering and construction journal* **2 (78)** 136-150

[11] Anisimova S V, Shurygina Yu N, Pavlikova S M, Korshunov A E 2018 Polymer water dispersions in technologies of application of dry building mixes *Construction materials* **3**

80-85

[12] Baskakov P S, Strokova V V, Maltseva K P 2015 Effect of alkaline action on the properties of acrylic and styrene-acrylic dispersions for water-based paint materials *Construction materials* **12** 81-84

[13] Schekina A Yu, Aziati Komlava Mawuko, Cherkasov R A 2019 Research of colloid-chemical processes on the border of the phases section waste-floto reagent *Vector of Geosciences* **2(4)** 5-8.

[14] Suraneni P, Ramanathan S, Wang Y, Hajibabaee A, Weiss J 2019 New insights from reactivity testing of supplementary cementitious materials *Cement and Concrete Composites* **103** 331-338

[15] Pandey A, Derakhshandeh M, Kedzior S A, Pilapil B, Shomrat N, Segal-Peretz T, Bryant S L, Trifkovic M 2018 Role of interparticle interactions on microstructural and rheological properties of cellulose nanocrystal stabilized emulsions *Journal of Colloid and Interface Science* **532(12)** 808-818

[16] Klyuev S V, Khezhev T A, Pukharenko Yu V, Klyuev A V 2018 To the question of fiber reinforcement of concrete *Materials Science Forum* **945** 25 – 29

[17] Lebedev M S. Kozhukhova N I 2018 Rheological characteristics of bitumen mastic depending on composition and filler dispersity *Journal of Physics: Conference Series* **1045(1)** 012026

[18] Dmitrieva T V, Markova I Yu, Strokova V V, Bezrodnykh A A, Kutsyna N P 2020 Effeciency of stabilizers of various compositions for strengthening soils with mineral binders *Construction Materials and Products* **3(1)** 30-38

[19] Rakhimbaev Sh M, Tolypina N M, Khakhaleva E N 2016 Influence of fine sand aggregate on the effectiveness of diluent additives *Bulletin of the Siberian State Automobile and Road Academy* **3(49)** 74-79

[20] Bryuzgina E B, Kolyaganova O V, Bryuzgin E V, Derbisher E V, Derbisher V E, Navrotsky A V 2020 Influence of an acrylamide-based copolymer on the properties of cement mortar *Proceedings of Volgograd State Technical University* **5(240)** 87-93

[21] Tarasova O I, Yurkin Yu V, Toroshchin A S 2020 The influence of several types of resins on the dynamic mechanical properties of a polymer mixture based on butyl rubber *Construction Materials and Products* **3(2)** 36-45

[22] Herrman D, Rickert D, Hecker A, Ludwig G M 2017 Interaction of methylcellulose and cements with mineral additives in dry building mixes *ALITinform: Cement. Concrete. Dry mixes* **4-5(48)** 86-90

Acknowledgments

This work was realized in the framework of the Program of flagship university development on the base of the Belgorod State Technological University named after V G Shukhov, using equipment of High Technology Center at BSTU named after V G Shukhov.

IOP Conf. Series: Materials Science and Engineering **945** (2020) 012032 doi:10.1088/1757-899X/945/1/012032

Impact of a pump unit and geothermal pump use on soil condition

K P Orlov[1*], V A Orlova[2], S V Kupriyanov[2] and Y N Bojkov[2]

[1] National Research University of Technology (NITU) «MISiS», 4 Leningradsky Avenue, Moscow, 119049, Russia
[2] Belgorod State Technological University named after V.G. Shukhov, 46 Kostyukova st., Belgorod, 308012, Russia

E-mail: orlov_kp@gmail.com

Abstract. The article considers the advantages of air heat pump use. Their positive impact on ecology compared to other types of similar pumps, in particular geothermal ones, is shown. The author aims to focus on the problems of gross interference with the soil structure. In the majority of sources, geothermal pumps are presented as environmentally friendly and recommended pumps. The task of the study is to prove analytically, as well as using general scientific methods, that these sources are wrong. The use of geothermal heat pumps causes damage to the environment, since well drilling in the vast majority of cases is followed by formation segregation, interstratal insulation violation, change in natural processes of water bodies self-purification, rate of mineralization flow, kinetics of microflora and fauna development and die-away. Significant areas of uninhabitable trees and plants are formed due to subsoil cooling and freezing. Various heat pump layouts were analyzed and conclusion on the most environmentally friendly integrated use of heat air pump and solar energy was made.

1. Introduction

The environment is of interest as a source of energy when its temperature level slightly differs from the temperature required by the consumer, which is specific to room heating. Indeed, depending on the season, indoor air temperatures and the temperature of an environmental source of energy differ by no more than 10-35°C. Environmental sources of energy for a heating pumping unit (hereinafter HPU) can involve soil, ground and surface waters, air, river water, well water, sea water.

With HPU, the existing environmental energy can be used, for room heating primarily. However, the efficient use of HPU involves consideration of appropriate operating conditions associated with the temperature pattern of energy sources.

To work with heat pumps, there is the need to study geology of a site where drilling or horizontal collector laying is expected. This issue is thoroughly discussed in the article written by the leading Alliance-Neva LLC engineer, K.P. Prunenko "Heat pump installation in Russia: operational experience" [1]. The companies that agree to develop budget for probe installation not having studied soil properties at a particular site of the customer should be cautiously dealt with.

In the 1980s, the European market also faced the problems of poor-quality installation and design of heat collection systems, but it has managed to overcome the increasing problems. Studies of soil

IOP Conf. Series: Materials Science and Engineering **945** (2020) 012032 doi:10.1088/1757-899X/945/1/012032

properties were carried out, maps were designed. The works which were carried out in Lund University, Sweden, by the expert Goran Hellstrom [2] are of a special interest.

Let us consider the types of heat sources for heat pumps.

Geothermal (using groundwater or top surface heat):

a) closed-type:

- horizontal – the collector is laid radially or wavily in horizontal trenches below the soil frost zone (usually below 1.2 m and more). This method is the most cost-effective for residential buildings, provided that there is no shortage of land for the circuit;
- vertical – the collector is laid vertically in wells of 50 - 200 m in depth. This method is used when the land plot area either does not fit to lay the circuit horizontally, or there is a risk to damage the landscape [3.4];
- water - the collector is laid wavily or radially in a reservoir (pond, lake, river) below the soil frost zone. Thus, it becomes the most cost-effective option, but there are restrictions on the minimum depth and water volume in the reservoir for a particular region;
- with direct heat exchange - unlike the previous types, there is no heat conductor (salt solution) in the external circuit. Here, the coolant supplied through copper tubes by the heat pump compressor is directly involved. Copper tubes are located vertically in 30m-long wells 80 mm in diameter, horizontally below the soil frost zone;

b) open-type: this system type uses water as a heat exchange liquid, pumping it directly through the geothermal heat pump system within an open cycle, which means that having flowed through the system water is fed back to the ground. This option can be implemented only if there is enough moderately clean water and provided that this approach to use groundwater is not prohibited by law.

2. Materials and methods

To write the article, general scientific, private and special methods were used to study technological phenomena and processes occurring in the field of geothermal pump drilling: retrospective, logical, analytical, system, structural, analysis and synthesis, deduction and induction, etc. The basic method used in the current research is a descriptive method involving observation, interpretation, comparison, generalization approaches. The study also used the methods of synchronous and diachronic scientific fact analysis.

To achieve the goal and test the hypothesis on the environmental damage caused by geothermal pumps, a general scientific method was used, including the literature analysis on the issue under research, generalization, comparison and systematization of empirical and theoretical data.

The influencing factors and major soil cover projected changes occurred during the well development were studied. The final-stage method of the study involved heat flow simulation. Visual methods were also used. Distribution of heat flows and temperatures in the source of energy for the jointly laid solar and heat pump collector is shown.

3. Results

Air heating system is a modern compact and environmentally friendly heating system that supplies hot water and heat to warm cottages by means of air heat use. Heating system that use waste heat (e.g. waste water pipeline heat, heat network pipeline) are also efficient. This option is most suitable for industrial facilities having waste heat sources (WHS) requiring recycling.

IOP Conf. Series: Materials Science and Engineering **945** (2020) 012032 doi:10.1088/1757-899X/945/1/012032

Table 1. Capacity of wastewater discharge, geothermal and low-temperature heat of Russian soils and reservoirs. Source: compiled by the author based on the source [3].

Parameter	Gross capacity	Technological capacity	Economic capacity	Production capacity
	Million tons of oil equivalent			
Totally across the Russian Federation, including:	Over 30000	12 058	216.3	20.08
- wastewater discharge	40.8	19.1	8.56	2.82
- soil and reservoir heat	-	26.4	13.22	4.36
- power plant condenser cooling systems	142.4	37.4	9.50	3.14
- water recycling systems	370.3	106.0	20.0	6.60
Geothermal heat	Over 30000	11869	165	3.16

Ambient air as a heat source depends on daily and seasonal temperature fluctuations. Other sources of energy may be less responsive to these factors, or may even be irresponsive to them. Thus, the temperature level of the environmental energy depends on different external conditions affecting each other. When using HPU for heating, the environmental energy is not a primary goal for using, but high-temperature sources are focused instead to gain a high energy conversion coefficient due to the temperature difference between a heat source and HPU heat carrier.

Soil heat exchangers are a system of pipelines, either laid horizontally in the ground, or installed in vertical wells (U-shaped pipes that reduce the required soil surface by 10÷20 times compared to horizontal ones). Non-freezing power fluid (water, calcium chloride, methanol, ethylene glycol (anti-freezing agent)) circulates through pipes made of copper, steel (or its alloys), polyethylene or polypropylene. The amount of heat supplied by HPU depends on certain conditions: solar radiation intensity, soil humidity, soil thermal and physical properties, laying depth, distance between pipes and heat exchanger temperature.

CT pipes are often laid in trenches during drainage work on the site allocated for construction, which increases harmful effects on the soil.

The drilling company shall be authorized in accordance with DVGW W 120 (for Germany, Austria and Switzerland) worksheet. Planning shall be carried out in cooperation with the customer. The drilling company shall develop a work plan in which all permits and restrictions are recorded.

The soil probe and its feeding and return lines are laid at a distance of not less than 70 cm from water supply and sewage pipes, as well as from other pipeline systems. When crossing feeding lines, the collection pipe shall be isolated in the intersection area. Soil probes are delivered to the construction site being assembled. To avoid the damage, probes should be handled with special care.

When planning a construction site, the following requirements shall be fulfilled:

- approximate aisle width for track-type drilling rig vehicles is not less than 1.5 m, 2.5 m for trucks;
- site size for small track-type drilling rig vehicles is not less than 6x5 m; 8x5 m for trucks;
- washing reservoir or basin, if required;
- 400V power supply;
- cold water supply;
- layout indicating electric cables, water supply and sewage pipes or other underground structures [5].

IOP Conf. Series: Materials Science and Engineering **945** (2020) 012032 doi:10.1088/1757-899X/945/1/012032

Figure 1. Soil probe. Source: made by the author.

The required information can vary considerably depending on the needs of the drilling company and equipment type and, therefore, should be considered only as an estimated starting point. Theoretically, drilling operations are performed during the pipe laying stage. When working near buildings under construction, their protection from dirt should be ensured.

Radial clearance clogging between the geothermal probe and the well walls (Figure 1) is very important to ensure heat exchange between soils and the geothermal probe and to seal permeable water-bearing underground layers at different depths.

A prerequisite is to prevent groundwater (from a depth of 20 m, for example) from flowing into the water-bearing layer (60 m, for example). This situation poses a real threat to water supply based on extracting water from an underground source.

There is also a major risk of damage to glycol underground pipeline integrity [6].

4. Discussion

Impacts on soils and projected changes. The main types of man-made effects on soils during drilling are mechanical damage to the soil cover and chemical pollution. The greatest mechanical impact on soils is produced by construction work (construction of drilling platforms, their infrastructure and utility systems) and remediation (fertile soil layer removal, excavation backfilling, land planning, etc.).

They are followed by the fertile layer removal and pedoturbations (turnover of soil genetic horizon substrates), which worsens the ecological properties of soils disturbed in the landscape [8]. Soil chemical pollution is associated with the impact of vehicles, household waste, and potential hydrocarbon and drilling fluid spills, as well as leads to negative changes in physical, chemical, ion-exchange soil properties and their biological activity. Though in general, this is not an inevitable consequence of drilling operations. In addition, manifestation of this impact type varies considerably depending on the properties of certain soils. Factors and types of impacts, as well as the major projected changes in the soil cover during well development are given in Table 2.

Table 2. Influencing factors and major projected changes in the soil cover during well development for geothermal pump use. Source: made by the author [7].

Influencing factors	Impact types	Projected changes
preparatory works: logging, site cleanup, site planning, trench work, drainage trenching, site backfilling, utility system construction, etc.	Mechanical predominantly	bedding damage and destruction, linear and area deformations of surface soil layers, soil profile rearrangement, partial and complete soil cover destruction or removal, antecedent soil lodging

construction and installation works: winch unit and other infrastructure facilities erection, foundation water isolation, construction unit mounting, mechanisms, etc.	Mechanical predominantly	bedding damage and destruction, linear and area deformations of surface soil layers, soil profile compaction and restructuring, partial or complete soil cover destruction, chemical soil pollution
well drilling and testing: water and heat supply, layered mineral water (salt water) inflows are possible	Mechanical and chemical	linear and area soil deformations, chemical soil pollution

Heat pumps using air masses are considered to have great potential compared to systems operating on the soil surface layer heat and groundwater heat. This is due to the fact that the use of the latter is limited to the availability of land plots having the required properties, while the use of groundwater as a heat source is determined only by availability of their appropriate properties in acceptable proximity to the consumer. However, the risk to disturb the ecological groundwater balance should be eliminated, which may occur if there is a groundwater scarcity and it is simultaneously used to supply water to people [9].

Though, such pumps, despite being relatively environmentally friendly, are low-powered. The efficiency of heat supply using a solar and geothermal source decreases during the winter season due to the increase in heat load and the decrease in solar activity; in this time of year additional potential is required to be used.

The decision to improve the air heat pump efficiency during the heating period will involve combining an air pump and a solar collector. Since solar radiation does not generally reach the ground in winter (due to the clouded sky, short daylight hours), we will use a solar collector installed on the roof of the building, and the heat exchanger circuit will be laid together with the air circuit, thereby accumulating solar energy directly in the air throughout the year. To prevent heat loss, we will use a thermal insulation layer at the top of the circuit.

By means of the method of static heat flow simulation, we will determine the most efficient circuit arrangement using graphic examples of heat flow and temperature air distribution [10]. The ELCUT simulation examples show (Figure 2) [11] that the heat flow is directed upward towards the cold surface.

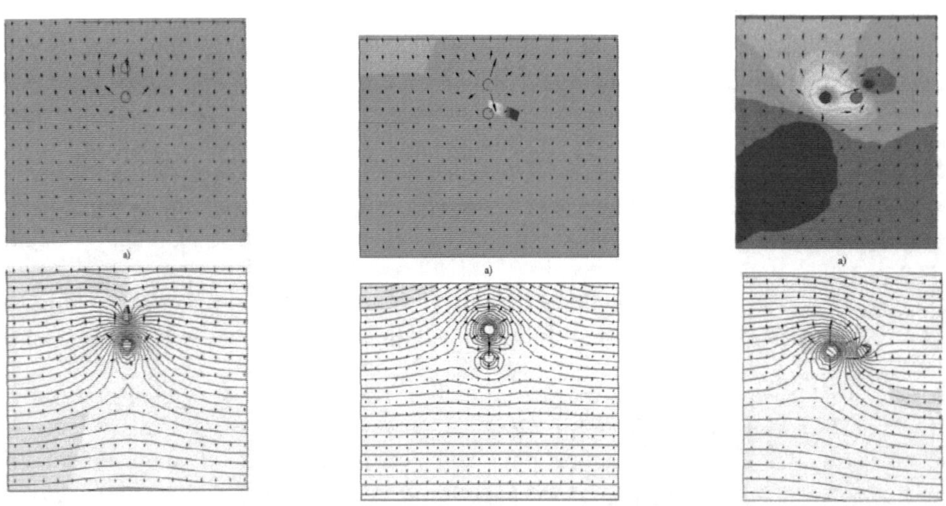

Lower heating circuit arrangement
a - heat flow distribution;
b - temperature distribution

Upper heating circuit arrangement
a - heat flow distribution;
b - temperature distribution

Parallel heating circuit arrangement
a - heat flow distribution;
b - temperature distribution

Figure 2. Heat flow simulation. Source: made by the author.

At the top of the graph, temperature of 0.6°C at the edge of a heat insulation layer is accepted, top circuit depth is at -1.5 m. The lower circuit boundary is at a depth of 2 m with a temperature of 1.3°C. Temperature of the heat exchanger operating from a solar collector is 15 °C [12].

Having arranged the circuit in the way shown in Figure 1 (left side), it is clear that the cold circuit is affected by the regular influence of heat flow and temperature, providing equal air heating at the level of the circuits laying in a radius of 300-350mm. Figure 2 (the middle part) with the upper heating circuit arrangement shows the heat flow "dips" and the temperature distribution throughout the circuit. Figure 1 with parallel heating circuit arrangement shows a chaotic distribution of heat flow and temperature, which results in heat loss. Thus, heat flow and temperature distribution in the source of heat combining solar and heat pump collectors is clearly shown.

5. Summary

The obtained results lead to the following conclusions:

– The integrated utilization option is the most efficient in terms of accumulating temperature in a certain range, since solar energy transfer to the source directly and using it as a heat accumulator makes it possible to smooth up the peak demand for heat supply;

– The use of geothermal heat pumps causes environmental damage, since in 96% of cases well drilling is followed by formation segregation, interstratal insulation violation, change in natural processes of water bodies self-purification, rate of mineralization flow, kinetics of microflora and fauna development and die-away. Significant areas of uninhabitable trees and plants are formed due to subsoil cooling and freezing. There is also a risk of glycol underground pipeline leakage;

– heat flow and temperature simulation when combining a heat air pump and a heat supply source (solar collector) circuits allows determining the optimal circuit arrangement. In addition, the search for materials more energy-intensive than sand is supposed to be extended in order to accumulate solar energy seasonally at a shallow depth.

6. References

[1] Prunenko K P 2013Heat pump installation in Russia: operational experience *S.O.K.* **11**

[2] Goran Hellstrom 1997Residential Vertical Geothermal Heat Pump System Models: Calibration to Data, Jeff W. Thornton, Timothy P. McDowell, John A Shonder, Jan.

[3] Sheremet E O, Seminenko A S, 2013. Heat pump application in centralized heat supply systems to improve heating network energy efficiency *Modern knowledge-intensive technologies* **8** 54-57

[4] Bondar E.S., 2011, Heat pumps: design, choice, installation *Plumbing, heating, air conditioning* **4** 62-37

[5] Reich 2011 *Geothermal heat pumps* (Plumbing, heating, air conditioning) **1** 80-83

[6] *Economics-Technology-Environment* (Regional Energy Service Company) http://eco-t-eco.ru/pumps (date of the address: 19.07.2020)

[7] *Solarsoul:* data resource on the solar energy use and energy saving. Detailed information on renewable energy technologies http://solarsoul.net/vodyanoj-teplovoj-nasos (date of the address: 19.07.2020)

[8] *Soil probes and water probes* (ALTAL production portal) http://www.altalgroup.com/ghp.htm (date of the address: 22.07.2020)

[9] Manzhilevskaya S E 2019 Organizational and economic issues of environmental safety in construction *Construction Materials and Products* **2** **(4)** 73-78 https://doi.org/10.34031/2618-7183-2019-2-4-73-78

IOP Conf. Series: Materials Science and Engineering **945** (2020) 012032 doi:10.1088/1757-899X/945/1/012032

[10] Karpov D F 2019 Algorithm for complex diagnostics of the technical state of building structures based on the heat pattern analysis *Construction Materials and Products* **2** **(2)** 23 – 28 https://doi.org/10.34031/2618-7183-2019-2-2-23-28

[11] *Geotherm*: Geothermal heat pumps http://www.geotherm.com.ua/about/closedloop/heatbasket.html (date of the address: 22.07.2020)

[12] *Thermal service*: Spiral and vertical geoprobes - innovative soil heat collection technologies for heat pumps http://teplo-v-dome.net/spiralnye-vertikalnye-geozondy-novejshie-texnologii-sbora-tepla-grunta-dlya-teplovyx-nasosov/ (date of the address: 23.07.2020)

IOP Conf. Series: Materials Science and Engineering **945** (2020) 012033 doi:10.1088/1757-899X/945/1/012033

Dynamic stress-strain state of earth dam considering an inhomogeneous base

K D Salyamova[1]

[1]Institute of Mechanics and Seismic Stability of Structures of the Academy of Sciences of the Republic of Uzbekistan named after M.T. Urazbayev, Durmon Yuli St., 33, 100125, Tashkent, Uzbekistan

E-mail: klara_51@mail.ru

Abstract. The design, construction and operation of important water retaining structures (high earth dams) in seismic regions, such as the territory of the Republic of Uzbekistan, require constant improvement of the methods to calculate their behavior under various loads, including seismic ones. A variational statement of the problem of natural vibrations of a plane elastic earth dam with and without an adjacent base is given. The problem is solved numerically by the finite element method. The aim of this study is to determine the effect of a piecewise-inhomogeneous section at the base of an earth dam on the stress-strain state of the dam itself. The result of solving the problem lies in the analysis of eigenfrequencies, vibration modes, stresses of a concrete earth dam.

1. Introduction

The consequences of the accidents or damage in high earth dams can lead to large human casualties [1, 2]. The methods for calculating earth dams under static and dynamic impacts (including seismic ones) are regulated by a number of regulatory documents [4-8], created on the basis of research conducted by leading experts in the field of hydro-technical engineering. These norms are usually based on a one-dimensional model that does not take into account the structure geometry and the piecewise inhomogeneity of physical and mechanical characteristics of the structure and its base; this results in applying numerical methods for solving the problems of the mechanics of rigid bodies [9-12].

Seismic impacts are resonant in nature. The magnitude of seismic impact on a structure at a given earthquake intensity is determined by the structure's own dynamic characteristics (frequencies and modes of natural vibrations) - a kind of a "passport" of the structure. These characteristics are determined both by the design (configuration, dimensions, weight and physical and mechanical properties of the structure materials) and (to no less extent) by the rigidity or flexibility of the base. Seismic resistance and strength of the structure is checked by various calculation methods (static, linear-spectral and dynamic analysis) and by experimental methods.

2. Materials and methods

Mathematical statement of the plane problem for an elastic plane-deformable structure includes [9-11]:

1) The variational principle of the minimum total energy of the system:

$$\delta\Pi - \delta W = 0 \qquad (1)$$

IOP Conf. Series: Materials Science and Engineering **945** (2020) 012033 doi:10.1088/1757-899X/945/1/012033

where $\delta\Pi$ is the increment of potential energy of the system, and $\delta' W$ is the sum of the work of external forces on virtual displacements.

2) The equation of state expressing the relationship between the components of stresses σ_{ij} and strains ε_{ij} in an elastic medium (the Hooke's law))

$$\sigma_{ij} = \lambda\varepsilon_{kk}\delta_{ij} + 2\mu\varepsilon_{ij} \tag{2}$$

3) Cauchy relations connecting strains with displacements

$$\varepsilon_{ij} = \frac{1}{2}\left(\frac{\partial u_i}{\partial x_j} + \frac{\partial u_j}{\partial x_i}\right) \tag{3}$$

4) Boundary conditions on a rigid base, meaning the absence of virtual displacements $\delta\overline{u}$

$$\delta\overline{u} = 0 \tag{4}$$

Here $\overline{u} = \{u_i, u_j\}$ - are the horizontal and vertical displacements of the point of a body with coordinates $\{x_i, x_j\}$; $\sigma_{ij}, \varepsilon_{ij}$ are the components of the stress and strain tensor; λ, μ are the Lamé constants.

For the numerical solution of the problem (1-4), the finite element method is used, which is invariant with respect to the geometry of the object under study and mechanical characteristics of soils; it allows accounting the structure interaction with surrounding medium, various mechanical (static and dynamic) loads, various types of boundary conditions.

To carry out calculations, the options were considered for a plane earth structure (of trapezoidal cross-section) with a rigid base and with account for the piecewise-inhomogeneous characteristics of the base soil. It was necessary to choose a bounded design model “structure-base”, which would adequately represent the base settlement under the structure.

The boundary conditions are:

- there are no virtual linear displacements (vertical and horizontal ones) on the lower (rigid) face of the model, i.e. for y=-10, δu=0, δv=0;

- the sliding hinges are installed on the side faces of the base, which do not impede vertical displacements, i.e. for x=0 and x=L, δu=0.

Such boundary conditions on the side faces of the base mean that only vertical displacements of the endless strip of base take place.

The first step in the dynamic analysis of a structure is to determine its dynamic characteristics – the frequencies and the modes of natural vibrations.

With the use of the developed methodology and a system of applied programs [9-11], the problem of natural vibrations has the form

$$[M]\{\ddot{q}(t)\} + [K]\{q(t)\} = 0, \tag{5}$$

and the solution to system (5) is sought in the form

$$\{q\} = \{q_0\}\sin\omega t.$$

As a result of substitution of solution into equation (5), a homogeneous system of algebraic equations is obtained; it has a unique solution only if the determinant of the system is equal to zero

$$([K] - \omega^2[M])\{q_0\} = 0 \tag{6}$$

Thus, with the application of the finite element method based on the statement of variational problem; introduction of a vector of generalized displacements (displacements of nodes of finite element discretization mesh); determination of stiffness matrices (matrices of response to unit displacements of nodes); mass matrices (matrices of response to unit accelerations of nodal points); and arrangement of separate matrices of the elements of overall matrices of mass and stiffness for the entire plane area (of a dam), we obtain an algebraic problem for eigen values (6), where the following designations are taken: [K], [M] are the stiffness and mass matrices of the structure; ω^2 – is the eigenvalue (square of natural vibration frequency); $\{q\}$ is the mode of natural vibrations corresponding to the frequency (the vector of displacements of mesh nodes).

IOP Conf. Series: Materials Science and Engineering **945** (2020) 012033 doi:10.1088/1757-899X/945/1/012033

The eigen frequencies ω and the modes of natural vibrations $\{q\}$, representing the two-component displacements of the mesh nodes, corresponding to one or another vibration frequency, are determined using the Muller numerical method.

3. Results

The reliability of the results obtained was validated by a test problem, carried out by the authors when determining the first three eigenfrequencies and modes of vibrations of the Nurek dam. The results were compared with the data obtained by the All-Russian Research Institute of Hydraulic Engineering named after B.E. Vedeneev and NIS Hydroproject named after S.E. Zhuk, and a good agreement was observed.

Calculations of dynamic characteristics of the operated earth dam of the Tupolang hydro power plant (HPP) were carried out, taking into account the design features and real physical and mechanical characteristics of soil of the structure and the underlying base. This study was carried out by order of JSC "Hydroproject" to assess dynamic characteristics and to predict the stress-strain state of earth dam and its inhomogeneous base under static and dynamic loads.

When calculating structures for seismic effects, the formula of seismic load was determined as [5]

$$S_{ik}=K_1K_2Q_kA\beta_iK_\psi\eta_{ik}.$$

The formula includes various coefficients, among which the coefficient of the calculated seismicity A, which depends on the intensity of expected earthquake; Qk is the weight acting on the k-th node of the structure; ηik are the coefficients of natural vibration modes of the structure, obtained by solving the problem on eigenvalues; dynamic factor β_i - inversely proportional to the i-th period of natural vibration taken in calculations.

Thus, the first step in dynamic design of a structure is to determine its dynamic characteristics – the frequencies, especially those that could cause intense vibrations of the structure due to their proximity to the frequency spectrum of seismic vibrations in a given region, and the corresponding vibration modes.

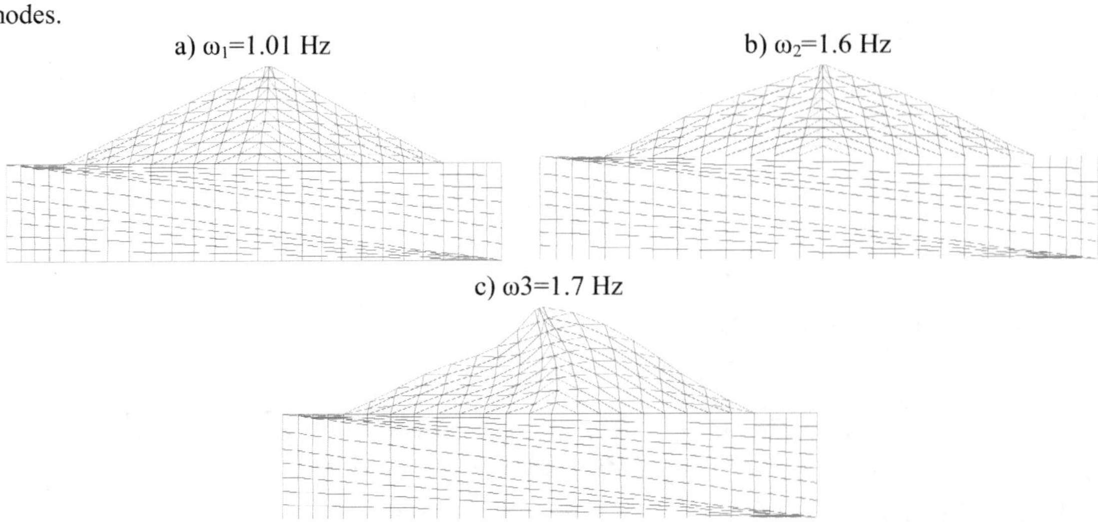

Figure 1. Fundamental frequencies and corresponding modes of natural vibrations of the dam on a homogeneous base.

When solving the eigenvalue problem, the dynamic characteristics of earth dam of the Tupolang HPP (H = 165m) were obtained in two options: on a homogeneous rock base without a weakened zone (E = 13000 MPa, ρ = 2.56 t/m3) and on a rocky base in the presence of a weakened zone – an anhydrite zone with the following physical and mechanical characteristics of soil E=16 MPa, ρ=1.58 t/m^3.

IOP Conf. Series: Materials Science and Engineering **945** (2020) 012033 doi:10.1088/1757-899X/945/1/012033

The parameters of the dam soil were selected as follows: slope ratios of 2 and 1.9 for the upstream and downstream slopes, respectively, E=70 MPa, ρ=2.04 t/m^3, Poisson's ratio for all options was taken equal to 0.3.

The resulting vibration modes with indication of frequencies are shown in Figs. 1 and 2. Here the deformed state by modes is shown against the finite element mesh. A weakened zone in the form of an anhydrite zone is located along the entire diagonal line of the base; it is uniform in thickness, of a height of 50 m in two finite elements. In the first option of calculations - in the case of a homogeneous base - the parameters of this strip are chosen similar to the parameters of surrounding soil. In the second option, the parameters of this strip correspond to the parameters of the anhydride zone.

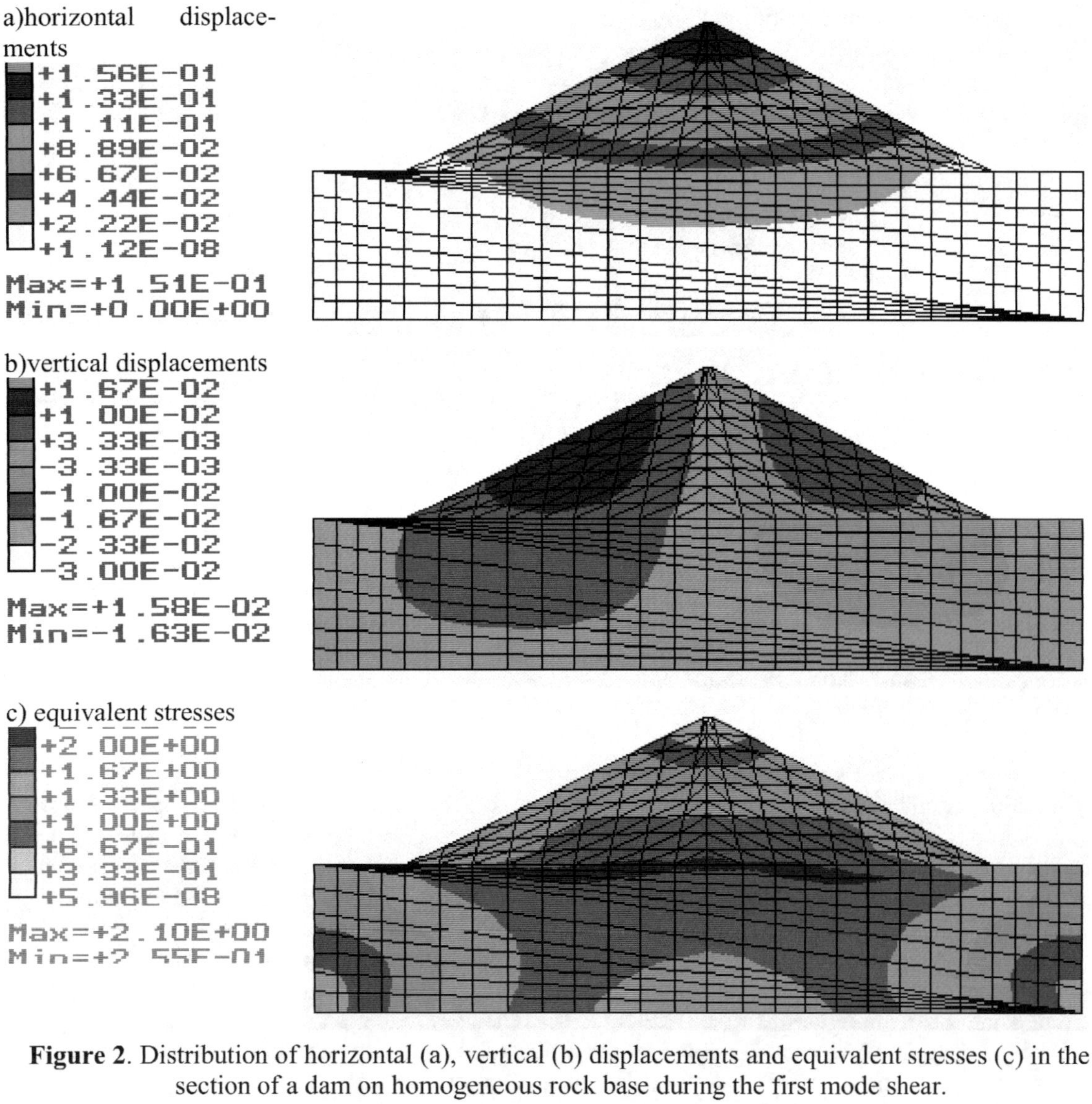

Figure 2. Distribution of horizontal (a), vertical (b) displacements and equivalent stresses (c) in the section of a dam on homogeneous rock base during the first mode shear.

The modes shown in the figures are: the shear in transverse direction (the 1st mode), a vertical mode (settlement) (the 2nd mode) and a complex mode of transverse vibrations (the 3rd mode). The rigid rock base remains undeformable.

4. Discussion

The horizontal displacements occurring in the cross section (Fig. 2,a) and the equivalent stresses (Fig. 2,b) are determined from the basic mode - the most energy-intensive one - the transverse displacement of the dam. The choice of these parameters is explained by the fact that they determine the strain mode (a horizontal shear) and the virtual transition of the model from elastic state in plastic.

Dynamic characteristics of the dam located on an inhomogeneous base (with a weakened zone) were determined in calculations: ω_1 = 0.148 Hz; ω_2= 0.21 Hz; ω_3= 0.46 Hz and the corresponding modes of natural vibrations were plotted. Comparison of fundamental frequencies and modes of vibrations of the "dam-base" plane system shows that the presence of a weakened zone at the base changes the spectrum of fundamental frequencies: it decreases noticeably (due to the weakened zone) along with the decrease in the base rigidity.

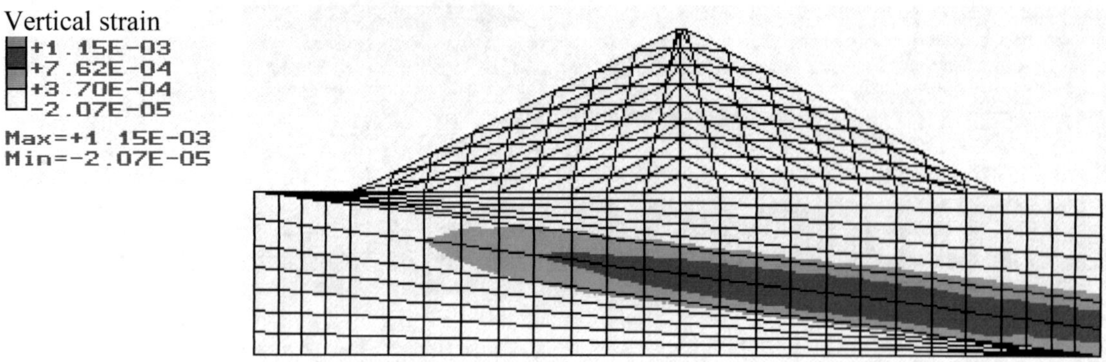

Figure 3. Vertical strain distribution of the "dam-inhomogeneous base" model during the first mode shear.

While the fundamental modes of vibration on a homogeneous rock base include the structure strain as a less rigid part of the model, then in the presence of a weakened zone, the fundamental modes of vibration include non-uniform vertical strain of the base. In this case, the structure tilts like a solid body, almost without deformation, as indicated in Fig. 3 by the vertical component of strain at the first mode of natural vibrations of the "dam-inhomogeneous base" model. The maximum value of vertical strain (dark color) falls on the right side of the anhydrite zone, and the "+" sign indicates vertical extension of this area.

5. Summary

A methodology and a complex of applied programs for the dynamic calculation of plane elastic system "earth dam-base" have been developed numerically by the finite element method. The analysis of the stress-strain state of this system indicates the need to account for physical and mechanical characteristics of soil of the structure itself and its base under various kinds of loads, not provided for in the standards. The influence of the weakened zone at the base on the SSS was investigated.

It was revealed that the first mode and the corresponding frequency reflect the oscillatory processes in the base, which do not affect the structure itself (the dam). Therefore, a comparison of the stress state in a dam on a homogeneous base (without a weakened zone) and on a inhomogeneous base (with a weakened zone) must be carried out using identical modes, which are the first mode for a dam on a homogeneous base and the third mode for a dam on a heterogeneous base.

6. References

[1] Akhmedov M A, Salyamova K D 2016 Analysis and damage assessment of hydro-technical structures *Science* 160
[2] Veksler A B 2002 Reliability, Social and Environmental Safety of Hydro-technical Facilities:

IOP Conf. Series: Materials Science and Engineering **945** (2020) 012033 doi:10.1088/1757-899X/945/1/012033

Risk Assessment and Decision Making, Saint Petersburg *Textbook* 120
[3] 1999 Law of the Republic of Uzbekistan "On the safety of hydro-technical structures"
[4] 1996 KMK 2.01.03-96 (Building Code) Construction in seismic regions Gosarchitekstroy **56**
[5] SHNK 2.06.11-04 2016 Town planning norms and rules. Construction in seismic areas. Hydro-technical structures Gosarhitekstroy 56
[6] 1998 KMK 2.0.05-98 Dams built of earth materials. Gosarkhitekstroy Tashkent 58
[7] Mirsaidov M M, Salyamova K D, Li A T , Shagrai M K 2003 Investigation of SSS and dynamic characteristics of earth dams Proceedings of the V Russian National Conf. On earthquake-resistant construction and seismic zoning Moskow 170-174
[8] Mirsaidov M M , Matkarimov P Zh , Yuldashev B Sh 2005 Investigation of dynamic behavior of soil structures taking into account various strain properties and water saturation of soils *Journal Earthquake-resistant construction. Safety of structures*. Moscow 5 25-30
[9] Zenkevich O 1975 The finite element method in technology Moscow *Mir* 541
[10] Salyamova K D , Rumi D F Dynamics of earth dams 2015 LAP LAMBERT Academic Publishing *Science*150
[11] Salyamova K D , Rumi D F 2015 Dynamics of an earth hydro-technical structure, taking into account the rheological properties of soil. *Scientific-theoretical journal Vestnik BGTU* **3** 16-35
[12] Yangiev A., Salyamova K., Turdikulov KH.,Fayziev X. Dynamics of an earth dam with account for rhological properties of soil under dynamics effect IOP Conf. series:materials science and engineering 869(2020) doi 10/1088 (175) 899x/869 17/72005

IOP Conf. Series: Materials Science and Engineering **945** (2020) 012034 doi:10.1088/1757-899X/945/1/012034

Method for justification options for housing stock reconstruction

Yu Biryukov[1], R Makarov[1] and E Dobryshkin[1*]

[1] Military Academy of Logistics named Army General A.V. Chrulev, St. Petersburg, Russia

E-mail: edobryshkin@mail.ru

Abstract. The main contradictions in the theory and practice of housing stock objects reconstruction are revealed. The technical condition of housing stock reconstruction facilities is analysed and a study of the organization and planning of housing stock reconstruction in the North-western federal district is performed. A method of substantiating options for reconstruction of housing stock in the North-Western federal district has been developed. Its practical significance lies in its application to justify the reconstruction option for technical re-equipment and modernization of housing facilities in Federal and municipal authorities implementing state policy in the field of providing the population with comfortable living conditions, as well as in regional departments of contractors, which perform housing facilities repair work, at the stage of design estimates development for the reconstruction of facilities. Thus, the purpose of the study (to improve the efficiency of planning the housing facilities reconstruction in the North-western federal district on the basis of the developed scientific and methodological apparatus) has been achieved.

1. Introduction

At the present stage, a significant number of buildings and structures in the North-western federal district of Russia have increased indicators of building structures physical wear, since most of them were built and put into operation before 1990 [1, 2].

Figure 1 shows information about the technical condition of buildings and structures in the North-western federal district by year of construction as of 2019. At the same time, in the article the technical condition of the object is understood as a set of building properties that characterize data on its serviceability and operability. 72.5 % of the buildings (mostly the objects, which were built in the period 1941-1990) were buildings of the second and third categories of technical condition [3, 4].

IOP Conf. Series: Materials Science and Engineering **945** (2020) 012034 doi:10.1088/1757-899X/945/1/012034

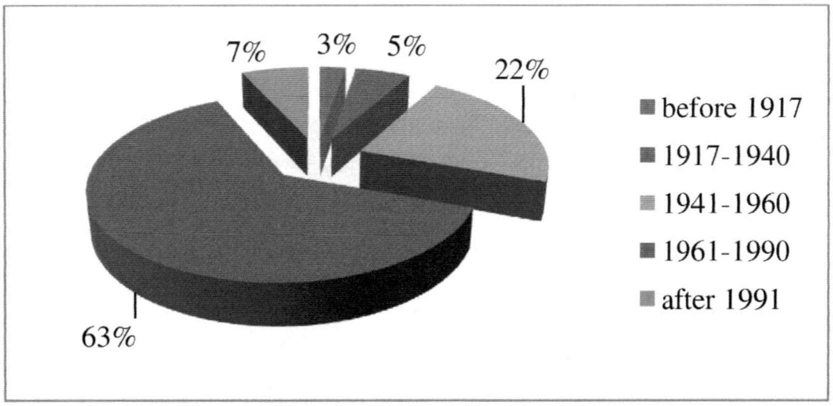

Figure 1. Characteristics of buildings and structures in the North-western federal district by year of construction (as of 2019).

The analysis of capital investments for housing reconstruction in the North-western federal district (table 1; figures 2 and 3) shows that a significant number of objects require repair work for the reconstruction and buildings and structures overhaul and increase the capacity indicators of existing urban infrastructure due to the reduced level of performance and the need for buildings technical and re-equipment [5, 6, 7].

Table 1. Capital investments implementation under the program part of the Federal targeted investment program (FTIP) in order to improve the technical condition of the North-western federal district housing stock from 2016 to 2018.

Number	Years	Allocated funds amount, thousand rubles	Realized funds volume, thousand rubles	*Realization level, %*
1	2016	5 844 020,0	5 115 258,0	87,0
2	2017	3 037 600,0	2 656 328,6	87,4
3	2018	2 911 539,6	2 568 650,9	88,2
Summary		11 793 159,6	10 340 237,5	87,7

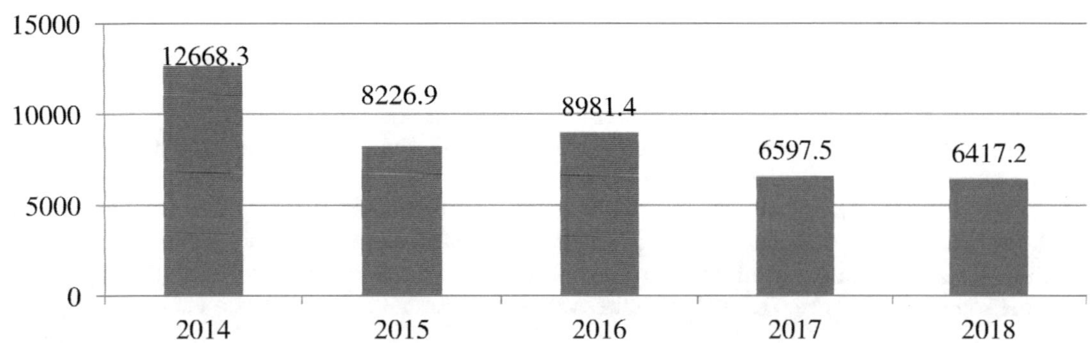

Figure 2. Analysis of FTIP capital investments for the construction of the North-western federal district housing stock from 2014 to 2018 (million rubles).

IOP Conf. Series: Materials Science and Engineering **945** (2020) 012034 doi:10.1088/1757-899X/945/1/012034

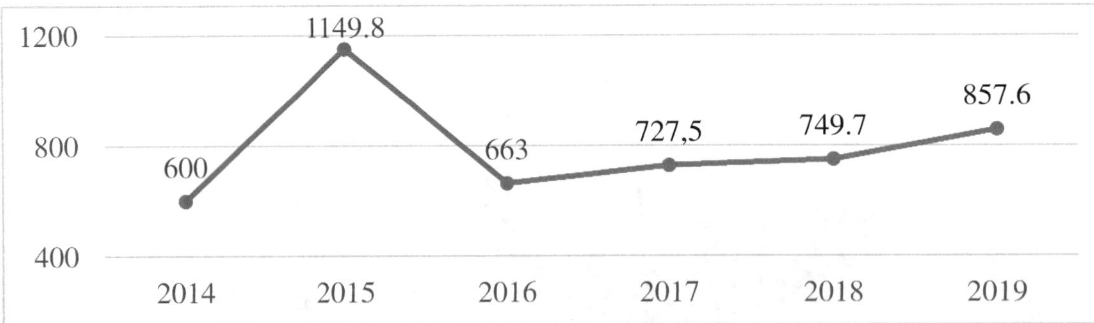

Figure 3. Analysis of capital investments for the reconstruction of the North-western federal district housing stock from 2014 to 2019, million rubles.

An analysis of the housing stock reconstruction and management in the North-western federal district showed that restoration and reconstruction predominate over some other forms of construction production. The capital investments volume in reconstruction in this direction exceeds the same indicators for current repairs in the last five years, which indicates a great demand for this method of improving the housing stock technical condition. At the same time, it is known that the specifics of the reconstruction object, the degree of work complexity, restrictions and features directly affect the capital investments volume and affect management decisions on the reconstruction project justification [8, 9, 10].

The capital investments analysis for infrastructure reconstruction in the North-western federal district also shows that this region has a large volume of reconstruction of existing buildings and structures, and there is a need to increase the buildings and structures capacity indicators. Funding for buildings and structures capital repairs in the North-western federal district amounted to 391.43 million rubles in 2017; 254.5 million rubles were allocated in 2018. Thus, in terms of allocated funds reducing, there is a growing role of housing stock reconstruction and organizational-technological solutions optimizing. Besides, improving the efficiency of investment is most urgent and requires scientific and methodological justification [11, 12, 13, 14 and 15].

In modern conditions the decision making task of the housing stock reconstruction is characterized by the presence of contradictions:

lack of a well-founded approach to the organization and planning of housing stock reconstruction, and as a result, non-compliance with the characteristics required for comfortable living of the population after the completion of buildings reconstruction;

increased physical and moral wear and tear, lack of reliable data on the buildings and structures technical condition, which leads to overstating the necessary financial resources for the objects reconstruction;

elimination of negative decisions consequences at the design stages, that increase capital investment during the buildings and structures operation and reduce the effectiveness of providing the population with comfortable living conditions;

insufficient theoretical developments in domestic technical science, lack of methodological documents on the need to optimize the reconstruction of facilities due to limited capital investment;

there is no scientifically based classification of housing stock objects according to the priority of reconstruction planning, which results in haphazard and inefficient capital investments spending [14].

2. Materials and methods

The authors' scientific task of the study was a methodology development for justifying the option and planning of housing stock reconstruction in the North-western federal district based on the generalized criterion method.

IOP Conf. Series: Materials Science and Engineering **945** (2020) 012034 doi:10.1088/1757-899X/945/1/012034

The purpose of the study was to improve the efficiency of planning the North-western federal district housing stock reconstruction based on the developed scientific and methodological apparatus.

The authors of the article performed the following tasks during the research: the technical condition of housing stock was analyzed and a study of the organization and planning of housing stock reconstruction in the North-western federal district was performed; a method of substantiating options for reconstruction of the North-western federal district housing stock has been developed; an assessment of the economic efficiency of the developed scientific and methodological apparatus is made.

The validity and reliability of the obtained scientific results is achieved by selecting real economic and statistical input data and using proven mathematical and methodological tools, which is confirmed by the economic efficiency of the customer organizations in making decisions on the facilities reconstruction in the North-Western Federal district. A computational experiment was performed to confirm the rationality of the developed method.

The research theoretical significance is determined by: conceptual approaches substantiation and contradictions' identification in the theory and practice of planning the reconstruction of housing facilities; indicators system justification for evaluating options for housing stock object reconstruction; using the generalized criteria method and modernization of previously developed methods based on analysis, the theory of the effectiveness of object management during technical re-equipment and reconstruction, which is a significant contribution to the development of the scientific basis for the organization and planning of restoration of the buildings and structures technical condition, investment evaluation during reconstruction, housing stock restoration, as well as improving the theory and methods of feasibility studies and time costs.

The practical significance of the study is: the justification of making reasoned decisions on capital investments for the housing facilities reconstruction was made; scientific and methodological apparatus was realized in organizations and enterprises that plan and perform housing facilities reconstruction; the prospects for the developed methodology practical use in universities and research organizations of federal executive authorities was determined.

3. Results

A study of the organization and planning of housing facilities reconstruction has been conducted, which showed that housing facilities have different levels of reliability of structural elements and terms of their failure-free operation, due to different design schemes and construction periods. It is revealed that there are four approaches to solve the problem of organizational and technological systems functioning reliability, what is based on the conducted research (table 2). The criteria for evaluating options for objects reconstruction are justified.

Table 2. Forms of matching functions for performance indicators in various approaches to the organizational and technological systems reliability problem.

Number	Desired result	Function	*Performance indicator*
1	Random event A	$p = \begin{cases} 1, & by\ A\ is\ occurred \\ 0, & by\ A\ is\ not\ occurred \end{cases}$	$P_u(A)$
2	Set by the parameter $y^{\text{тр}}$		
3	Fuzzy variable $y^{\text{тр}}$ with the membership function μ	$p = \begin{cases} 1, & by\ y(u) \geq y^{\text{тр}} \\ 0, & by\ \hat{y}_{\text{н}} \leq y^{\text{тр}} \end{cases}$	$P(\hat{y}(u) \geq y^{\text{тр}}) = 1 - F(y^{\text{тр}})$
4	$y^{\text{тр}}$ - random variable $\hat{y}_{\text{н}}$ with the distribution		$P(\hat{y}(u) \geq \hat{y}_{\text{н}}) = \int F_{\text{н}}(y)\ dF(y)$

IOP Conf. Series: Materials Science and Engineering **945** (2020) 012034 doi:10.1088/1757-899X/945/1/012034

	function $F_{н}(y)$		
5	Not defined	$p = y\,(u)$	$M[y(u)]$
6	Set by the parameter	$p = (y(u) - y^{тр})^2$	$M[(y(u) - y^{тр})^2]$
7	Not defined	$p = (\hat{y}(u) - M[y(u)])^2$	$D[\hat{y}(u)]$
8	The guarantee degree is set	$p = F^1(1-\alpha) = y_\alpha$	y_α

The method is applicable to substantiate the option of housing stock object reconstruction during technical re-equipment and modernization of objects in organizations and enterprises that perform repair work, as well as in Federal and municipal authorities that plan and organize the restoration of the housing stock technical condition.

It is necessary to define a number of key parameters within implementing the method. There is a housing facility in need of reconstruction, which can be characterized by the number of parameters *M*. It is necessary to take into account a number of external factors that can influence decisions within evaluating the reconstruction option for object *L*. It is necessary to make a decision on the justification of the reconstruction option.

Variants of reconstruction of the object *n* are defined. A number of General criteria is defined for these variants, that affect decision-making and by which it is possible to fully characterize each variant p_n.

For these options, a number of general criteria *n* have been identified that influence decision-making and by which it is possible to fully characterize each option p_n. The criteria should meet the general condition:

$$p_n \to max. \tag{1}$$

Comparison of reconstruction options of a building or structure is based on comparison of their vector estimates by preference. The vector estimation of a variant is calculated using the formula:

$$P(p/v) = v_1 \times \widehat{p_1} + \cdots + v_n \times \widehat{p_n}. \tag{2}$$

where: $P(p/v)$ – vector estimation of the housing stock reconstruction option; $v_1 \ldots v_n$ – the coefficients of criteria importance n; $\widehat{p_1} \ldots \widehat{p_n}$ – normalized values of the *n* criteria, that affect the decision to justify the reconstruction option for the object.

Define the normalized values of the criteria $\widehat{p_n}$ and assign a significance factor v_n to each criterion (table 3).

The weight coefficients are determined by the formula:

$$v_j = \frac{x_j}{\sum_{j=1}^{n} x_j}, \tag{3}$$

where: x_j – the sum of expert ratings m based on one of the criteria *n*; $\sum_{j=1}^{n} x_j$ – total amount of values x_j.

Table 3. Determining weight coefficients using the ranking method.

Experts	Criteria						
	y_1	y_2	y_3	y_4	…	y_n	
1	x_{11}	x_{12}	x_{13}	x_{14}	…	x_{1n}	4. $P_u(A)$
2	x_{21}	x_{22}	x_{23}	x_{24}	…	x_{2n}	5.
…	…	…	…	…	…	…	6.

m	x_{m1}	x_{m2}	x_{m3}	x_{m4}	...	x_{mn}	7.
Sum $x_j = \sum_{j=1}^{m} x_{ij}$	x_1	x_2	x_3	x_4	...	rx_n	8.
$v_j = \frac{x_j}{\sum_{j=1}^{n} x_j}$	v_1	v_2	v_3	v_4	...	v_n	9.

Thus, the coefficients v_n for the criteria n that affect the decision on the reconstruction of the object are obtained, which in total are equal to one.

Since the evaluation criteria are heterogeneous and have different scales due to their nature, it is necessary to bring the initial criteria to a comparable form - normalize them. The normalized criteria $\widehat{p_n}$ are dimensionless and their values range from 0 to 1. Values $\widehat{p_n}$ calculate using the formula:

$$\widehat{p_n} = \frac{p_n^{fac}}{p_n^{pr}}, \quad (4)$$

where: p_n^{fac} – the actual values of the criteria n; p_n^{pr} – limit values of criteria n (which are determined and based on the maximum values of the object's characteristics).

For the most complete representation of the methodology for housing stock reconstruction evaluating options, an algorithm is formed (figure 4).

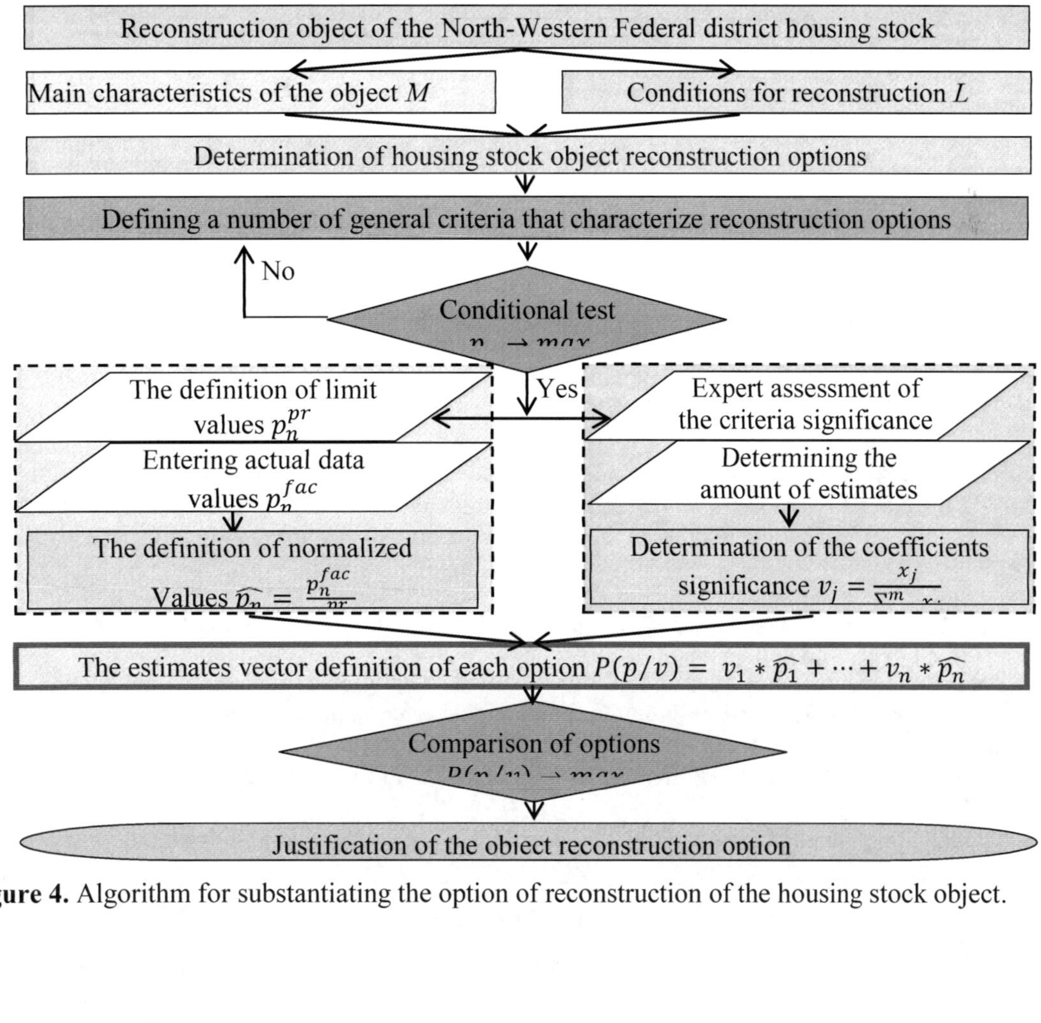

Figure 4. Algorithm for substantiating the option of reconstruction of the housing stock object.

According to the algorithm for justifying the housing stock reconstruction option, the calculation of significance coefficients, the determination of normalized values, and the determination of option vector estimates are performed. A comparison of vector estimates of each option for reconstruction of a housing facility shows which option is the most optimal.

In order to increase the automation of the process and reduce the time spent, the method is implemented using the received patent for the invention "Decision support system for building restoration" (figure 5). The technical result is automation and support of management decision - making in the conditions of several criteria, based on which reconstruction planning is carried out. The system use increases the efficiency of planning repairs in federal and municipal authorities that implement state policy in the field of providing the population with comfortable living conditions. Thus, the method forms a comprehensive approach, confirmed by the patent for the invention of the Russian Federation no. 2716351 "Decision support system for building restoration" [14].

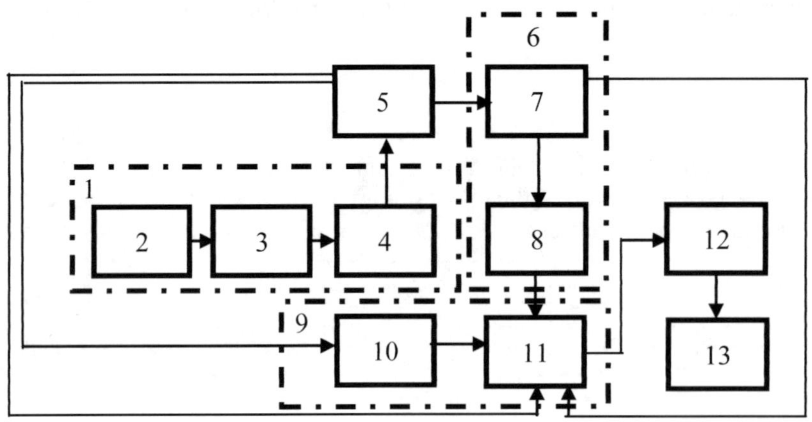

Figure 5. Decision support system for building restoration: 1 - module for monitoring the buildings.

technical condition; 2 - sensors; 3 -autonomous data collection units; 4 - control station; 5 - block for entering parameters; 6 - information analysis module; 7 - block for determining the building reconstruction cost; 8 - block for determining the building operation cost; 9 - information processing module; 10 - reference indicators block; 11 - ranking criteria and calculation block; 12 - memory block; 13 - displaying information block

Within developing the presented methodology the application' effectiveness was evaluated while analyzing options for restoring the residential building technical condition in Saint Petersburg.

The authors of the article expressed the comparative cost of reconstruction of objects by the ratio:

$$\frac{KV_{rec,i}}{T_{rec}} \leq \frac{KV_{new}}{T_{new}}. \quad (5)$$

The characteristics of residential house reconstruction options are shown in table 5.

The calculation performed by the authors allows us to conclude that option C of reconstruction of the housing stock is the best option for restoring the technical condition of the object.

The economic effect of the implementation of option C of the reconstruction will be:

$$E_v = E_n \times KV_{rec,i} \times (T_1 - T_2) = 0{,}15 \times 155(1{,}8 - 1{,}6) = 4{,}65\ million\ rubbles. \quad (6)$$

Table 4. Options for design solutions for reconstruction.

Indicators	Measure unit	Reconstruction options		
		A	B	C
The facility capacity W	m^2	2000	2050	1950
One-time costs for new construction of a similar facility KV_{new}	million rubles	500	500	500
One-time expenses for reconstruction of the object KV_{new}	million rubles	150	160	155
	thousand rubles/m^2	75	78	79,5
Annual operating expenses C_{epr}	thousand rubles/m^2·rubbles	1,55	1,36	1,53
Duration of construction of a new similar facility t_{new}	year	2	2	2
The duration of the reconstruction t_{rec}	year	1.7	1.8	1.6
Object service time after reconstruction T_{rec}	year	30	30	30
New object' service time T_{new}	year	50	50	50

The economic efficiency of the selected option is:

$$E = \frac{E_v}{KV_{rec,i}} \times 100\% = \frac{4,65}{155} \times 100\% = 3\%, \quad (7)$$

this shows the expediency of using the developed scientific and methodological apparatus in planning the reconstruction of a housing facility.

4. Discussion

As a result of the research, the scientific task of developing a methodology for justifying the option and planning of reconstruction of housing facilities in the North-Western federal district based on the generalized criterion method, which is essential for the of effective planning of housing facilities reconstruction in the North-Western Federal district, was solved.

The effectiveness of the research results was evaluated with the justification of using the method of options comparative analysis for project proposals for reconstruction, which showed that the use of the developed methodology, as well as practical recommendations for planning the reconstruction of housing facilities, allows you to get an economic effect of 10 to 15% of the reconstruction cost.

5. Summary

The scientific significance of the method lies in the fact that unlike the other, it enables the comparative assessment of the housing stock reconstruction variants on the basis of the generalized criterion taking into account weight coefficients and their normalized values to identify the most optimal variant. The method allows to justify organizational and economic decisions within reconstructing a housing facility.

The theoretical significance of the developed method is to justify the efficiency of capital investment distribution in the optimal case of object reconstruction based on the generalized criterion method.

Its practical significance lies in its application to justify the reconstruction option for technical re-equipment and modernization of housing facilities in Federal and municipal authorities implementing

IOP Conf. Series: Materials Science and Engineering **945** (2020) 012034 doi:10.1088/1757-899X/945/1/012034

state policy in the field of providing the population with comfortable living conditions, as well as in regional departments of contractors, which perform housing facilities repair work, at the stage of design estimates development for the reconstruction of facilities.

Thus, the purpose of the study (to improve the efficiency of planning the housing facilities reconstruction in the North-western federal district on the basis of the developed scientific and methodological apparatus) has been achieved.

6. References

[1] Biryukov A N, Bolotin S A 2013 Time management in drafting probability schedules for construction work *World Applied Sciences Journal 23 (Problems of Architecture and Construction)* 01-04
[2] Biryukov A, Dobryshkin E, Kravchenko I, Glinskiy M 2019 Optimization of management decisions for choosing the strategy of enterprises fixed assets reproduction *Engineering for rural development* **18** 1726-1736
[3] Biryukov A, Dobryshkin E, Biryukov Yu, Tishenko V 2020 Complex approach to organizations' capital assets reproduction *E3S Web of Conferences* 157, 04026
[4] Biryukov A, Dobryshkin E 2018 Optimization of supply of material and technical resources for repair and restoration works *Construction and road vehicles* **10** 52-56
[5] Klyuev S V, Khezhev T A, Pukharenko Yu V, Klyuev A V 2018 To the question of fiber reinforcement of concrete *Materials Science Forum* **945** 25 – 29
[6] Klyuev S V, Khezhev T A, Pukharenko Yu V, Klyuev A V 2018 Experimental study of fiber-reinforced concrete structures *Materials Science Forum* **945** 115 – 119
[7] Klyuev S V, Klyuev A V, Khezhev T A, Pukharenko Yu V 2018 Technogenic sands as effective filler for fine-grained fibre concrete *Journal of Physics: Conference Series* **1118** 012020
[8] Abolfazl Farahani, Holger Wallbaum, Jan-Olof Dalenbäck 2018 Optimized maintenance and renovation scheduling in multifamily buildings – a systematic approach based on condition state and life cycle cost of building components *Construction Management and Economics*
[9] Edmundas Kazimieras Zavadskas, Zenonas Turskis, Jolanta Tamošaitien 2010 Risk assessment of construction projects *Journal of civil engineering and management* **16(1)** 33-46
[10] Christen M, Schroeder J, and Wallbaum H 2014 Evaluation of strategic building maintenance and refurbishment budgeting method Schroeder *International journal of strategic property management* **18(4)** 393–406
[11] Kumar D, Setunge S, and Patnaikuni I 2010 Prediction of life-cycle expenditure for different categories of council buildings *Journal of performance of constructed facilities* **24(6)** 556–561
[12] Hellova K, Struharova A 2018 Optimization of thermal insulation of external walls of wooden frame constructions *IOP Conference Series: Materials Science and Engineering* **385** 012017
[13] Konikov A 2019 Use of satellite technology for monitoring buildings and structures *IOP Conference Series: Earth and Environmental Science* **403** 012223
[14] Kravchenko I, Biryukov A, Trostin V, Fedorov A 2019 Modernization of technological equipment for the enrichment of local building materials and recycling of destroyed structures *Construction and road vehicles* **12** 44-50
[15] Biryukov A, Dobrich B, Kazakov Yu, Tokarev N 2017 Application of composite materials in construction, repair and reconstruction of transport infrastructure facilities *Construction and road vehicles* **10** 46-52
[16] Biryukov A 2012 The concept of restoration and construction of facilities in response to emergency situations in the system of ensuring economic security of the country *Bulletin of civil engineers* **33(4)** 123-129
[17] Biryukov A N, Dobryshkin E O, Biryukov Yu A 2020 Decision support system for building restoration *Pat. 2716351 Russian Federation G06F 17/40* **2019129210**

IOP Conf. Series: Materials Science and Engineering **945** (2020) 012035 doi:10.1088/1757-899X/945/1/012035

Innovative structural and technological developments of pile foundation engineering

Vladimir Kocherzhenko[1], Lyudmila Suleymanova[1]* and Maksim Kolesnikov[1]

[1]Belgorod State Technological University named after V.G. Shukhov, Belgorod, Russian Federation

E-mail: ludmilasuleimanova@yandex.ru

Abstract. The paper considers innovative structural and technological solutions for precast piles, which significantly increase their load-bearing capacity: a transformable pile and a pile with "pockets" (recesses) on the side surface for use in loose soils. In the first case, in the manufacture of precast reinforced concrete piles in the center they set plastic cone-shaped cruciform brake detail with length 0.75 of the pile length, in which after driving the piles destructive non-explosive substance is poured, that increasing in volume in 2-3 times, transforms prismatic pile into a pyramid one. In the second case, when making precast concrete piles, they provide for the arrangement along the side faces of "pockets" (recesses). The liquid hardening compound located in the "pockets" penetrates into the soil of the pile during the driving process and secures it, which significantly increases the lateral friction force of the pile and its load-bearing capacity and reduces the energy of the pile sinking. The considered design and technological solutions allow reducing the energy of pile driving, while increasing their load-bearing capacity.

1. Introduction

Pile technologies in modern foundation engineering are the leading direction for improving their structural and technological solutions. The load-bearing capacity of pile foundations largely depends on the formation of the friction force of the soil on the side surface of the piles. Numerous works of scientists have been devoted to the study of soil friction during the sinking and operation of piles, as well as ways to increase it [1-12].

Currently, special attention is paid to the development and research of innovative structural and technological solutions for driving floating piles, which provide a significant increase in their bearing capacity. The authors conduct research in two areas:

– the transformation of prismatic to pyramidal piles with the use of non-explosive destructive substances;

– development of piles for use in loose soils.

The aim of this work is to develop and study innovative structural and technological solutions for driving floating piles, which provide a significant increase in their load-bearing capacity. To achieve this aim the following tasks were solved:

– development of precast concrete pile structures based on the requirements for them in a specific research area (transformable; for use in loose soils);

– choice: expanding substance for transforming piles; fixing compounds for piles used in loose soils;

– development of the technological sequence when applying the developed methods for increasing the load-bearing capacity of piles;

– experimental verification of the effectiveness of the developed methods for increasing the load-bearing capacity of floating piles in laboratory and field conditions.

2. Materials and methods

2.1. Transformable prismatic pile

The transformed pile (Fig. 1) increases its load-bearing capacity by wedging the prismatic (serial) pile into four equal-volume parts and turning it into a pyramid, which has a much greater load-bearing capacity. This is due to the tilt of the four faces of the pile and compaction of the soil around the pile when wedging. In addition, the use of these piles allows achieving energy savings on pile driving, as the prismatic pile is initially driven, and as a result of transformation, it exploits the pyramid pile [13].

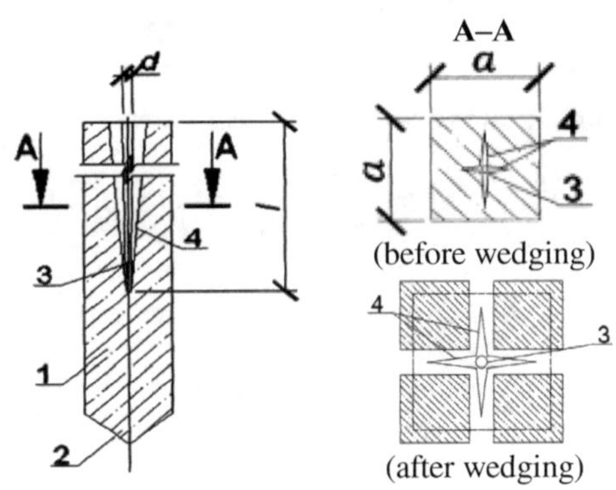

Figure 1. Scheme of a transformable driving pile: 1 – reinforced concrete pile body; 2 – sharpened end of the pile; 3 – plastic tube; 4 – plastic petals that divide the body of the pile into four parts.

However, to obtain a transformable pile in the manufacture of a conventional prismatic reinforced concrete pile, it is necessary to install a plastic cone-shaped cruciform embedded part with a 0.75 length of the pile length, dividing the pile body into four equal parts, in the center of the pile trunk.

Technological sequence of construction of the transformed pile:

– sinking to the design mark of a prismatic reinforced concrete pile with changes in its design made during manufacture;

– submission of a non-explosive destructive substance from above into a plastic cone-shaped cruciform embedded part;

– technological break: during 1.5...2 days, non-explosive destructive substance, increasing in volume by 2...3 times, wedges the body of the pile into four equal parts, thus creating a pyramid pile [14].

A non-explosive destructive substance is a powdery material that solidifies with increasing volume when interacting with water. They prepare a non-explosive destructive substance from crushed quicklime with various additives to improve performance: setting and hardening retarders, lime hydration accelerators, plasticizers and other additives.

2.2. Pile for use in loose soil

The authors solve the problem of increasing the load-bearing capacity of driving piles in loose soils, while the experiments of sinking piles with plaster were performed, given in the works [15-20].

IOP Conf. Series: Materials Science and Engineering **945** (2020) 012035 doi:10.1088/1757-899X/945/1/012035

The design of the developed and researched prismatic pile provides for the arrangement along the side faces of "pockets" (recesses) randomness, without affecting the working armature of the pile. The structure of the pile proposed by the authors is shown in Fig. 2.

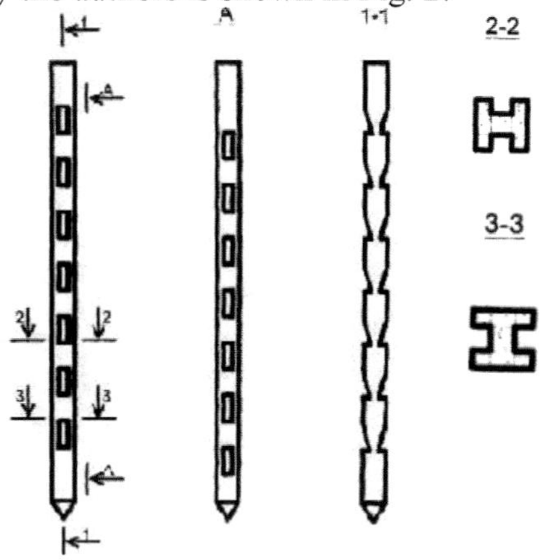

Figure 2. Construction of a reinforced concrete pile for use in loose soils.

When studying the load-bearing capacity of piles immersed in a hardening composition, the following assumptions are accepted:

– during operation the pile body and the surrounding soil perceive the load as a single structure;

– the liquid hardening compound located in the "pockets", during the hammering process, experiences hydrodynamic shocks, as a result of which it penetrates the soil of the pile and fixes it. A silicate-phosphoric formulation using orthophosphoric acid as a hardener and an aluminosilicate formulation using sodium aluminate as a hardener were used as a fixing compound for laboratory studies;

the depth of penetration of the chemical reagent into the soil depends on the porosity of the soil – *e*, the number of impacts of the moment – *n*, that is, in the upper part of the pile, it is much greater than in the lower, and the system "pile – fixed soil" has a slope of the side faces.

Transformable prismatic pile.

To determine the effectiveness of the transformed pile, the authors conducted laboratory tests of models of driving piles using a lever mechanism for static immersion of pile models in fine air-dry sand (Fig. 3, *a*).

a)

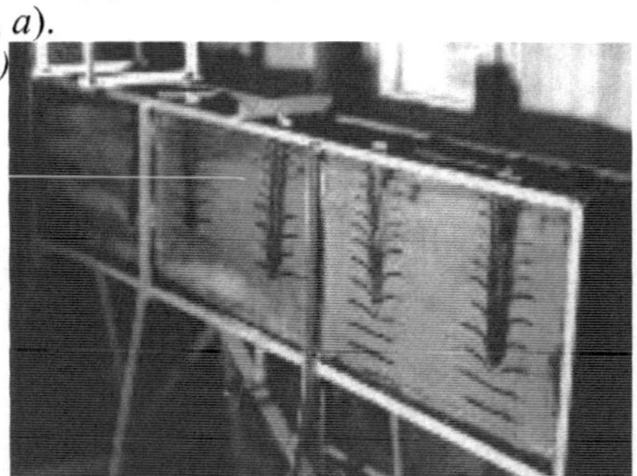

b)

Figure 3. Checking the effectiveness of transformed piles: a) the nature of deformation of the soil adjacent to the pile when it is driven and wedged; b) field tests of wedging four sections of a 1.5 m deep driving pile model using a non explosive destructive substance in fine medium density sand.

Measurements of the resulting deformations of the adjacent soil showed that when driving piles, the adjacent soil sinks together with the pile. In this case, the thickness of the soil layer involved on the surface of the tray reaches 1.0 *b* (where *b* is the cross-section of the pile), at a depth of 0.5 of the pile length it reaches up to 0.5 *b*, at the tip of the pile – equal to zero. When wedging models of piles, the adjacent soil is compacted in the horizontal direction. The thickness of the compacted soil varies within 0.5...1.0 *b* depending on the angle of inclination of the pile faces and depth.

In addition to laboratory tests, field experiments were conducted on models of piles submerged in fine sand of medium density to a depth of 1.5 m. Transformation of models of prismatic piles into pyramidal ones during field tests is shown in Fig. 3, *b*.

The results of studies of prismatic (basic) models and transformable (pyramidal) piles are shown in Fig. 4.

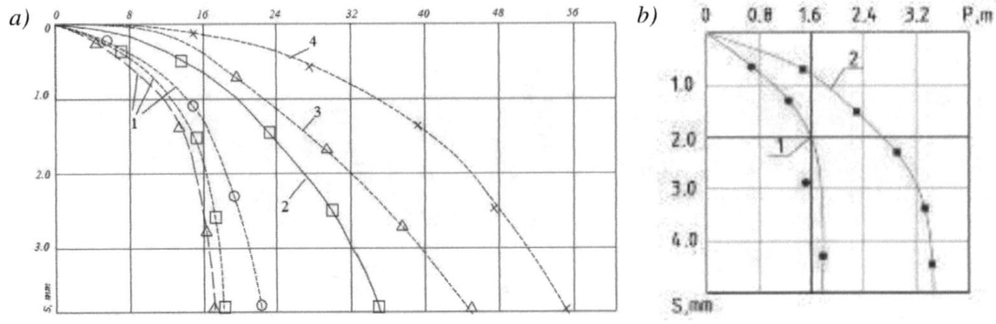

Figure 4. The results of the tests of model piles: *a)* laboratory data on the dependence of pile upset on load; *b)* field test data; *1* – for prismatic piles ($i_1 = 0.00$); *2, 3, 4* – for piles with the stress of the host soil medium by wedging in the same plane (respectively $i_2 = 0.02$; $i_3 = 0.04$; $i_4 = 0.074$).

The results of experimental studies allow concluding that the upset pattern of prismatic pile models and transformed pile models is qualitatively different. Landing a prismatic pile to a certain value is almost directly proportional to the load, after which the pile upset increases sharply, and it loses its bearing capacity.

When the load on the transformed piles increases, there is no sharp increase in upset. This is because, giving the angle of the side faces of the straining piles and the adjacent soil, with subsequent sediment pilings continues to solidify the adjacent soil mass and thus increases the bearing capacity of the pile.

Studies of the efficiency of transformable piles compared to prismatic ones have confirmed an increase in the load-bearing capacity of transformable piles by 40...45 %.

2.2. Piles for use in loose soils

Taking into account the accepted assumptions, models of piles with a size of 20×20×300 mm were tested in a soil tray with fine sand (Fig. 5, *a*).

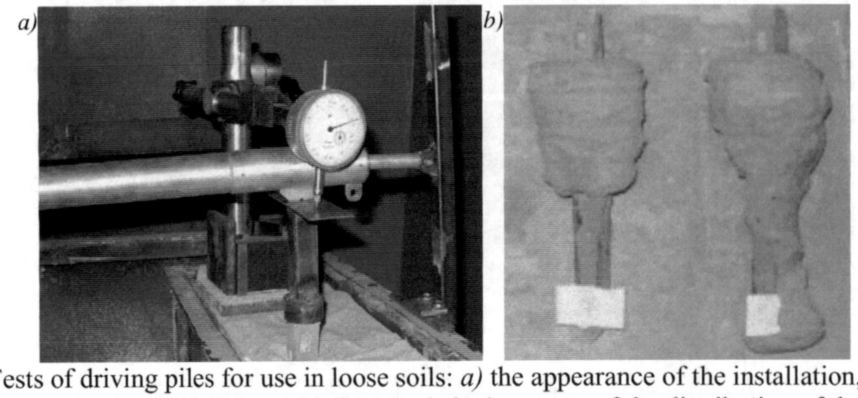

Figure 5. Tests of driving piles for use in loose soils: *a)* the appearance of the installation, devices and models of the pile in a soil tray with fine sand; *b)* the nature of the distribution of the fixing composition in the depth of models of piles with "pockets" during their driving and after their load testing.

Model of piles with “pockets” with a different location on the lateral surface and the base (without “pockets”) was loaded shock method, as piling with the “pockets” in the collar mounted on the surface of the sand, was added to the fixing composition (volume of structure for each pile – 150 g). When driving the reference (base) models of piles, the number of blows was 80 blows, driving models with “pockets” and in the presence of a fixing liquid composition around the pile, the number of blows was reduced to 30...35. Therefore, the presence of lubricants and greases when driving piles reduces the energy spent on sinking the piles. In our experiments, the energy of driving models using fixing compounds is reduced by more than 50 % compared to the immersion of reference piles.

The holding time of driven pile models before the start of testing is 7...8 days. Static load on the pile model is applied in steps, measuring the amount of their upset. As a result of tests, the bearing capacity of reference (base) piles and piles with “pockets” was established. The results are shown in Fig. 6. The appearance of excavated models of piles with “pockets” is shown in Fig. 5, *b*.

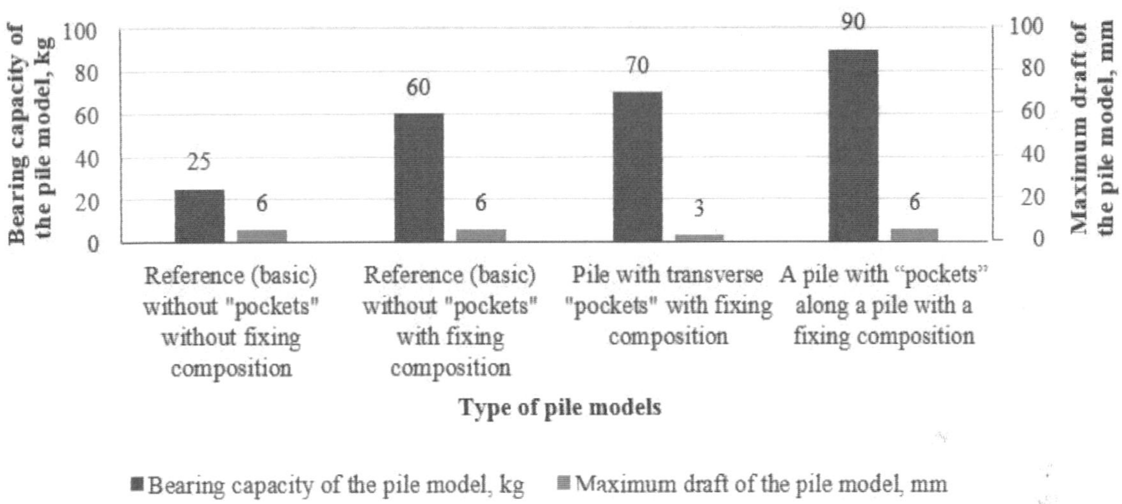

Figure 6. Results of laboratory tests of pile models in a ground tray with fine sand (air-dry humidity).

3. Summary

As a result of studies of transformable driving floating piles, the possibility of transforming prismatic driving piles into pyramidal ones with a slope of the side faces using a non-explosive destructive substance was proved. It was found that when driving a model of a prismatic pile, the adjacent thin layer of soil sinks together with the pile, and when the pile is wedged, the soil is compacted horizontally by 0.5 *b* at a depth of 0.5 of the pile length and 1.0 *b* at the ground surface, and depends on the angle of inclination of this pile during its transformation. A comparison of the hammering energy of pre-fabricated pyramidal precast concrete piles and prismatic transformable ones suggests that the transformation of piles reduces the energy of their hammering by more than 30 %.

Studies of the developed piles for use in loose soils using chemical fixing compounds allowed establishing that the body of the pile and the fixed soil around it under the influence of load work as a single structure, as when the pile is deposited, the cut plane occurs at the contact “fixed soil – soil”. The thickness of the fixed soil in the upper part of driven piles is considerably greater than in the bottom, while the depth of the pile it gradually decreases, resulting in a prismatic pile is transformed in the pyramid one, where the carrying capacity is much larger than the prismatic. The use of piles with “pockets” on the side surface, through which the chemical composition fixes the loose soil around the pile, allows increasing its load-bearing capacity by 2...3 times. The presence of liquid fixing compounds around the perimeter of the piles reduces the energy of immersion by more than 50% when driving them in comparison with the base piles.

4. References

[1] Saenko Y V and Nevzorov A L 2015 Deformation and Strength Characteristics of Piling Foundation Soil *Gornyi Zhurnal* 17-22

[2] Fakharian K and Khanmohammadi M R 2015 Numerical Modeling of Pile Installation Effects on Stress State in Clay *15th Asian Regional Conf. on Soil Mechanics and Geotechnical Engineering*

[3] Wulandari P S and Tjandra D 2015 Analysis of Piled Raft Foundation on Soft Soil Using PLAXIS 2D *Procedia Engineering*

[4] Baziar M H, Kashkooli A and Saeedi-Azizkandi A 2012 Prediction of Pile Shaft Resistance Using Cone Penetration Tests (Cpts) *Computers and Geotechnics* **45** 74-82

[5] Hokmabadi A S, Fatahi B and Samali B 2015 Physical Modeling of Seismic Soil-Pile-Structure Interaction for Buildings on Soft Soils *Int. J. of Geomechanics* **15(2)** 04014046

[6] Utkin V S 2018 Experiments and Design of Reinforced Concrete Friction Piles by Bearing Capacity of the Pile and Soil Base *Vestnik MGSU* **(8)** 952-958

[7] Basu P, Loukidis D, Prezzi M and Salgado R 2014 The Mechanics of Friction Fatigue in Jacked Piles Installed in Sand *From Soil Behavior Fundamentals to Innovations in Geotechnical Engineering*

[8] Gabibov F 2014 Workig out new technical solitions to eliminate forces of negative friction while setting pile fondations in subsuding soils *PNRPU Construction and Architecture Bulletin* **(3)** 167-78

[9] Reinforced-Concrete Driven Pile *Dictionary Geotechnical Engineering/Wörterbuch GeoTechnik* 1097-1097

[10] Prekop L 2017 Verification of the Vertical Bearing Capacity of a Reinforced Concrete Pile *Procedia Engineering* **190** 536-539

[11] Lundin L S and Rabinovich E A 1977 Performance of Driven Reinforced-Concrete Piles in a Pile Grillage Subject to Lateral Loading *Soil Mechanics and Foundation Engineering* **14(2)** 116-21

[12] Gaido A 2017 Development of Work Execution Design for Construction of Pile Foundations in Constrained City Conditions *PNRPU Construction and Architecture Bulletin* **8(4)** 74-85

[13] Kocherzhenko V V and Migrin DA 2017 Driven transportable pile *Utility Model Patent, №* 173756 RF Bul. 26.

[14] Sheremetyev Yu G and Udalov V V 1995 Brief excursion into the technology of obtaining and using environmentally friendly non explosive collapsible substances *Problems of construction materials science and new technologies* 113-120

[15] Kocherzhenko V V and Sokolov N G 2011 Development and research of a new design of a driving pile for use in loose sand *Construction industry: yesterday, today, tomorrow* 157-161

[16] Kocherzhenko V V and Sokolov N G 2013 Reinforced concrete driving pile for chemical fixing of loose sand and methods of fixing the soil surrounding the body of the pile *Utility Model Patent, № 2492293RF. Bul. 25*

[17] Kocherzhenko V V and Sokolov N G 2012 Selection of the optimal composition of fixing reagents to increase the load-bearing capacity of piles in loose soils *Efficient building structures, buildings and facilities* 112-116

[18] Dmitrieva T V, Markova I Yu, Strokova V V, Bezrodnykh A A and Kutsyna N P 2020 Effectiveness of stabilizers of various compositions when strengthening soils with mineral binders *Construction materials and products* **3(1)** 30-38

[19] Shaposhnikov N N, Sargsyan A E and Gerashchenko V S 2012 Calculation Model of Pile Foundations taking into account the Effect of their Interaction with the Ground Environment *Vestnik MGSU* **(4)** 69-72

[20] Bekarov I I 2019 On the Peculiarities of Modeling Process of Piles filling in Laboratory Conditions *Bulletin of LN Gumilyov Eurasian National University Technical Science and Technology Series* **127(1)** 19-27

IOP Conf. Series: Materials Science and Engineering **945** (2020) 012036 doi:10.1088/1757-899X/945/1/012036

Method for predicting the remaining life of the ball mill support axle based on changes in the design stresses of the external surface of the axle

J A Bondarenko[1], T A Duyun[2*] and O V Bestuzheva[3]

[1]Professor of Department of mechanical engineering, BSTU named after V.G. Shukhov, Belgorod, Russia,
[2]Head of Department of mechanical engineering, BSTU named after V.G. Shukhov, Belgorod, Russia,
[3]Docent of the Department of applied informatics and information technologies, Belgorod National Research University, Belgorod, Russia,

E-mail: kdsm2002@mail.ru

Abstract. The article considers a method for predicting the residual life of the ball mill support axle based on changes in the calculated stresses of the external surface of the axle. The main terms of forecasting the remaining life of large-sized equipment are considered. The task is set and controlled parameters are selected for the development of the methodology. An algorithm for implementing a method for predicting the residual life of the ball mill support axle based on external surface stresses is presented. A method has been developed for predicting the residual life of ball mill support axles based on changes in the design stresses of the external surface of the trunnion, which allows you to set parameters that affect the duration of the residual life without stopping the mill and to adequately calculate the residual life of the trunnion, including from the start of operation of the mill. The homogeneity of the obtained data is estimated. The parameters of the stress distribution of the external surface of the trunnion are determined, indicating that the obtained distribution corresponds to the theoretical Weibull distribution and the validity of using this method for predicting the resource of the ball mill support trunnion. The values of the guaranteed residual life of the ball mill trunnion are calculated according to the developed method. The analytical dependence of the residual life of the ball mill trunnion on the equivalent stresses occurring on the external surface of the trunnion is given.

1. Introduction

In the production of cement, grinding equipment is the main link of the production line [1]. A significant dynamic impact of the load on the mill leads to a loss of performance, which contributes to long downtime in repairs. Increasing the productivity of equipment is achieved either by replacing it with a more productive one, or by increasing the reliability of the existing one [2].

According to GOST 27.003-2016 [3], the operational reliability of the object of continuous long-term use with the possibility of recovery and maintenance, which is the ball mill, is characterized by indicators of durability. Durability is the property of an object to maintain its functional state until the

IOP Conf. Series: Materials Science and Engineering **945** (2020) 012036 doi:10.1088/1757-899X/945/1/012036

limit state is reached with the installed maintenance and repair system, according to GOST 27.002-2015 [4].

According to table B. 2 of GOST 27.003-2016 [3], the indicator of durability for an object whose transition to the limit state when used for its intended purpose can lead to consequences, with the main process determining the transition to the limit state – wear and aging simultaneously, is the average and gamma percent (guaranteed) residual resource and the average and gamma percent (guaranteed) service life. To assess the durability from the selected reference point to the onset of the limit state of the equipment, use the average and gamma percent (guaranteed) residual life.

The residual resource is the total operating time of object from the time of control of its technical condition up to the moment of reaching the limit state, with average life – expectation of the resource to a guaranteed resource is the total time during which the object reaches a limiting state with probability, expressed in percentage according to GOST 27.002–2015 [4].

Operational failures of equipment are manifested in changes in the parameters of the technical condition of the directly measured values of damage or output parameters of the equipment [4]. Their monitoring allows you to predict the moment when hardware failure occurs. According to GOST 27.002-89 [4], a failure is an event that results in a violation of the object's functional state. The technical condition of a ball mill, in which its further operation or restoration is impractical or unacceptable, is its limit state [3]. Restoration of worn-out equipment due to technical and economic indicators or violations of established safety requirements, according to GD 09-10-95 and GOST R 22.2.05-94 [5, 6], is impractical.

The limit state of the equipment, according to GOST 27.002-89 [4], is characterized by the criteria of the limit state. When providing operational and technical parameters, the operating time from monitoring the technical condition of the ball mill or its repair, the average gamma percentage and residual resource is an indicator of durability.

The complex stress-strain state of the ball mill support axle determines the dependence of stresses and deformations on the temperatures of the internal and external surface of the axle [7]. To determine deformations, measurements of deviations in the nominal diameter of the external surface of the trunnion are required. Based on the obtained data, it is possible to estimate the residual life of the ball mill trunnion by changing the output parameters, according to GD 26.260.004–91 [6].

The task is to develop a methodology for predicting the remaining life of the trunnion, which will allow you to adequately calculate the remaining life of the trunnion without stopping the mill, including from the beginning of its operation, and to establish parameters that affect its duration. The controlled parameter for the development of the method is the stress of the external surface of the trunnion, which occurs under the influence of gravity, rotation and temperature field.

2. Materials and methods

The algorithm for implementing the method for predicting the residual life of the ball mill support pin by external surface stresses is shown in figure 1. the Method for predicting the residual life of the pin by changing the controlled parameter-voltage includes the following steps:

1. Getting initial data: the temperature of the external surface of the unloading pin and the temperature of the material coming out of the mill.
2. Determination of surface areas for calculating the stresses of the external surface of the ball mill trunnion.
3. Calculation of equivalent stresses of the external surface of the ball mill trunnion by sections, according.
4. Statistical processing of the received data:

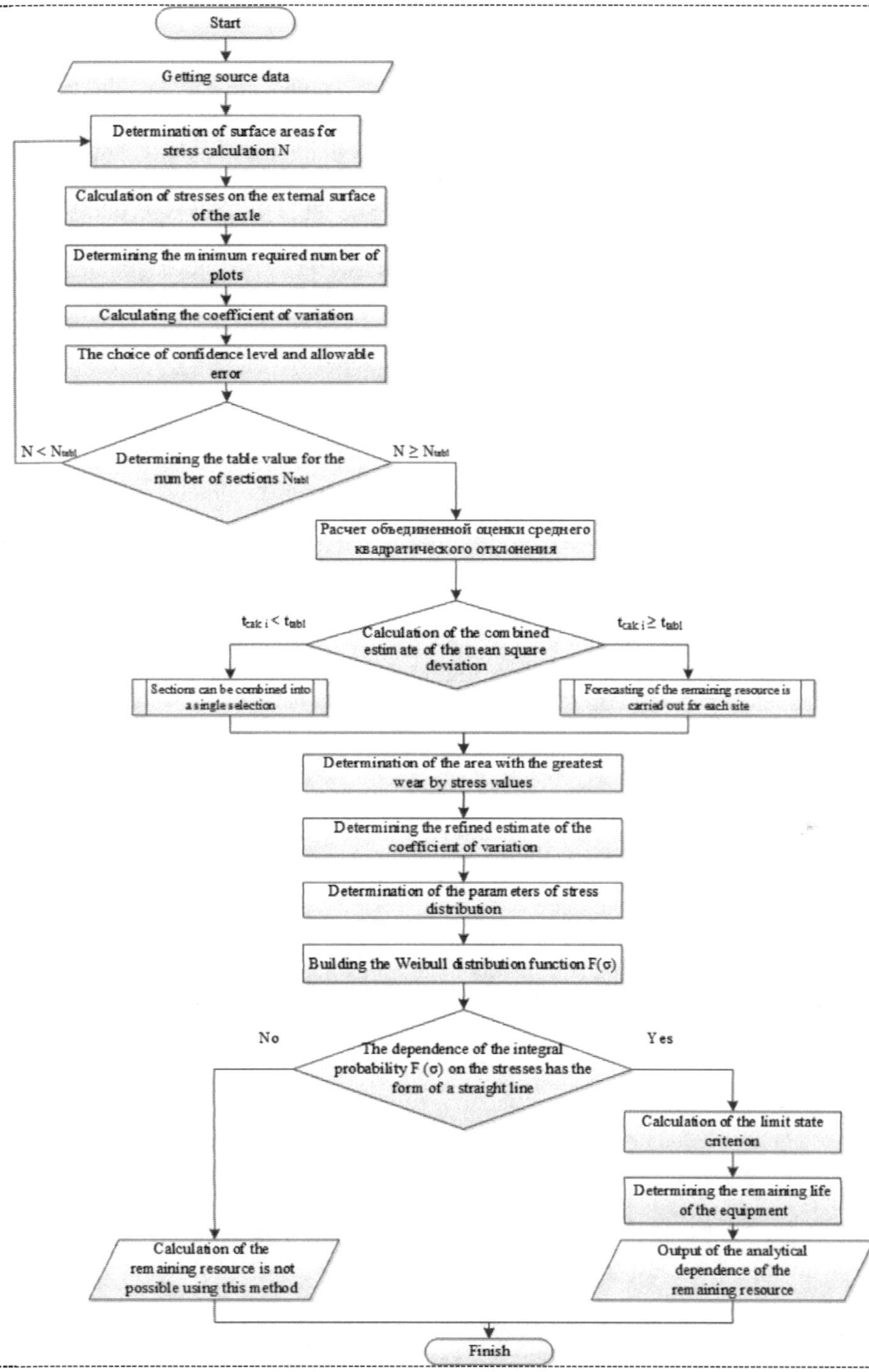

Figure 1. Algorithm for implementing the method for predicting the residual life of the trunnion by external surface stresses.

4.1. Determination of the minimum required number of plots for calculation:
4.1.1. Calculation of average values, average square deviation and coefficient of variation.
4.1.2. Selecting the confidence probability and maximum permissible relative error.
4.1.3. Comparison of the selected number of plots with the table value.

4.2. Assessment of the homogeneity of the obtained sample:
4.2.1. Calculation of the combined estimate of the mean square deviation.
4.2.2. Determining the calculated value of the student's criterion.
4.2.3. Comparison of the calculated and tabular values of the student's criterion.
4.2.4. Conclusion about the homogeneity of the sample and the possibility of combining the sample sections.
4.3. Check the sound application of the technique of forecasting of a residual resource of axle:
4.3.1. The choice of the site of greatest wear on the outer surface of the axle.
4.3.2. Determination of a refined estimate of the coefficient of variation for the area of greatest wear.
4.3.3. Selection of distribution parameters based on the updated estimation of the coefficient of variation.
4.3.4. Output of the Weibull probability distribution function with the construction of a dependency graph and its data to determine the reliability of the developed method.
5. Predicting the remaining life of the ball mill support axle:
5.1. Formulation of the criterion for the limit state of the ball mill axle.
5.2. Determination of the average wear rate and coefficient of variation of the resource.
5.3. Finding the average and guaranteed residual life of the ball mill trunnion.
6. Conclusion of the analytical dependence of the residual resource on the equivalent stresses occurring on the external surface of the trunnion.

A change in the technical condition parameter-the stresses that occur on the outer surface of the trunnion, indicates uneven wear [8], explained by the stochastic property of the system: temperature, composition, speed, and concentration of material particles [9].

3. Results

According to figures [7], the stresses on the outer surface of the ball mill pin change as the temperature of the inner surface increases. At the same time, it was found that the highest stress values are reached on the external surface of the unloading pin. Therefore, let's consider the method of predicting the residual life of the unloading trunnion on the example of a ball mill with a standard size with diameter of 3.2×15 m.

1. Determine the surface areas for calculating the stresses on the outer surface of the ball mill trunnion. It is impractical to estimate the degree of wear of the ball mill trunnion along the length of the working cylindrical surface for all sections under the same conditions. Selected 6 sections for determining the stress (figure 2). It is required to perform calculations on the sections of the external surface of the trunnion.

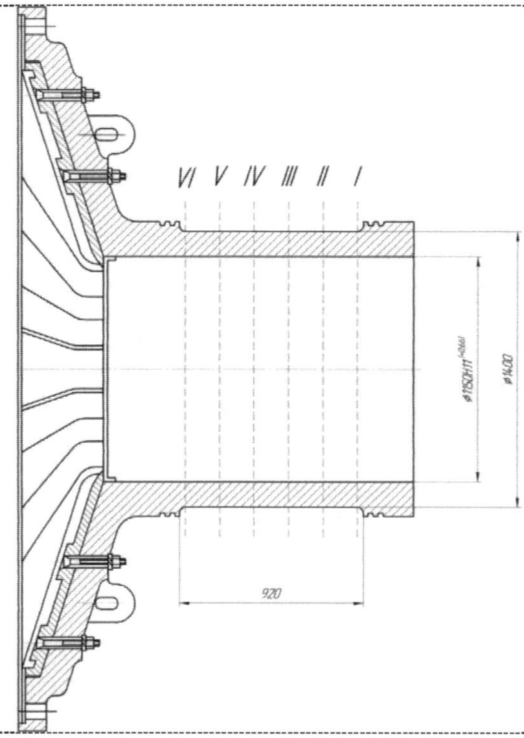

Figure 2. Layout of sections for calculating the stress on the surface of the trunnion.

Figure 2 shows the layout of the sections for calculating the stress on the surface of the trunnion.

2. Calculate the stresses on the outer surface of the trunnion at an average constant temperature of the outer surface of the unloading trunnion T_{R_2} = 58°C [7], varying the values of the inner surface temperature. The characteristics of the calculated stresses on the external surface of the trunnion are shown in table 1.

3. Produce statistical processing of the obtained data.

3.1. Determine the minimum required amount of sample size.

To do this, calculate the statistical characteristics: the average value of the external surface stress on the site, the average square deviation and the coefficient of variation (table 1). Table 1 shows and calculates the following characteristics:

- T_{R_1} – change in the temperature of the inner surface of the axle, °C;
- σ_i – calculated value of the equivalent voltage on the section, MPa;
- $\overline{\sigma_\iota}$ – average value of equivalent stresses on the i-th section, MPa:

$$\overline{\sigma_\iota} = \frac{\sum_{i=1}^{n} \sigma}{n}; \tag{1}$$

- S_i – mean square deviation on the I-th section, determined by the formula, according to [10]:

$$S_i = \sqrt{\frac{\sum(\sigma_i - \overline{\sigma_\iota})^2}{n_i - 1} - \left(\frac{\Delta}{3}\right)^2}, \tag{2}$$

where Δ – limit error of the method when determining the residual life of equipment by voltage;

- V_i – coefficient of variation on the i-th section, according to [10]:

$$V_i = \frac{S_i}{\overline{\sigma_i}}. \quad (3)$$

Table 1. Characteristics of the calculated voltages.

№ section	T_{R_1}, °C	σ_i, MPa	$\overline{\sigma_i}$, MPa	S_i	V_i
	115	137.33			
	116	139.73			
I	117	142.14	143.10	4.9761	0.0348
	119	146.95			
	120	149.35			
	121	151.76			
	122	154.16			
II	124	158.97	158.01	4.9761	0.0315
	125	161.37			
	126	163.78			
	127	166.18			
	129	170.99			
III	130	173.40	173.40	5.3674	0.0310
	131	175.80			
	133	180.61			
	134	183.02			
	135	185.42			
IV	137	190.23	189.75	5.7320	0.0302
	138	192.63			
	140	197.44			
	140	197.44			
	142	202.25			
V	142	202.25	202.73	3.5524	0.0175
	143	204.66			
	144	207.06			
	145	209.47			
	145	209.47			
VI	147	214.27	214.27	5.0909	0.0238
	148	216.68			
	150	221.49			

3. To select the required minimum number of plots for $\mathbf{N_{tabl}}$ calculations, you need to determine the values of confidence probability and acceptable error. Accept the confidence probability of the estimation, according to [11]:

$$\Delta = 0.95; \quad (4)$$

4. and the maximum permissible relative error, according to [11]:

$$\mathrm{D} = 0.10. \quad (5)$$

5. The calculated coefficients of variation (table 1) do not exceed the value of 0.1. According to table 4 [12], for the given values (4) – (5), we find the minimum:

6. $\mathbf{N_{tabl}} = 5$,

7. this confirms the validity of the selected number of sections, equal to N= 6.

8. 3.2. Perform an estimation of the homogeneity of the obtained data. Since different areas of the ball mill shaft surface experience different intensity of stress influence during operation [7], it is necessary to check the obtained data for uniformity. Checking the uniformity of the data obtained by the student's criterion of adjacent sections of the axle surface using the formula, according to [12]:

$$t = \frac{|\overline{\sigma_i} - \overline{\sigma_{i+1}}|}{S\sqrt{\frac{1}{n_i} + \frac{1}{n_{i+1}}}} < t_T, \qquad (6)$$

9. where $\overline{\sigma_i}$, $\overline{\sigma_{i+1}}$ – mean value of the voltage at the i-th and (i+1)-th portions of the surface; n_i, n_{i+1} – number of measurements on adjacent areas; S – joint estimation of mean-square deviation, calculated according to the formula [11]:

$$S = \sqrt{\frac{S_i^2(n_i - 1) + S_{i+1}^2(n_{i+1} - 1)}{n_i + n_{i+1} - 2}}, \qquad (7)$$

10. where S_i, S_{i+1} – RMS deviation at the i-th and (i+1)-th sections, defined by the formula (2) (table 1);

11. t_T – tabular value of the student's criterion for the calculated number of degrees of freedom (8) and significance level (3.4).

12. The number of degrees of freedom, according to [10], is determined by the formula:

$$f = n_i + n_{i+1} - 2 = 8. \qquad (8)$$

13. Let's calculate the values of the combined estimate of the mean square deviation for the sample sections and the student's criteria. The calculation results are shown in table 2.

14. Since condition (6) is met on all surface areas, the stress differences are not statistically significant and the calculation results can be combined, which will improve the accuracy of predicting the remaining life [12]. Thus, it is proved that the resource can be determined for the entire external surface of the trunnion by the values of the area of greatest wear [6].

15. 3.3. The reasonable application of the method of forecasting the residual life of the ball mill trunnion is verified. According to tables 1-2, the area of the greatest wear of the external surface is the VI section.

16.

Table 2. Assessment of the homogeneity data.

Sections		Statistical characteristic				Comparison with a table $t_T = 2.306$
		σ_i, MPa	S_i	S	t	
	I	143.10				
I – II			4.9761 4.9761	4.9761	1.8948	$t < t_T$
	II	158.01				
II – III			4.9761 5.3674	5.1754	1.8806	$t < t_T$
	III	173.40				
III – IV			5.3674 5.7320	5.5527	1.8624	$t < t_T$
	IV	189.75				
IV – V			5.7320 3.5524	4.7684	1.7222	$t < t_T$
	V	202.73				
V – VI			3.5524 5.0909	4.3896	1.6629	$t < t_T$
	VI	214.27				

17. A refined estimate of the stress variation coefficient was determined [11]:

$$V_\sigma = \sqrt{\frac{\sum_{i=1}^{N} V_i^2 \cdot n_i}{N}}, \qquad (9)$$

18. where $V_{\sigma i}$ – coefficient of variation on the surface area of the greatest wear; n_i – number of measurements on this area; N – number of surface areas.

19. Then

$$V_\sigma = 0.0642. \qquad (10)$$

20. 4. To describe failures caused by wear on the part surface, we will use the Weibull distribution function [13], which has flexibility in approximating the distribution and is widely used in the practical implementation of the reliability theory [14].

21. The distribution function is assumed in the form, according to [15]:

$$F(\sigma) = 1 - e^{-\left(\frac{\sigma_i}{a}\right)^b}, \quad (11)$$

22. where $\boldsymbol{\sigma_i}$ – equivalent stresses of the external surface of the axle; a - parameter of the distribution scale, determined by the formula, according to [16]:

$$a = \frac{\sigma_{imax}}{k_v}, \quad (12)$$

23. where $\boldsymbol{k_v}$ is the coefficient that depends on the coefficient of variation;
24. $\boldsymbol{b}$ – parameter of the distribution form, determined by the coefficient of variation.
25. For the calculated refined estimation of the stress variation coefficient (10), it is necessary to determine the values of the distribution parameters.
26. Distribution parameter values defined according to [17]:

$$k_v = 0{,}887;\ a = 274{,}5;\ b = 3{,}6;\ \gamma = 0{,}05. \quad (13)$$

27. According to the calculated values (13), the voltage distribution function (11) takes the form:

$$F(\sigma) = 1 - e^{-\left(\frac{\sigma}{274{,}5}\right)^{3{,}6}}. \quad (14)$$

28. The results of statistical processing were plotted on Weibull's probability paper (figure 3).
29.

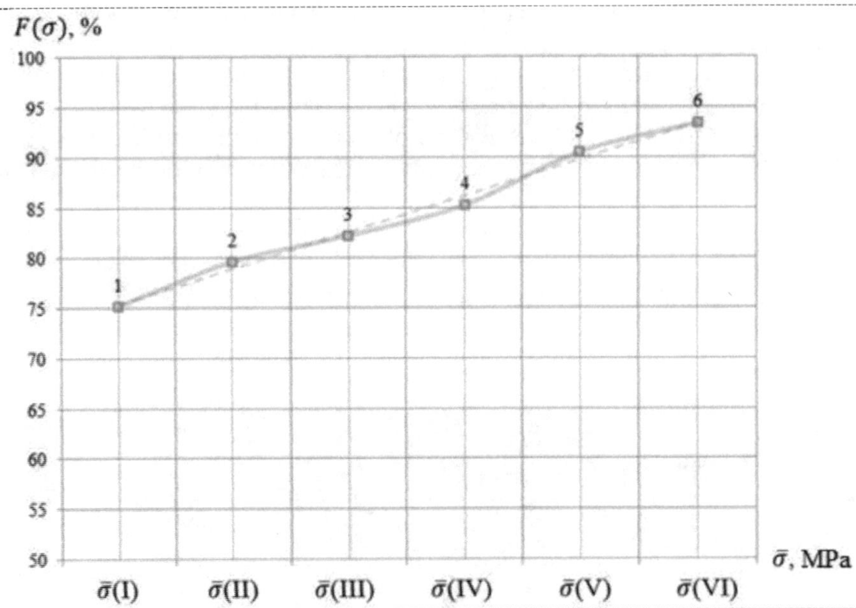

Figure 3. Stress distribution on the Weibull probability paper: 1 – probability $F(\sigma) = 75{,}1\%$ at $\bar{\sigma} = 214{,}27$ MPa; 2 – probability $F(\sigma) = 79{,}5\%$ at $\bar{\sigma} = 202{,}73$ MPa; 3 – probability $F(\sigma) = 82{,}4\%$ at $\bar{\sigma} = 189{,}75$ MPa; 4 probability $F(\sigma) = 85{,}1\%$ at $\bar{\sigma} = 173{,}4$ MPa; 5 – probability $F(\sigma) = 91{,}3\%$ at $\bar{\sigma} = 158{,}01$ MPa; 6 – probability $F(\sigma) = 93{,}5\%$ at $\bar{\sigma} = 143{,}1$ MPa.

30. The stress distribution graph indicates the probability of average stress values as a percentage: a decrease in stress leads to an increase in the distribution function (figure 3). In the graph shown in figure 3, the points are approximated in a single line, which corresponds to the obtained theoretical Weibull distribution and indicates the validity of using this technique to predict the resource of the ball mill trunnion.
31. 5. Determine the residual resource of the ball mill axle. To predict the remaining life, it is necessary to formulate a criterion for the limit state of the ball mill trunnion. The strength criterion for a complex stress-strain state of the ball mill trunnion with a standard size with diameter of 3.2×15 m, made of a material – 35L steel, is determined according to [18]:

$$\sigma_i < 490\ \text{МПа}. \quad (15)$$

32. Given [19], expression (15) will have the form:

$$\sigma_z + \frac{\tau_{-1}}{\sigma_{-1}}\sigma_\varphi < 490 \text{ МПа}, \quad (16)$$

33. To determine the average and guaranteed residual life, find the average wear rate and the coefficient of variation of the resource.

34. The average wear rate is determined by the formula:

$$c = \frac{\alpha}{\tau} = \frac{274{,}5}{4} = 68{,}61 \text{ МПа/г}, \quad (17)$$

35. where τ – average time to failure in years, according to [20].

36. The calculated average life of the ball mill trunnion is determined according to [6]:

$$\mathrm{T_p} = \frac{\sigma_\text{в} \cdot \mathrm{k_V}}{c \cdot (-\ln \gamma)^{\frac{1}{b}}} = \frac{490 \cdot 0{,}887}{68{,}61 \cdot (-\ln 0{,}05)^{\frac{1}{3{,}6}}} = 4{,}67 \text{ years}. \quad (18)$$

37. To determine the guaranteed resource, we calculate the coefficient of variation of the resource, according to [12]:

$$V_T = \frac{V_\sigma}{\sqrt{\sum_{i=1}^{N} n_i}} = \frac{0{,}0642}{\sqrt{30}} = 0{,}011. \quad (19)$$

38. The guaranteed (gamma-percent) resource is determined by the formula, according to [6]:

$$T_\gamma = T_p(1 - u_\gamma V_T), \quad (20)$$

39. where $\boldsymbol{u_\gamma}$ – quantile of the normal distribution with the level γ = 0.05: $\boldsymbol{u_\gamma} = 1.645$.

40. Thus, the value of the guaranteed residual life of the ball mill axle is calculated:

$$T_\gamma = 4.59 \text{ years}. \quad (21)$$

41. 6. Conclusion of the analytical dependence of the residual resource on the equivalent stresses occurring on the external surface of the trunnion.

42. A thorough technical diagnostics is required to assess the remaining service life of a ball mill depending on the complex stress-strain state of the trunnion. For diagnostics, devices are used that control the metal surface and determine the distribution and intensity of certain interference bands, which visually determine the concentration of mechanical damage, but the direction and values of deformations and stresses are ignored [21]. This method does not provide a complete assessment of the stress-strain state of the surface.

43. Complex stress-strain state of the axle considered in the second Chapter, involves the wear of the sliding surface, depending on many factors: accuracy of selection of materials of the bearing and the axle that the contact points of the insert bearing and axle, observance of technological process of recovery, compliance with performance standards, timeliness of diagnostics and preventive maintenance [22].

44. In view of the complexity of estimating the remaining life and the impossibility of conducting field tests to determine the durability of the operating equipment, depending on the above factors, it is of interest to predict the durability.

45. Tests of samples at various constant deformation rates [23] indicate that the influence of the duration of loading on the mechanical strength of parts is conditioned. In the course of durability studies conducted by many scientists [24, 22, 25], experimental and theoretical results have been obtained, which are the fundamental basis for the development of the kinetic concept of solid strength, which is currently generally recognized.

46. Wear is considered as an irreversible temporary process in the deformable volume of structural damage to the surface of the material. The destruction process occurs when the critical level of damage is reached, according to (15) [22].

47. Quantitative measure of damage is considered a partial gap corresponding to the relative duration of the voltage to the longevity of details in this load:

$$\sum_{i=1}^{k} \frac{t_i}{\tau_i(\sigma_i)} = 1, \quad (22)$$

48. where $\boldsymbol{t_i}$ – the time of action of the tension; $\boldsymbol{\tau_i(\sigma_i)}$ – durability parts for a given load.

49. If the voltage is determined by a continuous function of time, then the Bailey criterion is implemented instead of the condition (3.21) [25]:

IOP Conf. Series: Materials Science and Engineering **945** (2020) 012036 doi:10.1088/1757-899X/945/1/012036

$$\int_0^{\tau} \frac{dt}{\tau[\sigma(t)]} = 1, \tag{23}$$

50. which allows us to establish the time dependence of the strength for a given loading mode for an analytically defined function $\boldsymbol{\tau[\sigma(t)]}$ for $\boldsymbol{\sigma(t) = \sigma = const}$.

51. The linear law of damage accumulation in the form of equations (22) and (23) is approximate [23, 26] and is valid only under relatively unchanged material structure, that is, under conditions of brittle destruction in the absence of shape change. Nonlinear laws are more universal.

52. According to [25], the time dependence of the material strength under dynamic conditions is described by an exponent in terms of the ratio of the equivalent stress to the maximum allowable stress value in logarithmic coordinates [26].

53. In accordance with the biki criterion [27], the analytical dependence of the residual resource [30] is obtained:

$$\mathrm{T} = \alpha \cdot e^{-\left(\frac{\sigma}{\sigma_S}\right)^{\gamma}}, \tag{24}$$

54. where $\boldsymbol{\alpha}$ – the material parameter that depends on external temperature factors (coefficient of linear expansion, deg.$^{-1}$); $\boldsymbol{\sigma_s}$ – the material strength limit, MPa; $\boldsymbol{\sigma}$ – the stresses arising during operation of the trunnion, MPa; $\boldsymbol{\gamma}$ – the level of significance, according to (13).

55. Substituting the numerical values $\boldsymbol{\sigma_s}$ and $\boldsymbol{\gamma}$, the dependence (24) will have the form:

$$\mathrm{T} = \alpha \cdot e^{\left(\frac{\sigma}{490}\right)^{0,05}}. \tag{25}$$

56. Substituting the range of stress values and the value of the linear expansion coefficient, we construct a graphical dependence of the residual resource on the equivalent stresses that occur on the external surface of the trunnion (figure 4).

57.

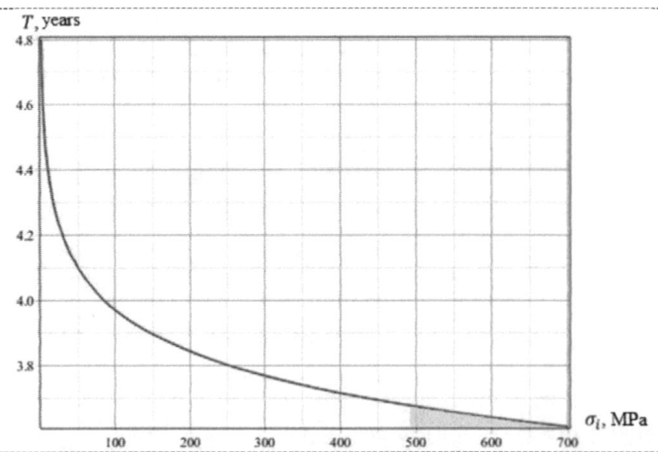

Figure 4. Graph of the dependence of the residual life of the ball mill trunnion on the equivalent stresses occurring on the external surface of the trunnion.

58. As follows from figure 4, the dependence has a monotonously decreasing character – with an increase in equivalent stresses, the residual resource decreases accordingly. The area of invalid values is highlighted in color on the graph, according to the limit state criterion (4).

59. 3. Results and discussions

60. Using a mathematical description and known analytical expressions, a method for predicting the residual life of ball mill support axles based on changes in the calculated stresses of the external surface of the trunnion is developed. The method allows you to set parameters that affect the duration of the residual life without stopping the mill and to adequately calculate the residual life of the trunnion, including from the start of operation of the mill. The homogeneity of the obtained data is estimated:

61. The parameters of the stress distribution of the external surface of the trunnion are determined, indicating that the obtained distribution corresponds to the theoretical Weibull distribution and the validity of using this method for predicting the resource of the ball mill support trunnion.
62. The values of the guaranteed residual resource of the ball mill trunnion are calculated according to the developed method: the average residual resource is $T_p = 4.67$ years, the guaranteed resource is $\mathbf{T_{p\gamma}} = 4.59$ years.
63. The analytical dependence of the residual life of the ball mill trunnion on the equivalent stresses occurring on the external surface of the trunnion is given.

4. Summary
Using a numerical method for a mill size with diameter of 3.2×15 m, the analysis of the dependence of stresses and deformations occurring on the external surface of the trunnion on the temperature difference of its internal and external surfaces was carried out. It was found that with an increase in the temperature difference in $\Delta T_R = 120°C$ ($T_{R_1} = 157°C$, $T_{R_2} = 37°C$), the stresses in both the circumferential and axial directions grow and amount to $\sigma_\varphi = \sigma_z = 172$ MPa, the deformations in the circumferential and axial directions monotonously increase and reach $\varepsilon_\varphi = 4.6$ mm, $\varepsilon_z = 5.8$ mm. The maximum strain value in the radial direction is $\varepsilon_r = 4.102$ mm. The unloading axle is subject to a more complex stress-strain state under the influence of a temperature field, and the external surface of the axle has the highest values of stresses and deformations.

5. References
[1] Boganov A 1961 *Mechanical equipment of cement plants* (Moscow: Mashgiz) 384
[2] Eremin N 1986 *Processes and apparatuses in the technology of building materials* (Moscow: Higher school) p 280
[3] GOST 27.003-2016 2017 Reliability in technology. Composition and General rules for setting reliability requirements *Moscow: standartinform* 23
[4] GOST 27.002-2015 2016 Reliability in technology. Terms and definitions *Moscow: standardinform* 28
[5] GOST R 22.2.05-94 1995 Safety in emergency situations *Moscow: publishing house of standards* 10
[6] GD 26.260.004-91 1992 Forecasting the residual life of equipment by changing the parameters of its technical condition during operation *Moscow: standardinform* 28
[7] Bestuzheva O, Bondarenko J and Khanin S 2019 Mathematical description of the stress-strainstate of the ball mill axle under the action of gravity and rotation *Bulletin of the Belgorod state technological University named after V. G. Shukhov* **3** 128-133
[8] Bestuzheva O 2018 Expert assessment of system reliability *European Scientific Conference: Collection of articles of the X International scientific and practical conference (Penza)* 69-71
[9] Abiev R and Strukov V 2018 *Reliability of mechanical equipment and complexes* (Saint Petersburg: Prospect Nauki) 288
[10] Lvovsky B 1988 *Statistical methods for constructing empirical formulas* (Moscow: Higher school) 239
[11] Voznesenskiy V 1974 *Statistical methods of experiment planning in technical and economic research* (Moscow: Statistics) 192
[12] Gmurman V 2000 *Theory and mathematical statistics: textbook for universities* (Moscow: Higher school*)* 479
[13] Karepov V, Bezverkhaya E and Chesnokov V 2018 *Reliability of mining machines and equipment* (Moscow: Infra-M) 133
[14] Pronikov A 1978 *Reliability of machines* (Moscow: Mashinostroenie) 592

[15] Pronikov A 1972 *Reliability and durability of machines and equipment. Experience and theoretical research* (Moscow: publishing house of standards) 314
[16] GOST 23942-80 1980 Evaluation of product quality indicators based on changes in the controlled parameter *Moscow: publishing house of standards* 27
[17] GD 50-490-84 1984 Methodical instructions. Technical diagnostics. Methods for predicting the residual life of machines and parts by indirect parameters *Moscow: standardinform* 38
[18] Filin A 1978 *Applied mechanics of a deformable solid* (Moscow: Nauka) 616
[19] Birger I, Shorr B and Iosilevich G *1979 Calculation on the strength of machine parts* (Moscow: Mashinostroenie) 702
[20] STO 70238424.27.060.01.011–2009 2009 Ball and drum mills. General technical conditions for major repairs. Norms and requirements *Moscow: NP INVEL* 138
[21] Gul V 1978 *Structure and strength of polymers* (Moscow: Chemistry) 344
[22] Gul V and Kuleznev V 1966 *Structure and mechanical properties of polymers* (Moscow: Higher school) 415
[23] Bokshitsky M 1978 *Long-term strength of polymers* (Moscow: Chemistry) 308
[24] Bartenev G and Zuev Yu 1964 *Strength and destruction of highly elastic materials* (Moscow: "Chemistry") 1964
[25] Zhurkov S and Kuksenko V 1974 Micromechanics of polymer destruction *Mechanics of polymers* **5** 792-801
[26] Kachanov L 1958 *Creep and long-term strength* (Novosibirsk: publishing house of Sib. Otd. As of the USSR OTN) 145
[27] Korotich V 1965 Analysis of body movement in a rotating cylindrical drum *Izvestiya vuzov. Gorny Zhurnal* **5** 20-22

IOP Conf. Series: Materials Science and Engineering **945** (2020) 012037 doi:10.1088/1757-899X/945/1/012037

Human resources for BIM jobs in the AEC industry in Vietnam: an investigation on job positions and requirements

The-Quan Nguyen[1,2,*], Nguyet-Anh Dau-Thi[2] and Thuy-Ninh Dao[1,3]

[1]Faculty of Construction Economics and Management, National University of Civil Engineering, 55 Giai Phong Road, Hai Ba Trung District, Hanoi, Vietnam
[2]Institute of Investment and Construction Management, National University of Civil Engineering, 55 Giai Phong Road, Hai Ba Trung District, Hanoi, Vietnam
[3]Ph.D. Candidate, Department of Civil Engineering, National Taiwan University, No. 1, Sec. 4, Roosevelt Rd., Taipei, 10617 Taiwan

Email: quannt@nuce.edu.vn

Abstract. The paper presents research results from 90 BIM jobs vacancies for the AEC industry in Vietnam which were advertised on popular recruitment channels. The results show that BIM jobs in the South are more popular than in the North and the Central. BIM-related job positions are different from company to company, but this study groups those positions into three basic positions according to the international best practice approach: BIM Manager, BIM Coordinator, BIM Modeler. For BIM Modelers, their main job is to develop 3D models, deploy design documents, and coordinate with stakeholders to perform the work. BIM Coordinator's on-demand job is to coordinate models across disciplines, check conflict and report, control models, support project management. The main task of a BIM Manager is to manage and develop BIM uses, to ensure the information to be communicated among the parties in the projects, and train BIM human resources in the company. There are many requirements for these positions, among them are the two basic requirements of "experience" and "software" (organisations may use different software for the BIM uses they apply). The article also emphasizes the need to have more involvement of professional associations, state management agencies, etc. to standardize job positions, then later they can award BIM licenses to the proper human resources appropriately.

1. Introduction

Building Information Modelling (BIM) has been widely introduced globally and it is claimed to be adopted in Vietnam since the last decade. The uses of BIM in the AEC industry (architecture, engineering and construction industry) have brought lots of benefits to the stakeholders of the construction projects [1] and have been claimed to enhance the productivity in the construction industry [2], and Vietnam may not be an exception. In order to implement BIM widely and effectively, the AEC industry needs to be provided with sufficient and qualified BIM human resources. In addition to attract the BIM specialists graduated from overseas, the local training centres in Vietnam must also take part in educating and training this type of human resources. However, the uses of BIM require multiple disciplines, a set of software and there are different BIM levels that each organisation may want to target; this fact leads to a great diverse in the requirements for this type of human resources.

IOP Conf. Series: Materials Science and Engineering **945** (2020) 012037 doi:10.1088/1757-899X/945/1/012037

Selected countries have officially categorised BIM jobs by issuing guidelines for BIM implementation, or training syllabus in which the BIM titles are clearly stated. Of course, roles and responsibilities of each position are different, but some titles have many common designated functions, i.e. BIM Managers and VDC Managers as in the classification of Lee and Borman [3].

There is a variety of BIM-related job titles that have been used in different countries. The study of Miyoung Uhm et al. [4] shows 35 types of BIM job titles. These are then divided into eight job categories by analyzing the relationship between job titles, namely BIM project manager, BIM director, BIM coordinator, BIM senior architect, BIM manager, BIM designer, BIM MEP coordinator, and BIM technician. Joseph's study classifies BIM job positions to 7 categories: BIM manager, BIM Modeler, BIM analyst, BIM application developer, BIM software developer, BIM modeling specialist, BIM facilitator, BIM consultant, BIM researcher [5]. In another approach, the United Kingdom CIC BIM Standard Protocol records the responsibilities of the BIM Manager, the BIM Coordinator, and the BIM Modeler [6], while other documents may refer to BIM job captain, BIM coordinator, BIM technician etc. [5] or BIM managers, BIM coordinators, and virtual design and coordination (VDC) managers [3]. More titles such as BIM Trainer, Director of BIM Technologies, BIM Consultant, Manager of BIM Marketing, BIM Software Applications Support Engineer have also been used in companies in North American countries, as Barison and Santos have summarized [7]. Mathews [8] also present a myriad of BIM titles in which there are some distinct positions.

BIM job titles are classified according to the theory analysis, task analysis, roles, and requirements for each position. This article adopts three indispensable positions in a BIM project: BIM Manager, BIM Coordinator, BIM Modeler (Modeling Technician BIM) from the UK AEC BIM Protocol [6], which are widely adopted by many countries, like Malaysia. The responsibilities of the BIM Manager, the BIM Coordinator, and the BIM Modeler are illustrated in Table 1.

Table 1. Responsibilities of the BIM Manager, BIM Coordinator, and BIM Modeler.

	Strategic						Management				Production	
Role	Corporate Objective	Research	Process+ Workflow	Standards	Implementation	Training	Execution Plan	Model Audit	Model Coordination	Content Creation	Modelling	Drawing Production
BIM Manager	Y	Y	Y	Y	Y	Y	Y	N	N	N	N	N
BIM Coordinator	N	N	N	N	N	Y	Y	Y	Y	Y	Y	N
BIM Modeler	N	N	N	N	N	N	N	N	N	Y	Y	Y

Source: [6]

As one of the followers in terms of BIM adoption, Vietnamese architecture, engineering, and construction (AEC) industry have been looking for good practices from overseas countries to learn and apply. However, since the AEC industry in Vietnam refers to different sources from different countries, they may have adopted different approaches for BIM; this leads to the fact that they perceive differently the BIM-related issues, including BIM jobs and positions.

Based on information collected recently from the recruitment advertisements for BIM human resources, this paper aims to explore the actual BIM job titles in Vietnam and the requirements that have been set for the advertised positions. This helps to standardize the BIM job titles and the level of competencies and skill set needed for each title, for the clients to have a framework to refer to when they select consultants and contractors to deliver their BIM projects. This can also be a reference for the educators and trainers to develop their training curricula regarding professional BIM courses or BIM degree programs.

2. Materials and methods

This research adopted a 4-stage approach (Figure 1). In Stage 1, a literature review was conducted to detect different job positions related to BIM in the global AEC industry, then find out the popular approaches for categorize BIM job titles and related roles. In Stage 2, the research team scanned the most favored recruitment channels in Vietnam to record advertisements for BIM jobs vacancies. The job descriptions and requirements in the collected advertisements were analyzed in Stage 3 for a thorough understanding of the BIM jobs in Vietnam. Results are then compared to the global findings and recommendations were proposed in Stage 4.

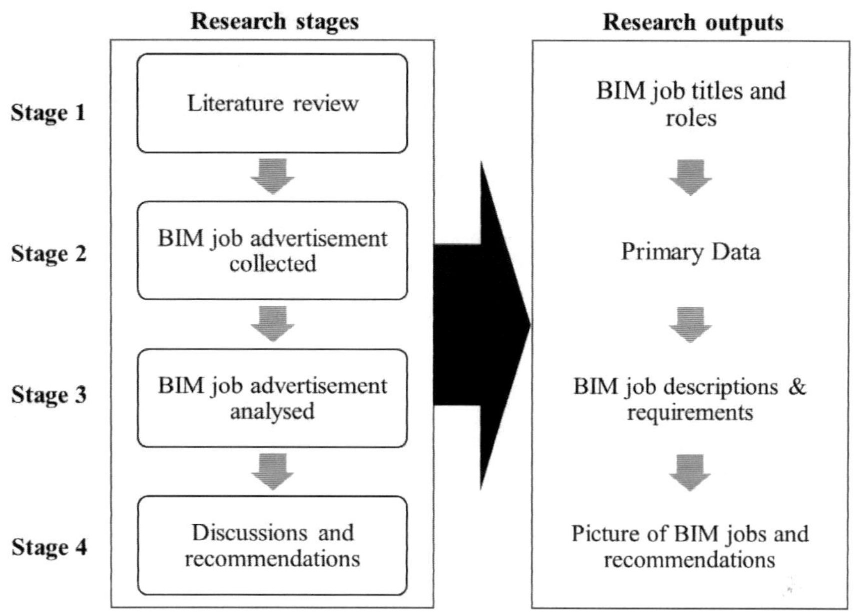

Figure 1. Research Approach.

In this research, advertisements for BIM job vacancies are collected from professional groups in social networks as Facebook, LinkedIn, construction online forums, and other reputable recruitment websites in Vietnam such as Indeed (https://vn.indeed.com), Vietnamworks, etc. The data collection process was carried out from October 2018 to August 2019, with 106 recruiting advertisements collected. After removing the duplicate recruitments and sketchy BIM job descriptions, 90 job postings remain for analysis (see Table 2).

Table 2: BIM job vacancies regarding job positions.

No	BIM job positions	Frequency	Percentage
1	BIM Modeler	33	36.67%
2	BIM Coordinator	37	41.11%
3	BIM Manager	20	22.22%
	Total	**90**	**100%**

Incidentally, the number of BIM Modeler recruitments is approximately equal to BIM Coordinator. In a posting with both BIM Modeler and BIM Coordinator recruitment, the number of BIM Modeler required is always greater than that of the BIM Coordinators. The number of BIM Manager vacancies found in advertisements is relatively smaller than that of the two other BIM-related positions. During the progress of the data collection, more and more BIM Manager recruitments have been being posted, revealing the increase in the demand for this type of human resources. Among the cities in Vietnam, only Hanoi (North), Da Nang (Central) and Ho Chi Minh City (HCMC - South) are the locations for the BIM jobs posted. Figure 2 shows that the BIM market in the South of Vietnam is much busier than in the North and the Central. In particular, the number of BIM recruiting jobs advertised in the South is 4 times higher than Hanoi, and up to 19 times higher than Da Nang.

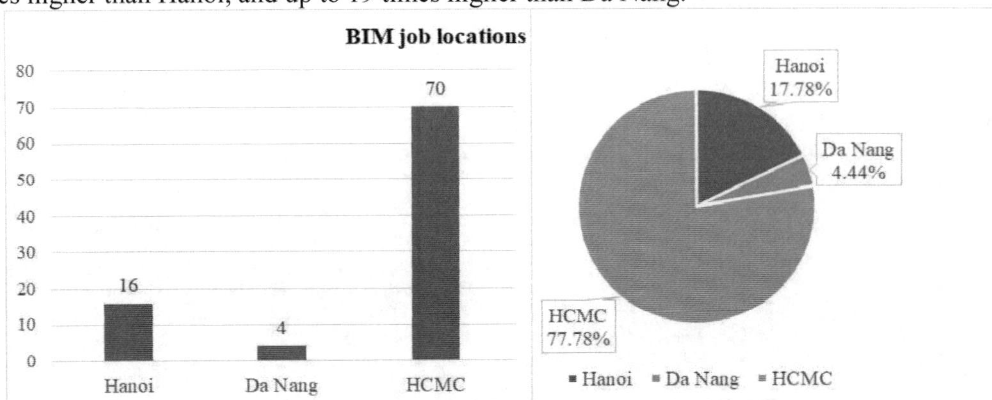

Figure 2. BIM job recruitments regarding locations of the posts.

Data collected were translated from Vietnamese into English to input into the NVivo software. Information generated from NVivo software will be analysed to describe the recent practice of BIM job descriptions and specifications in Vietnam.

3. Results and discussion

3.1. The required duties for BIM job positions

3.1.1. For BIM Modeler. In job postings, the title BIM Modeler is named differently, such as Revit Modeler, Construction Drawing Engineer, BIM designer etc. The main duties for this position in the 37 job recruitments posted include: developing 3D models, deploying design documents, and coordinating with the stakeholders to perform the work.

Table 3. The duties of BIM Modelers.

Duties	Frequency
Reporting	2
Controlling model quality	4
Controlling the schedule for model development	3
Developing 3D models	34
Clash detection	4
Developing project documents	3
Concept development	3
Works inspection	1
Developing and transferring models	6
Approving construction methods	1
Coordinating stakeholders	8
Coordinating models	2
Family creation	3
2D drawing design	5
Data integration and management	1
Manipulation construction workload	4
Deploying design documents	11
Exporting quantity	2

IOP Conf. Series: Materials Science and Engineering **945** (2020) 012037 doi:10.1088/1757-899X/945/1/012037

The full list of duties required for BIM Modeler, under the form of employers' job requirements, and the frequency of each duty required for this position are encrypted by Nvivo software, then are shown in Table 3.

As the title implies, modelling is the primary responsibility of a BIM Modeler. In addition, some BIM recruitments also put down clash detection, quantity take-off, exporting quantity from the models. Besides, other posts proposed other duties such as controlling schedule and quality for model development, checking conflicts, developing families, coordinating models among different disciplines. Besides 3D modelling, Table 3 shows that deploying design documents is also another popular duty of a BIM Modeler. Though each organisation may consider some extra duties for the position, the most popular ones are rather similar to what of the standard position, as presented in Section 3.

3.1.2. For BIM Coordinator. Table 3 illustrates the demanding duties advertised by the employers to BIM Coordinators. The most popular duties required for this position include controlling models, clash detection and report, coordinating multidisciplinary models, developing and updating the templates, standards, BIM guides, ensuring the quality of submittals and internal processes, developing 3D models, ensuring communication among stakeholders, project information updating, controlling the project progress and report and quantity surveying. The full list of duties for BIM Coordinators as advertised in BIM job vacancies are presented in Table 4.

Table 4. The duties of BIM Coordinators.

Duties	Frequency
Drawing updating	3
Project information updating	13
Setting up working environment and project standards	10
Ensuring the quality of submittals and internal processes	16
Ensuring communication among stakeholders	14
Team member evaluation	1
Assessing and checking design	3
Training BIM software for internal human resources	2
Developing 3D models	16
Supervising on construction sites	4
Supporting the managers	10
Controlling models	24
Controlling the project progress and report	12
Clash detection check and report	22
Working with clients	7
Works and project inspection and acceptance	3
BIM MEP department development	1
Developing and maintaining BIM Protocol	4
Coordinating multidisciplinary models	19
Quantity surveying	11
Monitoring monthly meetings	1
Designing project items	3
Shopdrawing	5
Developing component libraries	7
Developing and updating the templates, standards, BIM guides	18

IOP Conf. Series: Materials Science and Engineering **945** (2020) 012037 doi:10.1088/1757-899X/945/1/012037

It is noted that there are some duties required for BIM Coordinators but also appearing in the job descriptions for BIM Modelers. This proves the fact that there are some overlaps between the positions' duties, not only in the United Kingdom as in their CIC Standard Protocol [6] but also in Vietnam.

3.1.3. For BIM Manager. BIM Managers are among the key members of BIM crew of any organization with BIM adoption. They are in charge of implementing BIM in the organization, applying and developing BIM uses for the organization and its projects. A BIM Manager's duties also include ensuring the information flow among the organization's project stakeholders and delivering training to the internal human resources. According to the CIC Protocol [6], there are some overlapped responsibilities between BIM Coordinators and BIM Managers, such as developing and executing the BIM Execution Plan, delivering in-house BIM training. The data collected from BIM job advertisements in Vietnam also supports this fact. Table 5 below shows the full list of the duties advertised for a BIM Manager. Among the duties that have been assigned for BIM Managers, the most popular duties include managing BIM implementation and BIM uses development, managing information exchange among project stakeholders, then delivering BIM training to internal members of staff and model coordination and clash detection, document management. That is to say, main duties of the BIM Managers in the perception of the Vietnamese organization are quite similar to that from the United Kingdom, as stated in the CIC Standard BIM Protocol [6].

Table 5. The duties of BIM Managers.

Duties	Frequency
Quantity take-off	1
Controlling schedule	6
Managing information exchange among project stakeholders	24
Delivering BIM training to internal members of staff	14
Developing BIM objectives	9
Developing models	6
Controlling team performance	5
Controlling families and standards	6
Model checking	6
Model coordination and clash detection	13
Coordinating BIM with subcontractors	5
Document management	12
Subcontractors management	6
Managing BIM implementation and BIM uses development	33
Shopdrawing	6
Developing BIM application for project management	1

In these job vacancies advertisement, the overlapping duties are not only between BIM Managers and BIM Coordinators, but also between BIM Managers and BIM Modeler; this has not been articulated in any previous research or publications.

3.2. BIM jobs' specifications in Vietnam

A typical job specification, together with job description, makes up the two major job analysis outcomes. It identifies the kind of person who are qualified to do well in the job [9]. Job specification may include education, experience, skills, preferred qualifications [9].

The job specifications of the collected BIM job vacancies focus on 3 main aspects: experience, software skills and other competencies. Figure 4, 5, 6 show the information generated from NVivo

IOP Conf. Series: Materials Science and Engineering **945** (2020) 012037 doi:10.1088/1757-899X/945/1/012037

software for the competencies, skills and qualification requirements for the BIM Modelers, BIM Coordinators and BIM Managers accordingly.

Figure 3. BIM Modelers' required qualification and competencies.

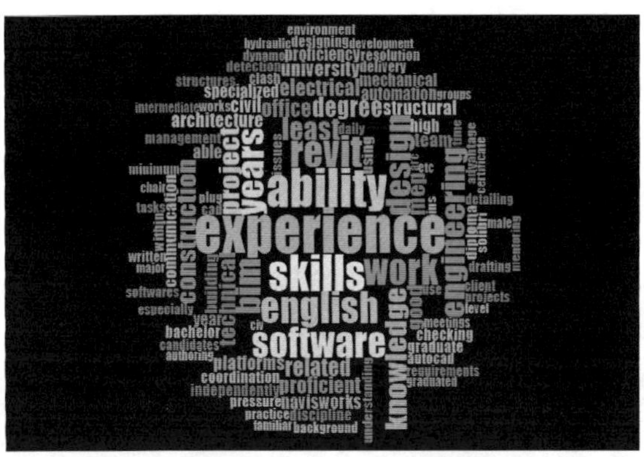

Figure 4. BIM Coordinators' job required qualification and competencies.

Figure 5. BIM Managers' required qualification and competencies.

"Experience" and "software" are the keywords that have been mentioned the most in the job advertisements. The number of years for similar experience and level of software proficiency are required differently for different positions and by different recruiters. As in Figure 4, 5, BIM Modelers and BIM Coordinators are required to use Autodesk Revit, which is claimed as the most popular BIM authoring software in Vietnam; while BIM Managers are expected to be proficient in Autodesk Navisworks, a coordinating tool.

Regarding the work experience, the average duration required for a BIM Modeler is 2.1 year, while that for a BIM Coordinator is 3.1 year and for a BIM Manager is 3.4 year, which are rather high for a new market regarding BIM like Vietnam. Other competencies required for BIM people include communication skills, teamworking skills, foreign language proficiency etc.; among which leadership and management skills are especially emphasized for the BIM Manager title. All of the three titles require at least a university degree; this fact reveals that in Vietnam, BIM people are considered as high-quality human resources.

4. Summary

The research results show that the AEC industry in Vietnam have started applying BIM and the organizations have been informed of the basic duties of the BIM people. However, the BIM titles are perceived differently by different employers, as well as the job descriptions for each BIM job position. As a results, apart from some common job requirements for each BIM position, the job specifications vary a lot from company to company. Therefore, there is a need for standardizing the BIM titles, as well as the basic requirements for each position. As the practice in Vietnam, the job standardization needs the involvement of not only the governmental agencies, but also professional associations in the AEC industry. The standards will support promoting BIM adoption widely and faster and facilitating quality management for the job performed by this type of BIM people. The standards also provide the basis for the clients in construction to assess and evaluate the contractors' bidding documents regarding BIM people. What is more, the standards can serve as the reference for universities to develop their education programs to ensure the future accreditation for the programs.

5. References

[1] Bryde D, Broquetas M and Volm J M 2013 The project benefits of building information modelling (BIM) *International journal of project management* **31(7)** 971-980
[2] Poirier E A, Staub-French S and Forgues D 2015 Measuring the impact of BIM on labor productivity in a small specialty contracting enterprise through action-research *Automation in construction* **58** 74-84
[3] Lee G and Borrmann A 2020 *BIM policy and management* (Taylor & Francis)
[4] Uhm M, Lee G and Jeon B 2017 An analysis of BIM jobs and competencies based on the use of terms in the industry *Automation in Construction* **81** 67-98
[5] Joseph J 2011 BIM titles and job descriptions: How do they fit in your organizational structure? (Autodesk University)
[6] Construction Industry Council 2018, Building Information Modeling (BIM) Protocol: Standard Protocol for use in projects using Building Information Models *Construction Industry Council: London, Great Britain*
[7] Barison M B and Santos E T 2011 The competencies of BIM specialists: a comparative analysis of the literature review and job ad descriptions, in *Computing in Civil Engineering 2011* 594-602.
[8] Mathews M 2015 Defining job titles and career paths in BIM, in *CITA BIM Gathering 2015 Technological University Dublin: Dublin*
[9] Lussier R N and Hendon J R 2017 Human resource management: Functions, applications, and skill development *Sage publications*

Acknowledgements
This research in funded by National University of Civil Engineering, Vietnam (NUCE) under grant number 223-2018/KHXD-TD.

IOP Conf. Series: Materials Science and Engineering **945** (2020) 012038 doi:10.1088/1757-899X/945/1/012038

Processes of forming protective and decorative coatings on concrete at plasma treatment

N I Bondarenko[1]* and D O Bondarenko[2]

[1]Department of Glass and Ceramics Technology, Belgorod State Technological University named after V.G. Shukhov, Kostyukov St., 46, Belgorod, 308012, Russia
[2]Department of Material Science and Materials Technology, Belgorod State Technological University named after V.G. Shukhov, Kostyukov St., 46, Belgorod, 308012, Russia

E-mail: bondarenko-71@mail.ru

Abstract. The development and improvement of technologies for obtaining coatings on various composites using low-temperature plasma is very relevant today. The paper considers the possibility of using broken glass of colored container glasses and other sanitary construction ceramics to obtain protective and decorative coatings on concrete, the selection of the ratio of which will allow getting different tints of the melted layer. To expand the color and decorative characteristics of coatings, mixtures for powder and further compaction were developed. The results of research on plasma treatment of concrete with developed protective and decorative coatings based on broken glass of colored containers, special sanitary construction ceramics and alumina cement are presented. An organoleptic assessment of the quality of coatings after plasma treatment is given. The optimal fractional composition for obtaining protective and decorative coatings on concrete was identified and the optimal speed of plasma treatment of the front layer of the product was established. For practical use, the fraction of colored container glasses and sanitary construction ceramics 0.63-0.8 mm is recommended. By adjusting the speed of the plasma jet passing over the front surface of concrete, they can get a coating with various decorative and relief effects. It was found that at the plasma treatment speed of 10 mm/s a solid coating is formed, and at the plasma treatment speed of 25 mm/s a coating with a bumpy surface is formed.

1. Introduction

Currently, the construction materials industry remains one of the most energy-intensive industries. The use of significant amounts of coal and other organic fuels can lead to even more intense environmental pollution and significantly worsen the given unhealthy environmental situation. In this regard, the development and implementation of environmentally friendly and energy-saving plasma technology in the construction materials industry is very relevant [1–8].

Plasma metallization of concrete and wall ceramics allows obtaining corrosion-resistant coatings, and new types of protective and decorative coatings on glass and ceramics are obtained with plasma spraying [9–11]. On the one hand, plasma and chemical modification increases the chemical stability of the coating and expands the possibilities for obtaining various decorative and color characteristics; on the other hand, it leads to thermal shock and dehydration of hydrosilicates in cement stone and, as a result, to its softening. This leads to deterioration in the performance of the protective and decorative coating on concrete products, namely, a decrease in strength and frost resistance. Therefore, the use of plasma treatment, which allows developing new technological solutions that improve the technical and operational characteristics of concrete with a protective and decorative coating, is relevant.

2. Materials and methods

Samples of 30×30×30 mm in size from the following raw materials were prepared for research: cement CEM I 42.5 N, quartz sand, water. When forming samples from fine-grained concrete, the mass ratio of Portland cement with quartz sand was 1:3, W/C = 0.4.

For the development of coatings, we used alumina cement of the VGC-1-35 brand, broken sanitary construction ceramics, and broken glass of colored containers. Plasma treatment was performed with an electric arc plasma torch of low-temperature plasma "Gorynych". The surface texture and color were determined organoleptically.

When developing protective and decorative coatings, the following factors were taken into account: the ratio of alumina cement and filler, filler fractions, and combinations of various types of fillers. Determination of the optimal fractional composition of colored container glasses and other sanitary construction ceramics was carried out experimentally. Fractions of blue and green container glass and sanitary construction ceramics 0.25-0.63 mm; 0.63–0.8 mm; 0.8–1.25 mm were obtained with sieving on laboratory sieves.

Then we prepared samples of fine-grained concrete with protective and decorative coatings. The prepared cement-sand solution was placed on ¾ in metal molds on the ground, which was solidified under usual conditions for 28 days. Then a layer of the developed protective and decorative coatings was placed on the ¼ of top layer and the front surface of the sample was sprinkled with colored container broken glass and sanitary construction ceramics. The powdered layer was compacted with a laboratory roller to obtain a good adhesion of the freshly formed coating to the concrete sample. After hardening for 3 days, plasma treatment of samples with a protective and decorative coating was performed on an electric arc laboratory plasma installation at different speeds of the plasma jet passing over the front surface.

3. Results

The following types of protective and decorative coatings were developed:

– alumina cement: broken glass at a mass ratio of 1:3 with a powder of colored container glass or a mixture of different fractions: 0.25–0.63 mm; 0.63–0.8 mm; 0.8–1.25 mm;

– alumina cement: sanitary construction broken ceramics at a mass ratio of 1:3 with a powder consisting of sanitary construction broken ceramics of various fractions: 0.25–0.63 mm; 0.63–0.8 mm; 0.8–1.25 mm;

– alumina cement: sanitary construction broken ceramics at a mass ratio of 1:3 with a powder mixture of colored container glasses of various fractions: 0.25–0.63 mm; 0.63–0.8 mm; 0.8–1.25 mm.

To expand the decorative and color characteristics, the compositions of mixtures for powdering and further compaction were developed:

– colored container glass and sanitary construction ceramics at the mass ratio of 1:1;

– a mixture of colored container glasses and other sanitary construction broken ceramics at a mass ratio of 2:1.

In order to determine the optimal fractional composition of fillers for protective and decorative coatings, the prepared samples were melted at the optimal speed of the plasma jet passing along the front surface of 10 mm/s. Below there are diagrams of the structure of fine-grained concrete with protective and decorative coating before plasma treatment (Figure 1) and after plasma treatment (Figure 2).

IOP Conf. Series: Materials Science and Engineering **945** (2020) 012038 doi:10.1088/1757-899X/945/1/012038

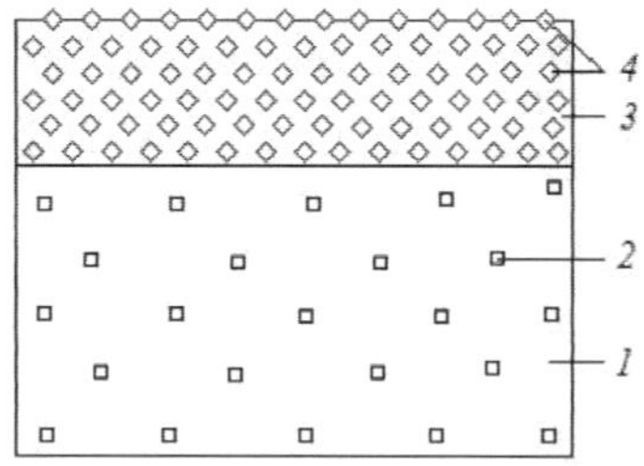

Figure 1. Structure of fine-grained concrete with protective and decorative coating before plasma treatment: *1* – fine-grained concrete; *2* – quartz sand; *3* – alumina cement; *4* – the broken ceramics or colored container glass.

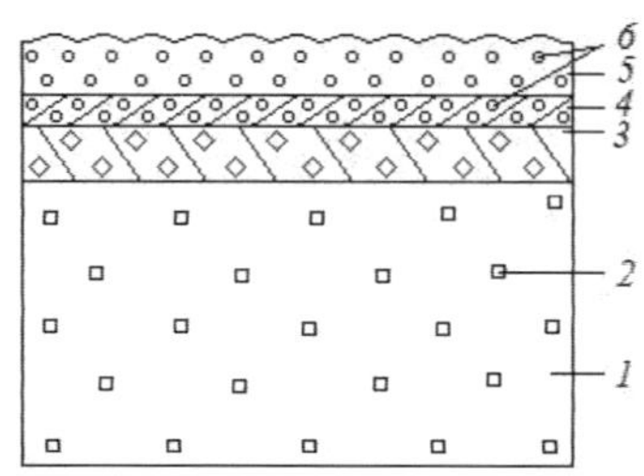

Figure 2. Structure of fine-grained concrete with protective and decorative coating after plasma treatment: *1* – fine-grained concrete; *2* – quartz sand; *3* – nonablated dehydration zone of protective and decorative coating; *4* – glass-crystal zone; *5* – amorphous zone; *6* – gaseous inclusions.

Samples with a fraction of 0.25-0.63 mm made of broken green container glass and sanitary construction ceramics had an uneven color, including black and brown zones due to melting of the surface part of alumina cement. This is because of small fractions of green container glass and sanitary construction ceramics were melted and partially mixed with the resulting melt of alumina cement.

Samples with a fraction of 0.63-0.8 mm from the broken blue container glass had a uniform blue color, from the battle of sanitary construction ceramics – a uniform light beige color.

Samples with a fraction of 0.8-1.25 mm made of broken green container glass and sanitary construction ceramics had a uniform color scheme, but the coating was foamed and bumpy, in some areas of the coating with drips and "melt assembly" (Figure 3).

Thus, for practical use, the fraction of colored container glasses and sanitary construction ceramics 0.63-0.8 mm is recommended. After determining the optimal fraction of the filler for coatings, these samples were subjected to plasma treatment with the regulation of the speed of passing the plasma jet from 2 mm/s to 25 mm/s (Figure 4, 5). Then an organoleptic assessment of the quality of the protective and decorative coating was carried out.

At the plasma treatment speed of 10 mm/s, a solid coating is formed (Figure 4). At a plasma treatment speed of 25 mm/s, the glassy coating is not solid, but similar to the decoration of "mound" glass products (Figure 5). This decor can be used as an independent type for industrial decoration of buildings and structures. The coating on the basis of container glass turned out to be more uniform than on the basis of sanitary construction ceramics, as the chemical composition of sanitary construction ceramics contains more refractory oxides, which do not have time to melt at a plasma treatment speed of more than 20 mm/s.

IOP Conf. Series: Materials Science and Engineering **945** (2020) 012038 doi:10.1088/1757-899X/945/1/012038

a *b*

Figure 3. Protective and decorative coatings with a fraction of 0.8–1.25 mm at plasma treatment speed of 25 mm/s: a) coating based on green container glass; b) coating based on sanitary construction ceramics.

a *b*

Figure 4. Protective and decorative coatings of 0.63–0.8 mm fraction at plasma treatment speed of 10 mm/s: a) coating based on blue container glass; b) coating based on sanitary construction ceramics.

a *b*

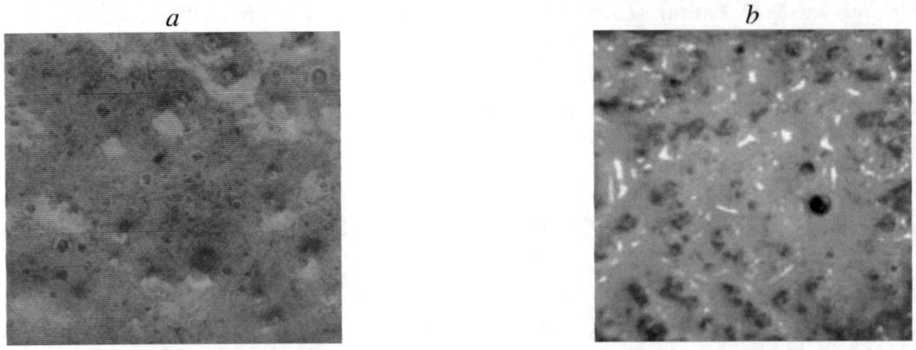

Figure 5. Protective and decorative coatings of 0.63–0.8 mm fraction at plasma treatment speed of 25 mm/s: a) coating based on green container glass; b) coating based on sanitary construction ceramics.

By adjusting the different colors of varietal glasses and other sanitary construction broken ceramics, they can get different shades of the fritted layer.

4. Summary

The fraction of colored container glasses and sanitary construction ceramics 0.63-0.8 mm is recommended for protective and decorative coating compositions. By adjusting the speed of plasma treat-

IOP Conf. Series: Materials Science and Engineering **945** (2020) 012038 doi:10.1088/1757-899X/945/1/012038

ment, they can get a coating with various decorative and relief effects. The selection of the ratio of different colors of glass and the composition of sanitary construction broken ceramics makes it possible to obtain different tints of the fritted layer.

5. References

[1] Volokitin G, Vlasov V, Skripnikova N, Volokitin O and Shekhovtsov V 2018 Plasma technologies in construction industry *Key Engineering Materials* **781** 143-148

[2] Shekhovtsov V V, Volokitin O G, Skripnikova N K, Volokitin G G, Vlasov V A 2019 Physical and technical processes for preparing silicate melts and silicate-melt products using low-temperature plasma *Journal of Physics: Conference Series* **1393** 012129

[3] Akulova M V, Shchepochkina Yu A 1999 New low-melting glazes for concrete *Glass and Ceramics* **56** 127-128

[4] Bondarenko D O, Bondarenko N I, Bessmertnyi V S, Strokova V V 2018 Plasma-chemical modification of concrete *Advances in Engineering Research* **157**, 105-110

[5] Volokitin O, Volokitin G, Skripnikova N, Shekhovtsov V 2016 Plasma technology for creation of protective and decorative coatings for building materials *AIP Conference Proceedings* **1698**. 070022

[6] Krokhin V P, Bessmertnyi V S, Panasenko V A, Drizhd N A, Nikiforov V M 1999 Decoration of glass and glass articles using the plasma-spraying method *Glass and Ceramics* **56** 78-80

[7] Bessmertnyi V S, Minko N I, Dyumina P S, Sokolova O N, Bakhmutskaya O N, Simachev A V 2008 Production of facing brick by plasma treatment using raw material from technogenic deposits *Glass and Ceramics* **65** 16-18.

[8] Bessmertnyi V S, Bondarenko N I, Bondarenko D O, Min'ko N I, Kochurin D V, Makarov A V 2019 Plasma technologies in glass production *Glass and Ceramics* **76** 241-245

[9] Bessmertnyi V S, Ilyina I A, Sokolova O N 2012 Obtaining protective and decorative coatings on wall construction materials of autoclave hardening *Bulletin of BSTU named after. V G Shukhov* **3** 155-157

[10] Bondarenko D O, Bondarenko N I, Bessmertnyi V S, Burlakov N M 2017 Plasma-chemical modification of concrete processed by colorific metal salts *Advances in Engineering Research* **133** 128-134

[11] Bessmertnyi V S, Krokhin V P, Panasenko V A, Nikiforov V M, Shvyrkina O N 1998 Glazed wall ceramics using KMA waste *Glass and ceramics* **55** 222-223.

[12] Bondarenko N I, Bondarenko D O, Kochurin D V, Bragina L L, Varfolomeeva S V 2018 Technology of plasma metallization of the wood and fibrous board *Construction Materials and Products* **1 (1)** 4-10

Acknowledgments

This work was realized in the framework of the Program of flagship university development on the base of the Belgorod State Technological University named after V G Shukhov, using equipment of High Technology Center at BSTU named after V G Shukhov.

IOP Conf. Series: Materials Science and Engineering **945** (2020) 012039 doi:10.1088/1757-899X/945/1/012039

Theoretical aspects of using fragments of destroyed buildings and structures of Iraq

Ahmed Ahmed Anees Ahmed[1]

[1]Department of Building Materials Science, Products and Structures, Belgorod State Technological University named after V.G. Shukhov, Kostyukov St., 46, Belgorod, 308012, Russia

E-mail: Civileng85@yahoo.com

Abstract. The relevance of the work is determined with the search for alternative sources of raw materials for the construction industry, associated with the utilization of technogenic waste. The novelty of the paper is to identify the scientific regularities of the influence of concrete scrap from fragments of destroyed buildings and structures on the formation of the microstructure of light and heavy concrete. Concrete scrap was prepared as both fillers of cement materials and small aggregates, based on which concretes with high mechanical properties were created. The compositions were designed from the point of view of geonics (geodaetica), in particular, taking into account the law of affinity of structures. The strength characteristics of concrete mixes were studied in accordance with EN 12390-3. In addition, microstructural, morphological and thermal properties of raw materials and concretes were determined during 28-day of curing. For the first time, a dense microstructure of the composite was provided, both by products of Portland cement and hydration, and, in part, by products of hydration of previously unreacted clinker, whose minerals are present in concrete waste and are activated when they are crushed. The use of fragments of destroyed buildings and structures as a filler of cementing material when replacing Portland cement up to 20% allows getting better compressive strength, both heavy and light concretes.

1. Introduction

Technical progress in the production of concrete and reinforced concrete requires additional sources of raw materials, in particular, the use of high-quality components [1,2]. The solution to this problem is hindered by the constantly growing shortage of mineral and energy resources, as well as increased environmental requirements[3,4]. An alternative source of raw materials for concrete production can be the waste of destroyed buildings [5].

States that have a large number of demolished or destroyed construction sites for various reasons have a great potential for this. The task of recycling these wastes is due to both the high cost of this event, and the need for early commissioning of land plots under these objects [6,7]. In this regard, it is advisable to use fragments of demolished buildings and structures as raw materials for the production of materials for the construction of new buildings, and especially "environment friendly" and "intelligent" composites.

A number of works are devoted to the study of the possibility of using the waste of destroyed buildings: in the EU countries [8], the Netherlands [9], Vienna (Austria) [10], China [11,12], South Korea [13], Lebanon [14], Brazil [15], the United States [11]. However, the results of these studies are very

IOP Conf. Series: Materials Science and Engineering **945** (2020) 012039 doi:10.1088/1757-899X/945/1/012039

contradictory, due to the different composition of waste from destroyed buildings in different countries. The general conclusion of these works is that the use of construction debris in newly created composites without reducing the required characteristics of these composites becomes possible only as a result of taking into account the chemical and mineralogical composition of the components, as well as the activation of the raw materials used in various ways. Currently, a number of studies were conducted on the use of concrete scrap in various building materials. Silva et al. [16] developed masonry solutions with aggregates from concrete scrap. Al-Bu-Ali U S et al. [17] used aggregate from concrete scrap to prepare high-density packaging. Amorim Beya and co-authors [18] studied the effect of demolition waste on the formation of microstructure of light and heavy concrete.

The theoretical basis for the design of composites using these raw materials is the transdisciplinary science of geonics (geodaetica), developed by V.S. Lesovik. This science uses the results of research on natural processes to create high-strength concrete and new-generation building structures. In particular, one of the aspects of this science is the law of affinity of structures, which consists in the selection of raw materials for a composite with similar physical and mechanical characteristics. The design of complex composites according to this law is possible only with the system analysis of macro -, micro- and nanoscale levels. These structures must correspond to the base composition as much as possible, in this case, the best adhesion and, consequently, the durability of the composite is ensured. The law is formulated based on the study of natural objects similar to construction composite materials. The results of this paper were obtained using the waste of destroyed buildings and structures of Iraq, which consist mainly of concrete, ceramic bricks and limestone wall blocks. In United Nations reports posted on the website www.un.org it is reported that between 80,000 and 100,000 industrial and residential buildings were completely destroyed as a result of military operations and terrorist acts in Iraq. Meanwhile, the country's natural resources are rapidly being depleted. Based on these facts, this paper analyzes the potential impact of fine mineral additives (concrete waste) as a filler and fine aggregate for light and heavy concrete.

2. Materials and methods

Portland cement CEM I 42.5 N (Belgorod cement, Russia) was used as a binder. The chemical and mineralogical composition as well as the main technical characteristics are given in tables 1 and 2. The specific surface area of Portland cement was 300 - 350 m^2/kg.

Table 1. Chemical and mineralogical composition of CEM I 42.5N concrete.

Chemical composition, % wt.							Mineralogical composition, % wt.			
CaO	SiO_2	Al_2O_3	Fe_2O_3	MgO	SO_3	Alkalis	C_3S	C_2S	C_3A	C_4AF
65.73	22.19	4.37	4.30	0.67	0.18	0.70	60.3	16.8	6.9	13.3

Table 2. Physical and mechanical properties of CEM I 42.5N concrete.

Setting time, h:min		Tensile strength in bending, MPa		Tensile compressive strength, MPa	
beginning	ending	3 day	28 day	3 day	28 day
2:30	3:35	5.8	8.1	33.6	46.6

The following raw materials were used as control natural aggregates: sand from the Kursk Deposit (Russia), granite crushed stone from the Erkilya Deposit (Russia), and expanded clay of the 10-20 fraction of the Aleksinsky plant (Russia). Characteristics of sand, crushed stone and expanded clay are presented in tables 3-5.

Table 3. Properties of used sand.

Characteristics	Values
Bulk density, kg/m^3	1400
Fineness modulus	1.8

IOP Conf. Series: Materials Science and Engineering **945** (2020) 012039 doi:10.1088/1757-899X/945/1/012039

Content of dust, clay and silt particles, % wt.	0.7
Specific weight, kg/m^3	2500
The loss on ignition, % wt.	<1
Residue on a sieve of 0.9 mm, % wt.	<1
Residue on a sieve of 0.5 mm, % wt.	>92

Table 4. Properties of used crushed stone.

	Characteristics	Value
Physical and mechanical properties	Specific weight, kg/m^3	2600
	Porosity, % vol.	4.1
	Compressive strength, MPa:	
	dry	168
	damp	156
Mineralogical composition	Microcline, % wt.	156
	Plagioclase, % wt.	15-20
	Quartz, % wt.	25-35
	Biotite, % wt.	2-5

Table 5. Properties of expanded clay.

Compressive strength, MPa	Frost resistance, cycles	Water absorption,%	Density, kg/m^3
0.3-6	15	10-25	800

Table 6 shows the composition of construction and demolition waste from Iraqi buildings and structures (figure 1). As it can be seen from table 6, construction and demolition waste resulting from military operations in Iraqi cities is 99.88% of crushed stone and cement. The appearance of crushed demolition waste is shown in Fig. 1, here they can see that the resulting technogenic large aggregate contains layers of cement mortar. Where such a layer is absent, film sections of hydrate phases are observed. There is every reason to believe that such inclusions will improve the adhesion of technogenic aggregate to the cement matrix. This assumption is based on the fact that the adhesion of various materials to the cement matrix decreases in the following sequence: clinker > limestone> granite > quartz.

Table 6. Composition of destroyed buildings.

Material	Portland cement	Crushed stone	Fibre	Wood	Metal	Plastic	Gypsum
Content, %	39.48	60.4	0.04	0.02	0.04	0.01	0.01

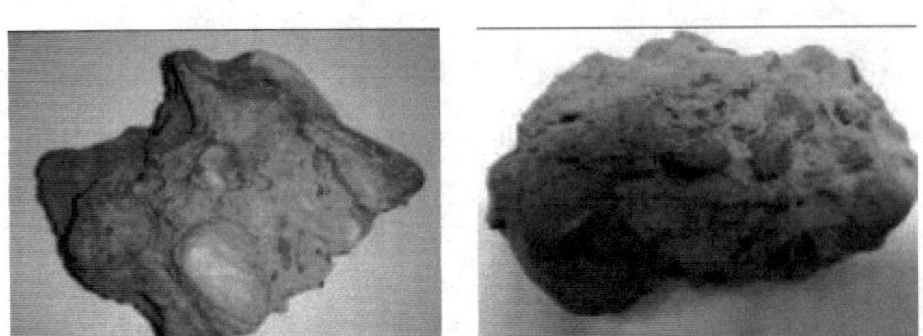

Figure 1. Appearance of crushed demolition waste.

Demolition waste was crushed in a jaw crusher to obtain fractions 2.5-5; 1.25-2.5; 0.63-1.25; 0.315-0.63; 0.16-0,315 and 0-0. 16. This raw material was used as a fine aggregate in further research. To study the demolition waste as a cement binder filler, they were further crushed at a VM-20 ball mill (Russia) to a size comparable to that of Portland cement.

The features of the microstructure and composition of raw materials and products used in the synthesis of Iraqi materials were studied by theoretical and empirical methods. The morphology of raw materials and small concrete samples with a size of 15 mm × 15 mm × 4 mm was studied using a scanning electron microscope (MIRA3 TESCAN, Brno, Czech Republic) operating at an accelerating voltage of 8.0 kV. The chemical composition of raw materials and small concrete samples was determined using an ARL 9900 WorkStation x-ray spectrometer (Thermo Fisher Scientific, Waltham, USA) with a built-in diffraction system. Differential thermal analysis (DTA) and thermal gravimetry (TG) were performed using a Shimadzu DTG-60H thermogravimetric analyzer (Japan).

The mixture was poured into a cone 30 cm high, then, after removing the cone, the sediment was measured. The average density of concrete was determined on samples-cubes 10 × 10 × 10 cm, dried to a constant weight at 105°C. The compressive strength of concrete samples was produced on 10 × 10 × 10 cm samples using a 200 kN Shimadzu test machine (Kyoto, Japan), in accordance with EN 12390-3.

During the experimental work, we used the system-structural foundations of the material science approach in "composition design – technology development – microstructure research – properties study".

3. Results

After crushing the demolition waste, the fractions were eliminated 2.5-5; 1.25-2.5; 0.63-1.25; 0.315-0.63; 0.16-0.315 and 0-0.16. Micrographs of the resulting dropouts are shown in Fig. 3. The mineral composition of various fractions of demolition waste is shown in table 7.

Figure 2. Micrographs of screening fractions: a) 0-0.16, b) 0.16-0.315, c) 0.315-0.63, d) 0.63-1.25, d) 1.25-2.5, f) 2.5-5.

Table 7. The mineral composition of various fractions of demolition wastes.

Fraction	Mineral composition, %					
	SiO_2	$Ca(OH)_2$	$CaCO_3$	CSH	C_3S	C_2S
0.00-0.16	48.4	11.5	10.0	5.8	12.0	12.3
0.16-0.315	55.2	7.4	11.0	4.4	11.0	11.0
0.315-0.63	56.4	11.0	3.9	12.0	6.7	10.0
0.63-1.25	65.1	12.0	6.0	5.9	5.0	6.0
1.25-2.5	64.4	10.5	6.9	3.0	7.6	7.6
2.5-5	62.8	11.0	0	6.0	9.2	11.0

IOP Conf. Series: Materials Science and Engineering **945** (2020) 012039 doi:10.1088/1757-899X/945/1/012039

Tests were carried out to replace natural sand with concrete scrap dropout in C8/10 class expanded clay concrete and C16/20 class heavy concrete. The tests were carried out with partial and complete replacement of sand with destruction waste with a particle size of 1.8 mm. In this case, the concrete mixtures were prepared on natural sand (control composition) and artificial sand with partial substitution (30% range of variation) to complete replacement (table 8-9).

Table 8. Compositions and properties of expanded clay concretes.

Number of content	Portland cement	Material content, kg/m³ Quartz sand	Claydite	Demolition waste	Water	Cone draft, cm	Average density, kg/m³	Compressive strength, MPa
LW-1	225	800	410	-	270	3-4	1670	16
LW-2	225	560	410	240	255	4-5	1600	17.2
LW-3	225	400	410	400	226	3-4	1625	19.8
LW-4	225	240	410	560	243	3-4	1562	17.9
LW-5	225	-	410	800	246	3-4	1564	17.7

Table 9. Compositions and properties of heavyweight concretes.

Number of content	Material content, kg/m³ Cement	Sand	Crushed stone	Demolition waste	Water	Cone draft, cm	Average density, kg/m³	Compressive strength, MPa
HW-1	475	600	1350	-	237	4	2360	27
HW-2	475	450	1350	150	237	4	2340	26.9
HW-3	475	300	1350	300	237	4	2365	26.5
HW-4	475	150	1350	450	237	4	2344	27.2
HW-5	475	-	1350	600	237	4	2350	26

As the research results showed (table 8), the compressive strength of expanded clay concrete using demolition waste increases, which is explained by an increase in the adhesion of the aggregate to the cement matrix. Table 9 shows the results of similar studies for heavy concrete. Tests to replace natural sand with artificial crushed sand in heavy concrete also showed that the strength of concrete does not decrease, but even exceeds the control composition. It is obvious that the artificial crushed sand obtained as a result of classification of crushed fragments of destroyed buildings and structures has a good potential for use as a fine aggregate in concrete.

The microphotographs shown in Fig. 3 show the interfacial transition zone of the technogenic aggregate and the cement matrix. The dense contacts of cement stone and large aggregate are clearly visible that form a single formed composite. Dense contact structure is formed due to the fact that in the early stages of hardening of the concrete mix crushed stone obtained from construction waste, absorb part of the closing water, and then during solidification of the cement stone this water is given away by the pores of technogenic crushed stone to bind clinker minerals into high-strength new formations. As a result, the activity of the binder increases, while the rheological characteristics of the mixture do not decrease. If the concrete waste is small, and the technology does not involve thermal water treatment, they are able to provide hydration of previously unreacted particles of clinker minerals and $Ca(OH)_2$, the total amount of which is approximately 15-25% in the concrete mix.

Figure 3. Contact zone between the cement matrix and the aggregate.

IOP Conf. Series: Materials Science and Engineering **945** (2020) 012039 doi:10.1088/1757-899X/945/1/012039

Thus, it was revealed that the composites obtained from active fillers of Porland cement binders and crushed stone from demolition waste have good adhesion between the components. Accordingly, this makes it possible to obtain building materials of a wide profile with reduced deformability, good crack resistance, durability, etc. These conclusions, based on the law of affinity of structures, have been proven earlier [19].

In Fig. 4 shows the compressive strength versus the amount of Portland cement replaced with crushed drift waste. At the same time, in accordance with the law of affinity of structures, concrete waste was crushed until the specific surface area of Portland cement (300-350 m2 / kg) was reached.

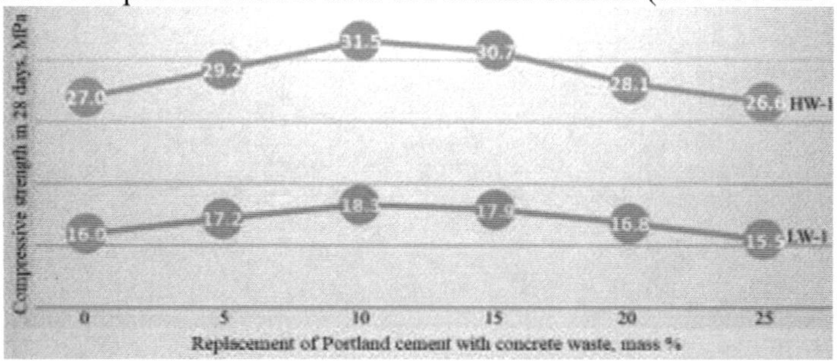

Figure 4. Effect of the amount of Portland cement replaced by crushed demolition waste on compressive strength of heavyweight and lightweight concrete.

The introduction of additives of various genesis into the binder entails a transformation of the mechanism and speed of hydration processes of clinker minerals in interaction with the closing water. It is possible that these additives can react with new formations of the Portland cement matrix, respectively, forming new phases that affect the characteristics of the concrete composite in various ways. Therefore, to achieve the goal of the paper, it was necessary to study the hydration mechanisms depending on the presence of waste in buildings. Differential thermal and x-ray phase analysis were used for this purpose.

Fig. 5 shows the results of DTA of 28-day-old solidified cement with the introduced concrete additives. There is a complex endothermic effect, reaching a maximum at a temperature of about 102°C. The peak at 448°C characterizes the endothermic effect resulting from the dehydroxylation of portlandite. This endoeffect shows that the secondary concrete, acting as an active filler of cement binders, did not react with all the calcium hydroxide, respectively, the liquid phase of the hardening mixture was saturated with this $Ca(OH)_2$. According to fundamental concepts, in a saturated solution of $Ca(OH)_2$, high-base C-S-H (II), as opposed to low-base C-S-H (I) remain stable. Accordingly, the most important component of cement materials modified with thin-ground concrete scrap is the C-S-H (II) fiber structure.

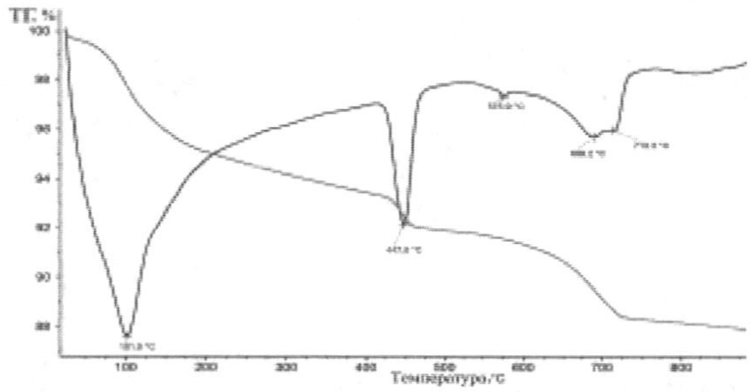

Figure 5. DTA patterns of HW-1specimen.

Small endothermic effects at temperatures of 689°C and 713°C are mainly characterized by decarbonization of weakly crystallized metastable forms of $CaCO_3$ formed due to partial carbonation $Ca(OH)_2$.

X-ray diffraction analysis of the HW-1 sample (Fig. 6) correlates with the above differential thermal studies. In particular, in addition to the reflections of non-hydrated clinker minerals (d = 4,343; 4,270; 3,357 and 3,198 A), a relatively small reflection of 2,637 A attracts attention as identified portlandite.

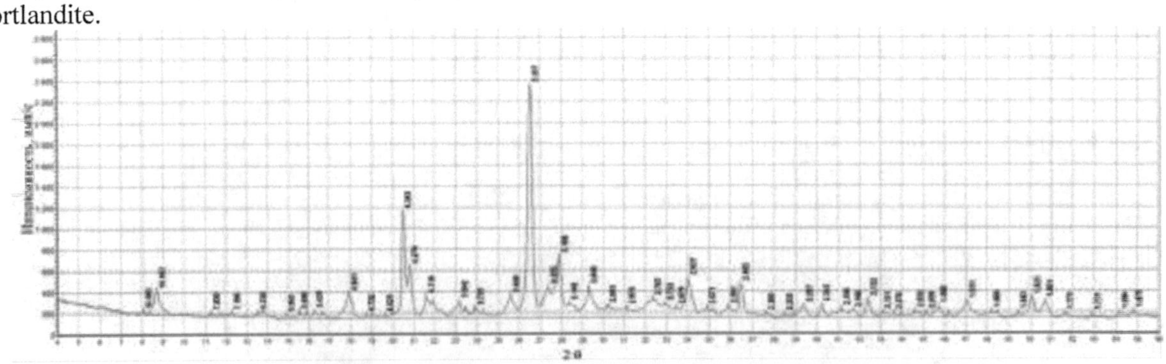

Figure 6. XRD results of specimen HW-1.

The same is confirmed by microphotography obtained with the help of SEM developed for concrete at the age of 28 days (Fig. 7). There is a dense microstructure provided by both Portland cement products and hydration products of previously unreacted clinker minerals present in concrete waste and activated during its grinding.

Figure 7. Microstructure of new growths in heavy concrete HW-1 at the age of 28 days.

4. Summary

This article presents the possibility of recycling construction waste, which leads to the solution of a number of problems: waste disposal and improving the physical and mechanical properties of concrete by adding recycled waste as a filler or fillers. Based on the results of a set of studies, the following conclusions are offered for further discussion:

The use of demolition waste as a fine aggregate gives similar results in the mechanical characteristics of heavy and light concrete, as when using natural sand. This is due to the compaction of the structure and the fusion of concrete waste with the surface of the cement matrix, as a result of which they look like a single whole.

Demolition waste as a cement binder filler when replacing Portland cement up to 20% allows getting a better compressive strength, both heavy and light concrete, due to the compaction of the microstructure by products of hydration of previously unreacted clinker minerals present in concrete waste, and activated during grinding.

5. References

[1] Szeląg M 2019 Properties of cracking patterns of multi-walled carbon nanotube-reinforced cement matrix *Materials (Basel)* **12(18)** 2942

[2] Klyuev S V, Khezhev T A, Pukharenko Yu V, Klyuev A V 2018 To the question of fiber reinforcement of concrete *Materials Science Forum* **945** 25 – 29

[3] Al-Fakih A 2020 Characteristic compressive strength correlation of rubberized concrete interlocking masonry wall *Structures* **26** 169-184.

[4] Klyuev S V, Khezhev T A, Pukharenko Yu V, Klyuev A V 2018 Experimental study of fiber-reinforced concrete structures *Materials Science Forum* **945** 115 – 119

[5] Danish A, Mosaberpanah M A 2020 Formation mechanism and applications of cenospheres: a review *Journal of Materials Science* 1-19

[6] Rajabi A M Omidi Moaf, F, Abdelgader H S 2020 Evaluation of Mechanical Properties of Two-Stage Concrete and Conventional Concrete Using Nondestructive Tests *Journal of Materials in Civil Engineering* **32(7)** 4020185

[7] Volodchenko A A, Lesovik V S, Zagorodnjuk L H, Volodchenko A N, Kuprina A A 2016 The control of building composite structure formation through the use of multifunctional modifiers *Research Journal of Applied Sciences, Engineering and Technology* **10(12)** 931-936

[8] Sáez P V, Osmani M 2019 A diagnosis of construction and demolition waste generation and recovery practice in the European Union *Journal of Cleaner Production* **241** 118400

[9] Zhang C 2020 Upgrading construction and demolition waste management from down cycling to recycling in the Netherlands *Journal of Cleaner Production* **266** 121718

[10] Lederer J, Gassner A, Kleemann F, Fellner J 2020 Potentials for a circular economy of mineral construction materials and demolition waste in urban areas: a case study from Vienna *Resources Conservation and Recycling* **161** 104942

[11] Aslam M S, Huang B, and Cui L 2020 Review of construction and demolition waste management in China and USA *Journal of Environmental Management* **264** 110445

[12] Yu B, Wang J, Li J, Lu W, Li C Z, Xu X 2020 Quantifying the potential of recycling demolition waste generated from urban renewal: A case study in Shenzhen, China *Journal of Cleaner Production* **247** 119127

[13] Cha G-W 2020 Evaluating recycling potential of demolition waste considering building structure types: A study in South Korea *Journal of Cleaner Production* **256** 120385

[14] Al Zaghrini N, Srour F J, Srour I 2019 Using GIS and optimization to manage construction and demolition waste: The case of abandoned quarries in Lebanon *Waste Management* **95** 139-149

[15] Silva R V, Brito J De, Dhir R K 2016 Performance of cementitious renderings and masonry mortars containing recycled aggregates from construction and demolition wastes *Construction and Building Material* **105** 400–415

[16] Beja I A, Motta R, Bernucci L B 2020 Application of recycled aggregates from construction and demolition waste with Portland cement and hydrated lime as pavement subbase in Brazil *Construction and Building Material* **258** 119520

[17] Al-Bu-Ali U S, Lesovik R V, Harhardin A N, Tolstoy A D, Ahmed A A A , Alaskhanov A kh, Aymenov Zh T 2020 Calculation and selection of grain composition of aggregate from high-density concrete scrap packaging *Bulletin of BSTU named after V G Shukhov* **5(6)** 18-28

[18] Ahmed A A, Fedyuk R S, Liseytsev Yu L, Timokhin R A, Murali G 2020 Use of Iraq concrete scrap as filler and aggregate of heavyweight and lightweight concrete *Construction Materials and Products* **3 (3)** 28 – 39

IOP Conf. Series: Materials Science and Engineering **945** (2020) 012039 doi:10.1088/1757-899X/945/1/012039

[19] Lesovik V S, Chulkova I L, Zagordnyuk L K, Volodchenko A A, Popov D Y 2014 The role of the law of affinity structures in the construction material science by performance of the restoration works *Research journal of applied sciences* **9(12)** 1100-1105

[20] Klyuev S V, Khezhev T A, Pukharenko Yu V, Klyuev A V 2018 Fiber concrete for industrial and civil construction *Materials Science Forum* **945** 120 – 124

Acknowledgments
This work was realized under the support of the President Scholarship; in the framework of the Program of flagship university development on the base of the Belgorod State Technological University named after V G Shukhov, using equipment of High Technology Center at BSTU named after V G Shukhov.

IOP Conf. Series: Materials Science and Engineering **945** (2020) 012040 doi:10.1088/1757-899X/945/1/012040

Design process of executive elements in stroke mechanisms based on rammer's deformations and stress modeling

M S Chepchurov[1*], V P Voronenko[2] and A V Chirkov[3]

[1]Department of "Mechanical equipment and machine building technology", Belgorod State Technological University named after V.G. Shukhov, Kostyukov St., 46, Belgorod, 308012, Russia

[2]Department of Technological Engineering, Moscow State Technological University "STANKIN" ("MSUT "STANKIN"),Russia, Moscow, GSP-4, Vadkovsky lane, 1, 127994

[3]Department of machine building technology Technical University of Liberec, Studentská 1402/2,461 17 Liberec 1,Czech Republic, EUROPE

E-mail: avtpost@mail.ru

Abstract. The article deals with the strength calculation issues of products by example of rammer under short-term loads in hammer drill. Based on finite element analysis' results of the rammer geometrical model, it is proposed to select a material with a strength limit significantly exceeding the allowable residual stresses inside the rammer. The harmonic analysis allows us to estimate the rammer's strength. The dynamic coefficient may be calculated by the use of Vibran 2.0 device (amplitude and signal frequency) and by calculation results of the maximum displacement. The FreeCAD software package was used for this. It allows to create internal scripts to automatically search for the optimal design parameters. A method of determining mechanism's static load and its dynamic coefficient under periodic impact loads has been developed. Static loads calculations and harmonic analysis according to the developed method allow to determine structural strength of the product, to select the rammer's material and to find the dynamic coefficient (which takes into account the stroke loads in calculations). The proposed calculation method may be used not only to solve the describing problem, but also for a whole class of problems related to stroke loading, where such parameters as application time of force or duration time of force's impulse are important.

1. Introduction

Designing process for machine components with rod elements under stroke loads is associated with additional time for choosing the material of rammer and its geometric form. It is not enough to perform only static strength calculations and endurance analysis for such elements. Because of this fact most of constructers have to order additional researches based on standard documents' requirements [7] with the use of special equipment [3], which demands additional costs. The result of such tests is not always sufficient for considering the right decision and leads to repeating of experiments. The whole cost of designing process and experimental work is shifted to the cost of the end product. That decreases the competitiveness of designed product and delays the mass production start-up date. Inte-

IOP Conf. Series: Materials Science and Engineering **945** (2020) 012040 doi:10.1088/1757-899X/945/1/012040

grated CAD / CAE automation systems for designing procedures with the FEM module may significantly reduce the costs of experiments and perform a great number of iterations in a short period of time [16]. In this case, the experiment itself moves to the virtual reality. Integration of the various auto-designing modules requires several things to organize in order to make a right decision in project lifecycle of product: to set automatic information exchange between all of such modules and to set process of outputting results at least. In this case, the designer or constructer should have a huge base of reference knowledge in order to choose correctly all of the physic and mechanical parameters for such part of hammer tool as a rammer.

Static problems may be solved by CAE applications rather simply, but dynamic tasks require additional qualification from a designer (theoretical physics, mathematic methods, material resistance etc.), as fully presented in [11], that may be problematic for a general constructor or designer. The authors propose a simplified, but effective way to solve the problem related to the stroke loads in axial direction for round rod part. This method also includes a process of selection rammer's material by the use of the licensed software GPL2.

2. Materials and methods

2.1. Materials

The authors faced with problems of material and geometric parameters, when developed a special device (for hammering dowels). The device consists of a rammer installed in a support, which is carrying by a spring fixed on the rammer with the help of stopper. The 3D model is presented in *Fig.1a*, and the really manufactured device in *Fig.1b*.

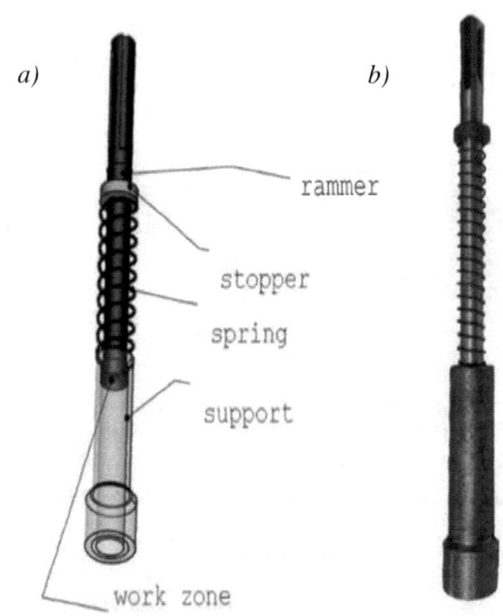

Figure 1. Device for hammering dowels in a hammer drill:
a- 3D model; b - photo of the real prototype.

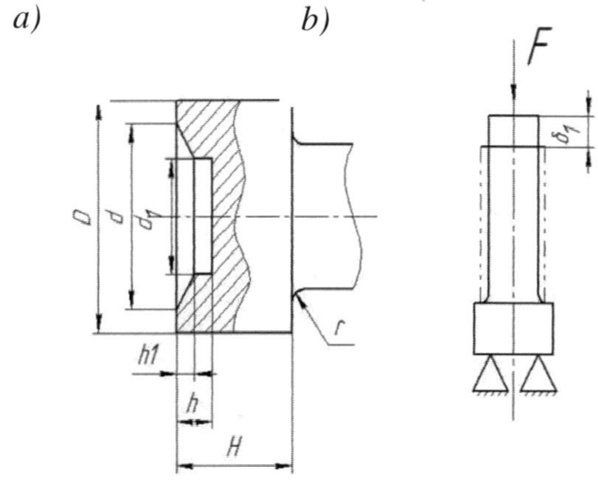

Figure 2. Hammer and loading scheme: a) the shape of the hammer's rammer, b) the scheme of loads.

The aim of calculations was to determine the best shape of the rammer's work zone (or rammer's impacting end) in accordance with the sketch shown in Fig.1a and selected axial loading scheme (Fig.2b), as the paper [18] recommends.

The purpose of researches is to determine rammer's diameter *D*, length *H*, which are in a certain range **from D_{min} to D_{max}, and from H_{min} to H_{max}** respectively. It is necessary to determine such values, which will not exceed limits of maximum stresses for selected material under short-term action of the *F* force. The general scheme of loads (*Fig.2b*) contains the force *F*, applied to the upper end of the rammer (rod), which has deformation δ_I. Its value depends on many factors: the applied force *F*, the cross-section of the rod and the material's elastic modulus.

The source of the force *F* is the hammer drill's mechanism (mechanical unit) of periodic rammer's displacements, i.e. it sets a certain frequency of a short-term applied force. In this case, the stroke energy, transmitted to the striker, expends on heating the elements and partly returns back to the executive element of the hammer drill's mechanism. Instrument's manufacturers claim different values of stroke energy, depending on the power of device. Meanwhile electro pneumatic rotary hammers provide maximum stroke energy [8] up to *20 J*. The strokes frequency from a heavy hammer drill reaches *1000 ... 2000 strokes / min*, which corresponds to oscillation frequency of the executive element *17 ... 35 Hz*.

Measurement results of rammer's vibrations, obtained by Vibran 2.0, were processed with the use of its special software. The software window with the results of harmonic analysis is presented on *Fig.3*.

The information about stroke frequency and types of hammer drill's movements in open sources is rather contradictory, so the authors made their own experiments and measured hammer drill frequency characteristics with Vibran 2.0 device. Measurements were based on the recommendations [15]. According to *Fig.3*, the movements of the executive element are periodic, but its movement over the period is irregular, because the graph of displacement is not sinusoidal. Therefore, we can't take such form of movement for further modeling as periodic movements of rammer [14]. The analysis of amplitude-frequency characteristic (*Fig.4*) shows maximum displacements *–0.9 mm* at a frequency of *29.9 Hz*, but the signal shows another additional harmonics at frequencies from *51.8 Hz* to *209.3 Hz*. The movement amplitude of the executive component at a maximum frequency of *209.3 Hz* is approximately 13% of the amplitude of the carrier frequency. Consequently, apart from calculations for static strength, it is necessary to perform a frequency analysis and to define critical values of mechanism's own frequencies. That also means to define whether the real investigated oscillations reach at least one of mechnism's own resonant frequencies.

IOP Conf. Series: Materials Science and Engineering **945** (2020) 012040 doi:10.1088/1757-899X/945/1/012040

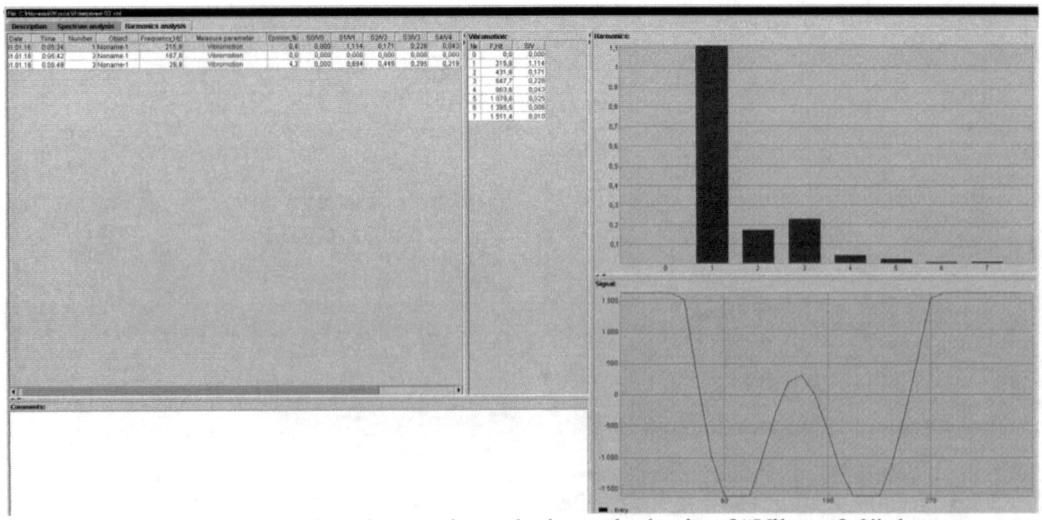

Figure. 3. Results of the harmonic analysis on the basis of "Vibran 2.0" data.

Rammer as an ordinary tool undergoes deformation. These deformations may exceed the maximum allowable value for selected material in some moment. Consequently, the material is selected according to the allowable yield strength. There are characteristics of some materials in *Table.1*, it also presents another parameter of materials - stroke viscosity or KCU (the stroke energy value at which the material will be destroyed). As it can be seen from *Table.1*: these values, as well as the yield strength, depend not only on the chemical composition, but also on the chosen heat treatment method. A material with a high yield strength may have a lower stroke viscosity. The table was compiled taking into account the requirements to loaded elements of mechanism with strength and elasticity properties, for which the authors used paper [19].

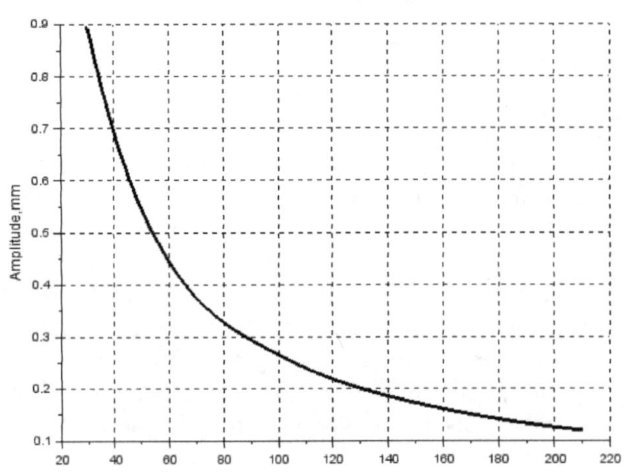

Figure. 4. Amplitude-frequency characteristic of movements of the rammer.

IOP Conf. Series: Materials Science and Engineering **945** (2020) 012040 doi:10.1088/1757-899X/945/1/012040

Table 1. Steel properties.

Steel grade DIN(EN)	Heat treatment: annealing, normalization			Heat treatment: quenching with oil cooling		
	Young's modulus, MPa	Strength's limit MPa	Stroke strength, J	Young's modulus, MPa	Strength's limit MPa	Stroke strength, J
C 20, Ck 20	210	420	2.2	-	-	-
C 45, Ck 45	209	610	1.2	203.6	1200	0.9
42Cr4	210	630	1,6	204	850	1.2
20MnCr5G	208	700	2.2	203.6	1000	2.0

2.2. Methods

The choice of CAE –application implies the use of an appropriate theoretical knowledge for solving the problem. Since the authors assume to use FEM analysis for simulation processes (rod under stroke loading), it is necessary to consider the recommendations for choosing equations of mathematical physics, for example [12].

During stroke process, the load is dynamic, and static strength calculations should be done by taking into account previously calculated dynamic coefficient.

Various sources recommend to use the same dynamic calculations methods as for static ones: $F_d = F_{st} \cdot k$; $\delta_{c\,max} = \delta_{max} \cdot k \leq [\delta]$, where k is the dynamic coefficient.

Let's consider, that stroke causes deformation of the elastic rod in the longitudinal direction. All of the energy is directed only to its compression. Its cross section in that case should be chosen in such a way as to eliminate bending in the transverse direction [2]. In the case of significant values of applied force, the rammer should be checked for the critical limiting values, according to one of the already known methods, for example:

$$F_{cr} = \frac{\pi^2 \cdot E \cdot I_{min}}{L^2} \quad (1)$$

where E, MPa — Young's modulus; I, mm^2 — axial inertia moment of the rod in the transverse direction; L, m— the length of the rod or the distance between the fixed rod's ends.

Another one thing should be noted. Authors used the distance between the supports as the distance from the hammer drill's chuck end to the rod's end in calculations of rammer. Actually, the critical force in form of production $E{\cdot}I$ may be considered as the rod bending stiffness. Calculations for such an element as a rammer may include just the previously selected values of the Young's modulus without further stability check. But deflection of the rod under the longitudinally applied force should be calculated in accordance with the recommendations [12]. For selected scheme of loading, for example, on *Fig.5*, you can use the differential equation:

$$\frac{d^2 \cdot y}{dx} = \frac{M}{E \cdot I}, \quad (2)$$

where $M = -F \cdot y$,N•m is the bending moment under the action of F force; x is the coordinate of maximum point of deflection, m.

$$T_0 = U_L + U_r + U_t \quad (3)$$

where U_L, J is the deformation energy of the rod, fixed according to the scheme on *Fig.6*; $U_r = \frac{m \cdot v^2}{2}$, J, is the energy required to return the power tool working element – striker (flying piston) with mass m, kg U_t, J]is the energy spent to heat the colliding parts, which can be neglected in case of elastic system.

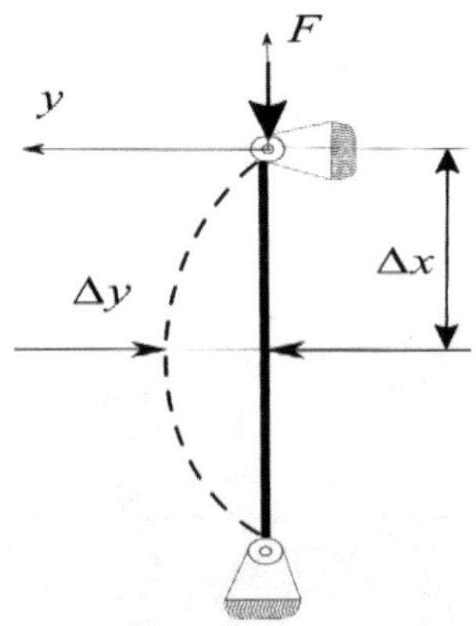

Figure. 5. Scheme of rod's loading.

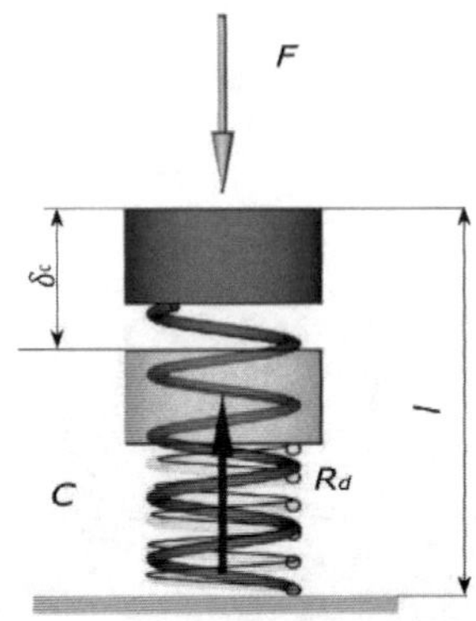

Figure. 6. Scheme of calculation for rod's defor-

The mass of a striker (flying piston) may be from *0,05 kg* to *0,8 kg*, depending on the power of hammer drill. For example, for a chosen one, the weight of the striker was *0.07 kg*, the frequency of the vibrations was *≈30 Hz*, the maximum movement was *30 mm*, i.e. $\frac{1}{30}$ movements were done for a one period, or taking into account the frequency $\frac{1}{60}$ *s*, the time of striker's movement, or its speed , h $v = 0.03 \cdot 60 = 1.8 m/s$ whence, the energy spent on striker's return is: $A_T = \frac{0.07 \cdot 1.3}{2} = 0.2268$, J. With the declared stroke energy of *2 J*, A_T means about 10% of this energy, thus

$$T_0 - A_T = U_L \Rightarrow 0.9\ T_0 = U_L \tag{4}$$

Let's calculate the deformation potential energy:

$$U_c = \frac{1}{2} \cdot F \cdot \delta_c, \tag{5}$$

J, where δ_c is the value of static deformation, mm.

Reaction of system *C* to the action of force *F*: $\delta_d = R_d \cdot c$,mm, where *c* is the proportionality coefficient, according to Hooke's law. Thus, it may be introduced as $U_c = \frac{c}{2} \cdot \delta_c$, J, a and $U_d = \frac{1}{2} \cdot R_d \cdot \delta_d = \frac{F}{2 \cdot \delta_c} \cdot \delta_d{}^2$, J. If we substitute the result expressions in the original expression (3), we will have: $0.9 \cdot T_0 = \frac{F}{2 \cdot \delta_c} \cdot \delta_d$, J, given that according to Hooke's law

$$c = F / \delta_c \tag{6}$$

Thus:

$$0.8\ T_0 - \frac{F}{2\ \delta_c}\ \delta_{d^2} = 0, \tag{7}$$

that gives us several unknown variables: F, δ_c, δ_d, which must be determined on the basis of calculations and results of experiments.

We will try to use similar calculations for some kind of situation with free fall (in our case, a striker of a power tool).

The load of mass m falls from the height *H*, so the work is $A = m \cdot g \cdot H$, J, i.e. it can be written as $0.9 \cdot T = mgH$, whence $H = 0.9 \cdot T_0 / mgH$.

In the dynamic coefficient calculations from various sources, it is H value that allows us to calculate the dynamic coefficient:

$$K_d = 1 + \sqrt{\frac{2 \cdot H}{\delta_c}}; \tag{8}$$

$$K_d = 1 + \sqrt{\frac{\frac{0.9.\ T_0}{m\ g}}{\delta_c}} = \sqrt{\frac{0.9\ T_0\ \delta_c}{m\ g}} \tag{9}$$

Let's equate the expressions of stroke energy: energy of speed and energy of height:
$\frac{m \cdot v^2}{2} = m \cdot g \cdot H$ or $H = \frac{g \cdot v^2}{2}$ (10), m.
Or the final dynamic coefficient:

$$K_d = \sqrt{\frac{g \cdot v^2}{\delta_c}} \tag{10}$$

It remains to calculate the force applied to the rod's end used in finite element calculations and modeling.

According to the «Vibran 2.0» data, the maximum movement amplitude of executive element will be $0.9 \cdot 10^{-3}$ meters, then the work force is $F = \frac{T}{A_{max}} \cdot H$, where A_{max} is the maximum movement amplitude or path, and $T_0 = 0.97 \cdot T$, J, as described above. Let's define: $F = \frac{0.97 \cdot 3}{0.9 \cdot 10^{-3}} \approx 3100$, N.

After creating the FEM analysis, we assign border conditions and loads. The border conditions are constant, but the force, applied to the rod's end, may vary depending on both factors: amplitude of movements and impulse energy of power tool as it is in our case (in general case - only energy of impulse).

The authors used FreeCAD as a software package for designing and calculating [1] with a GLP2 license, which complies the requirements about free software, for example [9]. The parametric modeling system of product's geometry allows to the use calculated data from the electronic table of the same project, as shown in *Fig.7*.

The spreadsheet and the window of project are shown in *Fig.6*. All the details of creation processes, modeling process and FEM analyze in FreeCAD with the help of CalculiX calculator [4] were done in accordance with the recommendations outlined in paper [13].

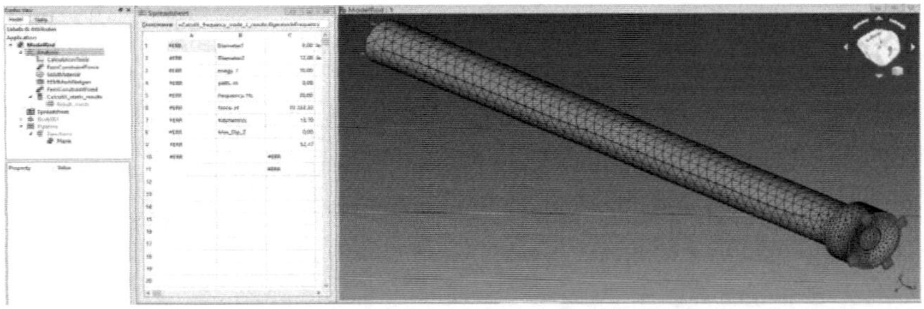

Figure. 7. Window of the electronic table and project.

IOP Conf. Series: Materials Science and Engineering **945** (2020) 012040 doi:10.1088/1757-899X/945/1/012040

The obtained calculation results should be analyzed for normal and maximum displacements in the model [17] and also should be determined the value of these displacements and deformations. According to the Vibran 2.0 data, the oscillatory process is not strictly sinusoidal, i.e. there are another additional harmonics, the frequency of which may coincide with the rammer oscillation frequency. For this, we will perform a mechanism's harmonic analysis by the use of CalculiX. The analysis result (*Fig.7b*) showed that in the range from *0* to *1000 Hz* there are only two resonance frequencies (*379.61* and *379.728 Hz*), which exceeds the limit values of the harmonics. Consequently, dangerous damage to the construction is not indicated. In this case, the maximum (by von Mises) stress is only *82.37 MPa*, that is less than with a static load only. The resonant frequencies are also recorded into the table for further analysis. At the same time, there are not specific resonant frequencies in the cells of the table (*Fig.6*), but there are parameters of the finite element model. That makes it possible to automate part of the calculations.

3. Results and discussion

The finite element analysis results make it possible to select a rammer's material with an acceptable strength's limit using the *table.1* of this paper.

The dynamic coefficient may be obtained with the help of calculation results based on the maximum displacements and Vibran 2.0 data (amplitude and frequency of signal) [12]. The calculated dynamic coefficient, according to expression (9) is *15.69* (for this case).

The FreeCAD package allows to create internal scripts for the automatic design calculations. This option is very useful for automatic searching of optimal design parameters [5]. The authors used Netgen 3.0 as a grid generator, where the grid changes automatically.

Automatic calculations should be organized with the direct access to project parameters from script: in order to make possible to change diameters, length, loads, to select the supporting surface. When all the initial parameters are set, the launch of the calculator will be performed also automatically. The results obtained by the FEM are processed by the FreeCAD integrated post-processor.

Thus, the authors obtained results allowing to determine the rational geometric form for the rammer. For example, if we change the type of support of rammer's end, the stresses arising into it will be also changed. *Fig.8* shows the restriction scheme of rammer's end movement (along two surfaces) (*Fig.8,a*) and the results of equivalent stresses calculations (by von Mises) (*Fig.8,b*).

According to the calculations, the maximum equivalent von Mises stress for the two-surface limiting scheme is *325 MPa*, which allows us to assign any material, for example, C20 without heat treatment, according to the data from *Table 1*.

During tests of the device [10] with a rammer from this material, the authors noticed the deformation of rammer's end surface (face of rammer), because at some moments the rammer relied only on this surface. Calculations were performed again, but with a different scheme for limiting movements, as shown in *Fig.8,a*.

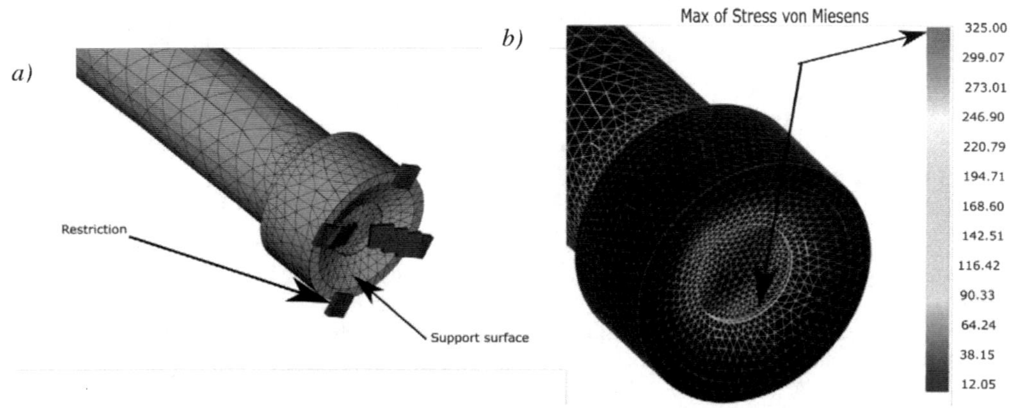

Figure 8. The calculation of the equivalent von Mises stresses for two-surfaces limitation: a - scheme for assigning the constraints; b - calculation results.

The results of the von Mises equivalent stress calculations (*Fig.8,b*) show, that the maximum stress value may reach *572 MPa*. As the width of the end surface's ring decreases, they will increase. So, to avoid catastrophic consequences for all the device, the authors chose the material C 45 DIN17200, which have a strength's limit (after quenching with oil cooling) - *1200 MPa*. The successful application of the authors' device may be found at https://cloud.mail.ru/public/2HBk%2F4q6iDpMmd.

Special steels with corresponding properties were described in *"Research and Development - 2016" (Springer, Cham)* and may also be assigned as a material for an stroke part, which allows to achieve mass production with minimum cost for such products.

4. Summary

Method for calculating the whole system's static load and its dynamic coefficient under the periodic stroke action has been developed. It makes possible to determine the magnitude of the maximum static deformation stress with the use of a finite element analysis. The technique can be used for calculating other displacements and stresses in the rods with periodic axial loads.

All the calculations (static loading and harmonic analysis) allow us to determine the mechanism's strength, to select the material for the blank (workpiece), to determine the dynamic coefficient for the rammer. The authors also proved the efficiency of using software packages for parametric modeling with the GLP2 license. There is a free program code in this software, that allows to use automatic searching for the best solution.

The proposed calculating method for both: the dynamic coefficient and the applied force magnitude is based on empirical calculations of stroke mechanisms. That makes possible to use it not only for solving the problem described in this article, but also for a whole class of problems with stroke loads, where exist properties, like time of force's application or duration of force's impulse.

5. References

[1] FreeCAD, URL: https://www.freecadweb.org, Accessed 20 August 2019

[2] Kruszka L, Vorobiov Y S, Ovcharova N Y 2014 FEM Analysis of Cylindrical Structural Elements under Local Shock Loading *Applied Mechanics and Materials* **566** 499-504 https://doi.org/10.4028/www.scientific.net/AMM.566.499

[3] Atapin V G, Rodionov A I, Rykov A A, Titorenko V P , Yur'ev G S Ivanov Y A 2009 Stands and devices for material tests and the specialist of products on impact resistance and shock stability *Scientific bulletin of Novosibirsk State Technical University* **3(36)** 87-98

[4] Wittig K 2018 CalculiX USER'S MANUAL - CalculiX GraphiX, Version 2.14 May 29 159 URL: http://www.dhondt.de/cgx_2.14.pdf

[5] Falck D, Collette, B 2012 Solid Modeling with the power of Python, Freecad [How-To] *Packt*

Publishing Ltd. 69

[6] Adhikari S, Murmu T, McCarthy M A 2013 Dynamic finite element analysis of axially vibrating nonlocal rods. *Finite Elements in Analysis and Design* **63** 42-50 doi: 10.1016/j.finel.2012.08.001

[7] GOST R 51371-99 2000 Mechanical environment stability test methods for machines, instruments and technical products. Test for influence of shocks *Moscow: Gosstandart Rossii: Izd-vo standartov*

[8] Chen H, Sun Y 2017 Development and application of reliability test platform for high-speed punch machine clutch brake system. *J Mech Sci Technol* **31** 53–61 doi.org/10.1007/s12206-016-1207-1

[9] Luca A, Francesco L, Paolo S 2016 Can Open-Source 3D Mechanical CAD Systems Effectively Support University Courses *International Journal of Engineering Education* 32 3(A) 1313-1324

[10] Zhukov E M, Chepchurov M S 2014 The device for driving in of expansion bolt shields *Patent RF* **142819**

[11] Polocoşer T, Kasal B, Li X J 2017 Design of Experiment and Pitfalls of Low-Velocity Pendulum Impact Testing *J. dynamic behavior mater* **3** 436-460 doi.org/10.1007/s40870-017-0123-5

[12] Nasdala L 2010 FEM-Formelsammlung Statik und Dynamik *Springer-Verlag* 230

[13] Bernd K 2007 FEM – Grundlagen und Anwendungen der Finite-Elemente-Methode im Maschinenund Fahrzeugbau *Springer-Verlag* 498

[14] Yang Y 2009 Understanding of Vibration Stress Relief with Computation Modeling *J. of Materi Eng and Perform* **18** 856–862 https://doi.org/10.1007/s11665-008-9310-9

[15] Buzdugan G, Mihăilescu E, Radeș M 1986 Instrumentation for vibration measurement. In: Vibration measurement. *Mechanics: Dynamical Systems Springer Dordrecht* **8** https://doi.org/10.1007/978-94-017-3645-9_5

[16] Louhichi B, Abenhaim G N, Tahan A 2015 CAD/CAE integration: updating the CAD model after a FEM analysis *Int J Adv Manuf Technol* **76** 391-400 doi.org/10.1007/s00170-014-6248-y

[17] Madenci E, Guven I 2006 The finite element method and applications in engineering using Ansys® *Springer US* 686

[18] Rajendran R, Sai K P, Basu S 2009 Axial impact studies on steel tubes and zircaloy rod Exp Tech **33(1) 17-22** doi: 10.1111/j.1747-1567.2008.00363.x

[19] Yamada Y 2007 Materials for Springs *Springer-Verlag Berlin Heidelberg* 377

[20] Lomaeva T V, Lukin L L, Maslov L N, Shavrin O I, Skvortsov A N 2016 Properties of Structural Steels with Nanoscale Substructure *In: Anisimov K et al. (eds) Proceedings of the Scientific-Practical Conference "Research and Development - " Springer Cham* 385-396

[21] Teterina I, Avdeeva A 2019 The determination of the structural parameters of mechanical engineering products by the finite element method *Bulletin of BSTU named after V G Shukhov* **2** 149-155 doi: 10.12737/article_5c73fc382e33e6.05913951

Acknowledgements

This work was realized under the support of the President Scholarship; in the framework of the Program of flagship university development on the base of the Belgorod State Technological University named after V G Shukhov, using equipment of High Technology Center at BSTU named after V G Shukhov.

IOP Conf. Series: Materials Science and Engineering **945** (2020) 012041 doi:10.1088/1757-899X/945/1/012041

Use of a supply ventilation system as an air curtain

V M Kireev[1*], A I Alifanova[1] and Yu G Ovsyannikov[1]

[1]Department of Heat and Gas Supply and Ventilation, Belgorod State Technological University named after V.G. Shukhov, Kostyukov St., 46, Belgorod, 308012, Russia

E-mail: vit31rus@mail.ru

Abstract. Recently, for a number of reasons, retail and office premises on the first floors of the building, which have a separate entrance to the street, have gained wide popularity. These rooms must have supply ventilation to maintain the required microclimate parameters. The article proposes, in addition to fulfilling its main function, to use general exchange ventilation as an air curtain in these types of premises. Due to the supply of air above the doorway with certain parameters, in some cases, it is possible to abandon the thermal air curtain as a separate energy-intensive device. The implementation of this task is supposed to be carried out by organizing the necessary air exchange scheme, without increasing its required performance. The article presents the results of the study, which make it possible to assess the possibility of using the proposed technical solution for the most standard case. In particular, an adapted calculation of the air distribution grille is presented to fulfill the stated purpose.

1. Introduction

All over the world, the level of requirements for the parameters of the air environment in rooms for various purposes is constantly increasing. Formation of comfortable parameters of the environment in habitable premises is the main function of life support systems. Open doors, windows and other openings lead to significant heat loss and are the main cause of drafts in the room. Air-thermal curtains help to solve several problems at once related to maintaining the microclimate and energy consumption of the building as a whole. They are designed to separate zones with different air temperatures on different sides of open openings of working windows, entrance doors and gates. By supplying a high-speed air flow, they form a kind of curtain that prevents warm air from escaping outside and does not let cold air into the room. When the curtain is used in a room with an open opening, the internal thermal comfort is improved, drafts disappear, heat losses are significantly reduced, and, consequently, heating costs.

In summer, in areas with a warm climate, the air curtain is also energy efficient, since it significantly reduces the cost of air conditioning.

In the modern world, the problem of energy conservation is very relevant, which makes scientific and technological progress require the introduction of new energy-efficient technologies, constant replenishment of scientific ideas, as well as fundamental research. Energy saving implies the efficient (rational) use of energy in all links: from the extraction of primary energy resources to the consumption of all types of energy by end users. Thus, the improvement of heating and ventilation systems aimed at reducing the energy consumption spent on creating and maintaining the required microclimate in the premises, is a very urgent task [1], [2].

Having carried out a comprehensive analysis of the operation of air-thermal curtains, we can confidently assert that they do not always meet modern energy saving requirements, due to the use of an electric heater for heating the supplied air [3].

Today, along with large retail and office buildings, small offices and trade organizations, often located on the ground floor of knowledge, have gained high popularity. These premises, as a rule, have a separate entrance and are equipped with a supply and exhaust general ventilation system [4], [5]. The main feature of these premises is not high attendance, therefore, the work of a separately installed heat curtain will be, although effective, but impractical. The operating mode of general ventilation is constant, since it is aimed at maintaining the required parameters of the microclimate [6].

Considering all of the above, a new promising technical solution appears, which consists in using a supply-side general ventilation system as an air curtain in small public spaces. By organizing the necessary air exchange in the room, general ventilation can function as a heat curtain, which will increase the energy efficiency of engineering systems and reduce the overhead costs of maintaining the premises.

2. Materials and methods

2.1. Materials

To use the proposed direction, it is first of all necessary to consider the existing methods for calculating the general ventilation system and air curtain.

Ventilation of premises is a prerequisite for work. The calculation of the required air exchange in the premises under consideration should be based on the type of room and the number of people working there [6]. The performance of the supply and exhaust ventilation system in this case is determined by the minimum amount of outdoor air per person. For one person, the air exchange rate ranges from 20-60 m^3/h. This average value varies markedly depending on the purpose of the room. In practice, as a rule, 60 m^3/h per person is sufficient for a permanent stay at the workplace and 30 m^3/h for each person for a temporary stay (less than 2 hours) [6], [7].

As for the calculation and design of air curtains, today there are a lot of works that are devoted to this issue [8-21]. However, most of them are considering industrial use cases that have a number of features.

2.2. Methods

To select the heat output of the air curtain, it is necessary to know the total time during which the entrance doors will be open, the specific flow of cold air, and the dimensions of the doorway.

The specific flow of cold air is determined by the formula:

$$j_{ex} = \mu_{ex} \cdot (2\rho_e \cdot \Delta p_{ex})^{0,5} \tag{1}$$

where μ_{ex}– the coefficient of air consumption at the inlet without taking into account the action of the air curtain and the influence of the figure of a person passing through the inlet;
ρ_e– he density of the outside air;
Δp_{ex} – the calculated pressure difference determined by the formula:

$$\Delta p_{ex} = 0,5 \cdot (H_{bu} + 2h_{fl} - h_d) \cdot (\gamma_{ex} - \gamma_{in}) \tag{2}$$

where H_{bu} – the height of the building from the ground surface to the top of the staircase, m;
γ_{ex} and γ_{in} – specific gravity of air, at the design temperature of outdoor and indoor air.

It has been experimentally established that the equivalent time z_{in} for a single passage of a person through single doors is 2 s, through double doors 1.5 s and through triple doors 1-1.2 s.

Knowing the number of people passing through the entrance for 1 hour, you can determine the total amount of cold air:

$$G_{ex} = 0,9 \cdot j_{ex} \cdot A_{ex} \cdot z_{in} \cdot N_{ex} \quad (3)$$

where j_{ex}– the specific flow of cold air, kg/(s m^2);
A_{ex} – door area, m^2;
z_{in} – equivalent time of door opening;
N_{ex} – the number of people passing through the entrance in 1 hour.

Heat consumption for heating cold air is found by the formula:

$$G_{ex} = 0,9 \cdot j_{ex} \cdot A_{ex} \cdot z_{in} \cdot N_{ex} \quad (4)$$

where c – the heat capacity of air, J/(kg °C);
t_{in} – internal temperature, °C;
t_{ex} – outside temperature, °C.

The amount of air G_z, kg/h, heated to t_z to create an air-thermal curtain, is determined by the formula:

$$G_z = \frac{Q_{ex}}{c \cdot (t_z - t_{ex})} \quad (5)$$

The volumetric flow rate of the supplied air at a temperature t_z of heated air [9]:

$$L_z = \frac{G_z}{\rho_z} \quad (6)$$

The above dependencies determine the amount of supply air required to compensate for heat losses due to opening the outer doors. However, when designing an air curtain, in our case, the determining parameter will be the air flow rate, which will prevent the flow of cold air.

In some works, [10] there is evidence that a sufficient condition for the operation of an air curtain is the passage of the outer boundary or axis of the jet through a certain point and finding the value of the average air velocity at the "end of the jet" within 2-3 m/s. At the same time, it is considered that such a balance of heat in the door opening is automatically observed, in which there is no heat loss with the air leaving outside, and with the incoming cold air. In our opinion, this approach is not entirely accurate, but we can use it, taking into account the specifics of the serviced object: a small area of the doorway, short-term opening of doors.

3. Results

A special place in the issue under consideration will be occupied by the method of supplying and distributing air to achieve the goal. The air distributor must fulfil the condition of maintaining the minimum required speed at the end of at least 2-3 m/s, and also fit harmoniously into the interior of the room. Since the doors in most cases for this type of premises have a small width (up to 1.5 meters, usually 0.8-1.1 m), it is most preferable to carry out air distribution using a ready-made standard air distribution device: a supply grille [23]. When calculating and selecting the most suitable grille, it is necessary to use the theory of movement of supply ventilation jets [24]. In our case, a flat jet will take place. Plane jets are formed when they flow out of elongated rectangular holes with a side ratio $a_0/b_0>5$, where a_0 and b_0 are respectively the sizes of the larger and smaller sides of the grating. For plane jets $x<6a_0$ we have:

IOP Conf. Series: Materials Science and Engineering **945** (2020) 012041 doi:10.1088/1757-899X/945/1/012041

$$V_x = \frac{m_1 \cdot V_0 \cdot \sqrt{b_0}}{\sqrt{x}} \cdot K_c \cdot K_b \cdot K_n \quad (7)$$

$$\Delta t_x^{\max} = \frac{n_1 \cdot \Delta t_0 \cdot \sqrt{b_0}}{\sqrt{x}} \cdot \frac{K_b}{K_c \cdot K_n} \quad (8)$$

where V_x– the maximum air velocity in the main section of the jet at a distance x from the grate, m/s;

$m_1 = m/2,45$;
m– lattice speed coefficient;
$n_1 = n/2,45$;
n– the temperature coefficient for a flat section of the jet;
F_0– area of the calculated section of the section;
b_0– the width of the design section of the lattice;
V_0– exit speed from the grid;
Δt_0– the temperature difference;
K_c– coefficient of constraint;
K_b– the coefficient of interaction;
K_n– non-isothermal coefficient;
x– the distance from the lattice to the point under consideration, m.

In our case, we will have a non-isothermal jet with a small value of $\Delta t_0 = 3$, according to [6], the supply air temperature in public premises should be no more than 3 °C above the calculated supply air temperature.

Since when designing a supply-side general-exchange ventilation system with the function of a heat curtain, it is first of all necessary to know the required speed at the outlet from the air distribution device, then from formula (7) we express the parameter V_0, setting the speed V_x = 2-3 m/s:

$$V_0 = \frac{V_x \sqrt{x}}{m_1 \cdot \sqrt{b_0} \cdot K_c \cdot K_b \cdot K_n} \quad (9)$$

However, this dependence is suitable only for a rough estimate, since it includes the non-isothermal coefficient K_n with the participation of the parameter V_0. In our and similar cases, it is quite possible to use this dependence, since there is an insignificant temperature difference between the ambient temperature and the supply air temperature, which leads to a minimal effect of gravitational forces.

Knowing the velocity in the section of the grating V_0, we can determine the air flow required to supply G, m^3/s:

$$G = F_0 \cdot V_0 \quad (10)$$

Let's make a calculation for the most typical conditions with the following initial data:
- room height h = 3 m;
-the width of the doorway c = 0,8 m;
- air temperature in the room t_{in}=19 ^{0}C;
- supply air temperature according to [6] 22 ^{0}C, temperature difference $\Delta t_0 = 3$;
- air speed at the end of the jet Vx = 2 m/s;
- let us set the air speed at the outlet from the grating V_0 = 5 m/s;
- the lattice is of the AMN type with dimensions 0.8x0.1 from the manufacturer's catalog [23] (Figure 1). The area of the calculated cross-section of the accepted lattice according to the catalog is F_0 = 0.073 m^2.

When calculating, we will take the scheme of supply air supply from top to bottom with conical compact jets (Figure 2).

There are two approaches to calculating non-isothermal jets. The first approach is based on the complex parameter *H*-geometric characteristic [13], which has the dimension of length and depends on the design parameters of the distributor (coefficients *m*, *n*, F_0) and the conditions at the outlet (V_0, *t*).

Figure 1. The appearance of the adopted ventilation grill.

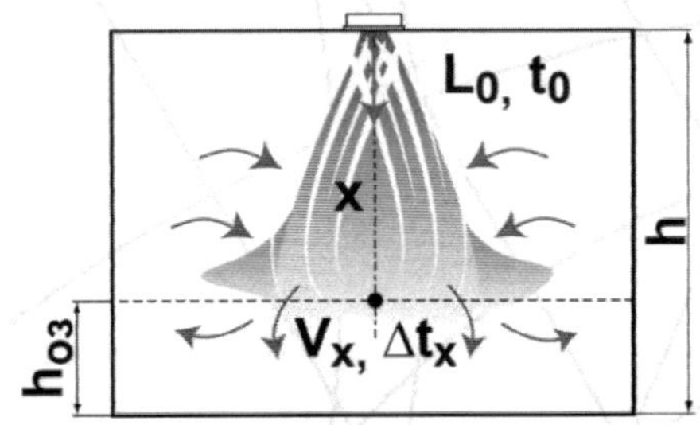

Figure 2. Air flow pattern.

The second approach is based on the dimensionless characteristic of the jet - the current Archimedes criterion [23], at a certain distance from the air distributor, which also depends on the above parameters and on the considered jet length.

The geometric characteristic H for a flat jet is determined by the formula:

$$H = \sqrt[3]{b_0 \cdot T_{in}^{\ 2} \frac{(m \cdot V_0)^4}{(n \cdot \Delta t_0 \cdot g)^2}} \qquad (11)$$

where T_{in} – the ambient temperature, K;
g– acceleration due to gravity 9.81 m/s^2;
for our case for the chosen lattice $m = 6$, $n = 5.1$.

When heated air is supplied, when its temperature is higher than the average air temperature, gravitational forces arise due to the difference in air densities, which reduce the range of the jet. This fact is taken into account by the non-isothermal coefficient K_n, which for our case is defined as:

$$K_n = \sqrt[3]{1 - 2 \cdot \sqrt{\left(\frac{x}{H}\right)^3}} \qquad (12)$$

Since in our case the temperature difference is insignificant $\Delta t_0 = 3\ ^0$C, and the velocity V_0 is sufficiently high, the value of the non-isothermal coefficient becomes insignificant due to its tendency in the unit $Kn = 0.99$ (Figure 3, 4), which allows using the dependence (9).

After making calculations, we got the value $V_x = 2.12$ m/s, which corresponds to the initially specified condition. We plotted the dependence of V_x on V_0 under the given conditions for the listed conditions (Figure 5).

For a speed V_0 = 5 m / s according to the diagram [23], the sound pressure level for this grille will be 46 dB, which is below the permissible 60 dB for this type of room.

4. Discussion

Consideration of the problem posed in the article is carried out using existing methods and data, however, some of them may be subject to criticism. In particular, there is evidence that the calculation of the required speed V_x is not entirely accurate [22], so the authors plan to continue research in this direction. However, today this direction is very promising and the use of this technical solution is very attractive from the point of view of energy saving in engineering systems.

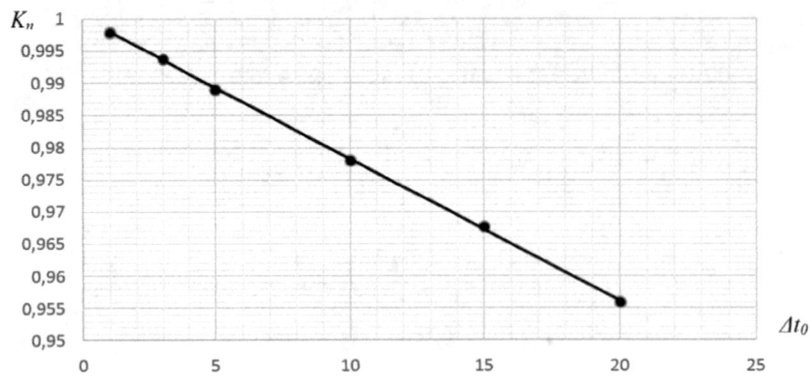

Figure 3. Dependence of the non-isothermal coefficient K_n on the parameter Δt_0.

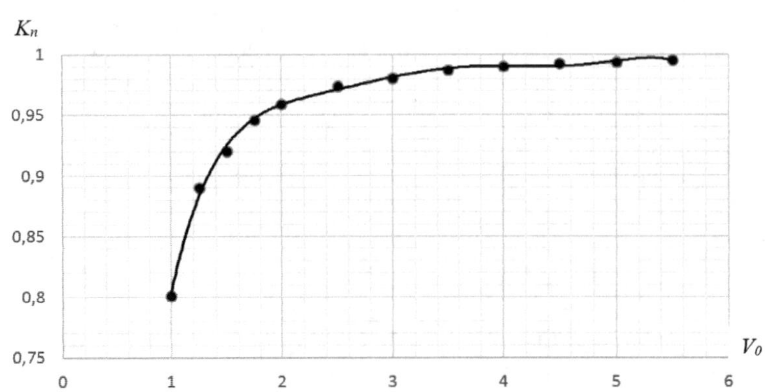

Figure 4. Dependence of the non-isothermal coefficient K_n on the velocity at the outlet from the grating V_0.

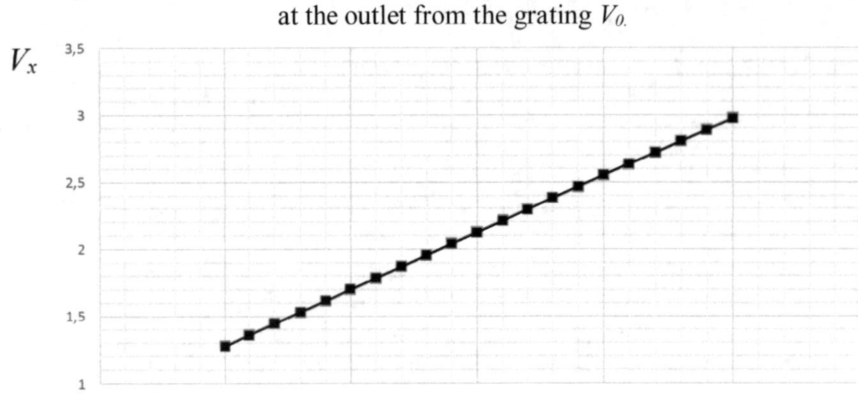

Figure 5. V_x versus V_0 plot under given conditions.

5. Summary

The article discusses the possibility of using general supply ventilation to prevent cold air from entering when opening doors in public premises. The implementation of this task is supposed to be carried out by organizing the necessary air exchange scheme, without increasing its required performance. Thus, general ventilation will simultaneously perform two functions: maintaining the required microclimate parameters and act as a heat curtain. The article deals with the most typical case for a public space of a small area, however, for a wider application of this technical solution, further research is needed.

6. References

[1] Overchenko M V, Belous A N 2019 Analysis of factors influencing the choice of insulation systems of educational institutions *Construction Materials and Products* **2** **(1)** 24 – 31
[2] Elistratova Y V, Kireev V M and Seminenko A S 2020 Contribution of heat loss by infiltration to energy saving and microclimate in multi-family residential buildings *IOP Conf. Series: Materials Science and Engineering 791*
[3] Skanavi A N 2002 *Heating* (Makhov-publishing ASV: Moscow)
[4] SNiP 31-06-2009 *Public buildings and structures* (M.: Gosstroy of the Russian Federation)
[5] SNiP 31-01-2003 *Residential apartment buildings* (M.: Gosstroy of the Russian Federation)
[6] SNiP-41-01-2003. *Heating, ventilation and air conditioning* (M.: Gosstroy of the Russian Federation)
[7] Ananiev V A, Balueva L N and Galperin A D 2001 *Ventilation and air conditioning systems, Theory and practice, Third edition* (Moscow: Euroclimate) p 416.
[8] Baturin V V 1941 *Air curtain design* (Moscow: Profizdat)
[9] Elterman V M 1961 *Air curtains* (Moscow: Mashgiz)
[10] Diskin M E 2003 To the question of calculating air curtains *AVOK* **. 7** pp 58-64.
[11] Bogoslovsky V N, Pirumov A I, Posokhin V N and others 1992 *Designer handbook: Internal sanitary devices. Part 3. Ventilation and air conditioning. Book. 1 / I. 4th ed., Revised. and add* (Moscow: Stroyizdat)
[12] Tatarchuk G T 1966 Refinement of the calculation method for air curtains *Heating and ventilation of industrial and agricultural buildings. Sat. scientific. tr.* **16**
[13] Shepelev I A 1950 *Fundamentals of calculation of air curtains, supply jets and porous filters* (Moscow: Stroyizdat)
[14] Baturin V V 1965 *Fundamentals of industrial ventilation* (Moscow: Profizdat)
[15] Abramovich G N 1948 *Turbulent free jets* (Moscow: Energoizdat)
[16] Butakov S E 1949 *Aerodynamics of industrial ventilation systems* (Moscow: Profizdat)
[17] Mott L F 1962 Design for protection by air curtain *Heating and Air Conditioning Journal* **№ 2**
[18] Lajos T, Preszler L 1975 Untersuchung von Torschleieranlagen *Heizung, Luftung, Haustechnik* **26** pp. 226-235
[19] Strongin A S, Nikulin M V 1991 A new approach to the calculation of air-thermal curtains *Construction and architecture: Ser. Izv. Universities* **1**
[20] *ASHRAE Handbook* 1999 (HVAC Applications)
[21] *Industrial Ventilation* 2001 (Design Guidebook. Academic press)
[22] Strongin A, Nikulin M 2004 On the issue of calculating air curtains *AVOK* **1**
[23] *Air distributors of the Akrtos company* Calculation and Application Notes 2008
[24] Grimitlin M I 2004 *Air distribution in premises* (St. Petersburg: Publishing house "AVOK North-West") 320

IOP Conf. Series: Materials Science and Engineering **945** (2020) 012042 doi:10.1088/1757-899X/945/1/012042

Modifying additive for concrete based on shungite processing waste

A S Yestemessova[1*], Z N Altayeva[2], B K Sarsenbayev[3], A M Budikova[4] and R O Karshygayev[4]

[1]KazGASA, Almaty, Kazakhstan
[2]Satbaev University, Almaty, Kazakhstan
[3]M.Auezov South Kazakhstan State University, Shymkent, Kazakhstan
[4]Korkyt Ata Kyzylorda University, 29a Ayteke bi Str., Kyzylorda, 12001, Kazakhstan

E-mail: stroitelstvo_ukgu@mail.ru

Abstract. The paper is devoted to the study of waste processing of shungite rocks, which are formed in the process of mining and technological operations. However, the market of national modifying additives based on local raw materials is not developed at the appropriate level in the presence of a rich resource base of the mining industry. In this regard, it is of great interest to obtain additives based on shungite rock processing waste, which allow expanding the range of mineral additives and providing qualitatively new properties for concrete and reinforced concrete products and structures. Such technologies for the production of additives are isolated and not developed at the proper level due to insufficient knowledge of shungite rocks.
Analysis of the state of research on concrete and the market for modifying materials showed that the development of production of energy-efficient and environmentally friendly materials due to increasing regulatory requirements for them, as well as to environmental ecology and the obvious need to recycle industrial waste initiated the progressive development of concrete science. One of its latest achievements is modified concrete with special properties, which marked a new stage in the development of concrete. In its production technology, significant progress has been made in improving its chemical additives, special modifiers, as well as active and inactive fillers. In Kazakhstan, the main volume of modifying additives is imported from abroad, which significantly increases their cost, so their choice is limited.
This paper is devoted to the research of waste processing of shungite rocks as a modifying mineral additive of concrete. Its influence on the main indicators of concrete is investigated and the features of the processes that occur in the hardening silicate system "cement - dispersed shungite particles"and contribute to increasing the strength of concrete, as well as reducing the duration of its heat-and-water treatment are studied.

1. Introduction

Currently, technologies related to the development of new methods for obtaining effective nanomodifying additives that allow to regulate the properties of composites based on mineral and organic binders are being intensively developed in the construction materials science [1-4].

The analysis of the issue showed that to improve the strength and other performance properties of concretes, solutions, polymers, the most acceptable are carbon nanoparticles, which include Kazakhstan shungite, which is a unique natural nanotechnological material [5-8].

Kazakhstan shungite is a natural nanostructured composite consisting mainly of a carbonaceous substance close to graphite and microcrystalline silica in mineralogical composition. The reserves of the shungite deposit, located only in the Almaty region, amount to more than 50 million tons. Therefore, the ecological and nanostructural potential of shungite, realized in its application as an effective mineral additive in concretes and solutions, seems promising and long-term [9-13].

Shungites of the Koksu Deposit are shungite rocks with a carbon content of up to 20% and have a wide ecological potential. The composition of shungite includes trace contaminants: Ti; K; Na; Mn; Ba; Zr; Sr; V; B; Zn; Ni; Y; Sc; P; Cr; Co; Mo; Li; Pb; Cu; Nb; Ga; Sn; Be; W; As; (the content of decreasing from 0.5 % to 0.5 ppm of the listed range of trace contaminants). Koksu shungites have an average chemical composition of macro components that do not contain toxic impurities or their dangerous concentrations [14-15].

Prerequisites for the use of shungite as a mineral additive are the development of the most effective and environmentally friendly technologies for obtaining nanomaterials for the modification of building materials [16-19].

2. Materials and methods

As a mineral additive of concrete, shungite waste was used under the brand "Taurite" of the mining company "Koksu". Taurite is a fine material with a size of 0-1 mm and up to 20 microns, which is based on the processes of mechanical activation and enrichment at a temperature of 400 to 600^{0}C. Properties of the mineral additive "TK-A", grain sizes 0-1 mm are shown in table 1.

Table1. Main properties of TK-A shale taurite.

Name of indicators	Values
Mass fraction of silicon dioxide (SiO_2), %	52.76
Content of water-soluble substances, %	0.73
pH of aqueous suspension	8.28
Bulk density, kg/m^3	674.3
Humidity, %	2.0

The properties of the mineral additive "TK-D-A" with grain sizes of 10 microns are shown in table 2.

Table 2. Main properties of TK-D-A shale taurite.

Name of indicators	Values
Mass fraction of silicon dioxide (SiO2), %	36.76
Content of water-soluble substances, %	0.7
pH of aqueous suspension	8.36
Bulk density, kg/m3	512
Humidity, %	0.7

Portland cement and all samples of cement stone were studied by a complex of physical and chemical analysis methods: x-ray phase and differential thermal.

Radiographs were obtained using a DRON-3M diffractometer with a copper anti-cathode in the angle range of 4-64 degrees. Samples in the form of fine-ground powder were pressed into a plexiglass cuvette.

Thermo grams were taken on a Q-1500D derivatograph from 10 to 1000^{0}C with a temperature rise rate of 100^{0}C per minute.

IOP Conf. Series: Materials Science and Engineering **945** (2020) 012042 doi:10.1088/1757-899X/945/1/012042

3. Results and discussions

Reducing the water demand of cement composite materials with mineral additives is achieved by the complex use of mineral additives with a superplasticizer, which has a significant water-reducing effect.

Table 3 shows the composition of concretes with mineral additives of taurite and polymethylene-naphthalene-sulfonate additive of the Kratasol brand.

Table 3. Concrete compositions with the mineral additive taurite.

Name of the material, kg							
№ of composition	Cement	Rubble	Sand	Mineral additive "TK-A"	Mineral additive "TK-D-A"	Superplasticizer Kratasol	W/C
1	380	990	875	19	-	-	0.64
2	380	990	875	19	-	3.8	0.56
3	380	990	875	-	11.4	-	0.68
4	380	990	875	-	11.4	3.8	0.64

The samples were solidified at room temperature for a day and after demolding, one series of concrete samples-cubes solidified for 3, 7, 14, 28 days under normal humidity conditions, and the second series of concrete samples-cubes after forming was placed in a heat-humidity treatment chamber after two hours (70^{0}C).

Analysis of table 3 data shows that the use of mineral additives, depending on their composition and content, affects the W/C ratio of the concrete mix, namely fine-grained additives without plasticizer on average 258l/m^3, with superplasticizer 243 l/m^3-247 l/m^3.

The effect of plasticization with different amounts of mineral and plasticizing additives is different: concrete mix №1 for 60 minutes with c grade on the mobility of PK5 decreased to PK4; concrete mix № 2 did not show a change in mobility for the same time; concrete mix №3 with PK5-PK3; concrete mix №4 with PK5-PK4.

The diluting effect of the superplasticizer "Kratasol" allows reducing the amount of closing water significantly and obtaining high-strength concretes with a given mobility of the concrete mix.

The results of research on concrete compressive strength are shown in table 4.

From the data given in table 4, it can be seen that the greatest strength of concrete was shown by compositions №3 and №4, containing mineral additives TK-D -A and superplasticizer "Kratasol".

Table 4. Strength of concrete with mineral additive Taurite and superplasticizer Kratasol in different conditions of hardening.

Numbers of compositions	W/C	Workability	Tensile strength at compression, MPa					
			Normal-humidity hardening, days			Hardening in the conditions of HHT, days		
		OK, cm	3	7	28	12 h	1	28
1	0,58	20	13.2	19.8	32.9	10.6	22.5	33.4
2	0,64	21	13.5	20.1	34.1	12.4	22.8	33.6
3	0,56	23	15.2	22.4	34.5	14.7	22.6	33.9
4	0,68	22	14.4	21.3	33.8	12.8	22.7	34.9
5	0,64	23	14.1	22.3	34.0	13.5	23.4	34.3

The increase in strength of compositions №2-5 in comparison with the control composition for the 3rd day is from 2.7-15.3%, for the 7th day from 1-27 %, for the 28th day 2.5-30% when solidified in normal humidity conditions.

The increase in the strength of concrete compositions №2-5 during heat and humidity treatment in comparison with the control composition in 12 hours is from 24.5-39.6%, after 1 day from 5.3-29.9 %, after 28 days 2.4-25 %.

The processes occurring in the cement-mineral additive system are studied.

Identification of products of hydration and hardening of silicate systems showed that mineral additives have a significant effect on the phase composition of hardening systems, so the microstructure of the solidified solution at the age of 3 days in natural conditions showed that the particles of the mineral additive have been already exposed to the alkaline components of the cement paste at the initial stage of hydration of the hardening silicate system. As a result of this action, the surface of the particles of the mineral additive is hydrated. The thickness of the hydration layer is 1 ... 5 microns. The hydrated part of the grains forms a "border" around them, without separating from the remaining main part. In some cases, the intermediate cementing phase, on the contrary, penetrates into the grain structure of the mineral component.

Analyzing the obtained data on the study of the microstructure of the solidified solution we can draw the following conclusions:

- initially, there is a mechanical interaction between the mineral additive and the clinker components due to the roughness of the surface of its particles;
- due to their amorphous nature, SiO_2 microparticles serve as centers of crystal formation for hydrates, such as calcium silicate hydrate (CSH), which explains the formation of an additional number of crystallites that initiate the setting process. In addition, SiO_2 silica reacts with portlandite [$Ca(OH)_2$] via the mechanism described in the reaction:

$$SiO_2 + Ca(OH)_2 \rightarrow CSH$$

- in the hardening silicate system, the grains of the micro-filler are mainly hydrated, creating a border around the residual part with a thickness of 1...5 microns;
- components of cement, hydrating and releasing the hydrate phase, actively affect the grain of the micro-filler;
- as a result of active hydration and surface hydration of the grains, an intermediate hydrate layer occurs between them, linking the cement stone and the aggregate (filler);
- this position of the intermediate hydrated layer between the grains of the micro-filler and the cement stone contributes to a significant strengthening and compaction of the solution.

Figures 1 and 2 show the obtained x-ray diffractograms of compositions after 28 days of natural hardening based on cement and mineral additives "TK-A", "TK-D-A".

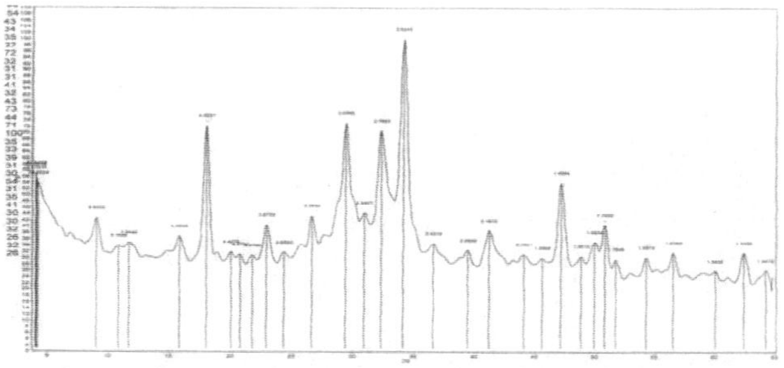

Figure 1. Diffractogram of cement stone with mineral additive "TK-A".

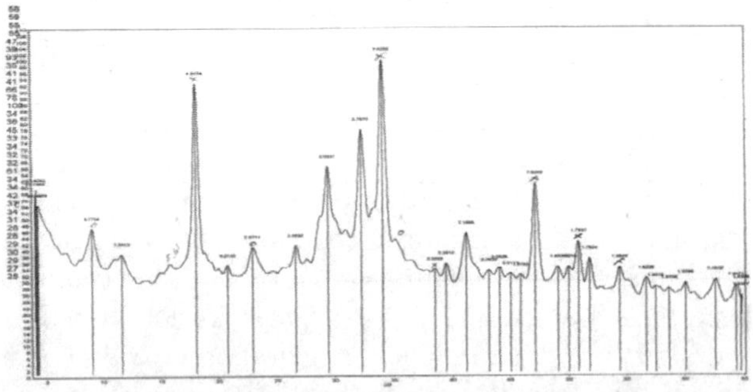

Figure 2. Diffractogram of cement stone with mineral additive "TK-D-A".

The lines of tricalcium silicate C_3S (3,035-2,76 Å), bicalcium silicate β-C_2S (2,89 Å) are fixed on the hardening diffractogram with "TK-A" (figure 1), there are quartz lines (4,27-3,346-2,451-2,283-1,988-1,540-1,447 Å) and hydration products: $Ca(OH)_2$ (4,923-2,624-1,926-1,795-1,687-1,485-1,447 Å), $CaCO_3$ (3,03-1,861 Å) and ettringite (9,803-5,59-3,874-2,20 Å). On the differential-thermal curve, endoeffects at 100^0C and 120^0C are associated with dehydration of ettringite and transition to CaO, at 500^0C with dehydration of $CA(OH)_2$, at 800^0C with calcite decomposition.

The lines of tricalcium silicate C_3S (3,037-2,76-2,32-2,188-1,853-1,762-1,48Å), bicalcium silicate β-C_2S (2,89 Å) are fixed on the hardening diffractogram with "TK-D-A" (figure 2), there are quartz lines (4,27-3,34-2,28-1,821-1,539 Å) and hydration products: $Ca(OH)_2$ (4,917-2,625-1,924-1,793-1,684 Å), $CaCO_3$ (3,037-2,28-2,093Å) and ettringite (9,73-3,871-2,77-2,18 Å). On the differential-thermal curve, the endoeffects at 120^0C and 150^0C are associated with the dehydration of ettringite and the transition to CaO, at 520^0C with the dehydration of $Ca(OH)_2$, at 800^0C with the decomposition of calcite.

According to the data of x-ray phase analysis, the positive effect of mineral additives "TK-A" and "TK-D-A" on the hydration of clinker minerals was noted, as evidenced by the increase in the intensity of the $Ca(OH)_2$ line with an inter-plane distance of 4.92 Å and the appearance of a line also related to $Ca(OH)_2$ of 2.63 Å. On thermograms endoeffects are recorded at 500°C and 520°C associated with the removal of water from the $Ca(OH)_2$.

4. Summary

The possibility of using a new mineral additive in concrete – "Taurite" of a fine-dispersed size from 0-1 mm and up to 10 microns, which is based on the processes of mechanical activation and enrichment at a temperature of 400 to 600°C is found. The research results showed that the introduction of the mineral additive "Taurite" and naphthalene superplasticizer "Kratosol" increases the strength of normal-humidity hardening concretes and in comparison with the control composition in 3 days is from 2.7-15.3%, 7 days from 1-27%, 28 days 2.5-30%, and concretes of heat-humidity hardening in comparison with the control composition for 12 hours is from 5.5-12.9%, after 1 day from 0.3-19.8%, 28 day 0.6-25 %. The optimal "soft" mode of heat-and-water treatment of concrete with an isothermal heating temperature of no more than 70 ^{0}C is established.

Features of dispersed particles of mineral additives from shungite waste processing in the hardening of silicate systems, namely that the surface of particles of mineral filler are susceptible of hydration that contributes to concrete and mortar with improved physical-mechanical indicators. Amorphous SiO_2 particles in the mineral filler contribute to the formation of foci of crystals of hydroaluminates and calcium hydrosilicates. The influence of mineral additives on the rate of hydration processes of hardening of silicate systems was studied and established. The main products of hydration of modified mixtures are ettringite, portlandite and gel-like silicate phase. It was found that in the presence of mineral additives, interdependent aluminosilicate and polymer phases are formed, and the additives "TK-A" and "TK-D-A""are inhibitors of structure formation.

IOP Conf. Series: Materials Science and Engineering **945** (2020) 012042 doi:10.1088/1757-899X/945/1/012042

5. References

[1] Sakharov G P 2009On the short-term perspective of nanotechnologies in the production of building materials and products *Technologies of concrete* **5** 13-15

[2] Aymenov A Zh, Sarsenbayev N B, Khudyakova T M, Sarsenbayev B K, Kopzhassarov B T 2016 Effect of additive of polymetallic ores' tailings on properties of composite cements// *Eurasian chemico-technological journal* 153-160

[3] Aymenov A Zh, Khudyakova T M, Sarsenbayev B K 2018 Studying the mineral additives effect on a composition and properties of a composite binding agent *Oriental journal of chemistry* **34 (4)** 1945-1955

[4] Bazhenov Yu M, Demyanova V S, Kalashnikov V I 2006 Modified high-quality concrete *Scientific edition. M.: Publishing house of the Association of construction universities* 368

[5] Khavari-Khorasani G, Murchison D G 1979 The nature of carbonaceous matter in the Karelian shungite *Chem. Geol.* **26** 165-82

[6] Mosin O V, Ignatov I 2012 Application of natural fullerene-containing mineral shungite in construction and construction technologies *Nanotechnologies in construction: scientific online journal. Moscow: CNT "Nano-Construction"* **4 (6)** 22-34

[7] Klyuev S V, Klyuev A V, Shorstova E S 2019 Fiber concrete for 3-d additive technologies *Materials Science Forum* **974** 367–372 DOI: 10.4028/www.scientific.net/MSF.974.367

[8] Klyuev S V, Klyuev A V, Shorstova E S 2019 The micro silicon additive effects on the fine-grassed concrete properties for 3-d additive technologies *Materials Science Forum* **974** 131–135 doi:10.4028/www.scientific.net/MSF.974.131

[9] Kovalevski V V, Buseck P R, Cowley J M 2001 Comparison of carbon in shungite rocks to other natural carbons: an X-ray and TEM study *Carbon* **39(2)** 243–256

[10] Mosin O V, Ignatov I (2013) Structural properties and composition of fullerene containing mineral shungite *Nano- i microsystemnaja technika* **1** 12–20

[11] Chernousov D I 2011 Application of asphalt binder with shungite in the construction of road surfaces: abstract dis. ... Candidate of Engineering Sciences. Voronezh 19

[12] Cascarini de Torre L E, Fertitta A E, Flores E S, Llanos J L, Bottani E J (2004) Characterization of shungite by physical adsorbtion of gases *J. Argent. Chem. Soc.* **92(4–6)** 51–58

[13] Kwiecinska B, Pusz S, Krzesinska M, Pilawa B 2007Physical properties of shungite *// International Journal of Coal Geology* **(71)** 455-461

[14] Zhurinov M Zh, Baeshov A B, Zhumabay I M, Serikbaev B A, Salaeva Z P 2006 Investigation of sorption properties of Koksu shungites: collection of scientific papers 6-12

[15] Yestemessova A S, Yesselbayeva A G, Yestemessova S M 2017 Shungite waste of mining production in modified concretes regional case stady *International Journal of Chemical Sciences* 16-26

[16] Rafienko V A 2008 Technology of processing of shungite rocks M.: GEOS 214

[17] Kalinin Yu K 2002 Carbon-containing shungite rocks and their practical use: autoref. dis. Doct. of Eng. Sciences. Moscow MM 50

[18] Lukuttsova N P, Pyshkin A A, Degtyarev E V, Shirko S V 2012 Structure and properties of cement stone and concrete with the addition of UKN modifier *Cement and its application* **3** 119-121

[19] Kalinin Yu K, Kalinina A I, Skorobogatova G A 2008 Karelian shungites for new building materials, in chemical synthesis, gas treatment, water treatment and medicine. SPb: Publishing House UNC SPbSU 219

Plane model of particle separation in disintegrators with internal circulation of the grinding material

T N Orekhova[1], A E Kachaev[2] and V S Prokopenko[1*]

[1]Department of Industrial Ecology, Belgorod State Technological University named after V.G. Shukhov, Kostyukov St., 46, Belgorod, 308012, Russia
[2]OOO «Resurrection plant «Mashinostroitel»»,1 Garage str., Voskresensk, 140200, Russia

E-mail: vlad.prokopenko1@yandex.ru

Abstract. The paper offers a method for evaluating the separation process of fine particles in the design of a classifier of a disintegrator with internal circulation of the dispersed material. Disintegrator type mills are fairly common in the construction materials industry and related industries. However, there are a number of technological problems with this type of equipment, which hinder their wider use in the production of dispersed powders with a high specific surface area. One of these problems is the presence of under-ground (coarse) particles in the volume of the finished product of a given fraction. The quality of classification is evaluated taking into account the velocity parameters of fine particles in the classifier device built into the disintegrator body. A new design of a gravity-type classifying device is considered, which contributes to the efficient return of large substandard particles to the finishes grinding in the disintegrator grinding chamber. The publication provides a method for determining the speed parameters of particle dynamics in the chamber of the classifier device of a disintegrator mill, and also presents a model for numerical calculation of the efficiency of the classifier device. This theoretical method of assessing the quality of separation of particles in the disintegrator is to define a measure of the degree of separation, to assess the effectiveness of recycle coarse fractions of grinding material powder: degree of operational separation shows the ability of disintegrator to operate in various process modes, and sets the granulometric boundaries for obtaining a particular fraction after grinding in the disintegrator.

1. Introduction

Along with the development of building materials science, it is necessary to solve the problem of obtaining dispersed materials of a narrow granulometric composition without the presence of particles with a diameter greater than the critical one in the finished product. Developed by a team of authors, the disintegrator [1] with internal circulation of the flow of grinding materials allows obtaining dispersed powders of a narrow granulometric composition with the absence of particles in their volume whose diameter is greater than the critical one.

Let us consider the design features of a gravitational type classifying device, which can be equipped with a disintegrator, or a rotary mill, of any design (figure 1).

IOP Conf. Series: Materials Science and Engineering **945** (2020) 012043 doi:10.1088/1757-899X/945/1/012043

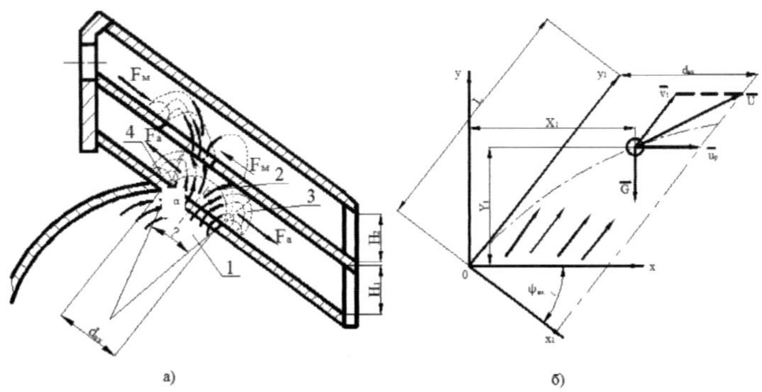

Figure 1. Scheme for calculating particle classification parameters in a disintegrator mill: a) scheme for calculating the speed mode of air-material flow in a conical multi-chamber classifier: 1 - core of the gushing layer; 2 - central part of the fountain; 3 –descending part of the fountain; 4 – area of bulk circular layer;
b) – scheme for determining the trajectories of movement and speeds of particles in the classifier.

To establish analytical expressions for determining the parameters of the process of dynamics, deposition and possible separation of particles from the conical surface of the classifier, we make the following assumptions: we assume that the particles move singly, have a spherical shape, the movement occurs in a limited volume of the classifier device, and the air (two-phase) flow is constant in speed and direction.

2. Materials and methods

In the weighed layer of a two-phase flow a fixed particle in the gushing layer is affected by the total force that can be formed from:

$$\sum F = G + F_a + F_{pod} + F_M. \quad (1)$$

Based on these components of the total force (1), we will keep in mind the model of particle classification in the disintegrator (figure 1b) in the self-similar region of the flow of a spherical particle. This means that the horizontal part of the particle velocity in the classifier in the projection on the X_1 axis will have the form:

$$m_ч \frac{du_p}{dt} = m_p a - 0{,}05\pi \cdot \mu \cdot d_{кp}^2 \cdot \rho_p \cdot u_p^2 + m_p g + V_p \cdot \rho_p \cdot g, \quad (2)$$

where a - particle acceleration in the air stream, m/s^2; u_p – the horizontal speed of the particle,m/s; ρ_p – density of particle material, kg/m^3; m_p - the density of the material particle, kg/m^3; m - mass of the material particle, kg; μ- dynamic viscosity of two-phase flow,Pa•s; V_p – volume of a particle as a spherical whole, m^3;g – acceleration of gravity, m/s^2; $d_{кp}$- the critical diameter of the particle, defined from the expression [2]:

$$d_{кp} \geq \sqrt{\frac{9 \cdot R_{vx}^2 \cdot S_{vx} \cdot j_o \cdot \mu}{\pi \cdot (\rho_p - \rho) \cdot \sin^2 \psi_{vx} \cdot R_{vix}^2 \cdot H \cdot v_{vx}}}, \quad (3)$$

here S_{vx}- the area of the hole for entering the two-phase flow into the classifier, m^2; R_{vx}- radius of the hole for entering the two-phase flow into the classifier, m; R_{vix}- radius of the classifier exit hole, m; H - the chamber height of the classifier, m; v_{vx} - the speed of air and material flow at the entrance to the classifier, m/s; ψ_{vx} - the angle between the axis of the hole and the direction of movement of the air-material flow, °C; ρ – air density, kg/m^3; jo - the ratio of windage of particle.
Let us represent a nonlinear quadratic expression (2) in letter form:
:

$$\frac{du_p}{dt} = A_1^2 - B_1^2 \cdot u_p^2, \quad (4)$$

where parameters can be defined like this

$$A_1^2 = a \cdot (1 - \bar{\rho}); \ B_1^2 = \frac{0{,}05\pi \cdot \mu \cdot d_{кр}^2 \cdot P}{m_p}, \qquad (5)$$

where P - full static pressure, Pa; $\bar{\rho}$ - the ratio of the particle density to the flow density of the carrier medium-air.

Let us integrate expression (4) under the initial condition $u_p(0) = 0$:

$$\int_0^{u_ч} \frac{du_p}{A_1^2 - B_1^2 \cdot u_p^2} = \int_0^t dt. \qquad 6)$$

Therefore, the equation for horizontal velocity will have the form:

$$u_p = \frac{A_1}{B_1} \frac{\exp(2A_1B_1t) - 1}{\exp(2A_1B_1t) + 1}. \qquad (7)$$

Substituting the expression (5) in (7), we obtain an equation for determining the horizontal velocity of the particle movement in the classifier chamber:

$$u_p = \frac{m_p a(1-\bar{\rho})}{0{,}05\pi \cdot \mu \cdot d_{кр}^2 \cdot P} \cdot \frac{\exp\left(2t\sqrt{a \cdot (1-\bar{\rho}) \frac{0{,}05\pi \cdot \mu \cdot d_{кр}^2 \cdot P}{m_p}}\right) - 1}{\exp\left(2t\sqrt{a \cdot (1-\bar{\rho}) \frac{0{,}05\pi \cdot \mu \cdot d_{кр}^2 \cdot P}{m_p}}\right) + 1}. \qquad (8)$$

We determine the movement of the particle from the central part of the two-phase flow to the periphery of the chamber along the X-axis - expression (8) is re-integrated. We will have with substitutions (5) for the initial condition X1 (t) = 0 for t = 0 the following expression for X_1:

$$X_1 = \frac{m_p}{0{,}05\pi \cdot \mu \cdot d_{кр}^2 . P} \cdot \left[\ln\left(1 + \exp 2d_{кр} \cdot t\sqrt{a \cdot (1-\bar{\rho}) \frac{0{,}05\pi \cdot \mu \cdot P}{m_p}}\right) - d_{кр} \cdot t\sqrt{a \cdot (1-\bar{\rho}) \frac{0{,}05\pi \cdot \mu \cdot P}{m_p}}\right]. \qquad (9)$$

Thus, an equation is obtained that allows the particles to move in the volume of the gushing layer at beginning of coordinates (figure 1 b). Along this trajectory, a heavy particle just needs to pass through and leave the active zone of the gushing (weighed) layer to settle on the surface of the conical classifier and return to re-grinding.

The obtained analytical dependences consider a flow evenly dusted with particles with a single value of their diameter and a certain concentration of C_{vx}. In this case, we use the condition that if large, heavy particles under the action of gravity G reach the lower cone surface of the classifier, we will assume that they differentiated, lagged behind the two-phase flow. When the two-phase flow exits the lower chamber of the conical classifier with a height $H_1 = f(L(y))$ and enters another chamber located above the first one, we observe that the dispersed two-phase flow with a thickness Y_1 leaves the solid particles of the crushed material. Therefore, it is necessary to install with the mass characteristics of a two-phase flow. Then the total mass flow rate of the particles separated in the classifier per unit height of each of the chambers will be determined by the expression:

$$M_Y = C_{vx} \cdot V_V \cdot L, \qquad (10)$$

where V_V- the speed of the particle along the axis of the hole of the conical classifier, which depends on the speed of the particle hovering $V_в$, m/s. However, it should be noted that the hovering speed $V_в$ can be calculated using the work formula [3]:

$$V_в = \sqrt{\frac{4 \cdot (\rho_p - \rho) \cdot g \cdot d_{sr}}{3\rho \cdot C_a}}, \qquad (11)$$

where d_{sr} - average diameter of the grinding material particles, m.

It should be noted that the dependence of the velocity of vertical movement of the particle along the axis of the hole of the conical classifier is experimentally confirmed, taking into account the gushing two-phase flow [4], which can be represented as:

$$V_V = V_в \sqrt{\frac{V_M^H}{V_M^K} \cdot \frac{S_{Mf}^K}{S_{Mf}^H}}, \qquad (12)$$

where V_M^H and V_M^K - the volume of material in the air flow at the entrance and exit of their classifier chambers, m^3; S_{Mf}^H and S_{Mf}^K - the area of the interphase boundary at the entrance and exit of the classifier chambers, m^2.

To calculate the vertical movement of a particle in a two-phase flow that is formed in the classifier along the Y axis, we integrate the expression (12) by time t. For initial conditions Y1 (t) = 0 for t = 0, we get:

$$L(y) = \tau \cdot \sqrt{\frac{4(\rho_ч-\rho)g\cdot d_{sr}}{3\rho\cdot C_a} \cdot \frac{V_M^н}{V_M^к} \cdot \frac{S_{Mf}^к}{S_{Mf}^н}}. \quad (13)$$

Expressions for u_p and X1, as well as V_V and $L(y)$ define the main parameters of particle dynamics in the conical classifier and help control the separation process in the device [5]. Using the obtained analytical equations it is possible to determine the degree of separation of particles crushed in a conical classifier of the disintegrator according to the method proposed below:

We determine the total mass flow rate of particles at the entrance to the cylindrical channel of the cone classifier:

$$\Sigma M_{vx} = 2C_{vx} \cdot V_V \cdot R_{vx}, \quad (14)$$

We calculate the degree of separation of particles with a critical diameter from a two-phase flow using the empirical expression:

$$\eta_d = \frac{M_y}{\Sigma M_{vx}} = \frac{H}{2R_{vx}} = \frac{H}{D_{vx}}. \quad (15)$$

We find the value of the degree of separation of particles in a two-phase flow entering the conical classifier, taking into account the substitution (6) and the time of movement in the channel $t = \frac{H(L(y))}{V_V}$:

$$\eta_d = K_{sep}\left[\frac{\ln(1+\exp 2S_p\cdot K_{sep})}{S_p\cdot K_{sep}} - 1\right], \quad (16)$$

here S_p - the flow parameter that affects the particle, determined from the works [6-7]:

$$S_p = \frac{0{,}05\pi\cdot d_{кр}^2\cdot\rho\cdot D_{vx}}{m_p}. \quad (17)$$

K_{sep} - separation coefficient for the quadratic law of particle flow, calculated by the results of the work [3]:

$$K_{sep} = \sqrt{\frac{(1-\bar{\rho})\cdot a\cdot H^2}{2S_ч\cdot R_{вх}\cdot V_V^2}}. \quad (18)$$

The distribution function (pass) can be expressed as follows, based on the fact that a two-phase flow in the volume of a conical classifier contains particles of different sizes:

$$\Phi(d_p) = \frac{M_d}{M}, \quad (19)$$

where M_d- mass flow rate of separated particles, kg, determined as $2r_p = d_p$; M - mass flow rate of all particles, kg.

The total mass flow rate of particles with a diameter dp within the deviation Δdp at a certain degree of separation η_d can be determined by their equations:

$$\Sigma M_S = M \cdot \eta_d \frac{d\Phi(d_p)}{dd_p}\Delta d_p. \quad (20)$$

The mass flow rate of separated particles ΣM_S, after integration, will be represented by the degree of separation of particles from the two-phase flow in the chambers of the conical classifier, which will look like this:

$$\eta = \frac{M_s}{M} = \int_0^{\infty} \eta_d \frac{d\Phi(d_p)}{dd_p} dd_p. \quad (21)$$

In the infinite set of volumes of dispersed dusts the logarithmic law of particle distribution is valid:

$$\Phi(d_p) = \frac{1}{\sqrt{2\pi}}\int_{-\infty}^{\tau_1} \exp\left(-\frac{1}{2}\tau_1^2\right) d\tau_1, \quad (22)$$

where $\tau_1 = \frac{\lg\left(\frac{d_p}{d_{50}}\right)}{\lg\sigma}$ - argument of the distribution function, and σ - variance that characterizes the standard deviation, d_{50} - the median diameter of the distribution at which the mass of particles with $d_p < d_{50}$ is equal to the mass of particles $d_p > d_{50}$.

Finally, the equation for the degree of separation of particles taking into account (21) will take the form:

IOP Conf. Series: Materials Science and Engineering **945** (2020) 012043 doi:10.1088/1757-899X/945/1/012043

$$\eta = 1 + \Phi(d_p)(\tau_{kr}) + \frac{1}{\sqrt{2\pi}} \int_{-\infty}^{\tau_{\kappa p}} \eta_d \cdot \exp\left(-\frac{1}{2}\tau_1^2\right) d\tau_{kr}, \tag{23}$$

where $\tau_{kr} = \tau_1$.

3. Results

The use of probabilistic expression (22), which is confirmed by experimental data, allowed obtaining theoretically equation (23), which establishes the dependence of the degree of separation of particles in the conical classifier of the disintegrator on the distribution parameters in the volume of the classifier chambers. Expression (23) includes all parameters that characterize this process. It should be noted that the obtained expression (23) is sufficiently correlated with experimental data obtained by electronic sieve analysis of particles after grinding and classification of various materials in the disintegrator.

Equations (16) and (23) allow determining the degree of separation under the influence of forces of different nature: gravitational, inertial, centrifugal, etc. in the working volume of the mill classifier of the developed design.

4. Summary

As a result of mathematical and experimental research, analytical dependences were obtained that allow determining the velocity parameters of particles in the working volume of the disintegrator chambers, which are based on the design and technological parameters of the conical classifier and the disintegrator separately. Taking into account the influence of processes occurring in the disintegrator, the effect of the disintegrator operation modes on the particle separation process in the classifier is established, and the relationship between the speed parameter of particle movement in the classifier and the degree of separation of fine particles in the two-phase circulation flows of the disintegrator mill is obtained.

5. References

[1] Pat. 2377070 Russian Federation, B02C 13/22 Disintegrator. Authors: Gridchin A.M., Sevostyanov V.S., Lesovik R.V., Glagolev E.S., Sevostyanov M.V., Kachaev A.E., Solopov N.V., Alfimova N.I., applicant and patent holder: BSTU named after V.G. Shukhov, decl. 06.05.2008-application №2008118079/03, publ. 27.12.2009 in bul. 36

[2] Kachaev A E 2013 Disintegrator with internal recycling of grinding materials: dissertation of Candidate of Engineering Sciences. Belgorod 210

[3] Krasnov N F 1976 Aerodynamics. P. 1 Fundamentals of the theory. Airfoil and wing aerodynamics.Higher school 384

[4] Timoshenko V I, Knyshenko Yu V, Lyashenko Yu G, Deshko A E 2008 Method of experimental substantiation of technological parameters of devices with the use of a gushing layer *Science and innovations* **4(2)** 21–32

[5] Kachaev A E, Sevostyanov V S 2012 Determination of particle velocities in the classifier of the disintegrator *Energy saving technological complexes and equipment for the production of construction materials* 196-201

[6] Smulsky I I 1983 Weighed layer of particles in a cylindrical vortex chamber *Journal of applied chemistry* **8** 1782–1789

[7] Blinichev V N 1975 Development of equipment and methods of its calculation for the intensification of processes of fine grinding of materials and chemical reaction in solids: Diss. for the academic degree of Dr. of Eng. Sciences. ICTI 422

[8] Bogdanov V S, Ilyin A S, Nesmeyanov N P 2004 Processes of grinding and classification in cement production. ASV 199

[9] Pugin K G 1994 Determination of parameters and productivity of rotary mills of intensifying action: autoref. of dissertation of Cand. of Eng. Sciences. Moscow 169

[10] Kachaev A E, Orekhova T N, Okushko V V, Shestakov Yu G 2020 Mechanics of a weighed layer with polydisperse particles in continuous air mixers *Bulletin of BSTU named after V G*

Shukhov **3** 121-127

[11] Romanovich A A, Romanovich L G, Chekhovskoy E I 2018 Determination of rational parameters for process of grinding materials pre-crushed by pressure in ball mill *IOP Conference Series: Materials Science and Engineering* **327** (042091)

[12] Orekhova T N, Prokopenko V S, Okushko V V, Kachaev A E 2019 The rotating layer mechanics of polydisperse particles in the continuous action pneumatic mixers *IOP Conference Series: Materials Science and Engineering* **698(2)** (022023)

Acknowledgments

This work was realized under the support of the President Scholarship; in the framework of the Program of flagship university development on the base of the Belgorod State Technological University named after V G Shukhov, using equipment of High Technology Center at BSTU named after V G Shukhov.

IOP Conf. Series: Materials Science and Engineering **945** (2020) 012044 doi:10.1088/1757-899X/945/1/012044

Using wastes of buckwheat processing as sorption materials for the removal of pollutants from aqueous media: a review

I G Shaikhiev[1], S V Sverguzova[2*], R Z Galimova[1] and A S Grechina[1]

[1]Department of Environmental Engineering, Kazan National Research Technological University, 41, st. Siberian tract, Kazan, 420074, Russia
[2]Department of industrial ecology, Belgorod State Technological University named after V G Shoukhov, 46, Kostukova Str., Belgorod, 308012, Russia

E-mail: fomina.katerina@mail.ru

Abstract. Based on data from various literature sources, information is given on the chemical composition of straw and husk of buckwheat grains. The information on buckwheat processing waste as a source for obtaining various valuable components is provided. Literature data on the use of ground buckwheat straw as a sorption material for the extraction of heavy metal ions and oil products from natural and sewage are presented. The possibility of increasing sorption characteristics by buckwheat straw treatment with various chemical reagents is shown. Determined that the husk of buckwheat grains contains cellulose and lignin most of all. Information is given on the use of the fruit shells of buckwheat grains as sorption materials for the removal of metal ions, oil and oil products from aqueous media. It was shown that to increase the sorption characteristics of these pollutants is possible by treating the sorption material with acidic chemical reagents and high-frequency low-pressure plasma. The parameters of acid treatment or plasma exposure are determined at which the highest adsorption indicators. Native straw and buckwheat husk are effective sorption materials for the removal of heavy metal ions and oil from aqueous media. It is possible to increase the sorption characteristics according to pollutants by a chemical or physical and chemical modification.

1. Introduction

In the world community, a new innovative direction in the field of environmental protection is currently developing rapidly such as the use of agricultural waste as reagents for the removal of pollutants from natural and sewage waters [1–10].

Now in the world, crops are grown most of all. Accordingly, when they are processed, a huge amount of various wastes such as straw, flour, flooring, grain husks, etc is formed. The latter are used in various industries as sorption materials for the extraction of harmful substances from aqueous media. Thus, previously published reviews showed that waste from the processing of wheat [11], barley [12] and oats [13] are effective sorption materials for the removal of metal ions, dyes, pesticides, oil and products of its processing from aqueous media.

The aim of this work is to analyze and summarize the literature on the use of waste from the processing of buckwheat as sorption materials for the removal of pollutants from aqueous media.

IOP Conf. Series: Materials Science and Engineering **945** (2020) 012044 doi:10.1088/1757-899X/945/1/012044

2. Materials and methods

The microstructure of the surface of the buckwheat husk was studied using a high-resolution scanning electron microscope TESCANMIRA 3 LMU. To study the specific surface area and porosity of the buckwheat husk at the micro level, we used the method of gas adsorption using the SoftSorbi – II ver.1.0 device.

3. Results and discussion

3.1. Basic information about buckwheat

Buckwheat (*Latin Fagopýrum*) is a genus of herbaceous plants of the Buckwheat family (*Polygonaceae*) (Figure 1, a). It is annual or perennial herbaceous plant, 10 – 100 cm high. Stems are bare, upright, branched, and evenly leafy. The leaves are alternate, arrow-triangular or arrow-ovate, 2.5–9 cm long, the tip is elongated. The flowers are bisexual, dimorphic, heterostyle. The perianth is five-parted, yellowish-green, white or light pink, 1.3–4 mm long. Fruits are trihedral, dull nuts, 4.5 –6.5 mm long. In practical terms, the most important plant of the genus is seed buckwheat (*Fagopyrum esculentum*), a bread and honey plant [14].

Figure 1. Buckwheat a), buckwheat straw b) and c) buckwheat husk.

It has been found that seed buckwheat contains a large amount of nutrients – 63 % carbohydrates, 11.7 % protein, 2.4 % fat, 9.9 % fiber, 11 % water and 2 % minerals [15]. Buckwheat contains a lot of iron, as well as calcium, potassium, phosphorus, iodine, zinc, fluorine, molybdenum, cobalt, as well as vitamins B1, B2, B9 (folic acid), PP, E. The flowering aerial part of buckwheat contains rutin, phagopyrine, pyrocatechol, gallic, chlorogenic and caffeic acids. Seeds have starch, protein, sugar, fatty oil, organic acids (maleic, menolenic, oxalic, apple and lemon), riboflavin, thiamine, phosphorus, iron. By the content of lysine and methionine, buckwheat protein is superior to all cereals; it has high digestibility – up to 78 % [16].

3.2. Sorption properties of buckwheat components

3.2.1. Straw

The most massive waste from the processing of buckwheat is straw (Figure 1, b). The composition of buckwheat straw is shown in Table 1 [17].

Table 1. The chemical composition of buckwheat straw [17].

Indicators	Value	Indicators	Value
Feed units	0.34	Calcium, g	9.3
Dry matter, g	906	Potassium, g	12.73
Crude protein, g	45	Phosphorus, g	1.3
Digestible protein (cattle), g	20.7	Magnesium, g	1.91
Lysine, g	1.22	Sodium, g	1.55
Methionine + cystine, g	1.35	Iron, mg	199.98

Crude fiber, g	349.5	Manganese, mg	34.54
Sugar, g	1.41	Zinc, mg	10.91
Biological extractive substances (BES), g	442	Copper, mg	3.45
Crude Fat, g	12.4	Cobalt, mg	0.18
		Iodine, mg	0.05
		Carotene, mg	3
		Vitamin D (calciferol), thousand ME	5

The presence of proteins in buckwheat straw can contribute to the adsorption of heavy metal ions (HMI) (Table 1). However, contrary to what was expected, information on the adsorption of HMI by buckwheat straw was very limited. In particular, the removal of Cu^{2+} and Cr^{6+} ions by a sorption material from a mixture of buckwheat straw and coal from iron agraterata (*Ageratina adenóphora*) was studied [18 – 20]. The influence of adsorption parameters, such as pH, dosage of SM, adsorption time and initial concentration of HMI, was determined. It was found that the maximum sorption capacity of the composite material on Cr^{6+} ions is from 46.2 to 55.2 mg / g in the temperature range 298 – 308 K under the following conditions: pH = 1, t = 4 h, 120 rpm of the mixer, – 45 mg / dm^3, dosage of sorption material – 0.4 g / dm^3. It was found that the adsorption isotherm of Cr^{6+} ions is most adequately described by the Langmuir model. Certain thermodynamic parameters ΔG = - 35.99 kJ / mol, ΔH = 143.7 kJ / mol and ΔS = 760.45 J / mol · K indicate that the process is spontaneous and exothermic [21]. At the same time, it was determined that the adsorption isotherm on Cu^{2+} ions is more accurately described by the Freundlich model, and the kinetics of the process obeys the pseudo-second order model. It was determined that the highest sorption characteristics of the studied SM on Cu^{2+} ions are achieved at pH = 3 [22].

The high fiber content (~ 350 g / kg dry weight) in buckwheat straw (Table 1) can contribute to the adsorption of oil and its processed products [23]. Initially, certain physicochemical parameters of crushed straw were determined: buoyancy – 77.2 %, bulk density – 0.111 g / cm^3, total pore volume on water – 2.95 cm^3 / g. It was determined that the maximum sorption capacity of buckwheat straw in relation to Devonian oil under static conditions is 6.26 g / g, and the maximum water absorption is 8.45 g / g. It is indicated that treatment of buckwheat straw with a weakly concentrated H_2SO_4 solution (0.5 %) contributes to a certain increase in oil intensity and a decrease in water absorption when removing the oil film from the water surface.

3.2.2. Husk

Another large-tonnage waste from the production of buckwheat is fruit shells (husk, shuck) (Figure 1, c). In the industrial buckwheat processing, the husk accounts for 14 to 30 % of the total grain weight [24].

Figure 2 shows a highly developed microrelief of the surface of the buckwheat husk; micropores up to 0.1 μm in size are noted.

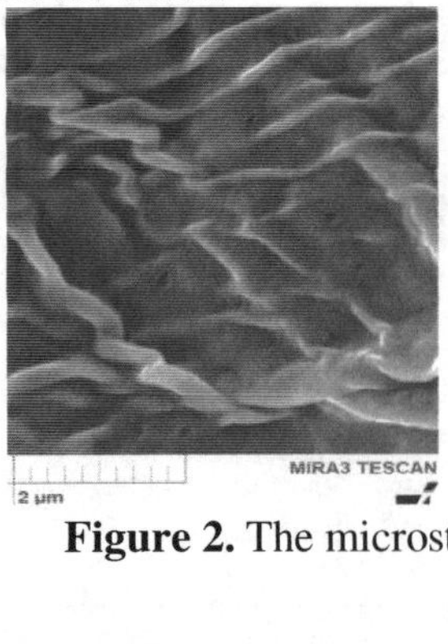

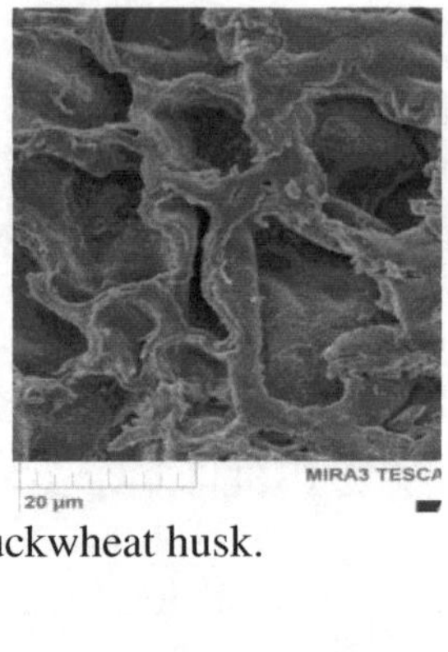

Figure 2. The microstructure of the surface of the buckwheat husk.

To study porosity at the microlevel and specific surface, we used the gas adsorption method. According to the results, it was found that the specific surface area of the husk by the BET method (SBET) was 21.5 m^2 / g (Table 2). The number of pores in the nanometric range from 1.7 to 300 nm is 0.63×10^{-3} cm^3 / g.

Table 2. Physical properties of buckwheat husk.

S_{BET} (m^2/g)	Total surface area of pores, $S_{p.}$ (m^2/g)	The total volume of pores with a diameter of 1,7 up to 300 nm, $V_{p.} \times 10^{-3}$ (cm^3/g)
21.5	0.32	0.63

Buckwheat husk contains fiber – up to 50 %, lipids – 4–5 %, polysaccharides –70 %, crude protein – 34 %, 0.2–0.3 % sugars, 9–10 % ash [24]. They contain such elements as potassium (3.1 g/kg), sodium (0.2 g/kg), copper (1.9 mg/kg), calcium (4.8 g / kg), magnesium (2.2 g/kg), zinc (5.4 mg/kg), iron (80.3 mg / kg), phosphorus (1.6 g/kg), sulfur (1.7 g/kg), manganese (21.8 mg/kg) and others in their composition [25]. The amino acid content in the composition of the shells of buckwheat grains is determined and shown in table 3.

Table 3. The amino acid content in the husk of buckwheat grains [26, 27].

Amino acid	Content (g/g)	Amino acid	Content (g/g)
Threonine	0.001	Glutamine	0.0002
Valine	0.002	Glutamic acid	0.003
Isoleucine	0.0009	Alanine	0.003
Leucine	0.008	β-alanine	0.0005
Lysine	0.0003	tyrosine	0.0008
Phenylalanine	0.0008	phosphoserine	0.01
Aspartic acid	0.005	Serine	0.005
Asparagine	0.0004	Taurine	0.0004

According to [28], the chemical composition of the shells of buckwheat grains in % of dry raw materials has alcohol-soluble substances – 4.9; total nitrogen – 0.63; ash – 1.9; easily hydrolyzable polysaccharides – 24.5; hardly hydrolyzable polysaccharides – 27.8.The remainder of the hardly hydrolyzable polysaccharides is mainly lignin – 30.7.

Considering the shuck composition of buckwheat grains, the latter is a promising raw material for the isolation of polysaccharides [29] and flavonoids [30, 31] from it. The fruit shells of buckwheat grains are more widely used in various sectors of the economy compared to buckwheat straw. In particular, it indicates the use of shuck of buckwheat grains as filler for pillows [32], concrete [33], composite materials [34, 35], as a feed additive [36], extracts as a dye [37, 38], etc.

The traditional scheme of the adsorption mechanism is to increase the concentration of substances on the phase interface (Figure 4).

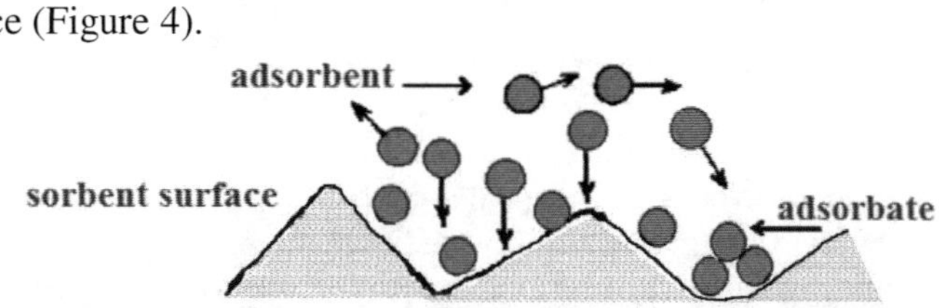

Figure 3. Scheme of the adsorption mechanism on a highly developed surface.

However, as it was indicated earlier (Table 3), the presence of proteins and amino acids in the composition of buckwheat shuck can contribute to the adsorption of HMI from aqueous media.

The adsorption of Au (III) ions from gold-containing sewage was studied [39]. It was found that the largest sorption capacity (1 mmol/g) is observed at pH = 2.5. With increasing temperature from 15

IOP Conf. Series: Materials Science and Engineering **945** (2020) 012044 doi:10.1088/1757-899X/945/1/012044

°C to 35 °C and contact time, the sorption capacity increases. It is also possible to increase the maximum sorption capacity to 2.84 mmol/g by modifying buckwheat shuck with an orthophosphoric acid solution [40]. The possibility of regeneration of the sorption material using 0.05% eluent solutions of thiourea in 0.1 mmol/dm^3 HCl solution is shown.

Buckwheat shuck was studied as a sorption material for the extraction of Cr^{6+} ions from model solutions with an initial concentration of 100 mg/dm^3. It was determined that the sorption capacity increased with decreasing pH and amounted to 99.87 % at pH = 2 and a dosage of sorption material of 5.0 g / dm^3 [41]. It was determined that the Langmuir model describes the adsorption isotherm of Cr (VI) ions on buckwheat husk best of all, and the maximum adsorption capacity at 45 °C, determined from the Langmuir equation, is 63.61 mg / g. An analysis of the thermodynamic parameters ($\Delta G°$, $\Delta H°$, and E) showed that the adsorption process was a spontaneous, endothermic, and chemisorption process. Besides, analysis of x-ray photoelectron spectroscopy showed that the adsorbed Cr (VI) ions are partially reduced to Cr (III) ions. IR Fourier spectra showed that carboxyl and amino groups are involved in the chemisorption process. It has been established that the most probable mechanisms of Cr (VI) removal using buckwheat husks are electrostatic attraction and chemical reduction [42].

Besides, buckwheat shuck was used to extract Hg^{2+} ions from aqueous solutions with an initial concentration of the last 802.36 mg /dm^3. It was found that at pH = 4 and a buckwheat husk dosage of 2 g / dm^3, the maximum sorption capacity was 110 mg / g [43]. The results showed that the kinetics of adsorption of buckwheat shuck by Hg (II) ions is well described by the pseudo-second order reaction model, and the calculated thermodynamic adsorption parameters ΔG, ΔH and ΔS were – 5.83 kJ / mol at 35 °C, 73.1 kJ / mol and 256 J / mol·K, respectively. It has been determined that the adsorption isotherm is described by the Langmuir equation best of all, and the maximum adsorption capacity on Hg (II) ions is 243.90 mg / g at 35 °C. In addition, a study of adsorption selectivity showed that buckwheat husk exhibits a strong affinity for mercury ions in aqueous solutions and exhibits 100% selectivity on Hg (II) ions in the presence of Zn (II) and Cd (II) ions [44].

The kinetics of adsorption of Zn (II), Cd (II), Co (II), Cu (II), Ni (II) ions from model solutions with an initial concentration of 10 – 50 mg / dm^3 of buckwheat shuck at a temperature of 25 °C in the pH range = 5 – 6 has been studied. Experiments determined that the lowest final concentrations are achieved with Cu^{2+} ions, and the highest ones with Co^{2+} ions (Figure 5) [45].

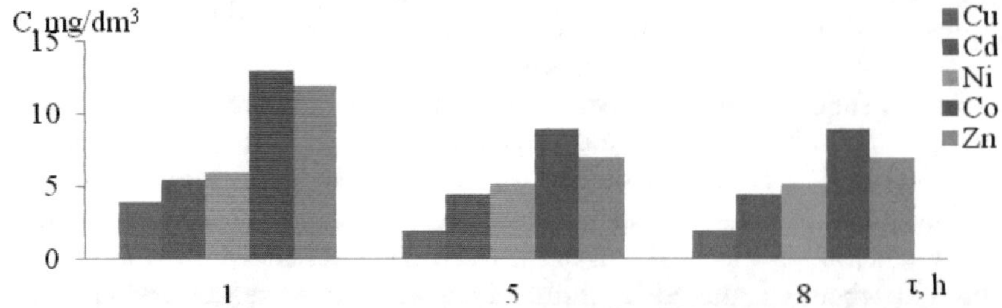

Figure 4. Diagram dependences of the adsorption kinetics of Zn (II), Cd (II), Co (II), Cu (II), Ni (II) ions by buckwheat shuck at an initial concentration of metal ions of 20 mg / dm^3 [45].

It has been determined that the adsorption kinetics of these metal ions is described by the equation with fractional derivatives, and adsorption is a physical process.

Sorption material was obtained by lignification of buckwheat shuck by successive hydrolysis of the latter with acids with a concentration of 0.5–10 % at a temperature of 80–200 °C and a water module of 1:10 for 20 – 180 minutes. Then, the obtained lignin was activated with an aqueous solution of an alkali metal carbonate with a concentration of 0.5–1.5 % at a temperature of 70–110 °C with a water module of 1:10 for 20–180 minutes. Then, the sorption material was neutralized [46]. The latter is an odorless dark brown powder, practically insoluble in water. The lignin content in the reagent was 79–88 %. The obtained sorption material was studied to extract metal ions, dyes, and some other pollutants from model solutions with an initial concentration of 5 or 50 mg / dm^3 (Table 4) [47].

Table 4. Sorption characteristics of buckwheat lignin on various pollutants [47].

Pollutant	Maximum sorption capacity (mg / g)	Efficiency extraction (%)
Ni (II) ions	2.3*/6.7**	76.9*/53.2**
Cu (II) ions	1.3/6.9	85.5/51.8
Sr (II) ions	4.4/7.8	88.1/53.8
Cd (II) ions	5.9/7.4	91.9/56.5
Pb (II) ions	36.4/42.5	90.2/76.4
Methylene blue	6.7/78.6	90.5/75.2
Congo Red	4.3/10.5	42.2/29.0
Uric acid	3.7/2.75	37.8/20.3
Ammonia nitrogen	1.5/6.9	81.6/18.1

* initial pollutant concentration of 5 mg / dm^3; ** initial concentration of pollutant 50 mg / dm^3

Native buckwheat shuck and processed during the extraction of polysaccharides of water, oxalate and alkaline extraction were studied to extract diesel fuel and toluene from aqueous media [16, 48]. It is indicated that the maximum degree of purification of dissolved and emulsified NPs with buckwheat shuck, in particular, is small and does not exceed 70% [49]. It is proposed to use sorbent obtained on the basis of buckwheat husks and polyurethane foam [50–52] for the purification of water from oil and oil products. The advantages of the latter include hydrophobicity and oleophilicity, versatility, a high percentage of profitability, ultra-fast adsorption, low density, non-toxicity, the ability to squeeze out absorbed oil up to 10 times. It is indicated that the oil capacity increases with increasing content of buckwheat husk in the composite material, reaching 16 g / g. It is shown that the oil capacity of the sorption material, called "PPU-45 SHG", depends on the volume of crumb grain and the type of oil. With a decrease in grain size and an increase in oil viscosity, the absorption capacity of the reagent increases. Among the advantages, it should also be noted the possibility of releasing the reagent in any geometric form, unlimited shelf life and speed of production (2 – 10 minutes). The latter circumstance allows the creation of mobile mini-plants for the production of adsorbent directly at the place of oil spill [51].

3.2.3. Composites

A composition consisting of polyurethane foam and buckwheat shuck called "PUF-OZG" was studied as a carrier for immobilizing oil-destroying microorganisms on its surface, which were used as a strain of *Rhodococcus erythropolis* O210, association V, consisting of a community of microorganisms (*Pseudomonas pickettii*, *Arthrobacter sp.*, *Aeromonas salmonicida*, *Vibrio sp.*), as well as activated silt, over silt fluid and the bacterial drug Destroyl, which is a powder consisting of microorganism cells isolated from a natural microbial culture [53]. It is indicated that it is possible to regenerate the absorbed oil products from saturated sorption material by squeezing or centrifugation [54]. The amount of regenerated oil is ~ 80–85 %.

3.2.4. Ash

In addition to native and modified buckwheat shuck samples used as sorption materials of heavy metal ions, dyes, and petroleum products, ash obtained by burning buckwheat husks also has sorption properties. The ash content in the buckwheat shuck varies from 1.5–2.3 %, in the straw – 2.8–11.1 % [55]. Nevertheless, multi-ton dumps of waste from buckwheat processing make it promising to use the latter as promising sorption materials after heat treatment of raw materials.

A thermal installation has been developed and is being used for burning buckwheat shuck in order to obtain a sorbent [54] at temperatures up to 1000 $^{\circ}$C with a productivity of 5 t/g. It was shown that the pyrolysis temperature of raw materials affects the yield of the target product and adsorption activity. The product of buckwheat husk pyrolysis (GS sorbent) is a black flake of size 1 – 5 mm, insoluble in water, not combustible and explosive [54], with a density of 0.08–0.12 g/cm^3 [55]. It is also possible to obtain sorbents based on buckwheat shuck by processing raw materials using a plasma arc setup [56] or low-pressure plasma [57].

It is proposed to produce sorption materials in the form of ash from buckwheat shuck for the extraction of heavy and rare - earth elements from sewage [58].

The GS sorbent also showed a high adsorption capacity with respect to oil and oil products (5 –10 g/g) [59]. It was found that the degree of removal of diesel fuel from the water surface by the sorbent GS exceeds 90 %, oil - more than 99 %. To increase the quality indicators of sorbents obtained as a result of heat treatment of buckwheat shuck, it was proposed to process them in the presence of a number of substances, such as sulfur, halides in an oxygen-free medium [60].

It is possible to isolate up to 44 % of sorbed oil [61] in the case of a saturated reagent squeezing, but at the same time, its structure is destroyed [62]. The possibility of using the used reagent in the production of asphalt, fuel briquettes is shown. The research results showed that the obtained samples of fuel briquettes have strength of 15–20 MPa and relatively low ash content (10–12 %). It was found that the specific heat of combustion of the fuel is 5400–5800 kcal/kg [67].

3.2.5. Extracts

As it was mentioned earlier, buckwheat shuck include compounds of the flavonoid group (including rutin and quercetin), lipids, polysaccharides, coumarins, amino acids, tannins, and other substances of organic origin. This circumstance suggests the use, in particular, of an alkaline extract from buckwheat shuck (ELG), which is used to precipitate ITM from sewage along with or instead of conventional methods for the extraction of metal ions [63]. Concentrated ELG represents a viscous product of dark brown color with a density of 1230 kg/m^3, pH = 11.5–12.3. The content of mineral and organic substances in the ELG is 137.84 g/dm^3 and 123.4 g/dm^3, respectively. The molar mass of organic substances contained in aqueous ELG is 300–600 g/mol.

The results showed that the sorption capacity of ELG has values, mEq / g: for Cu^{2+} ions – 7.56, Ni^{2+} – 10.05, Cd^{2+} – 5.25, Cr^{6+} – 6.53 and is at a level characteristic of cation exchange KS-1 or KB-4 resins.

A comparison of the results of experiments on the treatment of sewage from heavy metal ions using lime milk and ELG showed the expediency of using the latter due to the higher degree of purification. In addition, ELG, which is a solution, unlike lime milk, is not deposited on the walls of pipelines and does not clog shut-off valves.

4. Summary

Thus, the above data summarizes the literature on the use of waste from the processing of buckwheat as sorption materials for the extraction of pollutants from aqueous media. The chemical compounds present in the composition of straw and husks of buckwheat contribute to high chemical adsorption indicators on metal ions, oil and the products of the processing of the latter. It has been revealed that treatment of buckwheat shuck with chemical reagents and arc plasma, as well as with low-pressure plasma, contributes to an increase in the adsorption characteristics on the above pollutants.

5. References

[1] Anastopoulos I, Massas I, Ehaliotis C 2015 Use of residues and by-products of the olive-oil production chain for the removal of pollutants from environmental media: A review of batch biosorption approaches *Journal of Enviromental Science and Health, Part A. Toxic/Hazardous Substances and Enviromental Engineering* **50(7)** 677–718

[2] De Gisi S, Lofrano G, Grassi M, Notarnicola M 2016 Characteristics and adsorption capacities of low-cost sorbents for wastewater treatment: A review *Sustainable Materials and Technologies* **9** 10–40

[3] Hokkanen S, Bhatnagar A, Sillanpää M 2016 A review on modification methods to cellulose-

based adsorbents to improve adsorption capacity *Water Research* **91** 156–173

[4] Sulyman M, Namiesnik J, Gierak A 2017 Low-cost adsorbents derived from agricultural by-products/wastes for enhancing contaminant uptakes from wastewater: A review *Polish Journal of Environmental Studies* **26(2)** 479–510

[5] Bhatnagar A, Sillanpää M, Witek-Krowiak A 2015 Agricultural waste peels as versatile biomass for water purification – A review *Chemical Engineering Journal* **270** 244–271

[6] Sverguzova S V, Sapronova Zh A, Svyatchenko A V Tom Otiti 2018 Adsorption of spindle oil by native and thermal modified chestnut tree waste *Construction Materials and Products* **1(1)** 4–11

[7] Dai Y, Sun Q, Wang W, Lu L et al 2018 Utilizations of agricultural waste as adsorbent for the removal of contaminants: A review *Chemosphere* **211** 235–253

[8] Mo J, Yang Q, Zhang N, Zhang W et al 2018 A review on agro-industrial waste (AIW) derived adsorbents for water and wastewater treatment *Journal of Environmental Management* **227**

[9] Ahmad T, Danish M 2018 Prospects of banana waste utilization in wastewater treatment: A review *Journal of Environmental Management* **206** 330–348

[10] Omo-Okoro P N, Daso A P, Okonkwo J O 2018 A review of the application of agricultural wastes as precursor materials for the adsorption of per- and polyfluoroalkyl substances: A focus on current approaches and methodologies *Environmental Technology & Innovation* **9** 100–114

[11] Farooq U, Kozinski J A, Khan M A, Athar M 2010 Biosorption of heavy metal ions using wheat based biosorbents – A review of the recent literature *Bioresource Technology* **101** 5043–5053

[12] Shaikhiev I G, Galblaub O A, Grechina A S 2017 The use of waste from the processing of barley as a sorption material for the removal of pollutants from aquatic environments (literature review) *Bulletin of the University of Technology* **20(23)** 110–117

[13] Sverguzova S V, Shaikhiev I G, Grechina A S, Shaikhieva K I 2018 The use of waste from the processing of oat biomass as a sorption material for the removal of pollutants from aqueous media (literature review) *Economics of construction and environmental management* **2(67)** 51–60

[14] Ratan P, Kothiyal P 2011 Fagopyrum esculentum Moench (common buckwheat) edible plant of Himalayas: A Review *Asian Journal of Pharmacy and Life Science* **1(4)** 426–442

[15] Zemnuhova L A, Shkorina E D, Fedorischeva G A 2005 Study of the composition of inorganic components of buckwheat husk and straw *Journal of Applied Chemistry* **78(2)** 329–333

[16] Zemnuhova L A, Tomshich S V, Shkorina E D, Klyikov A G 2004 Polysaccharides from buckwheat production waste *Journal of Applied Chemistry* **77(7)** 1192–1196

[17] Zemnuhova L A, Makarenko N V, Tischenko L Ya, Kovalyova E V 2009 Investigation of amino acid composition in waste products of rice, buckwheat and sunflower *Chemistry of plant materials* **3** 147–149

[18] Chen J 2013 Study on biosorption of Cr(VI) and Cu(II) by a new low-cost adsorbent *Water Saving Irrigation* **11** 35–38

[19] Chen J, Yang P, Song D et al 2014 Biosorption of Cr(VI) by carbonized Eupatorium adenophorum and buckwheat straw: thermodynamics and mechanism *Frontiers of Environmental Science & Engineering* **8(6)** 960–966

[20] Chen J, Yang P 2015 Equilibrium kinetics studies on the biosorption of Cu(II) from aqueous solutions by a new adsorbent from a Eupatorium adenophorum Spreng/buckwheat straw mixture *Desalination and Water Treatment* **53(3)** 778–784

[21] Galblaub O A, Shaykhiev I G, Stepanova S V, Timirbaeva G R 2016 Oil spill cleanup of water surface by plant-based sorbents: Russian practices *Process Safety and Environmental Protection* **101** 88–92

[22] Devyatkin A I 1990 Rational use of feed *Rosagropromizdat Publ.* 256

[23] Gorshkov V V, Butskih O A 2016 Prospects for the use of non-traditional feed in livestock. Network edition Council of rectors of universities of the Big Altai *Western Mongolia on the way to sustainable development: conceptual design and project proposals* 124–133

[24] Quettier-Deleu C, Gressier B, Vasseur J et al 2000 Phenolic compounds and antioxidant activities of buckwheat (Fagopyrum esculentum Moench) hulls and flour *Journal of*

Ethnopharmacology **72(1-2)** 35–42

[25] Bhinder S, Kaur A, Singh B, Yadav M P, Singh N 2020 Proximate composition, amino acid profile, pasting and process characteristics of flour from different Tartary buckwheat varieties *Food Research International* **130** 108946

[26] Zhu F 2020 Dietary fiber polysaccharides of amaranth, buckwheat and quinoa grains: A review of chemical structure, biological functions and food uses *Carbohydrate Polymers* **248** 116819

[27] Hromádková Z, Ebringerová A 2003 Ultrasonic extraction of plant materials - investigation of hemicellulose release from buckwheat hulls *Ultrasonics Sonochemistry* **10(3)** 127–133

[28] Li J, Hossain M S, Hongchi M, Yang Q et al 2020 Comparative metabolomics reveals differences in flavonoid metabolites among different coloured buckwheat flowers *Journal of Food Composition and Analysis* **85** 103335

[29] Hong C S, Park H S, Oh S H 1987 Dermatophagoides Farinae, an Important Allergenic Substance in Buckwheat-Husk Pillows *Monsei Medical Journal* **28(4)** 274–281.

[30] Vaickelionis G, Valanciene V 2016 Lightweight concrete with an agricultural waste – buckwheat husk *Materials Science (Medziagotyra)* **22(1)** 98–104.

[31] Danchenko Yu, Andronov V, Kariev A et al 2017 Researh into surface properties of disperse fillers based on plant raw materials *Eastern-European Journal of Enterprise Technologies* **5(12)** 20–26

[32] Reutov V A, Lim L A, Zabolotnaya A M, Prischenko N A 2016 A method of manufacturing products from lignocellulosic polymer composite materials *Pat. 2582498 Russian Federation IPC C08L 97/02* **2015106744/13** 8

[33] Bhardwaj H, Hamama A A 2020 Yield and Composition of Buckwheat Biomass Relative to Forage Use *Journal of Agricultural Science* **12(8)** 129

[34] Shekurov V N, Tarenko B I, Shekurov K V 2014 Deep processing of buckwheat husks *Bulletin of Kazan Technological University* **17(7)** 205–207

[35] Ibragimov Sh N,Shekurov V N, Efremov B A, Lazko A S, Chechenev L A 1997 The method of obtaining food dye from buckwheat husk. *Pat.* 2086588 *Russian Federation IPC C09B 61/00* **95101265/13** 3

[36] Deng K, Yin P, Liu X, Tang Q, Qu R 2014 Modeling, analysis and optimization of adsorption parameters of Au(III) using low-cost agricultural residuals buckwheat hulls *Journal of Industrial and Engineering Chemistry* **20(4)** 2428–2438

[37] Yin P, Xu M, Qu R et al 2013 Uptake of gold (III) from waste gold solution onto biomass-based adsorbents organophosphonic acid functionalized spent buckwheat hulls *Bioresource Technology* **128** 36–43

[38] Li K, Wang Q, Dang Y et al 2012 Characteristic and mechanism of Cr(VI) biosorption by buckwheat hull from aqueous solutions *Acta Chimica Sinica* **7** 929–937

[39] Li K, Zhang Y, Dang Y et al 2014 Removal of Cr(VI) from aqueous solutions using buckwheat (Fagopyrum esculentum Moench) hull through adsorption–reduction: affecting factors, isotherm, and mechanisms *Clean Soil Air Water* **42(11)** 1549–1557

[40] Xu M, Yin P, Liu X et al 2013 Optimization of biosorption parameters of Hg(II) from aqueous solutions by the buckwheat hulls using respond surface methodology *Desalination and Water Treatment* **51(22-24)** 4546–4555

[41] Wang Z, Yin P, Qu R et al 2013 Adsorption kinetics, thermodynamics and isotherm of Hg(II) from aqueous solutions using buckwheat hulls from Jiaodong of China *Food Chemistry* **136**(3-4) 1508–1514

[42] Tomchak E, Kaminski W, Szczerkowska D 2013 Fractional derivatives for description of sorption kinetics in the plant sorbent – metal ions system *Ecological Chemistry and Engineering* **20(3)** 499–506

[43] Denisovich V A, Nikolaychuk A A, Kupchik L A, Kartel M T 2006 The method of obtaining lignin enterosorbent *Pat. 76835 Ukraine IPC A 61K 36/87*

[44] Kupchik L A, Denisovich V A, Nikolaychuk A A 2010 Obtaining environmental sorbents from buckwheat husks *Energy technologies and resource saving* **6** 48–50

[45] Shkorina E D 2007 The composition and complex processing of waste production buckwheat *Abstract of the diss. cand. chemical of sciences* 24

[46] Zemnuhova L A, Shkorina E D, Filippova I A 2005 The study of the sorption properties of rice husk and buckwheat in relation to petroleum products *Chemistry of plant materials* **2** 51–54

[47] Chikina N S, Muhamedshin A V, Zenitova L A 2008 The use of a sorbent based on polyurethane foam and buckwheat husk to reduce the environmental burden on aquatic areas *Environmental protection in the oil and gas complex* **10** 31–36

[48] Meng X, Liu F, Xiao Y et al 2019 Alterations in physicochemical and functional properties of buckwheat straw insoluble dietary fiber by alkaline hydrogen peroxide treatment *Food Chemistry: X* **3** 100029

[49] Chikina N S 2010 Liquidation of bottling oil and oil products using a sorbent based on polyurethane foam and waste of grain crops *Dis. cand. tech. of sciences* 20

[50] Chikina N S, Muhamedshin A V, Zenitova L A 2008 The use of sorbent based on polyurethane foam and buckwheat husks in the elimination of hydrocarbon spills *Life safety* **9(93)** 49–53

[51] Chikina N S, Muhamedshin A V, Zenitova L A 2007 Sorbent based on polyurethane foam and buckwheat husk to collect oil spills *Advances in chemistry and chemical technology* 21 **5(73)** 38–42

[52] Chikina N S, Muhamedshin A V, Ankudinova A V et al 2009 Reducing the environmental burden of oil and petroleum products spills using a sorbent based on polyurethane foam and waste grain crops *Bulletin of Kazan Technological University* **6** 184–192

[53] Ivanova M A, Murtazina R T, Chikina N S et al 2011 Regeneration of absorbed products by the sorbent PPU-OZG *Butlerovskie soobscheniya* **28(20)** 89–93

[54] AbdullinI Sh, Gafarov I G, Israfilov I H 2001 Unified ecological-technological complex for modifying the human habitat with the help of sorption cleaning of the hydrosphere *KGU Publ.* 419

[55] AbdullinI Sh, Gafarov I G, Israfilov I H 2001 Study of the properties of sorbents based on rice husk and buckwheat husk *Kama Polytechnic Institute Press* 12

[56] Gafarov I G, Mishulin G M, Israfilov I H, Schepakin M B 2001 Ecosorbent as a regional resource management product *Ecology and Industry of Russia* **12** 20–25

[57] Gafarov I G, Muhametzyanov M T, RastorguevYu I et al 2005 Sorbent for the removal of oil and oil products and method for its production from buckwheat husks. *Pat. 2259874 Russian Federation IPC C02F 1/28* **2003127907/15** 2

[58] AbdullinI Sh 2003 The use of plasma installations of low pressure arc discharge to obtain sorbents based on rice husk and buckwheat husk and modification of their properties. *Materials Science* **11** 45–53

[59] Elistratov G D, Volchanova M N, Malyigin N V et al 2004 The method of producing sorbent *Pat. 2222377Russian Federation IPC C01B 31/08* **2001130741/15** 7

[60] Samoylov N A, Hlestkin R N, Osipov M I, Chichirko O P 2004 Formation of a consolidated layer of the carbon sorbent-oil system while collecting oil from the site of an emergency spill *Journal of Applied Chemistry* **77(2)** 328–334

[61] Temirhanov B A, Temerdashev Z A, Eletskiy B D 2005 Investigation of the possibility of regeneration and reuse of some sorbents for collecting oil *Environmental protection in the oil and gas complex* **5** 19–21

[62] Shekurov V N, Kiseleva N V, Polovnyak V K 1998 The study of the sorption properties of the extract from the buckwheat husk with respect to heavy metal ions *News of universities. Chemistry and chemical technology* **41(3)** 1209–1212

IOP Conf. Series: Materials Science and Engineering **945** (2020) 012044 doi:10.1088/1757-899X/945/1/012044

[63] Kiseleva N V 1999 Reagent wastewater treatment of electroplating plants from heavy metal ions using buckwheat husk extract *Diss. cand. tech. of sciences* 113

Acknowledgments

The work is realized in the framework of the Program of flagship university development on the base of the Belgorod State Technological University named after V.G. Shukhov, using equipment of High Technology Center at BSTU named after V.G. Shukhov.

IOP Conf. Series: Materials Science and Engineering **945** (2020) 012045 doi:10.1088/1757-899X/945/1/012045

Energy characteristics of sands of deposits in the Ivanovo region

M V Morozova[1,2*], M V Akulova[1] and M A Frolova[2]

[1] Federal State Budget Educational Institution of Higher Education «Ivanovo State Polytech-nic University», Sheremetyevsky pr., 21, Ivanovo, 153000, Russia

[2] Federal State Autonomous Educational Institution of Higher Education «Northern (Arctic) Federal University named after M.V. Lomonosov», Severnaya Dvina Emb. 17, Arkhangelsk, 163002, Russia

E-mail: m.morozova@narfu.ru

Abstract. It is proposed to use the energy potential of rock-forming minerals obtained in the process of rock synthesis as an energy criterion for ranking raw materials for the production of construction materials. The energy of atomization and its specific values calculated based on the principles of crystal energy are used as a quantitative parameter. The estimation of two deposits of sands of the Ivanovo region on energy of atomization, specific mass energy of atomization, specific volume of energy of atomization (energy density) is carried out. It was found that, despite the difference in the chemical, mineral composition and origin of rocks, macro-energy parameters have similar values. The energy density value allows classifying the considered sands as high-energy-dense. The possible level of use of this energy reserve due to mechanical crushing of raw material is established. Quantifying potential energy of the system that fallen after crushing the material in the surface was carried out based on the values of critical surface tension and specific surface area of raw materials (formed after the process of mechanical grinding). The level of possible use of the potential energy reserve due to the formation of a new material surface was characterized by the activity of the surface of fine samples. The obtained values of surface tension and surface activity showed that the sands of the Nekhtskoe deposit, in comparison with the sands of the Khromtsovskoye deposit, are preferable to be used as an active fine-dispersed component for obtaining binding compositions.

1. Introduction

It is known that rocks have a reserve of free internal energy, which is determined by the method of their formation (igneous, sedimentary, metamorphic) [1]. The use of this energy reserve can reduce the work that must be performed during the technological transformation of raw materials in the production of building materials (mechanical disintegration of raw materials, destruction of crystal lattices of minerals, synthesis of new things, etc.) [2-4]. In addition, a significant interest for the construction materials industry is represented by fine-ground mineral powders [5-7], the addition of which in the production technology can improve the quality of building materials: strength, wear resistance, thermal and chemical resistance, fracture toughness, i.e., increase the operating period and life cycle of the product or structure as a whole [8 – 11]. Bringing powders to the nanoscale level of particles also leads to changes in a number of physical and chemical properties of the systems themselves [12-14].

IOP Conf. Series: Materials Science and Engineering **945** (2020) 012045 doi:10.1088/1757-899X/945/1/012045

When rocks are destroyed in the process of mechanical dispersion, according to the classical principles of thermodynamics, a certain amount of work is spent, because of which energy is released from the total energy potential of the rock, which is called free surface energy [3 – 4]. Moreover, this energy parameter is associated with the specific surface area (S_{sp}). A number of researchers have noted that when creating compositions of different composition, the greater S_{sp}, the better the interaction of the material particles with the composite matrix becomes (a more uniform distribution of nanoparticles in the material is achieved) [7, 8, 10, 15].

In addition, the dispersion of rocks to a micro-and nanoscale state is accompanied by the formation of an amorphous phase, which is an important component of increasing the reactivity of a highly dispersed material. For such systems, this capability is determined by surface activity (k_s). This parameter allows characterizing quantitatively the transition of potential energy accumulated by the rock during genesis to free surface energy due to activation of the surface of the raw material [6-7, 16].

The determination of the k_s value is based on the macro-energy indicators of the rock. The first is the energy of atomization (E_a, kJ/mol), which is defined as the sum of the standard enthalpy of crystal formation and the heat of the formation of constituent atoms (the values are given in the reference literature [17]). The derivatives of this value are the specific mass energy of atomization (E_m, kJ/g) and the specific volume energy of atomization (E_v, kJ/cm^3), which allow comparing correctly rocks by energy parameters. In this case, the E_v parameter, due to the specifics of its calculation, allows characterizing rocks by energy density [2 – 3].

Therefore, the calculation of the above parameters can help in ranking and selecting the most effective raw materials for the production of building materials (composition of composite binder, complex additive for concrete, active filler, etc.).

The aim of this work is to calculate the macro-energy indicators of polymineral sands of the most widely used deposits of the Ivanovo region, and a comparative assessment of fine-dispersed systems based on their surface activity.

The surface activity value was calculated using the working positions [3, 6, 11, 18], according to which this parameter is equal to the ratio $k_s = E_s / E_m$. The free surface energy (E_s, J/kg) was calculated by multiplying the surface tension ($\sigma_к$) of a highly dispersed system by its specific surface area. The OWRK method was used for experimental determination of $\sigma_к$ [13, 15 – 16, 18].

2. Materials and methods

2.1. Materials

Two deposits of polymineral quarry construction sands were selected as raw materials: Khromtsovskoye deposit (Khromtsovo village, Ivanovo region) and Nekhtskoye deposit (Teterinskoye village, Nerekhtsky district).

2.2. Methods

Before testing, the sand was washed and dried to a constant mass at a temperature of 105 °C. Determination of the main characteristics of the sands such as the size modulus and true density was carried out in accordance with GOST 8735-88.

The mineralogical composition of rocks was determined by recording x-ray diffractograms on a Shimadzu XRD-7000 S x-ray diffractometer (CCP "Arctic", Northern (Arctic) Federal University named after M.V. Lomonosov). The chemical composition of the sands was determined using a MetExpert x-ray fluorescence analyzer.

To calculate the activity value of the surface of the sand sample, the method of dry dispersion of raw materials was ground to a highly dispersed state in a planetary ball mill "Retsch PM100", whose grinding principle is based on the friction of particles and the impact of steel grinding bodies on them. The optimal dispersion parameters were selected experimentally to obtain the required particle size.

The size characteristics were determined using the submicron particle size analyzer Delsa Nano Series Zeta Potential and Submicron Particle Size Analyzers ("DelsaNano") by photon-correlation spectroscopy based on the principle of dynamic light scattering.

The specific surface of highly dispersed rock systems was determined by gas sorption using the "Autosorb-iQ-MP" analyzer, according to the BET theory.

To determine the edge angle of wetting on the "Easy Drop" installation at a temperature of 25±1 °C, we used samples-tablets made by pressing ground sand on a press "PLG-20" at a load of 20 kPa into a metal mold with a diameter of 20 mm.

The surface tension of the samples ($\sigma_к$, N/m) was calculated using the OWRK method (Ounce, Wendt, Rabel and Kjelble method), based on the measurement of the equilibrium wetting angle (θ) with liquids (distilled water, decane, glycerine and ethylene glycol) with known values of the surface tension (σ_j, N/m) and its dispersion (σ_L^D) and polarization (σ_L^P) components.

3. Results

The selected sand deposits do not differ significantly in size modulus (M_k), but they are medium-sized. Quarry polymineral sand deposits "Khromtsovskoe" (Kh) has a fineness modulus of 2.18 and for sand deposits "Nekhtskoe" (N) module size amounted to 2.43. The true densities (ρ_{tr}) of the studied rocks also have similar values and are for sand "X" ρ_{tr} = 2.50 g/cm^3; for sand "H" ρ_{tr} = 2.60 g/cm^3.

The studied rocks have the same composition of the main rock-forming minerals, but their quantitative content differs. So, the sand field "Khromtsovskoe" contains 40 % quartz, 11% albite; sand deposits "Nerekhtskoe" - 84 % quartz, and 14% albite.

When calculating the energy reserve of a rock, the atomization energy parameter was used as a criterion. Therefore, the initial data for calculating E_a are the results of chemical analysis of the rock presented in table 1.

Table 1. Chemical composition of sands in terms of oxides (%) and enthalpy of elements and oxides formations.

Defined component	Sand deposit		E_a, kJ/mol	Molar mass of oxide, g/mol
	Khromtsovskoe	Nerekhtskoe		
SiO_2	85.23	95.89	1861.34	60.084
Al_2O_3	1.97	2.32	3081.90	100.181
MgO	0.46	0.16	997.80	40.311
Fe_2O_3	5.75	0.56	2403.80	159.695
CaO	3.30	0.20	1062.10	56.079
TiO_2	0.05	0.01	1916.00	79.954
K_2O	0.01	0.75	789.00	94.203
SO_3	0.68	0.04	1463.60	80.061
P_2O_5	0.08	0.02	3385.80	141.943
Na_2O	2.41	0.04	879.04	61.979
MnO	0.06	0.01	918.80	70.937

Based on the results of chemical analysis of the samples, the enthalpy of formation of the corresponding elements and chemical compounds, the energy of atomization for each oxide was calculated, taking into account its quantitative content in the sample. The sum of these values allowed calculating the energy of atomization for samples "Kh" and "N".

Then the specific mass energy of atomization (E_m) was calculated based on the molar mass of the oxide and its percentage content in the analyzed sample. Table 2 summarizes all calculated energy parameters for sedimentary rocks of the studied fields.

Table 2. Calculation of atomization energy and specific mass energy of atomization.

Oxides	E_a oxide in the mixture, kJ/mol		E_m oxide, kJ/g	E_m oxide in the mixture, kJ/g	
	Khromtsovskoe	Nerekhtskoe		Khromtsovskoe	Nerekhtskoe
SiO_2	1586.42	1784.84	30.979	26.403	29.706
Al_2O_3	60.71	71.50	30.763	0.606	0.714
MgO	4.59	1.60	24.753	0.114	0.040
Fe_2O_3	138.22	13.46	15.052	0.866	0.084
CaO	35.05	2.12	18.939	0.625	0.038
TiO_2	0.96	0.19	23.964	0.012	0.002
K_2O	0.08	5.92	8.376	0.001	0.063
SO_3	9.95	0.59	18.281	0.124	0.007
P_2O_5	2.71	0.68	23.853	0.019	0.005
Na_2O	21.18	0.35	14.183	0.342	0.006
MnO	0.55	0.09	12.952	0.008	0.001

Based on experimentally determined values of rock density, the energy density values of the sands were calculated. All the obtained data on atomization energy, specific mass energy, and specific volume energy (energy density) are presented in table 3.

Table 3. Energy parameters of sedimentary rocks.

Energy parameter	Sand of the deposit	
	Khromtsovskoe	Nerekhtskoe
Energy of atomization, E_a, kJ/mol	1860.43	1881.34
Specific mass energy, $E_m \cdot 10^3$, kJ/kg	29.12	30.67
Specific volume energy, E_v, kJ/cm^3	72.80	79.73

Based on the data obtained (table 3), it is established that the energy parameters of the considered sands, despite the distinctive features of the processes of formation of these rocks, can have similar values due to similar chemical composition. At the same time, the calculated E_v allows characterizing materials by energy density. Based on the existing classification, all minerals and artificial compounds are divided into: super-dense (E_v=150 ... 230 kJ/cm^3), high-energy (E_v=60 ... 150 kJ/cm^3), medium-energy (E_v=30 ... 60 kJ/cm^3) and low-energy (E_v=1 ... 30 kJ/cm^3). Consequently, the considered deposits of sands of the Ivanovo region can be used as an active fine-dispersed additive in production, as the energy density of these rocks allows classifying them as high-energy dense.

The surface activity value of the studied sands was calculated for fractions with different dimensional characteristics and corresponding specific surface values (table 4).

Table 4. Grinding time and dispersion characteristics of sand fractions.

Parameter	Sand of the deposit							
	Khromtsovskoe				Khromtsovskoe			
Grinding time, min	5	10	20	30	5	10	20	30
Average particle value, nm	687 ± 5	549 ± 3	467 ± 1	341.7 ± 2.2	603.7 ± 6.1	510.6 ± 4.2	389.4 ± 2.3	313.9 ± 1.2
Specific surface area, m^2/kg	910 ± 11	1089 ± 13	1638 ± 11	2319 ± 14	778 ± 18	887 ± 17	2085 ± 15	3154 ± 16

To calculate the value of free surface energy (E_S, J/kg), the surface tension values (σ_K, N/m) were additionally calculated for samples with different values of S_{sp}. In this case, the obtained functional dependencies of the OWRK method:

$$\frac{\sigma_L \cdot (cos\theta + 1)}{2 \cdot \sqrt{\sigma_L^D}} = f\left(\frac{\sqrt{\sigma_L^P}}{\sqrt{\sigma_L^D}}\right) \quad (1)$$

are well described by linear equations with a high value of the approximation confidence coefficient (R^2), where θ is the edge angle of wetting of the material under study; σ_L, σ_L^D and σ_L^P are the total, dispersion, and polarization surface tensions of working fluids (respectively). A total surface tension of the system $\sigma_k = \sigma_S^P + \sigma_S^D$, where σ_S^P and σ_S^D – polar and dispersion components of the surface tension of the studied material.

So, for the sand of the Khromtsovskoe deposit, these equations have the following form: at 5 minute grinding: $y = 6.97 \cdot x + 4.35$ (R^2=0.98); at 10 minute grinding: $y = 6.99 \cdot x + 4.37$ (R^2=0.98); at 20 minute grinding: $y = 7.00 \cdot x + 4.35$ (R^2=0.98); at 30 minute grinding: $y = 7.05 \cdot x + 4.29$ (R^2=0.99). For the sand of the Nerekhtskoe deposits are the following: at 5 minute grinding: $y = 6.76 \cdot x + 4.41$ (R^2=0.98); at 10 minute grinding: $y = 6.83 \cdot x + 4.39$ (R^2=0.98); at 20 minute grinding: $y = 6.94 \cdot x + 4.39$ (R^2=0.98); at 30 minute grinding: $y = 6.95 \cdot x + 4.38$ (R^2=0.98).

The values of the surface tension of the sands and the characteristic of their surface activity are presented in table 5.

Table 5. Characteristics of the surface tension of sands.

Defined parameter	The sand of the deposit							
	Khromtsovskoe				Nerekhtskoe			
Polar component σ_s^P, mn/m	48.56	48.81	49.05	49.72	45.74	46.61	48.12	48.80
Dispersed component σ_s^D, mn/m	18.96	19.11	18.94	18.39	19.48	19.23	19.27	19.22
Surface tension σ_S, mn/m	67.51	67.91	67.98	68.11	65.22	65.84	67.39	67.51
Free surface energy E_S, J/kg	61.89	74.07	111.41	157.67	50.59	58.54	139.72	214.49
Surface activity $k_s \cdot 10^6$	2.13	2.54	3.83	5.41	1.65	1.91	4.56	6.99

4. Discussion

The obtained results show that the polar components of the surface tension of the samples prevail over the dispersion ones. This indicates the advantage of transformations associated with active surface centers. For highly dispersed samples "Kh" and "N", a fairly similar specific surface was obtained at the same grinding time. The free surface energy of the surface unit (surface tension) increases as the grinding time increases. It should be noted that the sands of the studied deposits, despite the difference in mineral composition, have similar values of surface activity of fine-dispersed systems obtained by mechanical grinding.

5. Summary

The calculated energy of atomization for the sands of the studied deposits in the Ivanovo region showed similar values and amounted to 1860.43 kJ/mol for the Khromtsovskoe deposit, and 1881.34 kJ/mol for the Nerekhtskoe deposit. The specific values of mass and volume atomization energy also have similar values. Thus, the energy density of these sands was: for the Khromtsovskoe deposit E_v = 72.80 kJ/cm^3, for the Nerekhtskoe deposit E_v = 79.73 kJ/cm^3. Based on this, the considered rocks can be classified as high-energy dense by their potential energy reserve.

The grinding ability of sands of Khromtsovskoe deposit lower than sands of Nerekhtskoe deposit, which is reflected in the increase in the activity of the surface with increasing time of grinding. Construction sands of this quarry (in comparison with the sands of the Khromtsovskoe deposit) are more active components for obtaining binding compositions.

6. References

[1] Lesovik V S, Fomina E V, Ayzenstadt A M 2019 Some aspects of technogenic metasomatosis in building materials science *Construction materials* **1-2** 100-106

[2] Abramovskaya I R, Ayzenstadt A M, Lesovik V S, Veshnyakova L A, Frolova M A, Kazlitin S A 2012 Calculation of the energy intensity of rocks as raw materials for the production of building materials *Industrial and civil construction* **10** 23-25

[3] Veshnyakova L A, Ayzenstadt A M, Frolova M A 2015 Assessment of surface activity of highly dispersed raw materials for composite construction materials *Physics and chemistry of materials processing* **2** 68-72

[4] Veshnyakova L A, Drozdyuk T A, Ayzenstadt A M, Frolova M A, Tutygin A S 2016 Surface activity of silicon-containing rocks *Material Science* **5** 45-48

[5] Tolstoy A D 2020 Fine-grained high-strength concrete *Construction Materials and Products* **3(1)** 39-43

[6] Drozdyuk T A, Ayzenshtadt A M, Frolova M A, Rama S V 2020 Mineral wool composite with the use of saponite-containing mining industry waste *Construction Materials and Products* **3(3)** 21-27

[7] Kozhuhova N I, Strokova V V, Chizhov R V, Kozhuhova M I 2019 Chemical reactivity assessment method of nanostructured low calcium aluminosilicates *Construction Materials and Products* **2(3)** 5-11

[8] Klyuev S V, Klyuev A V, Shorstova E S Fiber concrete for 3-D additive technologies *Construction Materials and Products* **2(4)** 14-20

[9] Fedyuk R S, Mochalov A V 2018 Issues related to management of the structure formation of a composite binder *AlitInform: Cement. Concrete. Dry mixes* **2(51)** 2-10

[10] Chernysheva N V, Shatalova S V, Evsyukova A S, Fisher H-B 2018 Features of the selection of the rational structure of the compositional gips binder *Construction Materials and Products* **1(2)** 45 – 52

[11] Shoshin E A, Strokova V V, Kozlov N A, Gribkov D S 2019 Increasing the efficiency of cement concretes by using a composite binder with silicate-calcium dispersion *ALITinform: Cement. Concrete. Dry mix* **4(57)** 22-29

[12] Boucedra A, Bederina M, Ghernouti Y 2020 Study of the acoustical and thermo-mechanical properties of dune and river sand concretes containing recycled plastic aggregates *Construction and Building Materials* **256** 119447

[13] Gavshina O V, Yashkina S Yu, Yashkin A N, Doroganov V A, Moreva I Yu 2018 Study of the effect of particulate additives on the setting time and microstructure of high-alumina cement *Construction Materials and Products* **1(4)** 30 – 37

[14] Danilov V E, Ayzenshtadt A M, Frolova M A, Tutygin A S 2018 Dispersion Interactions as Criterion of Optimization of Cementless Composite Binders *Inorganic Materials: Applied Research* **9(4)** 767–771

[15] Grishina A N, Korolev E V 2015 Efficiency of modifying cement composites with nanoscale barium hydrosilicates *Construction materials* **2** 72-76

[16] Sokolova Y V, Ayzenshtadt A M, Strokova V V 2017 Evaluation of dispersion interaction in glyoxal/silica organomineral system *Journal of Physics: Conference Series* **929(1)** 012110

[17] Binnewies M, Milke E 2002 Thermochemical Data of Elements and Compounds *Hannover. Germany: Wiley-VCH* 928

IOP Conf. Series: Materials Science and Engineering **945** (2020) 012045 doi:10.1088/1757-899X/945/1/012045

[18] Danilov V E, Korolev E V, Ayzenstadt A M, Strokova V V 2019 Features of the surface free energy calculation based on the Ounce–Wendt–Rabel–Kjelble interfacial interaction model *Construction materials* **11** 66–72

Acknowledgments
The reported study was funded by RFBR, project number 19-31-27001.

Nanodispersed additive for composite binders based on technogenic raw materials of Iraq

Ahmed Ahmed Anees Ahmed[1]

[1]Department of Building Materials Science, Products and Structures, Belgorod State Technological University named after V.G. Shukhov, Kostyukov St., 46, Belgorod, 308012, Russia

E-mail: Civileng85@yahoo.com

Abstract. The economic development of the Arab states is characterized by such trends as industrialization and urbanization, population concentration in large cities, saving land, increasing the number of storeys of buildings. Population growth in developing countries is causing acute housing needs. One of the most promising modern building materials is concrete and reinforced concrete. The main components of concrete, as it is known, are a binder (cement), large and small aggregates. The creation of comfortable conditions is provided by the optimization of the "man-material-environment" system. The theoretical basis for the development of materials science is knowledge about geological processes and the genesis of rocks. The evolution of the development of binders goes along the way of complicating their composition and structure. In the development of materials science there is the process of controlling the structure formation of binders using nanodispersed additives. Nanodispersed additive is produced by grinding in various grinding aggregates of the silica-containing component, the cement, the carbonate constituents, superplasticizers, air-entraining additives, etc. The result is a very complex system. After we introduce a nanodispersed modifier in cements in an amount of 3 to 12%, they absolutely change the processes of structure formation. Thus, it was found that when adding a superplasticizer and a nanodispersed modifier, the compressive strength increases to 47% and the bending strength increases to 27%.The nature of the influence of the nanodispersed modifier on the properties of binders is established. Maintaining a nanodispersed modifier in the composition of Portland cement allows optimizing the process of structure formation of the component of a nanodispersed modifier and has a multifunctional value.

1. Introduction

The economic development of the Arab states is characterized by such trends as industrialization and urbanization, population concentration in large cities, saving land, increasing the number of storeys of buildings. Population growth in developing countries is causing acute housing needs. The Republic of Iraq is one of the states in this region that has been in urgent need of accelerated development of the construction industry for many years. The territory of Iraq has rich deposits of raw materials that are used in the production of construction materials (limestone, sand, clay, gypsum), as well as deposits of iron, aluminum and other metals. In the southern region of Iraq, there are deposits of oil, natural gas, and mercury. The Western region is a rich source of limestone, gypsum, silica, phosphate rock, and natural gas. The Northern region is rich in oil, gas, basalt, and iron ore.

Numerous studies conducted on the use of technogenic raw materials as a filler for the manufacture of concrete products and structures have confirmed its high efficiency [1-7]. However, the question of the possibility of obtaining a nanodispersed additive for composite binders based on technogenic raw materials is not sufficiently studied, although the available data offer its high value in this quality. For example, the authors [8-11] investigated the effect of replacing secondary components, such as recycled concrete aggregate, as a partial replacement of coarse-grained aggregate and filler, respectively, along with the addition of polypropylene fiber to the concrete mix. The results showed that the optimal content of the filler is 22% at 150 min of activation and 0.5% of the organic impurity content. The standard consistency of the composite binder decreased from 26.5% to 22.4%. Mathematical models can be used in the design of self-compacting concrete mix.

The authors [12] investigated the properties of cement mortar developed for the production of environmentally friendly building material without carbon dioxide emissions and with various useful effects. The composite binder used in this mortar was developed by the reaction of a very small amount of inorganic minerals and activated hwangtoh (HB), which is rich in silica, SiO_2, and alumina, Al_2O_3. The test results reflecting the influence of various parameters on the properties of the HB solution confirmed that the developed HB has highly effective properties as a binder. Equations using nonlinear multiple regression analysis based on test results are presented to estimate the flow and 28-day compressive strength of the HB solution. Comparisons between predicted and measured flow and 28-day compressive strength show good agreement. The authors [13] investigated the development of a cement-free binder activated with sodium silicate powder. To evaluate the properties of alkaline-activated solutions, an alkali quality coefficient is proposed that combines the amounts of the main compositions of the starting materials and sodium oxide (Na_2O) in sodium silicate based on the hydration mechanism of alkaline-activated pastes. Fly ash and crushed granulated blast furnace slag were used as raw materials. The test results clearly showed that the loss of fluidity and the development of compressive strength of the alkali-activated solution largely depended on the proposed alkali.
Meanwhile, none of these studies reflects the production of a nanodispersed additive for composite binders based on technogenic raw materials. It is expected that this study will also provide more information to understand the interaction of new generation building materials and empirical fact to promote a green environment, and its overuse in large-scale concrete structures such as pedestrian walkways and blinding to the foundation of reinforced concrete structures among other applications.
Currently, it is very important to improve mineral binders on the basis of their modification with various highly dispersed including nano-additives based on local raw materials of the Republic of Iraq.

2. Materials and methods

To obtain and study a nanodispersed modifying additive, the following raw materials were used: Portland cement CEM1 42.5 GOST 31108-2003; basalt crushed stone; natural limestone; superplasticizer Polyplast SP-3 and quartz sand.

Due to the objective impossibility of using Portland cements produced by the Republic of Iraq in experiments, a comparison of Russian and Iraqi cements was made. The Cubais cement plant is the second most important plant, which is located 200 km northwest of Baghdad and produces conventional Portland cement. To conduct the research, we used Portland cement of JSC "Belgorod cement" similar in its properties and structure to the cement of the city of Cubais - Iraq, table 1.

Table 1. Material composition of Portland cement.

Clinker composition	Al-Qaim Plant Quantity, wt. %	Cubais plant Quantity, wt. %	JSC "Belgorod cement" Quantity, wt. %
CaO	60.72	62.65	67.22

SiO_2	21.82	21.90	22.5
Fe_2O_3	5.0	3.0	4.4
Al_2O_3	3.7	5.0	4.77
MgO	2.22	3.32	0.67
SO_3	2.18	2.60	0.18
Portland cement clinker minerals			
C3S	42.84	49.5	60.3
C2S	30.36	23.4	16.8
C3A	2.56	7.8	6.9
C4AF	15.41	9.4	13.3

Determination of the composition of raw materials of nanodispersed additives by x-ray phase analysis. Identification of different phases in their mixture based on the analysis of the diffraction pattern given to the samples under study is the main task of x-ray phase analysis (XPA). The samples were studied in the laboratory x-ray phase analysis. As a result of the research, the chemical composition of the basalt sample was obtained (table 2, figures 1,2, radiograph of basalt and limestone, respectively.

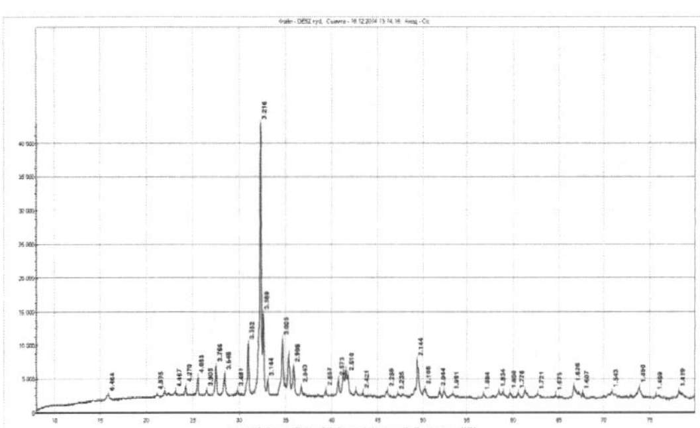
(a)

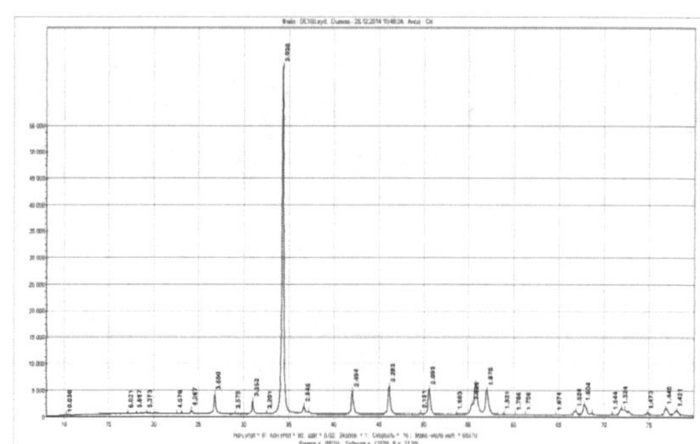
(b)

Figure 1. Radiograph of (a) basalt, b) limestone.

Table 2. Chemical composition of basalt.

Composition	m/m%	Std.Err%	Elements	m/m%	Std.Err%
SiO_2	51.7	0.25000	Si	24.19	0.12000
Al_2O_3	14.19	0.17000	Al	7.51	0.09000
MgO	12.46	0.17000	Mg	7.51	0.10000
Fe_2O_3	10.26	0.15000	Fe	7.18	0.11000
CaO	6.1	0.12000	Ca	4.36	0.09000
Na_2O	3.68	0.09000	Na	2.73	0.07000
TiO_2	0.6	0.03000	Ti	0.361	0.01800
P_2O_5	0.27	0.01400	Px	0.118	0.00600
K_2O	0.215	0.01100	K	0.178	0.00900
MnO	0.187	0.00900	Mn	0.145	0.00700
V_2O_5	0.0617	0.00310	V	0.0346	0.00170
SrO	0.0271	0.00140	Sr	0.0229	0.00110
SO_3	0.0233	0.00120	Sx	0.0093	0.00050
TeO_2	0.0153	0.00260	Te	0.0122	0.00210
ZrO_2	0.012	0.00060	Zr	0.0086	0.00040

SnO_2	0.0102	0.00200	Sn	0.008	0.00160
Au	0.0001	0.00047	Au	0.0001	0.00047

The study of limestone using x-ray phase analysis showed that the rock mainly consists of the mineral calcite, and the minerals $CaCO_3$ and SiO_2 are also present.

Thus, the composition of basalt and limestone has been studied, which will later be studied as multifunctional additives for optimizing the process of structure formation of cement stone.

The main characteristics of the grinding product are the specific surface area (in m^2/kg), the average size of the pieces (grains, particles), the change in dispersion, and the granulometric composition (in %). During the grinding process, the product dispersion was measured on the psh-12(sp) device until the desired result was achieved. The change in the specific surface of the nanodispersed modifier components depending on the grinding time is shown in table 3.

Table 3. Results of analysis of the granulometric composition of components.

Name of the material	Content of nanodispersed particles, %	Specific surface area, m^2/kg
Portland cement	6.63	649
Basalt	13.6	943
Limestone	17.11	1338

During the (time) of the grinding period, figure 2, the specific surface area of the components increases, depending on the grinding time, for about 3 hours they can see how the value of the specific surface fluctuates. The change in the specific surface of limestone depending on the grinding time is shown in figure 2. The figure shows how the specific surface of the test substance (limestone) changes over short periods of grinding time. This depends on the fact that limestone is a softer rock.

The specific surface area of basalt powder is slightly higher than that of Portland cement and was 943 m^2/kg in weight, with a particle diameter from 0.09 to 100 microns. The differential histogram has three peaks corresponding to the particle diameters of 2.15-2.45 microns, 11.87-13.42 microns, and 65.58-74.09 microns. They probably correspond to the dispersion of basalt minerals, plagioclase, clinopyroxene, magnetite, and volcanic glass, which have different grinding properties. According to the integral histogram, the content of nanodispersed basalt particles is 13.6%.

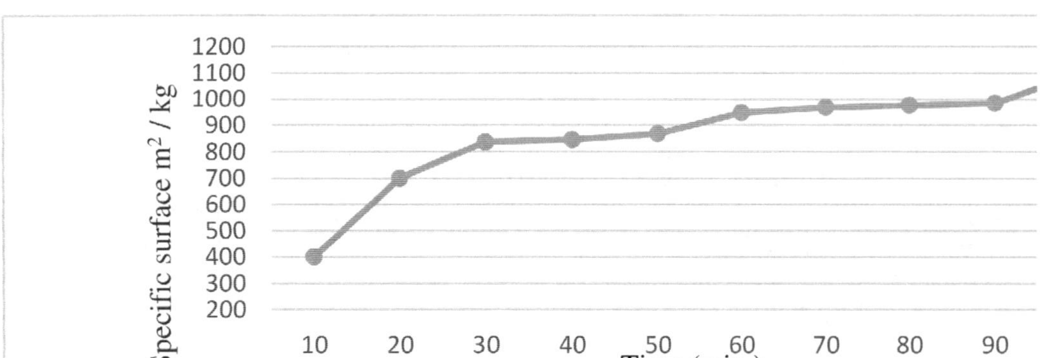

IOP Conf. Series: Materials Science and Engineering **945** (2020) 012046 doi:10.1088/1757-899X/945/1/012046

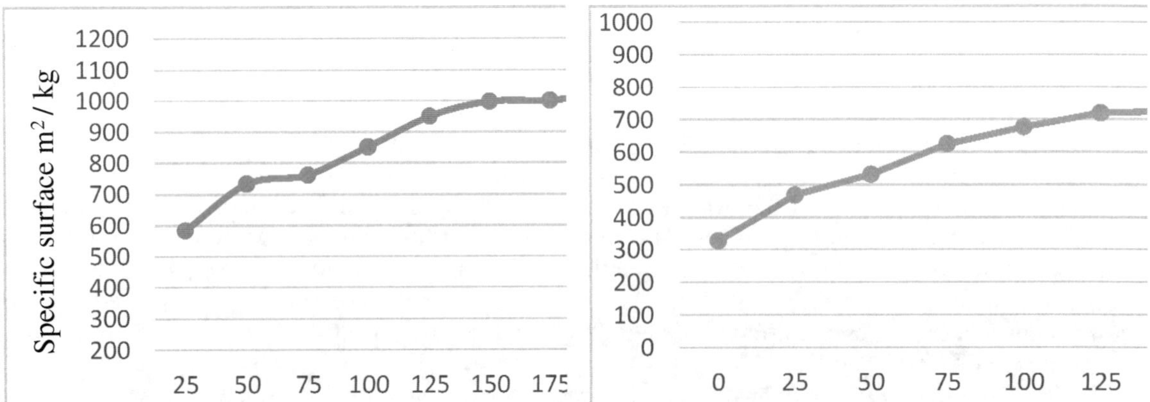

Figure 2. Change of the specific surface area of components during the grinding process.

As a result of the analysis of the granulometric composition of fine-ground limestone, it is shown that the specific surface mass of 1338 m^2/kg, the particle diameter from 0.09 to 100 microns. The differential diagram shows that the maximum percentage of particles contained in the powder corresponds to sizes from 0.01 to 65 microns. According to the integral histogram, the content of nanodispersed particles is 17.11%. The results of the analysis of fine-ground components of the nanodispersed modifier indicate the presence of nanoparticles in the amount of up to 13% of basalt, up to 17.11% of limestone, 6.63% of Portland cement from the mass of the analyzed samples.

Table 4. Results of analysis of the granulometric composition of nanodispersed modifier components.

Name of the material	Content of nanodispersed particles, %	Specific surface area m^2/m^3	Specific surface area by mass m^2/kg
Portland cement	6.63	20467	649
Basalt	13.6	2640	943
Limestone	17.11	3612	1338

To determine the optimal ratio of components of the nanodispersed modifier, the composition of a cement-sand solution of the necessary consistency was selected, at which the cone spread on the flow table was 106 mm. The composition of the cement-sand solution for the manufacture and testing of standard samples 4x4x16 cm was carried out in accordance with GOST 30744-2001. The optimal composition of the nanodispersed modifier was selected by changing the ratio of individual raw materials components. The modifier was introduced in the amount of 12% of the mass of cement required for the production of samples. The influence of different amounts of modifier on the properties of the composite binder was studied.

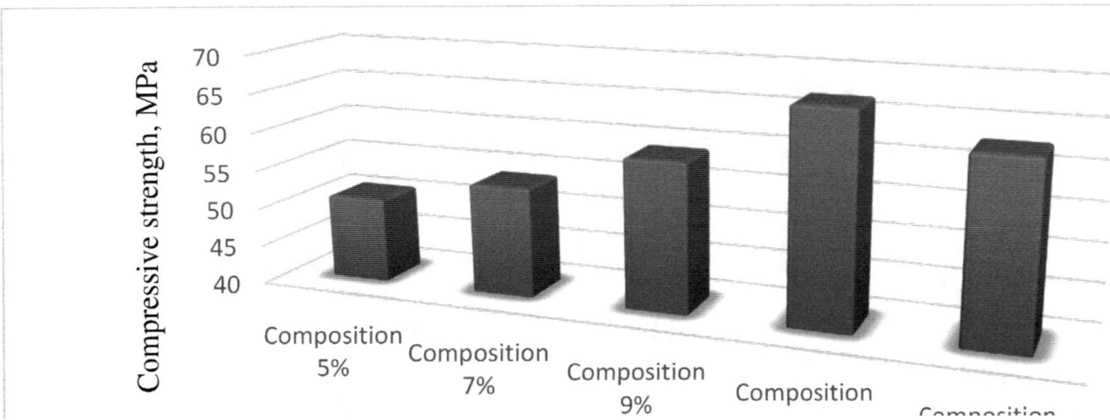

Figure 3. Changes in the compressive strength depending on the amount of nanodispersed modifier composition.

It was found that the optimal mix is a 12% nanodispersed modifier of the cement mass. In the future, we will use a composition with 12% nanodispersed modifier, figure 3. To increase the efficiency of the nanodispersed modifier, a superplasticizer SP-3 was introduced into its composition in an amount of 0.8% of the cement weight. The introduction of a superplasticizer allowed reducing the flow rate of the mixing water to obtain a cement-sand solution of a similar consistency.

The introduction of a nanodispersed modifier and a superplasticizer SP-3 allowed obtaining samples whose strength is 47% higher than the strength of control samples. When testing samples for flexural strength, the excess strength due to the introduction of a superplasticizer and a nanodispersed modifier is 27%, figure 4.

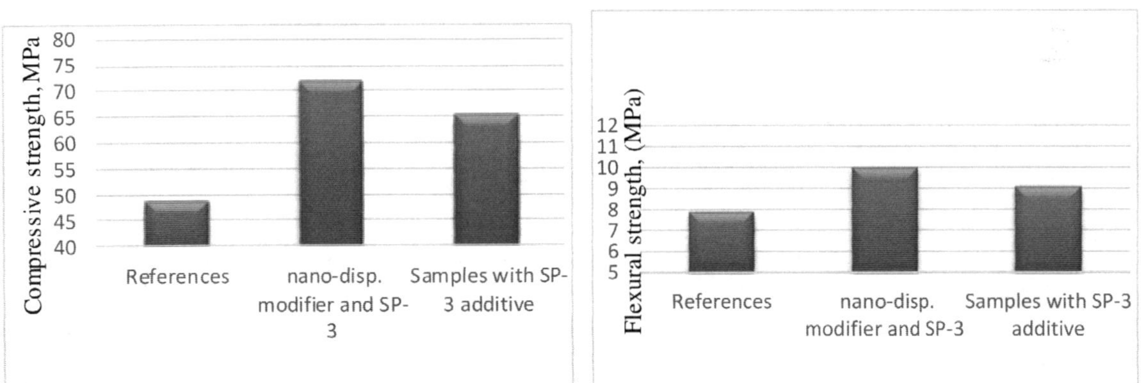

a) b)

Figure 4.(a) Change in compression strength, (b) Change in the flexural strength, depending on the introduction of superplasticizer and nanodispersed modifier.

To study the structure of samples with a nanodispersed additive and superplasticizer SP-3, a microscopic study was performed. For the study of samples on a microscope, compositions with different dispersion of the additive-modifier were made. The research results are shown in figures 5,6.

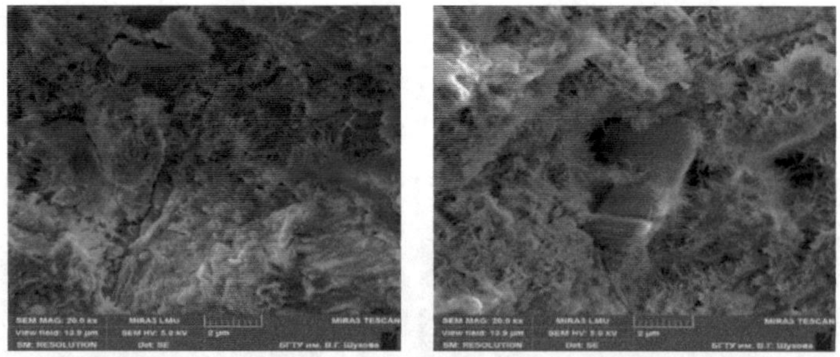

Fig.5. Images of samples without additives.

The images obtained on the microscope show that the structure of the cement stone in samples with nanodispersed additive is more uniform with fewer defects. During the dissolution of clinker minerals from the surface of cement grains to the formation of a saturated solution, in which the primary germs of new phases ($Ca(OH)_2$, ettringite and gel needles (C-S-H) begin to appear. The formation of low-base calcium hydrosilicates increases the strength of cement stone; when high-base hydrosilicates occur, its strength is less. It is possible that the introduction of up to 12% nanodispersed modifier creates additional conditions for the formation of tobermorite $5CaO\text{-}6SiO\text{-}5H_2O$, also characterized by well-formed crystals that strengthen the cement stone.

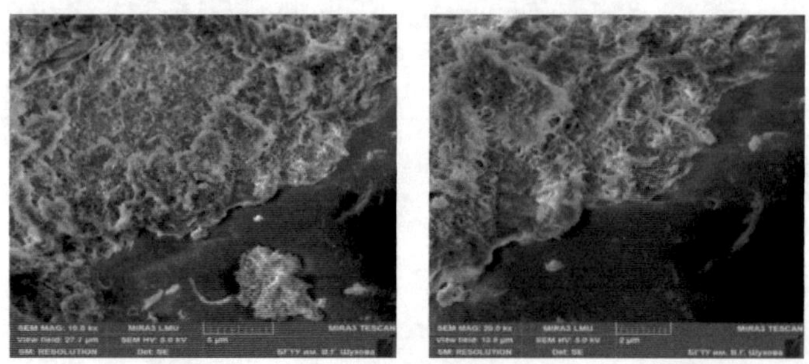

Fig.6. Images of microscopic analysis of samples with nanodispersed modifier with specific surfaces of components: cement - 727 m^2/kg, basalt- 1002 m^2/kg, limestone - 985 m^2/kg.

Thus, we can say that the proposed composition of concrete mixtures for a composite binder with nanodispersed modifier allows not only to reduce the consumption of cement or clinker, as the most expensive component of concrete, but also by controlling the process of structure formation, to create new system polyfunctional things, which not only increases strength and operating performance of concrete at this stage, but in the process of operation, they will contribute to the evolutionary development of operating performance, help to heal those cracks, which will be formed during the operation of structures made of these concretes.

3. Summary

It was found that the use of nanodispersed modifiers, which consists in the disintegration of basalt, limestone, cement and superplasticizer, with a specific surface area of 550-650 m^2/kg, thus forming a nanodispersed powder, the conduct of which will allow controlling the process of structure formation of binders. Cement stone with nanodispersed modifier has a denser structure, significantly changes the composition of new things and, in comparison with traditional cement stone, the amount of alite decreases, the number and variety of calcium hydrosilicates of various types increases, and calcium hydrocarboaluminates appear.

It is obvious that along with the decrease of porosity of material in the process of structure formation, on the one hand, nanodispersed particles of modifier are actively included, on the other hand, they also act as a sealing additive, thereby providing a significant improvement of physical and mechanical characteristics of the material. Fine-ground cement in a nanodispersed modifier is the center of crystallization, it primarily begins the processes of hydration. The basalt component also serves as the center of crystallization and absorbs the alite $Ca(OH)_2$ released during hydration in the calcium hydro-silicates of the second generation. The introduction of a nanodispersed limestone modifier makes it possible to synthesize calcium hydrocarboaluminates, which play an important role in the process of structuring and improving the performance of the composite.

4. References

[1] Adel Y, Usama E, Simon J, 2018 Life cycle cost analysis of structural concrete using seawater, recycled concrete aggregate, and GFRP reinforcement *Construction and Building Materials* **17530** 152-160

[2] Jianhe X, Chi F, Zhongyu L 2018 Effects of the addition of silica fume and rubber particles on the compressive behavior of recycled aggregate concrete with steel fibers *Journal of Cleaner Production* **197** 656-667

[3] Ahmed A A, Fedyuk R S, Liseytsev Yu L, Timokhin R A, Murali G 2020 Use of Iraq concrete scrap as filler and aggregate of heavyweight and lightweight concrete *Construction Materials and Products* **3 (3)** 28 – 39

[4] Al-Bu-Ali U S, Lesovik R V, Harhardin A N, Tolstoy A D, Ahmed A A A , Alaskhanov A Kh, Aymenov Zh T 2020 Calculation and selection of grain composition of aggregate from high-density concrete scrap packaging *Bulletin of BSTU named after V G Shukhov* **5(6)** 18-28

[5] Guoliang B, Chao Z, Chao L, Biao L 2020 An evaluation of the recycled aggregate characteristics and the recycled aggregate concrete mechanical properties *Construction and Building Materials* **24020** 117978.

[6] Larsen O, Naruts V, Aleksandrova O 2019 Self-compacting concrete with recycled aggregates *Materials Today: Proceedings* **19** 2023-2026.

[7] Volodchenko, A A, Lesovik, V S , Volodchenko A N, Zagorodnjuk L H 2015 Improving the efficiency of wall materials for «green» building through the use of aluminosilicate raw materials *International Journal of Applied Engineering Research* **10(24)** 45142-45149

[8] Klyuev S V, Khezhev T A, Pukharenko Yu V, Klyuev A V 2018 To the question of fiber reinforcement of concrete *Materials Science Forum* **945** 25 – 29

[9] Klyuev S V, Khezhev T A, Pukharenko Yu V, Klyuev A V 2018 Experimental study of fiber-reinforced concrete structures *Materials Science Forum* **945** 115 – 119

[10] Nakhaev M R, Salamanova M Sh, Ismailova Z Kh, 2020 Regularities of the processes of formation of the structure and strength of a clinker-free binder of alkaline activation *Construction materials and products* **3(1)** 21-29

[11] Klyuev S V, Khezhev T A, Pukharenko Yu V, Klyuev A V 2018 Fiber concrete for industrial and civil construction *Materials Science Forum* **945** 120 – 124

[12] Keun-H Y, Hey-Zoo H, Sun-Young K, Jin-Kyu S 2007 Development of a cementless mortar using hwangtoh binder *Building and Environment* **42(10)** 3717-3725

[13] Jeong-I C, Keum I S, Jin-K S, Bang Y L 2016 Composite properties of high-strength polyethylene fiber-reinforced cement and cementless composites *Composite Structures* **3815** 116-121

IOP Conf. Series: Materials Science and Engineering **945** (2020) 012047 doi:10.1088/1757-899X/945/1/012047

Analysis of power parameters of a rotary mill

T N Orekhova[1*], E O Sheremet[2] and A E Kachaev[3]

[1]Department of hoisting-and-transport and road machines, Belgorod State Technological University named after V.G. Shukhov, Kostyukov St., 46, Belgorod, 308012, Russia

[2] Department of Heat and Gas Supply and Ventilation, Belgorod State Technological University named after V.G. Shukhov, Kostyukov St., 46, Belgorod, 308012, Russia

[3]Department of hoisting-and-transport and road machines, Belgorod State Technological University named after V.G. Shukhov, Kostyukov St., 46, Belgorod, 308012, Russia

E-mail: nefact@mail.ru

Abstract. The paper (publication) presents a description of the developed design of a rotary mill with a complex mechanical and acoustic effect on the dispersed material, as well as analytical studies of the parameters of energy consumption of a rotary mill, based on the main physical processes occurring in it. The research was carried out by mathematical and simulation modeling of technological processes in the grinding chamber of the mill, as well as on various technological paths of the unit. To determine the main parameters that make up the total power consumption of the unit, an approach was chosen to differentiate the processes by their physical nature and contribution to the formation of the total power consumption of the rotary mill.

The obtained results of mathematical modeling of processes occurring inside the technological paths of the unit allowed getting expressions for determining the main power costs of the unit depending on the characteristics of the unit design for each component of the total power required for effective operation of the unit at the required parameters of the dispersion of the finished product.

1. Introduction

In view of the need for materials with high specific surface characteristics, mechanical engineering is faced with the task of developing and calculating effective technological equipment that could successfully meet these requirements in the production of construction materials and products. In accordance with this, the team of engineers formulated a goal in the need to design and calculate a special shredder that would be able to combine in its design various effects on the crushed material and solve various technological problems, which would increase the efficiency of the process of dispersion and homogenization of dispersed systems, increase the specific surface area of the finished product and reduce the specific energy consumption in industrial production.

To achieve this goal, the design of a rotary mill [1] was proposed for complex mechanical and acoustic effects on the dispersed material. The essence of the design presented in figure 1 is expressed in the fact that at one of the technological stages of pre-dispersion, the principle of shock-shear strains

IOP Conf. Series: Materials Science and Engineering **945** (2020) 012047 doi:10.1088/1757-899X/945/1/012047

is used; at another stage of the main grinding, the principle of vortex and acoustic action with a reflective effect on the mill stator; at other stages, the vortex-acoustic effect is used (to avoid agglomeration of fine particles); as well as in the mill design, it is possible to carry out dispersible mixing of dissimilar materials [2].

Using the resulting three-dimensional model of the projected unit, they can predict the behavior of the material at any stage. We used the FloXpress application in the SolidWorks 2018 program to simulate various processes that occur in the unit.

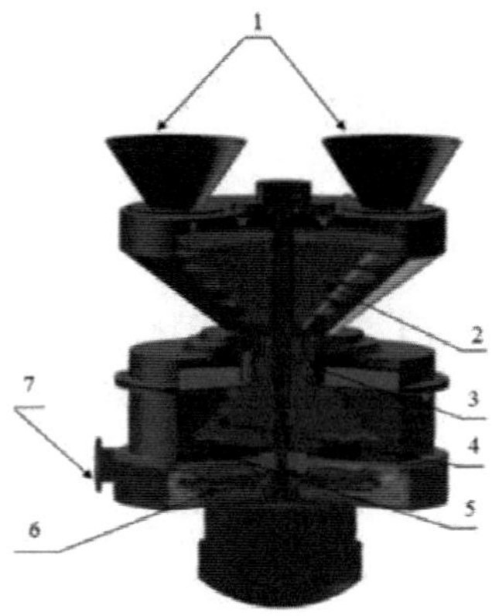

Figure 1. Rotary mill of complex mechanical and acoustic action: 1 – loading components; 2 – transportation or pre-shredding; 3 – the stage of grinding under the action of shock-shear strains; 4 – grinding with the use of the impact effect; 5 – processing of air-material flow in disk headsets (partial deagglomeration of particles); 6 - pneumohomogenization of dispersed powders, or acoustic dispersion; 7 – discharging the finished product.

The design of the rotary mill under consideration includes the possibility of various mechanical effects on the dispersed material. The combination of successive (or simultaneous) types of mechanical effects on the dispersed material, such as impact, abrasion, sound effects of acoustic wave vibrations, reduces the consideration of the grinding process in the mill to a complex approach, and the combination of such effects has every right to be called complex [2].

The works of Kirpichev, Kik, Rittinger, Rebinder, Bond, and Rundkvist are the most widely used in all the existing variety of developed theories for calculating energy costs for grinding [4], however, there are studies by modern authors that have significantly improved and clarified the classical understanding of this process [5,6] with respect to specific types of grinders.

2. Materials and methods

Let us express the necessary complete work spent by rotary mill in the dispersion process, as the sum of certain works: friction work - $W_и$; impact dispersion work - Wy; work of mechanical and acoustic destruction of particles from the influence of resonators - W_a. Then they can get an expression for the complete dispersion work, taking into account that the various stages of grinding the material in the mill will imply certain equations for the complete dispersion work:

$$W_{comp.} = \sum W = W_у + W_и + W_а. \quad (1)$$

The impact dispersion of material particles in a rotary mill is defined by a well-known expression:

$$W_у = \frac{\sigma_ч^2 \cdot V_ч}{2\mathrm{E}}, \quad (2)$$

where $V_ч$ – particle volume of the dispersed material, m^3;

$\sigma_ч^2$ - the strength of the material particles (compression), *Pa*;

E – modulus of longitudinal elasticity of the dispersed material, *Pa*.

We assume that the shape of the particle is spherical [4].

Taking into account [7] the energy parameter ξ plays an important role and indicates the ratio of the initial velocity of the particle before impact and the velocity of the particle after impact. This ratio is determined by the proportion of loss of movement speeds of the systems involved in the grinding process: the particle and the working body - the rotor:

$$\xi = \frac{v_{н}}{v_{к}}, \quad (3)$$

here $v_{н}$ - initial particle velocity of the dispersed material, *m/s*;
$v_{к}$ – the final velocity of a particle of the dispersed material when it collides with the working element of the rotor, *m/s*.

We describe the value of the energy required for dispersion by the impact of a single particle of material at the first stage of the grinding process:

$$W_{y1} = \sum_{i=0}^{z_1} \xi_i \frac{\sigma_{ч}^2 \cdot V_{ч}}{2E} = \sum_{i=0}^{z_1} \xi_i \frac{\sigma_{ч}^2 \cdot \pi \cdot d^3}{12E} \cong \sum_{i=0}^{z_1} \xi_i \frac{0{,}25\sigma_{ч}^2 \cdot k \cdot d^3}{E}, \quad (4)$$

where d – diameter size of the cross part of the particle, *m*;
k – the formal particle coefficient of the dispersed material.

The expression for calculating strength is represented as:

$$\sigma_i = \sigma_{ч} + \sigma_{theoret.}(1 - e^{-\beta i}), \quad (5)$$

here σ_i - ultimate compressive strength during impact at the *i*-th stage of dispersion, Pa;
$\sigma_{theoret.}$ – theoretical value of the material compressive strength under impact, *Pa*;
β - a scale coefficient that is determined empirically for each specific material;
i – stage of dispersion.

We find the amount of energy at the second stage of dispersion, which is necessary for grinding a single particle of raw material from the initial size $d_{кр}$ to the specified one:

$$W_{y2} = \frac{0{,}25k \cdot d_{кр}^3}{E} \sum_{i=0}^{z_2} \left[\xi_i \left(\sigma_{ч} + \sigma_{theoret.}(1 - e^{-\beta i})\right)^2\right], \quad (6)$$

where z_2 - dispersion stage.

The theoretical value of the material strength in compression during impact we calculate using Orowan-Kelly equation [8]:

$$\sigma_{theoret.} = \sqrt{\frac{E \cdot S_{пов.}}{r_0}}, \quad (7)$$

where $S_{пов.}$ - surface energy of a solid on an area of 1 cm^2; r_0 - the interatomic distance in the equilibrium state, cm.

Let us represent the notation of possible normal stresses from the shock pulse *S* as σ_4, from the influence of the back-up force $F_{под.}$ - σ_3.

Possible stresses σ_4 from the impact pulse *S* of the mill rotor on a particle of the dispersed material can be calculated from the dispersion criterion:

$$|\sigma_4| \geq [\sigma]_c, \quad (8)$$

where $[\sigma]_c$ - acceptable possible normal compression stresses of the dispersed material, *Pa*, which, in turn, can be set from the equation:

$$[\sigma]_c = \sigma_{theoret.}. \quad (9)$$

Taking into account equations (8) and (9), it can be concluded that:

$$\sigma_4 = \sigma_{theoret.} \quad (10)$$

The stresses in the material also have a value associated with the impact pulse *S*, so it can be written as follows:

$$\sigma_4 = \frac{F_{imp}}{S_{к}}, \quad (11)$$

IOP Conf. Series: Materials Science and Engineering **945** (2020) 012047 doi:10.1088/1757-899X/945/1/012047

where F_{imp} – impact force of the rotor on a particle of the dispersed material, *N*;
$S_к$ - contact area of the material particle with the rotor, m^2.

If we consider that in practice, for various geometric shapes of particles and its sizes, the pulse *S* from the impact force F_{imp} will be transmitted to the particle via an infinitesimal value of $S_к$, then:

$$\sigma_4 = \lim_{S_к \to 0} \frac{F_{imp}}{S_к} = sign\left(F_{imp}\right). \tag{12}$$

So expression (12) in expression (5) is represented as:

$$\sigma_i = \sigma_ч + sign\left(F_{imp}\right)\left(1 - e^{-\beta i}\right). \tag{13}$$

This expression (13) quantifies the boundary value of the compressive strength during the impact pulse *S* at the *i*-th stage of dispersion, depending on the geometric dimensions of the mill working elements and their speed during the impact period.

It is known that the possible normal stresses from the compression of particles on impact contribute to the appearance of tangential stresses τ in the particle, which, according to Mohr's theory of strength, are defined as:

$$\tau_{min}^{max} = \pm \frac{\sigma_{theoret}}{2}. \tag{14}$$

Taking into account the above and in accordance with the provisions of the Mohr strength theory, we make an assumption about shock-shear strains of particles during dispersion of the material. These strains contribute to the formation of new surfaces of the dispersed material, as well as to the production of powders with a lighter subsequent destruction of already thin particles at further stages of raw material grinding. Next, we get the expression (6) taking into account (13) in the form:

$$W_{у2} = \frac{0{,}25k \cdot d_{кр}^3}{E} \sum_{i=0}^{z_2} \left[\xi_i \left(\sigma_ч + sign\left(F_{imp}\right)\left(1 - e^{-\beta i}\right)\right)^2\right]. \tag{15}$$

The grinding of material particles during dispersion and impact interaction in an infinitesimal period of time occurs under the action of friction forces.

Let us note that each stage of dispersion due to attrition increases the number of fine particles by a (1 - ξ) times, while the total volume of material particles subjected to destruction is the following value - $k \cdot d_{кр}^3 (1 - \xi^z)$.

Based on the above, we will find the work of the friction forces spent on the abrasion of (1 - ξ) part of the material for parts with the initial size *d*:

- for the first dispersion stage

$$W_{и1} = \frac{0{,}25\sigma_{theoret.}^2}{E} k \cdot d^3 \left(1 - \xi^{z_1}\right), \tag{16}$$

- for the second dispersion stage:

$$W_{и2} = \frac{0{,}25\sigma_{theoret.}^2}{E} k \cdot d_{кр}^3 \left(1 - \xi^{z_2}\right), \tag{17}$$

We obtain the equation of the work required to disperse the material from the initial particle size *d* to the critical- $d_{кр}$ in z_1 stages, in accordance with expressions (4) and (16):

$$W_{у1} + W_{и1} = \frac{0{,}25 \cdot k \cdot d^3}{E} \cdot \sigma_ч^2 \sum_{i=0}^{z_1} \xi + \sigma_{theoret.}^2 \left(1 - \xi^{z_1}\right). \tag{18}$$

Then, in the same way, we add the components (15) and (17) for the material particles from the size of the beginning of the scale effect $d_{кр}$ to the final size $d_к$ at the z_2 stage:

$$W_{у2} + W_{и2} = \\ = \frac{0{,}25k \cdot d_{кр}^3}{E} \left(\sum_{i=0}^{z_2} \left[\xi_i \left(\sigma_ч + sign\left(F_{imp}\right)\left(1 - e^{-\beta i}\right)\right)^2\right] + \sigma_{theoret.}^2 \left(1 - \xi^{z_2}\right) \right). \tag{19}$$

To find the results of mathematical modeling, we obtained an expression for the dispersion of a material with a mass of $M_М$. We multiply the expression by the number of particles N contained in the material with a mass $M_М$:

$$W_у + W_и = \frac{M_М}{Е\rho_ч}\left(\left(\sigma_ч^2 \sum_{i=0}^{z} \xi_i + \sigma_{theoret.}^2 (1 - \xi^{z_1})\right) + \right.$$

$$\left. +2^{z_1}\frac{d_{кр}^3}{d^3}\left(\sum_{i=0}^{z_2}\left[\xi_i\left(\sigma_ч + sign\left(F_{imp}\right)\left(1 - e^{-\beta i}\right)\right)^2\right] + \sigma_{theoret.}^2 (1 - \xi^{z_2})\right)\right), \quad (20)$$

where $\rho_ч$ - particle density of the dispersed material, kg/m^3.

If with each subsequent stage the particle is destroyed into several - c - parts, we get:

$$c^z = \frac{d^3}{d_к^3} = i_i^3. \quad (21)$$

As it was assumed above $c = 3$, we will determine the number of stages for a certain size of the fraction of the dispersed material:

$$z = \frac{3 \lg i_i}{\lg 2} \approx 10 \lg \frac{d}{d_x}. \quad (22)$$

Based on (20), (21), (22) we conclude that the work that must be spent on dispersing the mass of the material with a volume of $M_М$ under complex mechanical action from the initial to the final particle sizes is determined by the equation of the form:

$$W_у + W_и = \frac{M_М}{Е\rho_ч}\left(\left(\sigma_ч^2 \sum_{i=0}^{10\lg\frac{d}{d_{кр}}} \xi_i + \sigma_{theoret.}^2 \left(1 - \xi^{10\lg\frac{d}{d_{кр}}}\right)\right) + \right.$$

$$\left. +2^{10\lg\frac{d_{кр}}{d_к}}\frac{d_{кр}^3}{d^3}\left(\sum_{i=0}^{10\lg\frac{d_{кр}}{d_к}}\left[\xi_i\left(\sigma_ч + sign\left(F_{imp}\right)\left(1 - e^{-\beta i}\right)\right)^2\right] + \sigma_{theor.}^2 \left(1 - \xi^{10\lg\frac{d_{кр}}{d_к}}\right)\right)\right). \quad (23)$$

In order to calculate the component equation (1) related to the work of destruction of dispersed particles under the influence of acoustic waves of resonators - W_a, we use the equation:

$$W_а = Р_{ак} \cdot V_x^3, \quad (24)$$

where $Р_{ак}$ - acoustic pressure in an elastic medium, Pa;
V_x^3 – volume of the particle affected by acoustic pressure, m^3.

The value of variable acoustic pressure can be determined by the amplitude of particle vibrations in the air by the formula:

$$Р_{ак} = 2\pi \cdot \omega \cdot \rho_{н.с} \cdot C \cdot А_ч, \quad (25)$$

here ω - frequency of oscillation of the wave at which the particle is destroyed, Hz;
$\rho_{н.с}$ – two-phase flow density, kg/m^3;
C – the speed of ultrasound distribution, m/s;
$А_ч$ – oscillation amplitude of dispersed particles, m.

Based on the above expressions included in equation (1), we finally get:

$$W_{comp.} = \text{Р}_{\text{ак}} \cdot V_x^3 + \frac{M_M}{\text{Е}\rho_{\text{ч}}}\left(\left(\sigma_{\text{ч}}^2 \sum_{i=0}^{10\lg\frac{d}{d_{\text{кр}}}} \xi_i + \sigma_{theoret.}^2\left(1 - \xi^{10\lg\frac{d}{d_{\text{кр}}}}\right)\right) + \right.$$

$$\left. +2^{10\lg\frac{d_{\text{кр}}}{d_{\text{к}}}}\frac{d_{\text{кр}}^3}{d^3}\left(\sum_{i=0}^{10\lg\frac{d_{\text{кр}}}{d_{\text{к}}}}\left[\xi_i\left(\sigma_{\text{ч}} + sign\,(F_{imp})(1 - \text{e}^{-\beta i})\right)^2\right] + \sigma_{theoret.}^2\left(1 - \xi^{10\lg\frac{d_{\text{кр}}}{d_{\text{к}}}}\right)\right)\right). \quad (26)$$

3. Results and summary

As a result of research on the design of a rotary mill using computer programs and mathematical modeling: analytical expressions are obtained to determine the main parameters of energy consumption of a rotary mill, based on the main physical processes occurring in it and its structural and technological features. As a result, it was found that the diameter of the final product particles exceeding the critical particle size is guaranteed by a more rational process of combined effect of impact with abrasion in combination with mechanical and acoustic impact. The obtained results of mathematical modeling of processes occurring inside the technological paths of the unit allowed obtaining expressions for determining the main power parameters.

4. References

[1] Glagolev S N, Sevostyanov V S, Gridchin A M, Kachaev A E, et al. Rotary mill *Applicant and patent holder Belgorod State Technological University named after V.G. Shukhov* Patent for the invention RU 2444407 C1, 10.03.2012. Application № 2010138884/13 dated 21.09.2010

[2] Sevostyanov V S, Kachaev A E, Boychuk I P, Kuznetsova I A 2017 Theoretical studies of processes in a rotary mill with a combined effect on the material being milled *Bulletin of the Belgorod State Technological University named after V.G. Shukhov,* **9** 122-129

[3] Uvarov V A, Sheremet E O, Orekhova T N 2016 Analysis of equipment for fine grinding *Scientific almanac* **12-2 (26)** 145-147

[4] Andreev S E, Petrov V A, Zverevich V V 1980 Crushing, grinding and screening of minerals *Textbook. 3rd edition. Nedra publishing house,* Moscow

[5] Bond F S Laws of crushing 1966 Proceedings of the European conference on crushing. *Stroyizdat, Moscow*

[6] Orekhova T N, Nosov O A, Prokopenko V S, Kachaev A E 2018 Complex modeling of technological processes in pneumatic mixers for the production of dry construction mixtures. *IOP Conf. Series: Materials Science and Engineering* **327** 022084

[7] Kachaev A E 2011 To determine the time of impact and the power of the disintegrator consumed for grinding under shock impact *Bulletin of BSTU named after V.G. Shukhov* **3** 60-64

[8] Prokopets V S Ivanova T L 2005 Mathematical model of efficiency of mechanical activation processes in building materials *Construction materials, equipment, technologies of the XXI century* **8** 71-73

[9] Kuznetsova, M M, Marakhovsky M B, Aleksina A A 2015 Determination of the energy consumption of the process of grinding solid materials *Journal of theoretical physics* **85 (5)**

[10] Landau, L D, Lifshitz, E M 2004 Mechanics *5th edition, stereotyped. M: Fizmatlit* 224-225

[11] Klyuev S V, Khezhev T A, Pukharenko Yu V, Klyuev A V 2018 To the question of fiber reinforcement of concrete *Materials Science Forum* **945** 25 – 29

[12] Menshikova, V K, Demina, L N 2020 Ceramic construction materials using non-traditional raw materials *Construction materials and products* **3 (3)** 40 – 46.
[13] Klyuev S V, Khezhev T A, Pukharenko Yu V, Klyuev A V 2018 Experimental study of fiber-reinforced concrete structures *Materials Science Forum* **945** 115 – 119

Acknowledgments
This work was realized in the framework of the Program of flagship university development on the base of the Belgorod State Technological University named after V G Shukhov, using equipment of High Technology Center at BSTU named after V G Shukhov.

IOP Conf. Series: Materials Science and Engineering **945** (2020) 012048 doi:10.1088/1757-899X/945/1/012048

The influence of additives of dehydrated clay on the properties of unburned binders and concrete on their basis

N B Sarsenbayev[1*], S K Kurbaniyazov[2], N A Saktaganova[3] and R O Karshygayev[3]

[1]M.Auezov South Kazakhstan State University, Shymkent, Kazakhstan
[2]International Kazakh-Turkish University named after Khoja Ahmed Yasavi
[3] Korkyt Ata Kyzylorda University, 29a Ayteke bi Str., Kyzylorda, 12001, Kazakhstan

E-mail: stroitelstvo_ukgu@mail.ru

Abstract: The purpose is development of technological bases of production of unburned alkaline binders and concrete on their basis from metallurgical and electro-thermo-phosphoric slags with an additive of the dehydrated clay with high operational properties.
In this regard the increased indicators of physical mechanical properties of unburned alkaline binders and concrete on their basis were researched with addition of 25% - dehydrated clay mixing alkalis.
The influences of fineness grinding, a type of slag and an alkaline component, curing conditions on properties unburned alkaline binders and concrete on their basis, namely on changes of average density, water absorption, strength characteristics were researched.
Thus, as a result of the conducted researches are received depending on a type of slag, an alkaline component, conditions of curing and fineness grinding of unburned alkaline binders and concrete on their basis with an additive of the ground dehydrated clay of grade from M300 to M1200.

1. Introduction

One of actual problems of production of cement in the Republic of Kazakhstan is lack of raw materials because the part of sources of the natural raw materials, being in available proximity, is exhausted and has no real replacement. In the foreseeable future, it will lead to inevitable partial or full replacement of natural raw materials technogenic which importance has to increase steadily including for nature protection reasons.

Now in Kazakhstan 10 million tons of electro-thermo-phosphoric slags and 85 million tons of metallurgical slags are accumulated, its yield makes from 700 thousand to 1 million tons per year [1-4].

Therefore research is directed on creation of industrial production of a new generation of hi-tech unburned alkaline binders of the Republic of Kazakhstan [5-7]. Its technology is based on a high grinding from metallurgical or electro-thermo-phosphoric slag with modifying additives and mixing solutions of alkaline components.

High physical and mechanical properties of concrete on the basis of unburned alkaline binders describing their durability and performance characteristics, are mainly connected with structural

IOP Conf. Series: Materials Science and Engineering **945** (2020) 012048 doi:10.1088/1757-899X/945/1/012048

characteristics of material which are basically determined with composition of new formation in binder and by contact with filler, also increased density and largely self-contained pores [8-12].

2. Methods and materials

In this regard, increased indicators of physical and mechanical properties of unburned alkaline binders and concretes based on them were researched with the addition of 25% - dehydrated clay mixing alkalis. The experiments were carried out in the laboratory. Used electrothermophosphoric slag of Limited Liability Partnership «Kazphosphate» plant (New Jambul phosphorous factor) in Taraz town and metallurgical slags of Limited Liability Partnership «Ispat Karmet» in Temirtau town had following chemical composition:

- electrothermophosphoric slag, in weight, %: SiO_2-42,0; CaO-45,1; MgO-2,4; Fe_2O_3-1,25; P_2O_5-2,60; Al_2O_3-2,65; SO_3-0,43; K_2O-0,15; F-2,20;
- metallurgical slag (blast-furnace slag), in weight, %: SiO_2-34,82; CaO-39,16; MgO-6,18; Fe_2O_3-3,45; Al_2O_3-12,7; FeO-0,89; BaO-0,17; SO_3-2,0; MnO-0,73; Na_2O-0,72; TiO_2-0,17.

Grinding of slag is carried in a mill Activator 4M. The samples of unburned alkaline binders was prepared in following composition: unburned alkaline binders: quartz to sand = 1:3.

For determination of degree of fineness of binder were used by PSH-12. Average density, water absorptions and strength characteristics were determined by the state of sectoral standard.

3. Results and discussions

The researching results of influence degree of fineness, type of slag and alkaline component, conditions of curing time on properties of unburned alkaline binders that changes in medium density, water absorption, strength characteristics of these factors are described with the dependences (Fig. 1,2).

The overall level of medium density of unburned alkaline binders on the base of electrothermophosphoric slag is lower on 0,8-1,2%, and water absorption is higher on 11,5-19% than on the basis of metallurgical slag independently on the kind of alkali component and hardening conditions, which is aligned with highly porous particles of dehydrated clay.

Depending on the type of alkaline component, hardening conditions and degree of fineness binder the average density of unburned alkaline binders on electrothermophosphoric slag ranges from 2,172 to 2,445 g/cm^3, water absorption 3,2-3-5,0%, and on metallurgical slag ranges from 2,160-2,429 g/cm^3 and water absorption - 2,9-4,5%.

After steam curing the density of unburned alkaline binders depending on the kind of slag, alkali component, degree of fineness of binder and presence of additives dehydrated powdered clay is higher on 0,3-0,5% than after uncured, water absorption is lower on 0,6-8%,

With increasing degree of fineness from 300 to 700 m^2/kg of unburned alkaline binders the compression strength depending on the kind of slag, alkali component, amount of additive dehydrated clay and hardening conditions, increases on 2,1-3,5%, water absorption is reduced on18,9-38,1%. Strength of UAB on the base of electrothermophosphoric slag depending on the kind of alkali component and hardening conditions is higher on compression - 3,5-21,5% and on flexing - 2,4-8,4%,.

With increasing degree of fineness from 300 to 700 m^2/kg of unburned alkaline binders the compression strength depending on the kind of slag alkali component and hardening conditions increases on 7-30,5%, and on flexing - 25-49,5%.

The researches of influence degree of fineness, kind of ash and alkaline components, hardening conditions at medium density, water absorption, strength of concretes based on unburned alkaline binders and also kinetics of curing time is similar to dependence given for unburned alkaline binders.

The analysis of obtained results of researches showed that the overall medium density of concretes based on CUAB unburned alkaline binders on the base of electrothermophosphoric slag is less on 0,15-0,9% and water absorption higher on - 8,1-17,2 % than on the basis of metallurgical slag independently on the kind of alkali component and hardening conditions.

IOP Conf. Series: Materials Science and Engineering **945** (2020) 012048 doi:10.1088/1757-899X/945/1/012048

Depending on the kind of slag, alkaline component, hardening conditions and degree of fineness of binder the medium density of concretes on the base of CUAB unburned alkaline binders on electrothermophosphoric slag ranges from 2,408 to 2,485 g/cm^3 and water absorption - 2 ,6-4, 3%, and on the basis of metallurgical slag in the ranges from 2,44 to 2,500 g/cm^3 and water absorption - 2,1-3,6%.

After steam curing the medium density of concrete on base of unburned alkaline binders is higher than after natural curing on 0,2-0,6%, water absorption lower on 0,3-5,0%, depending on the kind of slag, alkali component, degree of fineness of binder and additives presence of dehydrated clay, with increasing degree of fineness from 300 to 700 m^2/kg medium density of concretes on base of unburned alkaline binders, depending on the kind of slag alkali component, amount of addition of dehydrated clay and hardening conditions is increased on 0.3-1.2%, water absorption is reduced on 14,1-26%.

Depending on the kind of slag alkali component, hardening conditions and degree of fineness of binder strength of concretes of unburned alkaline binders ranges from 40 to 115 MPa - on electrothermophosphoric slag, and on metallurgical slag in the ranges from 32 to 105 MPa.

After heat and humidified treatment the strength on compression of CUAB is higher than after uncured up to 4,7%, depending on the kind of slag alkali component, tannin of grinding of binder and quantity of amount of additive dehydrated powdered clay.

With increasing of degree of fineness from 300 to 700 m^2/kg strength of concretes on the base of unburned alkaline binders depending on kind of slag alkali component, amount of additive dehydrated powdered clay and hardening conditions, increases on 27-32%.

The results of physico-chemical studies have shown that the addition of dehydrated clay reduces maintenance the free highly soluble alkali.

The presence of the optimal content of additive dehydrated clay - 25% the content of free alkali is reduced on 18.3%. Perhaps this is a result of interaction between clay minerals and active ingredients, dehydrated clay alkaline compounds mix to form soluble products of hydration.

4. Summary

The influences of additives of the dehydrated clay on operational properties of unburned alkaline binders and concrete on their basis depending on a type of slag, an alkaline component, and conditions of curing and grinding fineness are researched. Thus, the results of studies depending on the type of slag alkali component, hardening conditions and fineness binder of unburned alkaline binders and concretes on their basis with the addition of powdered dehydrated clay marks from M300 to M1200 are produced.

The research was carried out in accordance with the following scientific and technical programs:"Scientific supplying of chemical industry development of the Republic of Kazakhstan" and "Presentation of innovation grants in the framework of direction Productivity – 2020", UDK 666.974.6:666.972.16, state registration 0110RK00446, according to themes of the following agreements "Development of production technologies of the non-calcining binders and concretes on their base of natural hardening out of metallurgical industrial waste" and "Development of production technologies of concretes out of alkaline binders on the base of phosphorus and blast furnace slag".

IOP Conf. Series: Materials Science and Engineering **945** (2020) 012048 doi:10.1088/1757-899X/945/1/012048

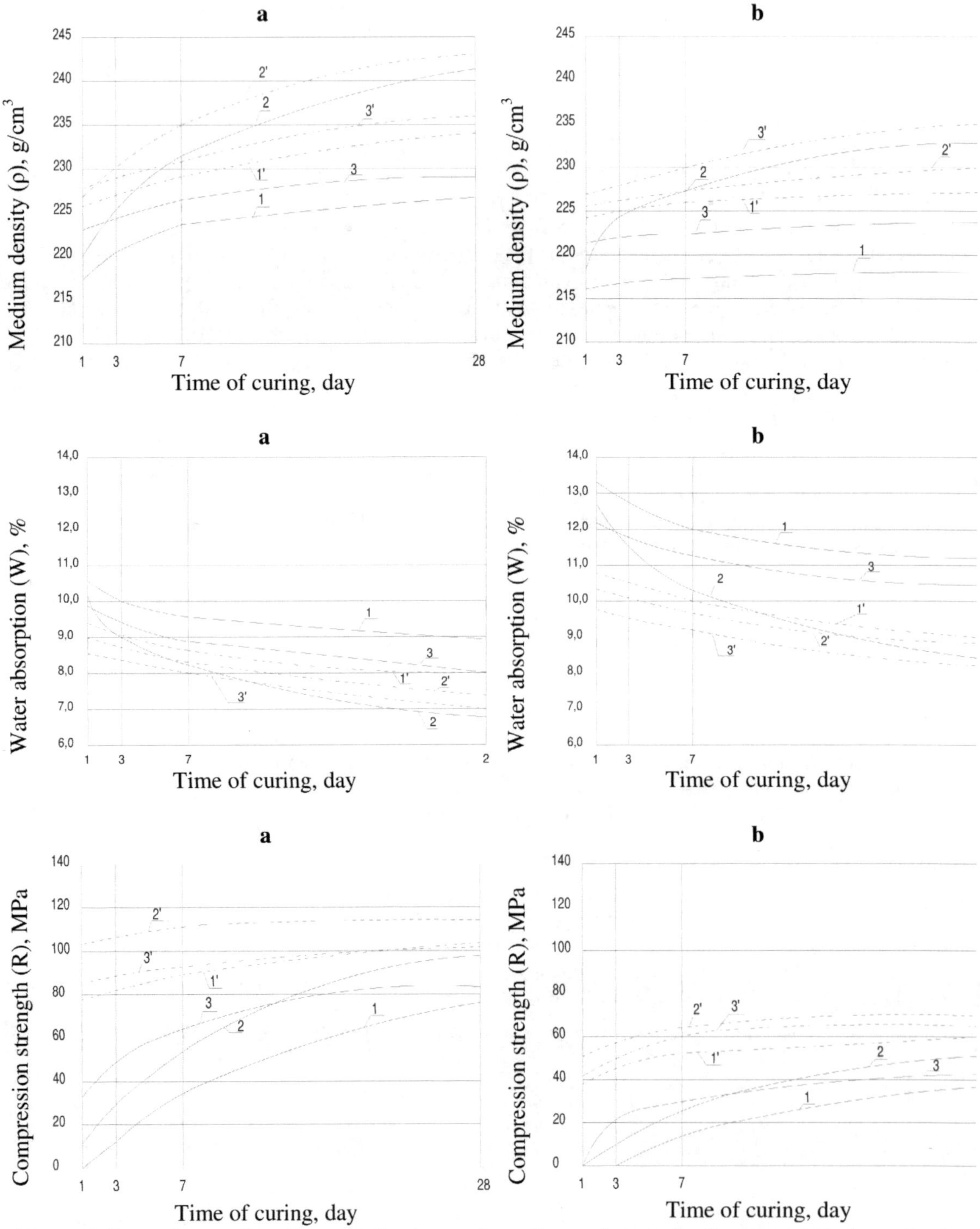

Figure 1. Dependence of medium density, water absorption and strength of unburned binder on basis of electrothermophosphoric slag with addition 25 % of dehydrated clay : a - liquid glass; b - caustic ash, composition: 1-st with S_{ss}= 300 m^2/kg; 2-nd with S_{ss}= 500 m^2/kg; 3-rd with S_{ss}= 700 m^2/kg.

------- after heat and humidified treatment; ——— after essential hardening;

IOP Conf. Series: Materials Science and Engineering **945** (2020) 012048 doi:10.1088/1757-899X/945/1/012048

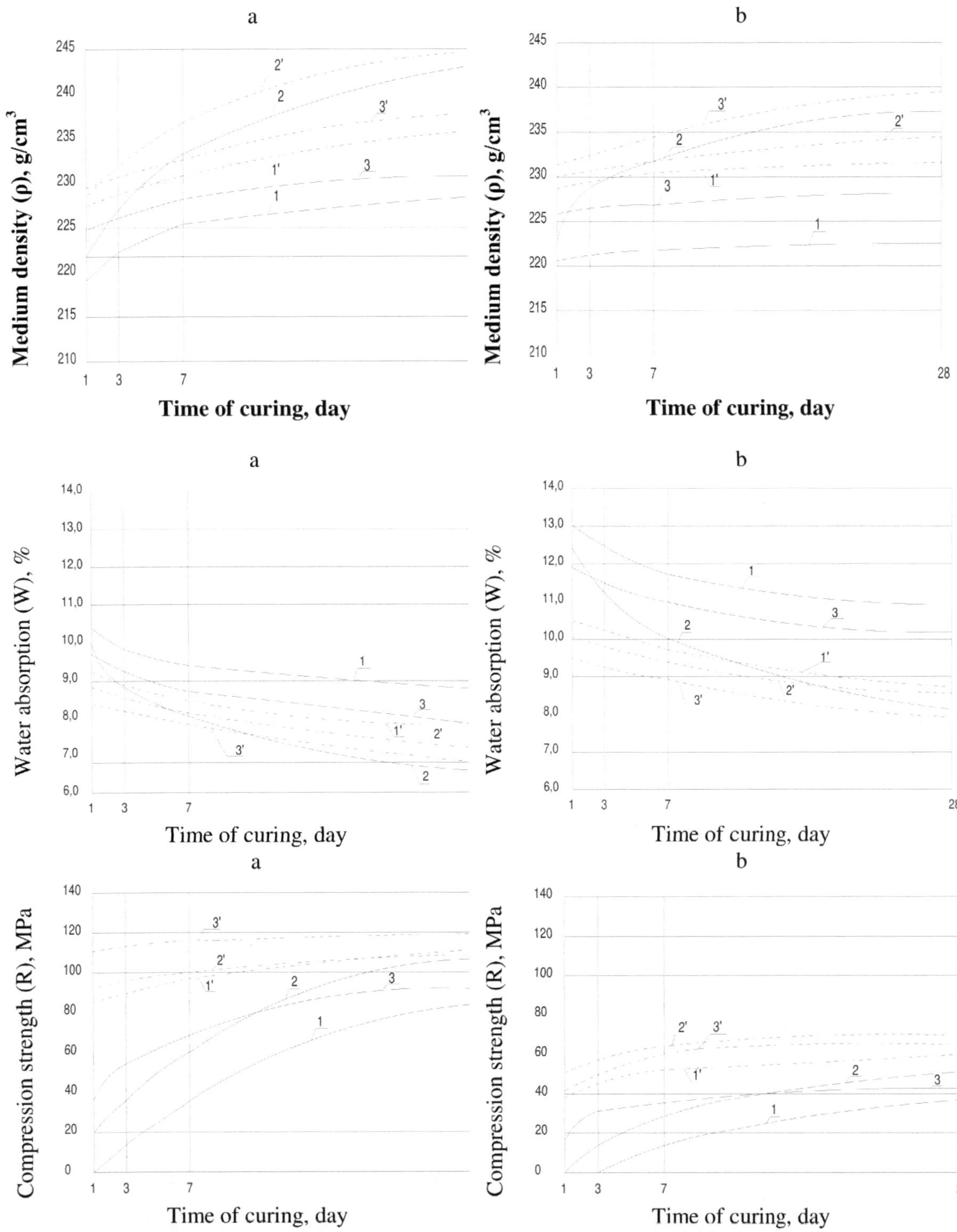

Figure 2. Dependence of medium density, water absorption and strength of unburned binder on basis of metallurgic slag with addition 25 % of dehydrated clay : a - liquid glass; b - caustic ash, composition: 1-st with Sss= 300 m2/kg; 2-nd with Sss= 500 m2/kg; 3-rd with Sss= 700 m2/kg.
----- after heat and humidified treatment; after essential hardening;

IOP Conf. Series: Materials Science and Engineering **945** (2020) 012048 doi:10.1088/1757-899X/945/1/012048

5. References

[1] Estemesov Z.A., Sadykov P.I., Barvinov A.V., Sarsenbaev N.B., Estemesov M.Z., Uralkhanova A.U., Tulaganov A.A. Physical and chemical processes occurring in the cranulated phosphorus slag dumps// NEWS Of the Academy of Sciences of the Republic of Kazakhstan JSC "D.V. Sokolsky institute of fuel, catalysis and electrochemistry". Series Chemistry and Technology. Almaty, Nas RK. 2 (440). march- april 2020. Pg.47-55.

[2] Estemesov Z.A., Barvinov A.V., Sarsenbaev B. K, Tulaganov A.A., Estemesov M.Z., Khaidarov A.M. New method for disposal of granulated phosphoric slag from hazardous gases// NEWS Of the Academy of Sciences of the Republic of Kazakhstan JSC "D.V. Sokolsky institute of fuel, catalysis and electrochemistry". Series Chemistry and Technology. Almaty, Nas RK. 3 (441). may- june 2020. Pg.6-14.

[3] Klyuev S V, Khezhev T A, Pukharenko Yu V, Klyuev A V 2018 To the question of fiber reinforcement of concrete *Materials Science Forum* **945** 25 – 29

[4] A.Zh. Aymenov, N.B. Sarsenbayev, T.M. Khudyakova, B.K. Sarsenbayev, B.T.Kopzhassarov// Effect of additive of polymetallic ores' tailings on properties of composite cements// «Eurasian chemico-technological journal» 2016 y. c.153-160.

[5] Aymenov A.Zh., Khudyakova T.M., Sarsenbayev B.K. Studying the mineral additives effect on a composition and properties of a composite binding agent/ Oriental journal of chemistry. G.A.Iqbal.- 2018. Vol.34, No.(4).Pg.1945-1955.

[6] Glukhovsky V.D. and etc. Slagalkaline concrete on fine-grained fillers. - Kiev: The higher school, 1981. - P. 115.

[7] Aubakirova T.S., Sarsenbayev B.K., Abdyramanova K.Sh., Sarsenbayev N.B. Phisico-himicheskye processi gidratasii i tverdenia slakochelochnyh vyzhychih i betonov na osnove electrotermophosphornyh slakov, Tehnika i tehnologia silicatov, Mejdunarodnyi jurnal po vyjuchim, keramike, steklu i emalym, Tom 20, № 2, 2013, 21-25, Moskwa;

[8] Aubakirova T.S., Sarsenbayev B.K., Iskakov T.U., Momymev T.A., Sarsenbayev N.B. Proizvodstvo slakochelochnyh vyzhychih i betonov na ih osnove, Nauchno-tehnicheskay I proizvodstvenyi jurnal Stroitelnye materiali №11, November 2012, 56-58 p;

[9] Aubakirova T.S., Sarsenbayev B.K., Iskakov T.U., Sarsenbayev N.B., The influence of various factors on the strength of slag-alkaline binders 18 Internationale Baustofftagung. 12-15. September 2012. Weimar, Bundesrepublik Deutschland.-Tagungsbeient.-Band 2. –0835-0840 p.

[10] Aubakirova T.S., Sarsenbayev B.K., Lesovic V.S., Aymenov A.Zh., Sarsenbayev N.B., Batyrkhanov A.T. Dehydrated clay additives entering influence on alkali-activated slag cement properties // International Conference of Industrial Technologies and Engineering (ICITE 2015). – Shymkent, 2015. – P.482-487.

[11] Klyuev S V, Khezhev T A, Pukharenko Yu V, Klyuev A V 2018 Experimental study of fiber-reinforced concrete structures *Materials Science Forum* **945** 115 – 119

[12] Klyuev S V, Khezhev T A, Pukharenko Yu V, Klyuev A V 2018 Fiber concrete for industrial and civil construction *Materials Science Forum* **945** 120 – 124

IOP Conf. Series: Materials Science and Engineering **945** (2020) 012049 doi:10.1088/1757-899X/945/1/012049

Movement of particles of concrete mixture aggregate subjected to vibration by cylindrical punches

I V Kirilov[1*], S I Khanin[1] and V P Voronov[1]

[1]Belgorod State Technological University named after V.G. Shukhov, 46 Kostyukova street, Belgorod, 308012, Russia

E-mail: igor_kirilov_1989@mail.ru

Abstract. For compaction of the concrete mixture in the manufacture of reinforced concrete products using the formless production method on long stands, as a universal and widespread method, vibration is used. The aim of the study was to establish the regularities of the process of spreading a cylindrical punch of a continuous formless machine of vibration oscillations on the aggregate particles of a viscous concrete mixture in the form of crushed stone and sand. The chaotic movement of filler particles of various sizes with an arbitrary number of degrees of freedom, due to their large number, movement, and exchange of impulses, is considered. The problems of the development of mathematical descriptions of the motion of spherical filler particles of different grain size in the vibration of a cylindrical punch with a viscous environment and determination of their average values of travel speed and distance through random motion were solved, the average values for these parameters at vibration of the punch with frequencies and amplitudes that are applied in industrial conditions were found. In the course of the work, the method of mathematical modeling was used. The result of this work is to establish the possibility, for particles of different sizes, of varying the average values of the speed of movement and the path traversed in a viscous environment, carried out by changing the characteristics of the vibration of the punch. The regularities of changes in the parameters under consideration for filler particles of various sizes depending on the frequency and amplitude of vibration oscillations of the punch surface are considered.

1. Introduction

It is known that the main properties of such a popular building material as concrete, reinforced concrete products and structures made of it in industrial conditions, are laid at the stage of forming its structure [1]. Various exposure conditions at the production stage can significantly increase these properties in order to form a dense and stable structure [2]. The effectiveness of these effects, mainly mechanical, provided by forming units, depends on the method in which the particles of aggregate and cement paste would form a uniform structure in the formed volume of the concrete product. Vibrating, as one of the most universal methods, has deservedly found wide distribution in various technological complexes. These include, in particular, the continuous production of prestressed long-length products without formwork, in which various functions are performed by aggregates relative to a stationary stand [3-5]. In Russia, the general use of forming machines of this type is over 80% [6].

The process of transmitting vibrations to a viscous concrete environment is provided by surface, deep or volumetric vibration, and when forming products of large cross sections, the effectiveness of the latter has been proven with unchanged characteristics and the content of the binder. Therefore, a significant part of the available scientific and technical literature is devoted to the description of vibration techniques of volumetric impact. However, scientists note its insufficiency in the field of scientific description of the process of transmitting vibration to a viscous concrete environment by the working bodies of a continuous formless forming machine in the form of a vibroform with fixed punches [7]. The complexity of the description is represented by the process of movement of concrete mixture particles, which is provided by rationally selected amplitude-frequency characteristics of vibration of the vibroform, allowing to provide the necessary characteristics of the formed product. In this regard, there is a need to develop a mathematical description of the speed of movement and displacement of particles in a viscous concrete environment, depending on the design of the vibroform punch and the rheological characteristics of the mixture.

2. Materials and methods

Let us consider the punch of a cylindrical shape. When vibrating, its surface has a cylindrical axial symmetry, so it is advisable to describe the action of vibrations in the coordinate system (r, φ, z) with the center on the axis (figure 1), taking into account the movement of the aggregate of filler particles in a volume equal to:

$$V = \pi L \cdot l \cdot (2R_0 + l), \tag{1}$$

where L – length of the punch; l – the size of the propagation zone of intense vibration in the environment; R_0 – the radius of the cylindrical surface of the punch, we take R_0 equal to 0.078 m in accordance with the reinforced concrete slabs manufactured by the industry of the Russian Federation.

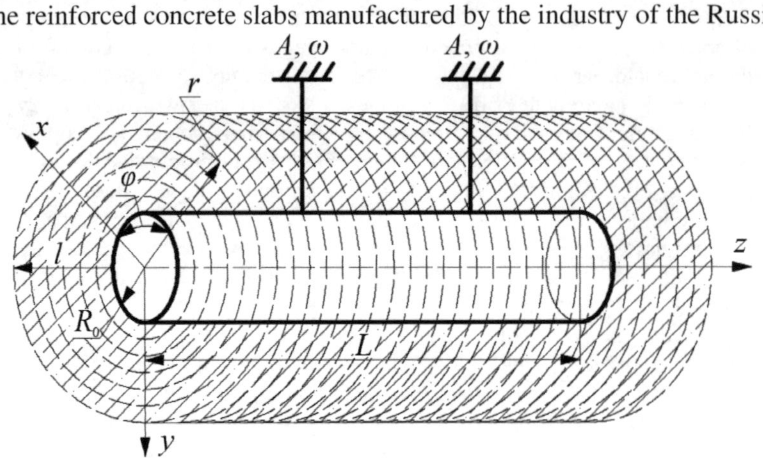

Figure 1. Scheme for determining the propagation of vibration in the volume of a viscous concrete medium, excited by the cylindrical surface of the punch.

As a result of vibration, the filler particles leave the equilibrium position they occupied in the viscous medium. The resulting movement of environmental particles leads to their interaction and exchange of impulses. As a result of a large number of material particles, the exchange of impulses contributes to their chaotic movement. Consequently, the material particles behave like ideal gas particles when the viscous environment vibrates [8]. Based on the above, we can conclude that the movement of material particles in a viscous environment can be described in the framework of statistical physics.

The velocity of a material particle in a viscous environment as a result of vibration is probabilistic and can be described using the velocity distribution function.

Due to the axial symmetry, it is convenient to present the velocity distribution function in the following form:

$$f(\upsilon_r) = B_0 \cdot \upsilon_r \cdot exp\left(-\frac{m\upsilon_r^2}{2 \cdot E_0}\right), \tag{2}$$

where B_0 – normalization constant; m – particle mass; υ_r – the value of the particle velocity in a plane perpendicular to the cylinder axis z; E_0 – the average value of the energy of motion of a viscous mixture as a result of vibration during the oscillation period.

Taking into account the average value of the square of the sine for a time equal to the period $<\sin^2(\omega_0 t+\varphi_0)>$, the average value of the motion energy of a viscous mixture takes the form:

$$E_0 = \frac{1}{2} \cdot \rho_{CM} \cdot V \cdot A_0^2 \cdot \omega_0^2 \cdot \psi^2(r), \tag{3}$$

here ρ_{CM} – mixture density; A_0 and ω_0 – amplitude and frequency of cylinder surface vibrations;

$$\psi(r) = \sqrt{\frac{R_0}{r}} \cdot exp\left(-\frac{r-R_0}{l}\right). \tag{4}$$

To find the maximum value of the velocity of a material particle in a viscous environment under the influence of vibration, the first derivative of (2) must be equal to zero:

$$B_0 \cdot exp\left(-\frac{m\upsilon_r^2}{2 \cdot E_0}\right) \cdot \left(1 - \frac{\upsilon_r^2 \cdot m}{E_0}\right) = 0. \tag{5}$$

Expression (5) turns to zero if:

$$\upsilon_r = \upsilon_{max} = \sqrt{\frac{E_0}{m}}. \tag{7}$$

Substituting (3) in (7) leads to the following result:

$$\upsilon_{max} = \sqrt{\frac{\rho_{CM} \cdot V}{2 \cdot m}} \cdot A_0 \cdot \omega_0 \cdot \psi(r). \tag{8}$$

Based on (2) the probability of different values of the velocities of the material particles in a viscous environment under the action of vibration can be written as:

$$d\omega = f(\upsilon_r) d\upsilon_r. \tag{9}$$

The normalizing constant B_0 can be found based on the normalization condition:

$$\int_0^{\upsilon_{max}} d\omega = 1. \tag{10}$$

Applying (10) to (9) with (2) in mind, we get the following expression:

$$B_0 \int_0^{\upsilon_{max}} \upsilon_r \, exp\left(-\frac{m\upsilon_r^2}{2 \cdot E_0}\right) \cdot d\upsilon_r = 1. \tag{11}$$

We transform:

$$I_1 = \int_0^{\upsilon_{max}} \upsilon_r \, exp\left(-\frac{m\upsilon_r^2}{2 \cdot E_0}\right) \cdot d\upsilon_r = \left[\xi = \upsilon_r \sqrt{\frac{m}{2 \cdot E_0}}\right] = \frac{2 \cdot E_0}{m} \int_0^{\xi_{max}} \xi \cdot e^{-\xi^2} d\xi = \frac{E_0}{m} \int_0^{\xi_{max}} e^{-\xi^2} d(\xi^2), \tag{12}$$

where ξ_{max} – coefficient of penetration of vibrations to depth l;

$$\xi_{max} = \upsilon_{max} \cdot \sqrt{\frac{m}{2 \cdot E_0}}. \tag{13}$$

Substituting (7) in (13) leads to the following result:

$$\xi_{max} = \frac{\sqrt{2}}{2}. \tag{14}$$

Calculating the integral in (12), taking into account (14), allows obtaining the following relation:

$$I_1 = \frac{E_0}{m} \cdot \left(1 - e^{-\frac{1}{2}}\right). \tag{15}$$

Based on (11) and (15), we find the value of the constant B_0:

$$B_0 = \frac{1}{I_1} = \frac{m}{E_0} \cdot \frac{1}{1 - e^{-\frac{1}{2}}}. \tag{16}$$

Substituting (16) in (2) allows getting the final form of the speed distribution function:

$$f(\upsilon_r) = \frac{m}{E_0} \cdot \frac{\upsilon_r \cdot exp\left(-\frac{m\upsilon_r^2}{2 \cdot E_0}\right)}{1 - e^{-\frac{1}{2}}}. \tag{18}$$

Based on the velocity distribution function (18), it is possible to find the average value of the velocity of a material particle in a viscous environment subject to vibration relative to the initial equilibrium position:

$$\upsilon_{cp} = \int_0^{\upsilon_{max}} \upsilon_r \cdot f(\upsilon_r) d\upsilon_r. \tag{19}$$

Substituting (18) in (19) gives the ratio:

$$\upsilon_{cp} = \frac{m}{E_0} \cdot \frac{1}{1 - e^{-\frac{1}{2}}} \cdot \int_0^{\upsilon_{max}} \upsilon_r^2 \cdot exp\left(-\frac{m\upsilon_r^2}{2 \cdot E_0}\right) \cdot d\upsilon_r. \tag{21}$$

Let us calculate the integral:

$$I_2 = \int_0^{\upsilon_{max}} \upsilon_r^2 \cdot exp\left(-\frac{m\upsilon_r^2}{2 \cdot E_0}\right) \cdot d\upsilon_r = \left[\xi = \upsilon_r \sqrt{\frac{m}{2 \cdot E_0}}\right] = \left(\frac{2 \cdot E_0}{m}\right)^{3/2} \int_0^{\xi_{max}} \xi^2 e^{-\xi^2} d\xi =$$
$$= \left(\frac{2 \cdot E_0}{m}\right)^{3/2} \cdot \left[-\frac{\xi}{2} \cdot e^{-\xi^2}\Big|_0^{\xi_{max}} + \frac{1}{2}\int_0^{\xi_{max}} e^{-\xi^2} \cdot d\xi\right] = \left(\frac{2 \cdot E_0}{m}\right)^{3/2} \cdot \left[-\frac{\xi_{max}}{2} \cdot e^{\xi_{max}^2} + \frac{\sqrt{\pi}}{4} \cdot erf(\xi_{max})\right], \tag{22}$$

here $erf(\xi)$ – function that defines the value of the probability integral [9].

Substituting (14) in (22) leads to the following relation:

$$I_2 = \left(\frac{2 \cdot E_0}{m}\right)^{3/2} \cdot \frac{1}{4} \cdot \left[\sqrt{\pi} \cdot erf\left(\frac{\sqrt{2}}{2}\right) - \sqrt{2} \cdot e^{-\frac{1}{2}}\right]. \tag{23}$$

Based on (21) – (23), we can get the following result:

$$\upsilon_{cp} \cong 0{,}633 \cdot \sqrt{\frac{E_0}{m}}. \tag{24}$$

If we consider the relation between the mass of a spherical material particle and its diameter, then substituting (3) with (1) in (24) allows obtaining the following relation:

$$\upsilon_{cp} = 0{,}633 \cdot \sqrt{\frac{3 \cdot \rho_{cм} \cdot L \cdot l \cdot (2 \cdot R_0 + l)}{2 \cdot \rho \cdot d^3}} \cdot A_0 \cdot \omega_0 \cdot \psi(r), \tag{25}$$

where ρ – particle mass; d – particle diameter.

IOP Conf. Series: Materials Science and Engineering **945** (2020) 012049 doi:10.1088/1757-899X/945/1/012049

Using the expression (25), we can find the average value of the displacement of a material particle over a time equal to the oscillation period:

$$r_{cp} = \upsilon_{cp} \cdot \tau, \tag{26}$$

where τ – the period of fluctuations of the viscous environment.

Based on (25), the formula for the average value of the particle displacement (26) is given as:

$$r_{cp} = 2\pi \cdot 0{,}633 \cdot \sqrt{\frac{3 \cdot \rho_{cM} \cdot L \cdot l \cdot (2 \cdot R_0 + l)}{2 \cdot \rho \cdot d^3}} \cdot A_0 \cdot \psi(r). \tag{27}$$

3. Results

Using the obtained expressions for constructing graphical dependencies of changes in the average values of the velocities of movement and displacement of particles subject to vibration, we set the frequency and amplitude of vibrations as variable parameters. For implementing the vibrations of the prototype we will take vibrioform machine of the continuous cold forming, consisting of two parts [10].

The density of the concrete mixture ρ_{cM}, with which the punch directly interacts during the operation of the vibroform, according to the manufactured factory products, is assumed to be equal to 2400 kg/m^3.

When selecting aggregate fractions, based on their largest and smallest sizes found in the formed product, the particle size for crushed stone is assumed to be 0.01 m in diameter, and for sand – 0.00125 m in diameter.

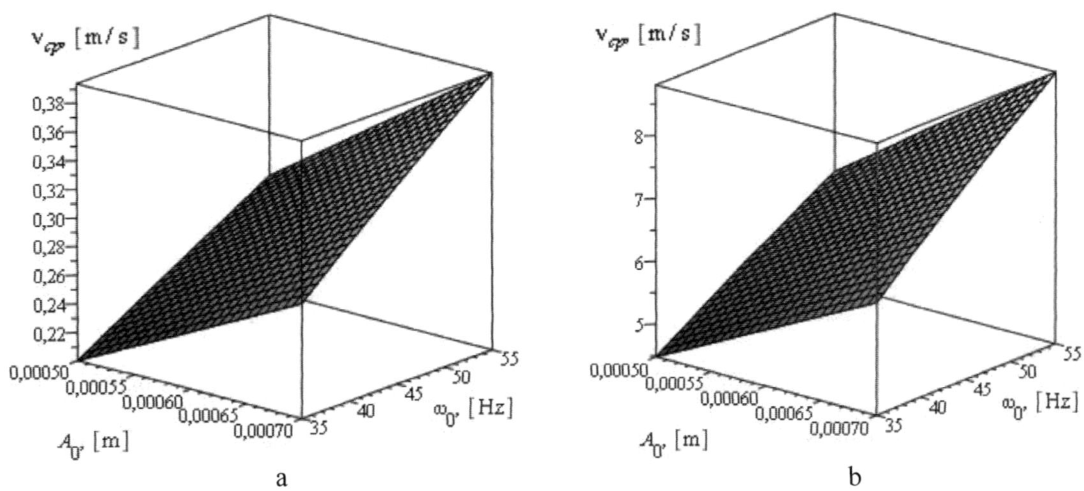

Figure 2. Dependences of the average values of the speed of movement of particles for a vibrating mold: a – for particles with a size of 0.01 m; b – for sand particles with a diameter of 0.00125 m.

The graphs of the functions υ_{cp} from ω_0 at A_0 = 0.0005...0.0007 m (figure 2) and A_0 = 0.0003...0.0004 m (figure 3) corresponding to the parts of the vibroform are monotonously increasing. At the same time, with a decrease in the diameter of the filler from 0.01 m to 0.00125 m, the υ_{cp} increases over the entire frequency range ω_0 of the first and second parts of the vibroform. For example, when ω_0 is equal to 55 Hz and A_0 = 0.0007 m, the average speed of movement of a crushed stone particle with a diameter of 0.01 m corresponds to a value of 0.393 m/s (figure 2, a), and for a sand particle with a diameter of 0.00125 m (figure 2, b) increases to a value of 8.82 m/s, i.e. by 22.44 times. Similarly, the increase of the velocity of the particle depending on its diameter can be observed when the frequency and amplitude of oscillation of the punch surface of the second part of the vibrating mold is 75 Hz and 0.0004 m, respectively, where with decreasing particle diameter from

0.01 to 0.00125 m increase in υ_{cp} occurs from 0.283 to 6.34 m/s (figure 3, a and b).

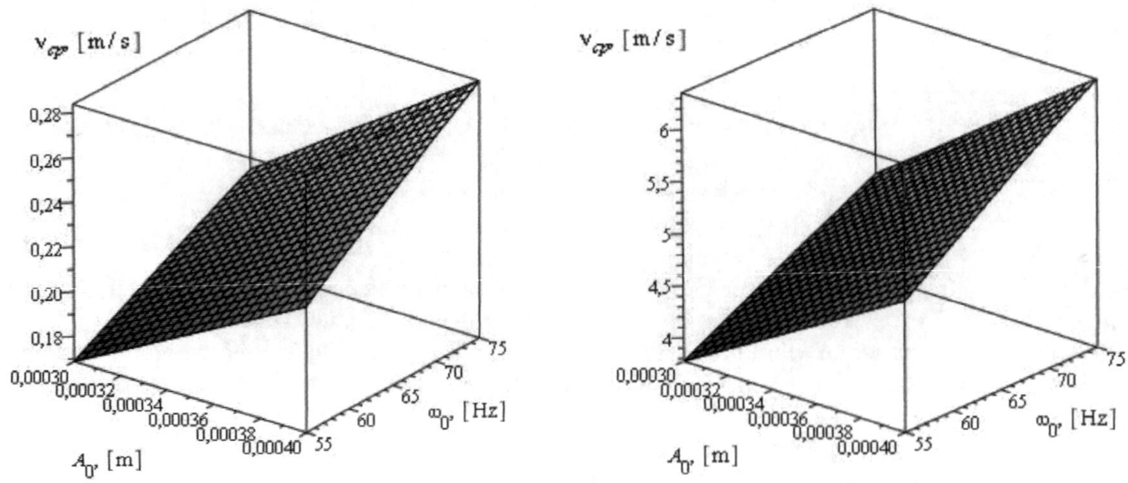

a b

Figure 3. Dependences of changes in the average values of the speed of movement of particles for the second part of the vibroform: a – for crushed stone particles with a diameter of 0.01 m; b – for sand particles with a diameter of 0.00125 m.

Let us consider the change in the υ_{cp} for given particle diameters and frequency ω_0. With a change in the amplitude A_0 of the first part of the vibroform (figure 2) in the range of 35...55 Hz, the average speed of movement of the crushed stone particle increases linearly from 0.200 to 0.280 m/s, i.e. by 40%. A larger increase in this parameter is observed for the diameter of sand particles equal to 0.00125 m. Thus, at ω_0 = 35 Hz and A_0 = 0.0005...0.0007 m, the υ_{cp} changes from 4.42 to 6.27 m/s, which corresponds to an increase of 41.85%.

In the section of the second part of the vibroform (figure 3), this phenomenon is observed, but with a slightly smaller percentage discrepancy, with a variable amplitude of vibrations of the punch surface in the range A_0 = 0.0003...0.0004 m and a given value ω_0 = 55 Hz. Thus, for crushed stone with a diameter of 0.01 m, the υ_{cp} increases from 0.169 to 0.235 m/s (an increase of 39.05%), and for sand particles with a diameter of 0.00125 m – from 3.78 to 5.03 m/s (an increase of 33.06%). Hence, the average speed of movement of particles of a viscous environmental material is significantly affected by the amplitude-frequency mode of vibrations of the surface of the punch, which can be adjusted based on their size by changing the technological mode of the unit that excites these vibrations.

The graphical dependences of the change in the average values of the displacement of the particles of the r_{cp} material from the frequency ω_0 in the considered range of amplitudes of the surface of the punch (figures 4 and 5) are nonlinear decreasing in nature. It follows from the dependencies that when $\omega_0$increasing in the considered range, the average value of the displacement of crushed stone particles in the viscous environment of the first part of the vibroform occurs by a greater 1.79 times, compared with the second for similar particles. So, a reduction in the average value of the displacement of the particles of crushed stone in the first part of the vibrating mold reaches 0.0450 m (at a frequency of 55 Hz) in comparison with the average displacement of the particles of crushed stone equal to 0.0504 m (at a frequency of 35 Hz), i.e. a decrease of 12%, and in the second part of the vibrating mold – with an average value of the displacement of the particles is equal to 0.0238 m (at a frequency of 75 Hz) in comparison with the average displacement is equal to 0.0254 m (at a frequency of 55 Hz), the decrease is 6.7%. environment of the first part of the vibroform, in comparison with the second one, is much higher and is 2.92 times. This is evidenced by a decrease in the average values of particle displacement in the first part of the vibroform from 1.128 to 1.007 m (by 12%) and in the second from 0.556 to 0.534 m (by 4.1%), respectively. Also, based on the above, we can conclude that with a decrease in the diameter of the spherical filler particles, the differences in the average values of their displacement increase significantly.

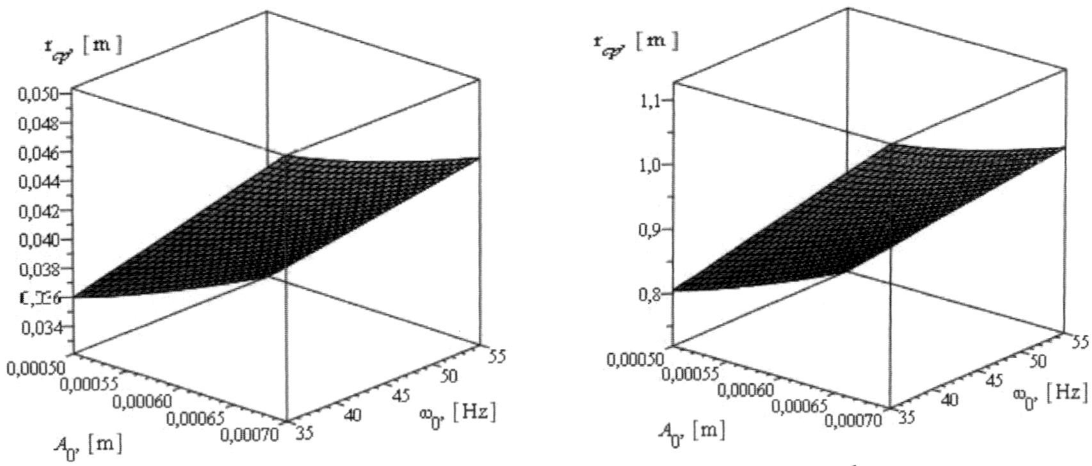

a b

Figure 4. Dependences of the average values of the displacement of particles subject to vibration in the first part of the vibroform: a – for crushed stone particles with a diameter of 0.01 m; b – for sand particles with a diameter of 0.00125 m.

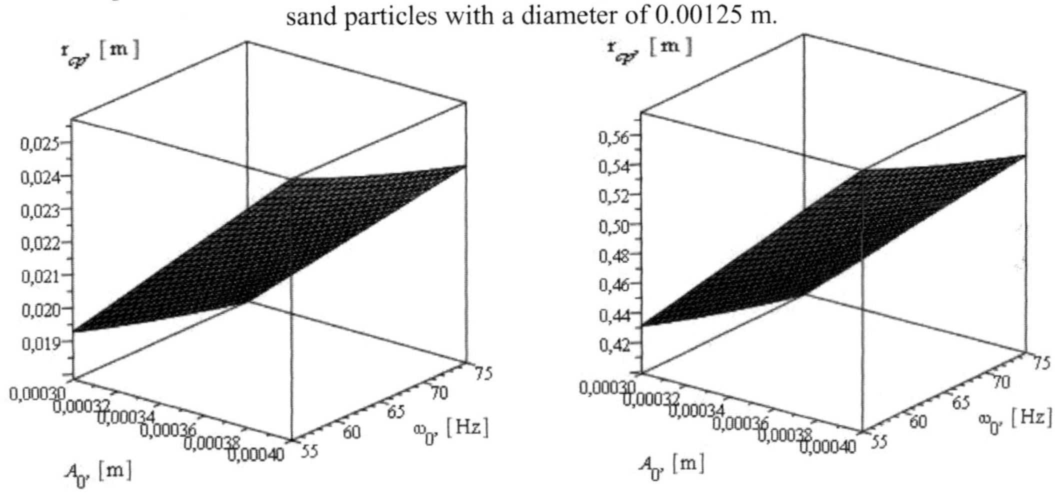

a b

Figure 5. Dependences of the average values of the displacement of particles subject to vibration in the second part of the vibration form: a – for crushed stone particles with a diameter of 0.01 m; b – for sand particles with a diameter of 0.00125 m.

4. Summary

The need to develop a mathematical description of the speed of movement and displacement of particles in a viscous concrete environment, depending on the structural and technological parameters of the vibroform and the characteristics of the mixture, was substantiated. Analytical expressions are obtained that allow describing the movement of spherical filler particles in a viscous environment subjected to a cylindrical punch and determining their average values of the speed of movement υ_{cp} and the displacement r_{cp}. Using the obtained relations, graphical dependences of changes in the studied parameters are constructed for a composite vibroform with punches of radius $R_0 = 0.078$ m, oscillating in its first part with a frequency of $\omega_0 = 35...55$ Hz, an amplitude of $A_0 = 0.0003...0.0004$ m, and in its second part – with $\omega_0 = 55...75$ Hz and $A_0 = 0.0005...0.0007$ m. It is possible to vary the average

IOP Conf. Series: Materials Science and Engineering **945** (2020) 012049 doi:10.1088/1757-899X/945/1/012049

values of the velocity of movement and displacement of particles, taking into account their size, by changing the amplitude-frequency characteristic of the punch vibrations. The regularities of changes in the parameters under consideration for filler particles of various sizes depending on the frequency and amplitude of vibration oscillations of the punch surface are considered.

5. References

[1] Pylnev V G 2004 Improving the efficiency of mixing and compaction of concrete mixture by vibration method *Dissertation of the candidate of technical sciences 05.23.05* 214

[2] Suleimanova L A, Malyukova M V, Slepukhin A S, Krushelnitskaya E A, Tolstoy A D 2019 Influence of a modifying additive with a hydrophobizing effect on improving the performance of vibro-pressed products *Bulletin of BSTU named after V.G. Shukhov* **9(4)** 8-13

[3] Utkin V V, Utkin V L, Utkin L V 2015 Formwork-free molding of reinforced concrete *Publishing house LLC "Papyrus"* 226

[4] Romanov E S, Kapyrin P D 2011 Modern technological lines for the production of floor slabs *Vestnik MGSU* **4** 499-508

[5] Li V A 1986 Production of reinforced concrete products by means of continuous molding *Study guide* 52

[6] Kopsha S P 2013 The form-free molding technology is a key to industrial upgrading and cost reduction *Concrete technologies* **11** 29-33

[7] Buswell R A, Soar R C, Gibb A G F, & Thorpe A 2007 Freeform construction: Mega-scale rapid manufacturing for construction *Automation in Construction* **16(2)** 224-231

[8] Ovchinnikov P F 1983 Vibrorheology *Chemical literature editorial office* 272

[9] Dwight G B 1966 Tables of integrals and other mathematical data *The macmillan company* 228

[10] Kirilov I V, Khanin S I 2019 Machine for continuous vibratory molding of concrete and reinforced concrete products *Pat. 186574 Russian Federation IPC B28B 1/08, B13B 1/02, B28B 15/00* **2018139138** 1-10

Acknowledgments

The work was carried out within the framework of the Development program of the flagship university on the basis of BSTU named after V G Shukhov.

IOP Conf. Series: Materials Science and Engineering **945** (2020) 012050 doi:10.1088/1757-899X/945/1/012050

The advantages of using polymer-bituminous concentrate for the production of polymer- bituminous binders

V P Denisov[1], A E Akimov[1*] and V V Yadykina[1]

[1]Belgorod State Technological University named after V G Shukhov, Kostyukov St., 46, Belgorod, 308012, Russia

E-mail: akimov548@gmail.com

Abstract. Construction of road surfaces with an extended service life, as well as the transition to methods of volumetric and functional design of asphalt concrete mixtures, leads to the need for the use of polymer- bituminous binders, as bitumen without polymer modifiers does not provide the required physical and mechanical characteristics and durability. Traditional PBB production technologies involve the use of complex technological processes and equipment. In addition, existing technological features lead to a significant decrease in the quality of the resulting polymer- bituminous binder. In this paper, we consider the possibility of obtaining PBB with high physical and mechanical parameters by using the PBC modifier (polymer-bituminous concentrate). The modifier under study is a polymer dispersed in an oil base and stabilized with active components. When added to bitumen, a colloid mill and complex processes of preparation and maturation of the binder are not required. Experiments have shown that the resulting binder has the characteristics of PBB, in particular, an extended temperature range of plasticity, elasticity and resistance to cyclic loads. At the same time, gentle heating modes provide increased resistance to aging during the preparation of asphalt concrete mixture and its operation. The ease of use of this modifier makes it possible to obtain PBB with a wide range of characteristics that correspond to the operating conditions.

1. Introduction

New requirements for the operational reliability of roads, as well as the transition to volumetric and functional methods of designing asphalt concrete mixtures leads to the need for expanded use of polymer-bituminous binders in road construction. Polymer asphalt concrete in the covering has a lower temperature dependence of strength and deformation characteristics compared to the use of unmodified asphalt concrete [1]. At high operating temperatures in the summer period, the use of PBB provides greater rigidity and lower creep of the material, as well as stability of the dynamic elastic modulus of the material. At low temperatures during the winter period of operation, the material remains in an elastic state longer, preventing the material from increasing its stiffness and preventing the appearance of temperature and fatigue cracks [2]. In addition, the use of PBB allows obtaining a covering that combines high resistance to the accumulation of plastic deformations with significant resistance to fatigue failure, due to high elasticity and the ability to relax internal stresses [3]. The use of polymer-bituminous binders produced using traditional technology in the production of asphalt concrete mixtures is associated with technological and economic difficulties [4]:

- complex equipment, including: colloid mills, a large capacity park for preparing and maturing the binder, a complex system for pumping bitumen;
- technological process that includes the selection of the composition, dosage and preparation of components, homogenization and subsequent maturation of PBB;
- the need for high-quality PBB components: bitumen, polymer, plasticizer.

These difficulties largely limit the production of PBB by road construction companies. Purchasing finished PBB from specialized manufacturing companies also has a number of negative aspects:

- high cost;
- tendency to delamination of PBB during transportation and storage, which leads to the need for additional preparation of the binder;
- variable quality of the binder and complicated input control.

In addition, long-term production of the binder at high process temperatures leads to increased aging, which significantly reduces the service life of such a material [5].

It is worth considering the use of polymer-bituminous binders in asphalt concrete mixtures developed by the method of volumetric and functional design. In the technical conditions for such mixtures, special requirements for binders are imposed. In particular, depending on the temperature conditions of operation and the frequency and intensity of load application, the binder must have different temperature and strength characteristics, including under dynamic influences. At the moment, manufacturers of PBB are not able to cover the required wide range of binders.

To solve the above problems, BSTU named after V.G. Shukhov together with the Research and Production Association "Selena" developed a polymer modifier for bitumen - polymer-bituminous concentrate (PBC). This modifier is a styrene-butadiene-styrene polymer dispersed in a high concentration in an oil base and stabilized with active additives. Due to the stable structure of the polymer in the additive, its distribution in the bitumen can occur without the use of a colloid mill, using a paddle stirrer. Multi-stage polymer grinding is excluded from technological processes and the maturation time is significantly reduced, which has a positive effect on the quality of the resulting product. In contrast to the traditional PBB formulation, the binder with PBB does not require the use of a plasticizer, which greatly facilitates the procedure for selecting the composition of various PBB brands. In this paper, we study the physical and mechanical characteristics of bitumen modified with the addition of PBC, in comparison with a polymer-bituminous binder of factory production.

2. Materials and methods

BND 60/90 bitumen was used as a binder for further modification. The concentration of the PBC modifier is 8% of the bitumen mass. Homogenization of the additive in bitumen was performed using a paddle stirrer. Uniformity was determined using an optical microscope and visually. As an object of comparison, the factory-produced PBB-60 was selected, corresponding to the actual brand PG 57-16.

The following research methods were used: determination of the dynamic viscosity of bitumen using a rotary viscometer with a coaxial spindle (ASTM D4402/D4402M) [6], determination of the fatigue strength of the binder under the action of a repeatedly applied load (AASHTO T350) [7], determination of high-temperature and low-temperature properties of the binder using a shear rheometer (AASHTO T 315) [8]. Aging of the binder in a thin layer for 5 hours was carried out at a temperature of 165^{0}C with constant air convection. The optical fluorescence microscope was used to assess the uniformity of PBB.

3. Results and discussions

The obtained data indicate the effectiveness of PBC as a polymer modifier for bitumen. The results of measurements of physical-mechanical and rheological characteristics are shown in table 1.

Table 1. Rheological properties of modifying binders.

Properties	Measurement unit	Bitumen + 8% additive	Polymer bitumen binder PBB 60
Dynamic viscosity at 165 ^{0}C	Pa*s	0.082	0.124
Dynamic viscosity at 135 ^{0}C	Pa*s	0.287	0.411
Shear stability	kPa	1.12	1.90
at temperature	^{0}C	58	70
Critical low temperature	^{0}C	-28.77	-19.01
Critical cracking temperature,	^{0}C	-31.16	-37.04
Fraas brittle point	^{0}C	-25	-20
Fatigue resistance	kPa	3832	1503
Elasticity	%	58	64

These results indicate a positive effect of the modifier on the properties of bitumen. Indicators of fatigue resistance and elasticity indicate the formation of a polymer spatial coagulation grid in the bitumen, so that the binder has acquired the ability to relax the internal stresses arising in it. Special attention should be paid to the low-temperature characteristics of the PBB under study. In comparison with a factory-produced binder, bitumen modified with PBC additive works much more effectively in the low temperature region, as evidenced by a decrease in the Fraas brittleness temperature and a significant decrease in the critical low temperature creep index. This is due to the absence of a plasticizer in the binder under study, which is usually made from light oil fractions. The balance of oil components in such a binder is not disturbed, the structure of the binder is stabilized by natural surfactants and introduced modifiers, as a result of which the glass transition processes of the light fraction occur at lower temperatures, allowing the formed structure “bitumen-polymer” to work in a viscoelastic state [9]. At the same time, it should be noted that the high-temperature properties of the resulting binder are not inferior to factory-prepared PBB. According to the technical requirements of GOST 58400.1-2019 binder, obtained with the use of PBC is of PG 58-28 brand (actual temperature limits of operability of the binder 58.7-28.7), tested the factory analogy is of PG 52-16 brand (the actual temperature limits of operability of the binder 57.7-19). Thus, the binder under study has a temperature range of the plastic state of 87.4^{0}C, while this parameter for the factory analog is 76.7^{0}C.

The method of fluorescence microscopy was used to study the homogeneity of the polymer distribution in the binder under study. Microphotographs are shown in figure 1.

Figure 1. Microphotographs of factory-produced PBB (left) and the binder under study (right).

The obtained data indicate a more uniform distribution of the polymer in the structure of the binder under study, compared with PBB produced using a colloid mill.

IOP Conf. Series: Materials Science and Engineering **945** (2020) 012050 doi:10.1088/1757-899X/945/1/012050

The simplified technology of PBB production with the use of polymer-bituminous concentrate reduces the time when the binder is at high temperatures. This technological feature leads to less aging of the binder, and, consequently, will have a positive impact on its durability [10-12].

For the analysis of the aging process, after the aging procedure in a thin layer for 5 hours at a temperature of 165 ^{0}C, the dynamic viscosity was measured for each of the samples at different temperatures in order to determine the degree of change in the structure of the binder by the stability of its rheological characteristics. Curves of changes in the dynamic viscosity of polymer-bituminous binders are shown in figure 2.

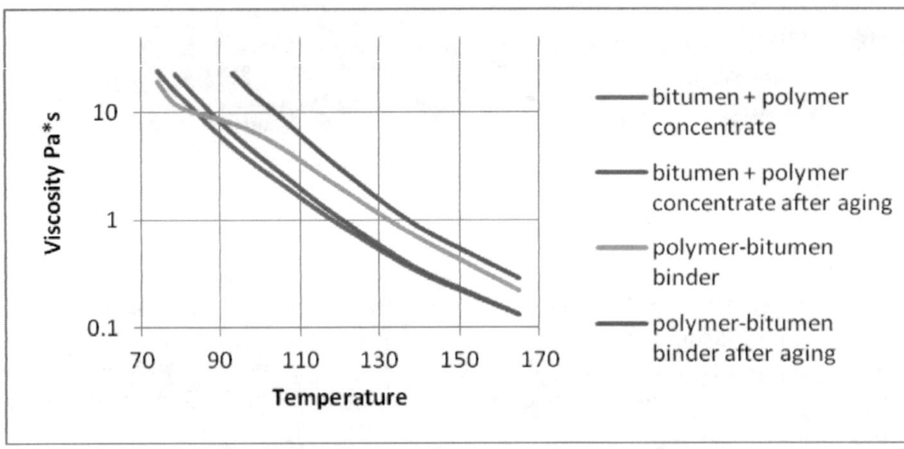

Figure 2. Dynamic viscosity of binder samples after aging.

These data demonstrate that modified PBB bitumen has more stable rheological characteristics compared to factory-produced PBB. These dependencies indirectly characterize smaller structural and qualitative changes in the binder under study under the influence of high temperatures and oxygen. To quantify the degree of aging, a solidification index can be introduced, which is the ratio of the dynamic viscosity before aging to the dynamic viscosity after aging. The results of calculating the solidification index are shown in table 2.

Table 2. Solidification index of modifying binders.

Bitumen + 8% additive	Polymer bitumen binder PBB 60
0.93	0.77

A higher value of the solidification index for the binder under study indicates a smaller change in the dynamic viscosity, and, consequently, a smaller influence of the aging processes of the binder on its structure and qualitative composition.

4. Summary

Research shows that the use of polymer-bituminous concentrate allows getting a high-quality binder without the use of complex equipment and multi-stage technology. The dissolution of the PBC additive in bitumen occurs evenly with mechanical mixing and requires less prolonged maturation. In the production of PBB using the modifier under study, the use of a plasticizer is not required. At the same time, the resulting polymer-bituminous binder is not inferior to industrial-made binder in terms of its main physical and mechanical characteristics, including key parameters: strength and elasticity. Due to the absence of a traditional plasticizer in the formulation, the PBB composition under study has a much larger temperature range of the viscoelastic state compared to its analog. The simplified technology for obtaining PBB based on polymer-bituminous concentrate reduces the duration of exposure to high temperature on the binder, which made it possible to learn a more aging-resistant composition.

5. References

[1] Polacco G, Filippi, S, Merusi F, & Stastna G 2015 A review of the fundamentals of polymer-modified asphalts: Asphalt/polymer interactions and principles of compatibility *Advances in colloid and interface science* **224** 72–112

[2] Odenbah I 2020 Some new nechnologies in road construction *Construction Materials and Products* **3(1)** 62–69

[3] Yue Y, Abdelsalam M, Luo D, Khater A, Musanyufu J, Chen T 2019 Evaluation of the Properties of Asphalt Mixes Modified with Diatomite and Lignin Fiber: A Review *Materials* **12(3)** 400

[4] Ghasemirad A, Bala N, Hashemian L 2020 High-Temperature Performance Evaluation of Asphaltenes-Modified Asphalt Binders *Molecules* **25(15)** E3326

[5] Zhang F, Yu J, Wu S 2010 Effect of ageing on rheological properties of storage-stable SBS/sulfur-modified asphalts *Journal of hazardous materials* **182(1-3)** 507–517

[6] ASTM International 2015 ASTM D4402/D4402M - 15: Standard Test Method for Viscosity Determination of Asphalt at Elevated Temperatures Using a Rotational Viscometer. In *Annual book of ASTM standards* West Conshohocken, PA: American Society for Testing and Materials

[7] AASHTO 2019 AASHTO T 350: Standard Method of Test for Multiple Stress Creep Recovery (MSCR) Test of Asphalt Binder Using a Dynamic Shear Rheometer (DSR). *AASHTO provisional standards.* Washington, D.C: American Association of State Highway and Transportation Officials

[8] AASHTO (2019) AASHTO T 315: Srandard Method of Test for Detetrmining the Reological Properties of Asphalt Binder Using a Dynamic Shear Rheometer. *AASHTO provisional standards.* Washington, D.C: American Association of State Highway and Transportation Officials

[9] Wang P, Shi F Q, Liu X Y, et al. 2019 Role of Aliphatic Chain Characteristics on the Anti-Cracking Properties of Polymer-Modified Asphalt at Low Temperatures *Polymers* **11(12)** 2025

[10] Cuadri A A, Delgado-Sánchez C, Navarro F J, Partal P 2020 Short- and Long-Term Epoxy Modification of Bitumen: Modification Kinetics, Rheological Properties, and Microstructure *Polymers* **12(3)** 508

[11] Ren S, Liu X, Fan W, Wang H, Erkens S 2019 Rheological Properties, Compatibility, and Storage Stability of SBS Latex-Modified Asphalt *Materials* **12(22)** 3683

[12] Yadykina V V, Gridchin A M, Trautvain A I, Verbkin V I 2015 Study of the influence of various polymers and plasticizers on the properties of BND 60/90 bitumen and asphalt concrete based on it *Bulletin of BSTU named after V G Shukhov* **6** 40-45

Acknowledgments

This work was realized in the framework of the Program of flagship university development on the base of the Belgorod State Technological University named after V G Shukhov, using equipment of High Technology Center at BSTU named after V G Shukhov.

IOP Conf. Series: Materials Science and Engineering **945** (2020) 012051 doi:10.1088/1757-899X/945/1/012051

The acceleration of the peat secondary consolidation due to the sorption of bound water with disperse clay

G Yu Ivakhnova[1*] and A L Nevzorov[1]

[1]Department of Engineering Geology, Bases and Foundation, Northern (Arctic) Federal University named after M.V. Lomonosov, Severnaya Dvina Emb. 17, Arkhangelsk, 163002, Russia

E-mail: g.ivakhnova@gmail.com

Abstract. According to observations at the experimental site, the primary consolidation of the 6-6.5 m peat layer completed in 3-5 years after the sandy embankment was filled on. Secondary consolidation of peat caused by the removal of bound water from micropores and channels within vegetation remains is the cause of settlement that has not stabilized for decades. The aim of laboratory research was to evaluate the effects of two methods for peat settlement stabilization at the stage of secondary consolidation – vacuuming and sorption of bound water of disperse clay. Tests were carried out with peat, which has a degree of decomposition of 40-45 %, a density of 0.98–1.12 g/cm^3, a void ratio of 11.8–13.8. Samples with a diameter of 87 mm and a height of 30, 50 and 70 mm were set into odometers and maintained under a load of 25, 50 and 100 kPa for 5-6 months, after that experiments to stabilize secondary consolidation were started. In contrast to vacuuming, the use of pre-dried clay with the following properties turned out to be very effective: maximum hygroscopic water content 43%, liquid limit 198%, plastic limit 58%, specific surface area 411 m^2/g. The volume of clay was about 20% of the volume of peat samples. Deformations of samples due to water sorption by clay increased within 2-4 days and reached 0.09-0.11 of their initial height. The penetration of clay into the peat can be performed by drilling of boreholes, vibration stamping and other methods.

1. Introduction

During the warm interglacial period that began 10-12 thousand years ago, the vast plains of North-West Russia and Western Siberia became swampy. The presence of few meters of peat deposits significantly complicates construction. If during the construction of buildings and laying of communication peat is pierced with pile foundations, and at arranging the transport routes it is replaced by mineral soil, then at the rest territory of the cities and enterprises carrying out continuous excavation of this organic soil is not possible. Here they use the method of peat loading, dumping a layer of sand several meters thick on the surface of the swamp. The example is Arkhangelsk, that was built on swampy peatland shows that the peat settlement under a sand embankment does not stabilized for several decades. In areas of the city where construction was carried out in the 70s of the last century, it is currently developing at a rate of 5-15 mm/year, which causes damage to roads, sidewalks etc. (Fig.1). Over the past two decades, the problem of stabilizing peat settlement has become even more urgent, as the city is undergoing intensive compaction of existing districts, where it is very difficult to complete the excavation of peat and replace it with mineral soil.

a) b)

Figure 1. Manifestations of long-term peat settlement at the base of structures: a) destruction of the blind area of the building, b) setting of the sidewalk around the well manhole.

A peat has specific two-level structure - closed micropores and channels are contained inside large plant residues. Primary consolidation is caused by the removal of free water from macropores between large particles, and secondary - bound moisture from micropores and channels [1;2;3;4;5]. Our observations at the experimental site showed that the change in the slope of the graphs shown in Fig.2, indicating the transition from primary to secondary consolidation, takes place 3...5 years after the filling of the embankment [6]. Therefore, long-term settlement of the city territory is caused by secondary consolidation of peat.

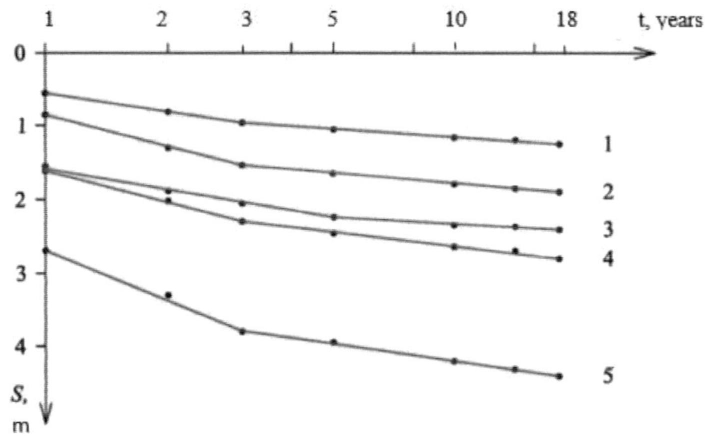

Figure 2. Surface setting at the test site with the initial thickness of the peat layer of 6-6.5 m at the thickness of the sand embankment: 1 – 0.9 m; 2 – 1.7 m; 3 – 2.2 m; 4 – 2.4 m; 5 – 3.7 m [6].

A very common way to accelerate the consolidation of peat is to create an additional load on it by dumping an additional temporary embankment. But for secondary consolidation, its effectiveness is low. As it can be seen from Fig. 2, the coefficient of secondary consolidation, which is equal to the tangent of the angle to the x-axis of the graphs plots for the stages of secondary consolidation, is increased only by 2 times from 0.025 to 0.050 - in the height of the embankment by four times, from 0.9 to 3.7 m. In addition, a temporary embankment delays the introduction of new objects into operation. The next way to stabilize the settlement is the usage of vertical drains, which have been supplemented by vacuuming. Vacuuming can be combined not only with vertical drains, but also with a temporary additional load [7;8]. It is clear that vertical drains that shorten the path of pore water filtration affect the course of primary consolidation, but vacuuming, apparently, can also contribute to the removal of bound moisture from micropores and channels.

Numerous studies on a peat stabilization with binders – cement, lime, gypsum – have shown that the required application of binder is very significant - from 100 to 400 kg per 1 m^3 of peat [1;9].

IOP Conf. Series: Materials Science and Engineering **945** (2020) 012051 doi:10.1088/1757-899X/945/1/012051

Solutions of binders penetrate the macropores, pushing the particles apart and compacting the peat, and create a spatial structure. On the contrary, the introduction of dry mixtures can contribute to the consolidation of peat due to the sorption of pore moisture. However, the same result at the stage of secondary consolidation is likely to be achieved when using inert highly dispersed materials, such as clay powders.

The aim of this research was to evaluate the effects of two methods for stabilization peat settlement at the secondary consolidation stage. Peat samples that had been exposed to constant compressive stresses in compression devices for a long time were vacuumed and treated with montmorillonite clay powder.

2. Materials and methods

We tested undisturbed samples taken at the neighborhood of Arkhangelsk on an undeveloped, undrained swamp from pits from a depth of 0.5-0.8 m. Initial properties of peat: decomposition degree 40-45 %, density 0.98-1.12 g/cm^3, water content 7.70–10.60, void ratio 11.8–13.8. Here and further, water content is given as the ratio of the mass of water to the mass of the solid phase. The structure of the peat is shown in a photo taken with a scanning electron microscope *Vegan Tescan 3* (Fig. 3).

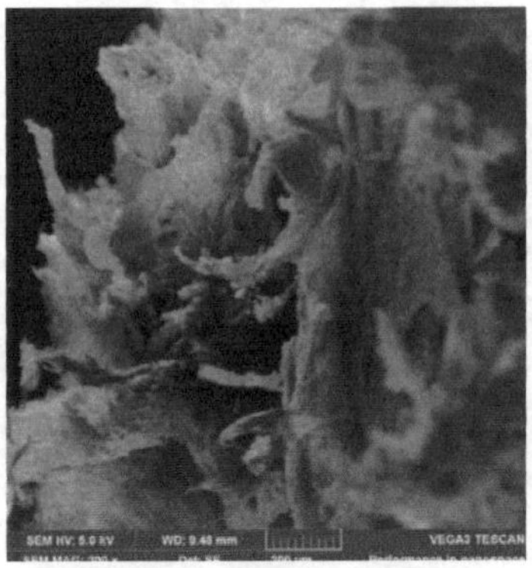

Figure 3. Structure of the studied peat.

The experiments were carried out in oedometers with a diameter of samples 87 mm, a height of 30, 50 and 70 mm. The load on the samples p was 25, 50 and 100 kPa and applied at once in one-step, as required by standards for determining consolidation coefficients. The devices were placed in a room with a temperature of 5-12°C, which slowed down the decomposition of organic matter of peat. The load was maintained for 156-195 days until the rate of deformation development decreased to 0.005 mm/day. After that, they started experiments to stabilization.

3. Research results

A characteristic graph of the dependence of deformations ε of the peat sample on time t is presented in Fig. 4a. As we can see, the primary consolidation of peat was completed 8 hours after the start of the experiment. When testing other samples, the duration of this stage varied from 8 to 22 hours.

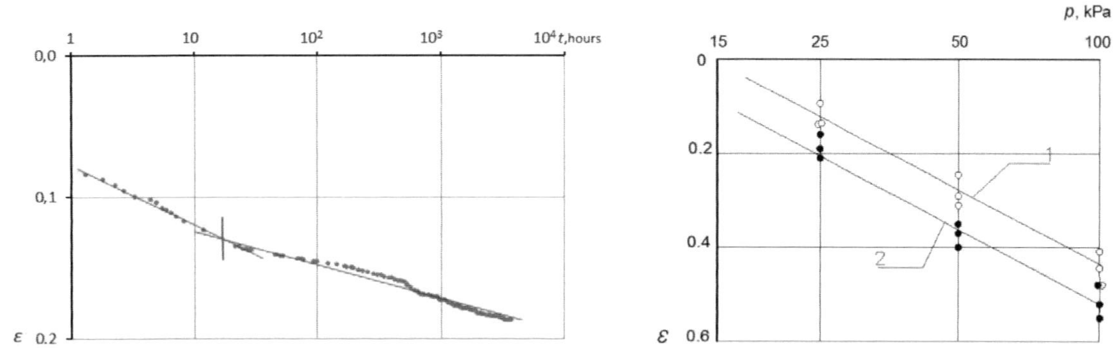

a) b)

Figure 4. The results of peat compression testing: a) development of deformations of one of the samples in time; b) the dependence of deformations of the load: 1 – at the end of primary consolidation, 2 – at the end of the test.

Figure 4b shows the compression dependencies $\varepsilon = f(p)$ for the stage of completion of primary consolidation and at the time of completion of compression tests, where a logarithmic scale is traditionally used on the abscissas axis. As we can see, the deformations of peat vary from 0.1 to 0.5 depending on the load.

Six samples were subjected to vacuum after standard compression tests were completed. To do this, the lower chambers of the devices were connected to the receiver of the vacuum pump (Fig. 5a). The pressure below atmospheric pressure by 45-50 kPa was maintained for 4 days. As it can be seen from the dependency shown in Fig. 5b with solid line, the effect of the vacuum, especially at a load of 50 and 100 kPa, was very small. The deformations of peat samples have not changed much either.

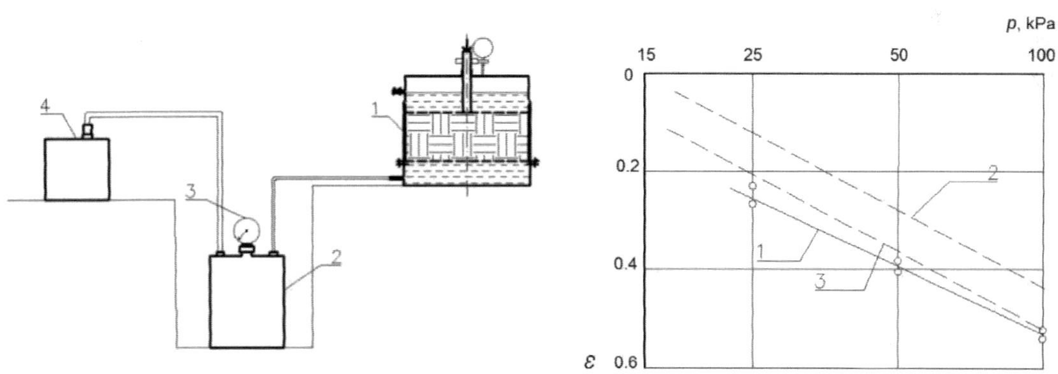

a) b)

Figure 5. Vacuuming of peat samples: a) scheme of the experiment; 1- odometer, 2 – receiver, 3 – pressure gauge, 4 – vacuum pump; b) the dependence of deformations of the load (1). The dotted line shows deformations during primary consolidation (2) and the completion of the compression test (3).

The other six samples were brought into contact with a powder of montmorillonite clay. Properties of clay: hygroscopic water content 6%, maximum hygroscopic water content 43%, liquid limit 198%, plastic limit 58%, specific surface area 411 m^2/g. Pre-dried clay was poured on top of the samples - into specially opened holes with a diameter of 14 mm in the pistons of oedometers. The volume of clay was about 20% of the volume of samples. Deformations of a peat due to water sorption by clay increased within 2-4 days and reached 0.09-0.11 from the initial height of the samples or 0.12-0.16 from the height of the samples after completion of compression (Fig. 6). The clay water content at the end of the experiments reached 0.58-0.60, the peat water content decreased on average from 5.81 to 4.80.

4. Discussion

The reason for the small effect when vacuuming samples is, apparently, the almost complete absence of macropores in the peat at the stage of secondary consolidation (especially at loads of 50 and 100 kPa). Bound moisture prevails in micropores, and to remove it, it is necessary to create a significantly greater underpressure than was the case in our experiments. We note that the execution of such works in production conditions is quite complex and expensive procedure.

On the contrary, the rapid increase in deformations was the case when exposed to disperse clay. When the thickness of the peat layer is 2-3 m under the sand embankment, the penetration of clay, for example, into the boreholes will provide additional peat draft by 20-30 cm, thereby eliminating the long-term development of deformations and damage to structures shown in Fig. 1.

The proposed method of a peat stabilization is very prospective, especially taking into account the presence in Arkhangelsk of large volumes of highly dispersed waste from one of the enterprises of the diamond industry – saponite clays.

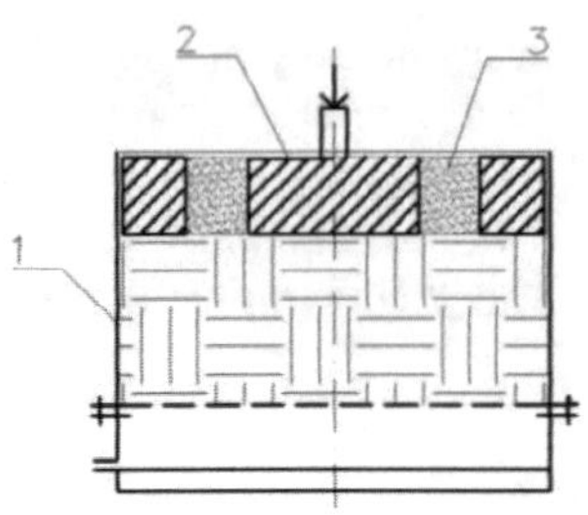

a)

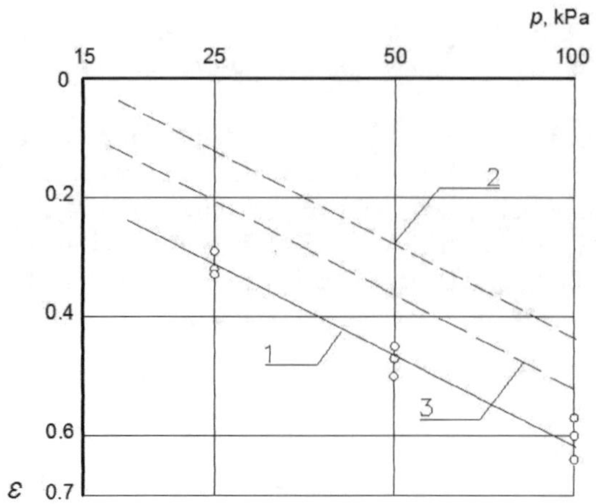

b)

Figure 6. Deformation of peat samples due to water sorption by clay: a) scheme of the experiment: 1-odometer, 2 – piston, 3 - holes filled with clay powder; b) the dependence of deformations of the load (1). The dotted line shows deformations during primary consolidation (2) and completion of compression tests (3).

5. Summary

1. Deformations of peat caused by its secondary consolidation have been developing for decades, and specific methods are required to settlement stabilizing, which differ from the methods of acceleration at the stage of primary consolidation.

2. Treatment of peat with clay powder contributes to its stabilization at the stage of secondary consolidation due to sorption of water contained in micropores.

6. References

[1] Amaryan L.S. 1990 Properties of weak soils and methods of their study 220

[2] Gofar N., SutejoY. 2007 Long-term compression behavior of fibrous peat. *Malaysian J. of Civil Engineering* **19** 104-116.

[3] Kogure K., Yamaguchi H., Shogaki T. 1993 Physical and pore properties of fibrous peat deposit. *Proceedings of the 11th Southeast Asian Geotechnical Conference, Singapore*, 135-139.

IOP Conf. Series: Materials Science and Engineering **945** (2020) 012051 doi:10.1088/1757-899X/945/1/012051

[4] Wong L.S., Hashim R., Ali F.H. 2009 A review on hydraulic conductivity and compressibility of peat. *Journal of Applied Sciences* **9 (18)** 3207-3218.

[5] Zhang L., O'Kelly, B.C. 2015 Effect of salt grain additions on fibrous peat consolidation. *Proceeding of the institution of civil engineers* **168** 14-21.

[6] Tyurin D.A., Nevzorov A.L. 2017 Numerical simulation of long-term peat settlement under the sand embankment. 1st International conference on the material point method. *Procedia engineering* **175** 51-56

[7] Zhong R., Indraratna B., Rujikiatkamjorn C., Kan M., Vincent P., Ryan G. 2017. A model for vacuum-assisted soft soil consolidation with PVDs. *Proceedings of the 19th International Conference on Soil Mechanics and Geotechnical Engineering, Seoul* 2685-2688

[8] Chai J.C., Carter J.P. 2013. Consolidation theory for combined vacuum pressure and surchange loading. *Proceedings of the 18th International Conference on Soil Mechanics and Geotechnical Engineering, Paris* 2449-2452

[9] Kazemian S., Huat B.B.K.,Moayedi H.2012. Undrained shear characteristics of tropical peat reinforced with cement stabilized soil column.*Geotechnical and Geological Engineering* **30** 753-759

IOP Conf. Series: Materials Science and Engineering **945** (2020) 012052 doi:10.1088/1757-899X/945/1/012052

Modeling the external business environments for the Vietnamese construction SMEs in a transition economy

The-Quan Nguyen[1*] and Christopher N Preece[2]

[1]National University of Civil Engineering, 55 Giai Phong Road, Hai Ba Trung District, Hanoi, Vietnam

[2]College of Engineering, Abu Dhabi University, United Arab Emirates

E-mail: quannt@nuce.edu.vn

Abstract. Vietnam has been shifting from a centrally planned economy with subsidy schemes, to a socialist-oriented market economy with multiple economic components. The internal dynamics and external pressures have caused many changes in the environment for small and medium size construction enterprises. This research models the external environment in order to aid understanding of the strategic management of such contractors in the construction industry of Vietnam. Semi-structured interviews of senior managers were combined with a mailed questionnaire to collect data. Seven major concepts were found to emerge including the economic administration of the country, the legacy of the old system, changes in society, nature of the workforce and the variety of business styles found. The research results show that the behaviour of key stakeholders has a major impact on the strategic/competitive behaviour of the concerned firms and their profitability.

1. Introduction

Since the adoption of the "renovation" or “reform” policy in 1986, Vietnam has been shifting from a centrally planned economy with subsidy schemes, to a socialist-oriented market economy with multiple economic components. Concurrently, the country has also embarked on a process of globalisation and integration into local and global economies with the significant milestones of joining the Association of South-East Asian Nations (ASEAN) in 1995, obtaining the World Trade Organization (WTO) membership on 11th January 2007, and becoming a member of the ASEAN Economic Community (AEC) when it established in 2015. Both the internal dynamics and the external pressures have caused the country many changes, especially in the business environment for firms, including contractors in the construction industry in general, and Vietnamese Construction Small and Medium-Sizes Enterperises (SMEs) in particular [9, 10].

Accounting for over 90% of the number of construction firms in Vietnam in recent years, the SMEs play a very important role in the industry. While the industry is carrying out its own transition, the business environment for them has changed significantly. State-owned firms cannot wait for a governmental authority to distribute the jobs to them anymore, but instead have to compete for construction contracts, among each other and the emerging private contractors. In addition, new sectors such as private and foreign direct investment have been accepted to join the market. As a result, the business environment for Vietnamese Construction SMEs becomes more and more competitive.

In an attempt to give a solution for the problems above, this research was designed to model the external environments which reflects the reality that the Vietnamese Construction SMEs have operated successfully in the new business environment created by the country's economic transition in order to aid understanding about the strategic management practice in this type of firm.

2. Materials and methods

Since no similar model has been found in the literature, both in Vietnam and worldwide at the time of the research, no existing model for the construction SMEs' environment can be used and tested using a hypothesis testing approach for this research. For the research convenience, there is a need of a knowledge base which can provide guidelines for the research questions, help with the data analysis and sensitise the emergent concepts. Therefore, a modified grounded theory approach was considered the best choice for this research.

Due to the unavailability of the necessary data on Vietnamese Construction SMEs - no sampling frame listing was available - snowball sampling was selected as the sampling method. In order to reduce the disadvantage of using a snowball sampling, a triangulation method was selected for data collection, using multiple data sources, which include two major sources: interviews, questionnaires and a secondary source of available internship-reports developed by final year students from National University of Civil Engineering, Hanoi, Vietnam (NUCE) (Figure 1).

Semi-structured interviews was selected for the interview survey; 77 top managers from Vietnamese Construction SMEs were contacted and 41 interviews were completed. For the survey with mail questionnaires, 500 questionnaires were sent and 99 completed, (a response rate of 20 percent). However, 50 questionnaires were ineligible and a further 23 of the collected questionnaires could not be used because the information was poor. Only 26 questionnaires could be used for data analysis. Respondents who answered the questionnaires were from Hanoi (20), Ho Chi Minh City (5) and Da Nang (1). Information from these mail questionnaires is not as rich as the interviews. That is why they were only used as a second source of information for the theory generation process. Secondary source information was also utilised from internship reports from final year students in NUCE. Additional 24 reports were selected as secondary data because they were from the firms that participated in either the interviewing survey (15) or the mail survey (9).

The interviews were transcribed to allow data analysis. Firms were coded for the ease of monitoring the data: interviewed firms were coded as Qi, with i=1-41 while firms in the questionnaire survey were coded as Ij, with j=1-26. Then data from each source was broken into codes using a line-by-line analysis. The main questions and some additional topics such as firm's code, date of the interview or questionnaire and recommender's names and contacts, etc. were listed in the initial columns in order to categorise the emerging codes more easily.

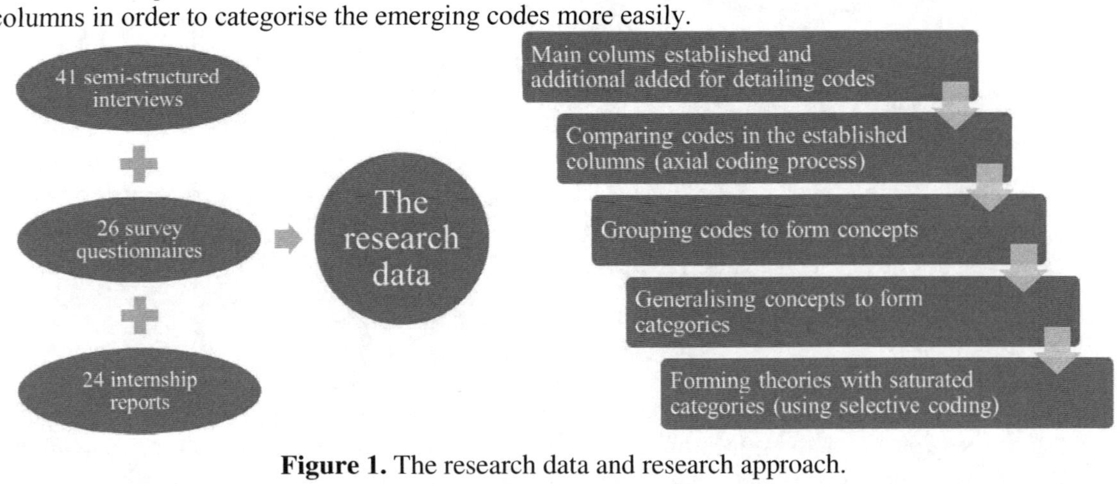

Figure 1. The research data and research approach.

While analysing the data, the emerging codes were put into the appropriate columns temporarily. Additional columns then were added when there was a need to break down an initial column into more detailed ones. For example, column "customers" was divided into "recognition of customer types", "returning customers", "role in contracts", "reasons for choosing the roles", "tendering methods". Later, another column was added to this group: "customer's criteria to choose a contractor". Using axial coding, codes in one column were then compared with another and also with other codes from other columns. Based on their common features, the codes then were grouped to form concepts, e.g. codes "asynchronous and incomplete legal system", "conflicts between regulations issued by different ministries", "discrimination between firms in the state-owned and private sectors", "discrimination between domestic investors and foreigners", etc. were grouped together to form a concept "economic administration mechanism". Concepts then were generalised to form categories. In the previous example, "economic administration mechanism" was generalised as "an influential factor from the general environment". Selective coding then was carried out to select the saturated categories in order to build up theories (Figure 1). Then, since there were several types of firms participating in the survey, categories were put back into several tables using different grouping methods to check the trend of the data in different types of firms. The validation was considered as not being important in this research because in grounded theory, no separate verification or validation is needed [1].

3. Emerging concepts

As a result of the data analysis process, there are 7 major concepts emerging from the data that underpin this category, as follow (Figure 2):

- the economic administration mechanism of the country;
- the economic integration of Vietnam into regional and global economies;
- the legacies of the old economic system;
- the variety of business styles in different regions of the country;
- the changes in society;
- the workforce of the country; and
- the behaviours of the organisations that are concerned with the targeting market segments of a firm.

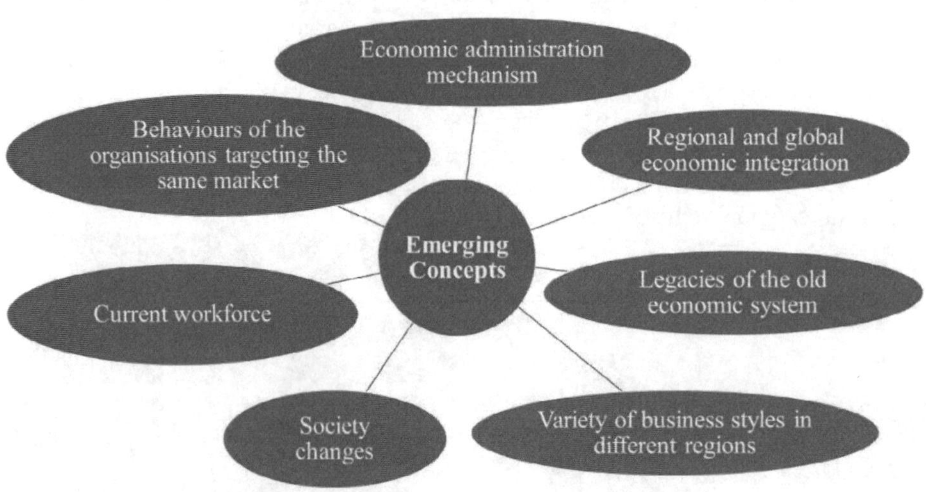

Figure 2. The emerging concepts from the data collected.

Besides the saturated phenomena that have formed the above conceptual codes, there are several individual factors that were mentioned by the respondents, but they either did not reoccur in the data or could not be grouped due to a lack of similar or relevant concepts. These factors, being regarded hereafter as "other/specialist factors", and will be discussed where relevant later.

IOP Conf. Series: Materials Science and Engineering **945** (2020) 012052 doi:10.1088/1757-899X/945/1/012052

4. Modelling the general environment

4.1. The model development

Looking at the concepts that have emerged from the data, one can easily recognise that the factors underpinning these concepts can be listed under appropriate categories of PEST analysis (see [2]) or one of its variants. However, the use of a PEST analysis or one of its variations in the case of the Vietnamese Construction SMEs may not be convenient, due to the following: (i) many factors in a standard PEST model and its variants do not appear as emerged concepts from the survey results; (ii) the use of a PEST framework or one of its variants to analyse the general environment of the firms is time-consuming as the user needs to go through the long lists under each heading of the framework being used, then evaluate the influence of each factor in the lists; (iii) there are some factors emerging from the data that cannot be put under a single heading in the PEST framework, for instance, the "mechanism" factor needs to be analysed under both PEST's economic and political headings.

The general environment of the Vietnamese Construction SMEs with the most important influential factors is illustrated in Figure 3, as a hexagon under the analogous form of a one-storey building. Six important influential factors are put at six vertices of the polygon, and the group of specialist factors occupies the roof. The firm (represented by letter C) is at the centre of the figure. The shape of the model implies that the general environment is like a solid, closed building which keeps the firm inside and constrains its business. But it has a keyhole, which leaves the firm with a chance to escape from the constraining environment (the building) to develop.

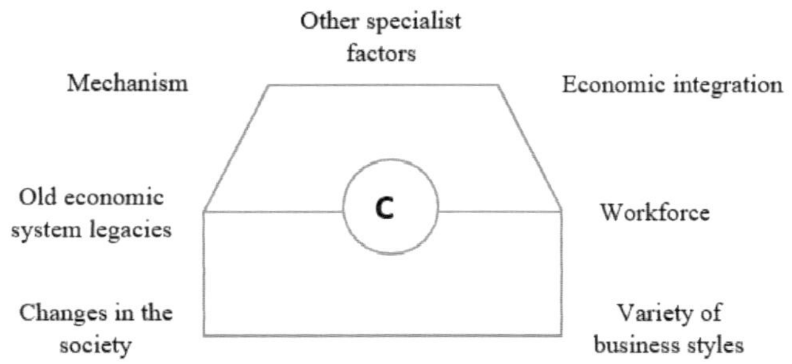

Figure 3. A Vietnamese Construction SME and the influential factors from the general environment – The House of the General Environment Model.

4.2. The significant influential factors from the general environment

4.2.1. The Economic Administration Mechanism

In research into SMEs in Vietnam, Hansen et al. [3] stated that the Government has a very important role in the development and survival of enterprises. A similar result has been discovered in this research. Talking about the external environment of their firms, informants kept mentioning the Government's "mechanism" (as a short form of the "economic administration mechanism") in the economy as a strong influential factor on their businesses. With 49 respondents showing interest in this factor, it can be regarded as the most important factor within the participating firms, regardless of type and sector. After the economic reform of the country began, a new economic mechanism was developed to replace the central-command and subsidized economic mechanism. Though the new mechanism has created a lot of favourable conditions for firms and has been improved a lot since the economic reform started, when the informants mentioned the word "mechanism", they often referred to its negative impact, considered it as a straight jacket constraining their business and that firms wanted to take it off. The situation however has been improving. WTO membership has required that the country amends its legislative and legal system to suit their legitimate requirements. Firms are expecting a clearer and more effective legal system and legislative system, which they believe could help to make a better administration mechanism in the near future.

4.2.2. Economic Integration into Regional and Global Economies
Respondents have also noted several factors which can be considered to be as a results of the economic integration of Vietnam into local, regional and global economies. They claim that the integration of the Vietnamese economy into the outside world (by joining AFTA, ASEAN and obtaining its WTO membership, etc.) has caused both positive and negative impacts on their business environment, and on their firms' businesses. Besides the advantage of an increase in foreign investment, the chance for learning and using modern technologies from the outside world, the construction markets have turned out to be more competitive because of the entrance of foreign contractors or foreign investment firms. There are overseas contractors that have signed partnership agreements with domestic firms and formed international joint-venture companies, which are often stronger competitors than the others. The country's economy is now connected closely with the global economy; therefore, it has become sensitive to the behaviour of the world's market.

4.2.3. The Legacies from the Old Economic System
In its transition, the economy in Vietnam is a mixed economy which is still on its way through a transition from a centrally-planned economy to a particular type of market economy, a socialist-oriented market economy. It is therefore understandable that there are still legacies from the old economic system and they have certain impacts on firms.

One significant legacy from the former economic system is the psychology of highly valued personal relationships over other issues. In the centrally-planned economy, the whole economy was run by orders from the top management bodies of the country, even the distribution of work to firms was done by the governmental authorities. Therefore, making relationships with key people in those bodies was very important to firms, even now, as forty three informants stated. This fact has created a custom of using personal relationships to deal with any problem in Vietnamese society. This custom is backed by the tradition of valuing the hierarchical social roles or the importance of hierarchy in society, which the Vietnamese adopted from Confucianism during the Chinese occupation in its history (around 1000 years in the first millennium) [4]. In the early stages of the market economy in Vietnam, those personal relationships tended to be corrupt. Those personal relationships are now considered more positive, and still very important.

4.2.4. The Variety of Business Styles in Different Regions of the Country
According to the survey, the respondents believed that the differences in regional markets within the country, especially between the Northern and the Southern markets due to cultural and historic issues have led to the differences in business styles.

4.2.5. Changes in Society
The next factor that the survey participants consider important in the external environment is the change in the Vietnamese people with regard to their awareness of the outside world and living styles. People now know more about their rights and are ready to fight for them, ask for better living conditions, and pay more attention to the environment. People now have higher requirements for living standards, including accommodation. These factors have contributed in raising the need for new types of construction products. The migration from rural areas to big cities is also a significant factor pushing up the development of many industries and services, which indirectly or directly creates more jobs for the construction industry. However, Vietnamese industries have been required to pay more attention to the environment and to avoid activities that may badly influence the natural environment and living conditions of society.

IOP Conf. Series: Materials Science and Engineering **945** (2020) 012052 doi:10.1088/1757-899X/945/1/012052

4.2.6. The Workforce

Respondents strongly expressed their concern about the existing workforce in the economy. Some of them claimed that the country was in an alarming situation of "having an excess of graduates but lack of skilled workers". [5]. Other respondents argued that there was an asynchrony in education and training. It was reported that the ratio of training among bachelors, vocational degree holders and skilled workers in Vietnam was 1: 1.16: 0.95, while in other countries this ratio was 1: 4: 10. Many respondents showed their disappointment regarding the quality of the newly-graduated applicants, especially new engineers (N=21). They claimed that new engineers did not have the relevant practical skills to start work immediately; therefore, firms had to provide them with retraining. Firms should invest in the human resources of the country instead of just waiting to get a benefit from it.

The agriculture-based economy of the country has been mentioned as one issue impacting on the workforce. Thirteen respondents reported that the majority of construction works (in quantity) in their firms have been carried out by temporary workers. Most of these workers have left the rice fields to migrate to the cities wishing to have a better life or only wanting a temporary job while waiting for the harvest. Obviously, firms cannot expect high quality work from this workforce.

4.2.7. Other/Specialist Factors

Besides the saturated phenomena that have formed the above data groups, there are several influential factors impacting on the general environment that respondents have mentioned individually, but do not have the same importance for every participating firm. Figure 3 shows these factors. That is to say, there are many factors from the general environment that may have an impact on the Vietnamese Construction SMEs, yet respondents did not mention them. Those factors may not be considered very important to the survey participants, as they did not recall them during the interview or when they completed the mailed questionnaire.

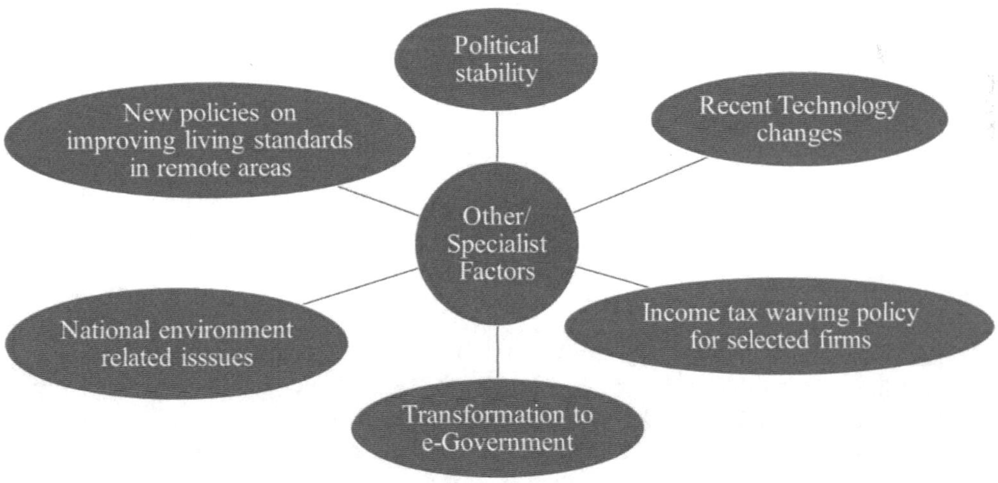

Figure 4. Phenomena forming the other/specialist factors.

IOP Conf. Series: Materials Science and Engineering **945** (2020) 012052 doi:10.1088/1757-899X/945/1/012052

5. Modelling the Competitive Environment

The data shows that the Vietnamese Construction SMEs share a set of seven types of stakeholders, which include direct customers, indirect customers, competitors, subcontractors, material suppliers, commercial banks and financial institutions, and specialist contractors. Besides, each group of firms also has a second set of important and common stakeholders, as in Table 1. There are also some more stakeholders that can influence some firms only that are omitted from Table 1.

Table 1. Major influential stakeholders in the competitive environment for different groups of Vietnamese Construction SMEs.

Stakeholder types	Firms operating in state-managed-capital market	Firms operating in foreign-invested-capital market	SOEs
1. The common stakeholders			
Direct customers	v	v	v
Indirect customers	v	v	v
Competitors	v	v	v
Subcontractors	v	v	v
Material suppliers	v	v	v
Commercial banks and financial institutions	v	v	v
Specialist contractors	v	v	v
2. The contingent stakeholders			
Governmental authorities	v		
Customers' consulting bodies	v		
Clients' societies		v	
Management body			v

In Table 1, the first seven types of stakeholder form the common set, which is the set that all of the Vietnamese Construction SMEs need to pay attention to. Each of the last four types is important to a certain group of Vietnamese Construction SMEs only – the 'contingent stakeholders'. It is noted that the firm categories in Table 1 are not mutually exclusive. State-owned firms (SOEs) can target the state-managed market as well as foreign-invested market and private customers and need to consider the influence of the stakeholders in the types listed under relevant columns.

Therefore, it is worth considering power and behaviour of these stakeholders when looking at the competitive environment of an SME. These issues can be used to describe and explain the competitive environment of the firm in order to reflect the complexity of the Vietnamese competitive environment for the Vietnamese Construction SMEs. If we use an "S" to indicate a stakeholder, the competitive environment of an Vietnamese Construction SMEs can be illustrated by a cloud of "S"s surrounding the firm (which is represented by letter C), as in Figure 4. This cloud represents the markets that the concerned firm operates in.

In this model, there are seven bold "S"s floating around the firm which represent the important stakeholders to all the firms (the common set in Table 1). Any SME doing business in the Vietnamese construction industry needs to pay attention to these parties' behaviours, though each stakeholder impacts in different degrees on different firms. The remaining normal "S"s represent other relevant stakeholders, belonging to the contingent set, whose behaviour may not affect all the firms in the market, but one or more individual groups of firms. Since each firm may be concerned about some particular stakeholders, the number of normal "S" in the model is only for reference. When looking at these stakeholders, an SME in construction should analyse the mutual influence with them as well as investigate the chance to cooperate with them.

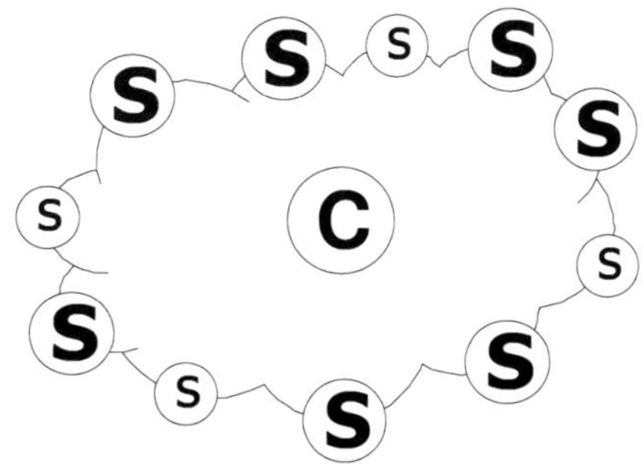

Figure 5. A Vietnamese construction SME and its stakeholders - the cloud of Ss model.

This model is similar when compared to the Five Forces Model (see [6]) in that it also tries to evaluate the impact of the forces in the Industry to the strategic behaviour, including the competition of Vietnamese Construction SMEs, and to their profitability. The major forces in the industry presented in the Five Forces model are included in the proposed model, except for the substitute providers. The "Cloud of Ss" model has proposed more groups of forces in order to emphasize their importance and different impacts on the Vietnamese Construction SMEs. Though they can be regarded as "customers", actually the indirect customers and direct customers can have opposite impacts on Vietnamese Construction SMEs' competition and behaviour.

One can classify subcontractors, material suppliers and commercial banks and financial institutions as "suppliers" in the Five Forces Model, but the proposed model separates them out into different groups to emphasize the difference in the impact they may have on the firms as well as to illustrate the complexity of the production chain in construction. Inheriting the idea from the resource-based framework, the proposed model is concerned more about competitors from both the current industry and competence-related industries. Like Freeman's model [7] and the modified model of Five Forces with "complementors", the proposed model looks for opportunities of cooperation of Vietnamese Construction SMEs with other forces in the market.

This model also emphasizes the importance of "specialist suppliers", who can either be complementors, subcontractors or forces that influence the substitute-selection of customers in the construction market. It also differentiates between forces for different groups of Vietnamese Construction SMEs. By looking at the market instead of the industry, it demonstrates the project-orientation of construction firms in their strategic management. In comparison to the topics suggested by Warszawski [8], it is noticeable that the proposed model focuses more on the competitive environment while Warszawski's model includes the general environments within the topics. Nevertheless, most of the topics that Warszawski mentioned which are related to the competitive environment of construction firms are included in the proposed model.

That is to say, the proposed model provides a fuller and more detailed view about the forces that make up the competitive environment of the Vietnamese Construction SMEs.

6. Summary

This paper has detailed the development of a model of the external business environment of small/medium construction enterprises in Vietnam. The results show that the behaviour of key stakeholders in the economy and industry in Vietnam has a major impact on the strategic/competitive behaviour of the concerned firms and their profitability. The influential factors identified are the economic administration mechanism of the country, the economic integration of Vietnam into regional and global economies, the legacies of the old economic system, the variety of business styles in different regions of the country, the changes in society, the workforce of the country, and the behaviours of the bodies/organisations that are concerned with the market segments that a firm targets.

Those bodies/organisations are classified into two groups: seven types of common stakeholders who have significant impacts on all of the concerned firms and the contingent stakeholders which include four groups of bodies.

The findings have confirmed and extended the western models of external business environments such as PEST and 5 forces and their variants, in the context of Vietnamese Construction SMEs. However, when applying to specific firms or firm types, more thorough empirical research may need to be carried out to reconfirm the significance of each of the external factors.

7. References

[1] Dey I 2012 *Grounding grounded theory: Guidelines for qualitative inquiry* (Crane Library at the University of British Columbia)

[2] Johnson G, Scholes K, and Whittington R 2008 *Exploring corporate strategy: text & cases* (Harlow, Financial Times Prentice Hall)

[3] Hansen H, Rand J, and Tarp F 2004 *SME Growth and Survival in Vietnam: Did Direct Government Support Matter?* (Institute of Economics University of Copenhagen: Copenhagen., Denmark)

[4] Tran T K 1919 *Vietnam: A brief history*

[5] Bowers-Brown T and Harvey L 2004 Are there too many graduates in the UK? A literature review and an analysis of graduate employability *Industry and Higher Education* **18(4)** 243-254

[6] Porter M E 2008 *Competitive strategy: Techniques for analyzing industries and competitors.* (New York, Free Press)

[7] Freeman E 1984 *Strategic management: A stakeholder approach*

[8] Warszawski A 1996 Strategic planning in construction companies, *Journal of construction engineering and management* **122(2)** 133-140

[9] Klyuev S V, Khezhev T A, Pukharenko Yu V, Klyuev A V 2018 To the question of fiber reinforcement of concrete *Materials Science Forum* **945** 25 – 29

[10] Klyuev S V, Khezhev T A, Pukharenko Yu V, Klyuev A V 2018 Experimental study of fiber-reinforced concrete structures *Materials Science Forum* **945** 115 – 119

Features of setting and strength gain of the foam concrete massif

A I Savenkov[1] and P S Gorbach[1*]

[1]Federal State Budgetary Educational Institution of Higher Education "Angarsk State Technical University", 665835 Angarsk, Russia

E-mail: pgs@angtu.ru

Abstract. The paper presents a study of the regularities of the processes occurring in the matrix and aerated concrete mass at the initial time after laying and later, during the strength gain. The colloidal properties that appear in the foam concrete mix begin at the stage of mixing the components containing binder and aggregate particles, and are activated by the mixer blades. In this case, the mixture is transformed by mass transfer of the matrix material onto the foam grid and a new spatial reorientation of the bubbles under conditions of physical and mechanical effects on the thin-film structure of the foam. In technology, it should be taken into account that the processes occurring in the foam concrete mixture can be attributed to two opposite actions occurring simultaneously. First, in a freshly laid foam concrete mixture, as in any thermodynamically unstable system, sedimentation and coalescence of air pores begins, which is especially manifested at forcedly high water-solid ratios. This can lead, if the technology is not followed, to complete stratification of the mixture. Secondly, in the cement slurry, the setting and further formation of the cement stone occurs. A necessary technological condition for product quality is compliance with the optimal water-solid ratio of the mixture and ensuring the beginning of the setting of the foam concrete mass before its stratification.
It was also experimentally determined that an increase in the hardening duration of foam-cement systems for more than 28 days is impractical, since it was revealed that an increase in the strength of foam concrete is similar to the same process in structural concrete. It is quite possible to achieve the strength required for foam concrete according to GOST 25485-89 within 28 days and even earlier due to the high technological discipline and the use of modern plasticizing additives.

1. Introduction

In Russia, foam concrete is used in construction mainly for the construction of enclosing structures due to its low thermal conductivity, environmental friendliness, and fire resistance. But in terms of manufacturability, the issues of accelerating the setting time and strength gain remain relevant [1,2,3]. In this regard, there is a need to investigate the processes occurring in foam concrete in the initial period after laying and later, during setting and aging to gain brand strength. The purpose of this article is to determine the specific features of the transition of a foam concrete mass from a colloidal solution to a porous stone.

2. Materials and methods

2.1. Materials
To assess the strength of the matrices, as well as the dynamics of strength gain, experiments were carried out on a cement matrix of foam concrete containing surfactants (surfactants). The surfactants used were cement grade PC-500-D0 produced by JSC Angarskcement, protein foaming agent Ompor and synthetic foaming agent PentaPAV (PPAV-430).

2.2. Methods
Studies of the dynamics of growth of the strength of the matrix were carried out on samples of beams 40x40x160. A foaming agent (surfactant) at a concentration of 0.5% to cement was introduced into the composition of the cement slurry without filler. In parallel, surfactant-free control beams were made. After stripping the formwork, the beams gained strength in the normal hardening chamber for 3, 7, 14, 28, 56 days. During these periods, further probes were taken from the batch of samples and the halves of the beams were tested for compressive strength in accordance with GOST 310.4-81.

3. Results
The tendency towards a decrease in strength and density with an increase in the concentration of the foaming agent was confirmed. Both protein and synthetic foaming agents caused a decrease in strength, density and loosening of the matrix stone. This effect clearly shows the difficulty of growing neoplasms and leads to a change in the microporosity of the cement stone towards an increase, and strength - towards a decrease. According to the Taube rule, the activity of the foaming agent in the cement system depends on the length of the molecular chain of the active substance. In protein surfactants, the molecule consists of thousands of atoms, which is probably why samples containing Ompor protein foaming agent lost their strength least of all.

According to the results of preliminary experiments obtained when determining the strength of the matrix, depending on the concentration of foaming agents in the samples with the content of surfactants of different nature, carried out earlier on the basis of the laboratory of the Department of PGS, it was shown that the steric effect in the cement matrix is expressed in a decrease in density and strength (Figure 1). To identify the further influence of negative factors, it is necessary to consider the hardening process as the foam concrete mass.

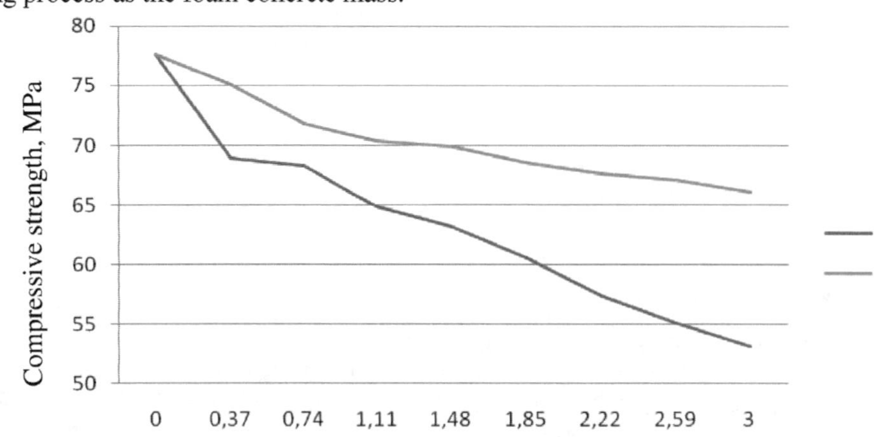

Figure 1. The strength of the cement matrix from the concentration of foaming agents
1 - matrix with Ompor foaming agent; 2 - matrix with PentaPAV-430.

In this regard, the question arises about the refinement of the hydration kinetics, which can be estimated from the strength of the matrix containing the surfactant during the standard hardening period and in the future.

Based on the results obtained, graphs of strength growth were plotted at 3,7,14,28 and 56 days of age of the matrix on M600 cement, which contains a surfactant and a control one (Figure 2).

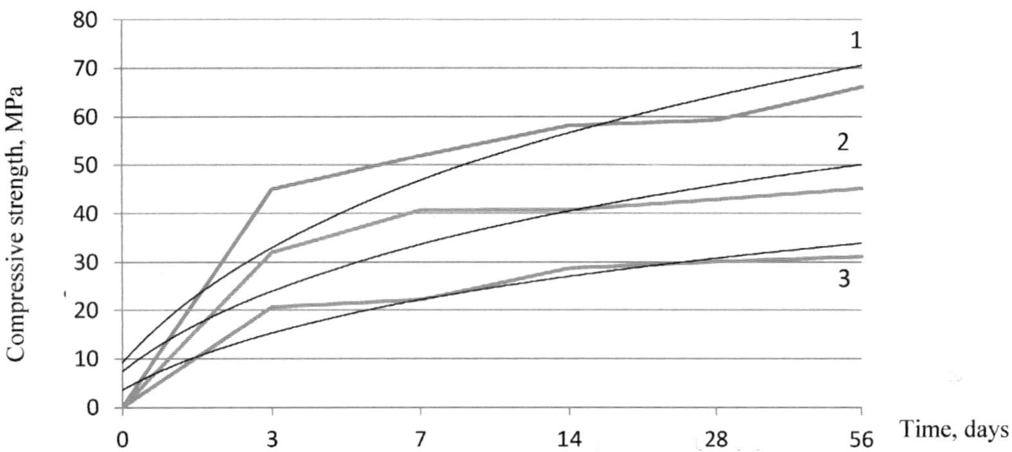

Figure 2. Comparative graph of the dynamics of growth of the strength of the matrix on cement M600, control and containing a foaming agent. 1 - control, the matrix does not contain surfactants; 2 - matrix with Ompor foaming agent; 3 - matrix with foaming agent PentaPAV-430.

In general, it can be argued that the value of the intermediate values of the strength of the cement stone of the matrices in the standard 28-day time range is adequately described by the logarithmic model:

$$R_n = R_{28}\frac{\lg n}{lg\,28} \qquad (1)$$

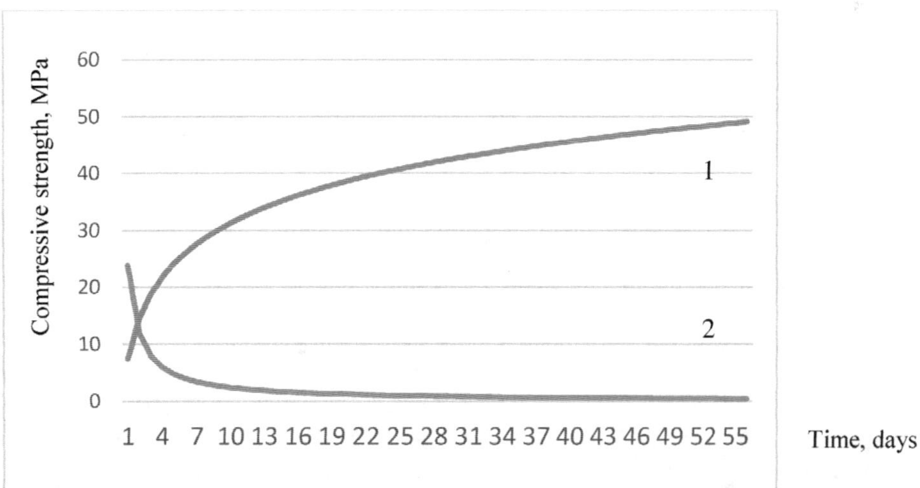

Figure 3. An approximated graph of the growth of the strength of the matrix with a synthetic foaming agent (1) and the differential curve of the growth of the strength of the sample on cement M600 with the OMPOR protein foaming agent (2).

This mathematical model of strength growth is consistent with the general laws [4].

The highest strength was obtained on matrices made on M600 cement. The model of the strength dynamics of the matrix containing the Ompor foaming agent is described by the equation:

$$Rn = 23{,}839 \ln n + 7{,}424 \quad (2)$$

where Rn is the compressive strength of the matrix for n - days.

Strength gain can be estimated from the first derivative of the function of equation (2). After the first differentiation of this model, there is the following expression:

$$Rn' = \frac{23{,}839}{n} \quad (3)$$

Analyzing equation (3), it can be argued that after 28 days of hardening, the increase is less than one percent, therefore, further holding of the samples is impractical. In foam concrete, a similar pattern of strength growth is observed (Figure 4).

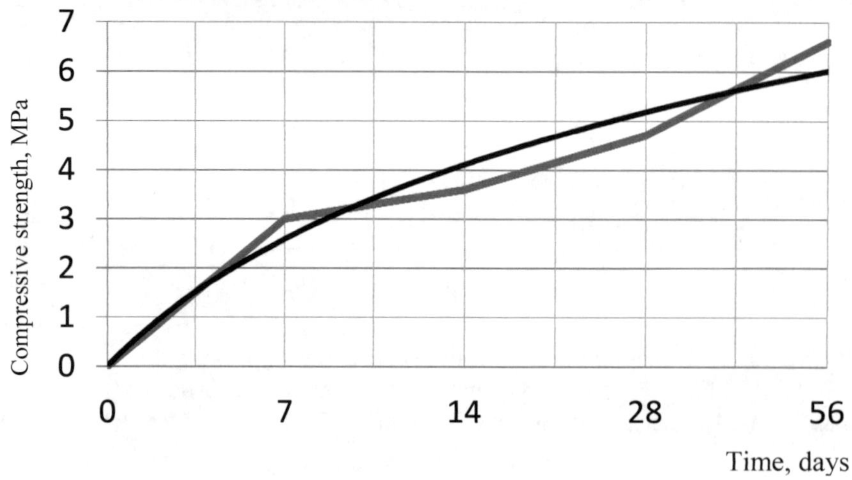

Figure 4. Graph of strength growth of non-autoclaved aerated concrete D700.

4. Discussion

As a result of ensuring all quality factors, it is possible to achieve the best structure of the material structure and maximum strength. It is known that in the presence of surfactants in a foamed cement mortar, the setting time of the massif slows down and its strength decreases. This deceleration is due to the fact that surfactants create a thin adsorption layer on the surface of the solvation shells of cement grains, which gives the same charges of particles and initiates the steric repulsion effect [5].

Foam concrete mixtures in their structure are multiphase and non-stationary. The stability of the structure in such mixtures is predetermined by the degree of decompaction of the solid phase. In foam concrete mixtures, its value is such that the stability of the mixture in the period of early structure formation can be provided only due to the optimal ratio between gravitational and electrostatic forces [5].

The colloidal properties that appear in the foam concrete mix begin at the stage of mixing the components containing binder and aggregate particles, and are activated by the mixer blades. In this case, the mixture is transformed by mass transfer of the matrix material onto the foam grid and a new spatial reorientation of the bubbles under conditions of physical and mechanical effects on the thin-film structure of the foam. This is accompanied by partial destruction of the foam, and when low-density foam concrete of grades D200-D300 is obtained - to the formation of a torn cellular structure and heterogeneous open cellular porosity.

Thermodynamic equilibrium formed in the foam concrete mixture during mixing is established from the equilibrium between gravitational, electrostatic and capillary forces immediately after the termination of the external influence.

In technology, it should be taken into account that the processes occurring in the foam concrete mixture can be attributed to two simultaneously acting and oppositely directed groups of actions. First-

IOP Conf. Series: Materials Science and Engineering **945** (2020) 012053 doi:10.1088/1757-899X/945/1/012053

ly, in a freshly laid foam concrete mixture, as in any thermodynamically unstable system, sedimentation manifests itself, especially when it is forced to have high water-solid relations. If the technology is not followed, this leads to complete stratification of the mixture (Figure 5). Secondly, setting begins in the cement slurry, the formation of solvation shells of cement grains, the nucleation of new formations from the solvation shells, the formation of clusters from particles and their accretion, leading to the setting of the entire foam concrete mass.

A necessary technological condition for product quality is to ensure the setting of the foam concrete mass before its stratification, in other words, the reduction of the "first" of these processes and the acceleration of the "second".

At the initial stage of the structure formation of foam concrete mixtures, when the mixing of the components is completed, but crystalline intergrowths have not yet formed, gravitational forces in combination with the energy capabilities of the surfactant films with changing properties of the binder control the redistribution of water within the dispersed system, which is carried out mainly through viscous contacts between them. The viscosity control in this system is carried out by the water-solid ratio or by plasticization of the mixture. Moisture on the surface of solid phase particles is physically firmly bound to them and does not contain surfactants. Therefore, the surfactant concentration in the bulk of the liquid phase of the mixture inevitably increases, which leads to an abrupt decrease in the viscosity of the liquid phase and a decrease in the elasticity of foam films [4]. As the process continues, this causes rupture of the foam films, initiating coalescence at the gas-liquid interface, leading to the coalescence of small gas bubbles into larger ones (Figure 5).

In this case, there is a redistribution of surfactants in the volume of the concrete mixture. Excess surfactants from the gas-liquid interface during coalescence are forced to move deep into the liquid phase of the concrete mixture. Then energy conditions arise for the formation of surfactant micelles, which reduce the viscosity of water. In the case when elastic bonds are formed between the particles of the solid phase and their strength is sufficiently developed to preserve the macrostructure of the concrete mixture, an increase in the concentration of surfactants in the liquid phase cannot lead to negative consequences. If only viscous bonds are formed between the components of the solid phase, with a high mobility of the foam mixture, then the development of coalescence naturally leads to the rupture of interpore partitions, stratification and sediment of the foam concrete mixture [5].

The first - the initial stage - occurs immediately after mixing and lasts from 30 minutes to 1 hour. The second - the main stage - begins approximately 5-6 hours after the start of mixing. There is a long induction period between the two stages of the process. The dissolution of the C3S mineral occurs simultaneously with the predominant transition of Ca2 + ions into a true solution. The silica remains in the solid phase as an orthosilicic acid gel. After reaching saturation of the liquid phase with respect to Ca (OH) 2, the process of formation of calcium hydrosilicate begins and proceeds rapidly by the reaction of Ca (OH) 2 in solution with orthosilicic acid gel in the solid phase. At the same time, the excess lime by crystallization precipitates into a solid phase. During hydration, there can be several (2 ... 4) such elementary hydration cycles. At a certain stage, due to the formation of hydrosilicate under the primary shell, this shell is mechanically destroyed, which creates the conditions for the main hydration period to proceed.

Since it can be conditionally considered that the steric effect partially inhibits the hydration reaction, then, as indicated above, the presence of surfactants near the reacting atoms can prevent the approach of these atoms and make the reaction difficult or make it impossible in a certain number of bonds. Almost all foaming agents slow down the reaction of interaction of clinker minerals with a grout, and change the rate of formation of hexagonal hydrate phases, especially for the most active mineral of cement clinker - C3A. Moreover, large dosages of additives have a more significant effect on the crystallization kinetics of hydrated neoplasms and to a lesser extent affect the phase composition. In solutions of foaming agents, the interaction of C3A with water does not proceed completely, and even after 28 days of the hydration process, the unreacted part of the original alite partially remains in the system.

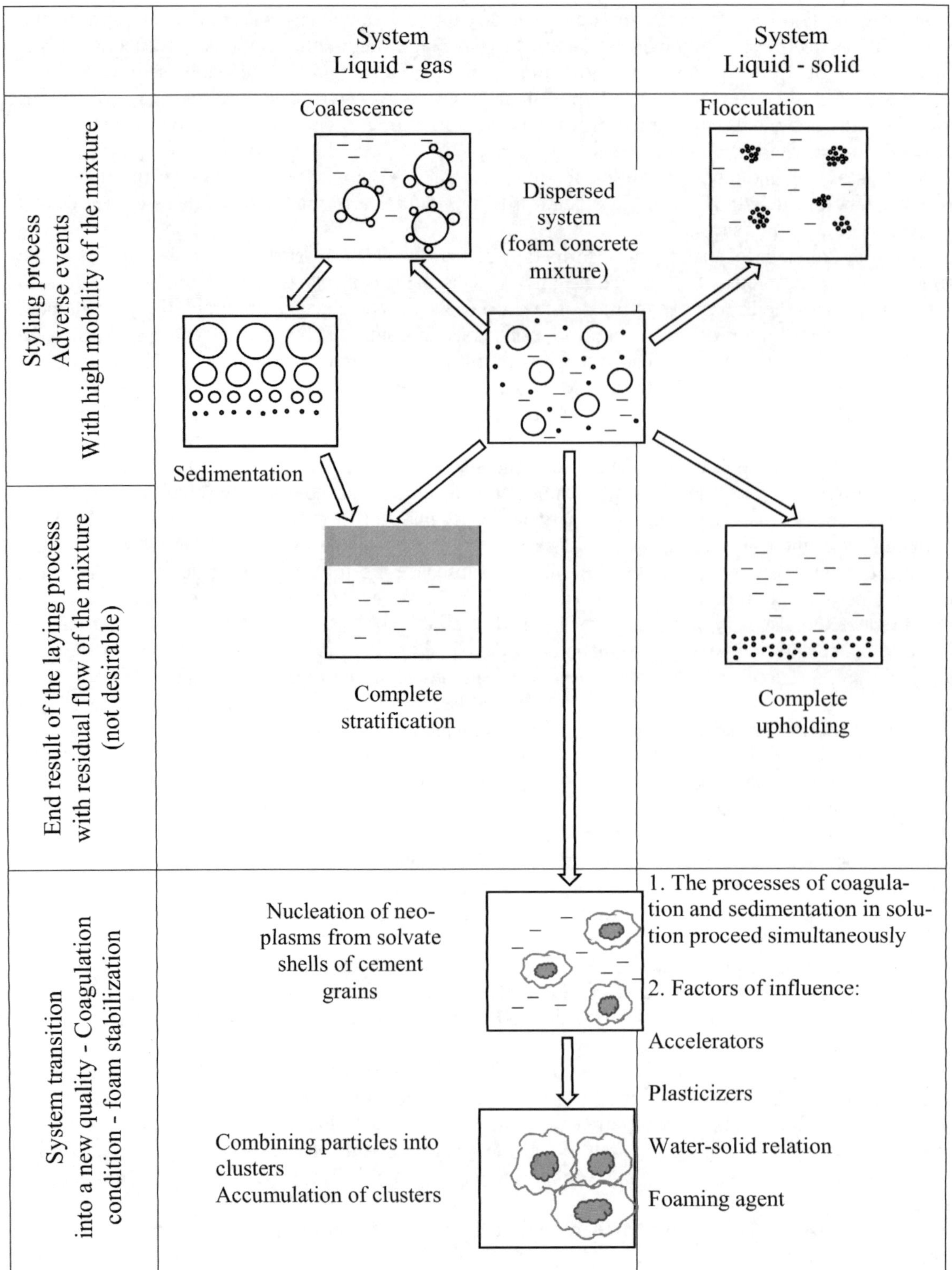

Figure 5. Process flow diagrams in foam concrete mix.

The elimination of these disadvantages can be achieved by a significant increase in W / T> 0.7, but this negatively affects the strength and quality of the material. It is also possible to use dry mineral compositions of reduced wettability and water demand. However, the resulting low-density foam concrete in this case has a non-uniform cellular structure with many enlarged contact pores, which means low strength. When these foams are combined with a binder solution, it is necessary to increase the water-solid ratio [6,7,8].

5. Summary

The difficulty of growing neoplasms caused by the presence of a foaming agent leads to loosening of the structure of the cement stone, the formation of communicating micropores and, accordingly, a clear decrease in density and strength. It is generally accepted that the regularities of hardening of non-autoclave foam concrete based on Portland cement are expressed in the lengthening of the setting time, low set speed and low final strength, and a tendency to crack formation [9,10,11,12,13]. This is the traditional approach, but it turns out that these negative properties are formed only during the mixing, laying and setting of the mixture. Subsequently, the course of hydration in foam concrete is identical to the process of hydration of structural concrete.

Based on the assumption that the growth dynamics of the strength of the control matrix, which does not contain surfactants, is reference (Figure 2), then to compare a similar process in matrices containing foaming agents, the ratio of matrix strengths was determined according to formula (4) at different times of hardening according to points 3,7,14,28,56.

$$\frac{Rn(matrix+Surfactant)}{Rn(matrix)} \quad (4).$$

It turned out that these relations remain close to constant during the hardening period. Consequently, the loss of strength in the presence of surfactants, formed in the first day of hardening, remains unchanged in the future. The steric repulsion effect, which hindered the setting and formation of crystalline intergrowths of the cement stone at the beginning of hardening, no longer manifests itself.

Thus, it was experimentally determined that an increase in the duration of hardening for more than 28 days is impractical, since the increase in the strength of both matrices and foam concrete after this period was very small.

At present, it has become quite possible to achieve a strength class value corresponding to the density grade for non-autoclave foam concrete according to GOST 25485-89 within 28 days and even earlier due to high technological discipline and the use of modern plasticizing additives [15,16].

6. References

[1] Shakhova L D, Balyasnikov V V, Black T I. 2000 Study of hydration processes of clinker minerals with additives of foaming agents of various nature *II International. meeting on the chemistry and technology of cement, St. Petersburg: Publishing House of the Center for Education Information*

[2] Klyuev S V, Khezhev T A, Pukharenko Yu V, Klyuev A V 2018 To the question of fiber reinforcement of concrete *Materials Science Forum* **945** 25 – 29

[3] Klyuev S V, Khezhev T A, Pukharenko Yu V, Klyuev A V 2018 Experimental study of fiber-reinforced concrete structures *Materials Science Forum* **945** 115 – 119

[4] Savenkov A I, Gorbach P S, Shcherbin A S 2013 Non-autoclaved foam concrete. Quality factors. Monograph. - Angarsk: Publishing house of the Angarsk State Technical Academy.

[5] Morgun L.V. 2008 Scientific principles of aggregate stability of foam concrete mixtures *Concrete technologies* **31**

[6] Kobidze T E, Kiselev A Yu, Listov S V 2005 The relationship between the structure of the foam, technology and properties of the resulting foam concrete *Construction Materials* **1**

[7] Hajili R A, Merkin A P 1981 Surfactants in construction. Baku: Azerbaijan State Publishing House

[8] Gleckel F L, Kopp R Z, Akhmedov K S 1986 Regulation of hydration structure formation by

surfactants. Tashkent: Publishing House "Fan", UzSSR
[9] Ruzhinsky S, Portico A, Savinykh A 2006 All about foam concrete. 2nd ed. -SPb: -OOO Stroybeton
[10] Roy D M, Gouda G R 1973 Porosity-strength relationship in high-strength cementitious materials. J. Am. Ceram. Soc. **56**
[11] Kirsley E P, Weinwright P J 2002 Influence of porosity on the strength of foam concrete [J]. Cem. Concr. Res. **32**
[12] Nambiar EKK, Ramamurthy K 2008 Models for Predicting the Strength of Foam Concrete *J. Mater. Struct.* **41** 247-254
[13] Chen, H., Wu, Sh., Zhou, J. 2013 Influence of porosity on the compressive and tensile strength of cement slurry *Construction and Building Materials* **40**
[14] Kharkhardin A N 2005. Structural topology of foam concrete *News of higher educational institutions. Building* **5** 18-25
[15] Peskov V I, Otsokov K A, 2004 The effectiveness of the use of cellular concrete in the construction of Russia *Building materials* **3**
[16] Sychev M M 1976 Patterns of manifestation of astringent properties *Sixth International Congress on Cement Chemistry.M .: Stroyizdat*

IOP Conf. Series: Materials Science and Engineering **945** (2020) 012054 doi:10.1088/1757-899X/945/1/012054

The research of different types of clays of the Kazakhstan for the production of wall ceramic products

B K Sarsenbayev[1*], G O Karshyga[2], A M Budikova[2] and A M Shakey[3]

[1]Doctor of Engineering Sciences (Advanced Doctor), Professor, M.Auezov South Kazakhstan State University, Shymkent, Kazakhstan
[2]Candidate of Engineering Sciences (Ph.D.), Korkyt Ata Kyzylorda University, 29a Ayteke bi Str., Kyzylorda, 12001, Kazakhstan
[3]Master of Arts (M.A.), M.Auezov South Kazakhstan State University, Shymkent, Kazakhstan

E-mail: stroitelstvo_ukgu@mail.ru

Abstract. The manuscript presents the results of analysis and synthesis of researches of scientists of the CIS and Kazakhstan on the use of various clays for the production of wall ceramic products. There was investigated the application of refractory and brown clays, loam to optimize the composition of the furnace charge and the choice of technological parameters of the production of wall ceramic products. The drying and burning properties of the samples burned at 850 °C, 900 °C, 950 °C, 1000 °C and 1050°C, 1100÷1150 °C with determination of strength, shrinkage, water absorption of the material at all specified burning temperatures were studied. The optimal burning temperatures for the study of the burning properties of the above three clays are determined. In many regions of the Republic of Kazakhstan the problem of production of high-quality wall ceramic manufacture is particularly acute. Most seasonal plants turns out low-branded products (the M75-100, B<25), often not conforming to requirements of normative documents, does not have the expanded range, the decorative brick necessary for construction of low houses (country, cottage buildings) as implementation of the similar program requires, first of all, detailed studying of technological and physical and chemical features of clay raw materials is not issued.

1. Introduction

Clays, being the main raw materials of ceramic industry, represent complex structures and demand serious approach to development of technologies taking into account specifics of raw materials. Since the most ancient times (China, the Middle East) paramount attention was paid to processing of raw materials. Different manufacturing techniques of pottery developed on the basis of practical experience which passed from generations to generations throughout centuries. Panel housing construction and excessive hobby for reinforced concrete led in the 50-60th years to decline of the domestic ceramic industry [1-2]. The experience and traditions developing within many decades was lost. As result, at the enterprises of ceramic industry less attention began to be paid to questions of high-quality preparation and processing of raw materials taking into account specifics of production. In this regard the research of a complex of technological and physical and chemical features of clay raw materials based on modern scientific concepts acquires special importance and relevance. In this regard development of structures and technology of receiving more quality front ceramic brick taking into account specifics of local raw materials is relevant [3-6].

The main technological requirements of production of wall ceramics are prevention of formation of cracks because of uneven shrinkage on thickness of a product and difference of humidity on an

IOP Conf. Series: Materials Science and Engineering **945** (2020) 012054 doi:10.1088/1757-899X/945/1/012054

internal and outer surface. And a condition of emergence of cracks and ways of fight against them the considerable number of works nevertheless is devoted to studying of the mechanism, the sharpness of a problem remains to this day [7-10].

Results of a research of the last years are directed generally to improvement of drying properties of masses by exhaustion and moistening steam of clays, their warming up, pumping out, additive in clay of plaster, ashes, surfactants, irrigation of a mouthpiece by the moisture detaining structures, introduction of electrolytes, creation of pre-depressed films.

There is theoretically and practically established that drying of stoneware of plastic formation differs in big complexity and it is bound to the considerable expenses of power supplies (gas, coal, fuel oil).

Therefore in the conditions of an energy crisis the special relevance is acquired by synthesis of structures of the ceramic masses, insensitive to drying, and the development of the intensive modes of drying of products allowing reducing duration of drying and will lower costs of energy carriers.

2. Materials and methods

For production of stoneware we in researches used the following raw materials: refractory and brown clay, loam, and as additive chamotte, $BaCO_3$ and sawdust.

For optimization of compositions of fusion mixture and the choice of process parameters of production of stoneware three types of clay of Kazakhstan, and conditionally designated as are picked up: refractory; brown; loam.

The next technological researches were in laboratory conditions:

Researches on processing were carried out on the laboratory hammer swing mill equipped with a bolter with openings in 3 mm, and the subsequent 100% processing with use of a sieve with openings <1 mm.

Researches of hashing and vacuum formation with determination of the humidity providing consistency of a bar at the exit from a mouthpiece in 1.3÷1.6 kg/cm^2.

The normal consistency of a bar at extrusion is in limits of 1.4 ÷ 1.8 kg/cm^2.

Researches of drying properties were studied drying properties of three mixes with determination of the following parameters:

- Durability of dry samples.
- Drying shrinkage.
- Crack resistance limit.
- A sensitivity indicator to drying according to Nosova.
- Curve Bigota.
- Percent of reabsorbed moisture.
- Extent of impact of reabsorbed moisture on durability of dry material.

The burning properties of the samples burned at 850°C, 900°C, 950°C, 1000°C and 1050 °C, 1100÷1150 of °C with determination of durability, shrinkage, water absorption of material at all specified burning temperatures. Depending on a type of clay and type of the final product, burning was begun at a temperature of 850 °C of 900 °C, and was complete at a temperature of 1100 °C of ÷1150 of °C.

Burning properties at the optimum temperature determined in the previous researches:

- Burning shrinkage.
- A lose of weight in burning (loss of ignition).
- Forming humidity concerning the burned weight.
- Water absorption of the burned material.
- Density of the burned material.
- Mechanical strength of the burned material.

Thermo-dilatometric researches: for identification of volume changes of material when burning and projections on this basis of an optimum temperature curve of burning.

3. Results and discussions

Drying properties and interaction with humidity of a surrounding medium are given in figures 1, 2, 3, 4.

Consistency of paste on escaping of a mouthpiece: 1.5 kg/cm².

Water for interfusing, necessary for samples with normal consistency, expressed as a percentage weights in the clay which is dried up at 110 °C: 27.76%.

Shrinkage at the drying after elimination of water for interfusing: 7.41 %.

Limit of cracks formation or the maximum length without cracks of sample 200x10x100 mm which is dried up in the following conditions: 20 cm.

Temperature: 75-120 °C.

Time: 90 minutes

Airflow speed: 6 m/s.

Residual humidity at the end of drying: 2.94%.

Sensitivity coefficient on nasal or a ratio between pressed volume during drying and the total pores volume of drying material: 0.67.

The tearing strain on a bend is value between 4 samples (deleting dispersion): 64.85 kg/cm^2.

The humidity of a surrounding medium absorbed by dry samples after finding of 24 h is in the saturated atmosphere (the relative humidity of 100%): 4.02%.

Flexural strength of samples from the surrounding medium absorbed by moisture: 17.95 kg/cm^2.

The strength lose of dry samples is due to effect of absorption of moisture of a surrounding medium: 72.32%;

The testing of burned samples:

The recommended temperature for burning: 1050 °C.

The shrinkage of manufactures when burning: 6.08%.

Lose of weight expressed as percentage weights of the clay which is dried up at 110 °C: 6.77%.

Water for interfusing per the mass of the burned products: 29.78%.

Water absorption after 2 h immersion of samples in the boiling water, per the mass of the burned and dry material: 5.58%.

Apparent specific gravity: 2.15 g/cm^3.

The tearing strain on a bend, test is carried out on 3 samples (deleting dispersion): 366.79 kg/cm^2.

Such studies were conducted using brown clay and loam and the following conclusions were made.

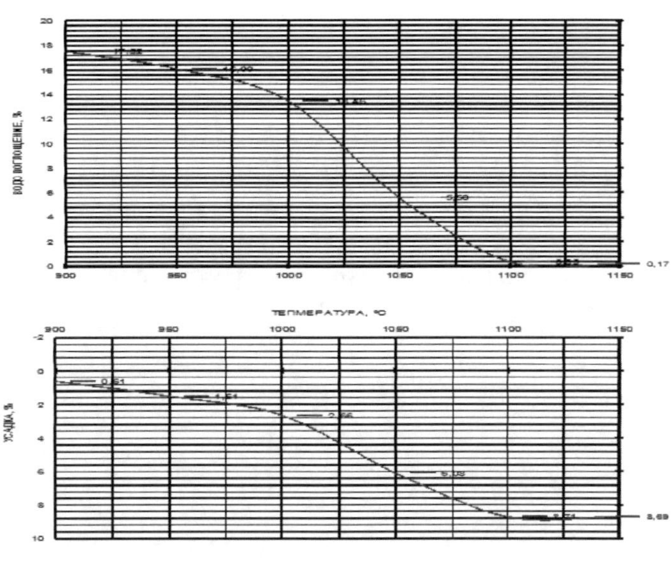

Figure 1. Determination of durability and shrinkage of refractory clays.

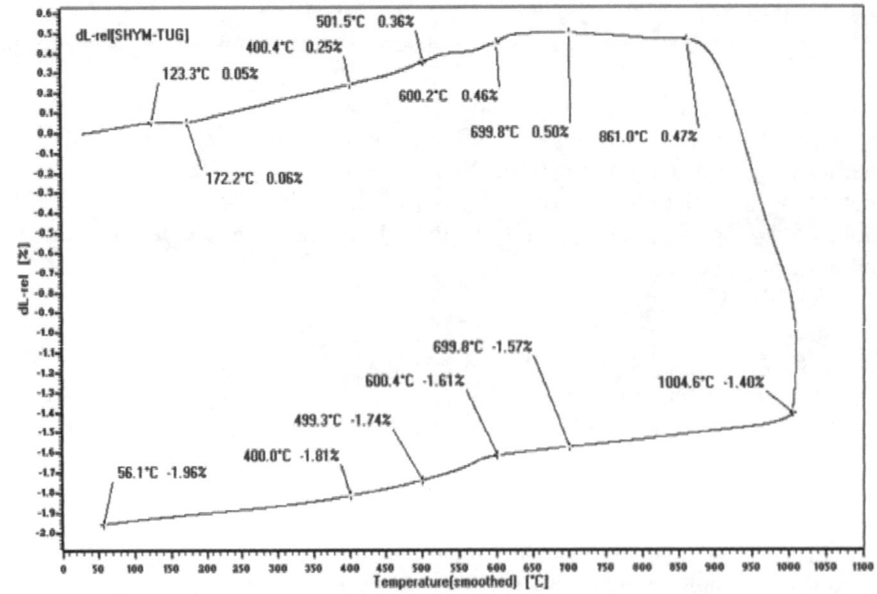

Figure 2. Dilatometric accumulative curve of % expansion/temperature of refractory clays.

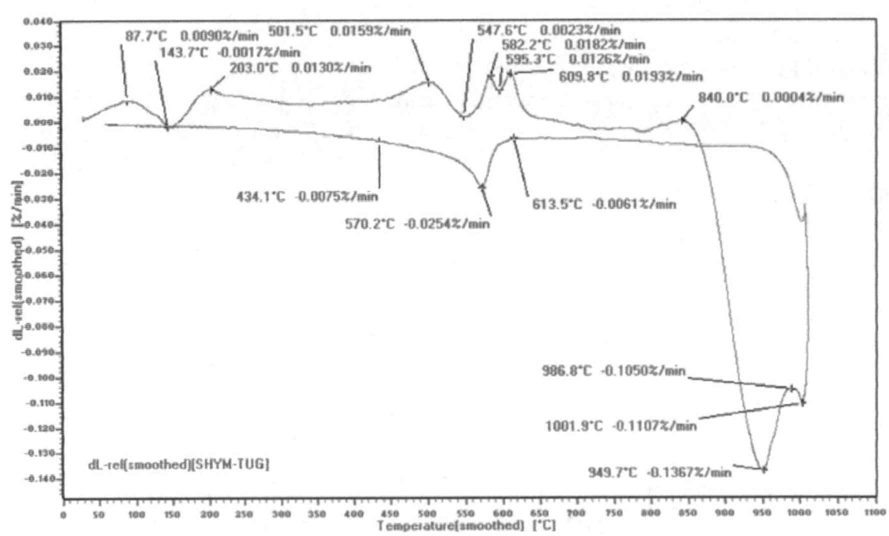

Figure 3. Differential dilatometric curve: % expansion speed on minutes / t° burning of refractory clays.

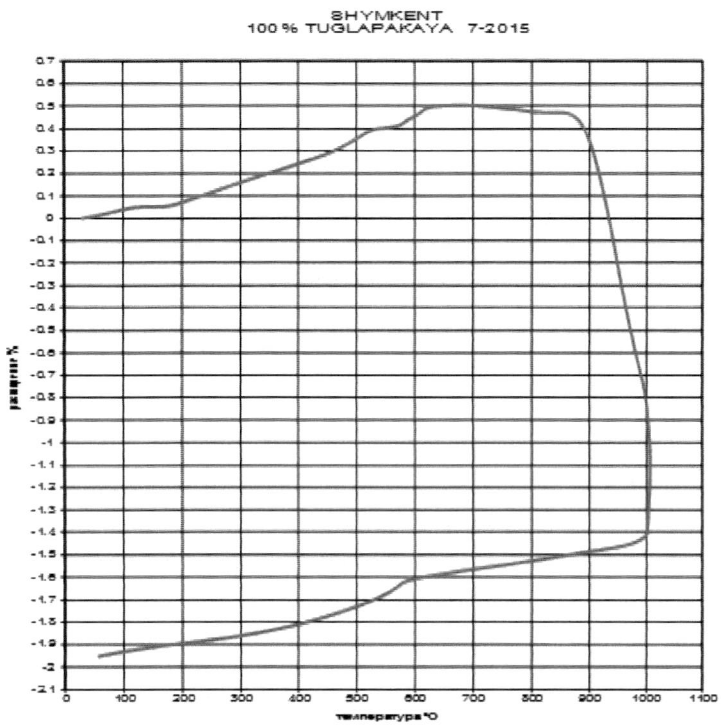

Figure 4. Variation of water absorption, shrinkage and durability depending on temperature of refractory clays.

4. Summary

It is established that percentage of moisture of refractory clay is high: composes 27.76%, also as well as an index of drying shrinkage: 7.41%.

It is revealed that percentage for manufacture of bricks moisture of formation usually fluctuates between 18 and 25%, and percent of drying shrinkage between 4 and 7% in clays.

It is established that indexes of capillary moisture are also high, as well as a ratio of capillary moisture of shrinkable humidity: 1.53 – this indicator favors to drying process. In clays for manufacture of bricks this ratio fluctuates within 1-1.5.

It is established that the mineralogical composition of brown clay has the same components, as refractory clay. The difference consists only in the relative concentration of each component and also the content of quartz (sand) which in refractory clay is very low, in brown clay is high within norm.

It is revealed, despite the high content of sand in brown clay, the volume of the remained capillaries in a dry product from brown clay: 11.4 in %, is lower, than in refractory clay: 16.80 in %, and the ratio of capillary and shrinkable moisture decreases from 1.53% in refractory clay to 1.50% in brown clay. However, effective diameter of the remained capillaries of brown clay is more, than clays refractory that provides faster process of drying of brown clay.

It is established that loam is less wax, than refractory and brown clays because of higher content of quartz (sand) in it and lime carbonate.

It is established that at loamy clay, besides a sandy quartz component, there is a limy component which isn't in two other clays that reduces its plasticity, compactness and resistance to tension created in material at the beginning of drying process.

It is revealed, the percent of shrinkable humidity which gives plasticity to material gives very low indicators in loam: 5.44 to % while this index is equal in refractory clay to 10.96%, and in brown clay - 7.60%. The percent of drying contraction also gives lower indicators: 4.32 to %.

IOP Conf. Series: Materials Science and Engineering **945** (2020) 012054 doi:10.1088/1757-899X/945/1/012054

5. References

[1] Ashmarin G D, Kurnosov V V, Lastochkin V G 2010 Energo - and resource-saving technology of ceramic wall materials *Structural materials* **4** 24-27

[2] Kondratenko V A, Peshkov V N 2002 Problems of brick production and ways of their decision *Structural materials* **3** 19-21

[3] Klyuev S V, Khezhev T A, Pukharenko Yu V, Klyuev A V 2018 To the question of fiber reinforcement of concrete *Materials Science Forum* **945** 25 – 29

[4] Klyuev S V, Khezhev T A, Pukharenko Yu V, Klyuev A V 2018 Experimental study of fiber-reinforced concrete structures *Materials Science Forum* **945** 115 – 119

[5] Klyuev S V, Khezhev T A, Pukharenko Yu V, Klyuev A V 2018 Fiber concrete for industrial and civil construction *Materials Science Forum* **945** 120 – 124

[6] Menshikova V K, Demina L N 2020 Ceramic building materials using non-traditional raw materials *Construction Materials and Products* **3 (3)** 40 – 46 DOI: 10.34031/2618-7183-2020-3-3-40-46

[7] Montayev S A, Almagambetova M Zh, Adilova N B, Adyrova G M 2016 Technology of Wall Ceramics *Textbook* 280

[8] Groshev I A, Sarsenbayev B K, Groshev A I 2013 About effectiveness of use of plasticizing additives by production of a ceramic brick from loessial raw materials of Central Asia *II scientific and practical seminar Production power and a resource of the preserving structural materials and products* **1** 117-121

[9] Groshev I A, Sarsenbayev B K, Verner V F, Groshev A I 2012 Receiving high-strength structural ceramics on the basis of the activated silica-alumina slate // *International conference "Modern Architecture and Innovations"* 120-122

[10] Groshev A I, Groshev I A, Ismailov A A 2015 Development of structures of masses for receiving front ceramics from raw materials of Kazakhstan *Glass and Ceramic* **6** 31-39

IOP Conf. Series: Materials Science and Engineering **945** (2020) 012055 doi:10.1088/1757-899X/945/1/012055

Investigation of ground and foundation vibrations of an existing building during impact pile driving

V A Veshnyakov[1]

[1]Department of geotechnics, Northern (Arctic) Federal University named after M.V. Lomonosov, Severnaya Dvina Emb. 17, Arkhangelsk, 163002, Russia

E-mail: v.veshnyakov@narfu.ru

Abstract. Pile driving technology remains the most common in Arkhangelsk where the presence of soft soils is an engineering-geological feature. New construction is mainly carried out in the conditions of existing urban development. During pile driving dynamic effects that propagate in the ground can have a negative impact on existing buildings. In common conditions, a distance of more than 25 m is considered safe. Sometimes driving piles by hammer is safe for neighboring buildings at a shorter distance. Often such buildings have pile foundations. The monitoring of pile driving by diesel hammer was carried out at one of the sites in Arkhangelsk. Vertical vibration velocities were measured by a vibration analyzer with independent channels. One vibration sensor was installed on the ground surface, the other one on the foundation of a 5-storey building. Continuous recordings of vibrations allowed a detailed analysis of the driving process of each pile. The level of dynamic effects propagating in the ground changed with the penetration of the driven pile. The highest values were observed when pile foots passed through a layer of glacial deposits. Change in soil resistance to pile driving also affected the operation of diesel hammer. The level of dynamic effects increased as the distance to the driven pile decreased both on the ground surface and on the foundation of the building, while the values on the ground surface were higher. The ratio of the highest vertical peak vibration velocities on the ground surface and on the building foundation was determined by field studies.

1. Introduction

Geotechnical conditions of Arkhangelsk are difficult due to the presence of soft layers such as peat or silt [1] and in most cases the impact method is still used for driving piles.

Brick and reinforced concrete buildings continue to replace old wooden apartment buildings built in the middle of the last century. New buildings are located next to them as well as to typical 5- and 9-storey buildings of the Soviet period construction.

Pile driving by hammer causes vibrations that can have a negative effect on existing buildings. A distance of more than 25 m is considered safe [2]. As practice shows, pile driving by hammer is acceptable for neighboring buildings at a shorter distance [3-6]. Often such buildings have pile foundations. In this case, the level of vibration of the ground and foundation of the building will be different. According to VSN 490-87 [7], ratio of amplitudes of foundation displacements to amplitudes of ground displacement during vibrations is in the range 0.5-1.0, the specific value depends on the number of stores and the foundation soil. Unfortunately, the type and parameters of the foundation of the existing building are not taken into account.

2. Materials and methods

This paper presents experimental data on determining the values of peak vibration velocities of the ground and pile foundation of a building that occur during impact pile driving on a construction site with typical geotechnical conditions for Arkhangelsk. The stratification of soils and their properties are shown in table 1.

Table 1. Geotechnical conditions of the building site.

№	Soil*	h m	ρ g/cm³	W %	E MPa	φ °	c kPa	I_P %	PI %	I_L %/%	LL %
(1)	Filled soil	0.7…1.4	–	–	–	–	–	–	–	–	–
(2)	Peat	1.7…4.0	1.01	372.0	0.3	–	–	–	–	–	–
(3)	Lean clay (CL)	0.6…1.1	1.94	30.1	8.0	16	16	11.5	20.0	0.57	43.6
(4)	Lean clay (CL)	5.9…9.5	2.13	17.5	20.0	23	34	7.8	10.8	0.23	26.5
(5)	Lean clay (CL)	0.8…3.6	2.01	22.0	16.0	18	24	7.6	10.9	0.72	27.4
(6)	Lean clay (CL)	0.6…5.1	2.13	17.5	20.0	23	34	7.8	10.8	0.23	26.5
(7)	Lean clay (CL)	2.4…7.2	2.08	20.1	26.0	24	34	11.5	19.2	-0.16	41.1
(8)	Lean clay (CL)	>5.7	2.08	21.6	20.0	22	34	9.5	15.4	0.18	35.5

* Soil formations from the pit bottom are shown in figure 2.a

A pit with a depth of 2.5-3.5 m was excavated for foundation work of the 12-storey section of a residential building. Part of the peat layer remained at the base of the pit, and fine sand (1') 0.5 m thick was filled at the pit bottom to allow construction machinery to work. Diesel hammer DD-35 with hammer striking mass 3.5 t and maximum impact energy 89.6 kJ was used for pile driving. Precast-concrete sectional piles 17 m long (bottom section 9 m, top section 8 m) with a cross section of 0.35×0.35 m were driven to the depth of 16.8 m.

A typical 5-storey brick building on the project 1-300-2m (figure 1. a), with plan dimension of 54.0×14.16 m is located next to the building site. This building has a pile foundation. Reinforced concrete piles 9.0 m long with a cross section of 0.30×0.30 m are spaced at intervals of 1.0-1.1 m. Pile foots are located in lean clay (4) of glacial deposits.

The nearest row of piles to the building is at a distance of 8.7 m. According to Russian building codes [2], if the distance from the nearest driving piles less than 25 m, the allowable safe distance is taken based on the condition that the vertical velocity of foundation vibrations does not exceed the maximum allowable value for building taking into account its design features, engineering status and base soil. For this building, the maximum allowable values of the foundation vibration velocity are 20 mm/s [2], 10.8 mm/s [8], [9]. The prediction of the velocity of ground and foundation vibrations was calculated according to VSN 490-87 [7]. The calculation takes into account the type and state of the most solid layer of soil lying up to a depth of 6-8 m from the pit bottom, the mass of the hammer striking part and the number of storeys of the existing building. For a 5-storey building the allowable velocity of foundation vibration of 20.0 and 10.8 mm/s occurs when driving piles with diesel hammer DD-35 at a distance of less than 1.2 and 3.5 m, respectively.

Vertical peak velocities of ground and building vibrations were measured with vibration analyzer Vibran-3 with the vibration velocity range 0.1...500 mm/s, the frequency range 2...100 Hz, the sampling frequency 256 Hz. The vibration sensor 1 is located on the ground surface at a distance of 0.5 m from the building wall, the vibration sensor 2 on the protruding base plate at the level of the brickwork base (figure 1. b). Let us assume that the vibrations of the building at this location will correspond to the vibrations of the building foundation. This assumption is possible because measurements were made on the lower part of the outside bearing wall of the building, at a height of no more than 1 m above grade [8]. It should be noted that the measurements were made on April 03, 2020, and the ground near the building at the location of the vibration sensor installation was thawed.

a)

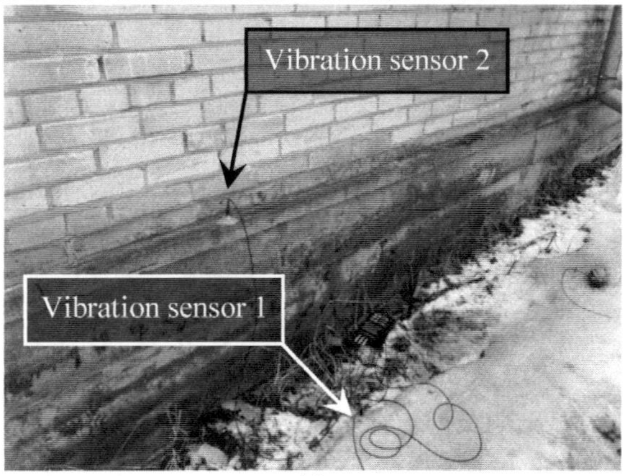

b)

Figure 1. a – view of the building site; b – the location of vibration sensors.

To obtain the dependence of ground and building foundation vibrations on the depth position of the pile foot position from the impact during driving the number of blows for every 0.5 m, and for the last meter – for every 0.1 m were counted. Vertical velocities of ground and foundation vibration were measured during driving of four piles located at different distances approximately on the same line perpendicular to the wall of the building. First the most distant pile was driven, gradually approaching the existing building. All bottom pile sections were driven before the top sections.

3. Results

The piles moved without impact to a depth of about 1.0 m under their own weight and the weight of the diesel hammer. Stable operation of the diesel hammer began only with a depth of 2.0-3.0 m. Driving the bottom pile sections to a depth of 8.3 m required 94-112 blows of the diesel hammer; the total number of blows for driving the piles to a depth of 16.8 m is 481-732. Changes in driving resistance influenced the operation of diesel hammer. Increase of the resistance resulted in rising of striking part of diesel hammer, thus increasing the time between successive hammer impacts Δt (figure 2.b).

Vertical vibration velocities of the building foundation when driving the bottom pile sections did not exceed 2.0 mm/s, which is less than 10.8 mm/s, so it was decided to drive the top sections. The peak vibration velocities at the beginning of driving the top sections of the pile were higher than the peak vibration velocities recorded at the end of driving the lower sections (figure 2. c). On the ground surface this difference was 45-77%, on the foundation of the building 15-78%. An increase in the level of propagating dynamic effects during a break in pile driving is associated with the restoration of the structure of clay soil [10]. When driving four piles, the break after the end of driving the bottom sections and the beginning of driving the corresponding top sections was 1.5-2.0 hours. The increase in vibrational velocities of the ground and building foundation can probably be explained by the fact that the foot of the piles could be at the border of layers (5) and (6).

The intensity of propagating ground vibrations induced by pile driving depends on the penetration of the driven pile. The highest values were observed on the ground and on the building foundation when the pile foots passed through lean clay (6) of glacial deposits (figure 2.c).

The level of dynamic effects decreases as the distance from the driving pile increases, both on the ground surface and on the building foundation (figure 3). For the distances considered in this paper, the highest peak vibration velocities of the building foundation were 2.0-2.1 times lower than the highest peak vibration velocities of the ground surface near the building. A similar ratio only for the

IOP Conf. Series: Materials Science and Engineering **945** (2020) 012055 doi:10.1088/1757-899X/945/1/012055

acceleration of ground vibration and building vibration at the first storey level was obtained for a 9-story brick building in the paper [11]. In figure 3, a dashed line shows the dependence of the ground

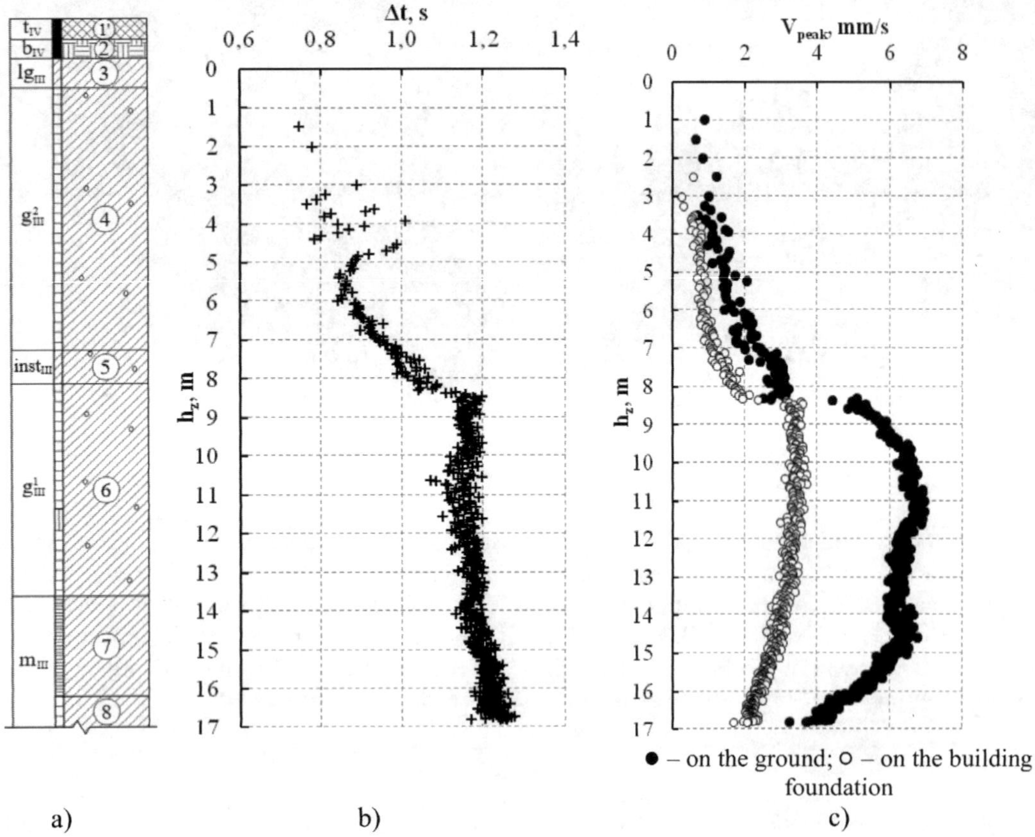

Figure 2. a – soil conditions; b – relation of the time between successive impact of the diesel hammer on the penetration of the driven pile; c – relation of the vertical peak velocity on the penetration of the driven pile.

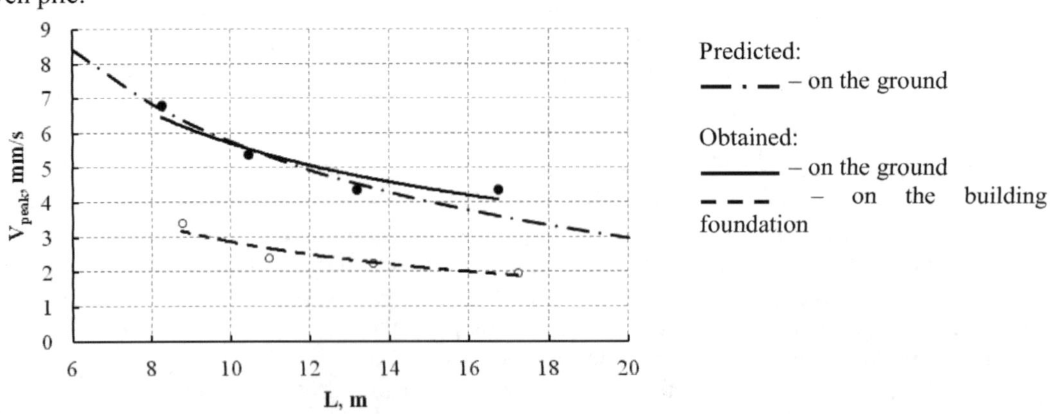

Figure 3. Graphics of the dependence vertical peak velocities on the distance from the driving piles

vibration velocities on the distance from the driving pile, as determined by VSN 490-87 [7]. In this particular case, the obtained level of dynamic effects on the ground surface is very well consistent with the predicted level determined by the calculation.

4. Discussion

As a result of field measurements, it was found that the vertical peak vibration velocities of the building foundation did not exceed the allowable values. The velocity of foundation vibrations was less than the velocity of ground surface vibrations, which is due to the load of 5 storeys, structural stiffness of the building and the fact that pile foots are restrained in a reliable clay soil.

A good agreement between the obtained and calculated values of the ground vibration velocities shows that the predicted level of dynamic effects is possible according to the method of VSN 490-87 [7] for building sites with similar geotechnical conditions and when driving piles with rod diesel hammers. Coefficient of 0.5 should be used to get the vibration velocity of the pile foundation of a 5-storey building. According to the VSN 490-87 [7] for this building and ground conditions this coefficient is 0.9, the difference can be explained by the fact that the type and parameters of the foundation of the existing building are not taken into account.

5. Summary

For buildings on pile foundations which are based on reliable clay soils the dynamic effects caused by pile driving are lower than on the ground surface. For a 5-storey building, the ratio of the highest peak vibration velocities of the pile foundation to the ground surface vibrations can be taken as 0.5 on average.

Based on the results presented in this work in the future the numerical modeling of dynamic effects in the soil during pile driving to study the vibration of buildings of different storeys, influence of pile spacing and pile length of this building and other parameters can be conducted.

6. References

[1] Nevzorov A L, Nikitin A V, Zaruchevnykh A V, Marko Ya Yu 2007 Prospects for the development of foundation construction in Arkhangelsk *Urban construction on soft soils* 87-89

[2] SP 24.13330.2011 Pile foundations. Updated version of SNiP 2.02.03-85

[3] Borisov E K., Usov A G 2007 Dynamics of soil and protected structures during pile driving *Bulletin of the Kamchatka State Technical University* **6** 86-91

[4] Nikitaeva G A, Stavnitser L R 2013 Determining the level of seismic impacts caused by pile driving *Earthquake engineering. Constructions safety* **4** 21-23

[5] Smolin Yu P, Beloborodov V N 2015 Effect of vibrations on nearby civil buildings at pile driving *Bulletin of the Siberian State University of Railway Engineering* **3 (34)** 27-31

[6] Evlakhova E Yu, Ivanova A V, Zheleznyakov V A 2019 Analysis of the results of monitoring of dynamic impacts during experimental immerson of driving piles at the construction site of the campus of SFU *Constrnction Materials and Products* **2 (6)** 12-19

[7] VSN 490-87 Design and construction of pile foundations and sheet pile walls in the conditions of reconstruction of industrial enterprises and urban development

[8] GOST R 52892-2007 Vibration and shock. Vibration of buildings. Measurement of vibration and evaluation of its effects on the structure

[9] NS 8141:2001 Vibration and shock – Measurement of vibration velocity and calculation of guideline limit values in order to avoid damage of constructions

[10] Simagin V G 2010 Design and construction of foundations near existing structures in conditions of dense development *Publishing House ASV* 126

[11] Kozmodemyansky V G, Scherba D V, Ginzburg A V, Bakhronov R R 2009 Investigation of the effect of ground vibrations during pile driving on buildings on soft soils by seismometric method *Vestnik MGSU* **3** 238-241

IOP Conf. Series: Materials Science and Engineering **945** (2020) 012056 doi:10.1088/1757-899X/945/1/012056

Effective risk management of transport facilities construction

E S Spiridonov[1*], A A Logvinenko[2] and A A Dukhov[3]

[1]Department of Design and Construction of Railways, Russian University of Transport (MIIT), Obrazcova St., 9b9, Moscow, 127994, Russia
[2]Department of Roads and Railways, Belgorod State Technological University named after V.G. Shukhov, Kostyukov St., 46, Belgorod, 308012, Russia
[3]Department of Design and Construction of Railways, Russian University of Transport (MIIT), Obrazcova St., 9b9, Moscow, 127994, Russia

E-mail: a.logvinenko534@gmail.com

Abstract. From the point of view of the resource intensity of creating communication tools (investment, labor, materials, energy, time), the leading place in production belongs to transport, and in it – railways and highways. This study examines various aspects of the application of the risk management system in the management of transport construction. Methods of risk minimization depending on the type of risk are presented: the risk of legal error, the risk of the customer's insolvency, the risk of a suboptimal loan portfolio, and the risk of nonfulfillment of financial indicators. A matrix form of comparison of identified risks is proposed for the development of the most effective methods of managing them. Using the example of the insolvency risk, which is one of the most characteristic for transport construction, the possibility of assessing the risk by a quantitative method is shown. The risk was quantified using the G. Markowitz model. To assess risks, it is proposed to use a coefficient method that allows determining the criteria that characterize their solvency based on an express assessment of the financial condition of customers, and justify recommendations for concluding contracts with them for the construction of a transport facility. The proposed matrix form of comparison of the identified risks allows formulating recommendations for reducing the impact of risks on the stability of the construction production process in a more reasonable way.

1. Introduction

A high degree of uncertainty and a risk factor are essential attributes of a market economy, especially for developing countries. Risk in the construction of a transport facility is the probability of losing part of the revenue of each organization and industry as a whole, or the possibility of obtaining part of the additional profit compared to the existing option because of activities related to overcoming uncertainty. Risk research in the construction industry is mainly divided into two categories: (1) risk identification and (2) risk assessment [1].

In the first category, the problem of research is related to risk recognition. The researchers proposed dividing risk factors into the following categories: financial risk, legal risk, management risk, market risk, social risk, political risk, technical risk, and climate risk [2-5]. When implementing international railway construction projects, the number of risks is even greater. So Liu and co-authors [6]

IOP Conf. Series: Materials Science and Engineering **945** (2020) 012056 doi:10.1088/1757-899X/945/1/012056

identified twenty important risks that affect the cost of an international project, including price fluctuations, contractor experience, and management skills. The research team looked at even more specific risk factors. In work [7] it was suggested to include the level of bureaucracy, the country's credit rating, security as risk factors, and in works [8-11] delays in the delivery of materials, rising prices for materials, construction regulations, and corruption. Social unrest, strict environmental requirements, and immaturity of the country's legal system are recognized as risk factors[10, 12]. These risk factors significantly affect the cost of construction and price indices[7, 13, 21].

In the second category of research, the risks of construction projects are assessed. Risk assessment inevitably depends on the decision makers, namely their individual experience, judgment, and ability to use different models and methods of risk measurement [14-15]. Lack of accurate and general methods for assessing the cost of risks leads to inconsistencies and uncertainties in the preparation of risk ratings. In research conducted by Dikmen I. with co-authors [16] it is pointed out that instead of a quantitative approach, it is necessary to use fuzzy approaches to risk assessment, using previously obtained knowledge about the risks that occurred. In a number of studies, researchers have combined some quantitative tools with qualitative methods for measuring risk. In a number of studies, researchers have combined some quantitative tools with qualitative methods for measuring risk. Lee K. P. and co-authors [17] applied a genetic algorithm to optimize the extracted cases and evaluate the estimated cost coefficients of construction projects. Attempts were made to assess project risks using agent-based modeling, combined with system dynamics [18] and fuzzy analytic hierarchy [19], Monte Carlo modeling [1]. An analysis of the literature suggests that the understanding of the importance of risk assessment has increased significantly over the past time. However, there is still no single and convenient method or model for risk assessment.

2. Materials and methods

2.1. Materials

To consider the risks arising from management in transport construction, the following methods were used: a systematic approach that takes into account the integrity and stability of the process; standardization and grouping by features that reflect the achievement of the final classification goals; comprehensive risk assessment. Figure 1 shows the main stages of risk management.

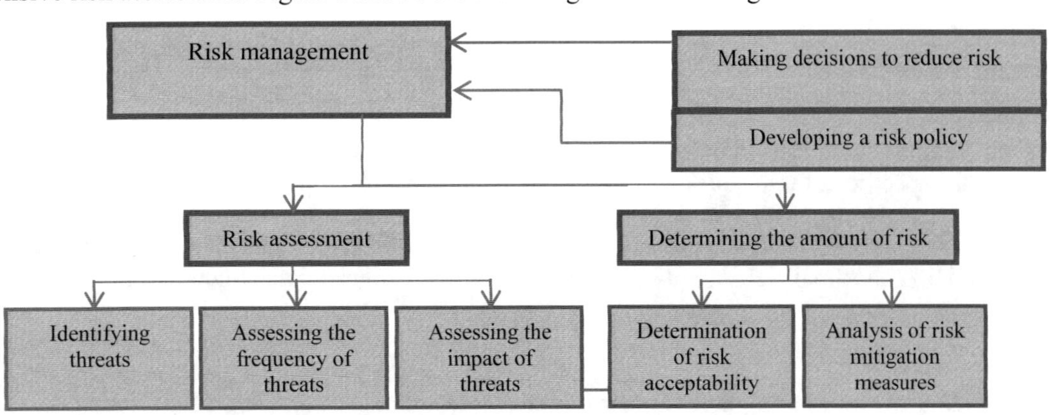

Figure 1. Main stages of risk management.

2.2. Methods

First of all, the totality of all tasks facing designers and builders is determined as a single system, and all threats are identified. After that, risks are assessed by analyzing the probability or frequency of their occurrence and their consequences are analyzed. Risk can be quantified using developed methods, such as the QRA method for quantifying risk.

IOP Conf. Series: Materials Science and Engineering **945** (2020) 012056 doi:10.1088/1757-899X/945/1/012056

There are 4 levels of risk assessment and management:

1. Global risk assessment - study of the scale of the problem and analysis of its components in order to formulate and develop risk management policies and principles.
2. Relative risk assessment in order to determine priorities and further actions in relation to the existing problem.
3. Risk assessment on a specific site to identify the main threats, risk levels, and make decisions to reduce the risk to an acceptable level.
4. Building risk maps for threat zoning and risk control planning in a given region.

Table 1 shows the risk minimization methods that can be applied when contractors work with customers.

Table 1. Methods for minimizing risks when contractors interact with customers.

Type of risk	Methods to minimize
Risk of legal error	Conclusion of the contract is possible only with legal entities registered in accordance with the current legislation; when making the contract, the employees involved carry out preparatory work with the potential customer, accepting from him a package of notarized copies of documents required for the conclusion of the contract
Risk of insolvency of the customer	Operational risk-analysis of the customer's financial condition using a coefficient method for determining the level of economic risk based on a copy of the balance sheet provided by the customer. Based on the results of this analysis, a decision is made to conclude a contract with him for the organization of construction
The risk of sub-optimal order portfolio	Optimization of the order portfolio, which is to determine how much of the order portfolio should be allocated to a particular customer so that the amount of expected accruals is distributed evenly over time
Risk of non-fulfillment of financial indicators	Improving the quality of revenue forecasts based on market research. Conclusion of an insurance contract

The risk of the customer's insolvency is understood as the possibility of losses of part of the income resulting from the provision of construction services to financially insolvent customers. Taking into account the specifics of transport construction and generalization of the practice of using methods of quantitative assessment of this risk in other areas of activity, a coefficient method for determining the level of the customer's insolvency risk was chosen. It is based on the calculation of indicators of solvency, liquidity, and financial stability used in the analysis of the financial condition of the enterprise.

3. Results

When concluding a contract with the customer for the construction organization, the following rules of operational risk analysis of the customer's solvency are proposed. The term "order portfolio of a railway transport organization implementing the construction" of a transport facility is understood as a set of customers of this division, acting as an integral management object. For example, the types of risks and the risk portfolio of Gazprom's promissory notes are shown in tables 2 and 3, respectively.

Table 2. Types of promissory note risks.

Type of risk	Risk characteristic
The credit risk of the issuer	Non-payment (delay in payment) on the bill at maturity date
Intermediary's credit risk	Non-payment or late payment of the bill. Refusal to transfer the purchased bill
Market risk	Loss of asset value because of fluctuations in the market prices of promissory notes. It occurs when purchasing promissory notes in a speculative portfolio or when urgent implementation of investment promissory notes is required to maintain liquidity
Legal risk	Purchase of a false promissory note
Operational risk	Arbitrage risk: • refusal of one of the parties to sign the contract arising from the transaction concluded over the phone; • buyer's refusal to purchase a promissity note purchased for an arbitration transaction. Risk of loss or theft of the bill отказ покупателя от покупки векселя, приобретенного для арбитражной сделки. Риск потери или хищения векселя

Table 3. Risk portfolio of promissory notes.

Risk portfolio	Operations
Exchange rate risk (includes foreign currency)	Buying an asset for resale: hard currencies, soft currencies. (Tod, tom, spot), securities, precious metals, forward contracts. The open foreign exchange position on assets and liabilities not related to the conduct of speculative operations
Interest rate risk	Transactions with assets and liabilities that are sensitive to changes in interest rates (interest-bearing assets and liabilities that do not match in terms of maturity). Transactions with different frequency of interest payment or receipt of other types of income and expenses
Counterparty risk (credit risk)	Interbank credit. Calculations (operations on counter payments, or prepayment). Pre-settlement risk (the risk of the counterparty's refusal to conduct the transaction, which will lead to exchange rate or currency risk). The issuance of the guarantee (the risk of the object of guarantee). Credit under guarantee (the risk of the guarantor). Purchase of promissory notes (risk of the promissory note issuer)
Liquidity risk	Exceeding the maturity of assets over the maturity of liabilities
Operational risk	Risk of loss of funds as a result of deficiencies in arranging the bank work

The cost of the "portfolio" is determined by the amount of cash receipts for all customers included in the portfolio and is determined by the formula (1):

$$C = \sum_{i=1}^{n} N^{\text{пол}_i} \qquad (1)$$

where C – the portfolio cost;
$N^{\text{пол}_i}$ - the amount of revenue received by the i-th customer for the time period under review.

IOP Conf. Series: Materials Science and Engineering **945** (2020) 012056 doi:10.1088/1757-899X/945/1/012056

Revenue for a particular customer is distributed unevenly throughout the year. As a result, there is a "risk of suboptimal order portfolio", which is understood as uneven cash receipts over time.

All calculations for quantitative risk assessment and optimization of the order portfolio were made using the G. Markowitz model. Despite the fact that the Markowitz model was proposed more than 60 years ago, it is still widely used by investors. A survey of 229 European portfolio managers, each with between €5 billion and €100 billion under management, found that half of those surveyed use the Markowitz optimal portfolio model when deciding whether to form an investment portfolio [20]

The value of the order portfolio is a random variable and is numerically characterized by the average monthly receipt (N) and the standard deviation (δ_p^2) calculated using formulas (2) and (3), respectively:

$$N = N^{\text{пол}_i} / n, \tag{2}$$

where $N^{\text{пол}_i}$ - amount of receipts of the i-th customer;
n – number of months.

$$\sigma_p^2 = \sum_{i=1}^{k} W_i^2 S_i^2 + \sum_{i=1}^{h}\sum_{j=1}^{h} W_i W_j \, cov_{ij}, \tag{3}$$

where S_i^2 - standard deviation of the i-th customer's receipts;
W_i - i-th client's share in the order portfolio;
W_j - j-th client's share in the order portfolio;
cov_{ij} - covariance between receipts for i-th and j-th clients.

4. Discussion

In the context of railway transport reform, the problem of ensuring the sustainable functioning of the industry and neutralizing emerging risk situations is very relevant. This problem is fully manifested in the construction of transport facilities, which makes it necessary to identify risks and develop the most effective methods of managing them.

The proposed matrix form of comparison of the identified types of risk and processes occurring in each subsystem of construction management of a transport facility allows formulating more reasonable recommendations for reducing the impact of risks on the stability of the construction production process.

5. Summary

To determine the integral point assessment of the potential customer's insolvency risk, it is proposed to use the coefficient method, which allows determining the criterion that characterizes their solvency based on an express assessment of the financial condition of customers, and justify recommendations for concluding contracts with them for the construction of a transport facility.

It should be emphasized that economic methods of risk management give a chance not only to maintain a stable position of the industry, but also to create certain prerequisites for economic development. Moreover, we will proceed from the fact that both builders and customers of the transport construction sector now understand this

6. References

[1] Yuan T, Xiang P, Li H, Zhang L 2020 Identification of the main risks for international rail construction projects based on the effects of cost-estimating risks *Journal of Cleaner Production* **274** 122904

[2] Shen L Y, Wu G W C, Ng C S K 2001 Risk assessment for construction joint ventures in China *Journal of Construction Engineering and Management ASCE* **127(1)** 76-81

[3] Loosemore M, Lim B T H, Ling F Y Y, Zeng H Y 2018 A comparison of corporate social responsibility practices in the Singapore, Australia and New Zealand construction industries *Journal of Cleaner Production* **190** 149-159

[4] Austin P, Groenhart L 2020 Improving public transport infrastructure in Auckland: Who will benefit? *2018 Joint Asia-Pacific Network for Housing Research and Australasian Housing Researchers Conference APNHR and AHRC* 2018 - Proceedings 296-305

[5] Atampugre G, Larbi M, Ojo T K, Liu G 2020 Transport system and climate change risks: Potential adaptation constraints and opportunities in Ghana *Transactions on Transport Sciences* **11(1)** 65-81

[6] Liu J, Zhao X., Yan P 2016 Risk Paths in International Construction Projects: Case Study from Chinese Contractors *Journal of Construction Engineering and Management* **142(6)** 05016002

[7] Zhang R, Xing J, Ye K, Lu W, Shan Y 2019 Synchronicity of global construction cost indexes Engineering *Construction and Architectural Management* **26(3)** 367-385

[8] Andrić J M, Wang J, Zou P X W, Zhang J 2019 Fuzzy Logic–Based Method for Risk Assessment of Belt and Road Infrastructure Projects *Journal of Construction Engineering and Management* **145(12)** 04019082

[9] Bhargava A, Labi S, Chen S, Saeed T U, Sinha K C 2017 Predicting Cost Escalation Pathways and Deviation Severities of Infrastructure Projects Using Risk-Based Econometric Models and Monte Carlo Simulation *Computer-Aided Civil and Infrastructure Engineering* **32** 620-640

[10] Saenz C, Brown H 2018 The disclosure of anticorruption aspects in companies of the construction sector: Main companies worldwide and in Latin America *Journal of Cleaner Production* **196** 259-272

[11] Owusu E K, Chan A P C, Hosseini M R 2020 Impacts of anti-corruption barriers on the efficacy of anti-corruption measures in infrastructure projects: Implications for sustainable development *Journal of Cleaner Production* **246** 119078

[12] Lingard H, Warmerdam A, Shooshtarian S 2019 Getting the balance right: Regulating occupational health and safety planning and coordination in the Australian construction industry *Engineering, Construction and Architectural Management* **26 (4)** 599-617

[13] Spiridonov E S, Dukhovny G S, Suprun E V 2015 Economic efficiency of cimulation of construction technology and organization *Bulletin of BSTU named after V G Shukhov* **6** 281-286

[14] Li H, Zhang X, Ng S T, Skitmore M 2018 Quantifying stakeholder influence in decision/evaluations relating to sustainable construction in China – A Delphi approach *Journal of Cleaner Production* **173** 160-170

[15] Hendiani S, Bagherpour M 2019 Developing an integrated index to assess social sustainability in construction industry using fuzzy logic *Journal of Cleaner Production* **230** 647-662

[16] Dikmen I, Birgonul MT, Anac C, Tah J H M, Aouad G 2008 Learning from risks: A tool for post-project risk assessment *Automation in Construction* **18(1)** 42-50

[17] Lee K P, Lee H S, Park M, Kim D Y 2017 Management-Reserve Estimation for International Construction Projects Based on Risk-Informed k-NN *Journal of Management in Engineering* **33(4)** 04017002

[18] Wu C, Chen C, Jiang R, Wu P, Xu B, Wang J 2019 Understanding laborers' behavioral diversities in multinational construction projects using integrated simulation approach *Engineering, Construction and Architectural Management* **26(9)** 2120-2146

[19] Ribas J R, Arce M E, Sohler F A, Suárez-Garcíab A 2019 Multi-criteria risk assessment: Case study of a large hydroelectric project *Journal of Cleaner Production* **227** 237-247

[20] Amenc N, Goltz F, Lioui A 2011 Practitioner portfolio construction and performance measurement: Evidence from Europe *Financial Analysts Journal* **67(3)** 39-50

[21] Glagolev S N, Bukhonova S M, Chikina E D 2020 Implementing Lean Production Management System (LPMS) in the Practice of Russian Organizations *Materials Science and Engineering* **753(7)** 072019

Acknowledgments
This work was realized in the framework of the Program of flagship university development on the base of the Belgorod State Technological University named after V G Shukhov, using equipment of High Technology Center at BSTU named after V G Shukhov.

IOP Conf. Series: Materials Science and Engineering **945** (2020) 012057 doi:10.1088/1757-899X/945/1/012057

Binders from gypsum-containing waste and products based on them

N I Alfimova[1*], S Yu Pirieva[1], M Yu Elistratkin[1], I S Nikulin[2] and A A Titenko[3]

[1]Belgorod State Technological University named after V.G. Shukhov, 46 Kostyukova str., Belgorod, 308012, Russia
[2]Belgorod National Research University, 85 Pobedy st., Belgorod, 308015, Russia
[3]Engineering Center NRU "BelSU", Belgorod, 2a/712, Koroleva st., Belgorod, 308015, Russia

E-mail: alfimovan@mail.ru

Abstract. Gypsum binders are among the most environmentally friendly from the point of view of their production. The raw material base for their production is represented not only by natural raw materials, but also by a wide variety of technogenic materials, phosphogypsum is the most multi-ton and studied of them. At the same time, citrogypsum is also a promising raw material for the production of gypsum binders and, in comparison with phosphogypsum, does not contain toxic impurities that require the development of additional measures for their removal. Synthesis of the binder from gypsum-containing waste is possible with the use of various technologies using autoclaves, chemical reagents and at elevated temperature and atmospheric pressure. Each method has its own advantages and disadvantages. The most acceptable from the point of view of industrial production is the method of calcination at atmospheric pressure, but it does not allow producing high-strength binders and products based on them. It is possible to increase the efficiency of using such binders for the production of small-piece products by applying special molding modes with increased pressing pressure. The possibility of obtaining a binder from the waste of citric acid – citrogypsum production by calcination in a semi-industrial plant is shown. Using standard methods, a complete analysis of the resulting binder was performed: granulometry, grain shape and morphology, normal density and setting time. A comparison was made with construction gypsum (G5) and high-strength gypsum (G16). The possibility of obtaining small – piece products from a synthesized binder with a dry compressive strength of 20 MPa is shown by using the semi-dry pressing method at high pressures.

1. Introduction

Gypsum binders and products based on them have always been very popular in the construction materials market, this is due to the simplicity and environmental friendliness of their production, low energy costs, capital investment and metal consumption of equipment for the production of this type of binders in comparison with cement. However, not all regions have a raw material base for their production. An alternative in this case can be gypsum-containing waste from various industrial productions.

To date, there is a wide range of research in the world aimed at considering ways to dispose of gypsum-containing industrial waste (GCW) [1-4, etc.], including as raw materials for the production of single- and multi-component binders [5-23]. Prerequisites for this research is the fact that the volume of natural gypsum stone extracted for the production of binders using traditional technology is significantly lower than the volume of GCW of various industries that are annually received in dumps. Currently, there are more than 50 types of technogenic gypsum-containing raw materials, which were considered as objects of research by various scientific groups [1].

The main difficulty in the way of large-scale and widespread use of gypsum-containing waste in the production of binders is the difficulty in applying standard technological modes and equipment used in the production of binders from natural raw materials. This is due to a number of features inherent in GCW, in particular their high humidity, dispersion, the presence of a large number of impurities in their composition, the amount and nature of which varies significantly and depends on the raw material and the technological cycle during which it was formed. Analysis of literature sources has shown the presence of a large number of variations in technologies for obtaining binders from gypsum-containing waste from various industries around the world.

Most of the research is focused on phosphogypsum, which is due to the large-scale nature of this raw material [2-4, 24], but this type of gypsum-containing waste is characterized by an increased content of impurities, including radioactive ones, which determines the need to search for methods of cleaning it before entering the production process. Another by-product of the industry containing calcium sulfate is citrogypsum – a waste of the biochemical production of citric acid. The volume of this raw material is much smaller (0.03-0.04 million tons per year.) [1], however, for a number of regions where there are accumulations of these raw materials, the problem of their processing is relevant not only from the point of view of solving environmental issues, but also from the point of view of expanding the raw material base.

There are studies aimed at considering the possibility of obtaining a binder from citrogypsum under conditions of high atmospheric pressure and temperature [25-27] and by the non-firing method using sulfuric acid as a chemical reagent [17-22]. The first method is characterized by high energy costs and the need for special equipment, the second – less energy, however, its disadvantages include the complexity of the technological process, gradual monitoring and sampling, creating unfavorable working conditions rapid wear of the equipment. Despite the fact that these studies have proved the possibility of obtaining α-modification gypsum binder with strength of up to 30 MPa, these disadvantages do not allow the organization of cost-effective production of gypsum binders by these methods.

As an alternative, we can consider the synthesis of gypsum binders from GCW by calcination, this method is the simplest from the point of view of the arrangement of the production process and equipment, but its disadvantages include the possibility of synthesizing only calcium β-semihydrate with low physical and mechanical parameters of the synthesized binder. First of all, this is due to the high specific surface area of the gypsum-containing waste itself and, as a result, the binder obtained on their basis, as well as the porosity of the particles, which determines increased water absorption and, as a result, a decrease in the compressive and bending strength of both the binder and products based on them.

This problem can be solved by using superplasticizers of different nature [28, 29], which, according to the results of various studies, can increase the strength of the binder obtained from phosphogypsum to an average of 15 MPa, and in some cases up to 29 MPa [29].

Another method of increasing the efficiency of use of such plaster is a cementitious molding with the use of elevated pressures. In particular, a group of researchers proved the possibility of obtaining plaster tiles – an analog of plasterboard. [30, 31] from a gypsum binder ($CaSO4•0.5H2O$) obtained from phosphogypsum. In the first case [30] phosphogypsum was washed and dewatered at a temperature of 150 °C, moistened to produce granules, then repeated pressing was performed in an aqueous medium at a pressure of 20 MPa with a frequency of once every 2 minutes for 24 hours. The bending strength of the resulting tile is 18.9 MPa. In the second [31], phosphogypsum was dehydrated at a temperature of 170 °C. Pressing time – 3 hours, the frequency of application of the load – every 10 minutes. The bending strength of the obtained samples was 14.7 MPa.

IOP Conf. Series: Materials Science and Engineering **945** (2020) 012057 doi:10.1088/1757-899X/945/1/012057

In this regard, this paper considers the possibility of obtaining gypsum binders from the waste of the biochemical production of citric acid – citrogypsum (Belgorod, Russia) by calcination and products based on them using the method of semidry molding at high pressures.

2. Materials and methods

The object of research was the waste of the biochemical production of citric acid – citrogypsum Belgorod, Russia.

The element and chemical composition of the studied raw materials is shown in Fig. 1 and table 1, respectively.

Visually citrogypsum is a gray powder with high specific surface area. The analysis of the morphology of citrogypsum particles made with the help of SEM showed that this raw material is mainly represented by large plate-shaped particles with a developed surface, against which the column-shaped particles are visible (Fig. 2).

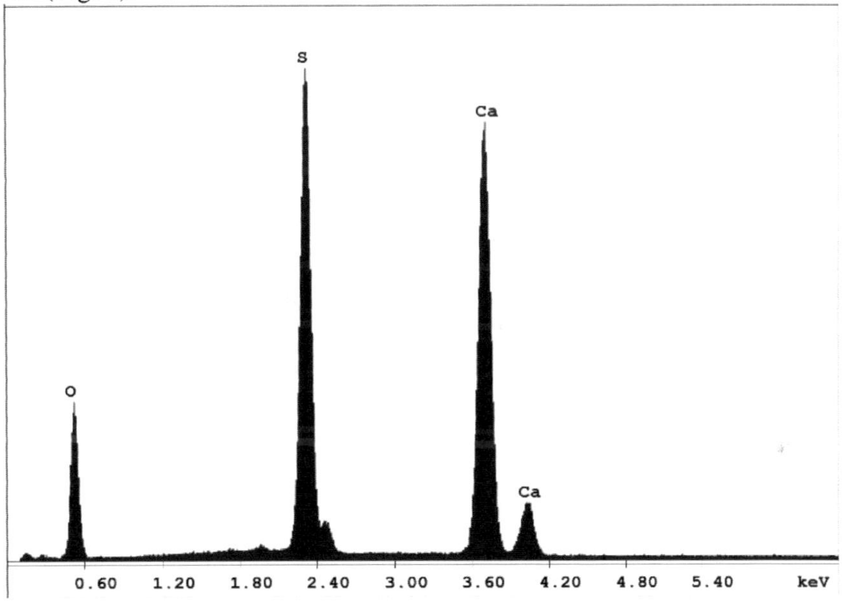

Figure 1. Elementwise composition citrogypsum, Belgorod, Russia.

Table 1. Chemical composition of citrogypsum.

Content of oxides, %				
CaO	SO_3	SiO_2	MgO	o.e.r
29.7–31.5	44.0–45.5	0.3–1.2	0.1–0.4	rest

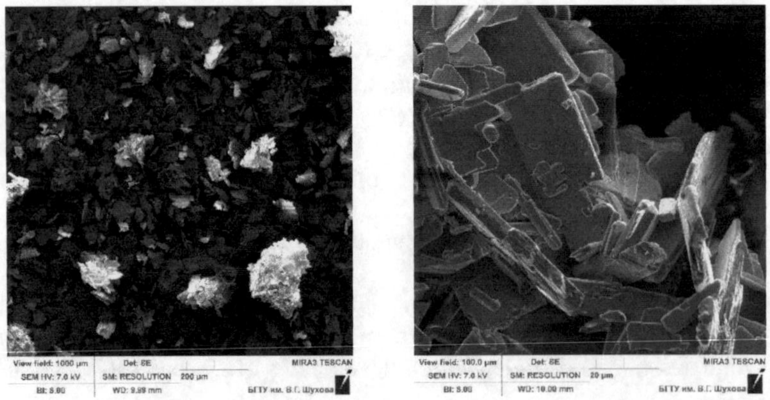

Figure 2. Morphology of the citrogypsum particles.

The process of waste accumulation in the waste storage facility has been carried out for 50 years. The total storage area is about 58.5 thousand m^2, the total reserves according to preliminary estimates – 351.2 thousand m^3. The humidity of waste in its natural state is about 80 % by weight. Their storage was collected by bucket excavators followed by mixing to average the material composition, drying to a humidity of 30 % by weight, carried out in natural conditions in the open air, by periodic mixing.

When producing gypsum stone by semidry pressing, perlite dust with a bulk density of 75 kg/m^3 was used as filler; with a granule size of 5 to 100 microns.

Test methods of gypsum binders were carried out according to GOST 23789-79.

The analysis of the particle surface morphology was performed using a high-resolution scanning electron microscope TESCAN MIRA 3 LMU, which includes an x-MAX 50 Oxford Instruments NanoAnalysis energy-dispersed spectrometer.

Qualitative assessment of the material composition of raw materials was carried out using x-ray phase analysis (RPA). The research was carried out in the range of double angles 2θ 4÷56° using the powder method by means of a general-purpose x-ray diffractometer (DRON-3M) operating in automated mode using the PELDos program.

The granulometric composition of powdered materials was determined using the laser granulometry method on the ANALYSETTE 22 NanoTec plus laser particle analyzer.

Production of the binder by firing was carried out in a semi-industrial installation. Based on the results of numerous studies in this field, a mode was chosen in which the temperature of the binder at the output was about 120 °C.

To obtain the binder in the laboratory, a heat chamber was used.

3. Results

The specific surface area of the resulting binder was 600-650 m^2/kg.

When determining the normal density, it was found that for a gypsum test with a spreading capacity of 180±5 mm, 370 water is needed for 300 g of gypsum binder, thus NG=123%.

According to the terms of setting, the resulting gypsum binder according to GOST 125-79 refers to a normally hardening one (index B, beginning no earlier than 6 min, ending no later than 30 min) (table 2).

Table 2. Setting time of the produced binder.

Water inflow time, h-min	Start of setting time, h-min	End of setting time, h-min	Start of setting, h-min	End of setting, h-min
9-37	9-55	10-07	0-18	0-30

IOP Conf. Series: Materials Science and Engineering **945** (2020) 012057 doi:10.1088/1757-899X/945/1/012057

The strength of samples made from gypsum paste of standard consistency is determined 2 hours after the contact of the gypsum binder with water. According to this indicator, the tested sample of gypsum binder does not meet the requirements of GOST 125-79.

The ultimate strength of samples-beams with dimensions of 40 × 40×160 mm at the age of 2 h, formed from gypsum paste with a humidity of 30 % on a vibrating platform (non-standard method), was 2.5 MPa (25 kgf/cm^2) at bending, 3.3 MPa (34 kgf/cm^2) at compression. Gypsum binder conditionally (due to non-standard molding methods and normal density) can be attributed to the G-3 brand.

The analysis of the particle size of the obtained binder and comparison with high-strength gypsum and citrogypsum was carried out (Fig. 3). The graphs show that high-strength gypsum is characterized by a polymodal distribution of particles in a wider range, while the graphs that characterize the granulometry of the original citrogypsum and the binder are single-modal with a distribution of particles in a narrower range. It should also be noted that after high-temperature exposure, the particle size decreases slightly, which is due to the settling of a larger fraction during the passage of citrogypsum through the turbine-type dryer.

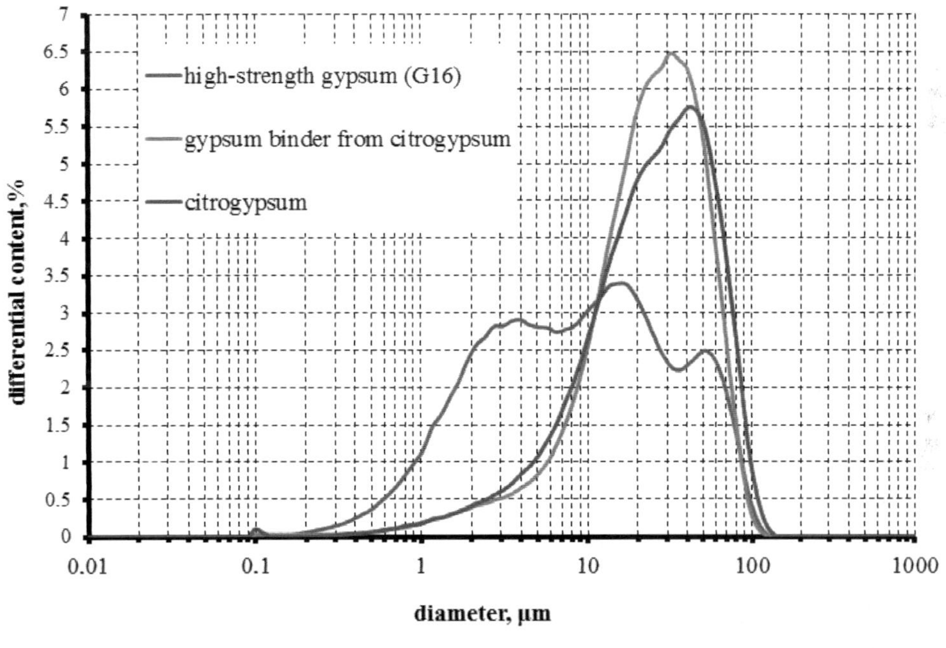

Figure 3. Granulometric composition of citrogypsum and gypsum binders.

Comparative analysis of the obtained binder particles with construction (G-5) and high-strength gypsum (G-16) revealed differences in their morphology and size. Thus, in photos of high-strength gypsum, enlarged columnar crystals stand out clearly against the background of small particles of polydisperse composition (Fig. 4. c). For building gypsum, there is also a significant spread of particles in size and the presence of elongated crystals, but their size is much smaller (Fig. 4, b). At the same time, the crystals of the binder obtained from citrogypsum have a lamellar shape and not a significant size range.

IOP Conf. Series: Materials Science and Engineering **945** (2020) 012057 doi:10.1088/1757-899X/945/1/012057

a

b

c

Figure 4. The morphology of the particles of gypsum binders
a – binder from citrogypsum; b – building gypsum (G-5) c – high-strength gypsum (G-16).

4. Discussion

According to the results, the gypsum binder obtained from citrogypsum corresponds to the setting time of normally hardening construction gypsum. However, the determination of the strength of samples made from gypsum paste of standard consistency at the age of 2 hours showed that they do not meet the requirements of the standard.

Production of products using injection molding technology from a synthesized binder is not possible due to its low strength. However, the fact that reducing the water demand of gypsum paste to 30 %

and the use of vibrocompression allowed obtaining samples with a compressive strength of 3.3 MPa, gave reason to believe that the use of non-standard molding methods, in particular, pressing at high pressures, will allow obtaining products with the necessary physical and mechanical characteristics.

In particular, a patent was obtained [32] for the composition of the raw mixture and a method for producing gypsum stone.

The raw mixture consisted of a binder obtained from citrogypsum and perlite dust. Perlite dust was water-saturated before mixing with a gypsum binder made of citrogypsum, which made it easier to mix and obtain a homogeneous mass without the formation of lumps and a uniform distribution of moisture by volume. After that, the molding mass was sent to the mold and held for 30 seconds under a pressing pressure of 10 MPa.

High specific surface of the binder of citrogypsum in conjunction with the proposed method of molding increases the number of contacts between particles, while low water-binder ratio (water is taken in an amount sufficient for the occurrence of a chemical reaction) minimizes the number of pores formed during evaporation of excess moisture and amount of the involved air, which in the aggregate contributes to the strength of the final product.

The composition of the raw material mixture and physical and mechanical characteristics are shown in table 5.

Table 3. Raw mix compositions and physical and mechanical characteristics of products.

Material consumption % by weight			Water-binder ratio	Compressive strength after stripping MPa	Compressive strength 2 hours after pressing, MPa	Compressive strength of dry products, MPa	Water absorption, %
Binder, obtained from citrogypsum	Perlite dust	Water					
83	1.5	15.5	0.18-0.19	2.5	15–16	19–20	No more than 5 %

5. Summary

The principal possibility of obtaining a gypsum binder from the waste of the biochemical production of citric acid – citrogypsum by calcination is shown.

The possibility of increasing the efficiency of the binder obtained from citrogypsum as a raw material for the production of small-piece products using semidry pressing technology is proved. The strength of the products obtained using this technology after stripping was 2.5 MPa, which will significantly reduce the number of defects formed at this stage of product production, and will also allow the production of multi-hollow products. The compressive strength of products in the dry state reached 20 MPa.

6. References

[1] 2004 Plaster materials and products (production and use) *Directory. Edited by A V Ferronskaya. Publishing House ASV* 488 (rus)

[2] Rashad A M 2017 Phosphogypsum as a construction material *Journal of Cleaner Production* **166** 732–743 doi:10.1016/j.jclepro.2017.08.049.

[3] Cánovas C R, Macías F, Pérez-López R, Basallote M D, Millán-Becerro R 2018 Valorization of wastes from the fertilizer industry: Current status and future trends *Journal of Cleaner Production* **174** 678–690 doi:10.1016/j.jclepro.2017.10.293

[4] Klimenko V G 2018 Influence of modifying composition of gypsum binders on the structure of composite materials *Journal of Physics: Conference Series* **1118 (1)** 43285 doi: 10.1088/1742-6596/1118/1/012019

[5] Yang J, Liu W, Zhang L, Xiao B 2009 Preparation of load-bearing building materials from au-

toclaved phosphogypsum *Construction and Building Materials* **23(2)** 687–693 doi:10.1016/j.conbuildmat.2008.02.011

[6] Myasnikov N F, Bershakov N G, Kozlov V P, Naumov E G, Shevchenko N N 2003 Gypsum binder preparation method *Patent RF* 2210540

[7] Semlyov V S, Reutov V A, Kondrikov N B 2008 Method of recycling gypsiferous raw materials *Patent RF* 2324654.

[8] Gubskaya A G, Podluzsky E Ya, Melenko V S 2018 Production of gypsum binder and products from natural and industrial raw materials in the Republic of Belarus *Building Materials* **3** 73–75 (rus)

[9] Ma B, Jin Z, Su Y, Lu W, Qi H, Hu P 2020 Utilization of hemihydrate phosphogypsum for the preparation of porous sound absorbing material *Construction and Building Materials* **234** 117346 doi:10.1016/j.conbuildmat.2019.117346

[10] Li X, Zhang Q, Ke B, Wang X, Li L, Li X, Mao S 2018 Insight into the effect of maleic acid on the preparation of α-hemihydrate gypsum from phosphogypsum in Na_2SO_4 solution, Journal of Crystal Growth **493** 34–40 doi:10.1016/j.jcrysgro.2018.04.025

[11] Li X, Zhang Q, Shen Z, Li L, Li X, Ma S 2019 L-aspartic acid: A crystal modifier for preparation of hemihydrate from phosphogypsum in $CaCl_2$ solution *Journal of Crystal Growth* **511** 48–55 doi:10.1016/j.jcrysgro.2019.01.027

[12] Lu W, Ma B, Su Y, He X, Jin Z, Qi H 2019 Preparation of a-hemihydrate gypsum from phosphogypsum in recycling $CaCl_2$ solution *Construction and Building Materials* **214** 399–412 doi:10.1016/j.conbuildmat.2019.04.148

[13] Ma B, Lu W, Su Y, Li Y, Gao C, He X 2018 Synthesis of α-hemihydrate gypsum from clean-er phosphogypsum *Journal of Cleaner Production* **195** 396e405 doi:10.1016/j.jclepro.2018.05.228

[14] Guan B, Yang L, Wu Z, Shen Z, Ma X, Ye Q 2009 Preparation of α-calcium sulfate hemihydrate from FGD gypsum in K, Mg-containing concentrated $CaCl_2$ solution under mild conditions *Fuel* **88** 1286–1293 doi:10.1016/j.fuel.2009.01.004

[15] Shen Z X, Guan B H, Fu H L, Yang L 2009 Effect of Potassium Sodium Tartrate and Sodium Citrate on the Preparation of α-Calcium Sulfate Hemihydrate from Flue Gas Desulfurization Gypsum in a Concentrated Electrolyte Solution *J. Am. Ceram. Soc.* **92** 2894–2899 doi:10.1111/j.1551-2916.2009.03330.x

[16] Guan B, Kong B, Fu H, Yu J, Jiang G, Yang L 2012 Pilot scale preparation of a-calcium sulfate hemihydrate from FGD gypsum in Ca–K–Mg aqueous solution under atmospheric pressure *Fuel* **98** 48–54 doi:10.1016/j.fuel.2012.03.032

[17] Tarasova G I, Sverguzova S V, Bubnova N Yu, Kozlov V P, Naumov E G 1999 Method of gypsum article making *Patent RF* 2132310.

[18] Sverguzova S V, Tarasova G I, Bubnova N Yu 1998 Promising technologies for the processing of citrogypsum *Ecology and Industry of Russia* **8** 20–24 (rus)

[19] Sverguzova S V, Bubnova N Yu, Tarasova G I 2001 Utilization of gypsum-containing waste by energy-saving technology *Science-production* **3** 41–43 (rus)

[20] Sverguzova S V, Chernysheva N V, Chernysh L I, Shamshurov A V 2010 The in-fluence of the processing conditions of citrogyp-sum on the composition of the obtained gypsum bind-er, *Building Materials* **7** 31–32 (rus)

[21] Sverguzova S V, Tarasova G I, Chernysheva N V, Chernysh L I 2010 Theoretical substantiation of the possibility of non-fired de-hydration of citrogypsum *Bulletin of BSTU named after V. G. Shukhov* **2** 117–121 (rus)

[22] Chernysheva N V, Sverguzova S V, Tarasova G I 2010 Obtaining a gypsum binder from phosphogypsum of Tunisia *Building Materials* **7** 28–30 (rus)

[23] Jin Z, Ma B, Su Y, Lu W, Qi H, Hu P 2020 Effect of calcium sulphoaluminate cement on mechanical strength and waterproof properties of beta-hemihydrate phosphogypsum *Construction and Building Materials* **242** 118198 doi:10.1016/j.conbuildmat.2020.118198

[24] 2020 Mineral Commodity Summaries *Department of the Interior, U S Geological Survey 122–*123 https://minerals.usgs.gov/minerals/pubs/mcs/2016/ mcs2020.pdf
[25] Lukyanova A N, Starostina I V 2013 Building composite materials based on modified gypsum binders obtained from industrial wastes *Fundamental Research* **4(4)** 818–822 (rus)
[26] Edamenko A S, Klimenko V G 2013 About the possibility of using technogenic raw materials in the production of building materials *Internet magazine "Technosphere safety technologies"* **1 (47)** (rus)
[27] Nesterova L L 2016 Possibility of obtaining gypsum binder from citrogypsum *Bulletin of the Yugorsk State University* **2 (41)** 29–33 (rus)
[28] Zheng C, Yu Q, Ning P, Ma J-B, Zheng Y-L 2015 Preparation of β-hemihydrate gypsum with phosphogypsum *Xiandai Huagong/Modern Chemical Industry* **35 (5)** 60–63
[29] Bumanis G, Zorica J, Bajare D, Korjakins A 2018 Technological properties of phosphogyp-sum binder obtained from fertilizer production waste *Energy Procedia* **147** 301–308 https://doi.org/10.1016/j.egypro.2018.07.096
[30] Zhou J., Sheng Z., Li T., Shu Z., Wang Y 2016 Preparation of hardened tiles from waste phosphogypsum by a new intermittent pressing hydration *Ceram. Int.* **42** 7237–7245 doi:10.1016/j.ceramint.2016.01.117
[31] Zhou J, Li X, Zhao Y, Shu Z, Wang Y, Zhang Y, Shen X 2020 Preparation of paper-free and fiber-free plasterboard with high strength using phosphogypsum *Construction and Building Materials* **243** 118091 doi:10.1016/j.conbuildmat.2020.118091
[32] Alfimova N I, Titenko A A, Nikulin I S, Galdun Yu V, Pirieva S Yu, Elistratkin M Yu 2018 Raw mixture for modified citrogypsum articles and method of their production *Patent RF* 2693808.

Acknowledgments
The work was carried out as part of the implementation of activities of the world-class scientific and educational center “Innovative Solutions in the Agricultural Sector”of Belgorod State University. Agreement #50 of 07.08. 2020; using equipment of High Technology Center at BSTU named after V G Shukhov.

IOP Conf. Series: Materials Science and Engineering **945** (2020) 012058 doi:10.1088/1757-899X/945/1/012058

Plasters with wood processing waste additives

T A Drozdyuk[1*], Yu V Sokolova[1] and T A Makhova[1]

[1]Department of composite materials and environmental engineering, Northern (Arctic) Federal University named after M.V. Lomonosov, Severnaya Dvina Emb., 17, Arkhangelsk, 163002, Russia

E-mail: t.drozdyuk@narfu.ru

Abstract. The paper assesses the influence of additives based on wood processing industry waste on the physical and mechanical properties of plasters. As additives to the plaster, the bark and sawdust of Scots pine were used, previously crushed and purified from extractive and easily hydrolysable substances. Extraction of extractive substances from wood processing waste was performed by hot water extraction. The content of additives was in the range from 1.5 to 6 % by weight of cement. The average density, compressive strength, flexural tensile strength, water absorption, and thermal conductivity were monitored for the obtained samples of plasters. As a result of the carried out experiments, it was found that the quantitative increase of the additive in the mixture reduces the density of the mortar, water absorption increases slightly, the coefficient of thermal conductivity and the tensile strength during bending decreases. The compressive strength increases with the addition of sawdust 1.5 %, in other cases it decreases with the increase in the amount of the additive. It is shown that mortars with the addition of fine wood processing waste can be used as plasters with reduced thermal conductivity for interior work.

1. Introduction

Construction is a large-scale industry, the development of which moves on. Mass construction of residential and non-residential buildings creates the need to re-evaluate the material and raw material base of the construction industry in order to rationalize its use and reduce net cost.

When building objects, major repair or refurbishing finishing works are necessary, in the production of which dry plaster mixtures are used. Cement plaster mixtures are common because they provide high strength, good adhesion to the surface, wear resistance, low thermal conductivity and water resistance.

The task of choosing raw materials for plaster mixtures that provide high quality and low cost of the solution becomes urgent, so currently there is an active search for effective additives that help reduce the cost and improve the strength and thermal insulation properties of plaster mixtures [1, 2].

One of the leading areas of production in the Arkhangelsk region is the woodworking industry. In this regard, a large amount of waste is generated, which includes wood bark, sawdust, wood chips, wood flour, and others that can be used as a replacement for part of the cement in mortars, including plasters ones. The use of such woodworking waste in composite materials [3-6] will reduce the consumption of cement, the production of which has a negative impact on the environment. In addition, replacing part of the cement with wood processing waste will reduce the cost of the final product and improve its quality characteristics.

However, when adding wood processing waste to plasters, it is necessary to take into account the fact that wood and Portland cement have a complex composition, and mixing them with water causes a significant number of competing chemical reactions [7, 8]. Organic matter (wood) and inorganic hydraulic binder (cement) are antagonistic in nature. Under the action of the highly alkaline liquid phase of cement (pH=12 ... 14), lignocarbon bonds are destroyed, thus violating the integrity of the material structure. The effect of water-soluble substances of wood on hardening cement is shown in the stabilizing effect [9]. "Cement poisons", i.e. extractive and easily hydrolysable substances, consist mainly of carbohydrate groups [10-12]. These substances are deposited on the surface of the cement mineral particles $3CaO \cdot SiO_2$ (tricalcium silicate) and $3CaO \cdot Al_2O_3$ (tricalcium aluminate) and form the thinnest shells that hinder the course of cement hydration processes. But monomeric sugars (glucose, xylose, derivatives of glucuronic and ascorbic acid) in a small amount (up to 0.125 wt. %) accelerate the setting processes and promote structure formation, and at a concentration of more than 0.25 wt. %, on the contrary, slow down the setting and hardening of cement, which completely eliminates the possibility of obtaining strong compositions. One of the easiest ways to reduce the negative impact of water-soluble extractive and easily hydrolysable substances on the strength of wood-cement compositions is pre-extractive wood processing [13, 14].

Based on the above, the aim of this study is to assess the impact of additives based on pre-prepared wood processing industry waste on the physical and mechanical properties of plasters.

2. Materials and methods

As an additive to the plaster mixture, the bark and sawdust of Scots pine were used, pre-extracted with hot water to extract the water-soluble components of the wood, and then dried to a constant mass at a temperature of 80 °C. The efficiency of the extraction process was controlled by reducing the optical density of the water extract taken from the reaction system at specified time intervals. The optical density was determined by recording the electron absorption spectra of experimental liquid samples on a SF-2000 spectrophotometer in the wavelength range of 270-480 nm. Extraction was stopped when the absorption spectrum of the water extract stabilized. The wavelength range was chosen based on the literature data, according to which the maximum absorption for most water-soluble extractives of wood (water-soluble phenolic compounds, monosaccharides, etc.) is located in this range. Then the bark and sawdust were ground by mechanical dry dispersion lasting 15 minutes at a rotor speed of 420 rpm with large grinding bodies made of stainless steel with a diameter of 20 mm in the amount of 18 pieces at the Retsch RM100 planetary ball mill. The grinding time was taken based on pre-conducted experiments. To obtain a uniform plaster mixture, to mix the dry components well, and to evaluate the effect of additives on the properties of mortars, the size of the bark and sawdust particles must correspond to the size of the cement particles. In order to meet this condition, it is necessary that the remaining 0.09 mm of crushed bark and sawdust on the screen does not exceed 0.8 %, which is achieved with a grinding time of 15 minutes.

As a binder, we used Portland cement of type CEM I normal hardening strength class 42.5 MPa. According to the quality certificate, this cement has a residue on the sieve of 0.09 mm-0.8 %, so it is not necessary to grind sawdust and bark to a size smaller than the cement particles, i.e. more than 15 minutes, and carries extra energy costs.

For the production of experimental samples of plasters, fine river polymineral sand of the Krasnoflotsky-Zapad Deposit was used, pre-ground and sifted through a sieve with round holes with a diameter of 2.5 mm. Production of prototypes began with the calculation of the ratio of components in the mortar. The ratio of cement and sand was assumed to be 1:4. For samples with additives, part of the cement was replaced with crushed sawdust or bark. These samples had sawdust or bark in the amount of 1.5 %, 3 %, 4.5 % and 6 % by weight of cement. Then the dry components of the mixture were mixed, after which the required amount of water was gradually added. Plaster mixtures were thoroughly mixed and stacked in layers in metal molds, tamping each layer. The samples were

IOP Conf. Series: Materials Science and Engineering **945** (2020) 012058 doi:10.1088/1757-899X/945/1/012058

decomposed after 24 hours. At the same time, during the production of prototypes, the mobility, density, and water-holding capacity of the mixture were monitored using standard methods. The average density, compressive strength, flexural tensile strength, water absorption, and thermal conductivity were determined for the obtained samples of plaster mortars using standard methods.

3. Results and discussion

Thus, samples of plaster mortars were made with the addition of sawdust and bark, the consumption of the mixture components for which is shown in table 1. This table also provides data on the density and water retention capacity of mixtures.

Table 1. The consumption of the plaster mixture components.

Additive	Amount of sand, g	Amount of cement, g	Amount of water on calculation, g	Actual amount of water, g	Amount of additive, g	Mixture density, g/cm^3	Water retention capacity of the mixture, %
Without additive	2000.0	500.0	250.0	407.0	0	1.89	95.55
Sawdust 1.5 %	2000.0	492.5	246.3	414.5	7.5	1.89	95.65
Sawdust 3 %	2000.0	485.0	242.5	419.5	15.0	1.86	95.69
Sawdust 4.5 %	2000.0	477.5	238.8	424.5	22.5	1.82	95.72
Sawdust 6 %	2000.0	470.0	235.0	428.5	30.0	1.79	96.14
Bark 1.5 %	2000.0	492.5	246.3	410.5	7.5	1.90	95.87
Bark 3 %	2000.0	485.0	242.5	415.5	15.0	1.90	95.97
Bark 4.5 %	2000.0	477.5	238.8	420.5	22.5	1.88	96.80
Bark 6 %	2000.0	470.0	235.0	425.5	30.0	1.87	97.07

From the obtained data, it can be seen that, as expected, the greater the amount of additive, the lower the density of the mixture. This is due to the fact that sawdust and bark are much lighter than cement, so the mass of the mixture is reduced. In accordance with clause 4.5.3 of GOST 33083-2014, the water-holding capacity of mixtures ready for use must be at least 95 %. This requirement for prepared mixtures is met. As it can be seen from the data presented in table 1, the water retention capacity increases with the increase of the additive. But with the addition of bark, the increase is more intense than with the addition of sawdust.

Figures 1 and 2 show the testing results on compression and flexural strength of samples of mortars after their 28-day of exposure.

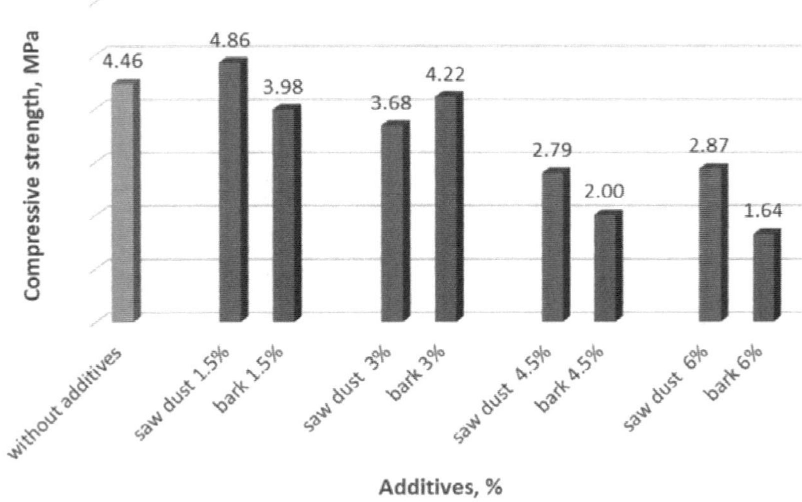

Figure 1. Change in the compressive strength of mortar samples with different additive content.

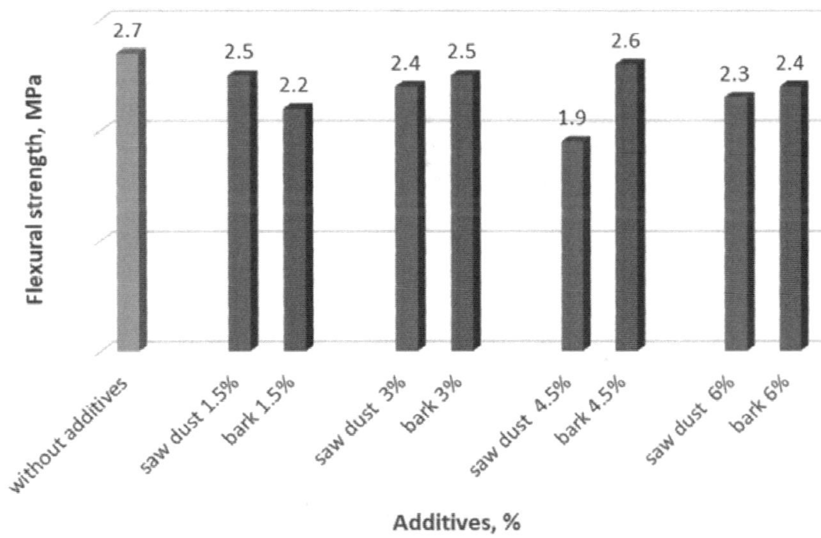

Figure 2. Change in the flexural strength of mortar samples with different additive content.

From the data presented in figure 1, it can be seen that the greatest strength of the samples was achieved with the addition of sawdust in the amount of 1.5 %. In our opinion, this is due to the residual amount of extractive substances contained in the sawdust. When extracted with water, tannins and sugary substances, organic acids, mineral salts and acids are extracted from wood, but other extractive substances – fats, resins, waxes, essential oils and others can only be isolated by organic solvents, so they are present in sawdust. It is known that extractive substances negatively affect the adhesion of cement to wood. But with a small amount of them, namely 0.02-0.04 % of the mass of the mortar, its strength increases. This is due to the fact that with small amounts of extractive substances, i.e. with a low degree of saturation with adsorption layers (these layers are formed around the cement grains due to the hydrophilicity of their surface), the dispersion effect prevails. The predominance of the dispersion effect (distribution of active fine particles in the bulk phase) over the stabilization effect (shielding of the surface of these particles) causes a marked increase in strength.

Also, the decrease in the strength of the plaster mixture, observed with an increase in the amount of the additive, can be explained by the properties of the additive itself, namely: significant volumetric humidity deformations and the development of swelling pressure; expressed anisotropy of properties and structure; high permeability; low adhesion to cement stone. It should also be noted that the strength of samples with the addition of bark is less than with the addition of sawdust. The bark is a dead layer of cells, has a low density and low strength. Sawdust has a much higher density and strength than bark, so the mortar with the addition of sawdust has better strength indicators than samples with bark.

The results given in figure 2 show that the flexural strength of the samples with plant additives is reduced compared to the control sample. For samples with added bark, the flexural strength is higher than for samples with added sawdust. However, mixtures with the addition of sawdust 1.5 % and with the addition of bark 3 % and 4.5 % have indicators close to the strength of the control. Crushed sawdust and bark are fibers that reinforce the hardened mixture. Also, sawdust and bark have a rough surface, so as the amount of additive increases, the adhesion of the wood additive to the cement matrix increases.

The results of measuring the coefficient of thermal conductivity of samples of plaster mortars are shown in figure 3.

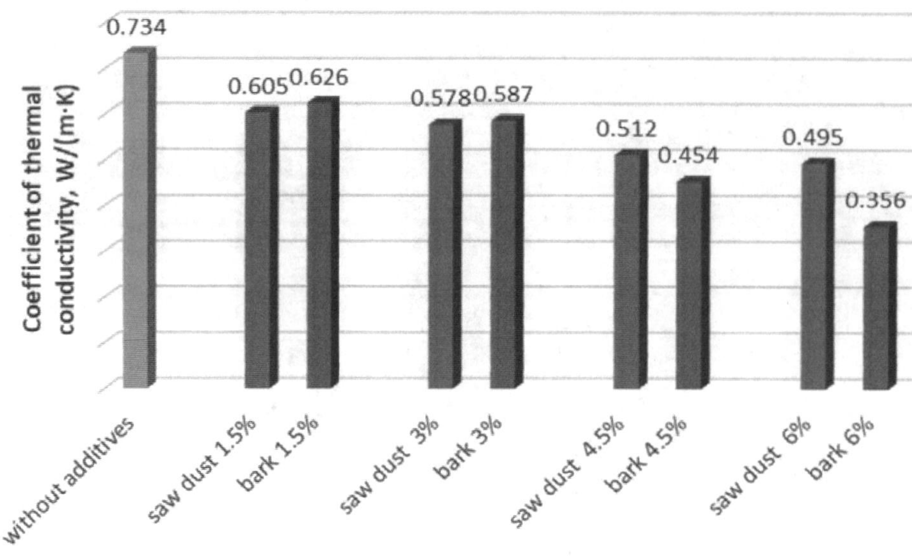

Figure 3. Change in the value of the thermal conductivity coefficient of experimental samples from the number of additives.

From the data presented in figure 3, it follows that with an increase in the amount of plant additives, the coefficient of thermal conductivity decreases. With the addition of 6 % bark, the coefficient of thermal conductivity is almost 2 times lower than that of a mortar without additives. Wood has a low coefficient of thermal conductivity due to its low density, so samples with an additive have a lower coefficient compared to the control sample without additives.

Table 2 shows the results of determining the average density and water absorption by weight and volume of samples of plasters.

Table 2. Average density and water absorption of plaster mortar samples.

Additive	Mortar density, g/cm^3	Water absorption of the mortar by weight, %	Water absorption of the mortar by volume, %
Without additive	1.89	0.11	0.21
Sawdust 1.5 %	1.80	0.12	0.22
Sawdust 3 %	1.78	0.14	0.24
Sawdust 4.5 %	1.76	0.14	0.24
Sawdust 6 %	1.75	0.14	0.24
Bark 1.5 %	1.84	0.10	0.18
Bark 3 %	1.82	0.12	0.23
Bark 4.5 %	1.79	0.13	0.24
Bark 6 %	1.77	0.14	0.26

The data given in table 2 show that as the additive content in the mixture increases, the density of the mortar, as well as the density of the mixture, decreases. As the amount of additive increases, water absorption increases by both weight and volume. This is due to the microporous structure of the wood matrix, which can be filled with water. But the increase in water absorption is insignificant, even with the addition of sawdust or bark 6 %.

4. Summary

It is shown that the waste of mechanical processing of plant raw materials (sawdust, bark) after preliminary preparation, consisting in the removal of extractive substances, can be used as an additive that replaces part of the cement in plasters.

It is experimentally established that the replacement of 1.5 % (by weight) of cement with sawdust allows to obtain plasters with improved thermal insulation characteristics, while not inferior in their physical and mechanical characteristics (compressive strength, flexural strength, water absorption) to mortars without additives used in this work. Mortars with the addition of fine wood processing waste can be used as plasters with reduced thermal conductivity for interior work.

5. References

[1] Elistratkin M Yu, Minakova A V, Jamil A N, Kukovitsky V V, Issa Jamal, Issa Eleyan 2018 Composite binders for finishing compositions *Construction Materials and Products* **1(2)** 37-44

[2] Elistratkin M Yu, Minakov S V, Shatalova S V 2019 Composite binding mineral additive influence on the plasticizer efficiency *Construction Materials and Products* **2(2)** 10-16

[3] Danilov V E, Shinkaruk A A, Ayzenstadt A M 2019 Technological features and prospects of production of innovative wood-mineral composite *Construction materials, equipment, technologies of the XXI century* **7-8** 24-27

[4] Danilov V E, Ayzenstadt A M, Makhova T A 2017 Structural thermal insulation based on waste from the woodworking and mining industries *Industrial and civil construction* **1** 97-100

[5] Kilyusheva N V, Danilov V E, Ayzenstadt A M 2016 Thermal insulation material from pine bark and its extract *Construction materials* **11** 48-50

[6] Garcez M R, Garcez E O, Machado A O, Gatto D A 2016 Cement-wood composites: effects of wood species, particle treatments and mix proportion *International journal of composite materials* **6(1)** 1-8

[7] Sushkov A S 2014 Formation of wood composite materials from wood processing waste *Actual directions of scientific research of the XXI century: Theory and practice* **2(2-1)** 160-169

[8] Posharnikov F V, Filichkina M V 2011 Research of regularities of cement hydration in wood-composite materials *Modern problems of science and education* **2** 21-21

[9] Juenger M C G, Jennings H M 2002 New insights into the effects of sugar on the hydration and

IOP Conf. Series: Materials Science and Engineering **945** (2020) 012058 doi:10.1088/1757-899X/945/1/012058

microstructure of cement pastes *Cement and concrete research* **32(3)** 393-399

[10] Polishchuk A I, Rubinskaya A V 2012 Chemical aggressiveness of plant-based aggregate in relation to cement *Actual problems of the forest complex* **34** 72-74

[11] Kondrashchenko V I, Krivoborodov Yu R, Kendyuk A V 2017 Influence of easily hydrolyzable wood substances on the properties of cement stone *Third interdisciplinary youth scientific forum with international participation "New materials"* 596-598

[12] Krivorotova A I, Usoltsev O A 2015 On the influence of extractive substances of wood on the formation of cement-wood composites *Scientific works SWorld* **13(1)** 97-101

[13] Tjeerdsma B F, Militz H 2005 Chemical changes in hydrothermal treated wood: FTIR analysis of combined hydrothermal and dry heat-treated wood *Holz als roh-und Werkstoff* **63(2)** 102-111

[14] Safin R R, Nazipova F V, Voronin A E, Ziatdinov R R 2015 Ultrasonic extraction of water-soluble sugars from wood in the production of composite materials *Bulletin of Kazan Technological University* **18(16)** 188-190

Acknowledgments

The research was carried out on the unique scientific equipment "Physical Chemistry of Surfaces of Nano-Dispersed Systems".

Improvement directions of technical regulations in the field of reinforced cellular concrete structures

P V Kladieva[1*] and Yu A Koshlich[2]

[1]Department of Construction Management and Real Estate, Belgorod State Technological University named after V.G. Shukhov, Kostyukov St., 46, Belgorod, 308012, Russia
[2]Department of Automation and Informatization, Belgorod State Technological University named after V.G. Shukhov, Kostyukov St., 46, Belgorod, 308012, Russia

E-mail: davidenkopolly@mail.ru

Abstract. Effective management of real estate complexes, which determines the rational capital intensity of ownership of buildings and structures, largely depends on the consumer qualities of building materials and products included in the investment and construction project at the design stage. The implementation of the principles of rational design, which are regulated in unique and technically complex and responsible structures, is still not properly reflected in the design of second-class objects of responsibility, which, however, make up more than 70% of the capital intensity of the construction market, mainly housing. Complex composite materials that combine innovative and energy-efficient components and technologies, as well as structures made from them that contain significant reserves of load-bearing capacity, which are opened and implemented by improving design standards and rational technical solutions, have the greatest potential for rationalization. Reinforced cellular concrete, which effectively combines load-bearing and enclosing functions, is one of the most popular construction technologies in the perspective practice of low-rise construction. Currently, however, the design of reinforced structures made of cellular concrete in Russia, abroad, mainly performed according to the normative documents, often not taking into account the significant differences in physical and mechanical properties, material structure, principles and working mechanisms of various types of concrete under load, as well as the uniqueness of plastic properties of cellular concrete when assigning the form and composition of discrete reinforcement. In this regard, there is a need to revise and systematize existing standards that contain requirements for the design of directly reinforced structures made of cellular concrete, generalize European experience and these standards, update information in them, as well as develop modern national standards in this area. In addition, the practice of designing bent cellular concrete structures has revealed a significant reserve for increasing the load-bearing capacity of these elements, associated with insufficient inclusion of the material of the traditionally used rod reinforcement due to the primary destruction of the contact layer (pulling the reinforcement). In order to increase the durability and reduce the resource intensity of national cellular concrete structures, the adopted methodology for their design needs to improve both the regulations that establish the principles and rules for calculating reinforcement elements of structures, and structural elements used in precast and in-situ practice of discrete reinforcement.

IOP Conf. Series: Materials Science and Engineering **945** (2020) 012059 doi:10.1088/1757-899X/945/1/012059

1. Introduction

Rational design of load-bearing and enclosing structures, which is an essential and integral part of effective management of real estate complexes, is supported by the principles and rules of rational design in the current technical regulations. Modern national standards for the design of cellular concrete structures are reflected in a variety of existing guidelines of the regulatory framework of the Russian Federation in this area. They are represented by codes of rules, state standards, technical requirements and instructions, as well as building codes and some other manuals that establish requirements for the calculation and design of concrete and reinforced concrete structures of industrial and civil buildings and structures.

2. Materials and methods

2.1. Materials

To date, the design of reinforced structures made of cellular concrete is mainly carried out according to regulatory documents that extend their effect and establish the same requirements for the design of concrete and reinforced concrete structures made from various types of concrete: heavy, fine-grained, light, cellular and straining [1]. However, it is worth noting that these rules do not take into account the significant differences in physical and mechanical properties, material structure, principles and working mechanisms of various types of concrete under load, as well as the uniqueness of plastic properties of cellular concrete, much different from the properties of heavy concrete. Thus, the basic requirements for finished cellular concrete products are similar to those for reinforced concrete structures made of heavy concrete, designed according to classical theories and models for calculating reinforced concrete, based on experimental data and laws of mechanics applicable to all types of concrete.

At the same time the requirements of SP 63.13330 do not apply to the design of structures made of concrete with an average density of less than 500 kg/m3, although many structures made of cellular concrete used in low-rise construction have a significantly lower density, and therefore cannot be calculated on the basis of this document, and the design of such structures must be carried out in accordance with relevant regulatory documents, such as SP 339.1325800.2017 “Structures made of cellular concrete. Design rules”, first introduced only in 2018 [2, 3]. This code applies to the design of concrete and reinforced concrete products made from precast cellular concrete, as well as to the design of reinforced monolithic structures intended for civil buildings. According to p. 4.2.1 of SP 339.1325800 “...basic design requirements for the design of stone, concrete and reinforced concrete structures made of cellular concrete are accepted in accordance with SP 63.13330, SP 15.13330...”, while this document provides a set of generalized requirements for the design of structures made of cellular concrete.

Also taking into account the periodic temporal extension of separate regulations to develop new or in connection with the cancellation, the introduction of relevant amendments to existing ones, it is often necessary to make use of outdated design guidelines, such as, a Manual for design of concrete and reinforced concrete structures made of cellular concrete (to BCR 2.03.01−84), approved by Gosstroy of the USSR on April 16, 1985, № 20, or Guidelines for the design of concrete and reinforced concrete structures made of cellular concrete, 1977 [4]. These documents contain the main design requirements, regulatory and design characteristics of materials, a complete list of methods for calculating elements of concrete and reinforced concrete structures made of cellular concrete according to the limit states of the first and second groups, while in SP 339.1325800.2017 “Structures made of cellular concrete. Design rules” only the requirements for these types of calculations are presented [5, 6, 7].

2.2. Methods

Standards for the design of cellular concrete structures are considered and presented in a variety of national documents, which often refer to each other, have incoherence and inconsistency, which causes a discrepancy in the understanding of the main design requirements [8, 9]. All this causes modern problems in the rational design of reinforced cellular concrete and, as a result, the need to integrate and harmonize the regulatory framework of technical regulation in this area.

3. Results

In order to update and adjust the main design characteristics, design requirements and recommendations related to the design of cellular concrete products, taking into account the current state of the production base and construction practice, MS standards of organizations are developed, which are generalized requirements and recommendations of modern regulatory documents, as well as experience gained in recent years both in Russia and abroad in the field of design of cellular concrete products.

The most frequently applied standards for the construction of structures made of autoclaved cellular concrete products are the standards of the National Association of autoclaved cellular concrete manufacturers (MS NAAG) and the National Association of builders (MS NOSTROY).

Rules for the design and construction of MS NAAG 3.1-2013, developed by a non-profit organization that represents more than half of the Russian cellular concrete market – the National Association of autoclaved cellular concrete manufacturers, which unites more than a dozen of the largest modern manufacturing enterprises in terms of their capacities, including such as JSC Bonolit-Construction solutions, LLC Eco, CJSC Interregional Production and Resource Corporation GRAS, LLC KSMK, LLC LSR. Cellular concrete, JSC Kcella-Airblock-Center and others. This standard provides guidance on the use of non-reinforced products and basic information on the use of reinforced bar lintels and panels, as well as recommendations for basic structural solutions [10].

Edition of MS NOSTROY 2.9.136-2013 is a standard for device structures with the use of goods and reinforced elements of cellular concrete autoclaved National Association of builders with its scope and partition wall blocks and panels, large unreinforced blocks and reinforced elements, such as blocks, load-bearing and curtain panels for exterior walls and load-bearing panels of the inner walls, panels and floor coverings, bearing and non-bearing lintel, and containing in the appendices guidelines for the design of structures made of reinforced products made of autoclaved cellular concrete, which are of a recommendatory nature [12].

It should be noted that in all national and European normative documents focused on the design and construction of walls made of small cellular concrete wall materials, the issues of reinforcement of masonry, the technical solutions of constructions and nodes of their mates with adjacent components, while the requirements to porous concrete structures, working on the curve, represented insufficiently. At the same time, the production of bendable reinforced products, which accounts for almost 1% of the total volume of cellular structures produced in the Russian market, which is mainly focused on the production of small-piece wall materials, is not of interest for large-scale research.

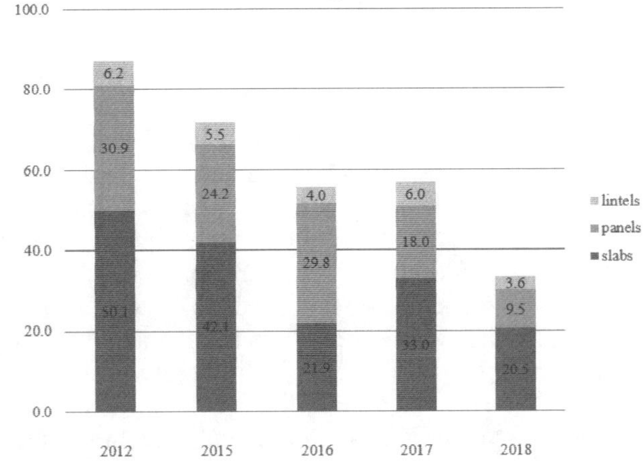

Figure 1. Structure of national production of cellular concrete products in 2012-2018.

IOP Conf. Series: Materials Science and Engineering **945** (2020) 012059 doi:10.1088/1757-899X/945/1/012059

As a result, leading manufacturers of bent cellular concrete structures need to develop their own standards for technical regulation of product quality. Thus, table 1 shows the author's analysis of the abundance of regulatory documents developed by leading Russian residents of the National Association of autoclaved aerated concrete manufacturers for their products on the example of an extensive range of various brands of reinforced cellular concrete jumpers, with an average density of D500 and a compressive strength class of at least B2,5, corresponding to all current national technological standards and requirements, and for individual manufacturers even European quality standards [2, 4].

Table 1. Nomenclature of manufactured lintels.

Manufacturer	Conditional Marking	Concrete	Reinforcement	Normative document
LLC DSK GRAS	PRS	B2.5 B3.5	B500C Bp-I	TC 5828-001-82651425-2014
LLC LSR. Wall			.	MS 5741-003-1522473-2016
LLC ECO	BPA	B3.5	AIII (A400) A500C B500C Bp-I A240	TC 5828-009-67236060-2015
CJSC Kcella-Airblock TM "YTONG"	PN	B2.5 B3.5	B500C	MS 73045594-004-2016
JSC BONOLIT-CONSTRUCTION SOLUTIONS;	BPA PB PORITEP	B3.5	A400 A240 AIII A500C B500C Bp-I	TC 5828-009-67236060-2015 TC 23.61.12.127-011-67236060-2018
CC AeroBel TM PORITEP				
LLC Glavstroy Ust-Labinsk	PG	B2.5		TC 5828-002-98818904-2016
JSC Glavnovosibirskstroy Plant "Sibit"	PB	B3.5	A400 A240 B-II	TC 5828-001-39136230-95 MS 39136230-02-2018 MS-39136230-2008

However, it is worth noting that the potential of standards developed by manufacturers taking into account the current state of capacities, current construction practices, as well as the critical analysis of numerous tests of reinforced cellular concrete structures, which contributes to a wider and correct application of cellular concrete structures on the Russian market, cannot be fully used by designers, as updated building codes and current codes of rules have priority, because they are official regulatory documents, mandatory for universal application, while the standards of organizations and technical specifications developed by manufacturers are only recommendations.

Successful and complete standardization of reinforced concrete elements, meanwhile, is obviously a derivative of the reliability, efficiency and prevalence of technical solutions used in its production, which are sufficiently deeply and fully justified by scientific research and production experience in the field of theory of operation and management of cellular concrete properties, which, alas, cannot be said about reinforcement products represented by traditional rods and frames, currently only partially adapted to use in cellular materials from the practice of reinforcing heavy concrete.

After analyzing the working drawings of the mass-produced bent cellular concrete products on the Russian market – bar lintels, we can conclude that all structures are the same type of traditional structural solutions of a porous array of cellular concrete reinforced with flat and three-dimensional reinforcement frames. Reinforcement of elements represents the following design decision: the placement of the joints concreting of the steel rods with subsequent grouting with concrete.

The author proposes to optimize design solutions when choosing a method for reinforcing cellular concrete bent structures by changing the design scheme and the principal approach to the perception, transmission and distribution of internal forces in the cellular concrete by studying and implementation of an innovative reinforcement product – a steel cogged tape (SCT, Fig. 2) [2]. The tape is a thin plate with a figured open perforation, the geometric dimensions and location of which are the subject of structural calculation depending on the physical, mechanical and deformative properties of cellular concrete and the accepted theory of strength of the contact layer of elements.

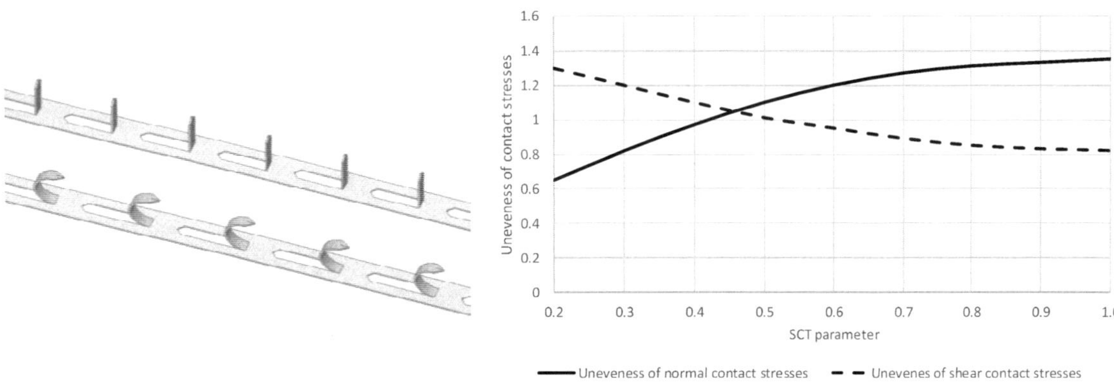

Figure 2. Structural options of SCT for cellular concrete bent elements: 90-degree cogged and 180-degree cogged.

The SCT is made of stainless steel with a yield strength of up to 300 MPa, and has a controlled side surface that most fully implements the principle of contact interaction of discrete reinforcement and cellular concrete surrounding array, which is the distribution and transmission of contact forces through a combination of tangential and normal stresses, the specific weights of which are determined by the SCT topology, selected according to general calculation algorithms for specified reinforcing materials and concrete. Practically significant for the purpose of proper codification of SCT cogged parameters and methodologically simple for the best formalization of calculation algorithms, the authors consider the convolution of the geometric parameters of the SCT cogged into a single indicator "SCT parameter", which functionally connects the key geometric characteristics of the element – the length, width, thickness of the plate, the angle of rotation of the cog (see Fig. 2). Numerical studies of information models of cellular concrete beams carried out by the authors (Fig. 3) demonstrated an expressed dependence of the calculated normal and tangential components of integral stresses in the contact zone (see Fig. 2). The unevenness index (deviation of the maximum from the average stress level), which determines the presence and distribution of stress concentrators and zones of potential destruction of the contact layer until the design bearing capacity of the object is reached (reinforcement pulling), is up to 40%, being leveled completely within the value of the SCT parameter 0.4-0.5, which allows evaluating this parameter's margins as a closest to optimal.

IOP Conf. Series: Materials Science and Engineering **945** (2020) 012059 doi:10.1088/1757-899X/945/1/012059

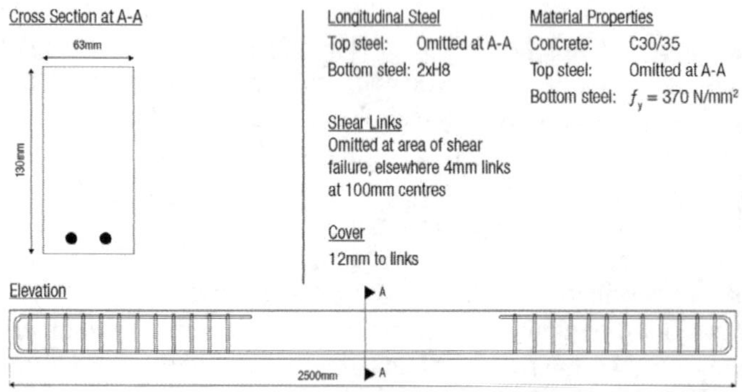

Costs	**Rubles**
The main costs of manufacturing PC-beams	
PC V2.5	85
Manufacturing	50%
Total	128
Additional costs of SGB	
SGB	20
Manufacturing, installation	50%
Total	158
Increase in production costs	**19%**
Increased load-bearing capacity	**50%**

Figure 3. Experimental cellular concrete bent structure analyzed in a numerical experiment to evaluate the efficiency of using a dispersed reinforcement element of the SCT.

Thus, author's proposal for use in the practice of discrete reinforcement of bent cellular concrete structures, that still retains significant unused bearing capacity with the regular rebars due to the insufficient contacting, offers innovative steel reinforcing element SCT with a side surface, better adapted to real working conditions of reinforcement in cellular concrete, as well as a high degree of controllability of the reinforcement efficiency by both normal and tangential stresses with the possibility of varying the specific weight of stresses in the integral contact resistance.

4. Discussion

The use of the innovative reinforcement product proposed by the author for cellular concrete structures in practice becomes possible only after technical regulation of the procedures for its selection, evaluation and calculation, which requires a detailed study of the actual operation of the reinforced bent structures under load, modelling their stress-strain states and developing an engineering applied methodology for selecting structural parameters of SCT that determine its cost and rationality.

5. Summary

Due to the abundance of regulatory documents for design of concrete and reinforced concrete structures and partially inconsistent requirements for reinforced products and structures from cellular concrete, there is a reasonable doubt about the admissibility of the application of the estimated requirements of certain normative documents in the design of bearing and enclosing reinforced cellular concrete structures for the purposes of the compliance characteristics of individual load-bearing and enclosing structures made of cellular concrete, as well as buildings and structures in general, the requirements of technical regulations on the safety of buildings and structures [11]. In this regard, there is a need to revise and systematize existing Russian and European standards that contain requirements for the design of directly reinforced structures made of cellular concrete, generalize data and update information in them, as well as develop modern national standards in this area in order to update the main design characteristics, specify design requirements and recommendations for the design of reliable and durable structures made of cellular concrete, they allow to improve the quality of manufactured construction products and increase the safety of buildings and structures with load-bearing and enclosing structures made of cellular concrete [6]. At the same time, the prospects for updating technical regulations in the design of reinforced cellular concrete structures are highlighted by the great potential for optimizing traditional discrete reinforcement products adopted in current practice, which prevent the implementation of rational design principles in this promising technology for the growth and development of resource - and energy-efficient individual housing construction.

Directions for improving the design principles and technical design solutions for bent cellular concrete structures also include improving the design of their reinforcement, which implements the expanded capabilities of complex contact resistance and the significant potential of the load-bearing

IOP Conf. Series: Materials Science and Engineering **945** (2020) 012059 doi:10.1088/1757-899X/945/1/012059

capacity of elements by more fully including the material of the reinforcement element in the work. The author's proposal in this matter is to develop structural design, design methodology, and applied engineering methods for calculating the innovative steel reinforcement discrete element of cellular concrete bent structures — SCT of controlled topology. Keeping the general principles of the computation algorithms and embedding the SCT in the cross-section of bent beams, provided the regulatory requirements of technical regulations, it's possible to enable up to 50% better use of the actual strength of the steel reinforcement to provide a suitable for mass use the plastic destruction of reinforced cellular concrete construction to ensure greater structural safety and performance of rebar in concrete and as a consequence, lower total cost of ownership of structures and the building made up of them by increasing the term, reducing the cost of maintenance and increasing of structural reliability [1]. Overall, the implementation of the proposed measures for the improvement of tools of technical regulation for designing of cellular concrete bending elements, with a proposal of use in practice and ensuring regulatory technique for the development of innovative reinforcing elements with controlled contact resistance, will in the short term allow to identify and implement potential additional resource efficiency and reduce the capital intensity of owning real estate property complexes constructed in reinforced cellular concrete structures.

6. References

[1] Kladieva P V Aerated concrete as a promising construction material 2019 *Collection of scientific articles of the 4th International scientific conference of promising developments of young scientists "Young Science – the future of Russia" [December 10-11, 2019] / Kursk: South-Western State University* 209-212

[2] Kladieva P V 2020 Directions for improving the technology of reinforcement of aerated concrete load-bearing structures *Design and construction: Collection of scientific papers of the 4th International scientific and practical conference of young scientists, postgraduates, master students and bachelors [March 13, 2020] / Kursk: South-Western State University* 143-146

[3] Grinfeld G I, Korkina E V, Pastushkov P P, Pavlenko N V, Erofeeva I V 2016 System of enclosing structures providing increased energy saving in buildings *Scientific Bulletin of the Voronezh State University of Architecture and Civil Engineering. Construction and architecture* **3 (43)** 25-35

[4] Naumov A E, Shevchenko A V, Dolzhenko A V, Bodyakov S N, Gvasalia Kh D 2019 Investigation of the contact strength of plastic pipes and concrete in the calculation of plastic-tube structures *Bulletin of the Belgorod State Technological University named after V.G. Shukhov* **9** 38-45

[5] Belykh T V, Davidenko P V, Zharikov I S 2018 Problems of methodological apparatus for conducting judicial construction and technical expertise *In the collection: Youth and the XXI century-2018 proceedings of the VIII international youth scientific conference: in 5 volumes.* 249-252

[6] Sotnikova E Yu, Abakumov R G, Naumov A E 2019 Construction materials and technologies used in housing construction in suburban areas of Belgorod *In the collection: Actual problems of development of socio-economic systems: theory and practice. collection of scientific articles of the 9th International scientific and practical conference. Southwestern State University*. P. 170-180.

[7] Fedyuk R S, Baranov A V, Mugahed Amran Y H 2020 Effect of porous structure on sound absorption of cellular concrete *Construction Materials and Products* **3 (2)** 5 – 18 DOI: 10.34031/2618-7183-2020-3-2-5-18

[8] Neyman K I, Abakumov R G, Naumov A E 2017 Theoretical and practical aspects of construction monitoring in post-structural operation *Innovative economy: prospects for development and improvement* **4 (22)** 167-170

[9] Otsokov K A 2020 Innovative technologies in construction and their use in organizational and technological events *Construction Materials and Products* **3** **(1)** 7 – 13 DOI: 10.34031/2618-7183-2020-3-1-7-13

[10] Martyanov E I 2020 Optimization of the construction of a two-bladed stirrer to increase the energy efficiency of the apparatus *In collection: The World of Science without Borders. Materials of the 7th International scientific and practical conference of young scientists* 105-109

[11] Klyuev S V, Klyuev A V, Shorstova E S 2019 Fiber concrete for 3-d additive technologies *Materials Science Forum* **974** 367–372 DOI: 10.4028/www.scientific.net/MSF.974.367

[12] Varlamov A A, Gavrilov B V, Davydova A M 2020 Destruction model of idealized concrete structure by sawing *Construction Materials and Products* **3** **(2)** 46 – 54 DOI: 10.34031/2618-7183-2020-3-2-46-54

Acknowledgments
The work is realized in the framework of the implementation of a comprehensive project to create high-tech production "Development of new methods and tools for management of property in the budget sector and their implementation in the software package of the information-analytical system for centralized management of property owned by the constituent entities of the Russian Federation, municipalities, as well as state property companies" (agreement No. 074-11-2018-026 of 07/11/2018).

IOP Conf. Series: Materials Science and Engineering **945** (2020) 012060 doi:10.1088/1757-899X/945/1/012060

Improving the efficiency of the material grinding process

A A Romanovich[1*], Ebrahim Amini[2] and M A Romanovich[1]

[1]Department of Hoist transport and road machines, Belgorod State Technological University named after V.G. Shukhov, Kostyukov St., 46, Belgorod, 308012, Russia
[2]K. N. Toosi University of Technology, Tehran, Iran

E-mail: AlexejRom@yandex.ru

Abstract. The paper provides an analysis of grinding systems used in the world practice, including the main technical solutions in the field of fine grinding of materials using a ball mill, and also presents indicators of their efficiency. It was found that the grinding systems of the partitioned grinding process with pre-grinding in a press-roll grinder showed the greatest efficiency from their use. However, today in various countries, free-standing units are used for deagglomeration of pressure-pressed semi-finished products, which require additional space for their placement and operation and maintenance costs. The authors describe the design of a press-roll unit with a device for breaking pressure-pressed materials, which allows reducing these costs, as it performs the following technological operations: volume-shear deformation of the material is carried out due to the conical profile of the main rolls; the compressed materials are disaggregated in the reverse direction of compression by means of additional rollers that have a reverse taper to the main rolls. A scientifically-based method for calculating the energy index of a new design of press roll unit is proposed, which allows for the energy calculation of these machines.

1. Introduction

Grinding materials using the sectional method, in which the coarse grinding stage is carried out outside the ball mill (BM) in a separate unit, is a very promising direction, while providing effective conditions for the destruction of materials. As, at the pre-grinding stage, aggregates are used that implement more efficient methods of destruction compared to the main grinding plant. Examples of such units are the press-roll type shredder (PRT), hammer mill (HM), cone inertia shredder (CIS), and others. Many prominent scientists Austin L, Klimpel R, Luckie P, Tavares L, de Carvalho R, Austin L., Shoji K., Bhatia V., Knieke C., Sommer M., Peukert W. are engaged in improving the efficiency of the partitioned grinding method [1-4].

Table 1 shows the energy indicators of grinding systems that are widely used in industry.

The analysis of specific indicators of these grinding systems indicates the feasibility of using the technology with pre-grinding in the press-roll unit. The use of this grinding technology reduces the specific power consumption by 30-40%, while increasing productivity by 20-45% [5-10, 11]. This effect can be explained by the results of studies on the grinding of materials of various strengths, which found that the energy spent on pre-grinding materials in the press-roll unit is 3-5 kW·h/t, and in a ball mill with a similar fineness of grinding is - q = 6-10 kW·h/t. For example, when grinding clinker to produce a product containing up to 30-35% of the fraction less than 80 microns, a press roll unit

IOP Conf. Series: Materials Science and Engineering **945** (2020) 012060 doi:10.1088/1757-899X/945/1/012060

under a pressure of 200 MPa consumes about 5 kW·h/t of electricity, and in a ball mill about 10 kW·h/t, which is 2 times less.

Table 1. Specific indicators of the main grinding complexes using a ball mill.

Indicator	Grinding complex				
	BM	PRT-BM		CIS - BM	
		PRU	BM	CIS	BM
Main characteristics of units					
Indexation	4.2 x 16	1 x 0.8	1.8 x 14	1.75	3.8 x 14
Specific power consumption, kW·h/t	38.8	4.6	23.2	5	23.2

This energy saving is achieved by the high-pressure compression of mineral materials between the working bodies, which causes the grinding of not only the particles interacting with the surface of the rolls, but also those in the layer and in contact with each other.

2. Materials and methods

However, the material after processing it by pressure between the working bodies of the unit has the form of pressed filter cakes with maximum strength in the direction of force application (figure 1), which imposes special conditions for its effective deagglomeration and grinding.

Figure 1. The shape of the material after pressure treatment in PRU.

Today, the world uses technological schemes for partitioned grinding of materials, including a press-roll unit for pre-grinding, a hammer mill or other unit for disaggregating compressed material, and a ball mill for final grinding, operating in an open or closed grinding cycle [6,7, 9].

However, when using these technological solutions, the use of a separate unit for deagglomeration of compressed filter cakes requires additional space for its placement and costs for operation and maintenance.

We developed a design of a press-roll unit that allows implementing the processes of pressure grinding and disaggregation of compressed materials in one unit (Fig. 2).

The unit includes two main rolls 1 mounted on the frame, which have a conical profile, and a hopper 2. Deagglomerating rollers 3, located under the main rolls, have a smaller diameter and reverse taper with respect to them. This allows for the disaggregation of compressed materials in the opposite direction to pressing, which leads to a reduction in energy consumption for their destruction. The main rolls are connected to the rollers by means of rods that regulate the size of the gap between them, in proportion to its change between the main rolls [6, 8, 10]. This design of the PRU eliminates the accumulation of product before the deagglomerating rollers.

IOP Conf. Series: Materials Science and Engineering **945** (2020) 012060 doi:10.1088/1757-899X/945/1/012060

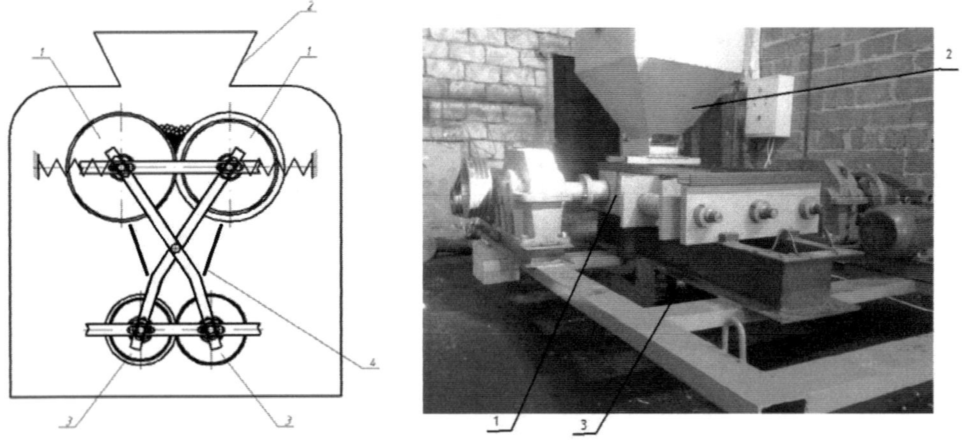

Figure 2. Press roll unit with deagglomeration device: a) scheme, b) pilot unit.

However, the lack of science-based recommendations for calculating the energy indicators of the unit for pre-grinding and deagglomeration hinders its widespread implementation in the industry.

When grinding materials in PRU, a number of technological operations are performed:

- volume-shear deformation of the crushed material - due to the use of a conical roll profile;
- disaggregation of pressed filter cakes in the direction of their lowest strength - due to additional rollers having the reverse taper of the main one.

For these operations, the drive power is consumed, which is determined by the following expression:

$$N = \frac{N_1 + N_2 + N_3}{\eta}, \tag{1}$$

where N_2 - power spent on creating volume-shear deformation of the material, W; N_3 - power required to overcome the friction forces in the bearing units of the unit, W,; N_1 - the power required for the disaggregation of the compacted filter cakes, W, η - efficiency.

The power spent on the destruction of the source material N_2 depends on the torque M_{kr} required for grinding the materials and the moment of friction M_2 that occurs in the roll supports.

The power spent on grinding the material is equal to:

$$N_2 = M_r \omega , \tag{2}$$

where M_r – the resulting moment that occurs between the conical rolls when grinding the material, Nm; ω - angular speed of rotation of the rolls, rad/s.

The value of the resulting moment, according to the calculation scheme Fig. 3, can be written as follows:

$$M_r = 2\frac{F}{2}R_{sr} = FR_{sr}, \tag{3}$$

where $F/2$ - circumferential force acting on the average roll radius R_{sr}.

The circumferential force is determined by the following expression:

$$F = 2P_{\Sigma} \sin \gamma_n , \tag{4}$$

where P_Σ – total spacer force acting between the rolls, N; γ_n - the value of the neutral angle, deg.

The total spacer force is determined by the equation:

$$P_\Sigma = qBR_{sr}\gamma_n \cos\alpha \ , \tag{5}$$

where q – specific grinding pressure,N/m^2, B- roll width, m, α - the angle of inclination of the working surface of the rolls.

Assuming that $\sin\gamma_n \approx \gamma_n$, as the angle is very small, we have the following expression:

$$N_2 = q\omega BR^2{}_{sr}\gamma_n{}^2 \cos\alpha \ . \tag{6}$$

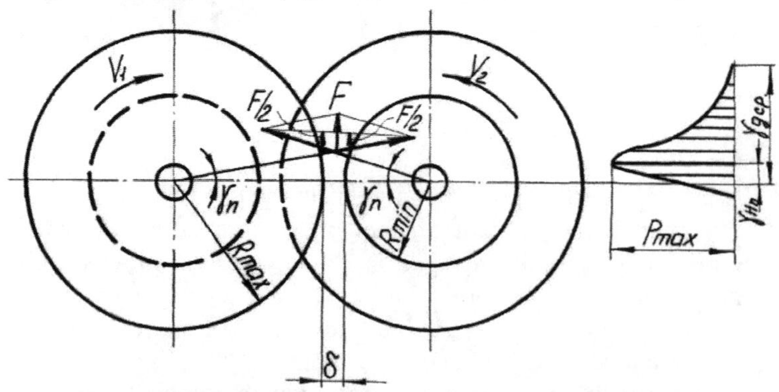

Figure 3. Scheme for determining the resulting moment.

The power required to overcome the friction forces in the roller supports can be determined by the following equation:

$$N_2 = 2\pi d\ \theta_r f_k n, \tag{7}$$

where d – the size of the diameter of the support part of the roll, m; θ_r – total force resulting from the pressure between the rolls and the weight of the roll, N, f_k - the coefficient of rolling friction, n - the number of the roll turns, s^{-1}.

The value of the total force is determined by the following equation:

$$\theta_r = \sqrt{P_\Sigma{}^2 + G_b{}^2}\ , \tag{8}$$

where G_b - force arising from the weight of the roll.

The drive power expended on the disaggregation of the pressed plates depends on the maximum deagglomeration force F, the value of which depends on the width of the rollers b, and the compressive strength of the pressed plates G, as well as the angular speed of rotation of the rolls, equal to $\omega = 2\pi n$.

$$F = \mu\, bG \sin\beta\ , \tag{9}$$

where μ - coefficient depending on the density of pressed filter cakes and material properties, β - maximum force angle between the rollers, deg.

Then the amount of torque is determined by the following expression:

$$M_1 = 2\frac{F}{2} r_{sr} = F r_{sr}\ , \tag{10}$$

where r_{sr} - the average radius of the conical rollers, m.

$$N_1 = Fr_{sr} . \omega \, .$$

Thus, the method proposed above allows calculating the engine power required for the implementation of the following technological operations in one unit:

- volume-shear deformation of materials due to the conical profile of the main rolls;
- disaggregation of compressed materials in the reverse direction of compression due to additional rollers that have an inverse taper with respect to the main rolls.

3. Summary

Thus, as a result of the analysis of the main technical solutions of grinding systems that work in conjunction with a ball mill and are widely used in world practice, it showed high efficiency from the use of a partitioned grinding process with pre-grinding in a press-roll unit. However, today in the world for the deagglomeration of pressure-pressed filter cakes between PRU rolls, free-standing units are used, the installation of which requires additional production space and operating and maintenance costs.

The developed new design of the press roll unit allows reducing these costs. When grinding materials in PRU with a device for deagglomeration of compressed materials, a number of technological operations are performed: volume-shear deformation is performed due to the conical profile of the main rolls and the compressed materials are disaggregated in the opposite direction to pressing due to additional rollers that have an inverse taper in relation to the main ones.

A scientifically-based method for calculating the energy index of a new design of a press roll unit is proposed. This method allows calculating and designing these machines, which will speed up their introduction into the industry.

4. References

[1] Tavares L, de Carvalho R 2009 Modeling breakage rates of coarse particles in ball mills *Minerals Engineering* **22 (7-8)** 650-659

[2] Austin L, Klimpel R, Luckie P 1984Process engineering of size reduction: ball milling. Process Eng of size reduct, Ball Milling *Published by Soc of Mining Engineers of AIME*

[3] Austin L, Shoji K, Bhatia V 1976 Et al Some Results on the Description of Size Reduction as a Rate Process in Various Mills *Industrial and Engineering Chemistry Process Design and Development* **15(1)** 187-196

[4] Knieke C, Sommer M, Peukert W 2009 Identifying the apparent and true grinding limit *Powder Technology* **195 (1)** 25-30

[5] Romanovich A A, Romanovich M A, Belov A I, Chekhovskoy E I 2018 Energy-saving technology of obtaining composite binders using technogenic wastes *Journal of Physics: Conf. Series* **1118** 12035

[6] Romanovich A A, Kolesnikov R S, Romanovich M A 2018 Study of device for precompaction and uniform supply of materials to working bodies of aggregate *Materials Science and Engineering* Conf. Series 42052

[7] Malov P N, Tarabanov N S 1968 On the effect of elastic consequences on the properties of pressed carbon materials *Chemistry of fuel* **1** 132-133

[8] Valkevich V L, Fmedlide G I, Fortunatov T V 1970 Dispersion and elastic expansion of oxide granular masses during pressing *Refractories* **8** 45-48

[9] Popilsky R Ya, Kondrashev F V 1988 Pressing ceramic powders *Metallurgy* 27-28

[10] Romanovich A A 2011 Study of the effect of the speed of rotation of the rolls on the output indicators of the grinding process and the development of recommendations for improving the wear resistance of their working surfaces *Bulletin of Belgorod State Technological University. V.G. Shukhov* **4** 71-73

[11] Pirotsky V Z 1986 Improving the technology and technology of grinding Portland cement clinker: evaluation of the effectiveness of grinding systems M. 3-23

IOP Conf. Series: Materials Science and Engineering **945** (2020) 012061 doi:10.1088/1757-899X/945/1/012061

Use of wood processing industry waste for bitumen modification

M A Turobova[1*], V E Danilov[1] and A M Ayzenshtadt[1]

[1]Department of composite materials and environmental engineering, Northern (Arctic) Federal University named after M.V. Lomonosov, Severnaya Dvina Emb., 17, Arkhangelsk, 163002, Russia

E-mail: turobowa.m@yandex.ru

Abstract. The paper provides information about the method of obtaining a modifying additive to bitumen, including wood filler. The pore space of the filler contains a basalt reinforcing filler with an average particle size of 150-300 nm. Scots pine bark with an average particle size of 0.5-1 mm was used as a wood filler. The modifying additive was obtained by mixing fine basalt (30 % by weight) and wood bark. After careful mixing for 5 minutes, water was added, the mass of which is twice more than the calculated mass of the reinforced bark. Then the mixture was homogenized. The use of an additive for the preparation of a bituminous binder improves the thermal and physical characteristics of the organic binder (softening temperature and brittleness temperature). In comparison with other traditional stabilizing additives, the use of the proposed composition for the preparation of a bituminous binder also improves the thermal and physical characteristics of the organic binder, such as the brittleness temperature. A decrease in the brittleness temperature of the mixture indicates an increase in the frost resistance of the modified organic binder. For other characteristics, the modifying additive has a comparable effect on the properties of the binder with traditionally used materials.

1. Introduction

The current level of development of road transport is associated with a constant increase in the volume of cargo transportation, with the intensity of movement of cargo vehicles, which in turn leads to an increase in dynamic loads on covering of roadway. These and other objective factors indicate that it is necessary to improve the quality of road bitumen.

Improving the quality of bitumen can be achieved by searching for new technological solutions that improve the physical and mechanical characteristics of asphalt covering [1]. The main way to solve this problem is currently the use of modifying additives in the bituminous binder. Moreover, the additive has a significant influence on the physical and mechanical properties of asphalt concrete [2, 3].

In practice and in the scientific literature, the necessary material has been collected on the effects of various substances on the properties of petroleum bitumen. For example, plasticizers, providing mobility of the dispersed bitumen system, have found their application in the production of roofing materials and polymer-bitumen, road bitumen with improved viscosity-temperature properties. Heavy gas-oil, selective oil purification extract, shale softener, used and fresh oils [4-7], etc., which are plasticizers, have also found their application. One of the types of modifying additives is surface-

IOP Conf. Series: Materials Science and Engineering **945** (2020) 012061 doi:10.1088/1757-899X/945/1/012061

active chemical compounds (surfactants). In [8-12], we consider the classification of surfactants and the mechanism of their action on the spatial dispersed structure of bitumen, the implementation of which leads to an increase in the contact area between the active centers of diphilic molecules of the additive and the organic substrate. This fact allows maintaining the necessary aggregate stability of the binder particles (the average particle size of the dispersed phase has decreased), which ultimately increases the adhesion interaction between the organic and mineral components of the reaction system.

Additives to bitumen that increase the adhesion of bitumen to the surface, such as alkyldiamines, amidoamines, diamidoamines, are widely used abroad and in Russia [13-15].

Polymer-bituminous binder (PBB), formed by the introduction of polymer additives for modifying bitumen, behaves as a material with a certain internal reinforcement when the internal structure is deformed. The use of PBB belongs to one of the dynamically introduced technologies of road covering construction and repair in the Russian Federation. The introduction of polymers into bitumen sets a primary goal – it is to reduce the temperature susceptibility of the binder, i.e. it is necessary that the hardness decreases in winter and increases in summer, in addition, the binder must have good elasticity. If this goal is achieved, the manufactured asphalt concrete containing PBB will have high shear resistance, low-temperature crack resistance, and fatigue life.

Polymer-based modifiers are distributed in a dispersion medium, and their own structural grid is formed in bitumen. The polymer frame compensates for the lack of fluidity when the temperature rises, as well as deformative properties when it decreases. Thus, the range of performance of various bituminous materials increases [16, 17].

As a possible modifier for bitumen, the use of lignin-containing plant components, such as waste from the woodworking industry, namely bark, is promising (in our opinion). In the structure of the lignin macromolecule, there are a significant number of functional groups that differ in their activity. In addition, the presence of some amino acids that are part of the protein structure and have an aromatic grouping in their composition is noted in native lignin [18].

However, it is known that the wood bark is a porous material that has a significant moisture capacity and anisotropy of properties. At the same time, there are currently works that allow obtaining an organic and mineral complex based on wood bark, which significantly reduces the above-mentioned negative effects. This organic and mineral composite can be very promising for use as an additive in bituminous mixtures.

In [19-22], the technology of filling the pore space of wood bark with highly dispersed basalt particles is considered, thus obtaining the material of a plant matrix reinforced with mineral particles. In the structure of the bark, there is an increase in the content of closed pores and reduce the open, due to the indicators of hygroscopicity and water absorption are greatly reduced, thermal characteristics are stabilized, even in case of high environmental humidity.

It should be noted that wood bark is a large-tonnage waste of enterprises associated with the mechanical processing of plant raw materials, which is currently stored in dumps or, at best, burned. Therefore, its use is also an important task in terms of environmental management [23]. One of the most common wood species that is well suited for the building materials industry is pine wood, so the use of the bark of this tree species is of primary interest. Being the most common and suitable for the production of building materials in the Russian Federation, pine bark is of the greatest interest in terms of reinforcement of the wood matrix.

Among the various reinforcing components, the most optimal, in our opinion, is basalt, as one of the most common natural stones. This rock is characterized by good strength and a fairly high density. It is also necessary to note the chemical resistance of the material to corrosion and aggressive environments and resistance to high temperatures. Withstanding temperatures of over 1500 °C, basalt has found use as a fire-resistant material for fire protection. In addition, it is necessary to note the low cost of basalt [24].

Based on the above, the aim of the research presented in this paper is to create a modifying additive that contains waste from the woodworking industry, namely bark reinforced with fine basalt, and

IOP Conf. Series: Materials Science and Engineering **945** (2020) 012061 doi:10.1088/1757-899X/945/1/012061

assess its impact on the thermal and physical (brittleness temperature, softening temperature, and flash point) characteristics of bitumen.

However, a necessary step in the preparation of wood raw material is the removal of water-soluble extractives. This action is necessary to increase the volume of free pore space of the bark, which will later be filled with an aqueous suspension of fine basalt particles. For these purposes, good results are shown by the use of the developed extractor, which allows for accelerated extraction of water-soluble substances from plant raw materials [25].

2. Materials and methods

One of the research objects was the bark of Scots pine, which was selected from the dump of a timber processing enterprise. Pine logs were cut on rotary machines and piled in one dump. The bark was evenly sampled along the perimeter of the dump from a depth of 30 cm to avoid boundary effects on the bark-air interphase. Previously, water-soluble extractives were removed from it. The bark for the study was dried to a constant mass in a drying cabinet at 60 °C for three days.

The reinforcing element has been sifting of crushing basalt sampled from the field Myandukha in the Arkhangelsk region. The basalt was pre-ground in a ball mill to a size of 4 ± 1 mm and then dried to a constant mass at a temperature of 110 °C.

BND-U 100/130 "Biturox" bitumen was used for experiments. It is manufactured using the patented Biturox technology, which is based on the process of concentrating oil oxidation by vacuum distillation. As a result, a material with a complex high-tech structure is formed: 75-80 % carbon, 10-15 % hydrogen, 5 % oxygen, metals, and nitrogen.

As the plant additive is a composite consisting of highly dispersed materials (bark and basalt), mechanical grinding of raw components was performed. Grinding was carried out using the Retsch PM 100 planetary ball mill. The bark was dispersed using stainless steel equipment. Basalt was crushed in the dry phase using carbide-tungsten equipment (grinding jars and balls).

Crushed fine wood bark and basalt were mixed in the required ratio (% by weight): bark – 70 and fine basalt – 30 for 5 minutes. Then water was added to the wood-mineral composition, the mass of which is twice more than the calculated mass of the reinforced bark, and the resulting mixture was homogenized. The components were homogenized using a vertical rotary agitator with digital control of the WiseStir model. After that, the resulting mixture was dried to a constant mass. Quality control of bark modification with fine basalt was performed by determining the specific surface area, the porosity of the composition (by gas sorption on the Autosorb analytical system), true and bulk density (by standard methods). Then the resulting additive (hereinafter D1) was introduced into dehydrated bitumen, which was preheated to 105 °C. Mixing was carried out using a two-bladed laboratory agitator.

Comparison of thermal and physical characteristics was carried out with respect to bitumen without additives (D0), and two main stabilizing additives were selected for comparison – "Stylobite" (hereinafter D2) and "Granulite-66" (hereinafter D3). The choice of these additives for bitumen is due to their wide use in road works.

The number of additives used in bitumen was assumed to be the same, based on the recommendations for the use of traditional components.

The following thermal and physical characteristics were selected: softening temperature (ring and ball method), brittleness temperature, and flash point. The choice of these characteristics is due to their influence on the quality properties of the final asphalt concrete – resistance to abrasion, cracking, and rutting. Samples for testing to determine the thermal and physical characteristics had a form that meets the necessary standards.

To determine the softening temperature of petroleum bitumen, a laboratory automatic Kish-20M4 device was used. The softening temperature of bitumen was taken as the arithmetic mean of two parallel measurements.

An automatic ATX-20 apparatus was used to determine the brittleness temperature of petroleum bitumen. The average arithmetic value of the three measurement results was taken as the result of the Fraas test of the bitumen brittleness temperature.

The flash point was determined in a closed crucible using the ATV-21 apparatus. The average arithmetic value of at least two consecutive measurements was taken as the test result.

All the obtained numerical values of the characteristics listed above were rounded up to integers.

3. Results and discussion

During the grinding of raw materials, the following dimensional characteristics of the initial components for developing the additive were obtained: bark – 0.5–1 mm; basalt – 150-390 nm. By mixing a highly dispersed fraction of the bark with an aqueous suspension of basalt, spontaneous filling of the free pore space of the bark occurs due to the process of its swelling. After separation of the solid phase from the aqueous suspension, the latter was brought to a constant mass in a drying cabinet at a temperature of 40 °C. The grayish tint of modified bark particles (figure 1) and changes in its physical and chemical characteristics (table 1) indicate a significant amount of fine basalt fixed on the surface of the plant material.

The results of determining the thermal and physical characteristics of bitumen with the considered additives and the control sample are presented in table 2.

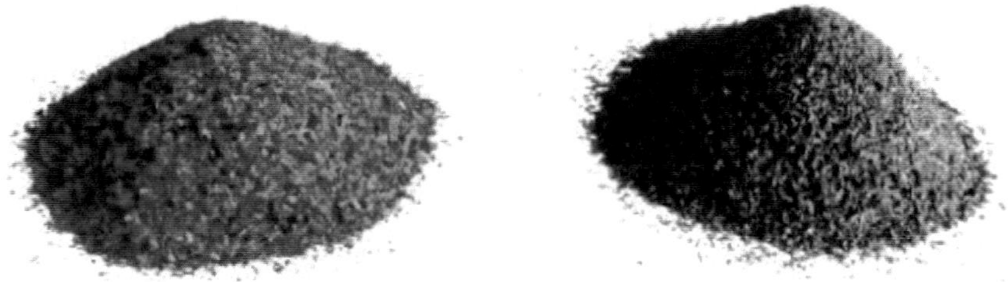

a b

Figure 1. The wood bark of the pine: a – initial; b – modified.

Table 1. Indicators of the quality of the modified bark.

Sample	Bulk density, kg/m^3	True density, kg/m^3	Specific surface area m^2/kg	Porosity, cm^3/g
Initial bark	185	1150	41887	0.257
Modified bark	313	1380	70311	0.093

The analysis of the obtained data presented in table 2 shows that the use of the additive (D1) has a positive effect on the determined thermal and physical characteristics of bitumen.

Comparing the obtained experimental data on the thermal and physical characteristics of bitumen with the quality requirements for road bitumen BND-U 100/130, it can be noted that when using an additive that includes wood filler, the softening temperature increased (compared to bitumen without additives) not significantly (2 °C). To a greater extent, the use of the D1 additive led to a significant reduction in the brittleness temperature. Thus, according to the data available in the literature, the softening temperature of organic binder BND-U 100/130 varies in the range of 40-48 °C, the brittleness temperature is -17 °C. Our experiments to determine these characteristics for the control sample fully correspond to the standard values. Thus, the softening temperature for the test sample D0 was 46 °C, and the brittleness temperature is -17 °C.

Table 2. Thermal and physical properties of bitumen.

Test name	Bitumen100/130			
	D0	D1	D2	D3
Softening temperature, ˚C	46	49	52	48
Brittleness temperature, ˚C	-17	-24	-20	-21
Flash point, ˚C	240	229	233	228

In comparison with other stabilizing additives, the use of the D1 additive for the preparation of the bituminous binder also improves the thermal and physical characteristics of the organic binder, such as the brittleness temperature. A decrease in the brittleness temperature of the mixture indicates an increase in the frost resistance of the modified organic binder. For other characteristics, the modifying additive has a comparable effect on the properties of the binder with traditionally used materials.

4. Summary

Thus, after analyzing the results obtained during the research, it can be assumed that a modifying additive, including a wood aggregate reinforced with fine basalt, can be used to produce asphalt concrete. However, it should be noted that the preliminary preparation of wood filler should be based on the removal of water-soluble extractives (increase the free pore space) from the plant material and bringing the size characteristics of particles of plant raw materials up to 1 mm (the practical absence of anisotropy properties of the bark). The use of additives based on a large-tonnage waste from the technology of mechanical processing of wood will help to solve issues of rational use of natural resources and reduce the environmental impact in the regions where the enterprises of the timber industry are concentrated.

5. References

[1] Odenbakh I A, Taurit E B 2020 Some new technologies in road construction *Construction Materials and Products* **3(1)** 62-69 DOI: 10.34031/2618-7183-2020-3-1-62-69
[2] Trautvain A I, Yadykina V V, Mulenko E S 2018 Study of physical-mechanical properties of asphalt-concrete specimens on activated mineral powders of various compositions *Construction Materials and Products* **1(4)** 44-50
[3] Trautvain A I 2019 Analysis of the influence of the qualitive composition of the asphalt-concrete mixture on the main performance characteristics of asphalt concrete pavement *Construction Materials and Products* **2(1)** 17-23
[4] Elshin A I, Kuzora I E, Moiseev V M et al. 2004 Industrial development of polymer-bituminous binders production *Materials of scientific and practical Conf. "Current state of oil refining processes"* 114-117
[5] Kotov S V, Timofeev G V, Levanova S V et al. 2003 Road bitumen with modifying additives *Chemistry and technology of fuels and oils* **3** 52-53
[6] Gokhman L M 2012 Effectiveness of the polymer. Advantages of using industrial oil for polymerasfalt concrete *Road surfaces* **1** 38-45
[7] Gokhman L M 2004 Complex organic binding materials based on block copolymers *ECON-INFORM* 510
[8] Korolev I V 1991 Technical surfactants from secondary resources in road construction *Transport* 144
[9] Basova S P, Leonenko V V, Safonov G A 2001 Influence of nitrogenous compounds on the rheological behavior of petroleum bitumen *Oil refining and petrochemistry* **2** 19-21
[10] Basova S P, Bembel V M, Leonenko V V, Safonov G A 1999 Influence of cyclic amines on the properties of petroleum bitumen *Oil refining and petrochemistry* **12** 26-29
[11] Aminov Sh Kh 2007 Modern bituminous binders and asphalt concretes based on them *Nedra* 336
[12] Shemonaeva D S 1975 Influence of surfactant concentration in bitumen on the properties of

bituminous mixtures *Improving the quality of road bitumen: works of SoyuzdorSRI* **80** 162-170

[13] Krut V V, Solomentsev A B, Kolodezny V P 1999 Adhesive additive for polyfunctional bitumen *Pat. 2130954 Russian Federation, IPC C08L 95/00, C08K 5/3415* **97100884/04** 1-6

[14] Kugele T G 1988 Asfalt antistrpping agents containing organic amines and Portland cement *Pat. 4743304 USA, C08L 95/00* **06/561382** 1-12

[15] Treybig Duane S., Chang Dane 1988 Bituminous compositions containing antistripping additives prepared from amines and substrituted heterocyclic compounds *Pat. 4765839 USA, C08L 95/00, C09D 3/24* **07/123009** 1-10

[16] Gureev A A, Gokhman L M, Gilyazetdinov L P 1986 Technology of organic binding materials *NUOG "Gubkin University"* 126

[17] Gokhman L M 2008 Bitumen, polymer-bitumen binders, asphalt concrete, polymerasfaltobeton *EKON-INFORM* 117

[18] Proskuryakov V A, Drabkin A E 1981 Chemistry of oil and gas studies. *manual for universities Chemistry* 359

[19] Stenin A A, Aizenshtadt A M, Shinkaruk A A, Makhova T A 2013 Formation of the flame retardant properties of building materials from wood using superfine basalt filler *Construction materials* **11** 47–50

[20] Stenin A A, Aizenshtadt A M, Shinkaruk A A, Demidov M L, Frolova M A 2014 Mineral surface modifier of building materials from wood *Construction materials* **10** 51–54

[21] Aizenshtadt A M, Makhova T A, Frolova M A, Tutygin A S, Stenin A A, Popova M A 2012 Designing of composition of nano- and microstructured construction composite materials *Industrial and civil engineering* **10** 14–18

[22] Danilov V E, Ayzenstadt A M, Frolova M A, Turobova M A, Karelsky A M 2015 Obtaining an organomineral filler based on wood bark and basalt for the development of composite materials *Construction materials* **7** 72-75

[23] Tsyvin M M 1973 Use of wood bark *Forest industry* 96

[24] Babaev V B, Strokova V V, Nelubova V V 2012 Basalt fiber as a component for micro reinforcing of cement composites *Bulletin of Belgorod State Technological University named after V.G. Shukhov* **4** 58–61

[25] Pogrebenkov V M, Sedelnikova M B, Vereshchagin V I 1997 Method for obtaining inorganic blue pigments *Pat. 172362 Poland, MKI 6C09 C 1/32*

Acknowledgments

The research was carried out on the unique scientific equipment "Physical Chemistry of Surfaces of Nano-Dispersed Systems" with the financial support of the Russian Foundation for Basic Research, Project No. 18-43-292002r_mk.

IOP Conf. Series: Materials Science and Engineering **945** (2020) 012062 doi:10.1088/1757-899X/945/1/012062

Synthesis of a mineralizing agent for Portland cement from aluminum production waste

I G Endzhievskaya[1*], A V Demina[1] and A A Lavorenko[1]

[1]Department of building materials and construction technology, Department of physical and inorganic chemistry, Siberian Federal University, 79 Svobodny Ave., Krasnoyarsk, 660041, Russia

E-mail: IEndzhievskaya@sfu-kras.ru

Abstract. Priority directions in innovative technologies of the construction materials industry are the expansion of the use of mineral and chemical additives in the production of cements, as well as alternative raw materials and fuels in production processes. During the production of aluminum at the enterprise JSC RUSAL Krasnoyarsk, the method of electrolytic reduction of aluminum oxide or alumina dissolved in a cryolite-based melt generates fluorocarbon-containing (FCC) waste, which currently accumulates more than 16,000 tons per year. Due to the increased year-round generation of waste with a high content of carbon, fluorine and aluminum, it is urgent to develop technologies that will effectively use them as a new mineralizer product for the production of Portland cement clinker - calcium fluoride. This paper describes the thermal method of waste processing FCC for receipt of the mineralizing agent (fluorite). The disadvantage of FCC as a raw material is an increased concentration of alkali metals, whose oxides negatively affect the process of hydration of clinker minerals, and lead to the formation of efflorescence on products. To reduce their content, the finished fluorite was washed with water. Calcium fluoride as a mineralizing agent in the production of Portland cement, reduces the clinkerization temperature, increases the content of alite, changes the stability, polymorphism and reactivity of phases. The resulting fluorite with a high mass concentration of 73.9% CaF_2 makes it possible to reduce the clinker formation temperature, which increases the energy efficiency of obtaining Portland cement. The technological process for removing alkaline components from the product will reduce the likelihood of destructive processes in concrete caused by the aggressive action of reactive silica aggregates and freezing and thawing processes.

1. Introduction

In the development strategy of building materials industry for the period until 2020 and further until 2030, approved by Decree of the Government of the Russian Federation from May 10, 2016 No. 868-r, it is stated that "the priority directions of innovative production technologies of building materials industry and new building materials are: the increased use of mineral and chemical admixtures in the production of cement, the use in the technological processes of production of building materials, alternative fuels and raw materials".

As the Siberian Federal district is redundant in the production of cement, the introduction of energy-efficient technologies for its production is very relevant.

The production of Portland cement is a very energy-intensive process [1]. Producing each ton releases a large amount of CO_2 into the atmosphere and requires 20 to 30% of thermal [2] and 12 to 15% of electrical energy [3] from total industrial consumption. Therefore, new solutions and approaches are required to implement energy-saving policies in the basic industry of building materials – the manufacture of cement binders. There is an urgent need to develop methods for producing Portland cement with reduced CO_2 emissions [4], and there is a need for energy savings, which can be achieved by reducing the firing temperature when using fluxes/mineralizers and other compounds. This solution has economic and environmental advantages [5, 6, 7].

Due to the complex mechanism of changing the crystallographic and other physical and chemical properties of the components of the oxide phases, mineralizers can significantly reduce the temperature of clinker formation. Such compounds as AlF_3, CaF_2, ZnO, MnO_2, SnO_2, CuO are considered to be effective mineralizers [7-10].

Calcium fluoride as a mineralizer reduces the clinkerization temperature, increases the content of alite in the clinker [8], and changes the stability, polymorphism, and reactivity of the phases [10, 12, 13]. However, it is mainly an imported product. The use of CaF_2 in multiphase systems can have both beneficial and adverse effects. As a rule, they were identified during the study of individual compositions and were not systematized to such an extent that it was possible to draw general conclusions and avoid or mitigate negative effects [6, 7, 8, 14-20].

The use of various types of alternative fuels, waste or secondary raw materials in the production of Portland cement allows saving energy and reducing greenhouse gas emissions (especially CO_2) [20]. Papers [8, 22] show that some industrial by-products contain a large amount of fluorine. During the production of aluminum at the enterprise JSC RUSAL Krasnoyarsk, waste with a high content of carbon, fluorine and aluminum is generated by the method of electrolytic reduction of aluminum oxide or alumina dissolved in a cryolite-based melt. By-products of the aluminum industry can be used as a mineralizer in the production of cement clinker [23]. Unlike traditionally used calcium fluoride-based mineralizers, these products have a high content of fluoride and aluminum [8]. Due to the increased circular generation of waste, it is urgent to develop technologies that will allow them to be used as new products [9].

The efficiency of using partially calcined or amorphous alternative raw materials in the production of Portland cement due to lower energy consumption has been studied and implemented well [24]. In the literature, there are examples of the use of fluorine-containing waste as a mineralizer, which differ in different fluorine content and different phase composition [5,8,9,23], and are also used without first improving the purity and quality of the used mineralizer. These studies show the suitability of using such products, when they act, the firing temperature is reduced by 100-150^{0}C, and the use of waste raw materials reduces production costs. Waste adsorbents after water treatment are also used as waste, for example, fluorine-containing sludge obtained during defluorination using Aluminum-PHA, which is obtained by applying hydroxyl aluminum oxide to the surface of PHA. Here it was found that fluoride works as a good mineralizer and is able to reduce the firing temperature for clinker preparation by 100^{0}C [22]. It is worth noting that there is no data in the literature on the production of high-purity CaF_2 from aluminum production waste, pre-purified from the amorphous component (most often C) and alkaline components. As a rule, manufacturers do not want to use direct waste in the production process, which does not have stability. This makes it difficult to select the weight ratio of raw materials and reduces the quality of the finished product.

We used a by-product of aluminum production of JSC RUSAL Krasnoyarsk, which currently accumulates more than 16,000 tons of FCC waste per year on its territory and landfills, to obtain a mineralizer for the production of Portland cement clinker - calcium fluoride. The disadvantage of this raw material is the increased concentration of alkali metals, whose oxides negatively affect the process of hydration of clinker minerals and lead to the formation of efflorescence on products. In addition, alkalis can interact with reactive SiO_2 included in the aggregate (opal, chalcedony), causing excessive expansion and even destruction of concrete, as well as reducing its frost resistance. Therefore, in some cements, for example for hydraulic engineering, the alkali content should not exceed 0.6%.

This study describes the optimal conditions for the synthesis of a high-purity fluorite mineralizer from FCC waste, with the preservation of useful components in the waste, such as aluminum and iron, and getting rid of impurities - alkali metal compounds. The composition, physical-mechanical and thermal properties of the product and raw materials were studied.

2. Materials and methods

2.1. Materials

The fluorocarbon-containing waste was provided by JSC RUSAL Krasnoyarsk.

Grade 2 quicklime is a national commercial product that contains 87% of active CaO+MgO according to chemical analysis results.

2.2. Methods

Chemical and x-ray phase analysis methods were used to determine the chemical composition of the FCC waste. The composition of the obtained fluorite was determined by x-ray phase analysis. The thermal stability, rolling boundary, lower yield boundary are considered, the number of plasticity, ash content, and bulk density of the mixture of FCC and quicklime waste are determined.

Powder radiographs of the FCC waste and the resulting fluorite were taken at room temperature using an x-ray diffractometer "X'pertpro" (PANanalytical). The wavelength of the radiation (CuKα) is 1.54 Å. The materials were identified using information and search engines (ISE FI).

The thermogram of the compound was obtained using a synchronous thermal analysis device SDT-Q600 TA Instruments Thermal Analyzer with an IR prefix Nikolet 380 in an argon atmosphere. Heating was performed in the temperature range from 273K to 1273K at a speed of 10K/min. The weight of the hitch was 55mg.

3. Results

3.1. Study of FCC waste

The FCC waste was studied to determine the composition and its stability. The radiograph of the FCC waste is shown in figure 1.

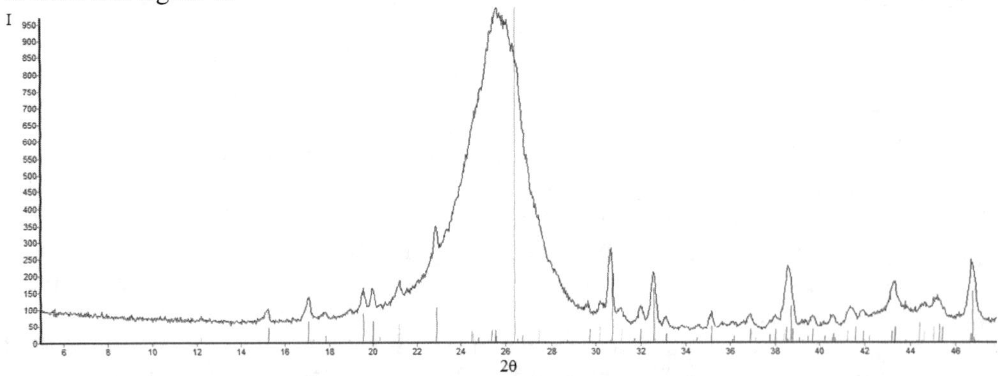

Figure 1. Radiograph of FCC waste.

Samples for analysis were taken in five parallels for several batches of waste, which were averaged by quartering. The results of x-ray phase and elemental analyses of FCC waste are presented in tables 1 and 2.

Table 1. Results of x-ray phase analysis of FCC waste.

Formula	Name of mineral	% wt.
Al_2O_3	Corundum	2.4
Na_3AlF_6	Cryolite	15.3
$Na_5Al_3F_{14}$	Chiolite	9
$Na_2Ca_3Al_2F_{14}$	Verneite	3.0
$NaCa(AlF_6)$	Thomsenolite	2.2
K_2NaAlF_6	Elpasolite	0.9
C	Graphite	67.1

Table 2. Elemental analysis of FCC waste.

Element concentration, % wt.						
O	C	Al	Na	Ca	F	K
1.1	67.1	5.5	7.9	1.2	16.8	0.3

The thermal behavior of the FCC waste was investigated. The thermogravimetric analysis curve (figure 2) shows that the substance is stable up to 550°C. At the stage of thermolysis in the temperature range 550-1000°C oxidative decomposition of the substance occurs with the release of CO_2, which is confirmed by the results of IR spectroscopic analysis of the exhaust gases. The mass loss at 725 and 948°C corresponds to exoeffects, and at 929°C - endoeffect.

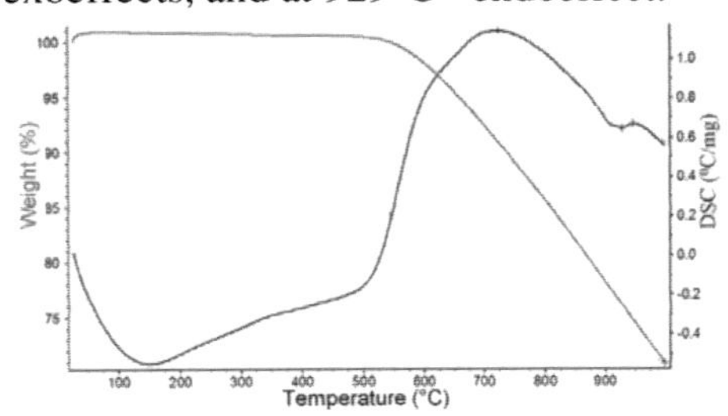

Figure 2. FCC waste thermogram.

3.2. Synthesis of CaF_2

Fluorite synthesis was carried out by mixing the components to a homogeneous mass, then the mixture was fired in a muffle furnace for 2 hours at t=1000^{0}C, and then cooled in air to room temperature.

A series of experiments were conducted with the excesses and drawbacks of FCC waste, as well as an experiment in accordance with the calculation of the stoichiometric ratio. Table 3 below shows the results of a chemical analysis of one of the experiments with a lack of FCC waste, at a ratio of 60/40 (FCC waste/quicklime).

Experiments with an excess, as well as with a lack, did not show a high content of fluorite. At these ratios, the mass fraction of fluorite is less than 35%, which is insufficient for an effective mineralization process in the production of Portland cement.

Table 3. Results of chemical analysis with a lack of FCC waste.

Concentration of the component, %											
poi	SiO_2	Al_2O_3	Fe_2O_3	CaO	MgO	SO_3	TiO_2	CaF_2	Na_2O	K_2O	F
2.23	4.41	18.34	4.77	26.73	0.73	10.08	0.83	33.58	15.00	1.09	20.87

IOP Conf. Series: Materials Science and Engineering **945** (2020) 012062 doi:10.1088/1757-899X/945/1/012062

The optimal ratio of the components of the FCC waste with quicklime 79/21 by weight was selected in accordance with the stoichiometric ratio. For example, the reactions of the interaction of CaO and cryolite/chiolite components (the main components of the waste mass, excluding carbon) are shown below):

- for the cryolite component:

$$2Na_3AlF_6 = 6NaF + 2AlF_3;$$
$$2AlF_3 + 3CaO = 3CaF_2 + Al_2O_3;$$
$$6NaF + 3CaO = 3CaF_2 + 3Na_2O;$$
$$2Na_3AlF_6 + 6CaO = 6CaF_2 + 3Na_2O + Al_2O_3.$$

- for the chiolite component:

$$3Na_5Al_3F_{14} = 5Na_3AlF_6 + 4AlF_3;$$
$$5Na_3AlF_6 = 15NaF + 5AlF_3;$$
$$9AlF_3 + 13.5CaO = 13.5CaF_2 + 4.5Al_2O_3;$$
$$15NaF + 7.5CaO = 7.5CaF_2 + 7.5Na_2O;$$
$$3Na_5Al_3F_{14} + 21CaO = 21CaF_2 + 4.5Al_2O_3 + 7.5Na_2O.$$

During laboratory tests, a number of experiments were conducted to increase the percentage of CaF_2 in the resulting mineralizer and to reduce the alkaline salts in the composition.

Chemical analysis of the product after the firing stage showed an increased content of by-products, about 47%, of which 15% is sodium salt, which is undesirable for the finished fluorite. According to qualitative x-ray phase analysis, the by-products are $CaSO_4{\cdot}(H_2O)_2$ (gypsum), Na_2SO_4, $Ca_2(SO_4)_2{\cdot}H_2O$ (bassanite), $MgAl_2O_4$ (spinel), $Ca_6(Al(OH)_6)_2(SO_4)_3(H_2O)_{26}$ (ettringite), $(CaO)_{11}(Al_2O_3)_7(CaF_2)$ (fluoromayenite), $CaSO_4$ (anhydrite). To minimize the presence of reaction byproducts in the finished mineralizer, it was washed with water.

According to x-ray phase analysis (XPA) (figure 3, table 4), the resulting fluorite consists mainly of calcium fluoride (73.9%), the rest of it consists of the minerals $MgAl_2O_4$ (spinel), $Ca_6(Al(OH)_6)_2(SO_4)_3(H_2O)_{26}$ (ettringite), $(CaO)_{11}(Al_2O_3)_7(CaF_2)$ (fluoromayenite), $CaSO_4$ (anhydrite).

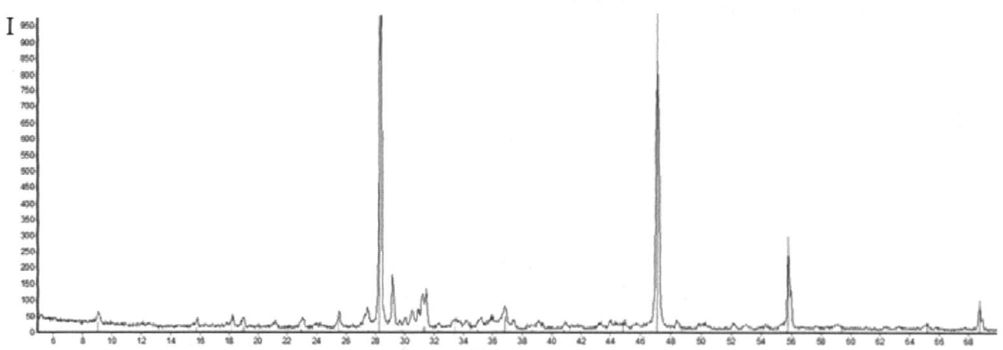

Figure 3. Radiograph of the obtained fluorite.

Table 4. Results of x-ray phase analysis of fluorite.

Formula	% wt.
CaF_2	73.9
$MgAl_2O_4$	8.50
$Ca_6(Al(OH)_6)_2(SO_4)_3(H_2O)_{26}$	4.78
$(CaO)_{11}(Al_2O_3)_7(CaF_2)$	4.65
$CaSO_4$	8.11

3.3. Analysis of physical and mechanical parameters

To assess the possibility of using a mixture of waste and quicklime in production as a precursor for the synthesis of fluorite, the physical, mechanical and thermal parameters of the mixture were analyzed. An analysis of the assessment of physical, mechanical and thermal parameters of the mixture of FCC waste and quicklime is shown in table 5.

Table 5. Results of analysis of physical and mechanical parameters of FCC and quicklime waste.

Indicator	Result
Humidity corresponding to the lower yield limit, %	61
Humidity corresponding to the rolling boundary, %	50
The number of plasticity, %	11
Bulk density, kg/m^3	925
Heat of combustion, J/g	31712
Ash content, %	32

4. Discussion

As a result of evaluating the process of obtaining and composition of the fluorite mineralizer, it is recommended to burn a mixture of FCC waste and quicklime in a ratio of 79/21 in the form of granules with a diameter of 10-20 mm, controlling the humidity of the mixture by adding water to 55% humidity. Washing of the firing product should be carried out at a low flow rate, no more than $8m^3/h$. The resulting fluorite with a high mass concentration of 73.9% CaF_2 will reduce the clinker formation temperature, which will increase the energy efficiency of obtaining Portland cement. The technological process for removing alkaline components from the product will reduce the likelihood of destructive processes in concrete caused by the aggressive action of reactive silica aggregates and freezing and thawing processes.

5. Summary

Thus, this paper describes a thermal method for processing FCC waste to obtain a mineralizer (fluorite), the mass fraction of CaF_2 in which is 73.9%. The obtained data confirm the possibility of using the product as a mineralizer in the production of Portland cement clinker. The increased content of calcium fluoride in the composition allows reducing its dosage when using. As there are currently no raw materials for the production of mineralizers in Russia, the production of fluorite will ensure that the manufacturer of the cement industry minimizes the risks of logistics costs and saves resources in the technological process.

6. References

[1] Bribian I Z, Capilla A V, Usón A A 2011 Life cycle assessment of building materials: Comparativ analysis of energy and environmental impacts and evaluation of the eco-efficiency improvement potential *Building and Environment* **46** 1133–1140
[2] Saidur R, Rahim N A, Hasanuzzaman M 2010 A review on compressed air energy use and energy savings *Renewable and Sustainable Energy Reviews* **14** 1135–1153
[3] Singhi M K, Bhargava R 2010 Sustainable Indian cement industry In: *Workshop on International comparison of Industrial Energy efficiency* **44** 27-28
[4] Elistratkin M Y, Kozhukhova M I 2018 Analysis of the factors of increasing the strength of the non-autoclave aerated concrete *Construction Materials and Products* **1(1)** 59 – 68
[5] Kani E N, Allahverdi A 2010 Fast set and high early strength cement from limestone, natural pozzolan, and fluorite *International Journal of Civil Engineering* **8(4)** 362–369

[6] Dominguez O, Torres A, Flores L M, Torres R 2010 Characterization using thermomechanical and differential thermal analysis of the sinterization of Portland clinker doped with CaF_2 *Materials Characterization* **61** 459-466
[7] Raina K, Janakiraman L K 1998 Use of mineralizer in black meal process for improved clinkerization and conservation of energy *Cement and Concrete Research* **28(8)** 1093-1099
[8] Menéndez E, Glasser F P, Aldea B, Andrade C, Zimmermann Y C 2016 Effects of the incorporation of aluminum fluoride mineralizers in Portlandcement clinker phases *National Council for Cement and Building materials*
[9] Maheswaran S, Kalaiselvam S, Karthikeyan S K, Kokila C, Palani G S 2016 β-belite cements (β-dicalcium silicate) obtained from calcined lime sludge and silica fume *Cement and Concrete Composites* **66** 57-65
[10] Misra K C, Borthakur P C 1985 Alite Formation in Rice Husk Ash-$CaCO_3$ – System: Influence of Mineralizers on formation *Transactions - Indian Ceramic Society* **44** 101-105
[11] Bordoloi D, Baruah A C, Barkakati P, Borthakur P C 1998 Influence of ZnO on clinkerization and properties of VSK cement *Cement and Concrete Research* **28** 329-333
[12] Odler I, Abdul-Maula S 1987 Hydration Reactions in the System CaO-SiO_2-CaF_2 *Journal of the American Ceramic Society* **70(1)** 39–42
[13] Klemm W A, Jawed I, Holub K J 1979 Effects of calcium fluoride mineralization on silicates and melt formation in portland cement clinker *Cement and Concrete Research* **9(4)** 489–496
[14] Altun I A 1999 Effect of CaF_2 and MgO on sintering of cement clinker *Cement and Concrete Research* **28** 1847-1850
[15] Blanco M T, Palomo A, Puertas F, Vázquez T 1995 Influence of the joint incorporation of CaF_2 and $CaSO_4$ in clinker process Obtainment of new cements *Materials Construction* **45** 239
[16] Chen Y L, Chang J E, Shih P H, KoM S, Chang Y K, Chiang L C 2010 Reusing pretreated desulfurization slag to improve clinkerization and clinker grindability for energy conservation in cement manufacture *Journal of Environmental Management* **91(9)** 1892–1897
[17] Emanuelson A, Landa A R, Hansen S 2003 A comparative study of ordinary and mineralised Portland cement clinker from two different production units Part I: Composition and hydration of the clinkers, *Cement and Concrete Research* **33** 1613-1621
[18] Emanuelson A, Landa A R, Hansen S 2003 A comparative study of ordinary and mineralised Portland cement clinker from two different production units Part II: Characteristics of the calcium silicates *Cement and Concrete Research* **33** 1633-1630
[19] Kacimi L, Simon A, Ghomari A, Derriche Z 2006 Influence of NaF, KF and CaF_2 addition on the clinker burning temperature and its properties Comptes Rendus Chimie **9(1)** 54-63
[20] Older I, Abdul-Maula S 1980 Structure and Properties of Portland Cement Clinker doped with CaF_2 *Journal of the American Ceramic Society* **63(11-12)** 654-659
[21] Bădănoiu A, Paceagiu J, Voicu G 2011 Hydration and hardening processes of Portland cements obtained from clinkers mineralized with fluoride and oxides *Journal of Thermal Analysis and Calorimetry* **103(3)** 879–888
[22] Sarmah S, Saikia J, Bordoloi D K, Kalita P J, Bora J J, Goswamee R L 2018 Immobilization of fluoride in cement clinkers using hydroxyl-alumina modified paddy husk ash based adsorbent *Journal of Chemical Technology and Biotechnology* **93(2)** 533–540
[23] Welch B J 1999 Aluminium production paths in the new millennium *JOM* **51(5)** 24-28
[24] Tsakiridis P E, Papadimitriou G D, Tsivilis S, Koroneos C 2008 Utilization of steel slag for Portland cement clinker production *Journal of Hazardous Materials* **152(2)** 805-811

Acknowledgments
This work was implemented using the equipment of the Collective Use Centre of IOG of SFU.

IOP Conf. Series: Materials Science and Engineering **945** (2020) 012063 doi:10.1088/1757-899X/945/1/012063

Method of forming technological parameters in the design of vibrating cone crushers

M D Gerasimov[1*], B S Chetverikov[1], N S Lubimyi[1] and I Ridvan[2]

[1]Department of Hoist transport and road machines, Belgorod State Technological University named after V G Shukhov, Kostyukov St., 46, Belgorod, 308012, Russia
[2]SanayiBölgesi, Lacivert Cad. No: 616065, Nilüfer, Bursa / TÜRKİYE

E-mail: nslubim@bk.ru

Abstract. The paper deals with issues related to expanding the possibility of using a promising direction for improving vibrating cone crushers of secondary and fine crushing. The paper deals with two major issues that need to be solved at designing and creating vibrating cone crushers. First question: the value of the ratio of the crushing force in the standard design of a cone crusher of secondary or fine crushing. The second question is the frequency of forced oscillations of the additional vibration device. The choice of the ratio of the crushing force and the value of the driving force of the vibrator is based on the output of the rational frequency of forced oscillations of the unbalanced vibration device. For the first time in this paper for this type of machine the method of dividing the circular crushing chamber into separate sections is used. Each section is a zone of wave displacement of the compressive stress in the material from the action of the crushing force. The length of the section is considered in the range of the current action of the crushing force, in which the stress in the crushed material changes from zero to maximum and again to zero. This section of the action of the crushing force is an elementary section of the force effect of the crushing force on the material. The paper contains some new terms that are caused by the need to apply a more accurate description of the kinetics of the process. The paper forms new approaches to the process of calculating the parameters of vibrating cone crushers, so it can be considered in the “in order of discussion” format.

1. Introduction

Issues of the state and development of design solutions in the field of design and calculation of vibrating cone crushers are fully covered in [1-9]. These works show a number of technological advantages of using vibrating cone crushers of secondary and fine crushing. Basically, there are three types of vibrating cone crushers: cone crushers of the traditional, classical scheme with additional vibration of the internal movable cone, cone crushers of the traditional, classical scheme with additional vibration of the external stationary cone, cone crushers that use only vibration forces. This paper discusses theoretical and technical problems related to cone crushers of the first type.

In [1], information about the kinetics of the process in classical cone structures is presented, and the features of their design solutions are revealed. Classic design solutions have successfully solved production problems for many years. However, in modern conditions, there is a need for further development of design solutions. The publication contains a large and useful amount of information both for expanding research work in the field of improving crushing equipment, and for use in the

IOP Conf. Series: Materials Science and Engineering **945** (2020) 012063 doi:10.1088/1757-899X/945/1/012063

practical application of machines of this type. This paper is an attempt to consider some issues of the problem of calculation and design of vibration crushers, formulated on the basis of the material presented in [1].

2. Materials and methods

Grinding in cone crushers by its nature “lies” in the boundary area between a continuous and cyclic process. This process can be considered both continuous and cyclic [10].

Most often, starting from the classic textbook of M.Ya. Sapozhnikov [11] this process is considered as a continuous process of material output from the output hole of the crusher. At the same time, the process is considered as the output of material in one turn of the eccentric shaft, that is, in one cycle, taking into account the number of cycles per unit of time, for example, in 1 hour.

Based on this approach, it is possible to separate the elementary cycle of force action of crushing cones in the process of grinding material in a cone crusher of secondary or fine crushing, which in time consists of the following components:

$$\mathrm{T}_C = t_1 + t_2 + t_3 + t_4 + t_5 + t_6 + t_7 \tag{1}$$

where t_1 – time of approach of the crushing cone surface to the grain (grains) of the material; t_2– the time of the capture material crushing surfaces; t_3– time during which fragments and crumbling of weak corners and sections of the lump material occur; t_4– capture and increase of compression stress in grains (pieces) of the crushed material; t_5 – reaching the ultimate strength, or temporary compression resistance, of grains (pieces) of brittle crushed material; t_6– formation of cracks and exposure of additional surface, destruction of grains (pieces) of the material and reduction of the size of the pieces; t_7– withdrawal of the crushing cone from the crushed material.

The area of active interaction between the crushing cone and the material can in principle be considered as an elementary crushing chamber.

In turn, the components $t_1 + t_2 + t_3 + t_4 + t_5$ within the cycle T_c represent the preparation time of the crushing conditions, t_{pcc}, and can be represented by the expression

$$t_{pcc} = t_1 + t_2 + t_3 + t_4 + t_5 \tag{2}$$

The components of the grinding process that occur within the time $t_1 \dots t_5$ can pass both sequentially and in parallel to each other. However, they meet all the conditions when the state of the ultimate compression stress has not been reached in the main material array yet [11]:

$$p_{sp} = \sigma_{comp} = E_0 \cdot \varepsilon \leq [\sigma_{comp}] \tag{3}$$

where p_{sp} – specific pressure of the crushing cone on the material.. N/m^2; σ_{comp} – current value of the compression stress in the material, N/m^2; E_0 – modulus of linear deformation of the material, N/m^2, table value [4]; ε – the relative linear deformation during compaction of the material; $[\sigma_{comp}]$ - temporary compression resistance of the source material, N/m^2.

At the time stage of destruction, grinding and crushing of pieces (grains) of the material, i.e. $t_{gr} = t_6$ the expression (3) takes the form:

$$\sigma_{comp} = E_0 \cdot \varepsilon \geq [\sigma_{comp}] \tag{4}$$

Finally, at the stage when the mobile cone is removed from the elementary volume of grinding, the stress in the material decreases from $\sigma_{comp\max}$ to 0. Denoting the time of the material unloading stage $t_{un} = t_7$,, we get the time of the elementary grinding cycle in the crushing chamber of a cone crusher CST or CFT:

$$\mathrm{T}_c = t_{pcc} + t_{gr} + t_{un} \tag{5}$$

In [1, p. 75], recommendations are given for “determining the crushing force in the mode of regular running-in of a movable cone over a stationary one” and it is shown that “the ratio between the forces developed by the cone and the unbalance determines the nature of the crusher operation:

$$N = v_b \cdot F_b + v_c \cdot F_c, \tag{6}$$

where F_b and F_b– forces developed by the cone and debalance, respectively; v_b and v_c – coefficients [1]".

It should be noted that this expression allows creating a condition that is close to resonance. Resonance is not so much mechanical, but to the resonance of the process.

The frequency of forced vibrations generated by the unbalance also plays an important role in achieving a condition close to the resonance of the grinding process in a secondary or fine cone crusher.

If the speed of the crushing cone is 246 rpm (CST-1750), or 4.1 swings/s, then the time of one turn of the crushing cone is $T_{C1\,revolution}$ = 0.244 s.

However, within the time of one turn, there may be several complete cycles described by the expression (1) or (5). Thus, if they know the time of one elementary cycle, T_c, within which the stress increases in the material from 0 to the maximum, grinding and reducing the stress again to 0, then they can determine how many elementary cycles of force action can be contained in one turn of the movable cone, i.e., within the time $T_{C1\,revolution}$ = 0.244 s.

The question is important for the process: "What is the frequency and period, T_c, of one elementary loading in the limit of the running period, $T_{C1\,revolution}$ = 0.244 s?".

We will conditionally set the number of such force actions, *n*, equal to four, namely:

$$n = \frac{T_{C1\,revolution}}{T_c} = 4.0 \qquad (7)$$

Then, the time of one elementary loading cycle will be: $T_c = \frac{0.244}{4} = 0.061$ s. At the same time, the beginning of each elementary cycle is shifted relative to the other by 90^0. Moreover, the reference point of the period $T_{C1\,revolution}$ can be taken anywhere in the circle from 360^0. The main condition remains that the period $T_{C1\,revolution}$ and the first elementary period T_c started simultaneously from one common point, for example, p. *b*.

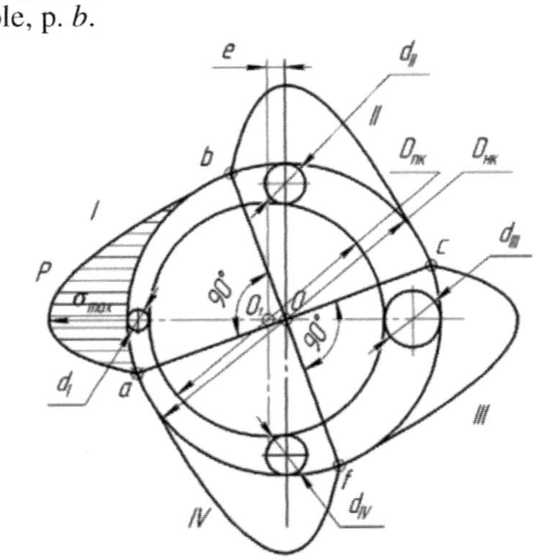

Figure 1. Scheme of elementary force actions of a moving cone on a material: I, II, III, IV – sequence of sections of elementary force action within a period, cycle, or one rolling of a moving cone; d_I – minimum size of a piece of material; d_{II} and d_{IV} – intermediate size of a piece of material; d_{III}– the size of the piece entering the grinding zone; D_{mc} – diameter of the movable cone b; D_{fc} – diameter of the fixed cone; *e* – eccentricity; σ_{max} – amplitude of the maximum stress from the crushing force of the movable cone; $\sigma_o = 0$ – minimum stress amplitude in the material; *O* and O_1 – the center of a fixed and movable cone, respectively; *b-a, c-b, f-c , a-f* – areas of elementary force impact of the crushing cone on the material; $\alpha = 90^0$ – central angle of the elementary force impact area, ω_c – angular speed of running-in of a movable cone.

IOP Conf. Series: Materials Science and Engineering **945** (2020) 012063 doi:10.1088/1757-899X/945/1/012063

The rolling period of the mobile cone is $T_{C1\,revolution}$= 0.244 s. The value of the amount of crushing force graphically can be depicted as a traveling wave with some asymmetry on frequency [12, 13], caused by the difference in time flow process consisting of three phases: increased tension ($t_{inc.ten.}$), grinding (t_{gr}) and the stress relief material ($t_{st.rel.}$): $t_{inc.ten.} > t_{gr} \cong t_{st.rel.}$.
The nature of such oscillations, regardless of the magnitude of the crushing force itself, can be represented by the sum of two oscillatory processes with a shift in the initial phases, table 1.

Table 1. Initial parameters for obtaining the value of the crushing force within one elementary loading cycle.

Parameter	№ of oscillatory process	
	1	2
Weight (kg)	1000	125
Radius (cm)	1.5	1.5
Init. phase (deg.)	180	270
Speed (rpm)	246	492
Ri (m)	0.015	0.015
φi (rad)	3.14	4.71
ωi (1/s)	25.76	51.52

Dividing the time of elementary exposure into 20 intervals by dt = 0.012 s. We obtain a change in the magnitude of the driving force of the first (F1) and second (F2) oscillations. The total calculated value of the driving force (Sum) forms the change in the crushing force in the area of elementary impact, table 2.

Table 2. Calculated values of the total driving force of two vibrators, displaying the form of the crushing force within the limits of the elementary force action.

№	t	F1	F2	Sum
0	0.000	-19.91	0.00	-19.91
1	0.012	-18.97	5.77	-13.20
2	0.024	-16.22	9.40	-6.82
3	0.036	-11.94	9.56	-2.39
4	0.048	-6.53	6.17	-0.36
5	0.060	-0.50	0.50	0.00
6	0.072	5.58	-5.36	0.22
7	0.084	11.13	-9.23	1.90
8	0.096	15.62	-9.68	5.94
9	0.108	18.64	-6.55	12.08
10	0.120	19.88	-1.00	18.89
11	0.132	19.25	4.93	24.17
12	0.144	16.78	9.03	25.81
13	0.156	12.73	9.79	22.52
14	0.168	7.47	6.92	14,39
15	0.180	1.50	1.50	2.99
16	0.192	-4.61	-4.49	-9.10
17	0.204	-10.29	-8.81	-19.09
18	0.216	-14.98	-9.87	-24.85
19	0.228	-18.26	-7.27	-25.54
20	0.240	-19.81	-1.99	-21.80
			Max	25.81
			Min	-25.81

The graph of changes in the size of the crushing force is plotted relative to a certain average line. The crushing force is $|Max| + |Min| = 51.62$ kN.

A graphical representation of the change in the crushing force within the limits of an elementary force action is shown in Fig. 2.

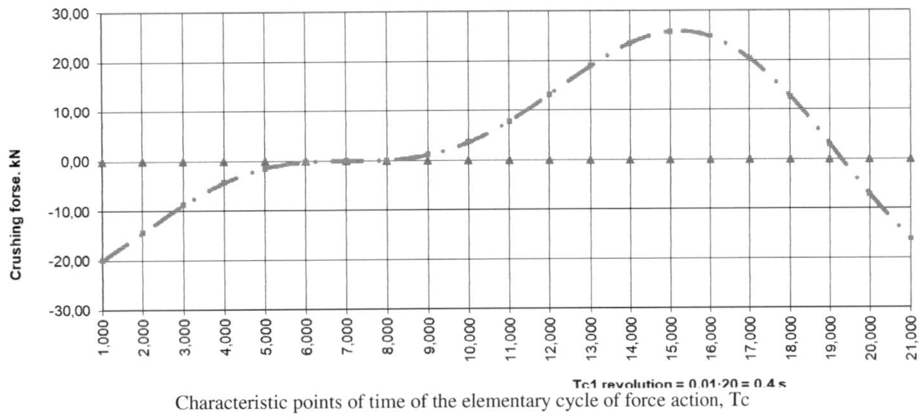

Figure 2. Graph of changes in the crushing force within the time of the elementary cycle of force action.

The time of the elementary cycle of the power impact of T_c consists of two components. In the first section: points 20-21-1-13, the material compression stress increases from 0 to the maximum value $\sigma_{comp} \leq [\sigma_{comp}]$. The duration of this section is 70% of the time T_c.

At the second time interval, 13-20, which is 30% of the T_c, the compression voltage is reset in the range from $\sigma_{comp} \geq [\sigma_{comp}]$ to zero.

This model and the form of changing the crushing force characterize the classic design of a cone crusher without vibration. Within the angle 90^0, the rotation of the movable cone, we have a fairly "large section" of time, which is characterized by a "slow" increase in voltage.

The ideas laid down in [1] allow taking advantage of this form and nature of the change in the crushing force. On the "stretched" section of the graph, they can embed additional force from the vibration mechanism.

Such a force effect can be created by a vibrating device with a multiple of the rotation frequency of the unbalanced mechanism in relation to the rotation frequency of the movable cone, table 3.

Table 3. Initial parameters for obtaining the value of the crushing force within one elementary loading cycle with a vibrating device.

Parameter	№ of vibrator		
	1	2	3
Weight (kg)	1000	125	50
Radius (cm)	1.5	1.5	1.5
Init. phase (deg.)	180	270	180
Speed (rpm)	246	492	738
Ri (m)	0.015	0.015	0.015
φi (rad)	3.14	4.71	3.14
ωi (1/s)	25.76	51.52	77.28

The calculation of the crushing force taking into account the vibrator is shown in table 4. Here F_1 and F_2 represent the crushing force of the cone of the basic cone crusher: $F_{cr} = F_1 + F_2 = 51.62$ kN. The value of the driving force of the vibrator F_{vibr} and F_{cr} form the total force.

IOP Conf. Series: Materials Science and Engineering **945** (2020) 012063 doi:10.1088/1757-899X/945/1/012063

Table 4. Calculation of the crushing force taking into account the vibrator.

№	t	F1	F2	F_{vibr}	Sum
0	0.000	-19.91	0.00	-8.96	-28.87
1	0.012	-18.97	5.77	-5.37	-18.57
2	0.024	-16.22	9.40	2.51	-4.31
3	0.036	-11.94	9.56	8.39	6.00
4	0.048	-6.53	6.17	7.55	7.19
5	0.060	-0.50	0.50	0.67	0.67
6	0.072	5.58	-5.36	-6.74	-6.52
7	0.084	11.13	-9.23	-8.76	-6.86
8	0.096	15.62	-9.68	-3.77	2.17
9	0.108	18.64	-6.55	4.24	16.32
10	0.120	19.88	-1.00	8.86	27.74
11	0.132	19.25	4.93	6.39	30.56
12	0.144	16.78	9.03	-1.19	24.62
13	0.156	12.73	9.79	-7.82	14.70
14	0.168	7.47	6.92	-8.19	6.20
15	0.180	1.50	1.50	-2.01	0.99
16	0.192	-4.61	-4.49	5.78	-3.32
17	0.204	-10.29	-8.81	8.94	-10.15
18	0.216	-14.98	-9.87	4.95	-19.90
19	0.228	-18.26	-7.27	-3.00	-28.54
20	0.240	-19.81	-1.99	-8.55	-30.35
	Max				30.56
	Min				-30.56

In table 4, in addition to the crushing force, F_{cr} we have the vibration force F_{vibr}. Together they form an analog of the expression (6):

$$N = F_{cr} + F_{vibr} \tag{8}$$

where F_{cr} – crushing force generated by the mobile cone of the crusher on the material, kN; F_{vibr} – the driving force of the vibrating device. In our case,

$$N = |30.56| + |-30.56| = 61.12\text{kN}.$$

3. Results and discussion

The graph of changes in the total crushing force is shown in Fig. 3.

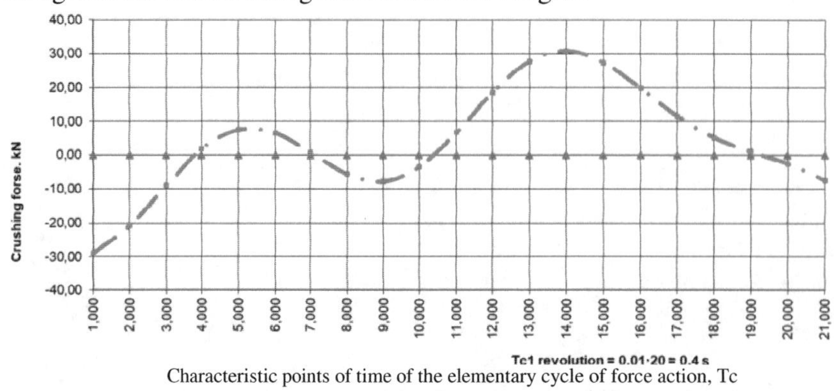

Figure 3. Graph of changes in the crushing force within the time of the elementary cycle of force action using a vibration mechanism at $\omega_{\partial} = 3 \cdot \omega_{\kappa} = 738\, rpm$.

IOP Conf. Series: Materials Science and Engineering **945** (2020) 012063 doi:10.1088/1757-899X/945/1/012063

The increase in the crushing force by the value of the driving force of the vibrator can be (0.4...0.5)· F_{cr} [1]. In this case, the total crushing force is increased by 18.4%.

In addition, during the crushing process, additional force pulses are formed within each elementary force action, which can favorably affect the efficiency of grinding.

The value of the additional vibration force must be determined by calculation. So, when using the frequency ratio: $\omega_{cr} = 3 \cdot \omega_c = 738\, rpm$, where ω_{cr} and ω_c– the angular velocity of rotation of the debalance of the vibrator and the movable cone, respectively, has two pulses within each elementary cycle of force action. Additional impulses of the crushing force are observed in p. 4-5 and p. 12.

If the rotation frequency of the unbalanced mechanism is increased, the picture of the force effect will take a different form. When $\omega_d = 4 \cdot \omega_c = 984$ rpm the total crushing force increases to 74.8 kN, which is 44.5% of the original. The graph of changes in the total driving force, at the same time, records an increase in the number of additional pulses, Fig. 4.

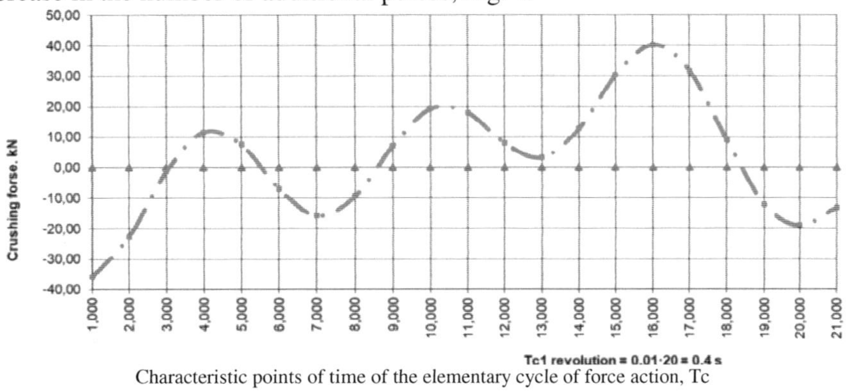

Figure 4. Graph of changes in the crushing force within the time of the elementary cycle of force action using a vibration mechanism at $\omega_d = 4 \cdot \omega_c = 984$ rpm.

The use of additional vibration impact on the crushed material is a reasonable means of increasing the speed of the grinding process in cone crushers of secondary and fine crushing significantly. A special feature of this effect is an attempt to create conditions for the process to approach the resonant effect [14,15,16,17,18]. When the effect is close to resonant, the speed of the process can increase not only by several times, but also by orders of magnitude. Attempts to solve such problems are given in [19]. However, solving such problems will require not only additional theoretical research and design development, but also financial support.

4. Summary

The use of a vibrating mechanism in the design of secondary and fine cone crushers can significantly increase the speed of the material crushing process in them. The purpose of using vibration mechanisms in the design of cone crushers is to create conditions close to the resonance of the process. An important condition for achieving the required conditions is the choice of a rational ratio between the size of the crushing force of the movable cone and the size of the driving force of the vibrating device. The paper presents the results of theoretical research in the development of methods for designing and calculating rational parameters of certain types of vibrating cone crushers.

5. References

[1] Vaysberg L A, Zarogatskiy L P, Turkin V Ya 2004 Vibrating crushers. Basics of calculation, design and technological application *SPb: VSEGEI* 306

[2] Zarogatskiy L P, Pirogova O A, Safronov A N 2009 Vibration Cone Crusher. Patent RF 87930

[3] Zarogatskiy L P, Pirogova O A, Safronov A N 2010 Vibration Cone Crusher. Patent RF 2401700

[4] Vaysberg L A, Zarogatskiy L P, Safronov A N, Cherkasskiy V A, Shuloyakov A D 2003 Inertial cone crusher. Patent RF 2212939
[5] Vaysberg L A, Zarogatskiy L P, Safronov A N, Turkin V Ya, 2004 Cherkasskiy V A Cone Vibratory Crusher. Patent RF 2234981
[6] Vaysberg L A, Turkin V Ya, Kazakov S V, Nagayev R F, Shishkin E V 2005 Vibrating crusher. Patent RF 2254929
[7] Zarogatskiy L P, Vaysberg L A, Kazakov S V, Turkin V Ya 2005 Vibration Cone Crusher. Patent RF 2257266
[8] Vaysberg L A, Turkin V Ya, Kazakov S V 2006 Cone Vibratory Crusher with Cross Axes. Patent RF 2292241
[9] Vaysberg L A, Zarogatskiy L P, Pirogova O A 2008 Inertial cone crusher. Patent RF 2314158
[10] Industrial vibrators and vibration equipment 2018 Product Catalog. Yaroslavl factory «KRASNYJ MAYAK» 41
[11] Hanns W Tappeiner, Roberta L Klatzky, Bert Unger, Ralph Hollis 2009 Good Vibrations: Asymmetric Vibrations for Directional Haptic Cues. *World Haptics (Third Joint Euro Haptics Conference and Symposium on Haptic Interfaces for Virtual Environments and Teleoperator Systems) SaltLakeCity* 18-20
[12] Rumyantsev S A, Astankov K Yu 2012 Increasing the efficiency of overcoming the frontal resistance of cohesive soils during vibration driving of piles *Bulletin of the Ural State Transport University* **4 (16)** 82-90
[13] Gerasimov M.D. 2016 Addition of vibrations in vibration absorbers *Bulletin of BSTU named after V G Shukhov* **3** 116-121
[14] Gerasimov M D, Mkrtychev O V, Glagolev S N, Gerasimov D M, Latyishev S S 2016 Method of determination of vibrating screens oscilation's amplitude in a characteristic point for plane motion *ARPN Journal of Engineering and Applied Sciences* **11(20)** 12295-12301
[15] Gerasimov M D, Glagolev S N, Gerasimov D M, Mkrtychev O V 2015 Determination of the driving force's asymmetry factor and the vibrostand's work's analysis *International Journal of Applied Engineering Research* **10(24)** 45392-45398
[16] Gerasimov M D, Romanovich A A, Vorobiev N D, Ebrahim Amini 2018 Results of research to improve efficiency of vibrating machines. International Conference "Complex equipment of quality control laboratories" *IOP Conf. Series: Journal of Physics: Conf. Series* 12015-12019
[17] Jiaguo Y, Ying P 2018 Research on Design of Vibration Platform of Molded Case Circuit Breaker and Reliability of Platform Frame *American Journal of Engineering and Technology Management* **3(2)** 35-41
[18] Liu Ye 2013 Analysis of Structural Design Method for Vibration System of Electrical Reliability Test Bench [J] *Science and Technology Guide* **26** 205-207
[19] Gerasimov M D, Lubimyi N S, Ryazantsev V G 2020 Methodology of vibro loader design with asymmetric oscillations *Bulletin of BSTU named after V G Shukhov* **5** 135–142.

Acknowledgments
The work was realized as part of performing service task in the framework of the Program of flagship university development on the base of Belgorod State Technological University named after V G Shukhov, using equipment of High Technology Center at BSTU named after V G Shukhov.

IOP Conf. Series: Materials Science and Engineering **945** (2020) 012064 doi:10.1088/1757-899X/945/1/012064

Method of calculation and selection of design parameters and operational characteristics of the spiral expander

S V Sergeev[1*], I S Nikulin[1], M Yu Saenko[1] and A A Titenko[2]

[1]Belgorod National Research University, 85 Pobedy st., Belgorod, 308015, Russia
[2]Engineering Center NRU "BelSU", Belgorod, 2a/712, Koroleva st., Belgorod, 308015, Russia

E-mail: svsergeev@gmail.com

Abstract. The paper proposes a method for calculating the main design parameters and operational characteristics of the HPEC using the Helmholtz equations of state. N-pentane, freon R11, and acetone were considered as the working fluid. The calculations were performed using the CoolProp 6.3.0 library. It was found that the most profitable for HPEC operating on the Rankin cycle with low-potential sources is n-pentane. L_D expander operation and mass flow rate of the working fluid are calculated. The calculations were performed using the freely distributed open source library CoolProp 6.3.0. The calculations show that the most profitable WF for HPEC running on the Rankin cycle with low-potential sources is n-pentane. It is determined that the increase in costs when the temperature decreases before the SD is due to a decrease in the level of operation of the Lд expander, which in turn depends on the value of the steam enthalpy at the entrance to the ex-pander.

1. Introduction

Taking into account the accumulated world experience in the design and operation of thermal machines operating on the organic Rankin cycle (ORC) [1-3], as well as the high level of development of some elements of such machines, it is appropriate and reasonable to use some ready-made solutions [4-6]. In this paper, the main attention will be paid to the method of calculating and selecting the design parameters of the drive of the generator of heat and power energy converters (HPEC), which is a spiral expander.

Spiral machines have recently gained popularity primarily as compressor units, including oil-free ones [7]. However, a number of studies have appeared on the use of spiral machines as expanders, including in ORC cycles [8-11]. Despite the significant similarity, expanders have completely different requirements, which must inevitably affect both the method of their calculation and design, as well as their design.

First of all, to do this, it is necessary to select and justify the applied structural materials and coverings with special properties. The solution of this issue will allow determining the design standards in terms of the permissible speeds of movement of parts and components of the expander, and the permissible loads on them, determine the rational pairs of friction materials, take into account the thermal impact of the environment on the deformation of elements, and create a basis for predicting the product life.

IOP Conf. Series: Materials Science and Engineering **945** (2020) 012064 doi:10.1088/1757-899X/945/1/012064

The second, no less important issue is the development of a method for selecting the optimal parameters of the thermodynamic cycle of HPEC, taking into account the possibility of its implementation in its individual elements. In addition, it is necessary to develop mathematical models of the main components of the HPEC, such as pump; recuperator; heater; expander; condenser; generator.

As a result of solving these problems, it is possible to create an optimal heat engine with the highest possible efficiency.

2. Materials and methods

2.1. Method for calculating the main design parameters of the HPEC

As the optimal WF for low-potential energy sources, liquids should be selected that, on the one hand, should boil at the minimum temperature, but not below the possible ambient temperature, and on the other hand, it is necessary to ensure effective condensation of WF. Three liquids meet these requirements: n-pentane (boiling point 36.074 °C), Trichlorofluoromethane (freon R11, boiling point 23.77 °C), and acetone (boiling point 56.1 °C).

The calculations used fundamental equations of state of substances in terms of Helmholtz energy [12]. A similar approach to the description is also presented in [13-14], devoted to the development of the fundamental equation of the normal pentane state.

These equations are implemented in the freely distributed open-source library CoolProp, where fluid models are based on Helmholtz energy formulas. The equations of state used to calculate the thermodynamic properties of substances are presented in [15].

Let us consider two variants of the Rankin cycle: a simple cycle and a cycle with heat recovery in order to determine the possible efficiency of the process in both variants and then compare them.

Comparison of cycles with and without recuperation, as well as comparison of application options for different WF was carried out by the value of the cycle efficiency:

$$\eta_t = \frac{q_1 - q_2}{q_1}, \tag{1}$$

where q_1 – heat supplied in the cycle; q_2 – the heat rejection in the cycle.

As heating and heat removal in this cycle are Isobaric processes, the input heat is the difference between the enthalpy at the output and input to the heater, and the output heat is the difference between the enthalpy at the input and output of the condenser [16]:

$$q_1 = h_{\text{н}\,out} - h_{\text{н}\,in}\ ; \qquad q_2 = h_{\text{к}\,in} - h_{\text{к}\,out}\ ; \tag{2}$$

$$\eta_t = \frac{(h_{\text{н}\,out} - h_{\text{к}\,in}) - (h_{\text{н}\,in} - h_{\text{к}\,out})}{(h_{\text{н}\,out} - h_{\text{н}\,in})} = \frac{L_{\text{д}} - L_P}{Q_H}, \tag{3}$$

where $L_{\text{д}} = (h_{\text{н}\,out} - h_{\text{к}\,in})$ – the work of expander; $L_P = (h_{\text{н}\,in} - h_{\text{к}\,out})$ – work performed by the pump; Q_H – energy supplied in the cycle by the heater.

The upper limit of the $T_{\text{г}}$ cycle temperature for all WF was assumed to be the same in order to ensure comparison of efficiency under equal conditions with respect to the heating source. The operating pressure at the inlet to the expander was assumed to be 10 bar, which according to preliminary calculations is an acceptable value for a spiral expander, and is not the limit for other elements of the HPEC. The calculation scheme is shown in figure 1, a.

IOP Conf. Series: Materials Science and Engineering **945** (2020) 012064 doi:10.1088/1757-899X/945/1/012064

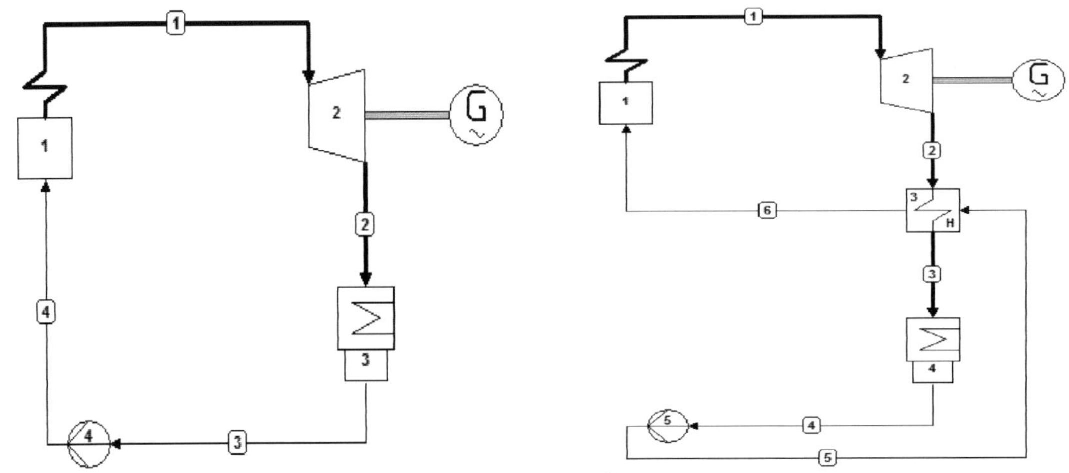

(a) (b)

Figure 1. A design scheme of HPEC, (a) – working on a simple Rankin cycle, (b) – working on the Rankine cycle with energy recovery.

The simple Rankin cycle is characterized by the following successive stages: isentropic compression – Isobaric heating – polytropic expansion – Isobaric condensation.

3. Results

Results of calculation of the main design parameters of the HPEC.

The calculations were performed using the CoolProp 6.3.0 library. Temperatures, pressures, density, specific volume, enthalpy, and entropy were obtained at the calculated points of the cycle. Then we performed calculations of Rankin cycles with heat recovery for the same WF under equal conditions of heating and condensation. The diagram of the simulated device is shown in figure 1, b.

The calculations assumed that the temperatures of both flows at the outlet of the recuperative heat exchanger will be equal, i.e. the maximum possible heat exchange characteristic of a shell-and-tube heat exchanger will be provided. Figures 2-4 show the results of calculations of the Rankine cycle with energy recovery for n-pentane, acetone, and freon R11.

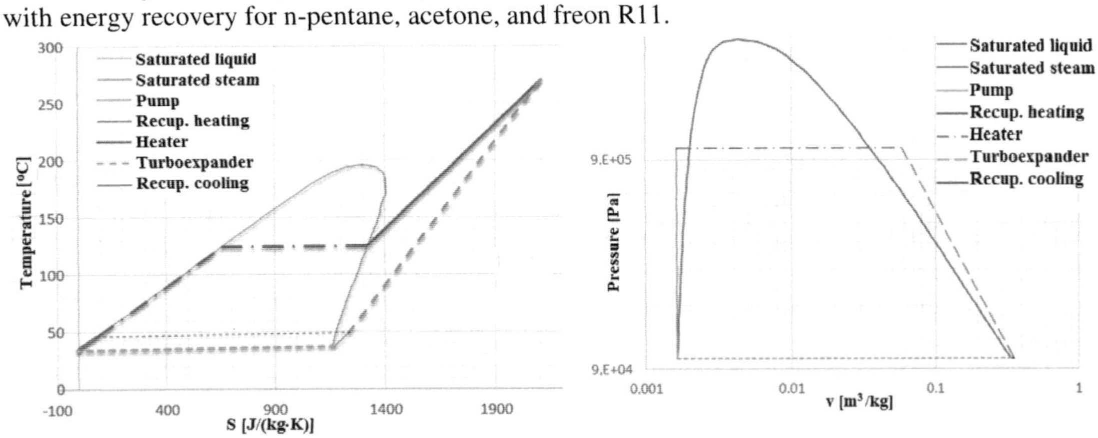

(a) (b)

Figure 2. T-s (a) and P-v (b) diagrams of the Rankine cycle with recuperation (n-pentane).

IOP Conf. Series: Materials Science and Engineering **945** (2020) 012064 doi:10.1088/1757-899X/945/1/012064

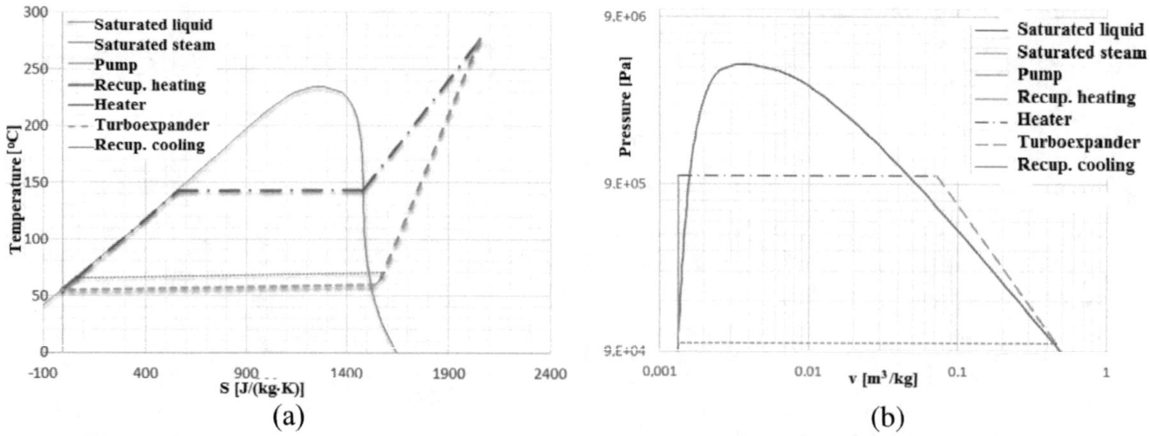

Figure 3. T-s (a) and P-v (b) diagrams of the Rankine cycle with recuperation (acetone).

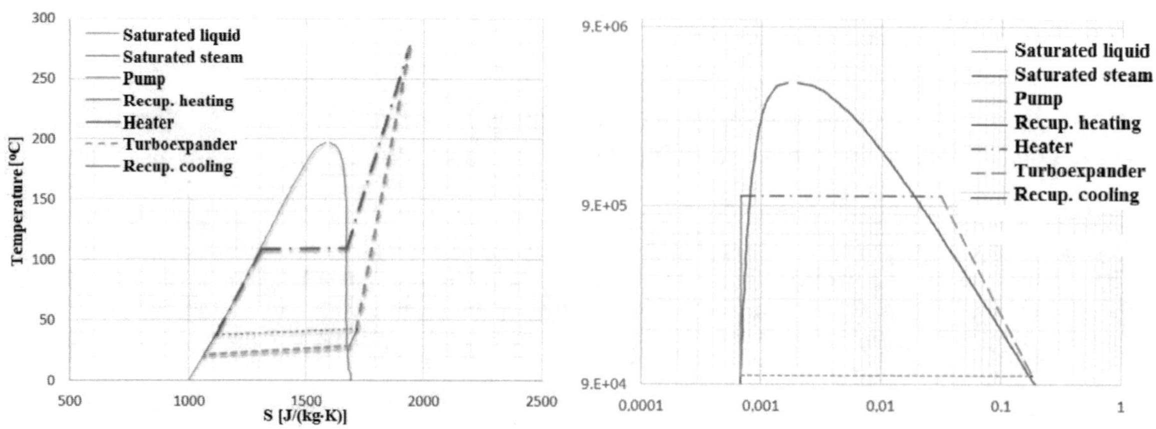

(a) (b)

Figure 4. T-s (a) and P-v (b) diagrams of the Rankine cycle with recuperation (freon R11).

The results of calculations show that almost half of the heat previously spent on cooling the condenser (direct heat loss in the system) during heat recovery is returned back to the circuit, and the heat loss in the refrigerator reduced by almost half.

To compare the effectiveness of the three listed WF, the maximum temperature for acetone is selected as the upper limit. In the temperature range from 150 °C to the maximum temperature for acetone, the thermal efficiency of the process was calculated for three types of WF for modifications of Rankin cycles with and without regeneration. The results of the calculations are shown in figure 5.

Calculations show that n-pentane is the most profitable WF for HPEC operating on the Rankin cycle with low-potential sources. The theoretical efficiency of the cycle on n-pentane with heat recovery (without taking into account irreversibility) at the temperature before the expander 573 K reaches 64%, and at the maximum WF temperature of 650 K reaches the value of 70.8%. Based on the performed calculations, the main parameters of cycles with and without heat recovery are determined.

The calculations assumed that the indicator efficiency of the spiral expander η_i is 85%, and the conversion efficiency of the synchronous generator η_e is 98%. According to these assumptions, when the rated electrical power N_e =9000 W, the rated mechanical power of the SD should be $N_{д} = \frac{N_e}{\eta_e} = 9183.67\, W$.

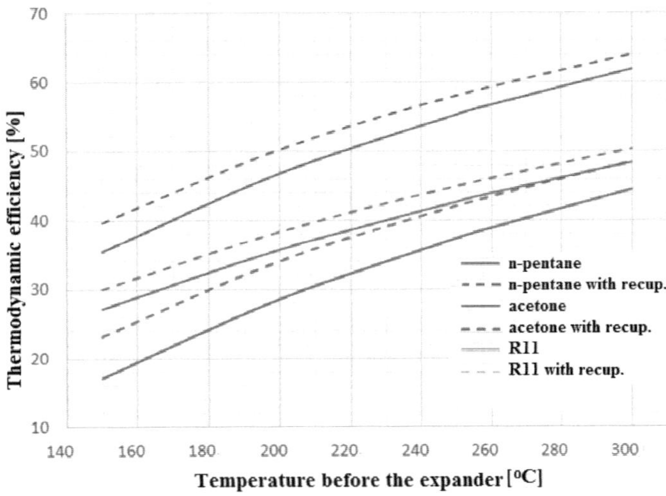

Figure 5. Results of comparative calculations of the thermal efficiency of the Rankin cycle for n-pentane, acetone and freon R11 with and without regeneration.

Knowing the difference between the enthalpy at the input and output of the SD, and thus determining the value of the expander $L_Д$, it is possible to determine the mass flow of the working fluid, which provides the mechanical and electrical power of the HPEC under specified conditions:

$$G_{m\mathrm{PT}} = \frac{N_t}{L_Д} = \frac{N_e}{\eta_e \eta_i \eta_t L_Д}, \tag{4}$$

where η_t – thermal efficiency of the process.

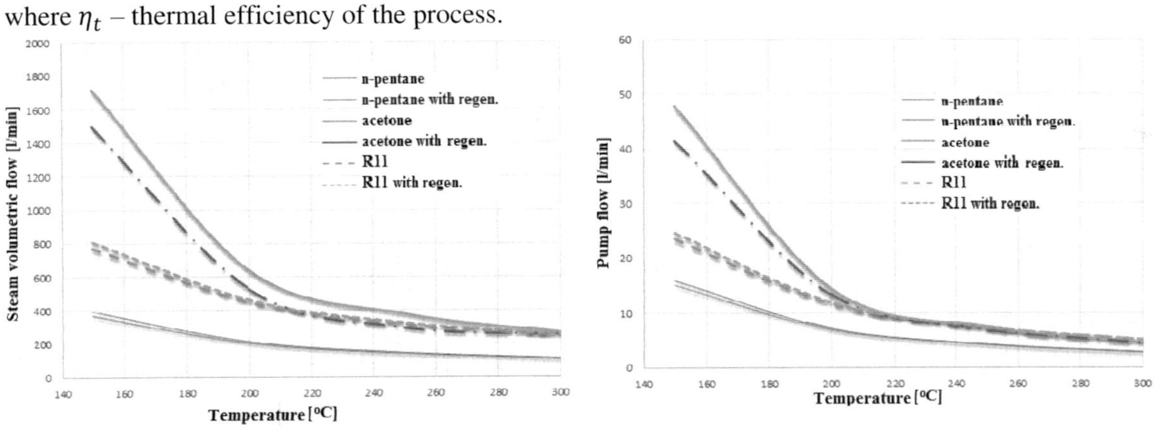

(a) (b)

Figure 6. The dependence of the volume flow of various WF in ORC with and without recuperation, provided that a constant power is reached: before the SD (a); through the pump (b).

At a known mass flow rate, using the fundamental equation of state [12], it is possible to calculate important parameters for design: the volume flow rate of vapors at the entrance to the SD, as well as the volume capacity of the feed pump. Such calculations were performed for the three WF types under consideration and for ORC variants with and without recuperation. Figures 6 show the dependences of the volume flow of the working fluid through the SD and the feed pump.

It is obvious that the increase in costs when the temperature decreases before the SD is due to a decrease in the level of operation of the $L_Д$ expander, which in turn depends on the value of the steam enthalpy at the entrance to the expander. Taking into account the constancy of the indicator efficiency

IOP Conf. Series: Materials Science and Engineering **945** (2020) 012064 doi:10.1088/1757-899X/945/1/012064

of the SD and the efficiency of the generator, it can be argued that the mass flow rate of WF is inversely proportional to the thermal efficiency of the cycle, provided that the constant electric power of the HPEC is achieved.

4. Summary

The method of calculation and selection of the main design parameters and operational characteristics of HPEC is developed. The given method of calculation using equations of state of real substances allows determining the main design parameters of the HPEC. Three liquids were considered as optimal WF: n-pentane, freon R11, and acetone. The calculations were performed using the freely distributed open source library CoolProp 6.3.0. The calculations show that the most profitable WF for HPEC running on the Rankin cycle with low-potential sources is n-pentane. It is determined that the increase in costs when the temperature decreases before the SD is due to a decrease in the level of operation of the Lд expander, which in turn depends on the value of the steam enthalpy at the entrance to the expander.

5. References

[1] Bianchi M, De Pascale A 2011 Bottoming cycles for electric energy generation: Parametric investigation of available and innovative solutions for the exploitation of low and medium temperature heat sources *Applied Energy* **88** 1500-1509

[2] Quoilin S, Van Den Broek M, Declaye S, Dewallef P, Lemort V 2013 Techno economic survey of Organic Rankine Cycle (ORC) systems *Renewable and Sustainable Energy Reviews* **22** 168-186

[3] Vankeirsbilck I, Vanslambrouck B, Gusev S, De Paepe M 2011 Organic Rankine cycle as efficient alternative to steam cycle for small scale power generation *8th international conference on heat transfer, fluid mechanics and thermodynamics, proceedings*785-792

[4] Ohman H 2012 Implementation and evaluation of a low temperature waste heat recovery power cycle using NH3 in an Organic Rankine *Cycle Energy* **2** 74-80

[5] Lecompte S, Huisseune H, Van den Broek M, Vanslambrouck B, De Paepe M 2015 Review of organic Rankine cycle (ORC) architectures for waste heat recovery. *Renewable and Sustainable Energy Reviews* **47** 448-461

[6] Schuster A, Karellas S, Kakaras E, Spliethoff H 2008 Energetic and economic investigation of Organic Rankine Cycle applications *Applied Thermal Engineering* **29** 1809–1817

[7] Low-temperature machinery. 2011 Cryogenic machines and tools *Publishing House of Bauman Moscow State Technical University* 584

[8] Quoilin S, Lemort V, Lebrun J 2010 Experimental study and modeling of an Organic Rankine Cycle using scroll expander *Applied Energy* **87(4)** 1260-1268

[9] Ali Tarique M, Dincer I, Zamfirescu C 2014 Experimental investigation of a scroll expander for an organic Rankine cycle *International Journal of Energy Research* **38(14)** 1825–1834

[10] Lemort V, Quoilin S, Pire C 2009 Experimental investigation on a hermetic scroll expander *7th International IIR Conference on Compressors 2009: Papiernička, Slovakia* 56-63

[11] Morozyuk L I, Morozyuk T V, Gayduk S V 2014 Thermodynamic analysis of a heat-using refrigerating machine with carbon dioxide. *Eastern European journal of advanced technologies* **2/8** 36-44

[12] Lemmon E W, Span R 2006 Short Fundamental Equations of State for 20 Industrial Fluids. *J. Chem. Eng. Data* **51** 785–850

[13] Grigoriev B A, Gerasimov A A, Alexandrova I S 2013 Tables of standard reference data GSSSD 291-2013. n-Peptane. Thermodynamic properties in the temperature range from triple point to 700 K at pressures up to 100 MPa *Russian research center for information on standardization, metrology and conformity assessment «STANDARTINFORM»* 54

[14] Alexandrov I S, Gerasimov A A, Grigoriev B A 2016. New fundamental equation of normal pentane state *Actual issues of reservoir systems research of hydrocarbon deposits* **4 (28)** 87-

95
[15] http://www.coolprop.org/fluid_properties/PurePseudoPure.html#introduction/
[16] Kirillin V A, Sychev V V, Sheyndlin A E 1983 Technical thermodynamics *Moscow "Energoatomizdat"* 416

Acknowledgments
The work was carried out with the financial support of the Russian Ministry of Education and Science. The agreement to grant № 05.608.21.0280 (RFMEFI60819X0280).

IOP Conf. Series: Materials Science and Engineering **945** (2020) 012065 doi:10.1088/1757-899X/945/1/012065

The use of augmented reality technology in the reconstruction of a lost cultural heritage site

E M Tomilina[1]*, E V Gontar[1] and M A Frolova[1]

[1]Department of Cultural Heritage Objects of the Northern and Arctic Territories, Northern (Arctic) Federal University named after M.V. Lomonosov, Severnaya Dvina Emb., 17, Arkhangelsk, 163002, Russia

E-mail: m.aizenstadt@narfu.ru

Abstract. The paper is devoted to the study of the possibility of using augmented reality technology in the process of hypothetical (virtual) restoration of the authentic appearance of lost cultural heritage structures on the example of an object of regional significance - the cinema "Sever". Based on the collected archival data and field studies, a historical and retrospective analysis of the cultural heritage object and its territory is presented, the construction history of the monument site is developed, dimensional drawings are restored and three-dimensional models of the object in various historical periods of the monument life are created. A detailed description of actions for creating an augmented reality object and exporting its animated model to an accessible digital platform (Instagram) was developed. The expediency of using the augmented reality approach in the reconstruction of lost cultural heritage objects is shown. This technology allows recreating the appearance of the lost monument in conjunction with the actual architectural component of the urban space. Successful implementation of this approach will allow for more detailed cultural, historical, archaeological, ethnographic, anthropological and other research.

1. Introduction

In the context of global digitalization, modern technologies, as a complex of information resources and equipment, are one of the main elements of any type of activity, including the preservation and promotion of cultural heritage objects [1].

Currently, a large number of unique historical buildings due to various reasons (anthropogenic impact, technological progress, imperfect legislation on the preservation of cultural heritage, etc.) are destroyed and irretrievably lost [2]. The issue of searching for modern, informative and at the same time cost-effective solutions to restore the appearance of lost monuments is becoming more and more urgent [3-6].

One of the possible methods of oriented work with historical and architectural heritage is the use of augmented reality technology (AR - augmented reality) - one of the fastest growing technologies of the XXI century.

Augmented reality is often confused with virtual reality (VR-virtual reality), but this technology is not formed only from digital objects located in 3D space. In augmented reality, objects are part of the existing environment, and the mobile device display shows the user the real physical world with added virtual objects.

In the context of the reconstruction of cultural heritage sites, the use of augmented reality technology is preferable, as it allows recreating the appearance of the lost monument in relation to the actual architectural component of the urban space [7, 8]. Successful implementation of this approach will allow conducting cultural, historical, archaeological, ethnographic, anthropological and other research in the future.

Thus, the aim of this work was to evaluate the possibility of using augmented reality technology to recreate the authentic appearance of lost cultural heritage structures on the example of an object of regional significance – the cinema "Sever".

The implementation of this technology involves a number of research activities: archival and bibliographic research; field research; historical and retrospective analysis of the territory and the object; development of a 3D model of the object that illustrates the appearance of the object in its life periods; creation of augmented reality of the object and export of its animated model to an accessible digital platform (Instagram) for further use and popularization.

2. Materials and methods

2.1. Materials

The object of research was selected as the object of cultural heritage of regional significance "E.K. Plotnikova's city estate in Arkhangelsk "Cinema" 1903-1915. The building of the cinema "Sever" is located on the line of Teatralny lane and is part of the complex of development of the city merchant's estate, which is a monument of history and culture of the late XIX-early XX centuries [9].

As of June 2020, the facility is almost completely dismantled. In its place there is now a vacant lot enclosed by a metal fence. Work on the restoration of the monument is not underway (figure 1).

a) b)

Figure 1. View of the main facade of the cinema "Sever" (June 2020):
a – photo-fixing point 1; b – photo-fixing point 2.

Archival and field studies, as well as existing project documentation for the restoration of this object served as the basis for the development of three-dimensional models, followed by their animation and integration into augmented reality technology.

2.2. Methods

To create realistic three-dimensional models of the Sever cinema in various historical periods, the following licensed application software was selected [10]:

IOP Conf. Series: Materials Science and Engineering **945** (2020) 012065 doi:10.1088/1757-899X/945/1/012065

1. Autodesk AutoCAD (the program created drawings of the main structure of the cinema in *.dwg format, which were then imported into the corresponding package for three-dimensional modeling in the form of two-dimensional curves (spline);
2. Autodesk 3ds Max Design 2009 (realistic three-dimensional models of the cinema were created in the image editor);
3. Adobe Photoshop CS3 (required for creating realistic images (textures) that are then superimposed on the corresponding parts of the modeled object exterior);
4. The program Spark AR Studio was used to work with augmented reality.

3. Results

Based on the historical and archival research of the object, six main periods of its existence were identified:

I period. 1895-1908. Market stalls.

When comparing the design drawings of stalls with subsequent drawings of reconstructions, it becomes obvious that only a small part of the masonry of the street facade along Teatralny lane survived fragmentally from the original building to the present day.

II period. 1908-1915. Power station.

Nothing remains of the power station. All its walls and ceilings were dismantled or redone during repairs.

III period. 1915-1929. Edison Electric Theatre.

The electric theatre, founded by E.K. Plotnikova stood for fourteen years. However, the private "Edison" was too small for a city movie theatre. After the expansion and reconstruction of the building, three facades were preserved in fragments: North, East and South.

IV period. 1929-1950. Edison cinema.

The principle of the internal layout of the 1929 cinema, oriented to the East (on the screen), did not change significantly. Unfortunately, it is little known about the design of facades and interiors. During this period, the main entrance to the building was arranged from the side of Teatralny lane.

V period. 1950-1991. Cinema "Sever".

Reconstructions of this period were mainly concerned with the interior design of the cinema, giving the interiors more grandeur and convenience in the style of the 1950s.

VI period. 1996 - to the present time. Cinema "Sever"

In 1996, the building was transferred to the State Museum Association "Art culture of the Russian North". During this period, restoration work begin at the monument to adapt to the modern use of the cinema building. Since its construction, the building has repeatedly changed not only its size and architectural appearance, but also its purpose – from "commercial" to cultural and entertainment.

By scaling archived plans, sections, and building facades, they can determine their true dimensions in different time periods. The study of project documentation for the restoration of the cinema, namely floor plans, facades and sections, as well as photos of the building and its individual parts, allowed developing three-dimensional models of the object in the main historical periods of its existence (figures 2-5).

Figure 2. Solid model of an extended extension to retail stores, 1903-1908. Facade of the yard.

Figure 3. Solid model of power station, 1903-1909. Facade of the yard.

Figure 4. Solid model of the Edison electric theatre, by 1916. Facade of the yard.

Figure 5. Solid model of cinema "Sever", late 1950s. Courtyard and main facade.

The next step after modeling an object is texturing and its rendering; and the final step is creating an augmented reality scene from the object and the planar tracker. Figures 6-13 show the results of this work sequentially.

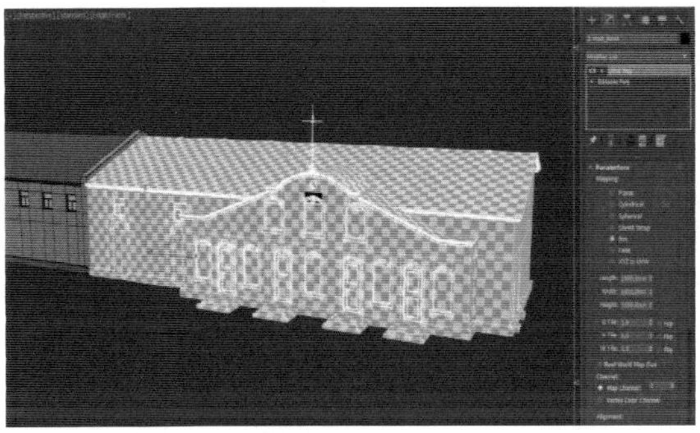

Figure 6. Configuring object texturing.

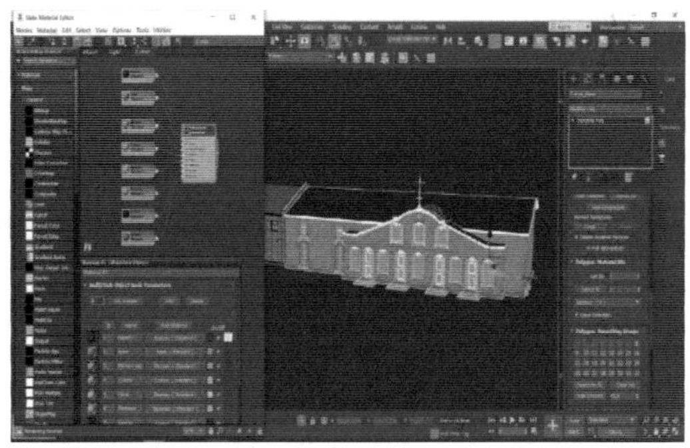

Figure 7. Applying materials to an object (texturing).

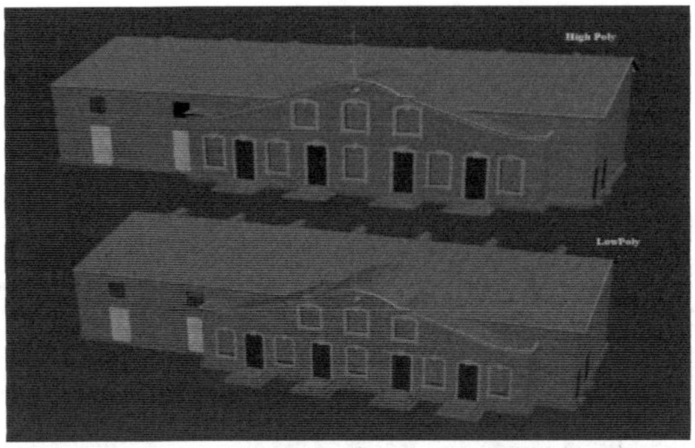

Figure 8. Rendering textures on the object.

Figure 9. Movement, rotation, and zooming of an object in Spark AR Player.

IOP Conf. Series: Materials Science and Engineering **945** (2020) 012065 doi:10.1088/1757-899X/945/1/012065

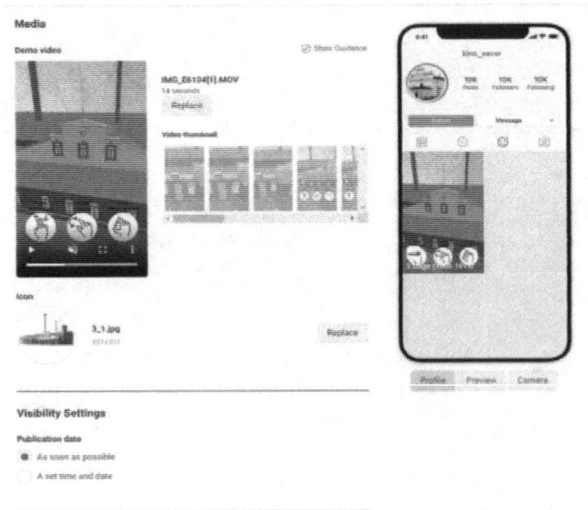

Figure 10. Publish an Effect page in Spark AR Hub. Media and Visibility settings groups.

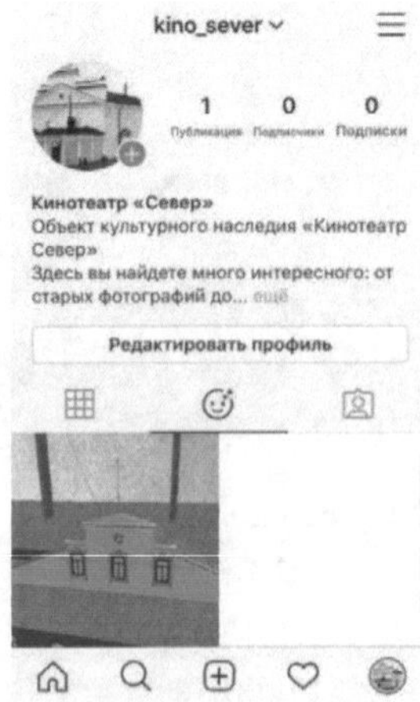

Figure 11. An Instagram page with accepted interactive work.

Figure 12. The backyard power plant, 1908-1913
Augmented reality on mobile devices.

Figure 13. Cinema "Sever", 1951
Augmented reality on mobile devices.

IOP Conf. Series: Materials Science and Engineering **945** (2020) 012065 doi:10.1088/1757-899X/945/1/012065

4. Discussion

Thus, the obtained practical results indicate the feasibility of using the augmented reality approach to solve the following tasks:

1) virtual reconstruction of lost cultural monuments based on historical documents;

2) virtual reconstruction of existing cultural monuments with the possibility of virtual visits to them.

5. Summary

The study was carried out as a practical recommendation for recreating cultural heritage objects by three-dimensional modeling and embedding in the existing architectural landscape using augmented reality technology. Using this approach allows conducting cultural, historical, archaeological, ethnographic, and anthropological research. In addition, it should be noted that the implementation of the proposed technology does not require special expensive technical support.

6. References

[1] 2004 UNESCO on the information society *main documents and materials St. Petersburg* 120

[2] Manzhilevskaya S E 2019 Environmental monitoring of ecological safety in areas of construction, reconstruction and operation of objects *Construction Materials and Products* **2(3)** 78-84

[3] Subbotin O S 2019 Features of using building materials in the restoration of architectural and town-planning heritage *Construction Materials and Products* **2(3)** 85-89

[4] Polyakova O M 2019 Architecture and design support for the implementation of the socio-economic development strategy of the city district *Construction Materials and Products* **2(3)** 90-95

[5] Pronina L A 2008 Information technologies in the preservation of cultural heritage *Analytics of cultural studies* **3(12)** 73-78

[6] Teryagova A N 2015 Architectural and urban planning solutions to the problems of accessibility of monuments of architecture, history and culture for people with disabilities *Traditions and innovations in construction and architecture. Urban planning: collection of articles Samara State University of Architecture and Civil Engineering Samara* 90-95

[7] Babkina K A, Kalugina E I 2019 Methods of restoration work for cultural heritage objects *Modern technologies in construction. Theory and practice: collection of articles Perm national Research Polytechnic University Perm* 111-116

[8] Isaev A O, Ginzburg A V 2019 Application of information modeling technologies in the restoration of historical buildings *Science and business: ways of development: collection of articles Foundation for the development of science and culture Tambov* 56-58

[9] Efremova S A 2009 History of the revival of the E.K. Plotnikova's estate in Arkhangelsk. *History, art, architecture, notes of the restorer Arkhangelsk: Vera* 191

[10] Lisitsyna A V 2011 Study of the historical and cultural heritage of the historical settlement *Method. guidelines for master students and postgraduates in the specialty 05.23.20 "Theory and history of architecture, restoration and reconstruction of historical and architectural heritage" Institute of Architecture and Urban Planning, Department of Architectural Design Nizhny Novgorod* 32

IOP Conf. Series: Materials Science and Engineering **945** (2020) 012066 doi:10.1088/1757-899X/945/1/012066

Employment of vacation queueing models for maintenance of engineering systems of residential buildings

G A Afanasyev[1]

[1]Moscow State University of Civil Engineering,129337, 26, Yaroslavskoye Shosse, Moscow, Russia

E-mail: AfanasievGA@mgsu.ru

Abstract We consider two main functions for servicing the housing stock-scheduled prophylactic inspection and repair of technical objects and the elimination of sudden malfunctions of technical equipment. An approach to analysis based on vacation queueing systems is proposed. We introduce two vacation queueing models in which during vacations the team relizes the prophylactic inspection and repair of technical objects. In the first model during vacation the elimination of sudden malfunctions of technical equipment stops and the second model has a working vacation scheme. For each model the characteristics which determine the quality of the work of the server (team), as well as the boundaries of the change of parameters at which the team copes with the work from the standpoint of a particular criterion are defined.

1. Introduction

Maintenance of residential buildings is a set of measures that ensure the highest reliability of all elements and systems of a building (see, for example, [1]-[5]). The main element of the technical operation of residential buildings is a system of scheduled prophylactic inspection and repairs. However even with its rational organization, there is always a positive probability of failure of building elements. Therefore, we consider two main functions for servicing the housing stock - scheduled prophylactic inspection and repair of technical objects and the elimination of sudden malfunctions of technical equipment. In this paper we propose an approach to analysis of technical operation of residential buildings, based on vacation queueing models. A vacation queueing system is one in which a server may become unavaible for a random period of the time from a primary service center. The time a way from a primary service center is called a vacation. There are various types of behavior of the server on the vacation period. In classical models the server completely stops service or is switched off when he is on a vacation. Servi and Finn [6] introduced the working vacation scheme, in which the server works in different rate rather than complexly stopping service during a vacation. We consider two vacation models, in which during vacations the team realizes the prophylactic inspection and repair of technical objects. In the first model during vacation the elimination of sudden malfunctions of technical equipment stops and the second model has the working vacation scheme.

IOP Conf. Series: Materials Science and Engineering **945** (2020) 012066 doi:10.1088/1757-899X/945/1/012066

The simplest vacation models for the technical operations of residential buildings were proposed in the articles [7,8]. The models considered in this paper generate obtained there results to the case of working vacation scheme and arbitrary distributied of the emergency repair times for breakdowns.

2. Methods: mathematical models based on queueing theory

Suppose that the management company (MC) of housing and communal services (HCS) has one or more teams of specialists to ensure the functioning of technical equipment's (heat supply, water supply, ventilation, etc) of residential buildings. These teams have two main tasks – elimination sudden equipment failures and conducting prophylactic inspections and repairs to ensure the necessary level of reliability of the relevant technical systems.

To simplicity, this article assumes that there is only one work team, i.e. one device in the queueing system. The transition to the multichannel case has no fundamental obstacles, but is quite complicated technically.

We consider two vacation queueing models M_1 and M_2. For the both models we make the following assumptions.

- Intervals $\{a_n\}_{n=1}^{\infty}$ between emergency repair requests are independent exponentially distributed random variables with rate λ, i.e. $P(a_n > x) = e^{-\lambda x}$. This means that the flow of customers incoming to the service system is Poisson (see, for example, [9]) and mathematical expectation $Ea_n = \lambda^{-1}$.
- Emergency repair times for breakdowns are independent random variables with distribution function $B(x) = P(\xi_n \leq x)$, $b = E\xi_n$, $b_2 = E\xi_n^2$, $\beta(s) = Ee^{-s\xi_n}$, $(Re\ s \geq 0)$.
- There is one service team that can be busy either with repairing sudden breakdowns, or with prophylactic inspection or equipment repair. This means that there is one server at the service system.
- The team may begin prophylactic inspection and repair only if there are no request for emergency service. We will call such requests (customers or requirements) of the first type.

At moments when the team is exempted from the customers of the first type, the team proceeds to a prophylactic inspection and repair of the next facility, wich lasts a random time η, which is called a vacation in queueing theory. Distribution function of the vacation period is $G(x) = P(\eta \leq x)$and $g(s) = \int_0^{\infty} e^{-sx} dG(x)$. We assume that $G(x) = 1 - e^{\mu x}$ for the modelM_2. Also denote $\bar{\eta} = E\eta$, so that $\bar{\eta} = \frac{1}{\mu}$ for the model M_2.

For the model M_1 we assume that customers of the first type are not served during vacation periods. For the model M_2 we have working vacation scheme. Namely, during vacation periods customers of the first type are served by unique server and service times have an exponential distribution with rate ν.

Suppose that the MC wants to organize the maintenance of buildings so that the average number of calls $\bar{q}$ for urgent repairs in the system does not exceed a certain $\delta > 0$ and the average number $\bar{n}(T)$ of facilities preventive inspection and repair of which is completed in time *T* is not less than *N(T)*, i.e.

$$\bar{q} < \delta, \quad \bar{n}(T) \geq N(T). \tag{1}$$

3. Results: formulars for operational characteristics

For the model M_i $(i = 1,2)$ let $T_n^{(i)}$ be the moment when the nth vacation starts and $\tau_n^{(i)} = T_{n+1}^{(i)} - T_n^{(i)} (n = 0,1,2, \ldots)$. Then $\left\{\tau_n^{(i)}\right\}_{n=1}^{\infty}$ – sequence of independent identically distributed random variables. One may easy obtain from results for a queueing system M|G|1 (see, for example [9]) the following formula for the mathematical expectation

$$\bar{\tau}_i = E\tau_n^{(i)} = \frac{\lambda\bar{\eta}_i(1-\rho) + \rho Y_1^{(i)}}{\lambda(1-\rho)} \quad (i = 1,2)\,. \tag{2}$$

For the model M_i here $\bar{\eta}_i = E\eta_n^{(i)}$ is the average of the vacation duration, $Y_1^{(i)} = EY_n\left(\eta_n^{(i)}\right)$ and $Y_n\left(\eta_n^{(i)}\right)$ – the number of the first type customers in the system at the end of the nth vacation period, $\rho = \lambda b$.

We introduce a random process $q_i(t)$ which is the number of emergency calls (the first type customers) at time t in the system M_i.From queueing theory (see, for example [9,10]) it follows that there are limits

$$\lim_{t\to\infty} P(q_i(t) = j) = P_j^{(i)} \quad (j = 0,1,2,\dots)$$

and $P_j^{(i)} > 0, \sum_{j=0}^{\infty} P_j^{(i)} = 1$ if and only if $\rho = \lambda b < 1$. Further, we assume that stability condition $\rho < 1$ is satisfied. We define the generating functions $P_i(z) = \sum_{j=0}^{\infty} z^j P_j^{(i)}$ for the limit distribution of $q_i(t)(i = 1,2)$. Then the avaverage number of emergency calls $\bar{q}_i = P_i'(1)$ in the system M_i. Using well-known results from queueing theory [9] one can obtain the following formula for $P_1(z)$

$$P_1(z) = \lim_{t\to\infty} Ez^{q_1(t)} = \frac{(1-\rho)(1-z)\beta(\lambda-\lambda z)}{\beta(\lambda-\lambda z)-z} \cdot \frac{1-g(\lambda-\lambda z)}{\lambda\bar{\eta}_1(1-z)}. \tag{3}$$

Differentiating $P_1(z)$ with respect to z and assuming $z = 1$ we obtain from (3) the average value $\bar{q}_1$ of the number of emergency calls

$$\bar{q}_1 = Eq_1 = \rho + \frac{\lambda^2 b_2}{2(1-\rho)} + \frac{\lambda E(\eta^{(1)})^2}{2\bar{\eta}_1}. \tag{4}$$

Since $Y_1^{(1)} = \lambda\bar{\eta}_1$ for the model M_1 it follows from (2) that

$$\bar{\tau}_1 = \frac{\bar{\eta}_1}{1-\rho}. \tag{5}$$

Now consider the model M_2. We define $Y_n^{(2)}(t)$ as the number of emergency calls which are situated in the system M_2 at time $T_n^{(2)} + t$. Then $Y_n^{(2)}(t)$ is the number of customers in the classical queueing system M|M|1 at time t. This system has a Poisson input flow with rate λ, exponentially distributed service time with parameter ν and initial condition $Y_n^{(2)}(0) = 0$. It is well known [9] that

$$P(z,s) = \int_0^\infty e^{-st} Ez^{Y_n^{(2)}(t)} dt = \frac{1}{\lambda(\alpha_1(s)-z)(1-\alpha_2(s))} \quad (Re\, s \geq 0). \tag{6}$$

Here $\alpha_{1,2}(s) = \frac{s+\lambda+\nu\pm\sqrt{(s+\lambda+\nu)^2-4\lambda\nu}}{2\lambda}$. Since $Ez^{Y_n^{(2)}\left(\eta_n^{(2)}\right)} = P(z,\mu)$ then

$$Y_1^{(2)} = EY_n^{(2)}\left(\eta_n^{(2)}\right) = (\alpha_1(\mu)-1)^{-1}.$$

Hence for model M_1 we have from (2)

$$\bar{\tau}_2 = E\tau_n^{(2)} = \frac{1}{\mu} + \frac{\rho}{\lambda(1-\rho)(\alpha_1(\mu)-1)}. \tag{7}$$

The stationary distribution of $q_2(t)$ is given by formula

$$P_2(z) = \lim_{t\to\infty} Ez^{q_2(t)} = \frac{(1-\rho)(\alpha_1(\mu)-1)}{\lambda(1-\rho)(\alpha_1(\mu)-1)+\rho\mu} \times$$
$$\times \frac{\beta(\lambda-\lambda z)(\lambda\alpha_1(\mu)-1)-\mu z+z(\mu-\lambda(\alpha_1(\mu)-1))}{(\alpha_1(\mu)-z)(\beta(\lambda-\lambda z)-z)}. \tag{8}$$

IOP Conf. Series: Materials Science and Engineering **945** (2020) 012066 doi:10.1088/1757-899X/945/1/012066

The proof is enough complicated and we omit it here. Differentiating $P_2(z)$ with respect to z and putting $= 1$, we have from (10)

$$\bar{q}_2 = Eq_2 = \frac{\mu}{\lambda(1-\rho)(\alpha_1(\mu)-1)+\rho\mu}\left(\rho + \frac{\lambda^2 b_2}{2(1-\rho)} + (1-\rho)\frac{\lambda}{\mu} + \frac{\rho}{\alpha_1(\mu)-1}\right). \quad (9)$$

4. Discussion

Suppose that the MC wants to organize the maintenance of buildings so that conditions (1) will be fulfilled. From the renewal theorem [11] we obtain that the average number $\bar{n}(T)$ of completed vacations for large T has asymptotics $\bar{n}(T) \sim \frac{T}{\bar{\tau}}$.

When $N(T) = \gamma T$ for the model M_1 conditions (1) have the form

$$\rho + \frac{\lambda^2 b_2}{2(1-\rho)} + \frac{\lambda E(\eta^{(1)})^2}{2\bar{\eta}_1} \leq \delta, \qquad \frac{1-\rho}{\bar{\eta}_1} \geq \gamma \ . \quad (10)$$

If the system parameters λ, b, b_2, $E(\eta^{(2)})^2$, $E\eta^{(1)}$ satisfy the inequalities (10), then we can assume that the team satisfactory copes with the tasks. If even one of the inequalities is not satisfied, managing actions should be taken.

We assume that model M_2 instead of the Model M_1 is employed.

Then we have the inequalities

$$\bar{q}_2 \leq \delta, \quad \bar{\tau}_2 \leq \gamma^{-1} \quad (11)$$

where $\bar{q}_2$ and $\bar{\tau}_2$ are defined by (11) and (9) respectively. We note that $\bar{q}_2$ and $\bar{\tau}_2$ are decreasing functions of ν and for $\nu = 0$

$$\bar{q}_2 = \bar{q}_1 , \qquad \bar{\tau}_2 = \bar{\tau}_1 \ .$$

When $\nu \to \infty$ the following asymptotics take place

$$\tilde{q}_2 \sim \frac{\mu}{(1-\rho)\nu}\left(\bar{q}_1 - \rho\frac{\lambda}{\mu}\right), \qquad \bar{\tau}_2 \sim \frac{\rho}{\nu(1-\rho)} .$$

Therefore, there exist ν_0 such that for model M_2 with $\nu = \nu_0$ inequalities (11) are fulfilled. So MC may guarantee the feasibility conditions (11) providing the service of emergency calls with rate ν_0 during vacation period.

5. Summary

In the present paper we propose two mathematical models with the help of which MC may solve two main tasks – eliminating sudden equipment failures and conducting prophylactic inspections and repairs to ensure the necessary level of reliability of the relevant technical systems. Our analysis relies on queueing theory, which is part of the probability theory that emerged from applications and widely used in solving applied problems in various fields.

6. References

[1] Notenko S N, Rimshin V I, Roytman A G and others 2008 Technical exploitation of buildings *Textbook editors Rimshin V.I. and Strazhnikov A.M., Vyshayashkola, Moskow* 639
[2] Kuzin N Y, Mishchenko V N, Mishenko S A 2017 Management with technical exploitation of buildings and structures *INFRA-M* **3** 156
[3] Kuzin N Y 1995 Technical exploitation of buildings *Penza* 90.
[4] Notenko S N and others 2000 Technical exploitation of buildings *Vyshayashkola, Moskow* 497
[5] Roshcina S M, Lukin M V, Lisyatnikov M S, Timokhova N S 2018 Technical exploitation of buildings and structures *M:* 232

[6] Servi L.D. and Finn S.G. 2002 M|M|1 Queueing with working vacations (M|M1|WV). *Performance Evalution*, 50, http://dx.doi.org/10.1016/S0166-5316(02)00057-3, 41-52
[7] Afanasyev G A, Shreiber K A 2019 Sheduling prophylactic maintenance of engineering systems of residential buildings *Journal of Physics: Conference Series* V.1425, art.N.012045. doi:10.1088/1742-6596/1425/012045
[8] Afanasyev G A 2019 Employment of queueing theory for maintenance of engineering systems of residential buildings *Vestnik TvSU, Prikladnaya matematika* **4** http:/dx.doi.org./10.26456/vtpmk546 52-64
[9] Saaty T L 1961 Elements of queueing theory with applications *Mc Graw Hill Book Company* 520
[10] Borovkov A A 1978 Stochastic processes in Queueing Theory *Springer-Verlag* 370
[11] Feller W 1966 An introduction to probability theory and its applications *John Willy & Sons Inc.* 750

Acknowledgments
This work is supported by Russian Foundation for Basic research project 20-01-00-487.

IOP Conf. Series: Materials Science and Engineering **945** (2020) 012067 doi:10.1088/1757-899X/945/1/012067

Properties of phosphogypsum as technogenic soil

Yu V Saenko[1*], A M Shiranov[1] and A L Nevzorov[1]

[1]Department of Geotechnics, Northern (Arctic) Federal University named after M.V. Lomonosov, Severnaya Dvina Emb., 17, Arkhangelsk, 163002, Russia

E-mail: yuri_saenko@mail.ru

Abstract. Phosphogypsum ($CaSO_4 \cdot 2H_2O$) is heavy tonnage waste generated by binding calcium in the process of sulfuric acid decomposition of phosphate raw materials when obtaining wet-process phosphoric acid in the production of phosphate fertilizers. In appearance, it is solid fine-crystalline substance of white or gray color with inclusions of large aggregates. The paper presents the results of laboratory studies of more than 390 samples of phosphogypsum neutralized with construction lime. The deformation and strength properties as well as the water permeability were determined on specially prepared samples with the relative compaction of 0.95, 0.90 and 0.80 by methods used in engineering and geological surveys. While the angle of internal friction, the cohesion and the coefficient of permeability of the phosphogypsum have the values typical for fine or silty sands, the compressibility characteristics are significantly worse due to the solubility of the particles. Phosphogypsum is characterized by subsidence and suffusion settling, which are proper for carbonate rocks (dolomite, limestone). The use of the phosphogypsum as ground is possible for subbase of roads with low traffic intensity with a gravel pavement, for dams at industrial solid waste storage facilities, as well as for engineering reclamation of those storage facilities, but only if implementing water protection measures.

1. Introduction

Phosphogypsum is dehydrate of calcium sulfate ($CaSO_4 \cdot 2H_2O$) generated by binding calcium in the process of sulfuric acid decomposition of phosphate raw materials when obtaining wet-process phosphoric acid in the production of phosphate fertilizers. Its annual output in the Russian enterprises exceeds 14 million tons and most of this by-product goes to dumps or gypsum storage [1].

There is certain experience of using phosphogypsum as ground material in road construction [2-4] and hydroengineering construction [5-8] as well as in the construction of dams at industrial solid waste storage facilities, but the deformation and strength characteristics of this material have not been sufficiently studied.

This paper presents the results of laboratory studies of phosphogypsum generated by the company LLC "Industrial group "Phosphorit" in the city of Kingisepp. We have tested the phosphogypsum neutralized with construction lime.

2. Materials and methods

The properties of the phosphogypsum were determined by methods used in engineering and geological surveys (GOST 5180, 12248, 22733, 25584). The samples were taken from the dump a month after it was released from the processing line.

The degree of solubility of the phosphogypsum in water under normal conditions was 3.8 g/l, the total content of easily soluble salts was 2.4% and the content of impurities insoluble in hydrochloric acid was 2.0 %. The average value of pH is 6.4. The corrosion activity against carbon and low alloy steel, to the aluminum cable sheath, to the concrete of all water resistance grades is high and against steel reinforcing rods in reinforced concrete structures is low.

The grain–size analysis performed by rubbing the samples on sieves and washing with a solution of gypsum with a concentration of 3-4 g/l. They showed that the composition of the examined phosphogypsum contains 47.8 – 82.4% of particles smaller than 0.1 mm, 6.8 – 21.2% of particles with size from 0.1 to 0.25 mm and 10.8 – 31.0% of particles larger than 0.25 mm.

The free water content of the samples determined by drying at the temperature of (60±2) °C was 19.8-29.5%. The content of hydrated water determined by calcination at the temperature of 400^{0}C of pre-dried samples was 25.2 %.

The density of the soil solids was 2.35 g/cm^3.

The deformation and strength properties were determined on specially prepared samples in modern automated devices manufactured by Geocomp Corporation, USA. Three types of samples were prepared – with the relative compaction (RC) of 0.95; 0.90 and 0.80, or with the average dry density of 1.31; 1.24 and 1.10 g/cm^3. Their initial water content was close to the optimal one that is from 21 to 24%. Approximately half of the total number of the samples were tested at the specified water content value and the remaining samples were saturated with water after loading into the devices.

The compressibility of the phosphogypsum was tested in the oedometer (GOST 12248). The load to the samples was increased by steps: 12.5, 50, 100, 200 kPa. 36 samples were tested – 6 for each value of RC and water content.

The consolidation tests were performed in oedometer too. The load to the samples was applied in one step and was equal 25, 50, 100, 200 and 300 kPa. All samples were saturated with water before testing. 90 samples were tested – 6 for each value of RC and load factor.

The strength characteristics (cohesion c and internal friction angle φ) were determined in the direct shear test according to the types of consolidated-drained (CD) and unconsolidated–undrained (UU) tests under four values of normal stresses – 50, 100, 200 and 300 kPa. The unconfined compression strength c_u was determined in triaxial consolidometers under three values of the p_{rc} reconsolidation pressure – 100, 200 and 300 kPa. A total of 126 samples were tested.

The water conductivity, which is characterized by the coefficient of permeability k, was tested in falling head permeameter (GOST 25587). The load of 25, 50, 100, 200 and 300 kPa was applied to the samples. A total of 108 samples were tested.

The shrinkage was determined by drying of water-saturated samples in three steps: in a box at room temperature 20-24 ^{0}C, in the open air at 20–24^{0}C and in a drying chamber at 60^{0}C. 6 samples were tested for each value of RC.

The swelling was determined in special devices without vertical pressure. 6 samples were tested for each value of RC.

The experiments to determine the subsidence were performed in special oedometers in accordance with the "two curves" diagram and under the load on samples of 25, 50, 100, 200 and 300 kPa. The water-soaked subsidence values: relative subsidence ε_{sl}, initial pressure p_{sl} subsidence and relative suffusion compression ε_{sf} due to dissolution and removal of components of the test material under water filtration were obtained for 6 samples for each pressure value.

3. Results

The tests with the Proctor device showed that under dynamic compaction, it is possible to achieve the maximum dry density of 1.37 g/cm^3 at the optimum water content of 23%, where the water content was found by drying the samples at the temperature of (60±2)^{0}C (Fig. 1).

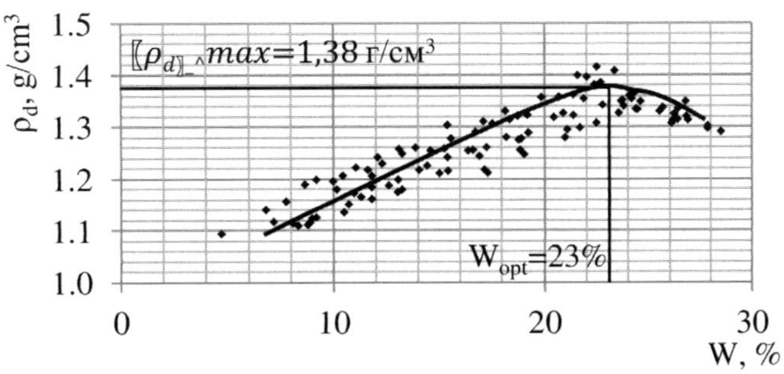

Figure 1. Compaction curve of phosphogypsum.

The average values of the oedometric modulus of phosphogypsum in the stress range from 100 to 200 kPa are shown in Table 1. As you can see, the tested material can be attributed to medium-deformable soils (GOST 25100) at RC of 0.95 and to highly and extremely deformable at RC from 0.90 to 0.80. Moreover, the water saturation of the samples with RC of 0.95 practically did not affect the modulus of deformation, but it decreased by 1.2 and 2 times for the samples with a lower initial density.

Table 1. Oedometric modulus.

Water content of samples	E_{oed}, MPa, at RC of samples:		
	0.95	0.90	0.80
Optimal	10.9	8.6	4.3
Full water saturation	11.3	4.2	3.6

The average values of primary and secondary consolidation coefficients – c_v and c_α are shown in Table 2. The Table shows the overall values of the secondary consolidation coefficient for the three sample types due to its insignificant variation. The obtained values of the consolidation coefficients are typical for highly deformable (weak) soils and it is most likely due to the dissolution of the particles and not to the slow removal of water from the pores nor to the creep of the particles as in the specified types of soils.

Table 2. Consolidation coefficients.

Load, kPa	c_v m^2/year, at RC of samples:			$c_\alpha \cdot 10^{-3}$
	0.95	0.90	0.80	
25	5.0	4.9	2.5	1.3
50	6.3	5.6	3.3	1.6
100	9.0	7.8	4.6	2.3
200	12.9	8.3	5.9	2.9
300	11.2	7.4	7.2	2.9

The average values of strength characteristics are shown in Table 3. As it turned out, the phosphogypsum has values of cohesion and internal friction angle, which are typical for sands. According to the value of unconfined compression strength, the phosphogypsum with RC of 0.95, 0.90 and 0.80 refers to very dense, dense and loose soils, respectively [9].

Table 3. Strength properties.

Type of test	Water content	Strength properties	RC of samples:		
			0.95	0.90	0.80
Direct shear test (CD)	Optimal	φ (degree) / c (kPa)	35.2 / 6.7	34.1 / 7.2	32.4 / 7.6
	Full water saturation	φ (degree) / c (kPa)	32.3 / 6.3	30.4 / 4.7	29.0 / 5.6
Direct shear test (UU)	Optimal	φ (degree) / c (kPa)	32.9 / 6.0	32.0 / 5.8	29.0 / 7.4
	Full water saturation	φ (degree) / c (kPa)	29.6 / 5.9	28.9 / 5.7	28.4 / 5.7
Triaxial test (UU)	Optimal	c_u, at p_{rc}: 100 kPa	110.8	53.0	24.4
		200 kPa	147.4	74.0	24.6
		300 kPa	185.6	88.1	39.5

The average values of the coefficient of permeability shown in Table 4, as well as its strength characteristics, are typical for fine or silty sands. When the relative compaction increases from 0.80 to 0.95, the coefficient of permeability decreases from 2.5 to 4.5 times depending on the vertical load acting on the samples.

Table 4. Coefficient of permeability.

Load, kPa	k, m/day, at RC of samples:		
	0.95	0.90	0.80
no	0.091	0.165	0.431
25	0.063	0.119	0.182
50	0.056	0.096	0.146
100	0.056	0.089	0.138
200	0.056	0.074	0.112
300	0.039	0.055	0.104

The tests have shown that the samples with RC of 0.95 tend to swell weakly and the phosphogypsum with a smaller compaction does not swell. When dryed, the samples do not shrink.

The values of subsidence in the process of soaking with water: relative subsidence ε_{sl},, initial pressure of the subsidence p_{sl} and relative suffusion compression ε_{sf} due to the dissolution and removal of the components of the test material during water filtration are shown in Table 5.

Table 5. Indicators of subsidence and suffusion compression.

RC of samples	ε_{sl} under load, kPa					p_{sl}, kPa	ε_{sf}, under load, kPa				
	25	50	100	200	300		25	50	100	200	300
0.95	0.001	0.001	0.003	0.003	0.003	>300	0.004	0.007	0.013	0.016	0.017
0.90	0.002	0.003	0.007	0.020	0.027	30	0.006	0.011	0.018	0.020	0.021
0.80	0.035	0.037	0.043	0.044	0.049	<25	0.022	0.028	0.035	0.046	0.058

The phosphogypsum has values of relative subsidence and suffusion compression typical for carbonate rocks (dolomite, limestone). When RC increases from 0.80 to 0.95, the relative subsidence and suffusion compression decrease by 14-37 and 2.7–4.5 times, respectively, depending on the vertical load acting on the samples. Due to the solubility of the phosphogypsum, any long-term filtration of water through the phosphogypsum body will lead to its leaching followed by development of suffusion settling.

IOP Conf. Series: Materials Science and Engineering **945** (2020) 012067 doi:10.1088/1757-899X/945/1/012067

4. Summary

The studied phosphogypsum can be used as ground for subbase of roads with low traffic intensity with a gravel pavement, for dams at industrial solid waste storage facilities, as well as for engineering reclamation of those storage facilities, but only if the following conditions are met:
–filling with compaction up to relative compaction 0.95 at least;
–applying a layer of soil that is not subject to water erosion on top of the phosphogypsum;
–surface levelling, installation of impervious blanket and lateral drainage to prevent infiltration of rain and meltwater, infiltration of ground water through the phosphogypsum layer;
–arrangement of geotechnical monitoring for timely detection of deformations and for eliminating defects.

Compliance with these conditions will reduce the risk of structures shifting into a limited operational or emergency state due to its inherent properties because of its particle solubility, subsidence and exposure to chemical suffusion.

5. References

[1] Kochetkov A.V., Shchegoleva N.V., Korotkovsky S.A., Talalay V.V., Vasiliev Yu.E., Shashkov I.G. 2019 *Conditions for obtaining phosphogypsum as a waste-by-product of nitrogen–phosphorus fertilizers production* (Transport facilities) **6-2** 10
[2] 1983 *Production and use in the construction of binders and products based on phosphogypsum.* Col. of papers, Kaunas, USSR, SRICA of the State Construction Committee of the Lithuanian SSR
[3] 1982 *New methods of processing and application of stone materials and industrial waste in road construction.* Moscow, USSR: SoyuzdorSRI
[4] 1985 *Problems of extraction phosphoric acid production and nature protection.* Abstracts of papers, Moscow, USSR: SRIUF
[5] Pavchich M.P. 1981 *Design and construction of embankment dams of a special type.* Moscow, Russia: Energoizdat
[6] Trunova I.A., Sidorenko R.V., Vakal S.V., Karpovich E. . 2010 *Analysis of the main directions of utilization of phosphogypsum-waste of phosphoric acid production* (Ecological safety) *2* 31-35
[7] Fayziev H.M., Sagdiev T.G. 2008 *Enclosing dams of hydraulic dumps constructed from production waste-phosphogypses* (Proceedings of RRSIG named after B.E. Vedeneev) **250** 100–107
[8] Kudashov E.S. 2015 *Engineering and geological substantiation of stability of alluvial gypsum storage* (abstract of the dissertation Saint Petersburg)
[9] 1997 *Uniform Building Code*. California, USA: International Conference of Building Official

IOP Conf. Series: Materials Science and Engineering **945** (2020) 012068 doi:10.1088/1757-899X/945/1/012068

The newly laboratory equipment for exploring of freezing and thawing soils

A A Korshunov[1*], S V Churkin[1] and A L Nevzorov[1]

[1]Department of Geotechnics, Northern (Arctic) Federal University named after M.V. Lomonosov, Severnaya Dvina Emb., 17, Arkhangelsk, 163002, Russia

E-mail: a.a.korshunov@yandex.ru

Abstract. Frost heaving is a challenge for geotechnical engineers in cold climate. Frost heave depends on two factors: the freezing rate and the velocity of water migration from unfrozen to frozen soil. The article deals with the existing apparatus for exploring frost heave which are systematized according to their purpose (measured parameters), design features and experimental conditions. The authors proposed the ways for improving laboratory equipment and presented newly developed apparatus for studying both the frost heaving and its consequences. The article presents apparatuses that allow controlling the freezing rate of soils by step-by-step movement of the freezing cell along the sample and providing even movement of the frost front in center and edges of the soil sample. Control of water migration to the frost front is carried out by making a vacuum in a base plate or using a porous item with a given water permeability. To study the impact of cyclic freezing-thawing on water permeability both along and across the cryogenic texture the authors developed a special apparatus in which horizontal permeability test is performed through the perforated double walls of multi-ring cell, vertical test- through bottom and top porous discs.

1. Introduction

Frost heaving is a challenge for geotechnical engineers in cold climate. Numerous studies showed that the primary frost heave caused by freezing of water contained in the soil is small. The main reason for the frost heave is the formation of ice lenses during freezing of water which are migrating to the frost front from the unfrozen soil. Now more than a dozen theories have been proposed to explain the mechanism of water migration, including capillary, osmotic, hydrodynamic and adsorption force theory. The latter, which is accepted by many researchers, is based on the concept of the adsorption-film migration mechanism. According to this theory at a negative temperature a layer of unfrozen bound water remains on the surface of the particles. Freezing soil seeks to maintain thermodynamic equilibrium by water absorption from the unfrozen soil. That is, the driving force of water migration is difference of the suction potential of unfrozen and freezing soil. This mechanism should be taken into account when designing laboratory equipment for studying the frost heave phenomenon.

2. Review of freezing test equipment

Many apparatus were developed for the frost heave studying of [1-16]. They can be systematized according to their purpose (measured parameters), design features, and experimental conditions. Principal schematic of apparatus for frost heave test are shown in Fig. 1.

The devices have the following design features:

- frost heave test is carried out on cylindrical soil samples with a diameter of 5 to 15 cm, the ratio of their height to diameter usually varies from 0.5 to 3;
- freezing of soil samples is performed, as a rule, from top to down;
- the samples are placed in a cylinder covered with a layer of thermal insulation, which provides steady-state one-dimensional heat flow along the sample;
- the overburden pressure on the samples is applied using a piston, its displacements are measured when freezing;
- free water supply is provided.

It should be noted that large laboratory apparatus are also known for testing specimens with dimension 40×40 cm and 45 cm height [17], 70×80 cm and 40 cm in height [18].

In order to minimize side friction between surface of the frozen sample and inner surface of cylinder, the latter is made from separate rings with a movable joint. Lubricants and membranes are also used. In apparatus proposed by K.Kujala [10, 11] the bottom plate is possible to move along the cylinder (Fig.1b).

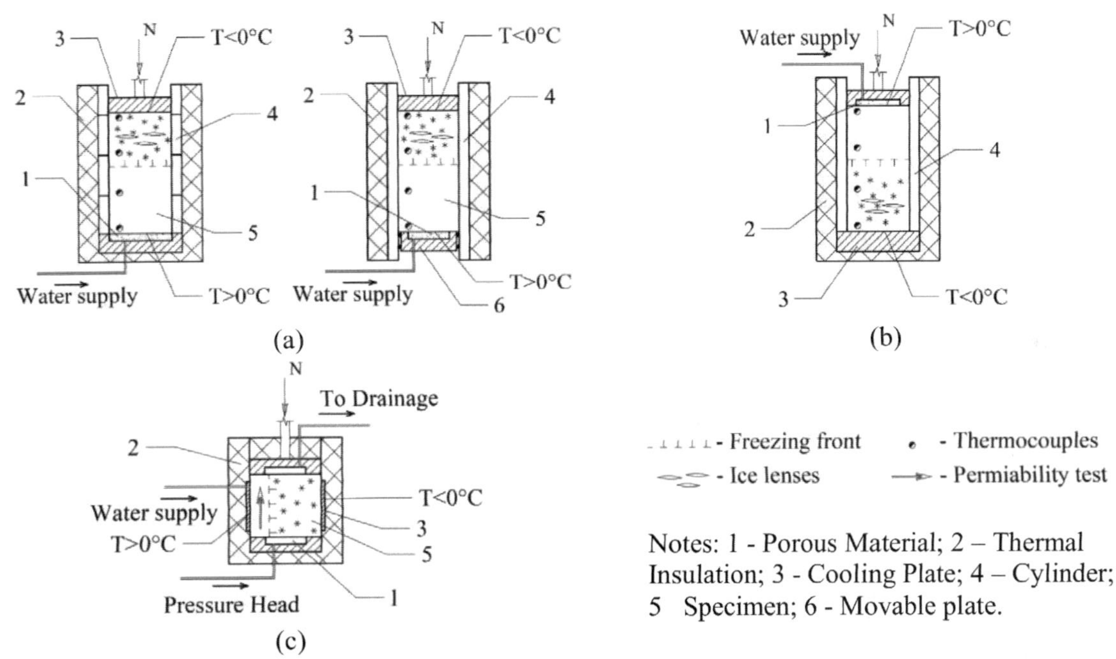

Figure 1. Principal schematic of apparatus for frost heave test in subject to direction of frost penetration: (a) – top-down; (b) – down-top; (c) - horizontal.

In most laboratory apparatus, as well as in-situ, the soil sample is frozen from top to down, and free water supply is provided from bottom. But T.Ishizaki, J.Kondrad, S.Freden studied frost heave phenomenon on apparatuses [9, 19-21], in which soil sample is frozen from down to top. Due to the frost heave deformations that are directed to unfrozen part of soil sample the influence of side friction was minimized.

Apparatus with horizontal movement of the frost front were used for assessment of frost heave around the columns during artificial ground freezing. Such devices can also be used for evaluation of frost heave on retaining walls and wells. Design of such apparatus can be one of the ways of development of the laboratory equipment.

Frost heave is mainly occurred due to the water migration from unfrozen to frozen soil, therefore, the slower the soil freezes, the greater frost heave phenomenon is appeared. That is why in laboratory experiments the freezing rate in frost action is to be as close as possible to field conditions. It is impossible to fulfill this requirement at constant temperatures at the top and bottom ends of the soil sample. However, in advanced apparatus thermocouples are placed into the soil sample. Due to collecting and processing data the temperature at the "cold" end plate is controlled by a PC.

The water intake flux from unfrozen to frozen soil is usually determined by the flow rate of water from the supply tanks, less often - using water suction sensors placed into the sample. There were also attempts to use fluorescent substances to color the water migrating to frost front. Layer-by-layer sampling for water content after completed of the experiment is mandatory.

Taking into account the mechanism of the frost heave, the advantages and disadvantages of the existing laboratory equipment, authors formulate the main requirements for the design of apparatus for studying frost heave phenomenon:

- ability to control the freezing rate and the intensity of water migration flux;
- minimization of influence of side friction on the frost heave and cryogenic texture;
- minimization of manipulations with the soil sample before water permeability testing and defining the mechanical properties of soil after freezing-thawing.

According to the above mentioned requirements the authors designed new apparatus for studying frost heave phenomenon.

3. Enhancement of laboratory apparatus

An apparatus was created in 2012 that used a very simple way to provide a given freezing rate of the sample (Fig. 2) [22]. Moreover, the device allows to stop the frost front to observe the growth of ice lenses. It includes a cylinder mounted from several HDPE rings, freezing and cooling chambers, a processor designed to control the temperature in the chambers, register the temperature inside the sample and frost heave. The cylinder with the sample is placed in the cooling chamber, where the temperature is maintained at 0 ... + 2°C. Free water supply is carried out through the bottom end of the sample. The freezing chamber, in which a negative temperature is maintained using Peltier elements, moves along the sample from top to down. The freezing rate is determined by the velocity of movement of this chamber. LoadTrac-II frame by Geocomp (USA) is used to apply an overburden pressure on the sample. LoadTrac-II frame also provides measurement of frost heave with an accuracy of 0.001 mm.

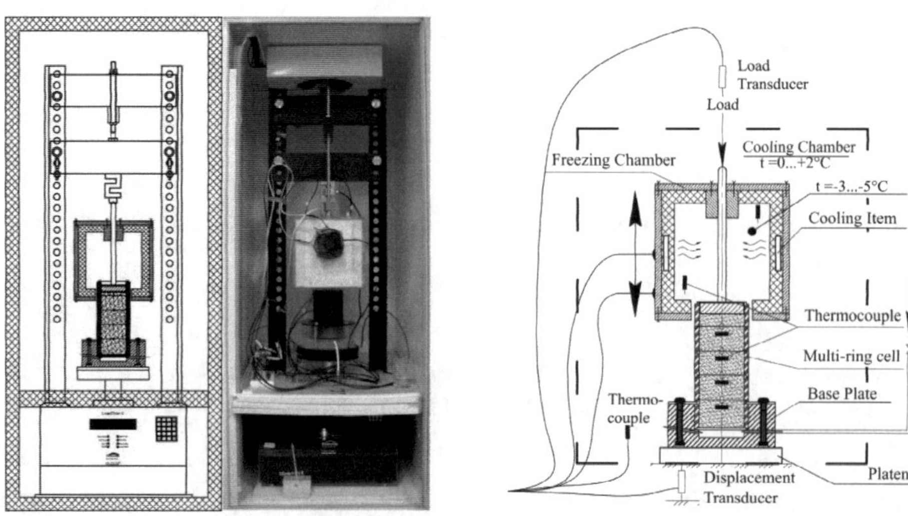

(a) (b)

Figure 2. Experimental apparatus for frost heaving test with controlled frost front penetration: (a) – photo of apparatus; (b) – schematic of apparatus.

The authors improved an original apparatus in which to ensure even movement of the frost front in center and edges of the soil sample dry mixture of special selected composition instead of thermal insulation are used (Fig. 3a). The thermal properties of the composition are priory defined by numerically simulation of laboratory apparatus when freezing of the soil sample in a special software (Fig. 3b) [23].

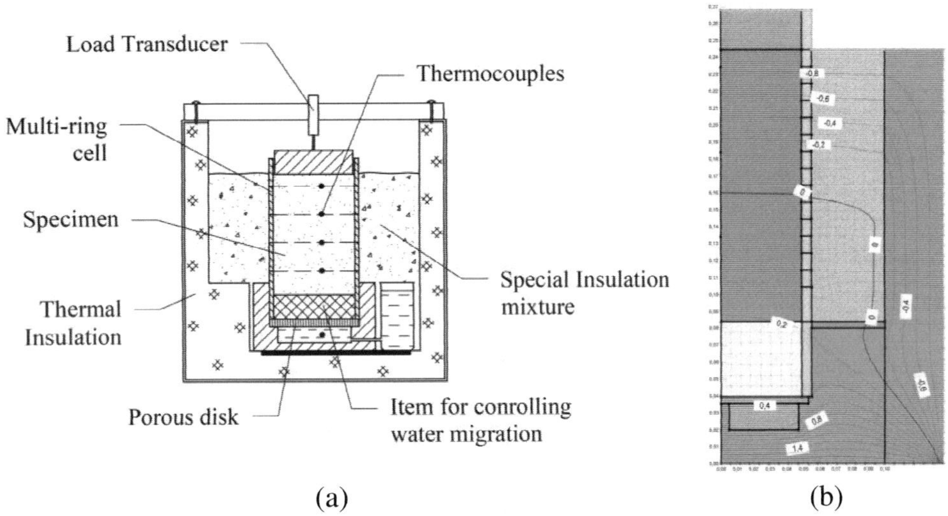

Figure 3. Experimental apparatus for frost heaving test: (a) - schematic of apparatus; (b) – numerical simulation.

The next step for improving apparatus is associated with the water migration control to the frost front. The presence of such a possibility, firstly, allows a deeper study of the frost heave phenomenon, and, secondly, take into account the groundwater level in soil stratum under building. In should be noted that standard apparatus provide only a free water supply to the bottom end of the soil sample with a height of about 15 cm, that is, the most unfavorable freezing conditions at the soil stratum are simulated.

The cylinders in the apparatus (Fig. 4a) filled with the soil are telescopic, that is, they are inserted into each other. It makes possible to mount a sample for laboratory testing with different heights and, in addition, to measure deformation below the frost front [24].

In the next apparatus, the aim is achieved by making a vacuum in a base plate (Fig. 4b) [25]. Freezing of the soil at a given groundwater level from the frost front is simulated by maintaining the negative pressure (below atmospheric pressure).

In the third apparatus (Fig. 3a), water migration is controlled by using special item placed under the bottom end of the soil sample with water permeability lower than the soil.

IOP Conf. Series: Materials Science and Engineering **945** (2020) 012068 doi:10.1088/1757-899X/945/1/012068

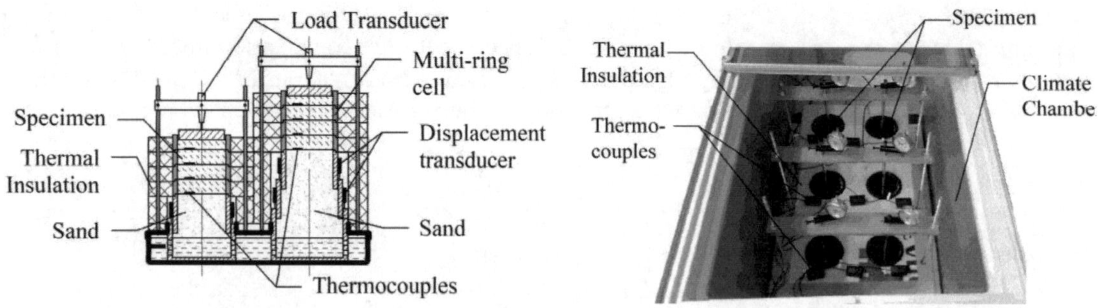

(a) (b)

Figure 4. Laboratory apparatus for frost heaving test with controlled water flow.

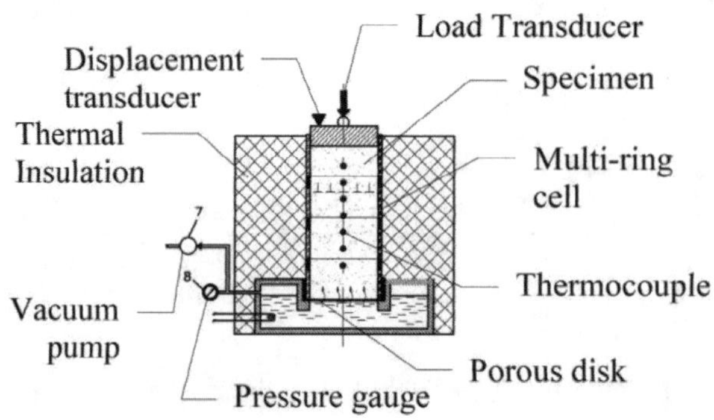

Figure 5. Laboratory apparatus for frost heaving test with controlled water flow.

To study the effect of cyclic freezing - thawing on the water permeability of soils, the authors created an apparatus that allows to determine the frost heave deformations and water permeability of soils both along and across the ice lenses (Fig. 6) [26]. A special feature of the apparatus is a cell mounted from double-walled rings. The space between the walls in each of the rings are divided into four sectors. Two sectors with perforated inner walls are used for water drainage when water permeability test is carried out in the horizontal direction, and the other two - with non-perforated walls - for placing tubes for water drainage. After a given number of freezing-thawing cycles, water permeability tests are carried out without removing the sample from the cell. It allows to keep post-cryogenic texture of soil. The water flow is directed either horizontally through the perforated inner walls, or vertically through porous disks at the ends of the sample.

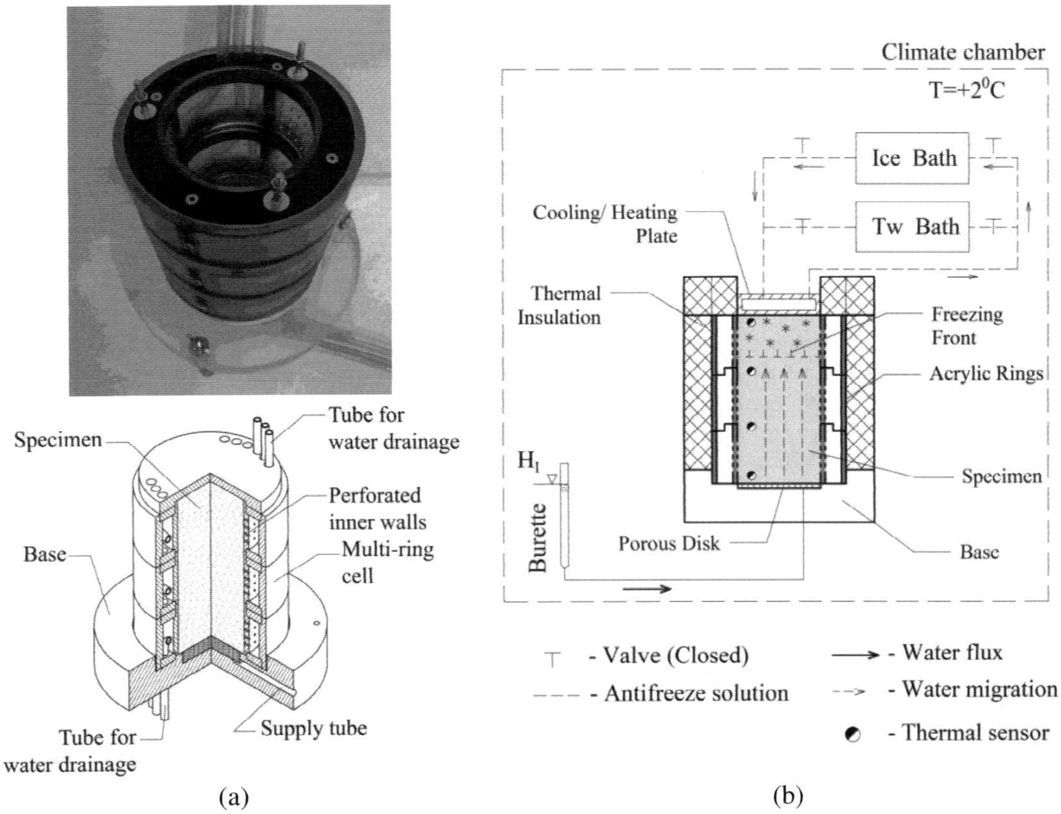

Figure 6. Laboratory apparatus for water permeability and frost heaving testing.

Thus, the newly laboratory equipment makes it possible to simulate the field conditions of soil freezing in the laboratory, to study the frost heave with given boundary conditions when testing and also to study the impact of cyclic freezing-thawing on the soil properties. The further enhancement of laboratory apparatus will improve the accuracy and reliability of assessment of frost-susceptibility of soils.

4. Summary

To obtain reliable results in the study of the frost heave phenomenon of soils laboratory equipment should provide the ability to control the freezing rate and the intensity of water migration flux, to minimize influence of side friction on the frost heave and cryogenic texture and exclude mechanical impact on the soil sample before water permeability testing and defining the mechanical properties of soil after freezing-thawing.

5. References

[1] Taber S 1930 The Mechanics of Frost Heaving *The Journal of Geology* **38(4)** 303-317
[2] Benson C H, Othman M A 1993 Hydraulic conductivity of compacted clay frozen and thawed in situ. *J GeotechEng*. https://doi.org/10.1061/(ASCE)0733-9410(1993)119:2(276)
[3] Bescow G. 1935 Soil Freezing and Frost Heaving with Special Applications to Roads and Railroads *Swedish Road Institute*
[4] Chamberlain E J, Gow A J. 1979 Effect of freezing and thawing on the permeability and structure of soils *Dev. Geotech. Eng.* https://doi.org/10.1016/B978-0-444-41782-4.50012-9
[5] Ershov, E D 1977 Water migration and cryogenic textures in dispersed soils. *PhD thesis* 452
[6] Feldman G M 1977 Forecast of the temperature regime of soils and the development of cryogenic processes *Science* 186

[7] Jessberger H L 1974 Bearing strength of frost-susceptible soils after thawing as a parameter for dimensioning roads and a measure for evaluating frost criteria *Int J Rock Mech Min Sci Geomech Abstr.* doi:10.1016/0148-9062(74)91170-X
[8] Johnson T C, Berg R L, Chamberlain E J, Cole D M 1986 Frost Action Predictive Techniques for Roads and Airfields: A Comprehensive Survey of Research Findings. *U.S. Army Cold Regions Research and Engineering Laboratory CRREL* 86-18
[9] Konrad J M, Morgenstern N R 1980 A mechanistic theory of ice lens formation in fine-grained soils. *Can Geotech J* https://doi.org/10.1139/t80-056
[10] Kujala K. 1993 Evaluation of factors affecting frost susceptibility in soils. *Frost Geotech Eng Proc 2nd Symp Anchorage*
[11] Kujala K, Ravaska O. 1989 Influence of test conditions and equipment on the frost heave test. *Frost Geotech Eng Int Symp Saariselka* **2**
[12] Luo T, Chen H, Yao Y P et al. 2019 Development of pot-cover effect apparatus with freezing-thawing cycles. *Sci China Tech Sci* **62** 1636–1648 https://doi.org/10.1007/s11431-018-9482-1
[13] Nevzorov A L 1997 Foundation on seasonally frozen soils. Handbook / Moscow *ASV* 152
[14] Orlov V O et al 1992 Frost-susceptible soils as a ground for structures. Novosibirsk, *Science,* 168
[15] Penner E. 1972 The Importance of freezing rate in frost action in soils. *Proc Am Soc Test Mater*
[16] Puskov V I, Krickij M.Ya., Melnikov I.A. *Frost heaving of consolidated soils Geology and ground engineering of transport structures* 76-84
[17] Pchelintsev A M 1964 Cmpoition and mechanical properties of frozen soils. *Science* 260
[18] Yakhtenfeld I P 1988 The reasons of frost heave phenomenon. *Melioration and water management* **11** 27-30
[19] Freden S, Stenberg L. 1980 Frost heave tests on tills with an apparatus for constant heat flow. *Comput. Chem. Eng.*
[20] Ishizaki T, Fukuda M 1994 Temperature dependence of unfrozen water film thickness of frozen soils https://doi.org/10.5331/seppyo.56.3
[21] Ishizaki T, Nishio N 1988 Experimental study of frost heaving of a saturated soil. *Gr Freez 88 Proc 5th Symp Nottingham* **1**
[22] Nevzorov A L, Korshunov A A 2013 Apparatus for determining frost heave deformations and forces. *Pat. 2473080 Russian Federation, IPC G01N 33/24* **2011135148/15**
[23] Nevzorov A L, Churkin S V, Korshunov A A 2014 Apparatus for determining frost heave deformations. *Pat. 156395 Russian Federation, IPC E02D 1/00* **2014126545/03**
[24] Nevzorov A L, Churkin S V, Korshunov A A 2015 Apparatus for determining frost heave deformations. *Pat. 2556681 Russian Federation, IPC G01N 33/24* **2014121241/15**
[25] Nevzorov A L, Korshunov A A 2011 Apparatus for determining frost heave deformations. *Pat. 2474650 Russian Federation, IPC E02D 1/00* **2011135096/03**
[26] Doroshenko S P, Nevzorov A L, Boldyrev G G, Korshunov A A 2019 Device for water permeability and frost heave of soils determination *Geotechnics* **4** 18–28 https://doi.org/10.25296/2221-5514-2019-11-4-18-28

Acknowledgments
The work was funded by RFBR according to the research project № 18-45-292004.

IOP Conf. Series: Materials Science and Engineering **945** (2020) 012069 doi:10.1088/1757-899X/945/1/012069

Influence of design and technological parameters of a high-speed paddle mixer on the quality of dry building mixes

Yu V Brazhnik[1*], N P Nesmeyanov[1] and A A Brazhnik[1]

[1]Department of mechanical equipment, Belgorod State Technological University named after V.G. Shukhov, Kostyukov St., 46, Belgorod, 308012, Russia

E-mail: Brazhnik_yuliay@mail.ru

Abstract. Today, one of the most promising building materials is dry building mixes. Their production is not only a large segment of the construction market, but also a kind of testing ground for promising developments of modern technological equipment. The paper discusses the design of a high-speed fixed drum paddle mixer for producing modified dry building mixes with a complex recipe. The question is raised about the complexity of the uniform distribution of components in the total volume of the mix. A number of existing factors affecting the distribution of components of the prepared mix are determined. Criteria for evaluating the quality of the finished mix are given. The method for determining the influence of design and technological parameters on the quality indicators of the finished product is described. The method of experimental research that allows evaluating the influence of the design and technological parameters of the mixer on the inhomogeneity coefficient of the resulting dry construction mix is presented. The obtained results confirm the effectiveness of the proposed design of a high speed paddle mixer to obtain high-quality modified dry building mixes at a minimum rate of heterogeneity of the mixture, as well as other indicators to measure the quality of the final product, diagrams of dependence of inhomogeneity coefficient of structural and technological parameters of the mixer are given. Rational values of the main design and technological parameters of the mixer for the production of dry building mixes are given.

1. Introduction

The use of dry building mixes in the world and national experience has shown their effectiveness and great advantage over classical methods of construction work: increased labor productivity from 1.5 to 5 times, reduced material consumption in comparison with traditional technologies by 3-10 times, uniformity of compositions and, as a result, improved quality of construction work, long shelf life without changing properties and the ability to transport and store at a negative temperature [1,3,5].

It is known that the quality of the resulting construction mix directly depends, first of all, on the mixing process in production. Therefore, in order to obtain high-quality mixes with a complex recipe, it is necessary to equip enterprises that produce construction mixes with modern mixing equipment that allows them to produce mixtures with the declared characteristics [1,4,6].

IOP Conf. Series: Materials Science and Engineering **945** (2020) 012069 doi:10.1088/1757-899X/945/1/012069

2. Materials and methods

2.1. Materials

To date, there is a wide variety of design versions of the working bodies of mixers for the production of building mixes. As practice shows, the most effective among them are considered to be high-speed forced mixers.

In order to increase the efficiency of the mixing process of dry components and, first of all, to improve the quality of uniformity of the resulting mixture, the design of a fixed drum paddle mixer with a vertical arrangement of the blade shaft operating in a high-speed mixing mode is proposed [2,9,10].

The mixer design change was based on the following:

– firstly, the organization (creation) of the "fluidized layer", i.e. the pseudo-fluidized state of the powdery material;

– secondly, creating additional circulation flows of the mixture components in the vertical direction (along the mixer body);

– thirdly, removing material from the stagnant areas of the mixer.

Paddle mixer (Fig. 1) consists of the following main components: a frame on which the drive and mixing drum are installed, a vertical paddle shaft (Fig. 2) mounted cantilevered inside a cylindrical mixing drum (Fig. 3). Several rows of blades are installed on the shaft, and in each row there are 3 blades that are turned to 120° relative to each other. A three-way helical spiral with breaks in the planes of rotation of the vertical shaft blades is fixed on the inner surface of the drum walls. It has a perforated surface. The size of the holes in this spiral exceeds 5..7 times the size of the part of the mixed components, and the value of the "live section" of the helical surface is at least 30-50% of its entire area.

Figure 1. Experimental installation of a paddle mixer.

Figure 2. Blade shaft.

Figure 3. Mixing drum.

A distinctive feature of this mixer is that when the blades of each row of the vertical shaft rotate, the mixed components are thrown onto the perforated helical spiral of the drum, which, in turn, intensifies the mixing process by moving the components of the mixture not only in the horizontal direction, but also in the vertical direction. This, in turn, leads to the destruction of stagnant zones in the drum and the creation of additional circulation flows of the mixture components in their total volume, which helps to improve the quality and reduce the inhomogeneity coefficient of the mixture.

2.2. Methods

As a result of the mixing process in the mixing drum, there is a mutual redistribution of particles of various materials that are separately or in a combined chaotic state before mixing. Ideally, you should get a mixture that preserves the ratio of the components to be mixed at any point in its volume. In fact, this ideal ratio is not observed due to a number of existing factors that affect the distribution of components in the mixture. These factors are divided into three groups [4,5,6,8]:

- mixing methods (pouring, shoveling, layering components, mixing components in a "fluidized" layer, etc.);
- design and technological features of mixers and their operating modes (the degree of filling of the mixing drum, the speed and nature of material circulation inside the mixing chamber, the design, speed and frequency of rotation of the mixing body);
- physical and mechanical properties of the mixture components (ratio of components, their granulometric composition, bulk mass and bulk density, coefficient of internal friction, etc.).

In the mixed volume of the material, an infinite variety of relative positions of the components of the mixed materials is possible, and under these conditions, the ratio of components at arbitrary points in the volume of the mixture is a random variable, so most modern methods of quality assessment are based on statistical analysis methods.

The most common criterion for evaluating the quality of the finished mixture is the value of the average square deviation S of the content of the key component in the mixture samples [1,4,7], but the most common criterion for evaluating the quality of mixing dry mixtures is the coefficient of heterogeneity (variation) V_c.

For obtaining high-quality dry mixes in a mixer of this type is necessary to ensure the pseudo-fluidization of the mixture components, which is typical for high-speed blending mode (frequency of rotation of the blade shaft more than 5 s^{-1}). It is necessary to consider the influence of the quality of the obtained mixtures (the heterogeneity coefficient of the mixture, Vc%) on the physical and mechanical properties of the mixture (tensile strength σ_o, MPa) and specific energy consumptions (q, kW·h/t) in the mixing process. Therefore, it is advisable to choose the named indicators (Vc, σ_o, q) as the response functions of the regression equations obtained when studying the influence of the main geometric and technological parameters on the mixing process of dry components.

The cycle of the mixing process, when using the laboratory setup consists of:

- the loading time of the drum with the components of the mixture, which depends on the volume of the body (near 8-10 s);
- the time required to get a high-quality finished product (16-20 s);
- the time of unloading the mixture (near 10 s).

The following geometric and technological factors influence the indicators of the response functions of the regression equations in this case:

$Vc, \sigma_o, q=f(D,H,d,i,j,b,l,\alpha,h,\beta,C,k,\varphi,n_p,t)$,

where D – diameter of the mixer drum, m; H – the height of the drum mixer, m; d – the diameter of blade shaft, m; i – number of rows of blades on the shaft, Pcs; j – number of blades in a row on the shaft, Pcs; b – the width of the paddle vertical shaft, m; l – vertical shaft blade length, m; α – angle of attack of the vertical shaft blade, deg..; h – width of the helical surface of the drum blades, m; Lφ – length of the forming helical surface of the drum blade m; β – angle of lift of the helical surface of the drum blades, deg.; C – the value of the live cross section of the helical surface of the drum blades, %;

k – number of turns of the helical surface of the drum blades, PCs; φ – material loading ratio of the drum; np – speed of rotation of the blade shaft, min^{-1}; t – mixing time, s.

To conduct research to determine the influence of design and technological parameters on the degree of heterogeneity of the mixture, a mixer with the following geometric parameters was studied [1]:

– the parameters of the drum of the laboratory installation are accepted D =300 mm, H =310 mm;

– the angle of lift of the helical surface of the drum blades is not more than 30° and not less than 15°;

– the number of rows of blades is determined by the ratio: $i=H/(h_0+b)$;

– geometric parameters of the mixer shaft blades: blade length l=(D-d)/2, blade width b and blade angle of attack α (varies in the range of 19÷51°);

– the number of blades in one row on the shaft is assumed to be 3 and are positioned relative to each other after 120°;

– the width of the helical surface of the drum blades is 40 mm, and the value of the “live section” C changes in the range from 14 to 46%.

The capacity of the mixer directly depends on the load factor (φ), so we assume its maximum value is 0.7. The mixing time (t) was chosen based on additional experiments, which determined that the mixing time is 16-20 s, and the full cycle time is 40÷50 s.

The interval for varying the speed of the blade shaft was adopted 470÷630 min^{-1} based on the fact that when the speed decreases, the load is not pseudo-fluidized, and with a further increase, the energy consumption increases significantly.

3. Results

As a result of the study, we obtained graphs of the dependence of the inhomogeneity coefficient V_C on the angle of attack of the blades α for fixed values of the “live section” values C =14%, 20%, 30%, 40%, 46% and the speed of the blade shaft n_p=470, 500, 550, 600, 630 min^{-1}. Analysis of these graphs showed that the minimum values of the inhomogeneity coefficient of the mixture equal to 2.7% and 2% are observed at the angle of attack in 35°, the speed of rotation of the blade shaft 550 min^{-1}, and the values of the live section equal to 14% and 46%, respectively (Fig. 4) [1].

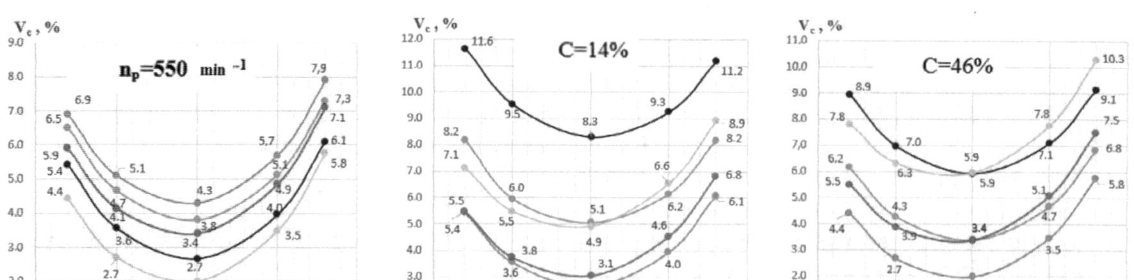

Figure 4. Graphs of the dependence of the inhomogeneity coefficient of the mixture V_C on the angle of attack α of the shaft blades at fixed values of the values of the live section C from the screw surfaces and the shaft rotation frequency.

4. Discussion

The results obtained are explained by the fact that with a minimum value of the “live section” equal to 14% due to the larger area of the helical surface, the bulk of the material, supported by its throwing blades, rises along the helical surface and at the end of it, passes from layer to layer under the influence of gravity, which contributes to mixing the components of the mixture. When the value of the

IOP Conf. Series: Materials Science and Engineering **945** (2020) 012069 doi:10.1088/1757-899X/945/1/012069

"live section" is equal to 46%, when the blades of the shaft throw the components of the mixture on the helical surface of the drum blades, the material almost immediately wakes up on the layer below, which intensifies the mixing process.

5. Summary

As a result of the study, rational values of the main design and technological parameters of the mixer for the production of dry building mixes were determined, namely: the frequency of rotation of the vertical shaft n_p, the angle of attack of the blade α, and the value of the "live section" of the helical surface of the drum blades ***C*** of the mixer, taking into account that $q \rightarrow min$, $V_C \rightarrow$ min and $\sigma_0 \rightarrow max$. So, $\alpha = 35^o$, $n_p = 470$ min^{-1} and *C*=46%.

6. References

[1] Brazhnik Yu V 2017 Improving the design and mixing process in a paddle mixer with high-speed operation **194** 10-112

[2] U1 154251 RU, MPK B01F 9/10. Mixer for mixing bulk materials/ Bogdanov V S, Nesmeyanov N P, Brazhnik Yu V, Gorshkov P S (BSTU named after V.G. Shukhov). - No. 20115107616/05; Application form 04.03.2015 //Utility model (Applications and patents) 2015 2 2.

[3] Makarov Yu I 1973 Devices for mixing bulk materials **216** 180-199

[4] Stewart R 2009 Effect of blade angle and gap on particle mixing in a vertical mixer **193 (8)** 303-311

[5] Bogdanov V S, Gorshkov P S, Nesmeyanov N P, Brazhnik Yu V 2013 *Bulletin of BSTU* **4** 62-64

[6] Gorshkov P S, Nesmeyanov N P Description of the process of changing the concentration of a key component in the production of dry building mixes 2013 *Bulletin of BSTU* **1** 45-47

[7] Gorshkov P S 2013 Rotary spiral-blade batch mixer **169** 10-112

[8] Bullard J W The Virtual Cement and Concrete Testing Laboratory Consortium 2003*Annual Report* 1-39

[9] Lozovaya S Yu, Lozovoy N M, Peshkov O G, Antsiferov S I 2017 Modeling of the process of obtaining high-quality dry building mixes in a continuous mixer *Fundamental research* **10-3** 471-476

[10] U1 112643 RU, MPK B01F 9/00. Mixer for mixing bulk materials / Bogdanov V S, Nesmeyanov N P, Gorshkov P S (BSTU named after V. G. Shukhov). - No. 2011122970/05; Application form 07.06.2011 //Utility model (Applications and patents) 2012 2

Acknowledgments

The paper was prepared within development program of the Flagship Regional University on the basis of Belgorod State Technological University named after V.G. Shukhov, using equipment of High Technology Center at BSTU named after V.G. Shukhov.

IOP Conf. Series: Materials Science and Engineering **945** (2020) 012070 doi:10.1088/1757-899X/945/1/012070

Numerical simulation of perimeter drainage system considering peat water permeability change over time

A V Nikitin[1*] and O M Zaborskaya[1]

[1]Northern (Arctic) Federal University named after M.V. Lomonosov (SAFU), 17 Severnaya Dvina Emb., Arkhangelsk, 163002, Russian Federation

E-mail: nikitinavsafu@yandex.ru

Abstract. A reliable forecast of water depression in waterlogged areas depends on the due consideration of specific peat properties: a considerable change in the water permeability of peat once water hydrostatic uplift is removed, clogging over time, decomposition and other factors. The article presents an example of numerical simulation in Plaxis software of a perimeter drainage system, where groundwater level has been monitored for 1 year. Piezometers were installed on site at the distance of 7, 17 and 30 m from the drain pipe of a building under construction. Deformation-strength properties and filtration properties of peat were determined on samples taken from excavated pits. Besides, the samples were tested using filtration tubes to identify changes in water permeability of peat over time and depending on the amount of water filtered through peat. The survey took 90 days at the ambient temperature of 6 to 12 °C, which allowed avoiding rapid peat decomposition. The laboratory test results have shown that water permeability of natural-state peat samples has decreased by a factor of 2.7 to 3.1 with 450 to 580 liters of water filtered. The deviation of the design groundwater level calculated in Plaxis from its actual position did not exceed 0.07 m. Numerical simulation with due consideration of decreased water permeability of peat over time allows obtaining forecast results closer to the actual groundwater level.

1. Introduction

The essential construction site preparation method for waterlogged areas includes creating a layer of sand over peat by filling or jetting.

Peat is highly porous and compressible, with prolonged consolidation under load and variable water permeability when compacted. In non-compacted condition, the natural-state peat is water permeable, but its permeability decreases hundreds of times as it is compacted under load [1-3]. Many types of peat feature anisotropic permeability both in vertical and horizontal directions [4, 5].

In addition, experimental results show that water permeability of natural-state peat without load can vary over time. Studies indicate that this process is related to peat clogging over time with fine particle transfer; swelling; adsorption; pore clogging with air bubbles; and peat decomposition [6, 7]. Microbiological processes at temperatures above 20 °C in full water saturation conditions lead to a considerable increase in peat decomposition rate [8, 9]. Under natural conditions, peat temperature over a year is 5-10 °C, and the decomposition process is slow.

This article presents the observation data and the results of perimeter drainage calculations with due account for peat permeability variation over time.

2. Materials and methods

A perimeter drainage system is installed in the basement of a high-rise building under construction at the depth of 2.7 m from the grade elevation (Figure 1). The layer of peat 2.5 to 3.5 m thick is surcharged with a layer of man-made soil 0.7 to 1.2 m thick. The peat is underlain by impermeable clay soil. The drainage system includes a tubular drain with inspection wells meant for pipe cleaning. Drain pipes are perforated corrugated polyvinyl chloride pipes with the inside diameter of 140 mm, wrapped in geotextile with the density of 100 g/m^2. A 100 mm thick layer of foamed plastic pellets is used as aggregate around the pipe. The aggregate is wrapped in geotextile with the density of 250 g/m^2. The drain pipes are backfilled with sand.

There is a four-storey building 43 m away from the high-rise building, which has a perimeter drainage system buried to the depth of 3 m. The drainage system includes asbestos-cement pipes 150 mm in diameter with crushed stone filling.

Three piezometers were installed to monitor the water level: at the distance of 7, 17, and 30 m from the high-rise building drainage system (Figure 1). The piezometers were made of perforated plastic pipes 50 mm in diameter and 3 m long. The water level was measured every 60 days during one year.

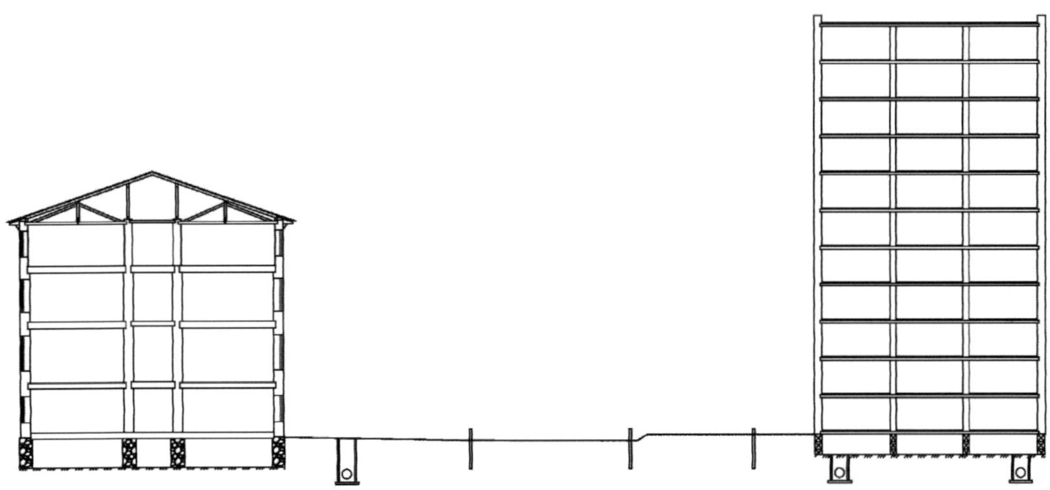

Figure 1. Perimeter drain design.

Samples of undisturbed structure were taken from excavated 1-m deep pits in the area without a surcharge sand layer. The density of sphagnum peat was 1.02 to 1.03 g/cm^3, moisture content was 11.6 to 12.6, and porosity coefficient was 17.4 to 18.9.

A Mohr-Coulomb model was used for peat when running a numerical simulation of the drainage system in PLAXIS software. The following parameters were set for peat:

- Modulus of deformation, E
- Specific cohesion, c
- Internal friction angle, φ
- Permeability factor, k_f
- Coefficient accounting for the change in the permeability factor during compaction, c_k

Numerical simulation of water depression in the drainage system area was performed using Plaxis 2D software package. The application of numerical techniques allows accounting for the change in water permeability of peat over time.

For numerical simulation purposes, permeability factor for natural-state peat was assumed. The c_k factor of 4.2 was assumed to account for the change in the permeability factor with compaction in the pressure range of 0-25 kPa.

The assumed PLAXIS calculation model takes into account the variation of peat water permeability after it is surcharged with sand. In addition, the permeability factor was reduced by a factor of 3 based on the findings of laboratory studies of peat permeability change. Infiltration was calculated on the basis of the actual amount of precipitation during the observation period considering partial evaporation. The amount of evaporation was determined on the basis of the ambient air temperature using Mayer data [10].

Peat compression tests were carried out using samples 8.7 cm in diameter and 70 mm in height. The sample height was adopted to ensure the integrity of natural peat structure was preserved when taking samples. The tests were conducted using compression and filtration test units with bottom-to-top water flow. The samples were loaded in increments of 5, 12.5, 25, 50, and 100 kPa. Each load stage was held until the deformation was assumed to have been stabilized.

Permeability factor was calculated after each load stage under variable head gradient conditions. Water permeability of sphagnum peat was studied in vertical direction only. According to V.V. Kramarenko [5], sphagnum peat is not anisotropic in terms of permeability.

Peat strength properties were checked using a shear test unit. Shear resistance limit was determined at the vertical pressure of 12.5, 25, and 50 kPa. Strength parameters were assumed in accordance with the results of peat sample shearing tests as follows: $c = 11.4$ kPa, $\varphi = 14.7°$.

Laboratory tests were conducted to identify any variation in water permeability of compacted peat over time. Figure 2 shows sample test setup. Filtration tubes were made of polypropylene pipes 105 mm in diameter. The adopted peat sample height was 300 mm. Water permeability was determined at the head gradients of 2.5 and 4. Constant water level in the tubes was maintained by make-up water supply from a tank. The samples were covered with a 30 mm thick layer of gravel.

The tests were conducted at the ambient temperature of 6 to 12 °C. This temperature corresponds to the soil temperature in natural conditions and prevents quick decomposition of peat.

The total duration of tests was 90 days. Permeability factor was determined before the test start and then after 15 to 150 liters of water filtered through the peat. The total amount of water filtered through the samples was 450 to 580 liters.

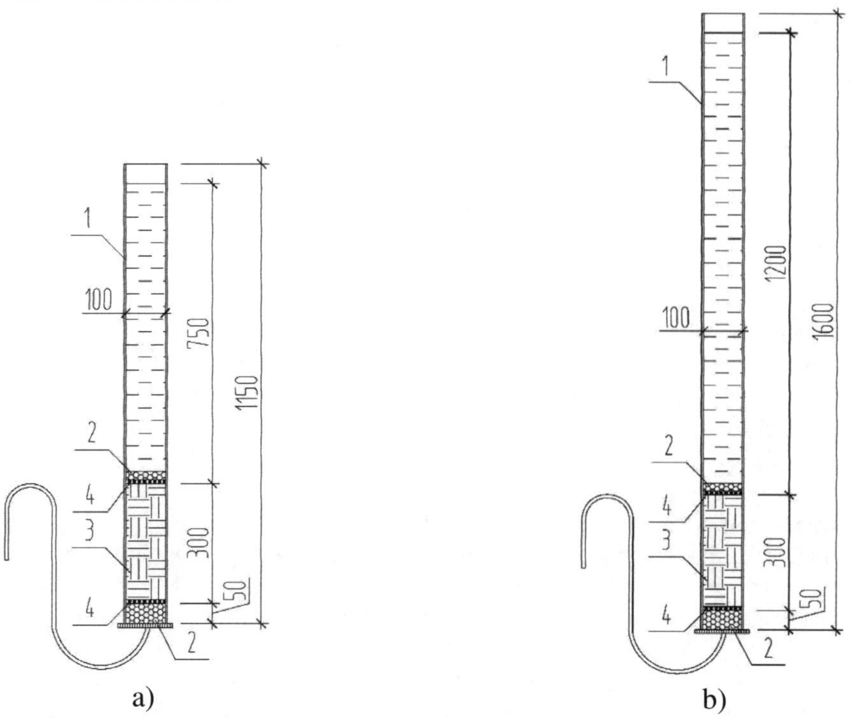

Figure 2. Test setup to determine peat permeability change over time.

a – head gradient 2.5; b – head gradient 4: 1 – filtration tube, 2 - gravel, 3 – peat sample, 4 - perforated piston

3. Results

Relative peat deformation against pressure is plotted in Figure 3, and permeability test results are summarized in Table 1.

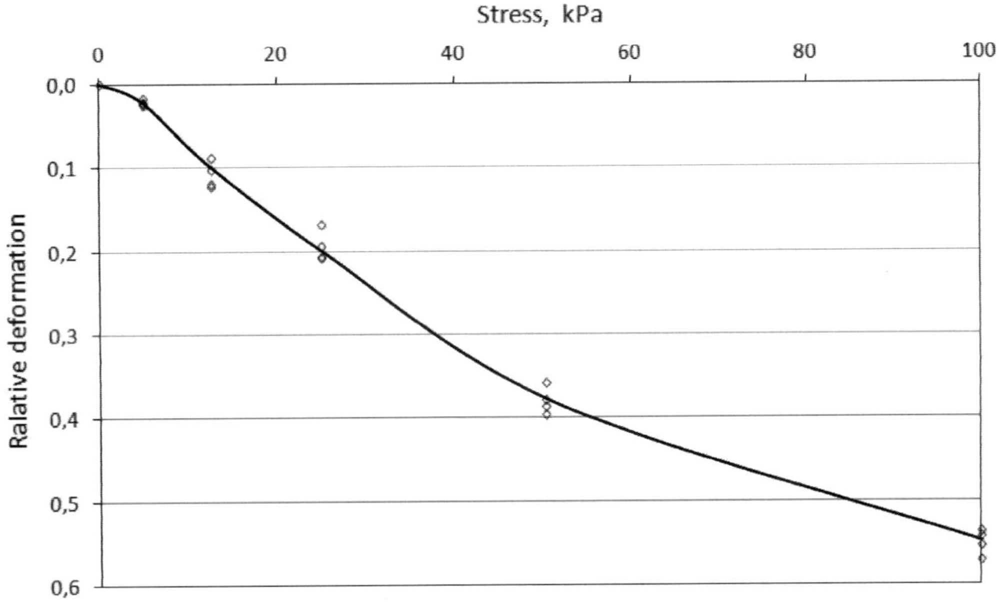

Figure 3. Peat compression curve.

The tests have shown that peat can be compressed to 55 % of its original height under the pressure of 100 kPa.

Table 1. Water permeability change with peat compaction.

Pressure on sample, kPa	Porosity factor	Permeability factor, m/day
0	17.4 to 18.9	0.30 to 0.37
5	17.0 to 18.4	0.23 to 0.33
12.5	15.5 to 16.7	0.10 to 0.14
25	13.8 to 15.1	0.041 to 0.046
50	10.1 to 11.8	0.002 to 0.008
100	6.9 to 8.0	0.0002 to 0.0008

The data obtained demonstrate that the permeability factor reduced by a factor of 450 to 1800 at the pressure of 100 kPa, and the peat turned into impervious soil. Reduced water permeability of peat is attributed to the decrease in the pore size with the increasing pressure.

The coefficient accounting for the variation of permeability factor due to compaction was determined in accordance with the results of compression and filtration testing using formula:

$$\log\left(\frac{k}{k_0}\right) = \frac{\Delta e}{c_k},$$

where Δe is the porosity coefficient change,
k_0 is the initial permeability factor,
k is the permeability factor at the pressure stage.

The test results are given in Figure 4.

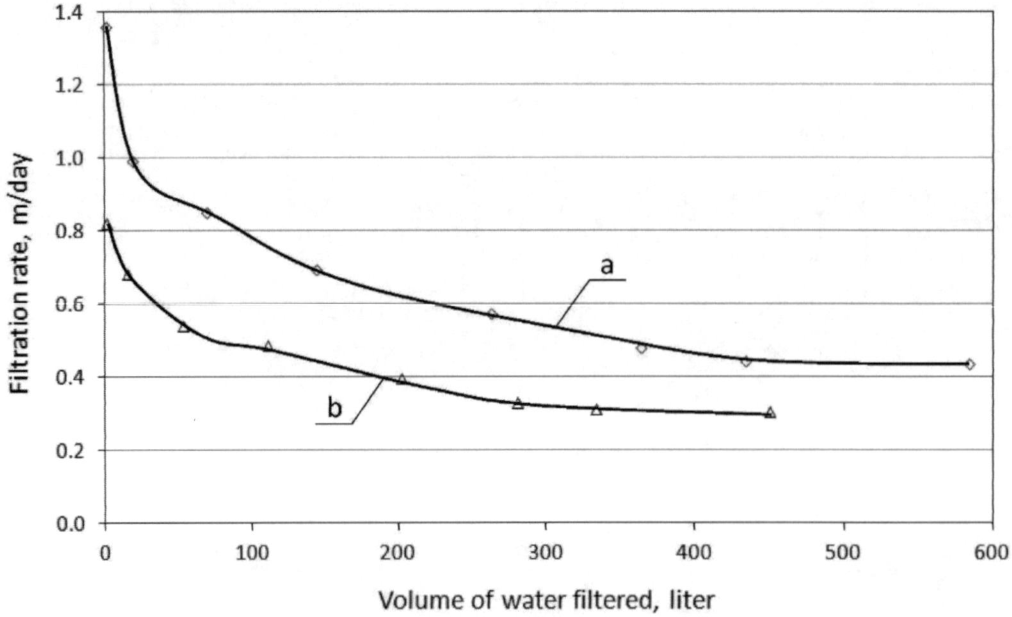

Figure 4. Variation of peat water permeability over time: a – head gradient 4; b – head gradient 2.5.

The test results indicate that the permeability factor reduced by a factor of 2.7 to 3.1. The most rapid change was observed at the initial stage of filtration. As filtration went on, permeability rate slowed down considerably.

Water level measurement results in piezometers are summarized in Table 2. Groundwater table depth in the table is given against the grade elevation in the immediate vicinity to the building under construction.

Table 2. Groundwater table monitoring results.

Date of measurement	Water table depth, m, at the distance from drain pipes, m		
	7	17	30
07.19	1.92	1.59	1.73
09.19	1.85	1.41	1.61
11.19	1.74	1.35	1.59
01.20	1.99	1.53	1.66
03.20	1.98	1.43	1.64
05.20	1.79	1.38	1.54
07.20	1.96	1.62	1.75

Table 3 provides the results of numerical simulation of the depression curve in the perimeter drainage operating area.

Table 3. Groundwater table position relative to grade elevation above the drainage system following the results of calculations and by observation data.

Water table position determination method	Water table depth from grade elevation above drainage system, m, at the distance to building under construction, m		
	7	17	30
Monitoring	1.89	1.47	1.65
Numerical simulation	1.95	1.44	1.58

The deviation of the design groundwater table depth from its actual position does not exceed 0.07 m. The best repeatability of the design depression curve and the actual groundwater table depth was achieved through the application of the reduction factor accounting for the change of permeability factor over time.

4. Summary

A reliable forecast of water depression in the perimeter drainage system operating area requires due consideration of peat water permeability variation over time caused by surcharge load, clogging processes, removal of hydrostatic uplift, decomposition in the zone of aeration, and other factors.

Laboratory test results have established that water permeability of peat can decrease several times if there is a filtering flow in the drainage system operating area.

Consideration of water permeability change over time brings the forecast water depression values closer to the actual depression curve.

5. References

[1] Wong LS, Hashim R, Ali FH 2009 A review on hydraulic conductivity and compressibility of peat*Journal of Applied Sciences***9(18)** 3207–3218
[2] Kovalenko NP, Hydyakov AB, Gorelyakov VS 1971 Preliminary loading of a peat 97
[3] Sutejoa Y, Saggaffb A, Rahayuc W, Hanafiah 2019 Hydraulic conductivity and compressibility characteristics of fibrous peat *IOP Conf. Series: Materials Science and Engineering* **620 012053**
[4] Ronkanen AK Klove B 2005 Hydraulic soil properties of peatlands treating municipal wastewater and peat harvesting runoff *Peat and Mires***56(2)** 43-56
[5] NevzorovAL,TelminovIV 2012 Filtration anisotropy of high-moor peat *Geoecology. engineering geology, hydrogeology, geocryology* **3** 152-154
[6] KramarenkoVV, Emelyanova TY2009 Characteristics of the physical properties of high-moor peat in the Tomsk region *Tomsk State University Bulletin* **322** 265-272
[7] Emelyanova TY, KramarenkoVV 2004 Substantiation of the methodology for studying the deformation properties of peat, taking into account the change in the degree of its decomposition *Bulletin of the Tomsk Polytechnic University* **5** 54-57
[8] KramarenkoVV 2004 Formation of the composition and physical and mechanical properties of peats in the Tomsk region *Abstract of a dissertation for the degree of candidate of geological and mineralogical sciences* Tomsk 22
[9] Konovalov PA1980 Construction of foundations on peaty soils *Stroyizdat* 160
[10] Rethati L 1989 Groundwater in construction *Stroyizdat* 432

IOP Conf. Series: Materials Science and Engineering **945** (2020) 012071 doi:10.1088/1757-899X/945/1/012071

Determination of thermal conductivity of backfill

S V Churkin[1*] and A A Korshunov[1]

[1]Geotechnical Department, Northern (Arctic) Federal University named after M.V. Lomonosov, Severnaya Dvina Emb. 17, Arkhangelsk, Russia; 163002

E-mail: s.churkin@narfu.ru

Abstract. Shallow foundations with thermal insulation can be used in the construction of low-rise buildings. Thermal insulation prevents freezing of soils under the foundation. The article presents the results of determining the thermal conductivity of sand with the addition of crushed brick and concrete rubble. The research was carried out in a specially designed unit, which consists of two climatic chambers connected to each other. The soil sample is maintained into the special cell. Heat flow and temperature sensors are placed in the soil sample. Before soil testing a numerical simulation of laboratory test was implemented in GeoStudio. Due to numerical modelling optimal dimensions of testing cell was defined. Results of laboratory testing of sand with a known thermal conductivity with using designed parameters of cell has a good agreement with calculated value. The test procedure was extended to soils with construction waste. The addition of 20% crushed brick to the sand reduces its thermal conductivity coefficient by 20.6% in the frozen state and by 10.1% in the thawed state. At the same time sand with concrete rubble showed a decrease in the thermal conductivity coefficient in the frozen state by 6.4% and an increase by 13.4% in thaw. The effect is occurred because of the addition of particles of different mineralogical composition with different thermal conductivity. The results of testing can be used in making forecast of the thermal regime of soil stratum when designing shallow foundations.

1. Introduction

Now shallow foundations are spreading in low-rise building. As a rule the foot of shallow foundations is placed below frost front but in some cases when non frost-susceptible soils are bedded under foundations or special conditions are provided to prevent soil freezing below foundations. Special conditions include thermal insulation that keeps the soil temperature above 0 C throughout the year and heating the soil stratum under the foundation. Foundations with thermal insulation made from expanded polystyrene are widely used around the world as a simple and cost-effective solution [1]. Shallow foundations are designed by numerical modeling. Thermal characteristics of insulation and base soils are input as initial data [2].

Backfill soils may include brick and concrete rubble, which can affect the thermal characteristics of the soils [3]. The heat capacity of materials can be determined by calculation when the composition and volume of the brick and concrete rubble are known. Heat transfer in dispersed multicomponent materials is a more difficult process. The purpose of research was the experimental determination of the thermal conductivity coefficient of sand and its mixtures with construction waste.

2. Materials and methods

The thermal conductivity coefficient was determined for the following soils: sand, a mixture of sand with 20% broken ceramic bricks and sand with 20% crushed concrete. The water content of samples was on average 15%, the dry density of soil was 1.68–1.70 g/cm^3, the size of the inclusions was no more than 10 mm.

The steady-state method was chosen to determine the thermal conductivity coefficient, which provides for the measurement of the heat flux through the sample and the determination of the temperature difference between the two planes [4]. Traditional devices are designed for soil samples in the form of a disk with a diameter of 120 to 250 mm and a height of 30 mm [5] or in the form of a parallelepiped, the dimensions of which are taken based on the size of inclusions in the soil [6], [7]. It should be noted that the heat flux sensors are placed in these devices at the ends of the samples. The authors offer to place the sensors inside the sample (figure 1a). To check the impact of the sensors on the heat flow and to define the dimensions cell for soil testing numerical simulation was carried out in the GeoStudio software.

The laboratory unit consists of two climatic chambers connected to each other. The soil sample is maintained into the special cell. Heat flux (HFP01) and temperature sensor are placed into the sample (figure 1b). The data from sensors are transmitted in the datalogger. Soil sample is placed between two climatic chambers. Each of the two the ends of the soil sample goes into one of the climatic chambers, in which the set the temperature in chambers is maintained with an accuracy of 0.1..0.3 K. The temperature gradient was 15..20 K/m.

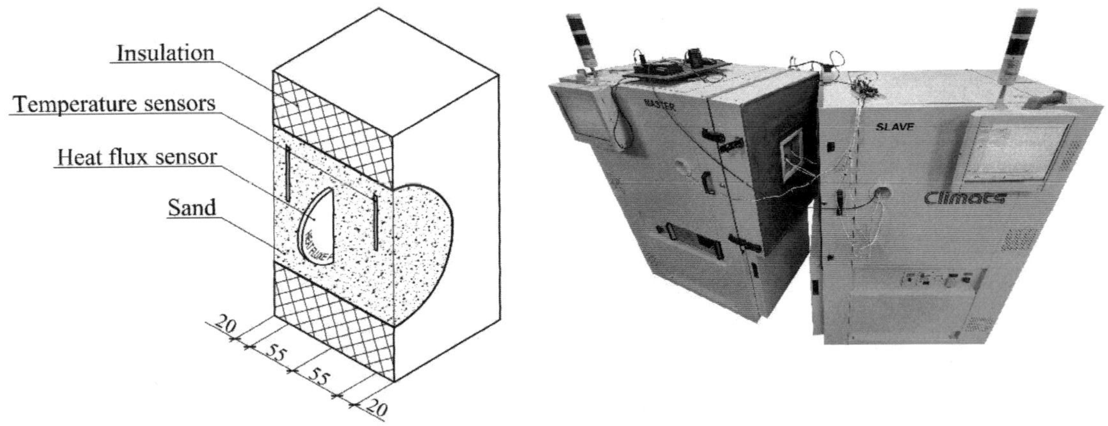

a) b)

Figure 1. a – scheme of cell (dimensions in millimeters); b – testing equipment.

The TEMP / W module of the GeoStudio 2012 program was used to simulate the processes of heat transfer and freezing by solving the steady-state heat problem.

The simulated cell has dimensions of 250 × 250 mm and is made of expanded polystyrene with a cylindrical soil sample inside. The diameter and height of the sample were selected to minimize the effect of the heat flow sensor and to get the smallest sample volume. Boundary conditions are applied to the ends of the cell. The heat flux sensor HFP01 has a diameter of 80 mm and a thickness of 5 mm and is placed in equal distance from both ends [7]. The heat flux follow the Fourier law in the software package [9]. Thermal properties of soil and materials are given in Table 1.

IOP Conf. Series: Materials Science and Engineering **945** (2020) 012071 doi:10.1088/1757-899X/945/1/012071

Table 1. Thermal properties of sand and materials.

Parameter	Units	Sand	Insulation (expanded polystyrene)	Sensor (heat flux plate)
Material model	-	Full Thermal	Simplified Thermal	Simplified Thermal
Thermal conductivity of material in:	KJ/m×K×day (W/m×K)			
frozen state, λ_f		207.4 (2.40)	2.6 (0.03)	69.1 (0.80)
thawing state, λ_{th}		172.8 (2.00)	2.6 (0.03)	69.1 (0.80)
Volumetric heat capability of material in:	KJ/K×m^3			
frozen state, c_{vf}		1864	48	1500
thawing state, c_{vth}		2400	48	1500
Volumetric water content, θ	m^3/m^3	0.25	0	0

3. Results

The numerical model of the cell and the steady-state heat flux through the sample are presented in Figure 2. The results of modeling the cell with sand showed that the optimal diameter and height of the sample are 150 mm (Figure 3). The impact of the sensor on the results is minimal at designed dimensions. Numerical modeling of laboratory test has significantly reduced the duration of laboratory testing.

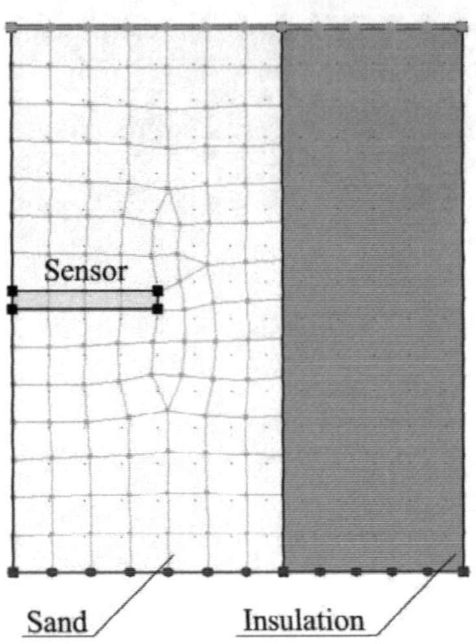

a)

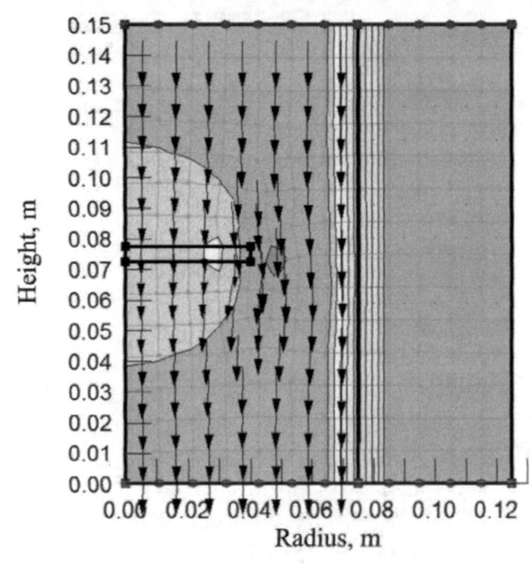

b)

Figure 2. a – model of cell; b – heat flow.

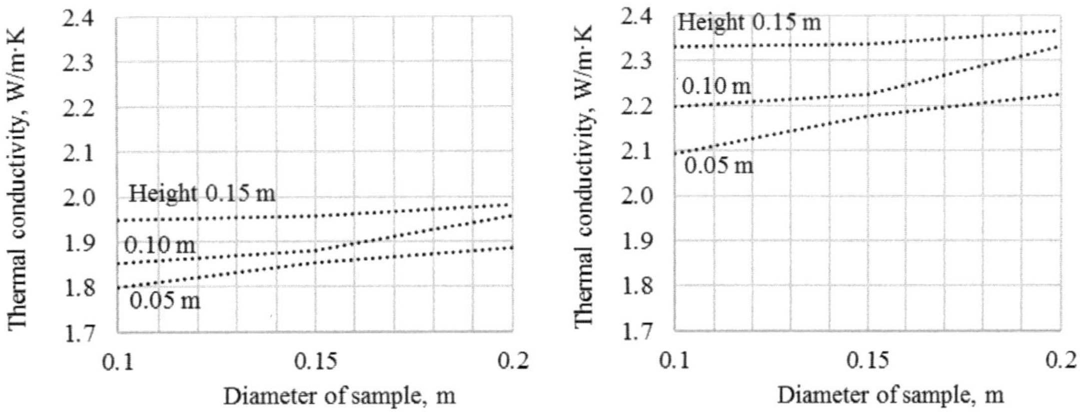

a) b)

Figure 3. a – thermal conductivity of thawing sand; b – thermal conductivity of frozen sand.

The cell for determining the coefficient of thermal conductivity was assembled considering the results of the numerical simulation. A quartz sand sample with a water content of 15% and a dry density of 1.68 g/cm^3 was tested to evaluate the laboratory results. The initial temperature of the sample was +20 °C. Then the sample was cooled to temperatures of +1 and + 5 °C at the ends. Next the sample was frozen to -3 and -10 °C. The sample was kept at positive and negative temperatures for at least 5 hours after the heat flow had steady state. Heat flux and temperature in the sample at two points were measured every ten minutes during the experiment. The distance between the temperature sensors is 110 mm (Figure 4, the plots show measurements every hour).

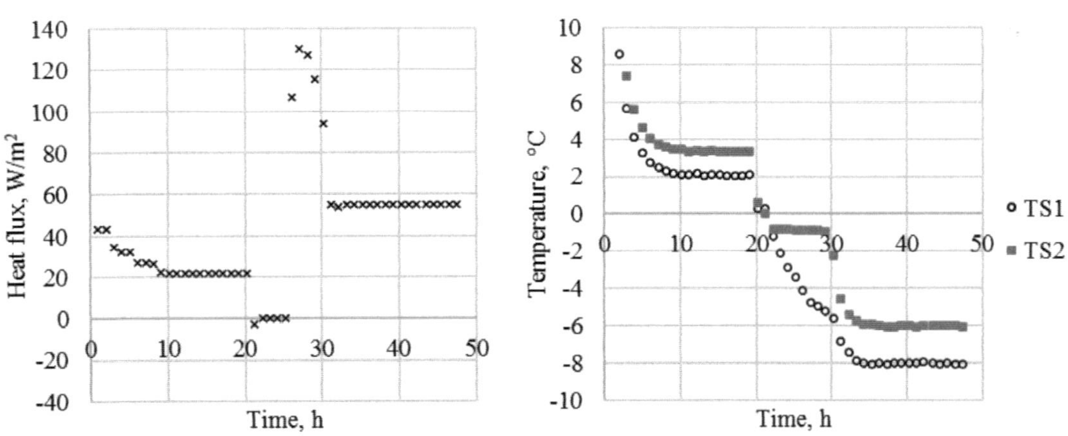

a) b)

Figure 4. a – time-heat flux curve; b – time-temperature curve.

The thermal conductivity coefficient was calculated as the ratio of the heat flux to the temperature gradient. The results were evaluated by comparing the test data with the calculated values obtained by various methods. Table 2 shows the test results in comparison with the values obtained by calculation methods. The experimental data are in the ranges of the calculated values.

Table 2. Thermal conductivity of quartz sand.

Method	Thermal conductivity, KJ/m×K×day (W/m×K)	
	frozen sand, λ_f	thawing sand, λ_{th}
Laboratory test	259.0 (3.00)	171.7 (1.99)
M.Kersten [9]	222.5 (2.57)	168.5 (1.95)
O.Johansen [11]	273.4 (3.16)	215.8 (2.50)
Russian Code [12]	203.9 (2.36)	171.4 (1.98)

The thermal conductivity coefficients for samples of different compositions were measured in a similar way. The impact of the cell design on the experimental data was clarified by implementing backward analysis in the GeoStudio program. The thermal conductivity coefficient of the simulated soil was acquired so that the ratio of the heat flux passing through the soil sample with the sensor to the temperature gradient corresponded to the thermal conductivity coefficient determined in laboratory conditions. It was necessary to increase the coefficients by 0.9% for thawed soil and by 1.6% for frozen soil average. The results of laboratory and numerical experiments are presented in Table 3.

Table 3. Thermal conductivity of composite soils according to the results of experiments.

Parameter	Units	Sand	80% sand + 20% ceramic brick	80% sand + 20% concrete
Laboratory test:				
frozen material, λ_f		259.0 (3.00)	206.7 (2.39)	243.2 (2.81)
thawing material, λ_{th}		171.7 (1.99)	154.8 (1.79)	194.7 (2.25)
	KJ/m×K×day (W/m×K)			
Improvement by numerical modelling:				
frozen material, λ_f		263.8 (3.05)	209.3 (2.42)	246.8 (2.86)
thawing material, λ_{th}		173.4 (2.01)	155.8 (1.80)	196.6 (2.28)

The results of research showed that the presence of construction waste in the sand affects its thermal conductivity. The addition of 20% crushed brick to the sand reduces its thermal conductivity coefficient by 20.6% in the frozen state and by 10.1% in the thawed state. At the same time sand with concrete rubble showed a decrease in the thermal conductivity coefficient in the frozen state by 6.4% and an increase by 13.4% in thaw. The effect is occurred because of particles of different mineralogical composition with different thermal conductivity.

The results of testing can be used in making forecast of the thermal regime of soil stratum when designing shallow foundations

4. Summary

Numerical simulation made it possible to design a cell for a laboratory apparatus. Results of numerical models have a good agreement with laboratory results and empirical expressions. Numerical modeling of laboratory test devices has significantly reduced the duration of laboratory testing.

The research has confirmed that the presence of foreign inclusions in the sand affects its thermal conductivity. The addition of 20% crushed brick to the sand reduces its thermal conductivity coefficient by 20.6% in the frozen state and by 10.1% in the thawed state. At the same time, sand with the addition of concrete rubble showed a decrease in the thermal conductivity coefficient in the frozen state by 6.4% and an increase by 13.4% in thaw. The effect is occurred because of particles of different mineralogical composition with different thermal conductivity.

IOP Conf. Series: Materials Science and Engineering **945** (2020) 012071 doi:10.1088/1757-899X/945/1/012071

The results of the research can be used in the design of shallow foundations, in calculating the depth of soil freezing near the foundations and underground utilities, which are backfilled with soil mixed with construction waste.

References

[1] Lushnikov V V 2013 Insulated foundation of an detached house 95
[2] Nevzorov A L 2000 Foundations on seasonally frozen soils *Textbook for universities* 152
[3] Lykov A V 1967 Heat conduction theory *Textbook for universities* 599
[4] Chudnovsky A F 1962 Thermal characteristics of dispersed materials *Fizmatgiz* 456
[5] 1985 GOST 26263-84 Soils. Laboratory method for determining thermal conductivity of frozen soils *Standards publishing house* 12
[6] 2001 BS EN 12667:2001 Thermal performance of building materials and products Determination of thermal resistance by means of guarded hot plate and heat flow meter methods Products of high and medium thermal resistance *British Standards Institution* 58
[7] 2004 ASTM C177-04 Standard test method for steady-state heat flux measurements and thermal transmission properties by means of the guarded-hot-plate apparatus *American society for testing and materials* 23
[8] 2012 Soil heat flux plate model HFP01 Instruction manual *Campbell Scientific* 18
[9] 2014 Thermal Modeling with TEMP/W. An Engineering methodology *GEO-SLOPE International* 172
[10] Farouki O T 1981 Thermal properties of soils *CCREL* 150
[11] Johansen O 1977 Thermal conductivity of soils *CRREL* 322
[12] 2012 Code 25.13330.2012 Soil bases and foundations on permafrost soils 123

Acknowledgments

The reported study was funded by RFBR and the government of Arkhangelskaya region according to the research project № 18-45-292004.

IOP Conf. Series: Materials Science and Engineering **945** (2020) 012072 doi:10.1088/1757-899X/945/1/012072

Use of chestnut sheel (Castánea) as adsorption material for removing pollutants from natural and sewage waters: a review

S V Sverguzova[1*], I G Shaikhiev[2], E V Fomina[1] and R Z Galimova[2]

[1]Department of industrial ecology, Belgorod State Technological University named after VG Shoukhov, 46, Kostukova Str., Belgorod, 308012, Russia
[2] Department of Environmental Engineering, Kazan National Research Technological University, 41, st. Siberian tract, Kazan, 420074, Russia

E-mail: fomina.katerina@mail.ru

Abstract. The literature data on the use of grinded chestnut shells as a sorption material for removing heavy metal ions, dyes and some organic compounds from aqueous media are summarized. It is shown that a large number of publications are devoted to the study of the adsorption of ions of heavy metals with the shell of chestnut fruits. Information on the chemical composition of the chestnut shell is given. It was shown that the chestnut fruit contains cellulose, lignin and sugars in the shell most of all. It is indicated that the adsorption of many metal ions is carried out by functional groups that is a part of the biopolymers contained in the shell of chestnut fruits. It was revealed that the acidity of the medium influences the degree of extraction of metal ions from aqueous solutions, while the temperature has little effect on the sorption capacity of the shell of chestnut fruits. It is shown that it is possible to increase the sorption characteristics of pollutants by chemical modification of the chestnut shells. It has been determined that sorption isotherms in most cases are more fully described by the Langmuir model, and the kinetics of the process in all cases obeys the pseudo-second order model.

1. Introduction

The aim of this work is to analyze and summarize the literature on the use of waste from the processing of buckwheat as sorption materials for the removal of pollutants from aqueous media. Nowadays, global scientific and industrial community intensively studies a new innovative direction in the field of environmental protection - the use of industrial waste [1, 2], lignocellulosic waste from agricultural production, as well as components and waste from the processing of wood coniferous and deciduous tree species as reagents to remove pollutants from aqueous media [3]. Ligno-, cellulose- and tannin-containing wastes from the processing of wood raw materials are of particular interest for the removal of pollutants of various origins from water bodies [4].

2. Materials and methods

Grinded bark, sawdust and leaves are effective adsorption ma-terials (AM) for removing pollutants of various origins from natural and waste waters (WW). During the processing of tree fruits in industrial production, waste is also generated in the form of seeds, fruit peel and nutshells.

During the processing of chestnut fruits, waste forms in the shape of shells and internal partitions occur. Up to 20% of shell accounts for 100 kg of chestnuts, and from 2.5 to 4% of the inner shells adjacent to the fruit accounts for inner shells. Thus, from 100 kg of raw materials, the yield of peeled chestnuts is 75–78 kg.

Nutshells are studied as AM for the extraction of metal ions, dyes, oil products, and a number of other compounds from aqueous media.

This article summarizes information from literary sources on the use of chestnut shells as AM for the removal of pollutants of various natures from wastewater and natural waters.

In order to collect information, the authors used publications in leading Russian and foreign issues. During the research such methods as the methods of system analysis, comparative criteria for the effectiveness of scientific results were used.

The microstructure of the surface of the sheelwas studied using a high-resolution scanning electron microscope TESCANMIRA 3 LMU.

3. Results and discussion

3.1. Basic information about chestnut

Chestnut (Lat. *Castánea*) is a small kind of trees of the Beech family (*Fagaceae*). Wood, bark and leaves contain from 6 to 14% of tannins. Chestnut has long been used as a fruit tree. Chestnut fruits are rich in carbohydrates and proteins, they are fried and baked, high-quality additives for flour. In addition confectionery products are obtained from them.

During the processing of chestnut fruits, waste forms in the shape of shells and internal partitions occur. The chestnut shell has no significant industrial and commercial use, which contributes to the emergence of a serious environmental problem.

3.2. Composition and possible uses of the chestnut shell

In the research work [5], it is indicated that during the processing of the chestnut a large amount of waste is generated, namely the inner shell, outer shell and leaves. The studies of chestnut by-products revealed a good profile of biologically active substances with antioxidant, anticarcinogenic, and cardioprotective properties. After valorization these agro-industrial wastes can be used by other industries, such as pharmaceutical, food or cosmetic.

The composition of the shells of chestnut fruits is determined in the research work [6], which is shown in Table 1.

Table 1. Dry matter composition of chestnut fruit shell [6].

Ingredients	Content, %
Cellulose	19.2
Lignin	29.15
Soluble lignin	**2.83**
Total sugar content	33.82
Of which monosaccharides are :	32.70
- glucose	19.23
- galactose	2.98
- xylose	6.45
- arabinose	2.52
- mannose	1.52
Ash	0.83
Extracts:	

- cold water	3.85
- hot wayer	26.87
- 1 % solution of NaOH	71.32
- a mixture of ethanol, benzene and water	34.52

3.3. Sorption properties of chestnut nutshell components

One of the possible ways to use the chestnut nutshell is to use them as an AM for extracting pollutants from natural and waste waters. The analysis of literature shows that the largest number of publications is devoted to the adsorption of heavy metal ions (HMI). In this review, the information on the adsorption of HMI by the shell of the chestnut fruit is presented in the alphabetical order of metal ion.

A large number of publications are devoted to the study of adsorption of Cd^{2+} ions by the shell of chestnut fruits [7]. The experimental conditions and the obtained adsorption characteristics are shown in Table 2.

Table 2. Experimental conditions and adsorption characteristics for the adsorption of Cd (II) ions by the shells of chestnut fruits.

Experimental conditions	Obtained adsorption characteristics	Note	Source
$C_{initial}$= 100–600 mg/dm^3, pH = 1.0–7.0, sorbent dosage – 1–8 g/dm^3, T = 30–70 °C, t = 0.5-7.0 h.	The maximum adsorption capacity is 51.28 mg/g. The maximum degree of removal of Cd ions is 98.9%	Discolored shell processed formaldehyde by cis-butenedioic acid, and irradiation. The adsorption isotherm is described by the Langmuir model (R^2> 0.9812). Kinetics of pseudo-second order.	[7]
$C_{initial}$= 20–100 mg/dm^3, pH = 5.5–8.5, T = 10–40 °C, dosage–10 g/dm^3, t = 24 h.	The maximum sorption capacity is 4.75 mg/g.	Incomplete 3^3 factoriallevel. The adsorption isotherm is described by the Freindlich model.	[8]
$C_{initial}$= 250 mg/dm^3, pH = 3–6, dosage – 10 g/dm^3, t = 15 minutes	The maximum adsorption capacity is 14.7 mg/g. The degree of removal of Cd ions is 95%.	The adsorption isotherm is described by the Langmuir model (R^2>0.998). Adsorption kinetics corresponds to a pseudo-second order model.	[9]

It is indicated that the adsorption of Cd^{2+} ions is carried out according to functional groups included in the biopolymers contained in the shells of chestnut fruits (-OH, C (O) OH, -NH and others).

3.3.1. Heavy metal ion adsorption

There are several publications on the adsorption of chromium ions by the fruit shells of chestnuts. In particular, the influence of the initial concentration of Cr (VI) ions – 100 or 150 mg/dm^3, contact time – 2 and 6 hours and the pH of the medium – pH = 2 or 5 on the adsorption characteristics was studied. It was found that the best conditions for the adsorption of Cr (VI) ions were pH = 2 and a contact time of 6 h for both initial concentrations.It was established that adsorption isotherms are more accurately described by the Langmuir model [10].

Research work [11] also indicated that the removal efficiency of Cr^{6+} ions reaches 99.7% at pH = 2. It was also determined that the removal efficiency of Cr(VI) ions increased with increasing temperature in the range from 15 °C to 35 °C.

The model drains were cleaned of chromium ions using chrome chips impregnated with tannins from the chestnut shell as AMs.It was determined that the optimal pH of the medium for the efficient extraction of Cr (VI) ions was pH = 4, and the maximum adsorption capacity of sawdust was 42 mg/g.It was determined that the equilibrium conditions were reached after 2 hours of contact. It was revealed that the adsorption isotherm obeyed the Langmuir model, and the adsorption kinetics obeyed the pseudo-second order model [12]

There are several publications on the study of adsorption of Cu^{2+} ions by the shells of chestnut fruits [13–15]. The experimental conditions and the obtained adsorption characteristics are shown in Table 3.

Table 3. Experimental conditions and adsorption characteristics for the adsorption of Cu(II) ions by the shells of chestnut fruits.

Experimental conditions	Obtained adsorption characteristics	Note	Source
Castaneamolissima, $C_{initial}$= 25–200 mg/dm^3, pH = 2–6,dosage– 10 g/dm^3, T = 20–40 °C, t = 96 h, 120 rot/min	The adsorption capacity is 12.42 mg/g	Adsorption isotherm obeys Langmuir isotherm models. The kinetic data were found to obey the pseudo-second-order model. ΔGo = -5.875, -5.714 and 5.442 kJ/mol at T = 293, 303 and 313 K, ΔHo = -12.296 kJ/mol, ΔSo = 21.534 J/mol· K.	[13]
Dynamic conditions. $C_{initial}$ = 20 mg/dm^3, pH = 6, particle size – 0.25 mm.	The maximum adsorptioncapacity is 10.94 mg/g	The adsorption process was endothermic. The process obeys the Thomas model. 96.1% was desorbed with 0.1 M HCl.	[14]
Insoluble pigment containing melanin from the chestnut shell. $C_{initial}$ = 10 mg/dm^3, t = 240 min	The maximum sorption capacity is 33.2 mg/g	The kinetic adsorption behavior obeyed the pseudo-second-order rate law, and the equilibrium adsorption data were well described with both the Langmuir and the Freundlich isotherms	[15]

In global literature there is little research works on the adsorption of Ni^{2+} ions in the shells of chestnut fruits. The removal of Ni^{2+} ions by some cheap AMs, including the shells of chestnut fruits, was studied [16]. The experiments were carried out at 25 °C with an initial concentration of a metal ion of 15 mg/dm^3 and the reaction time of 10–180 minutes at pH = 7 and a dosage of AM of 2–20 g/dm^3. It was found that the maximum adsorption capacity was 5.6 mg/g and the degree of extraction Ni (II) ions is 37.3%

The adsorption of Pb(II) ions by Korean researchers was also studied [17]. The process was carried out at pH = 2–6 for 24 hours at a temperature of 20–40 °C under static conditions. The dosage of the shell was 5 g/dm^3.The constructed adsorption isotherms at temperatures of 20–40 °C were well described by the Langmuir model (R^2= 0.99), and the kinetic dependences at the indicated temperatures at an initial concentration of Pb(II) ions from 10 to 40 mg/dm^{3}obeyed a pseudo-second order model. It was determined that the maximum adsorption capacity of the chestnut shell was 31.3 mg/g.

There are practically no studies on the adsorption of Zn^{2+} ions by the shells of chestnut fruits. Chinese authors conducted experiments on the use of chestnut shells as the AM of Zn (II) ions. The experimental results showed that when the concentration of the solution was 20 mg/dm^3, the pH was 6.03, the contact time was 120 minutes, and the dosage of the chestnut shell was 0.7 g, the degree of removal of zinc ions could reach 98.78% [18].

Several studies are carried out on the adsorption of the chestnut shell of several HMIs to compare adsorption activity. In particular, the influence of initial cation concentration, temperature and pH was investigated to optimize Pb^{2+}, Cu^{2+} and Zn^{2+} removal from aqueous solutions using acid formaldehyde pre-treated chestnut shell as adsorbent.The experiments were planned according to an incomplete 33 factorial experimental design. Under the optimal conditions selected, the metal ion adsorption equilibrium was satisfactorily described by the Langmuir isotherm model. The maximum pre-treated chestnut shell adsorption capacity was obtained for Pb^{2+} ions 8.5 mg/g, for Cu^{2+} ions 4.44 mg/g and for Zn^{2+} ions 3.35 mg/g. A model that considered the effect of axial dispersion was successfully used to describe the fixed-bed adsorption behavior of Pb^{2+}, Cu^{2+} and Zn^{2+} ions at the flow rates essayed [19].

Korean authors studied the adsorption of Cd^{2+}, Cu^{2+} and Pb^{2+} ions with the skin of chestnut fruits. It was determined that the maximum adsorption capacity for these ions was in the following row: Pb^{2+} (31.25 mg/g)> Cu^{2+} (7.87 mg/g)> Cd^{2+} (6.85 mg/g). It was shown that adsorption isotherms for the aforementioned ions of the chestnut shell were described by the Langmuir model ($R^2 = 0.99$) [20].

There are much more research works mainlypublished by Chinese researchers, in which the adsorption characteristics of the chestnut shell with respect to 4 metal ions are studied. In particular, the adsorption of $Cd^{2+,}$ Cu^{2+}, Pb^{2+} and Zn^{2+} ions was studied. It was found that the maximum adsorption by ions for Cd^{2+} ions occurred within 0.5 h, and for other ions after 1 h. It was determined that the temperature had little effect on the adsorption rate of Zn^{2+} ions and increased with respect to Pb^{2+} and Cd^{2+} ions at 40 °C [31]. It was noted in the research work that the highest removal rate of Pb^{2+}, Cu^{2+}, and Zn^{2+} ions was observed at pH = 5, and Cr^{6+} ions at pH = 2 [21].

The adsorption of Cd^{2+}, Pb^{2+}, Cu^{2+} and Zn^{2+} ions from acidic solutions (pH = 2) by chestnut shells under static and dynamic conditions was studied. At an initial concentration of the mentioned HMIs of 30 mg/dm^3, the degree of their removal under static conditions after 1 h of contacting was 98.9% for Cd^{2+} ions, 99.6% for Pb^{2+} ions, 98.9% for Zn^{2+} and 98.7 ions % – for Cu^{2+} ions. The values of maximum adsorption capacity for these metal ions were: 3.2 mg/g for Cd^{2+} ions, 90.8 mg/g for Pb^{2+} ions, 27.3 mg/g for Zn^{2+} ions and 52.4 mg/g by Cu^{2+} ions.

Fourier transform infrared (FTIR) and X-ray photoelectron(XPS) spectroscopies showed that the functional groups involved in metal ions binding included carboxyl,hydroxyl, ether, alcoholic and amino groups.

Literary sources contain references to the work of Korean scientists on the adsorption of gold ions by the shells of chestnut fruits [22]. It is indicated that under the conditions of the experiment: pH = 2, temperature – 20 °C, when the initial concentration of Au (III) ions is 300 mg/dm^3, the maximum sorption capacity is 189 mg/g. It is shown that this value of the maximum capacity for gold ions is higher than that obtained by other researchers. The authors explain this fact by the presence of a large number of tannins in the composition of the chestnut fruit shell.

The adsorption of platinum ions by the chestnut shell was studied under the above mentioned conditions: pH = 2, temperature – 20 °C, stirrer speed – 120 rpm, dosage of sorption material – 2.5 g/dm^3. It was indicated that the maximum adsorption capacity, determined by the adsorption isotherm, was low and did not exceed 1.5 mg/g. The adsorption isotherm was adequately described by the Freindlich equation ($R^2 = 0.931$) [23].

3.3.2. Adsorption of organic compounds and dyes

Using the shell of chestnut fruits as an adsorption material to reduce the dyeability of solutions is not entirely advisable because the shell itself produces tannins and is used for dyeing, in particular, of fabrics and wool. Nevertheless, the possibility of extracting dyes of the Reactive Brilliant Red K-2G brand, Methylene blue [24] is indicated. It was determined that the maximum sorption capacity for the dye "Methylene blue" of the chestnut shell at 25 °C is 126.5 mg/g, and the adsorption isotherm was most fully described by the Langmuir model.

The adsorption of Auramine O, Basic Violet 14, Basic Blue 11, Rhodamine B, and Methylene Blue dyes with tannins isolated from the chestnut shell was studied. It was found that natural tannins were effective in removing dyes of the Basic Violet 14, Basic Blue 11, and Methylene Blue brands, and were ineffective in the other two dyes [25]. It was determined that the adsorption isotherms were well described by the Langmuir model.

IOP Conf. Series: Materials Science and Engineering **945** (2020) 012072 doi:10.1088/1757-899X/945/1/012072

The chestnut shell was also studied as an adsorption material for the removal of pesticides of the trade names Pirimicarb, Imidacloprid, Acetamiprid and Thiamethoxam under dynamic conditions [26]. It was determined that the adsorption capacity for these pesticides was: according to Thiamethoxam – 14.3 mg/g, Acetamiprid – 4.70 mg/g, Imidacloprid – 8.51 mg/g. It was found that treatment of the chestnut shell with citric acid increased the adsorption characteristics of these pesticides by 15%.

3.3.3. Horse-chestnut shell

In fact, besides species of the Chestnut kind, other plants from other families can be called "chestnuts". For example, *Aesculus* - Horse Chestnut is a plant of the Soapberry family *(Sapindaceae)*, common horse chestnut (*Aesculushippocastanum* L.). Figure 1 shows a highly developed microrelief of the surface of the horse-chestnut shell.

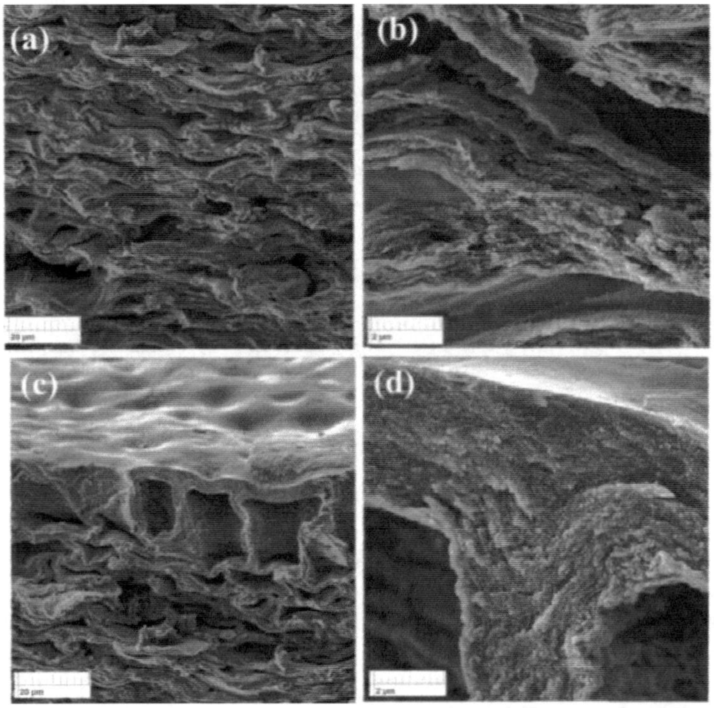

Figure 1. Microstructure: of chestnut skin surface (a, b); of the faults of peeled chestnut (c, d).

The shells of horse chestnut fruit were also investigated as AM HMIs. Thus, in particular, the adsorption of Cu^{2+} ions by the green shell of an ordinary horse chestnut was studied. It was determined that the greatest extraction of copper ions at the initial concentration of the last 10–100 mg/dm^3 was observed at pH = 5, and the time to achieve adsorption equilibrium was 5 minutes. It was determined that the degree of extraction of Cu^{2+} ions was 72.37–98.09%, depending on the process parameters. It was found that the adsorption isotherms were adequately described by the Freindlich model, and the kinetics of the process was described by a pseudo-second order model.

The adsorption of Cr (VI) ions with an initial concentration of 500–4000 mg/dm^3 was studied with varying process parameters of the shells of horse chestnut fruits. It was found that the highest sorption capacity (142.9 mg/g) was achieved at pH = 2 and contact time of 2 hours. It was found that the adsorption isotherm was most accurately described by the Langmuir model, and the kinetics of the process was described by a pseudo-second order model. The thermodynamic parameters of the process were determined, which indicate that the adsorption process was exothermic and was spontaneously.

Under dynamic conditions, the adsorption of Zn^{2+} ions was studied using various AMs, including horse chestnut shells. It was found that the maximum adsorption capacity was 12.25 mg/g with an initial concentration of zinc ions of 97.5 mg/dm^3 at 25 °C.

4. Summary

Thus, the above mentioned materials summarize the literature on the use of grinded shells of chestnut fruits as an AM for the removal of metal ions, dyes and some organic compounds from aqueous media. It was determined that adsorption isotherms, in most cases, were more ade-quately described by the Langmuir model, and the kinetics of the process in all cases obeyed the pseudo-second order model. The values of adsorption indicators for the studied pollutants were given. It was shown that it was possible to increase the adsorption characteristics of the chestnut shell by chemical modification.

5. References

[1] Rangabhashiyam S, Jayabalan R, Asok Rajkumar M, Balasubramanian P2019Elimination of toxic heavy metals from aqueous systems using potential biosorbents: A review*Green Buildings and Sustainable Engineering* 291–311

[2] Sverguzova S V, Bomba I.V.,Voronina Yu.S. 2018 Purification of oily emulsions with leaf litter of cherry and rowan *Chemical Bulletin* **1(4)** 4–19

[3] Sapronova Z A,Svyatchenko A V, Fomina E V, Voitovich E V 2019 Obtaining sorption material from the leaves of *Aésculus hippocastanum* L*Atlantis Highlights in Material Sciences and Technology***1** 311–315

[4] Sverguzova S V, Sapronova Zh A, Svyatchenko A V, Otiti T 2018 Adsorption of spindle oil by native and thermal modified chestnut tree waste *Construction Materials and Products* **1(1)** 4–11

[5] Braga N, Rodrigues F, Oliveira MBPP2015Castanea sativa by-products: a review on added value and sustainable application *Natural Product Resesrch* **29(1)**1–18

[6] Vázquez G, Freire MS, González-Álvarez J, Antorrena G2007Valorisation of lignocellulosic waste materials: Tannins as a source of new products*Proceedings of European Congress of Chemical Engineering (ECCE-6)* Copenhagen 1–7

[7] Huang Y, Zhao R, Liu Wet al 2018Using pretreated chestnut endothelium to adsorb lead and cadmium ions from water *Saudi Journal of Biological Sciences* **25** 1154–1162

[8] Vazquez G, Gonzalez-Alvarez J, Freire MS, Calvo M, Antorrena G2009Determination of the optimal conditions for the adsorption of cadmium ions and phenol on chestnut (*Castanea sativa*) shell *Global NEST Journal* **11(2)** 196–204

[9] Ding Y, Wang D, Jing D, Huang Z, Wang Z2013Adsorption of aquatic cadmium (II) by chestnut inner shell *Third International Conference on Intelligent System Design and Engineering Applications* 222–226

[10] Ertaş R, Öztürk N 2013Removal of Cr(VI) from aqueous solution onto chestnut shell: application of full factorial design and equilibrium studies *Desalination and Water Treatment* **51(13–15)** 2909–2914

[11] Wang G 2009 Study on adsorptive performance of heavy metal chromium(VI) by chestnut shell *Chinese Journal of Environmental Engineering* **3(5)** 791–794

[12] Chabaane L, Tahiri S, Albizane A et al 2011Immobilization of vegetable tannins on tannery chrome shavings and their use for the removal of hexavalent chromium from contaminated water *Chemical Engineering Journal* **174** 310–317

[13] Qi J, Liang Z, Deng X, Yao Z2009Adsorption of Cu^{2+} on to chestnut (*Castanea mollissima*) shells: Equilibrium, kinetics and process design*Acta Scientiae Circumstantiae* **29(10)** 2141–2147

[14] Qi J, Yao Z, Deng X, Liang Z, Hu C2009Absorption capability of chestnut shells for copper removal from aqueous solution *Journal of Northwest A & F University (Natural Science Edition)* **37(4)** 197–202

[15] Yao Z, Qi J, Hu Y, Wang Y 2016 Insolubilization of chestnut shell pigment for Cu(II)

adsorption from water *Molecules* **21(4)** 405

[16] Ince M 2014 Comparision of Low-Cost and Eco-Friendly Adsorbent for Adsorption of Ni(II) *Atomic Spectroscopy* **35(5)** 223–233

[17] Lee HY, Lim JE, Lee SS, Ok YS2012Lead biosorption by chestnut shell and oak sawdust in aqueous solution using thermodynamic equilibrium model *Joumal of the Environment* **8(1)** 15–31

[18] Li N, Liu R, Zhang Y 2014 Study on Adsorption Conditions of Waste Chestnut Shell and Walnut Shell for Zinc Ions in Aqueous Solution *Environmental Science and Technology* **S1** 129–131

[19] Vazquez G, Calvo M, Freire MS, Gonzalez-Alvarez J, Antorrena G 2009 Chestnut shell as heavy metal adsorbent: Optimization study of lead, copper and zinc cations removal *Journal of Hazardous Materials* **172** 1402–1414 DOI 10.1016/j.jhazmat.2009.08.006

[20] Lee H Y, Hong K C, Lim J E 2009 Adsorption of heavy metal ions from aqueous solution by chestnut shell*Korean Journal of Environmental Agriculture* **28(1)** 69–74

[21] WuY, HuangX2012Add to favoriteget latest updatestudy on adsorptive performance of Cr^{6+} and other heavy metals by chestnut shell *Journal of Anhui Institute of Architecture & Industry (Natural Science)* **6** 76–78

[22] Hong K C, Lee H Y, Lim J E 2008 Development of precious metal recovery process using. Agricultural by-product: Gold adsorption in aqueous solution by chestnut shell*Journal of Agriculture and Life Science* **19** 105–113

[23] Hong K C, Lee H Y, Lim J E, Yang J E, Ok Y S 2009 Development of precious metal recovery process using agricultural by-product: Platinum adsorption in aqueous solution by chestnut shell *Journal of Agriculture and Life Sciences* **21** 27–34

[24] Ozdemir C S 2018 Adsorptive removal of methylene blue by fruit shell: Isotherm studies. Fullerenes *Nanotubes and Carbon Nanostructures* **26(9)** 570–577

[25] Sanchez-Martin J, Beltran-Heredia J, Gragera-Carvajal J 2011 *Caesalpinia spinosa* and *Castanea sativa* tannins: A new source of biopolymers with adsorbent capacity. Preliminary assessment on cationic dye removal *Industrial Crops and Products* **34** 1238–1240

[26] Cobas M, Meijide J, Sanromán MA, PazosM2016Chestnut shells to mitigate pesticide contamination *Journal of the Taiwan Institute of Chemical Engineers* **61** 166–173

Acknowledgments
The work is realized in the framework of the Program of flagship university development on the base of the Belgorod State Technological University named after V.G. Shukhov, using equipment of High Technology Center at BSTU named after V.G. Shukhov.

IOP Conf. Series: Materials Science and Engineering **945** (2020) 012073 doi:10.1088/1757-899X/945/1/012073

Improvement of the design of the aeration device of the pneumatic chamber pump

S Yu Lozovaya[1*], Yu M Fadin[1], A V Gavrilenko[1] and Yu V Gavrilenko[1]

[1]Department of mechanical equipment, Belgorod State Technological University named after V.G. Shukhov, Kostyukov St., 46, Belgorod, 308012, Russia

E-mail: Andrej.402@mail.ru

Abstract. The production of building materials is a complex technological process involving the processing of raw materials with various physical and mechanical properties, as well as using a variety of degrees of complexity of technological equipment and auxiliary mechanisms. High-pressure pneumatic transport has found application in construction, refractory, chemical and other industries. It is widely used for transportation of fine-grained bulk materials such as cement, alumina, soda, apatite, ash, clay, synthetic products, mineral fertilizers. The process of pneumatic transportation occurs due to the action of compressed air on the material being moved. However, at the same time, as a rule, there are through channels in the material layer, which leads to an increased consumption of compressed air, part of the unloaded material remains at the bottom of the pump chamber, which reduces productivity. Due to the fact that some fluidizing devices can form through channels in the material layer and given the large volumes of material being transported, it is necessary to solve issues to improve the efficiency of pneumatic chamber pumps and reduce the consumption of compressed air required for transporting bulk materials. This can be achieved by organizing effective fluidization of the material, that is, uniform mixing of the transported material with air.

1. Introduction

Pseudo-fluidizing is a transition state between a stationary layer of bulk material and the entrainment of the material by the compressed air flow, which occurs in a certain range of air flow rates through the material layer, which depends on the design of aerating devices. It should be noted here that an increase in the concentration of the cement-air mixture leads to an increase in the performance of the pneumatic chamber pump.

Pneumatic transportation has a number of advantages over traditional means of transporting bulk materials: high lifting height and transportation distance (up to 3000 m), the use of areas unsuitable for other transportation methods, the complete absence of residues and losses of the transported product in pipelines, the exclusion of violations of technological and hygienic conditions of the air environment in industrial premises due to the lack of dusting, ease of installation, the possibility of full automation of management, compactness. However, pneumatic transportation consumes a large volume of compressed air. Therefore, for some materials, the air consumption reaches 250 m^3 per ton [1-4].

Depending on the location of the chamber, pneumatic chamber pumps are available with vertical, horizontal or inclined chambers. However, vertical pneumatic chamber pumps are widely used due to better material flow conditions and, consequently, more uniform discharge [5-7].

IOP Conf. Series: Materials Science and Engineering **945** (2020) 012073 doi:10.1088/1757-899X/945/1/012073

2. Materials and methods

A pneumatic chamber pump with upper material discharge and aeration device is a metal tank, the upper part of which is made in the form of a hemisphere, and the lower part is made in the form of a hemispherical bottom. The loading spout is closed with a conical valve.

Pump operation is periodic. They open the loading spout, and the chamber is filled with the transported material to a certain level. Then the fuelling port is hermetically closed with a valve and a tap is opened to supply compressed air to the space between the porous partition and the bottom of the chamber. Air penetrates through the porous partition into the chamber, brings the material to the fluidized state, and the chamber is discharged through the discharge pipe. After emptying the chamber, the air supply stops and the operation cycle repeats.

To solve the problem of reducing the consumption of compressed air for cement transportation, a new design of the aeration device of the pneumatic chamber pump was developed.

The proposed design of a multi-nozzle aeration device has a number of features that increase its efficiency in comparison with other used multi-nozzle systems. First of all, the number of aero elements that are evenly spaced across the entire cross-section of the chamber and have certain bending and turning angles has been significantly increased. As a result of a closer arrangement of the nozzles, the volume of stagnant zones located between them decreases and the uniformity of cement distribution increases.

The pneumatic chamber pump contains a chamber with a loading valve located in the upper part for supplying the material to be transported and a spout for releasing compressed air. At the bottom of the pump chamber, a multi-nozzle aeration device is placed, which is a pipe made in concentric circles, in the lower part of which are welded nozzles directed towards the bottom, the output ends of which have a bending angle of 50-70 ° relative to their vertical axis, and relative to the radius drawn through the nozzle axis from the center of the aeration device, have a rotation angle of 20-25°.

The discharge pipe located in the center of the pump chamber is a vertical pipe equipped with a confuser in the lower part. In the center of the bottom of the pump chamber there is a central nozzle, which is necessary to increase the speed of the cement-air mixture in the discharge pipe. A pipe is welded inside the upper part of the chamber to supply compressed air above the layer of loaded material.

The nozzles of the aeration device are oriented in such a way that the outgoing jets have velocity components directed downwards and tangentially to the air rings. This allows creating a vortex field in the lower part of the pump chamber that narrows to the entrance to the discharge pipe, preventing the formation of through channels in the material layer. As a result of air jets flowing out of the nozzles, the cement is intensively mixed with air and the cement layer is agitated. A number of authors claim in their works that the best pseudo-fluidizing properties have aeration devices that have flat aeration elements using porous partitions. Their main advantage is that they create a uniform pseudo-fluidized layer. As porous partitions of aeration devices, technical felt, fiberglass, and polystyrene are used, but they have a big drawback – clogging of pores as a result of humidification. Ceramic partitions are also used, which have good anti-corrosion properties and can withstand high temperatures, but quickly fail when the temperature changes. Metal partitions do not have the above disadvantages, but they are prone to corrosion. In general, gas distribution grilles tend to clog the pores and nozzles, which can lead to complete sticking of the holes, and, consequently, a decrease in the live section of the partition, which leads to a deterioration of the pseudo-fluidization of the material.

If the flow of the conveying agent (compressed air) passes through the material layer at different speeds, the state of the pseudo-fluidized bed is different depending on the speed of this flow. In order to achieve efficient pseudo-fluidization with an optimal flow rate of compressed air, it is necessary to maintain a constant airflow rate, which depends on the pseudo-fluidization rate of the material.

All these disadvantages can be eliminated by rational selection of the aeration device for a certain type of transported material. In particular, for cement with its physical and mechanical properties, pseado-fluidization using aeration devices in the form of porous partitions is difficult. The use of nozzle devices with low hydraulic resistance does not always have the necessary effect on the cement, which, following an increase in efficiency, can lead to an increase in the consumption of compressed air. Therefore, for effective operation of pneumatic chamber pumps with reduced compressed air consumption, it is necessary to use a rational design of aeration devices [6, 8, 9].

The proposed design of a multi-nozzle aeration device has a number of features that increase its efficiency in comparison with other used multi-nozzle systems. First of all, the number of aero elements that are evenly spaced across the entire cross-section of the chamber and have certain bending and turning angles has been significantly increased. As a result of a closer arrangement of the nozzles, the volume of stagnant zones located between them decreases and the uniformity of cement distribution increases.

The pneumatic chamber pump contains a chamber with a loading valve located in the upper part for supplying the material to be transported and a branch pipe for releasing compressed air. At the bottom of the pump chamber, a multi-nozzle aeration device is placed, which is a pipe made in concentric circles, in the lower part of which are welded nozzles directed towards the bottom, the output ends of which have a bending angle of 50-70° relative to their vertical axis, and relative to the radius drawn through the nozzle axis from the center of the aeration device, have a rotation angle of 20-25°.

The nozzles of the aeration device are oriented in such a way that the outgoing jets have velocity components directed downwards and tangentially to the air rings. This allows creating a vortex field in the lower part of the pump chamber that narrows to the entrance to the discharge pipe, preventing the formation of through channels in the material layer. As a result of air jets flowing out of the nozzles, the cement is intensively mixed with air and the cement layer is agitated.

3. Results

Thus, due to the specific location of the aeration device nozzles, the pseudo-fluidization of the material is improved and a homogeneous pseudo-fluidized layer is formed at the entrance to the discharge pipe with no empty channels, which reduces the consumption of compressed air and increases the performance of the pneumatic chamber pump.

This is confirmed by the results of simulation modelling of the creation of a pseudo-fluidized layer depending on the shape and location of the nozzles (figure 1) using the software product Solid Works.

Figure 1, a shows that the air stream coming out of the nozzle tends to the discharge pipe in an almost straight trajectory, and only at the entrance to the discharge pipe, due to the meeting of air flows coming out of neighboring nozzles, a pseudo-fluidized layer is created, where an active process of mixing air with cement occurs. However, it is possible to create undesirable pore channels through which air passes without affecting the cement, which is negative in the pseudo-fluidization process. On the other hand, there are so-called dead zones at the bottom of the pump chamber, in which the air jets practically do not have a useful effect on the loaded material, which leads to an increase in the unloading time and compressed air consumption.

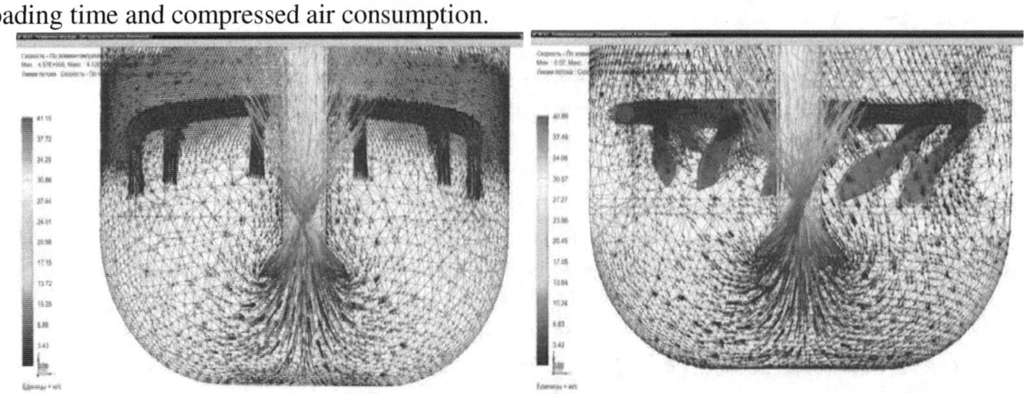

a) b)

Figure 1. Results of simulation modeling of pseudo-fluidized layer creation depending on the shape and location of the nozzles:
a – straight nozzles, *b* – nozzles with turning and bending angles.

IOP Conf. Series: Materials Science and Engineering **945** (2020) 012073 doi:10.1088/1757-899X/945/1/012073

Figure 1. b clearly shows the flows that are created by air jets coming out of the nozzles with angles of rotation and bending. As they can see, air flows are more active on the cement in the area of air supply from the nozzles most remote from the discharge pipe, as well as at the bottom of the pump chamber. At the bottom of the chamber, a vortex field is created closer to the center, which has a more effective effect on the material and prevents the formation of through channels, which increases the uniformity of the cement distribution in the pseudo-fluidized layer. Due to the action of the downward and tangential components of the jet velocity, stagnant zones and pore channels are eliminated, which significantly affects the flow of compressed air. At the same time, the created pseudo-fluidization zone (figure 1. b) is larger than when using direct nozzles (figure 1. a), which facilitates the process of involving the cement-air mixture in the discharge pipe, and, consequently, reduces the time of unloading the pump chamber.

Analyzing the results obtained using simulation modeling and based on the fact that the aeration device with nozzles having rotation and bending angles pseudo-fluidizes bulk material more effectively, the main factors affecting the process of pneumatic transportation were determined. The main parameters that characterize this process are the productivity and specific consumption of compressed air. The capacity is directly proportional to the discharge time of the pump chamber, therefore, for convenience, it is necessary to study the dependence of the discharge time [10, 11].

Using the analytical package Maple, three-dimensional figures were constructed showing the dependence of the unloading time on changes in the main factors at fixed time values τ_r = 6, 12, 18 seconds (figure 2). If necessary, they can build these figures for any value of not only time, but also other functions (if there is an appropriate regression equation).

Thus, it is possible to determine the geometric and technological parameters of the pump, that is, at what values of the main factors one or another value of the unloading time, productivity, and air consumption is obtained.

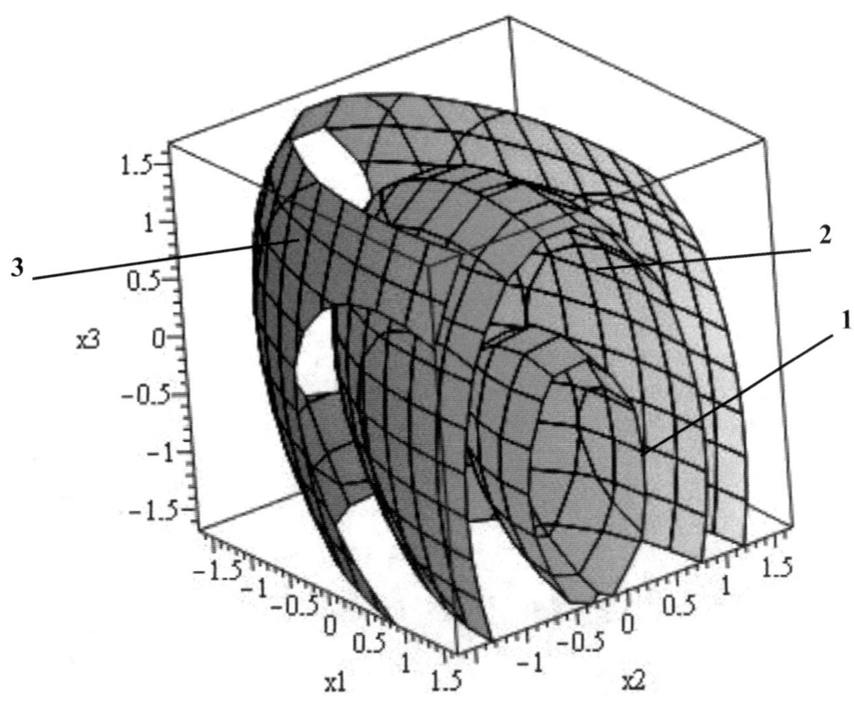

Figure 2. Graphical structures that display fixed values of unloading time depending on the main factors: 1 – τ_r = 6 sec.; 2 – τ_r = 12 sec.; 3 – τ_r = 18 sec.

Figure 2 shows surface *1*, where any point shows at what values of factors the minimum unloading time τ_r = 6 seconds can be obtained, namely, when the values of the main factors are combined in a

IOP Conf. Series: Materials Science and Engineering **945** (2020) 012073 doi:10.1088/1757-899X/945/1/012073

certain range. Geometric parameters are regulated structurally, because this is inherent in the design and manufacture of the laboratory installation, and the excess pressure affects the air flow rate, and, consequently, the energy consumption of the transportation process, it can be assumed that it is advisable to use the values of factors on the surface *1*, choosing the minimum pressure value in the pump chamber. With certain parameters, it is possible to get the maximum performance of the pneumatic chamber pump. In this regard, we can conclude that the second performance is one of the main technical and economic indicators of the pneumatic chamber pump.

4. Discussion

Under certain conditions, for example, when aggregating cement, its bulk density increases, so often when air is supplied from the nozzles, so-called pore or through channels occur, through which air passes through the cement without mixing with it and without creating a fluidized layer in this area.

5. Summary

Thus, the proposed design of the multi-nozzle device has a simple design in comparison with known aeration devices and provides effective aeration of cement in the discharge part of the pump chamber, as well as due to the geometric arrangement of the nozzles, a homogeneous pseudo-fluidized layer is formed in the pump chamber, which reduces the time of its unloading and reduces the consumption of compressed air.

6. References

[1] Mills D, Jones M G, Agarwal V K 2004 Handbook of Pneumatic Conveying Engineering. Marcel Dekker Inc 695

[2] Evtyukov S A 2005 Pneumotransport equipment in the construction industry and construction, Shapuno M M. S-Pb.: DNA 360

[3] Zarnitsyna E G, Terekhova O N 2011 Ventilation systems and pneumatic transport: a tutorial Barnaul: Publishing house of Altai State Technical University 228

[4] Bogdanov V S, Fadin Yu M, Shaptala V V, Gavrilenko A V 2015 Calculation of pseudo-fluidized layer parameters in a pneumatic chamber pump *Bulletin of BSTU named after V G Shukhov* **2** 65–67

[5] Bogdanov V S, Fadin Yu M, Shaptala V V, Gavrilenko A V 2016 Characteristics of flows of cement-air mixture during pneumatic transportation of cement *Bulletin of BSTU named after V G Shukhov* **2** 110–112

[6] Brox M, Zimmer D 2015 Pneumatic transport technologies for the cement industry *Cement and its application* **5** 1–4

[7] Bogdanov V S, Lozovaya S Yu, Fadin Yu M, Gavrilenko A V 2016 Research of influence of the main parameters on the capability of the pneumatic chamber pump with multijet aeration unit *International Journal of Pharmacy & Technology* **8 - 4** 24669-24680

[8] Shaptala V G, Shaptala V V, Gavrilenko A V 2015 Numerical simulation of the movement of a cement-air mixture in a pneumatic chamber pump *Bulletin of BSTU named after V G Shukhov* **2** 159–161

[9] Bukhmirov V V, Rodionov G A 2013 Mathematical modeling of two-phase flows in installations of pneumatic transport of fine materials *Bulletin of the Ivanovo State University* **5** 24–28

[10] Gavrilenko A V 2015 Determination of pressure losses in the installation of pneumatic transport of materials *Bulletin of IrSTU* **4** 23–26

[11] Shishkin S F, Gavrilyuk D N 2009 Calculation of high-pressure pneumatic transport *Bulletin of BSTU named after V G Shukhov* 3 114–117

Acknowledgements
The work is realized in the framework of the Program of flagship university development on the base of the Belgorod State Technological University named after V.G. Shukhov, using equipment of High Technology Center at BSTU named after V.G. Shukhov.

IOP Conf. Series: Materials Science and Engineering **945** (2020) 012074 doi:10.1088/1757-899X/945/1/012074

Calculation of energy parameters of fixed drum concrete mixer with helicoid-type blades

N P Nesmyanov[1]* and M G Matusov[1]

[1]Department of mechanical equipment, Belgorod State Technological University named after V.G. Shukhov, Kostyukov St., 46, Belgorod, 308012, Russia

E-mail: beren1989@yandex.ru

Abstract. Currently, one of the most important areas in the construction industry is the production of high-quality concrete and construction mixes used in the manufacture of reinforced concrete products and structures, road, foundation, monolithic works, as well as used in the laying of bricks, blocks, etc. Concrete mixes are able to ensure high quality of the resulting products and structures when solidified. The scope of their use is the most extensive of all building materials, and production volumes are among the largest. The practice of using concrete in world and domestic construction has shown that it is effective, i.e. it has a number of advantages over new types of building materials, provides high productivity during construction work, is universal and easy to obtain. Initial components for the production of concrete are very common, easily accessible and transportable over long distances, which allows for the production of concrete mix in remote locations using mobile plants. To date, the enterprises of building industry is the production of equipment of various designs and principles of operation used for mixing. In the construction industry, a variety of mixing machines with both advantages and disadvantages have been used in the production of concrete mixes.

1. Introduction

Every year, the volume of production of concrete and construction mixes, as well as the requirements for them, increases. Recently, both in Russia and in other countries [1], various promising developments of mixing equipment used in the production of concrete and construction mixtures have found their application. In modern economic conditions, the issue of efficiency of technological equipment for producing concrete and building mixes, reducing the energy intensity of production, as well as improving the technological properties of finished concrete is very important. The result of the analysis of long-term theoretical and experimental studies has shown that it is possible to increase the efficiency of mixing equipment and increase the uniformity of the resulting construction mixes, while reducing energy consumption and technical and economic costs, due to the rational mode of loading large aggregate into the mixer and creating an intensive circulation of the mixed components under the influence of the blade device. At the present stage of development of construction materials production, when solving problems to improve the mixing efficiency, it is necessary to take into account the properties and behavior of the components to be mixed, as well as to apply modeling of processes occurring in the mixer. To date, the enterprises of building industry is the production of equipment of various designs and principles of operation used for mixing. In the construction industry, a variety of mixing machines with both advantages and disadvantages have been used in the production of concrete mixes. Despite the large number and variety of different designs of mixers used in the production of concrete mixes, still the most common in enterprises are fixed drum mixers with a vertical rotor. The existing mixing equipment of foreign and national manufacturers does not meet the increasing requirements for improving the uniformity of concrete and construction mixes and, consequently, the quality of finished concrete. Mixers that have become widespread in the construction industry have low energy efficiency, a long mixing time and do not provide the required level of quality of the finished product in the modern world. Various methods of intensification used in existing mixers can improve the quality of the finished concrete mix, but at the same time complicate the design of the equipment, reduce its reliability and increase the level of energy consumption. [1, 2, 3].

2. Materials and methods

Analysis of fixed drum rotary mixers for concrete production, as well as their technological and design parameters showed that increasing the efficiency of the mixing process of the initial components can be achieved by improving the design of the mixing apparatus of the machine (blades) and increasing the efficiency of the loading process of the initial components [4, 5, 9, 10, 11, 12].

In order to increase the efficiency of the mixing process, which consists in reducing energy consumption and improving the quality of the finished concrete (strength, uniformity, etc.), the design of a fixed drum rotary mixer with a modified shape of the mixing device blades is proposed.

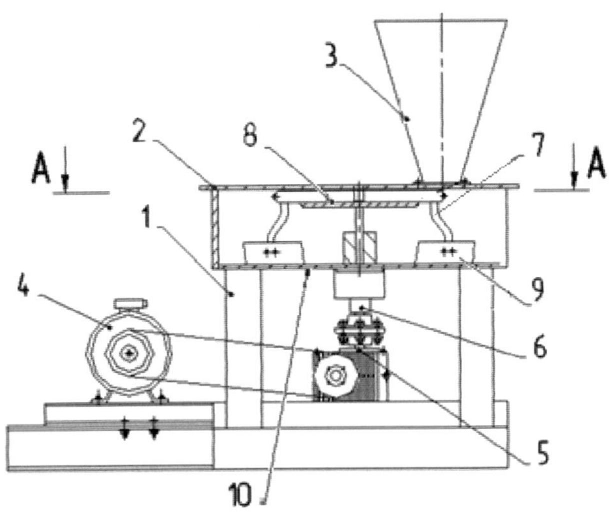

Figure 1. Concrete mixer:
1 - case; 2 - mixing bowl; 3 - boot device; 4 - electric motor; 5 - reducer; 6 - vertical shaft; 7 - bracket; 8 - rotary sprocket; 9 - blade; 10 - slide gate of the unloading device.

In a mixing bowl 2 console, in the bearing assembly, a vertical shaft 6 is mounted. On it there is a rotary sprocket 8, on which, with the help of adjustable brackets 7, the blades of the helicoid type 9 are fixed. The blades have a curved surface, which when moving in the mass of the blend components allows to give the motion of the particles of materials (gravel, sand, cement and water) a spiral trajectory [3, 4]. The angle of the blades varies from 23° to 69°. To load the initial components, a loading device 3 is installed on the mixing bowl. The device is attached to the cover on guides that allow changing the loading location of the source components. In the lower part of the mixing bowl for unloading the finished concrete mix, there is a discharge hatch, which is closed by a slide gate 10.

IOP Conf. Series: Materials Science and Engineering **945** (2020) 012074 doi:10.1088/1757-899X/945/1/012074

The concrete mixer (figure 1) works as follows: the initial components are fed through the loading device 3 installed on the lid of the mixing bowl 2 and get inside the installation, where they fall down under the influence of gravity and accumulate on the bottom of the mixing bowl.

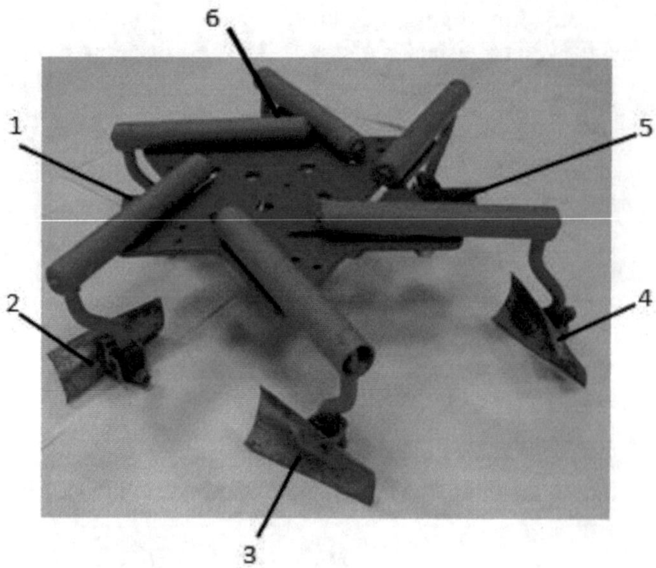

Figuire 2. Concrete mixer mixer: 1, 2, 3, 4, 5, 6 – mixing blades of the helicoid type.

After the end of loading the initial components into the bowl, an electric motor 4 is turned on, which, through a belt transmission, transmits torque to the gearbox 5, which rotates the vertical shaft 6 with a rotary sprocket 8, brackets 7 fixed to it and blades of the helicoid type 9. As a result of rotation of the vertical shaft 6, the blades 9 act on the mixed components, giving them a helical trajectory in the vertical plane, and lift solid and dense particles from the bottom of the mixing bowl. As the blades are installed at different angles to their trajectory, the components to be mixed start moving not only in the vertical plane, but also in the horizontal plane, providing radial-axial circulation from the outer wall of the mixing bowl and back. This allows intensifying the mixing process when getting the finished concrete mix.

Thus, the developed design of the fixed drum rotary mixer will improve the quality of the resulting concrete and construction mixes by giving the mixed components multiple convective trajectories of material particles that intensify mixing of the mixture components in both horizontal and vertical planes inside the mixing bowl of the installation [6, 7, 8].

To determine the power of the mixer, we use the dependency:

$$N_0 = N_{\text{тр}} + N_{\text{вр}} + N_{\text{c}}, W. \tag{1}$$

We calculate the power value $N_{i,mp}$, which must be spent on overcoming the friction force when the concrete mixture moves along the helicoid surface of the i blade:

$$N_{i,\text{тр}} = A_{i,\text{тр}} \cdot \omega_i, W. \tag{2}$$

There $A_{i,mp}$ the amount of work spent on overcoming the friction force when moving along the helicoid surface of the i-th blade. The value of this work is determined by the following ratio:

IOP Conf. Series: Materials Science and Engineering **945** (2020) 012074 doi:10.1088/1757-899X/945/1/012074

$$A_{i,\text{тр}} = \int_{0}^{\pi-\varphi_k} \cdot F_{i,\text{тр}} \cdot dl, J, \tag{3}$$

where dl - element of length equal to:

$$dl = R_0 \cdot d\varphi. \tag{4}$$

The value of the friction force is determined by the ratio:

$$F_{i,\text{тр}} = P_i \cdot f, N, \tag{5}$$

where f – coefficient of friction of the concrete mix on the surface of the blade; P_i – the value of the normal pressure force on the surface of the i-th blade, equal to:

$$P_i = f_{\text{ц},i} + P_{\text{д}}(\varphi), N, \tag{6}$$

$$P_i = M \cdot \omega_i^2 \cdot R_0 + M \cdot g \cdot \cos\varphi, N. \tag{7}$$

Substituting (7) in (5) gives the following expression for the friction force:

$$F_{i,\text{тр}} = f \cdot M \cdot \left(\omega_i^2 \cdot R_0 + g \cdot \cos\varphi\right), N, \tag{8}$$

then:

$$N_{i,\text{тр}} = f \cdot M \cdot R_0 \cdot \omega_i \cdot \left[\omega_i^2 \cdot R_0(\pi - \varphi_k) + g \cdot \sin\varphi_k\right], W. \tag{9}$$

Based on the ratio (9), we find that the amount of power spent on overcoming the friction force of the mixture on the helicoid surface of all the mixer blades will be determined by the ratio of the form:

$$N_{\text{тр}} = \sum_{i=1}^{z} N_{i,\text{тр}}, \tag{10}$$

where z – number of mixer blades.

Substituting (9) into formula (10) allows getting the following results:

$$N_{\text{тр}} = f \cdot \gamma \cdot R_0^3 \cdot l \cdot \left(\pi - \arccos\frac{H_0 \cdot \psi - R_0}{R_0}\right) \cdot \frac{\pi}{S} \cdot \Omega \cdot$$
$$\cdot \sum_{i=1}^{z} R_i \cdot \left[\frac{4 \cdot \pi^2}{S^2} \cdot R_i^2 \cdot \Omega^2 \cdot R_0 \cdot \left(\pi - \arccos\frac{H_0 \cdot \psi - R_0}{R_0}\right) + \right.$$
$$\left. + g \cdot \sqrt{1 - \left(\frac{H_0 \cdot \psi - R_0}{R_0}\right)^2}\right], W. \tag{11}$$

Thus, the obtained ratio (11) shows the amount of power spent on overcoming the friction force of the concrete mixture on the surface of all the mixer blades (figure 3), taking into account their length and the speed of rotation of the rotor.

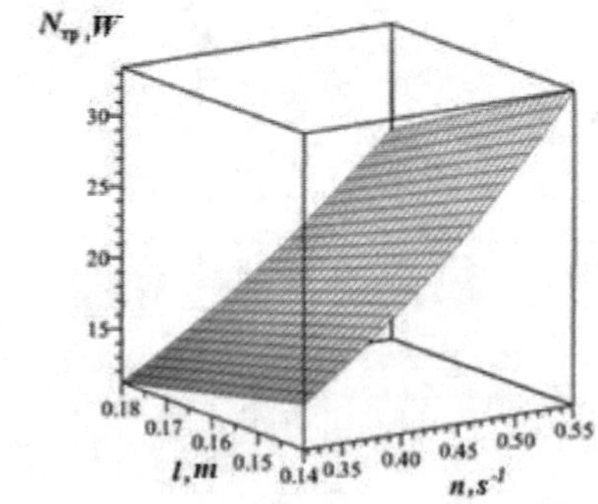

Figuire 3. The graph of the dependence of the power N_{TP}, which must be spent on overcoming the friction force when the concrete mixture moves along the surface of the blade on the blade length l and the rotor speed n.

3. Results

The analysis of the obtained dependence showed that when the rotor speed increases by 1.7 times, the power spent on overcoming the friction force when using blades of the minimum length (0.14 m) increases by 2.6 times. And when using blades of the maximum length (0.18 m) increases by 2.4 times, which is 1.27 times less than when using blades of the minimum length at the maximum speed of the rotor, and for the minimum frequency by 1.4 times.

To calculate the amount of power spent on bringing the concrete mixture into rotational motion on the surface of the i-th blade, we use the following ratio:

$$N_{i,\text{вр}} = T_i \cdot \omega_i, W, \tag{12}$$

where T_i – kinetic energy of rotational motion, the value of which is equal to:

$$T_i = \frac{m \cdot U_i^2}{2}, J, \tag{13}$$

here U_i - the speed modulus, the value of which is determined by the formula.

The total power spent on bringing the concrete mixture into rotational motion along the helicoid surfaces of all the blades will be determined by the ratio:

$$N_{\text{вр}} = \sum_{i=1}^{z} N_{i,tp}. \tag{14}$$

Converting expression (14) results in the following:

$$N_{\text{вр}} = \frac{4 \cdot \pi^4 \cdot R_0^4 \cdot l \cdot \gamma \cdot \Omega^3}{S^3} \left[1 + \left(\frac{S}{2 \cdot \pi \cdot R_0}\right)^2\right] \sum_{i=1}^{z} R_i^3, W. \tag{15}$$

The resulting ratio (15) determines the power spent on bringing the concrete mixture into rotational motion on the surfaces of all the blades, depending on the design and technological parameters. The analysis of the obtained dependence (figure 4) showed that with an increase in the rotor speed from 0.32 s^{-1} to 0.55 s^{-1}, the power spent on bringing the mixture into rotational motion will increase by 1.2 times when using blades with a length of 0.14 m compared to blades with a length of 0.18 m.

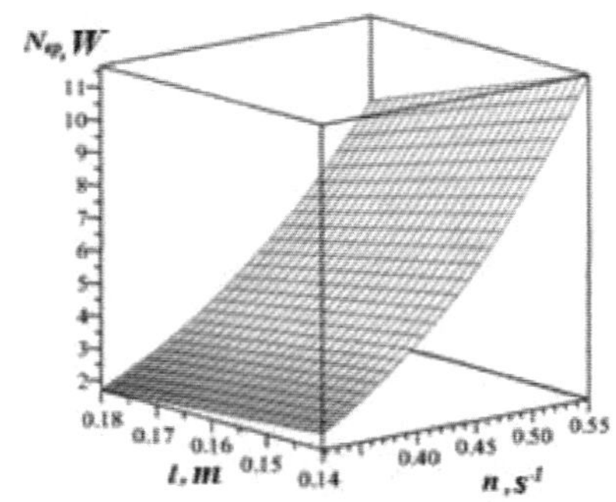

Figuire 4. The graph of the dependence of the power spent on bringing the mixture into rotational motion $N_{вр}$ on the length of the blade l and the rotor speed n.

The amount of power spent on overcoming the shear resistance of the i-th blade of the mixer will be equal to:

$$N_{c,i} = A_i \cdot \Omega, W, \tag{16}$$

$$N_{c,i} = 2 \cdot \pi \cdot \tau \cdot l \cdot \Omega \cdot (2R_{1,i} + l) \cdot R_i, W. \tag{17}$$

Based on (17), the total amount of power spent on overcoming the shear resistance of all blades is determined by:

$$N_c = \sum_{i=1}^{n_0} N_{c,i}, \tag{18}$$

Substituting (17) in (18) we finally can get the following expression:

$$N_c = 2 \cdot \pi^2 \cdot \tau \cdot l \cdot \Omega \cdot \sum_{i=1}^{n_0} (2R_{1,i} + l) \cdot R_i, W. \tag{19}$$

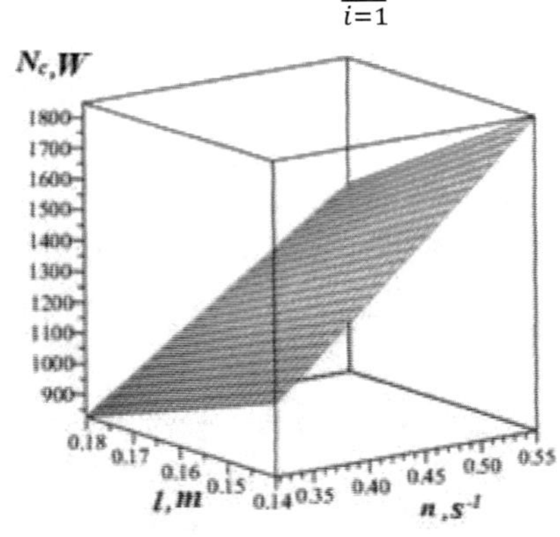

Figure 5. The graph of the dependence of the power N_c, spent on overcoming the force of resistance to shear of the material along the lateral edges of the blade on the blade length l and the rotor speed n.

Analyzing the dependence of the power consumed by the mixer (figure 5) to overcome the shear resistance of the material along the side faces of the blade on the blade length and rotor speed, it can be concluded that when the rotor speed increases from 0.32 s^{-1} to 0.55 s^{-1} when using blades with a length of 0.14 m, the power increases by 1.7 times, reaching 1.85 kW, and by 1.7 times reaching 1.4

kW when using blades with a length of 0.18 m, which is 1.3 times less than with blades with a length of 0.14 m.

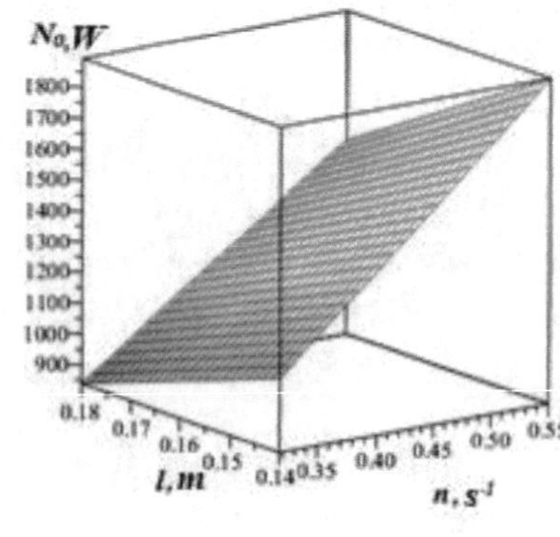

Figure 6. The graph of the dependence of the total power N_0 on the blade on the blade length l and rotor speed n.

Analysis of formula (1) allows concluding that the first two terms are proportional to the mixer speed n in the third power, and the last one is directly proportional to n.

4. Discussion

The obtained dependence (1) allows determining the total power consumed by the mixer depending on its technological and design parameters. Analysis of the dependence of the total power consumed by the rotary mixer when changing the values of the rotor speed and the length of the used blades showed that:

- the minimum power value of 850 W is achieved when using blades with a length of 0.18 m and a rotor speed 0.32 s^{-1};
- the maximum power value of 1.86 kW is achieved when using blades with a length of 0.14 m and a speed of the rotor rotation 0.55 s^{-1};
- at increasing the length of the blades by 1.2 times consumption of the mixer capacity reduced by 1.3 times, while increasing the rotor speed by 1.7 times consumed by the mixer capacity increased by 1.8 times.

5. Summary

Thus, a ratio is obtained that allows determining the total power consumed by the mixer depending on its technological and design parameters. Its analysis shows that:

- the maximum power value of 1.86 kW is achieved when using blades with a length of 0.14 m and a speed of the rotor rotation 0.55 s^{-1};
- the minimum power value of 0.85 kW is achieved when using blades with a length of 0.18 m and a rotor speed 0.32 s^{-1};
- at increasing the length of the blades by 1.2 times consumption of the mixer capacity reduced by 1.3 times, while increasing the rotor speed by 1.7 times consumed by the mixer capacity increased by 1.8 times.

6. References

[1] Bazhenov Yu M, Demyanov V S, Kalashnikov V I 2006 Modified high-quality concrete: scientific publication 368

[2] Bazhenov Yu M, Alimov L A, Voronin V V, Magdeev U Kh 2004 Technology of concrete, building products and structures: textbook 256

[3] Bogomolov A A 2007 Modern mixing equipment for the production of building mixes *Bulletin of the Kharkiv National Automobile and Road University* **38** 31-35

[4] Gustov Yu I, Kaytukov B A, Mishchagin D P 2016 Increasing the wear resistance of the working body of a rotary concrete mixer *Mechanization of construction* **77(4)** 39-41

[5] Gustov Yu I, Kaytukov B A, Grigoryev M N. 2016 Improving the efficiency of a fixed drum rotary concrete mixer *Mechanization of construction* **77(11)** 26-29

[6] Matusov M G, Nesmeyanov N P, Florinsky V V. 2016 Calculation of the capacity of a concrete mixer with a new form of blades *Bulletin of BSTU named after V.G. Shukhov* **7** 100-105

[7] Nesmeyanov N P, Starchik Yu Yu, Matusov M G. 2017 Analytical studies of changes in the concentration of crushed stone in a rotary concrete mixer taking into account the design features of the blades and the loading device *Construction and road machines* **11** 22-28

[8] Nesmeyanov N P, Pochupaylo B I, Dmitrienko V G, Brazhnik Yu V, Matusov M G. 2016 Methods for calculating the rational parameters of the working bodies of forced mixers *Fundamental research* **2** 503-508

[9] Selivanov Yu T. 2005 Methods of calculation and improvement of designs of circulating mixers that provide a given quality of the mixture: dis. ... Cand. of Eng. Sciences: 05.02.13 336

[10] Telichenko V I, Kaytukov B A, Skel V I. 2017 On the issue of productivity of rotary concrete mixers *Bulletin of the Belgorod State Technological University named after V.G. Shukhov* **2** 178-182

[11] Telichenko V I, Kaytukov B A, Skel V I. 2017 Determination of optimal speeds of movement of rotor concrete mixer blades *Bulletin of the Belgorod State Technological University named after V.G. Shukhov* **3** 80-84

[12] Moreno Juez J. 2017 Monitoring of concrete mixing evolution using image analysis *Powder technology* **305** 477-487

Acknowledgments
The work is realized in the framework of the Program of flagship university development on the base of the Belgorod State Technological University named after V.G. Shukhov, using equipment of High Technology Center at BSTU named after V.G. Shukhov.

IOP Conf. Series: Materials Science and Engineering **945** (2020) 012075 doi:10.1088/1757-899X/945/1/012075

A novel approach for coverings transformation to define a robot workspace

L A Rybak[1*], L Behera[2], M A Averbukh[1] and A V Sapryka[1]

[1]BSTU named after V.G. Shukhov, 46, Kostyukova street, Belgorod, 308012, Russia
[2]Indian Institute of Technology Kanpur, Kanpur- 208016, India

E-mail: rl_bgtu@intbel.ru

Abstract. The article discusses the application of optimization algorithms to solve the problem of determining the workspace of robots. The development of a numerical method for approximating the set of solutions of a system of nonlinear inequalities that describe the restrictions on the geometric parameters of a robot based on the concept of non-uniform coverings is considered. An approach is proposed that allows you to reduce the number of boxes and numbers describing each of the boxes. Reducing the number of boxes is achieved by combining the boundary boxes of the covering set and the area between them along one of the axes. The dimensions along the other axes are equal for all boxes, which allows them to be described by one number. The transition to the integer space is described. Thus, converting non-uniform covering sets to a partially ordered set of integers reduces computational complexity. An algorithm for converting boxes of a covering set is presented. The approach has been tested for a 3-RPS robot.

1. Introduction

Workspace is one of the most important characteristic of robots. The issues of structural synthesis, methods for studying the workspace, movement trajectory optimization of parallel robots are discussed in detail in [1-4]. The task of the workspace determination can be solved using various methods, in particular, deterministic ones. The actual problem for applying these methods is often significant computational complexity. One of the most effective deterministic methods is the non-uniform coverings method. The application of this method to determine the workspace is considered in works [5-7]. In [8], a comparison of two approaches is considered, one of which is based on the use of a system of inequalities to describe the structural constraints of the robot, and the other on the use of a system of equations.

One of the tools for implementing the method is iterative bisection. With each division, the box decreases by 2 times, respectively, the ratio of the sizes of the initial box and one of the boxes forming an approximation of the workspace can be estimated by the degree of division d, and the ratio itself is 2^d. With an increase in the degree of division d, the number of boxes, the combination of which describes the workspace, increases. Due to the increase in computational complexity for processing an approximated workspace of higher accuracy, the problem arises of reducing it by converting the resulting set of boxes describing the workspace. As part of this work, an approach is proposed for converting a covering set obtained using the method of non-uniform coverings into a partially ordered set of integers. It includes two components: reducing the number of boxes and the transition from the space of real to the space of integers. Application of the proposed approach and assessment of its effectiveness is considered on the problem of determining the workspaces of the 3-RPS robot.

2. Materials and methods

Consider the set of boxes which described the workspace. Using the system of inequalities $g_j(x) = 0, j \in 1, m$, the workspace is described by the union of two sets $Q_E = Q_I \cup Q_J$, where Q_I is an internal approximation set that is included in the set of solutions to the system of inequalities, Q_J is the boundary set. For each P_i from Q_I, the condition

$$\max_{j=1,..,m} \max_{x \in P_i} g_j(x) \leq 0. \quad (1)$$

For each P_i from Q_J, the system of inequalities is fulfilled:

$$\begin{cases} \max_{j=1,..,m} \max_{x \in P_i} g_j(x) > 0 \\ \max_{j=1,..,m} \min_{x \in P_i} g_j(x) \leq 0. \\ d(P_i) \leq \delta \end{cases} \quad (2)$$

We introduce the following notation: ∂Q_E is the boundary of the set Q_E, $Q_A = \partial Q_E \cup (\mathbb{R}^n \backslash Q_E)$ is the outer region for which the condition

$$\max_{j=1,..,m} \min_{x \in P_i} g_j(x) > 0. \quad (3)$$

The elements of the set Q_J are necessarily located between the elements of the sets Q_I and Q_A. Therefore, it is possible to describe the workspace as a set of boxes using the following approach. We introduce the following notation for the boundaries of the box:

$$\underline{x_i} \leq x_i \leq \overline{x_i}, i \in 1, n. \quad (4)$$

We denote two subsets in the set Q_J as Q_{J1} and Q_{J2}. The subset Q_{J1} includes only those boxes at the boundary of which there is a point $x = (x_1, .., x_n)$ for which the system is satisfied:

$$\begin{cases} x \in \partial Q_E \\ x_1 = \underline{x_1} \\ \underline{x_i} < x_i < \overline{x_i}, i \in 2, n \end{cases} \quad (5)$$

The condition for the subset Q_{J2} is similar, but for it $c_1 = \overline{x_1}$.

It should be noted that the number m of boxes in the subsets Q_{J1} and Q_{J2} is equal to each box $P_{k,J1}, k \in 1, m$ from Q_{J1} corresponds to the box $P_{k,J2}, k \in 1, m$ from Q_{J2} with equal values $\underline{x_i}, i \in 2, n$ and $\overline{x_i}, i \in 2, n$, while for all points $x = (x_1, .., x_n)$ for which the condition $\underline{x_{i,J1}} < x_i < \underline{x_{i,J2}}$ is satisfied, $x \in Q_E$ is true. Q_F is the set of boxes and $P_{k,F} = \left\{x: \underline{x_{i,J1,k}} \leq \underline{x_i} \leq \overline{x_{i,J2,k}}, i \in 1, n\right\}, k \in 1, m$.

The n-dimensional box is described by 2n real numbers. The proposed approach to the conversion of boxes allows us to describe boxes with a smaller quantity of numbers, while integers. First, we consider this concept in the general case of converting the set of real numbers Y into the set of integers Z with the approximation accuracy δ (Fig. 1). For each of the set points, its coordinates in the space of integers are calculated:

$$x_i^{(j)\prime} = \left[\frac{x_i^{(j)}}{\delta}\right], x^{(j)} \in \mathbb{R}, x^{(j)\prime} \in \mathbb{Z} \quad (6)$$

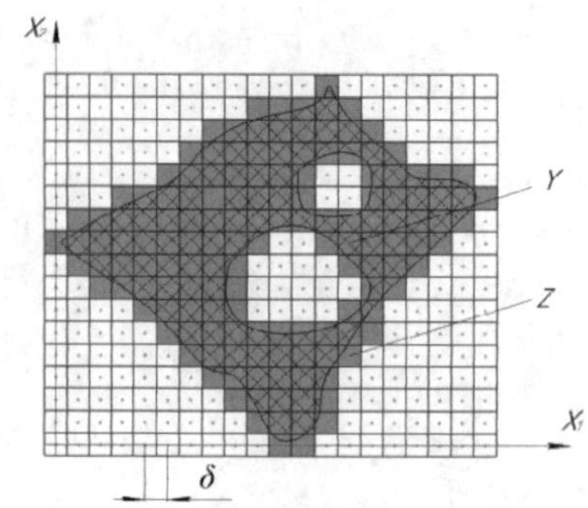

Figure 1. The transformation of the set into the space of integers.

Moreover, the Hausdorff distance between the sets depends on δ:

$$0 < h(Y,Z) \leq \sqrt{\sum_i |\delta_i|^2} \tag{7}$$

The integers of one of the coordinates are likewise combined into intervals, for each of which the values of the remaining coordinates are equal. Let us consider the application of this concept to the transformation of the set of boxes Q_F to Q_Z. For Q_F and Q_Z the following holds:

$$h_{max}(Q_F, Q_Z) = \sqrt{\sum_i |\delta_i|^2} \tag{8}$$

The accuracy of determining the workspace allows us to estimate the Hausdorff distance between the sets Q_Z and Y (Fig. 2a). The maximum distance $h_{max}(Y, Q_Z)$ is defined as

$$h_{max}(Y, Q_Z) = h_{max}(Y, Q_F) + h_{max}(Q_F, Q_Z) \tag{9}$$

In order to reduce the Hausdorff distance, we modify the formula (7):

$$x_i' = \left[\frac{2^d x_i + k_i}{\overline{x}_i^{(0)} - \underline{x}_i^{(0)}}\right], x_i \in \mathbb{R}, x_i' \in \mathbb{Z} \tag{10}$$

where k_i- bias coefficients, which are determined by the formula:

$$k_i = \left(\left[\frac{2^d \underline{x}_i^{(k)}}{\overline{x}_i^{(0)} - \underline{x}_i^{(0)}} + 0.5\right] - 0.5\right)\left(\frac{\overline{x}_i^{(0)} - \underline{x}_i^{(0)}}{2^d}\right) - \underline{x}_i^{(k)}, i \in \overline{1, n} \tag{11}$$

In this case, $h_{max}(Q_F, Q_Z) = 0$ (Fig. 2b).

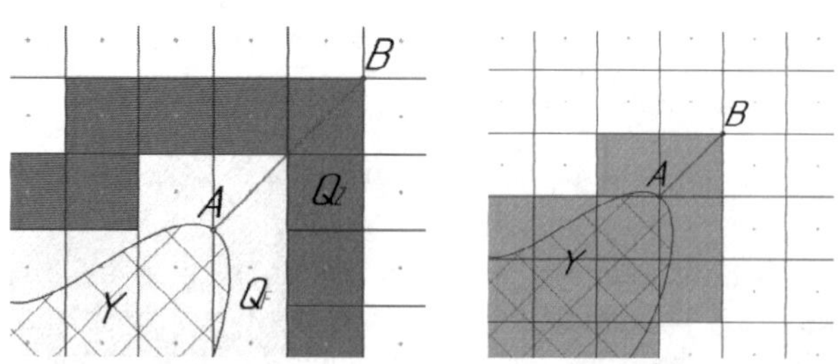

a b

Figure 2. The Hausdorff distance: a) between the sets Y and Q_Z, b) between the sets Y and Q_Z taking into account the displacement coefficients.

We introduce the following variables:

$$\underline{a_i} = \left[\frac{2^d \underline{x_i} + k_i}{\overline{x}_i^{(0)} - \underline{x}_i^{(0)}}\right], \overline{a_i} = \left[\frac{2^d \overline{x_i} + k_i}{\overline{x}_i^{(0)} - \underline{x}_i^{(0)}} - 0.5\right] \tag{12}$$

It is worth noting that an additional coefficient of 0.5 is added to exclude rounding of the upper boundary value to the next integer.

We synthesize a box conversion algorithm for n = 2, taking into account formulas (10) - (12):

IOP Conf. Series: Materials Science and Engineering **945** (2020) 012075 doi:10.1088/1757-899X/945/1/012075

Algorithm 1.
Input: Q_E.
$N_1 = 0$
for $k = 1, 2, \dots, N$:
Compute $k_1, k_2, \underline{a_1}, \overline{a_1}, \underline{a_2}, \overline{a_2}$,
for $a_1 = \underline{a_1}, \underline{a_1} + 1, \dots, \overline{a_1}$:
if ($N_1 = 0$) **then**:
$x_1^{(1)'} = a_1$, $\overline{x_2^{(1)(1)'}} = \overline{a_2}$, $\underline{x_2^{(1)(1)'}} = \underline{a_2}$ $N_1 = 1$, $N_2^{(1)} = 1$
else
for $j_1 = 1, 2, \dots, N_1$:
if ($a_1 = x_1^{(j1)'}$) **then**:
for $j_2 = 1, 2, \dots, N_2^{(j_1)}$:
if ($\overline{a_2} < \underline{x_2^{(j_1)(j_2)'}}$) **then**
if ($j_2 = 0$) **then**
if ($\underline{x_2^{(j_1)(j_2)'}} = \overline{a_2} + 1$) **then**
$\underline{x_2^{(j_1)(j_2)'}} = \underline{a_2}$
else
for $i = 0, 1, \dots, N_2^{(j_1)}$:
$\overline{x_2^{(j_1)\left(N_2^{(j_1)}-i+1\right)'}} = \overline{x_2^{(j_1)\left(N_2^{(j_1)}-i\right)'}}$
$\underline{x_2^{(j_1)\left(N_2^{(j_1)}-i+1\right)'}} = \underline{x_2^{(j_1)\left(N_2^{(j_1)}-i\right)'}}$
end for
$\underline{x_2^{(j_1)(j_2)'}} = \underline{a_2}$, $\overline{x_2^{(j_1)(j_2)'}} = \overline{a_2}$
$N_2^{(j_1)} = N_2^{(j_1)} + 1$
end if
else
if ($\overline{x_2^{(j_1)(j_2-1)'}} = \underline{a_2} - 1$) and ($\underline{x_2^{(j_1)(j_2)'}} = \overline{a_2} + 1$) **then**
$\overline{x_2^{(j_1)(j_2-1)'}} = \overline{x_2^{(j_1)(j_2)'}}$
for $i = j_2, 1, \dots, N_2^{(j_1)} - 1$:
$\overline{x_2^{(j_1)(i)'}} = \overline{x_2^{(j_1)(i+1)'}}$,$\underline{x_2^{(j_1)(i)'}} = \underline{x_2^{(j_1)(i+1)'}}$
end for
$N_2^{(j_1)} = N_2^{(j_1)} - 1$
else if ($\overline{x_2^{(j_1)(j_2-1)'}} = \underline{a_2} - 1$) **then**
$\overline{x_2^{(j_1)(j_2-1)'}} = \overline{a_2}$
else if ($\underline{x_2^{(j_1)(j_2)'}} = \overline{a_2} + 1$) **then**
$\underline{x_2^{(j_1)(j_2)'}} = \underline{a_2}$
else
for $i = 0, 1, \dots, N_2^{(j_1)} - j_2$:

IOP Conf. Series: Materials Science and Engineering **945** (2020) 012075 doi:10.1088/1757-899X/945/1/012075

$\overline{x_2^{(j_1)(N_2^{(j_1)}-i+1)'}} = \overline{x_2^{(j_1)(N_2^{(j_1)}-i)'}}$
$\underline{x_2^{(j_1)(N_2^{(j_1)}-i+1)'}} = \underline{x_2^{(j_1)(N_2^{(j_1)}-i)'}}$
end for
$\underline{x_2^{(j_1)(j_2)'}} = \underline{a_2}, \overline{x_2^{(j_1)(j_2)'}} = \overline{a_2}$
$N_2^{(j_1)} = N_2^{(j_1)} + 1$
end if
end if
go to NEXT
else if $(\underline{a_2} > \overline{x_2^{(j_1)(j_2)'}})$ and $(j_2 = N_2^{(j_1)})$ **then**
$\underline{x_2^{(j_1)(j_2+1)'}} = \underline{a_2}, \overline{x_2^{(j_1)(j_2+1)'}} = \overline{a_2}$
$N_2^{(j_1)} = N_2^{(j_1)} + 1$
go to NEXT
end if
end for
else if $(a_1 < x_1^{(j1)'})$ **then**:
for $i = 0, 1, \ldots, N_1 - j_1$:
$x_1^{(N_1-i+1)'} = x_1^{(N_1-i)'}$,
for $j_2 = 1, 2, \ldots, N_2^{(N_1-i)}$:
$x_2^{(N_1-i+1)(j_2)'} = x_2^{(N_1-i)(j_2)'}$,
end for
$N_2^{(N_1-i+1)} = N_2^{(N_1-i)}$
end for
$x_1^{(j_1)'} = a_1, \overline{x_2^{(j_1)(1)'}} = \overline{a_2}, \underline{x_2^{(j_1)(1)'}} = \underline{a_2}$
$N_1 = N_1 + 1, N_2^{(j_1)} = 1$
else if $(a_1 > x_1^{(j1)'})$ and $(j_1 = N_1)$ **then**:
$x_1^{(N_1+1)'} = a_1, \overline{x_2^{(N_1+1)(1)'}} = \overline{a_2}, \underline{x_2^{(N_1+1)(1)'}} = \underline{a_2}$
$N_1 = N_1 + 1, N_2^{(N_1+1)} = 1$
end if
end for
end if
end for
NEXT
end for

As the dimension increases, we add external loops and similar to actions with the x_1 coordinate.

3. Results and discussion

Consider the application of the method of non-un iform coverings to determine the workspace of one of the robot modules - a planar 3-RPS mechanism (Fig. 5), which consists of three kinematic chains containing variable-length rods pivotally attached to a fixed base at the vertices of an equilateral triangle (Fig. 3) The other ends of the rods are pivotally mounted at the vertices of an equilateral triangle on a movable platform. The input coordinates are the rod lengths (l_1, l_2, l_3), the output coordinates are the position of the geometric center of the moving platform in Cartesian coordinates (x, y) associated with the center of the base of the mechanism, and its rotation angle (φ) relative to the axis perpendicular to the plane of the base. R and r are the radii of circles describing triangles and, respectively.

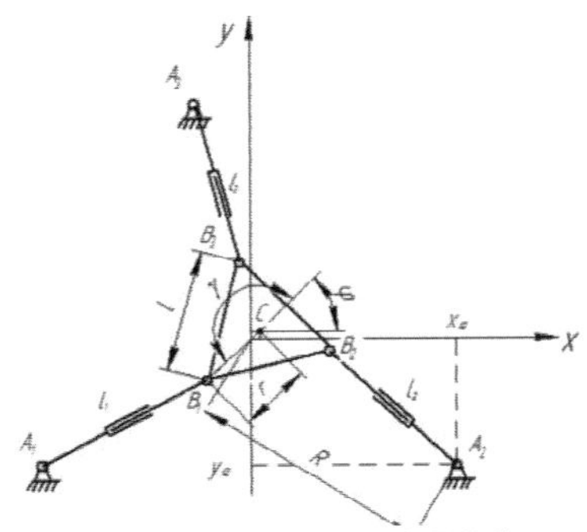

Figure 3. Scheme of a planar 3-RPS mechanism.

Define the workspace of the 3-RPS mechanism. To do this, we introduce restrictions on the geometric parameters of the mechanism

$$l_{min} \le l_i \le l_{max} \tag{13}$$

where l_{min}, l_{max} are determined by the design parameters of the mechanism, l_i is the current length of the i-th rod,

If the points A_i and B_i are located at the vertices of equilateral triangles, then the change in the length of the rods is determined by the formulas

$$l_1^2 = \left(x + 0.5r(\sin\varphi - \sqrt{3}\cos\varphi) + 0.5\sqrt{3}R\right)^2 + \left(y - 0.5r(\sqrt{3}\sin\varphi + \cos\varphi) + 0.5R\right)^2 \tag{14}$$

$$l_2^2 = \left(x + 0.5r(\sin\varphi + \sqrt{3}\cos\varphi) - 0.5\sqrt{3}R\right)^2 + \left(y + 0.5r(\sqrt{3}\sin\varphi - \cos\varphi) + 0.5R\right)^2 \tag{15}$$

$$l_3^2 = (x - r\sin\varphi)^2 + (y + r\cos\varphi - R)^2 \tag{16}$$

Algorithms for approximating the set of solutions of nonlinear inequalities were considered earlier in the authors' work [7]. To speed up the calculations, multithreaded calculations using the OpenMP library are used. This is considered in more detail in [8].

We use the proposed approach to transforming the covering set. In fig. 4a shows a visualization of the workspace symmetrical about the Y axis. The left half is described by the sets Q_I and Q_J, the right half is described by the set Q_F. The simulation results for $R = 400$ mm, $r = 50$ mm, $l_{1,2,3} \in [200\ \text{мм},\ 500\ \text{мм}]$, $\varphi = 0°$ are presented in Fig. 4b. The calculation time for approximation accuracy $\delta = 0.01$ mm, the grid dimension for calculating 8x8 functions using parallelization of computations into 8 flows on a personal computer was 6 seconds.

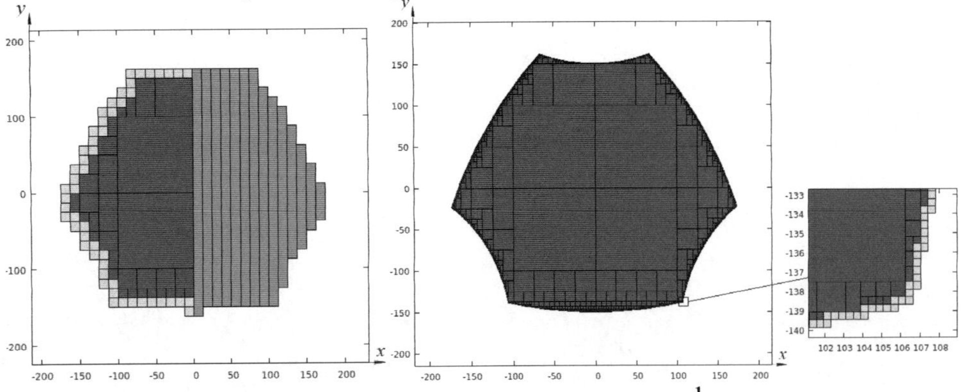

a b

Figure 4. The 3-RPS mechanism workspace: a) before and after the transformation of the covering set, b) fixed angle φ = 0 °: the blue area is the internal approximation, the yellow is the boundary.

IOP Conf. Series: Materials Science and Engineering **945** (2020) 012075 doi:10.1088/1757-899X/945/1/012075

With a degree of division of 2^{12}, the number of boxes for two-dimensional space decreased by 86.7% from 3010 to 400, for three-dimensional - by 80.5% from 6 991 738 to 1 363 683. At the same time, boxes of the covering set are described by fewer numbers. For three-dimensional space, the set of numbers describing the workspace decreased by 90.2% from 41 950 428 to 4 099 241.

4. Summary

The proposed approach to the transformation of sets to reduce the number of covering boxes describing the workspace has shown its effectiveness. The application of the approach allowed reducing the quantity of numbers describing the 3-RPS mechanism workspace by 90.2%. It is shown that the approach can effectively optimize the workspace of various dimensions. It can be applied to optimize the workspace covering of various robots. It will reduce the computational complexity of visualization and other calculations that use the workspace, including trajectory planning, the safe relative positioning of robots as part of a multi-robot system, determination of a joint workspace for relative manipulation devices, and many others.

5. References

[1] Kong H, Gosselin C M 2007 Type Synthesis of Parallel Mechanisms *Springer* 272
[2] Merlet J-P 2007 Parallel Robots *Springer* 269-285
[3] Aleshin A K, Glazunov V A, Rashoyan G V, Shai O 2016 Analysis of kinematic screws that determine the topology of singular zones of parallel-structure robots *Journal of Machinery Manufacture and Reliability* **45(4)** 291-296
[4] Srinivasa Rao P, Mohan Rao N 2013 Position Analysis of Spatial 3-RPS Parallel Manipulator *International Journal of Mechanical Engineering and Robotics Research* **2(2)** 80-90
[5] Zeng Q, Ehmann F K, Cao J 2016 Tri-pyramid Robot: stiffness modeling of a 3-DOF translational parallel Manipulator *Robotica* **34(2)** 383-402
[6] Virabyan L G, Khalapyan S Y, Kuzmina V S 2018 Optimization of the positioning trajectory of planar parallel robot output link *Bulletin of BSTU named after V.G. Shukhov* **9** 106–113
[7] Posypkin M 2019 Automated Robot's Workspace Approximation *Journal of Physics: Conference Series* **1163(1)** 012050
[8] Evtushenko Y, Posypkin M, Rybak L, Turkin A 2018 Approximating a solution set of nonlinear inequalities *Journal of Global Optimization* **71(1)** 129-145
[9] Malyshev D, Posypkin M, Rybak L, Usov A 2019 Approaches to the Determination of the Working Area of Parallel Robots and the Analysis of Their Geometric Characteristics *Engineering Transactions* **67(3)** 333-345
[10] Malyshev D I, Posypkin M A, Gorchakov A Yu, Ignatov A D 2019 Parallel algorithm for approximating the work space of a robot *International Journal of Open Information Technologies* **7(1)** 1–7

Acknowledgments
The research was funded by RFBR within the research project No 18-57-45014 and Science Comittee, the Ministry of Education and Science, Republic of Kazakhstan, project no. AP05133190.

IOP Conf. Series: Materials Science and Engineering **945** (2020) 012076 doi:10.1088/1757-899X/945/1/012076

Technopark and its role in improving the competitiveness of the country's economy: current issues in Russia and Iran

Nasser Safaie[1], M A Romanovich[1,2*], L G Romanovich[3] and I V Yarmolenko[3]

[1]Faculy of Industrial Engineering Dep. in K. N. Toosi University of Technology, Iran
[2]Department of Hoist transport and road machines, Belgorod State Technological University named after V.G. Shukhov, Kostyukov St., 46, Belgorod, 308012, Russia
[3]Department of Economics and organization of production, Belgorod State Technological University named after V.G. Shukhov, Kostyukov St., 46, Belgorod, 308012, Russia

E-mail: romanovich.ma@bstu.ru

Abstract. This article presents the recent problems of infrastructure support for innovative enterprises that topical for scientists, public authorities and business representatives. The authors consider modern innovation policy as a regulator of innovation processes, consider the experience of organizing technoparks in foreign countries and the features of Russian technoparks, give the dynamics of their creation in Russia, and analyze the structure of domestic technoparks depending on their specialization and affiliation. The structure of technoparks is analyzed depending on its specialization, as well as depending on the owners. Actual problems of the organization and activity of technoparks in Russia, including on the basis of higher education institutions, are highlighted and generalized. Noting the important role of new-generation technoparks in improving the competitiveness of the country's economy, practical recommendations for their development in modern conditions are given. The recommendations are universal and can be applied in the practical activities of authority, representatives of technoparks, universities and subjects of innovative business in Russia, Iran and other states, aimed at developing their own developments and their introduction into production. It can contribute to increase the competitiveness of the country's economy.

1. Introduction

The problems of technoparks organization and improving the efficiency of its existing are more widely used. In accordance with the operational strategies in the field of innovation, this implies that Russian Federation, in fact, like the other countries, is aimed at repeating successful international experience.

However, the research results have shown that successful technology parks in economically developed countries were created, as a rule, at large research centers or corporations, gradually turning into mega technoparks and agglomerations of innovative cities.

Due to various features such as: historical, socio-economic, regional, legal and others, domestic technology parks are characterized by low efficiency.

Recently, in the context of the economic crisis and the increasing role of innovative business in ensuring the socio-economic development of the country, attention of scientists, government bodies and

IOP Conf. Series: Materials Science and Engineering **945** (2020) 012076 doi:10.1088/1757-899X/945/1/012076

business representatives is being increasingly focused on the study of infrastructure support for innovative enterprises problems.

The study the certain matters of the development of the national innovation system, including technoparks in different periods until the exacerbation of the global economic crisis, the works of Romanovich L, Romanovich M [7,9,10,21] and other authors focus on the problems of University innovative companies and University technoparks, the work by Sizova Yu[8] studies the matters of technoparks in the regional aspect, Lytaev N [12], conducted the approach analysis to defining success of a tech park and the main factors influencing it.

Meanwhile, the authors' analysis of the technoparks development problems in Russia under modern conditions of the global economic crisis and their solutions will help to fill the existing gap in terms of scientific-practical issues of strategic development of the Russian economy. Based on the above, the relevance of this research is being formed

At the present time, a technopark represents an element of the national innovation system in developed countries and is a kind of "barometer" of innovative activity in regions and countries.

In these countries, various systems of innovative development have been formed and significant capital is invested in innovative business. The top priority is to realize a profit on the introduction of advance developments and technologies. According to research by Khaikina E, Akmaeva R, Marenkov N [1-3] the government has an urgent need to choose such development systems where small innovative enterprises that bring significant revenues to the budget of countries act as guides.

It is arguable that in these countries that have chosen an innovative path of development there are effective partnerships between the three leading institutional sectors: science – business – government.

Technoparks served as infrastructure elements and promoted the transition to a knowledge economy as the most effective form of integrating science and business and stimulating the country's economic development.

Considering the foreign experience of organizing technoparks and the special aspects of Russian technoparks, it is worth distinguishing stages of their organization and development.

2. Materials and methods

2.1. Materials

The first technoparks were organized in the late 1950s at the universities of Stanford, Cambridge, and other cities in the United States and Great Britain. This period is characterized by the formation of university-based platforms that involve inventors and researchers in solving practical problems. Many international companies in the field of high technologies were set up on the basis of the first technoparks. Then the first concepts of founding Science Cities and Technopolises were born.

Around the first technoparks there was the forming agglomeration of innovative cities such as Silicon Valley. As for our country, the USSR successfully implemented the concept of developing Science Cities.

As an example, Novosibirsk Akademic town, founded in 1959, has focused on its base the scientific-research and design institutes of various industry areas. Iran's first technopark was founded in 2001.

In our opinion, we can distinguish the following differences and common features of scientific centers of the Soviet period and western technoparks:

- The differences are that domestic research centers were not focused on market demand, as Western technology parks;
- A common feature - performance result of the first technoparks is innovative products;
- It is also common that technoparks are aimed at introducing accumulated scientific knowledge into production (business), the owners of which were scientists, researchers, universities or various innovative companies. At the present time, these parks remain the predominant form of industrial parks.

Technoparks of the 70s and 80s (the second stage) are characterized by a focus on commercialization of scientific and research developments by providing space for rent and providing services to innovative enterprises.

At that time, the concept of creation mega technoparks in countries with expanding economy (for example, Asia) also began to be implemented. The special aspect of this period is the large scale of the technoparks construction and the intensive growth in the number of specialized technoparks.

Since the early 1990s (the third stage), technoparks are gradually transformed into platforms for communication. Various professional associations of researchers, industry experts, and venture financiers have been actively developing. During this period, technoparks expand their communication and virtual opportunities and increasingly become organizational structures. One of the main performance indicators of technoparks is the number of events held on its basis and with its participation.

Figure 1 shows that the number of operating technoparks in Russia and Iran increased over the period from 2001 to 2019. Russia has seen a significant increase in technoparks since 2010, while in Iran the growth in the number of technology parks is not so fast. In Russia, the maximum number of operating technoparks fell on 2016. During the period from 2016 to 2019 the number of technoparks in Russia decreased to 157, while in Iran it increased from 32 to 42.

It should be emphasized that the number of 157 Russian technoparks includes 22 technoparks organized in 16 regions of the country with the volume of state support of 4.4 billion rubles (according to The Ministry of Economic Development and Trade data).

The results of the study show that in our opinion, this does not solve the current needs of the national economy, which is developing along the innovation-based development and setting ambitious goals.

In addition can be summarized as the present day the planned funding of 30 billion rubles until 2024 in Russia within the context of the national project to support Small and medium-sized enterprises for the technoparks development is insufficient to ensure a technological breakthrough in the conditions of the global crisis, technological shifts and competition. The number of technoparks created in Russia is at the level of 2011, however in Iran there is a steady trend in the growth of technoparks.

2.2. Methods

Our analysis of the economic literature [1-4, 6, 9-10, 16-20], shows that is established that only 2% of technoparks that were created more than 10 years ago are currently operating effectively and successfully in Russia. Thus, remaining 98% continue to be "at an early stage of activity", and it should be noted that only a small number of them have the necessary infrastructure, laboratories, high-tech equipment, and qualified personnel. We consider it important to note the fact that only 10% of resident companies from among the functioning technoparks received a set of necessary services in the field of business acceleration, assistance in attracting investment, mentoring support, etc.

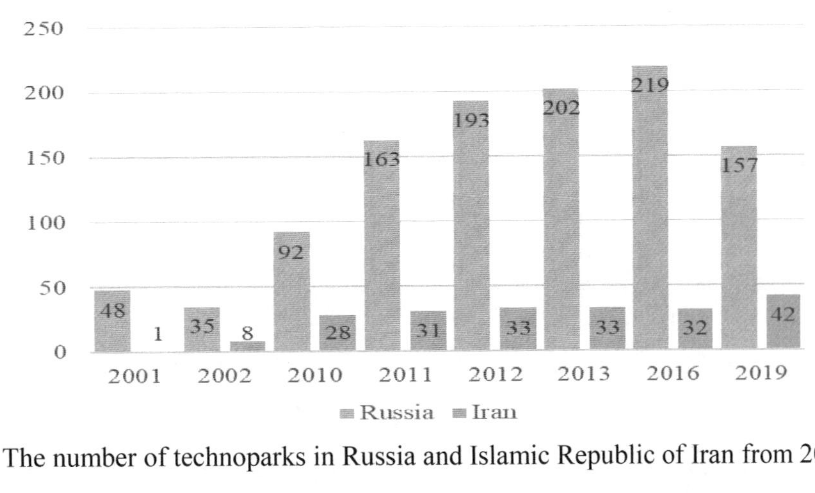

Figure 1. The number of technoparks in Russia and Islamic Republic of Iran from 2001 to 2019.

IOP Conf. Series: Materials Science and Engineering **945** (2020) 012076 doi:10.1088/1757-899X/945/1/012076

If we consider the structure of Russian technoparks depending on their specialization (figure 2), it is obvious that the majority of them are multi-industry technoparks (31%) and technoparks in the sphere of information technologies (29%).

Minimal activity of technoparks was found in the field of research and development (8%) and agriculture (1%) in the general structure of technoparks. In our opinion, the referred structure of technoparks, depending on their specialization, cannot be called rational and adequate to the public needs.

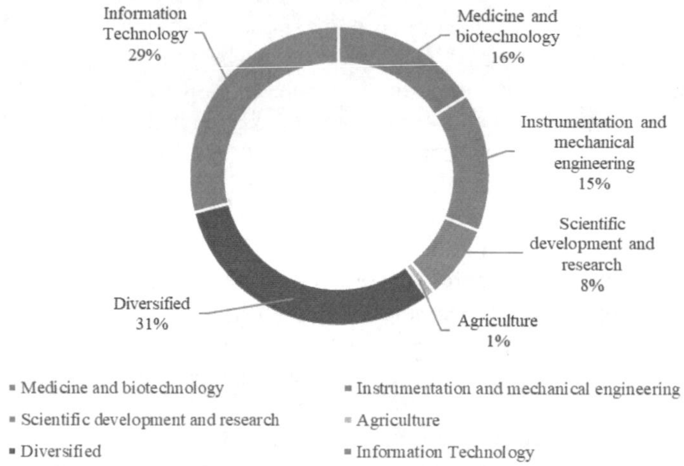

Figure 2. Specialization of Russian technoparks.

Thus, the US accounts for more than 20% of the innovation market, while Russian manufacturers account for less than 1 %. The measures carried out on the basis of domestic technoparks are extremely insufficient to achieve a synergistic effect in strengthening the relationship between the Russian academic community and company officials.

The founders mainly determine the potential opportunities and funding level for all the project stages of technoparks implementation. If we consider the ownership structure of technoparks (figure 3), it may be concluded that the regional administrations dominate as founders, which account for 39% of technoparks. The administrations of universities that have organized and maintain infrastructure facilities, usually on their own basis, account for 26% of technoparks. Individuals and companies 22% of technoparks. And the minimum amount (13%) falls on municipal administrations.

IOP Conf. Series: Materials Science and Engineering **945** (2020) 012076 doi:10.1088/1757-899X/945/1/012076

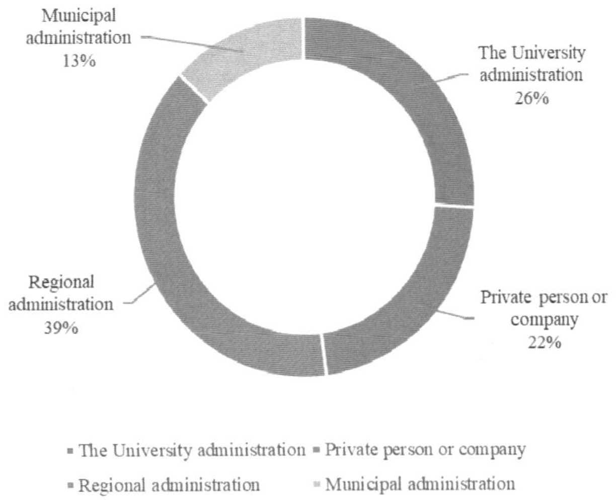

Figure 3. Owners of Russian technoparks.

In our opinion, the existing structure of technoparks depending on the owners is not exactly rational, since in a market economy based on knowledge, private individuals or companies and University administrations should prevail in the structure of owners.

As a result, in our prospects in this sense, if the present practice does not go beyond its current narrow framework of "servicing the founders needs", then they seem quite modest.

The results of our study show that modern conditions there are changing trends of technopark movement development: rapid development of innovative technologies against global competition, the cost escalation of construction and maintenance of parks, decrease in intangible assets and financial opportunities of small innovative enterprises in global environment.

3. Results

From the results of the analysis, it follows:

1. The domestic technoparks development is a necessary and major element in the formation and development of the national innovation system in order to ensure an innovative breakthrough under conditions of economic crisis and globalization.

2. Currently, a certain base of support for domestic science and education has been created. However, a systematically developed integrated state scientific and technical policy covering them, as well as the development and transfer of technologies, scientific-technological production modernization within the Russian national scale has not yet been finalized.

3. In our opinion, the system of state support should be aimed at stimulating scientific research, developing domestic high-tech and knowledge-intensive technologies, spreading technology transfer, strengthening scientific and technical human resources, increasing the prestige value of science, forming scientific elites, ensuring priority growth of science financing by attracting various sources, creating and developing new organizational forms and mechanisms of interaction with production (technopolises, tech cities, academic and technological valleys, technology and innovation zones, industrial-and-scientific park, etc.).

4. The problem of technoparks resides in experience reproduction of traditional technoparks organized in the past in other economic and social conditions and market opportunities. The value and success of modern technoparks of the new generation directly depends on the diversity of scientific potential. Technopark becomes a place for free communication and interaction of the "creative community" of technology developers and consumers working in the markets of technology free exchange.

IOP Conf. Series: Materials Science and Engineering **945** (2020) 012076 doi:10.1088/1757-899X/945/1/012076

4. Discussion

The information was gathered from literature, newspapers, Internet and by interviewing scientists and researchers [1-4, 6, 11, 16-20] we have established that various studies conducted in this area by Russian and foreign scientists in recent years have shown that infrastructure facilities created on the basis of higher education institutions are considered as the most important elements of stimulating the growth of the regional economy.

In our opinion after analysis shows that currently about 2% of technoparks that are more than 10 years old effectively operate in Russia while the remaining 98% continue to be "at an early stage of activity", and only a few of them have the necessary infrastructure, laboratory, high-tech equipment, and qualified staff. Only 10% of resident companies from the number of functioning technoparks received a set of necessary services for business acceleration, assistance in attracting investments and mentoring support. If we consider the structure of Russian technoparks depending on their specialization (figure 2), it is obvious that the majority of them are multi-industry technoparks (31%) and technoparks in the sphere of information technologies (29%). Minimal activity of technoparks was found in the field of research and development (8%) and agriculture (1%) in the general structure of technoparks. In our opinion, the referred structure of technoparks, depending on their specialization, cannot be called rational and adequate to the public needs.

5. Summary

The research revealed that the number of technoparks in Russia is extremely small (below the level of 2011). There is an increase in the number of technoparks in Iran, but the growth rate is not high (since 2010, the increase has amounted to 14 technoparks).

In our opinion, in the modern conditions of Russia and Iran, the current number of technoparks does not meet the needs of the economies.

The structure of russian technoparks, depending on their specialization and affiliation, is not rational because of the problems in the technoparks work.

In this context, the current issues of the organization and technoparks activities are highlighted and summarized, which hold back the growth of innovative companies. Based on the identified problems, the authors proposed recommendations that will contribute to the development of technoparks.

The authors focus on the fact that the modern technopark of a new generation is becoming a place for free communication and interaction of the "creative community" of developers and consumers. The process of developing new-generation technology parks is significant for Russia and Iran under conditions of global economic crisis. One of the ways out of the current crisis is to rely on domestic science and high-tech production. In this regard, new-generation technology parks, including those based on universities can play one of the most important roles in this process.

6. References

[1] Khaykina E M, Sizova Yu S 2016 Development of technoparks in the Russian Federation *Young scientist* **14** 409-412
[2] Akmaeva R I 2012 Innovative management of a small enterprise operating in the scientific and technical sphere *Phoenix* 541
[3] Marenkov N L, Zolotareva M A 2005 Innovations in Russia 315
[4] Vilenskiy A V 2011 Small business support paradox: preliminary results of the crisis *Economic issues* **6** 152-153
[5] Tekhnoparki. Tekhnogrady. Tekhnozhizn'. Available at https://мойбизнес.рф/novosti/news/tekhnoparki-tekhnogrady-tekhnozhizn
[6] Alekseeva N 2014 Why Invest in Russian Technoparks? *Social and political journal* - Public Russia. Available at http://politrussia.com/ekonomika/razvitie-innovatsiy-673/

[7] Romanovich L G, Evtushenko E I, Romanovich M A, Kudinov D V 2015 Innovation activity and business incubators based on institution of higher education. The experience of Russia *Journal of applied engineering science* **13(3)** 161-166
[8] Sizova Yu S 2016 The development of technopark structures in the Moscow region: dynamics, factors, prospects *Economics and Entrepreneurship* **3(1)** 270–272
[9] Romanovich L, Sevostyanov V, Romanovich M, Sevostyanov M, Arkatov A Y 2014 Innovation activity and technology transfer of higher education *Journal of Applied Engineering Science* **12(4)** 273-276.
[10] Romanovich L G, Romanovich M A, Vybornova V V, Riapukhina V N 2014 Small businesses is a sphere of innovation in the age of globalization *Journal of applied engineering science* **12(4)** 297-301
[11] Fedorova D R, Tarverdiev V Ch 2011 Features of functioning of Technopark structures at the present stage of the transition economy *Science* 191–194
[12] Lytaeva N N, Lipatnikov V S 2018 Factors of success for Russian technoparks: an economic model *Innovation* **4 (234)** 85-95
[13] Kotsyubinskiy V A, Mukhlisova A R, Rybalkin V V 2012 Technology parks around the world: organization of their activities and comparison *Science* 182
[14] Ministry of Science, Research and Technology of the Islamic Republic of Iran. Available at https://www.msrt.ir/en/page/25/science-and-technology-parks-and-incubators
[15] Pelkonen A 2006 The problem of integrated innovation policy: Analyzing the governing role Analyzing the governing role of the Science and Technology Policy Council of Finland *Science and Public Policy, Oxford University Press* **33(9)** 669-680
[16] Pérez-Canto S, Albahari A, Barge-Gil A 2017 Technology Parks versus Science Parks: Does the university make the difference? *Technological Forecasting and Social Change* **116** 13-28 https://www.mendeley.com/catalogue/8fc5e1af-e84b-3171-ab84-ae372c7f35ec/
[17] Fernández-Olmos M, Diez-Vial I 2017 The effect of science and technology parks on a firm's performance: a dynamic approach over time Journal of Evolutionary Economics **27(3)** 413-434
[18] Link A, Scott J, Hobbs K 2017 Science and technology parks: an annotated and analytical literature review *Journal of Technology Transfer* **42(4)** 957-976
[19] Florén H 2006 Managerial work in small firms: summarising what we know and sketching a research agenda *International Journal of Entrepreneurial Behaviour and Research* **12(5)** 272-288
[20] Assudani R, Mroczkowski T, Muñoz-Fernández A 2017 Entrepreneurial support systems: Role of the Czech accelerator *International Journal of Entrepreneurship and Innovation Management* **21(6)** 530-552
[21] Romanovich L G, Yevtushenko E I, Romanovich M A 2015 Small business of innovation sphere in the conditions of globalization *Bulletin of BSTU named after Shukhov V G* **6** 287-290

Acknowledgments
This work was realized under the support of the President Scholarship; in the framework of the Program of flagship university development on the base of the Belgorod State Technological University named after V. G. Shukhov, using equipment of High Technology Center at BSTU named after V.G. Shukhov.

IOP Conf. Series: Materials Science and Engineering **945** (2020) 012077 doi:10.1088/1757-899X/945/1/012077

Method of designing robotic complexes with relative manipulation modules

E V Gaponenko[1*], S I Anciferov[1], L R Kholoshevskaya[1] and V S Kuzmina[1]

[1]Belgorod State Technological University named after V. G. Shukhov, 46, Kostyukova Str., Belgorod, 308012, Russia

E-mail: gaponenkobel@gmail.com

Abstract. The paper offers a method for designing robotic complexes with six degrees of freedom for processing complex parts containing a base and a module for installing tools and a module for processing parts placed on it, made in the form of parallel structure mechanisms, using CAD/CAM/CAE of the NX system under the control of the Teamcenter PLM system. Based on the terms of reference and draft documentation, a preliminary composition of the complex for components of the upper level of the assembly was developed. The stages of designing an electronic-digital model of a robotic complex are presented. A fully parameterized control structure was developed, which is the basis for building robotic complexes of various classes. The use of this approach allows the designer to reduce the complexity of designing similar complexes and simplify the process of making changes, as all changes are made to one part of the product, which has many versions. The application of the developed design methodology for various versions of robotic systems is shown. An algorithm for strength calculation of the assembly is presented. It is shown that the most effective result can be achieved only by using a combined design and calculation methodology.

1. Introduction

Currently, due to the increasing complexity and cost of technical devices, as well as more stringent customer requirements for the quality of the designed product, modern computer-aided design systems are increasingly used, which allow designing high-tech, competitive products, minimizing the cost of purchasing materials, manufacturing and testing [1-3]. The NX CAD/CAM/CAE system, which works in integration mode with the Teamcenter PLM system, is widely used for design and development work [4].

2. Materials and methods

CAD/CAM / CAE-NX system is a computer-aided design system that has a large set of tools designed for performing design and technological preparation of production, as well as conducting engineering analysis of designed products. The NX CAD/CAM/CAE system achieves maximum efficiency when running the Teamcenter PLM system, which unites all participants in the product lifecycle from the moment the idea of a future product appears to its disposal, thereby reducing the time to market for a new product and the final cost, as well as improving its quality. NX is widely used in industries that produce products with a high density of layout, a large number of parts and produce products with complex shapes, including in the mechanical engineering production of robotic systems.

IOP Conf. Series: Materials Science and Engineering **945** (2020) 012077 doi:10.1088/1757-899X/945/1/012077

This paper discusses an approach to the design of robotic complexes of a robotic complex with the possibility of relative manipulation of the part and tool along six coordinates during the processing of complex parts (Fig. 1). The mechanism contains a module for installing the tool and a module for installing the processed part based on parallel structure mechanisms [5-10].

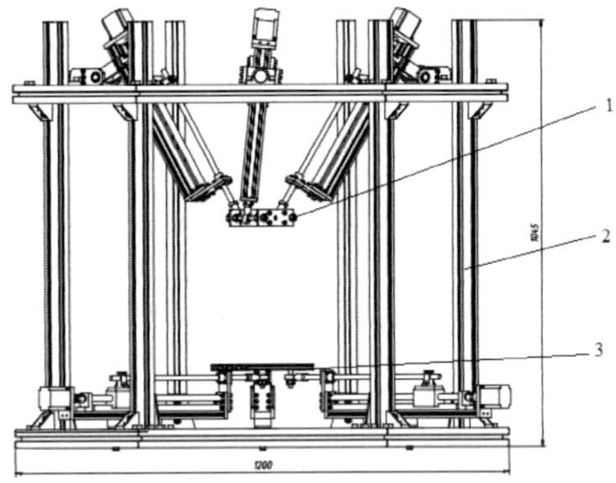

Figure 1. Robotic complex: 1 – The upper module for tool installation; 2 – Base frame; 3 – The lower module for installation details.

The use of such mechanisms will increase the technical and operational efficiency of devices for manipulating space in six coordinates. An important advantage of this complex is the lack of dynamic coupling of drives, which complicates the control task.

3. Results and discussion

Based on the terms of reference and draft documentation, the preliminary composition of the robotic complex is being developed.

The upper level of the robotic complex assembly includes such components as the lower module, the upper module, and the frame.

Then the product composition is formed in the "structure manager" application of the Teamcenter PLM system. The Teamcenter structure manager application allows creating a single product structure, making changes to it, and managing variant build conditions. They can only work with object revisions in the structure manager.

It is very important to understand that depending on the level of entry of products into the assembly, the complexity of the design depends. A well-formed composition of both the assembly units and the main assembly makes it easy to manage the parameters of the future product.

The process of designing an electronic-digital model can be presented in a simplified form by the following stages (Fig. 2): building a control structure, preliminary design, development of an electronic-digital model, checking the correctness of assembly constructions, strength calculation of the structure, design optimization, and release of design documentation.

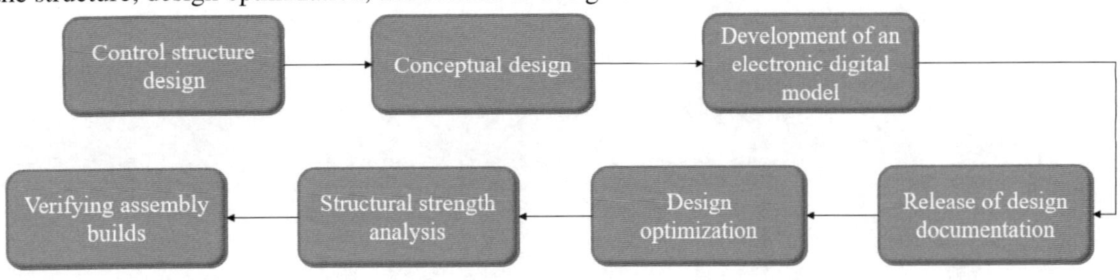

Figure 2. A simplified scheme of the design process.

IOP Conf. Series: Materials Science and Engineering **945** (2020) 012077 doi:10.1088/1757-899X/945/1/012077

The “Top-down” method is used in the design process [11]. This choice is justified by the fact that the geometry of most elements is completely unknown, and only their location in space is known. To facilitate the design process in accordance with the product composition, a control structure was developed (Fig. 3). The presence of a control structure in the assembly reduces the time for creating a model and provides the opportunity to work together on the product. A control structure is a set of different coordinate elements that occupy a certain position in the workspace. It is used for building geometry in the planes of the control structure, as well as for interfacing with assembly components. At the same time, according to the draft design, the planes are placed in the places where the main elements of the structure meet. The planes that limit the width of the frame are located relative to each other at a distance *d*, the plane of the desktop is raised from the plane of the base to a height *a*. The distance between the base plane and the top plane of the frame is *H*. The plane of the working table is located at a distance *b* from the upper plane of the frame. The values of parameters a, b, d, and H are set using the expressions command in the Modeling application. Figure 4 shows the parameters set for the robotic system.

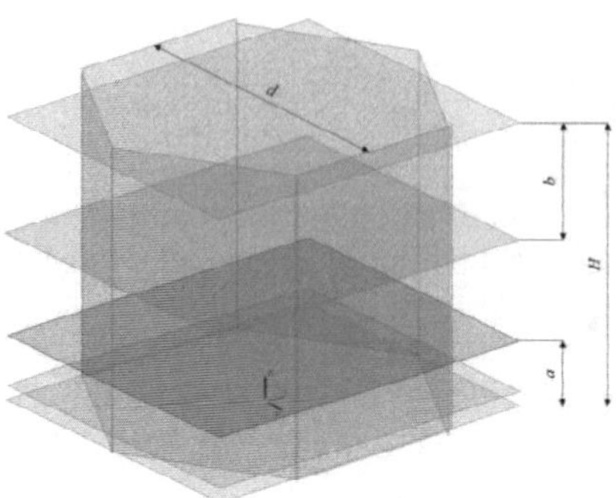

Figure 3. Control structure.

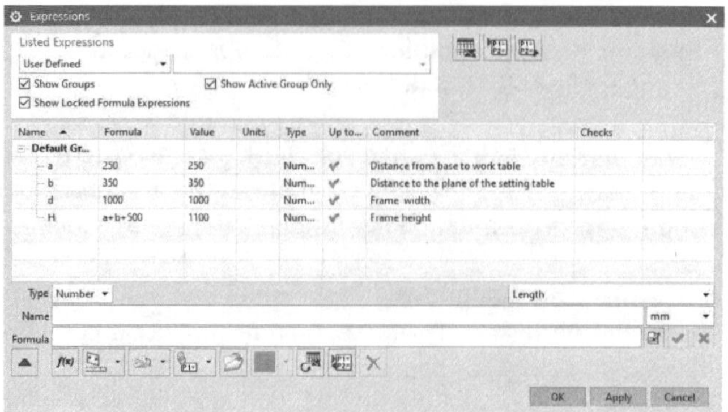

Figure 4. Dialog box “Expressions”.

When the control structure is built, a preliminary design process is carried out, during which the required configuration of the robotic complex is determined. Also, by setting the location of coordinate elements of the control structure through variables, they can create a fully parameterized control structure, which is the basis for building robotic complexes of various classes.

This approach makes it possible to manage the variant composition of the product, which can have a large number of different versions that are simultaneously in production and can be delivered to the customer. The use of this approach allows the designer to reduce the complexity of designing similar complexes and simplify the process of making changes, as all changes are made to one part of the product, which has many versions.

For example, if the customer needs to process large-sized parts relative to the vertical axis, in the “Structure Manager” application, a version of the robotic complex with an increased frame height can be selected (Fig. 5), or, if the customer does not need 5-coordinate processing of products, then you can make the active version of the product composition, in which the desktop is statically fixed (Fig.6).

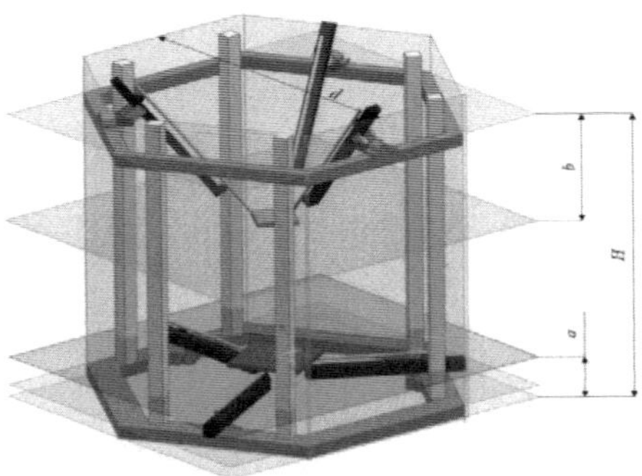

Figure 5. Layout of a robotic complex for processing large-sized products.

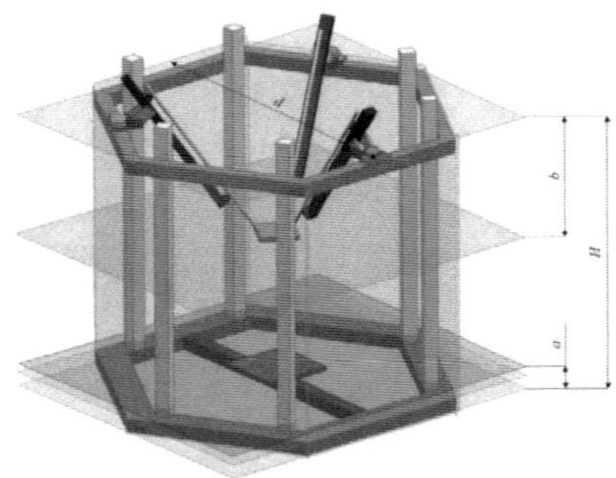

Figure 6. Layout of a 3-axis robotic processing complex.

In addition, this robotic system can have a version with a modification for the possibility of its use in the field of additive manufacturing (Fig. 7) due to its characteristics, which include high accuracy, speed of movement of the upper platform. Variants of a robotic complex with a completely different kinematic scheme of the device can also be developed (Fig. 8).

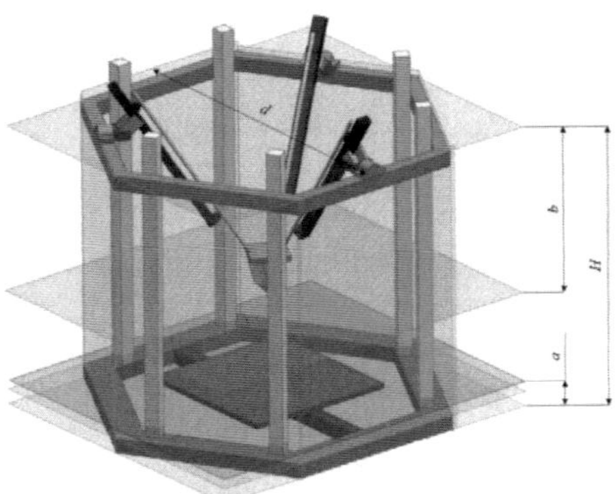

Figure 7. Layout of a robotic complex for working with plastics.

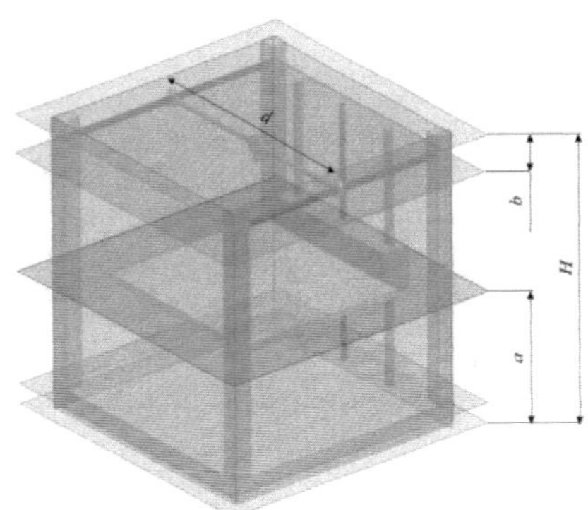

Figure 8. Ultimayker Platform.

After the required configuration of the robotic system has been selected, a rational profile is selected from which the frame should be made. For design reasons, the selected profile must provide the necessary strength, rigidity of the finished product and manufacturability of assembly with minimal time.

The next step is to build an electronic-digital model of the robotic complex in the NX CAD/CAM/CAE system.

The frame is constructed using a control structure, with reference to the base planes, the upper plane of the frame, and the planes that limit the width. When the frame of the robotic system is ready, they need to create the geometry of the desktop located in the assembly of the "Lower module".

This is followed by building level 2 assemblies, in particular the upper and lower modules. A sketch is created in the desktop plane, and then the "Pull" command creates a geometric model of the table. The next step is to fix the geometric model of the desktop for easy construction and positioning of subsequent components of the assembly relative to the main part. The rest of the assembly components are built in the same way. Figure 9 shows the completed model of the lower module.

Then the rest of the components are completed and the correct construction of the robotic complex assembly is checked, i.e. the gap analysis is performed. This command allows setting whether the geometry of the assembly components intersects with each other. If there is intersection geometry,

IOP Conf. Series: Materials Science and Engineering **945** (2020) 012077 doi:10.1088/1757-899X/945/1/012077

then the geometry of these components is either changed if it is an intersection between parts or subassemblies, or intersections are added to exceptions if it is an intersection geometry in threaded connections.

The other models in the assembly are also positioned relative to the control structure and the existing geometry. Figure 10 shows the completed model of the robotic complex.

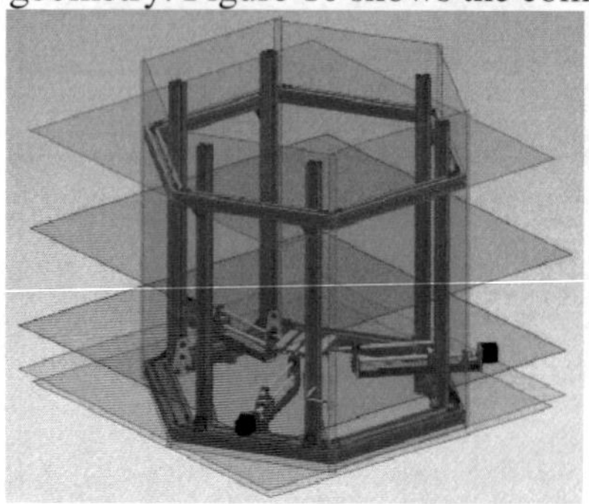

Figure 9. 2 ready build level.

Figure 10. Electronic-digital model of a robotic complex.

The next stage of designing a robotic complex is the strength calculation of the assembly, the stages of which are presented in a simplified form in Fig. 11.

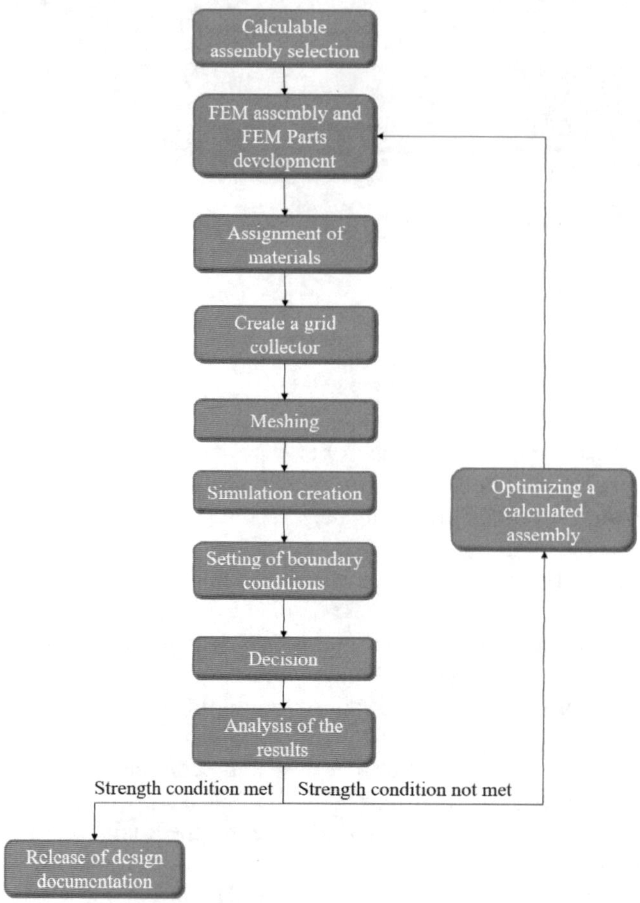

Figure 11. Algorithm for strength calculation of the assembly.

1) Select the assembly for which they want to perform the strength calculation.
2) Creating FEM assembly and FEM parts files.
3) The purpose of the materials. Materials are assigned to each part, this is necessary in order to correctly calculate the weight of the nodes that rest on the frame.
4) Creating a grid collector..
5) Creating a grid. At this stage, a 3D finite element grid is created. Creating a finite element model is a key step in the preprocessor, as the accuracy of the results depends on the quality of the finite element grid.
6) Creating a simulation file.
7) Setting boundary conditions. At this stage, all unconnected 3D solid-state finite elements are connected to a single finite element model using the "Surface-to-surface bonding" command. This is necessary to describe correctly the interaction of structural elements. After that, loads are applied to the frame elements. The weight of the lower and upper modules is applied as the load. The load from the weight of the upper module is applied to the middle sections of the finite element grid obtained when dividing the body of the upper profile model using the "Force" command with the selected "Total per object" distribution method. Gravity must also be taken into account. The "gravity" command takes into account the self-weight of the frame structure. As the frame is based on the lower faces of the "Lower Profile" pipes, they are fixed using the "Seal" command. This command denies the selected objects six degrees of freedom.
8) Solution. The obtained results during the solution are located in the postprocessor Navigator.
9) Analysis of the results obtained. In order to obtain a conclusion about the performance of the structure under the imposed restrictions and loads, it is necessary to check the conditions of strength and rigidity of the product. To do this, the load-bearing capacity of the frame is evaluated using the criteria of plasticity (fluidity).
10) If the strength conditions are met, the design documentation is issued; if not, the calculated design is optimized.

4. Summary
Thus, having studied past achievements in the field of design and calculation of robotic systems, we can say that the most effective result can be achieved only by using the design and calculation methods together. Using this method, they can develop a variety of variants of structures of this class, depending on the customer's requirements, with minimal time spent.

5. References
[1] Kyratsis P, Kakoulis K, Markopoulos A 2020 Advances in CAD/CAM/CAE Technologies *Machines* **8(1)** 13
[2] Milchev G, Miltchev R 2018 Development of Information Technologies, Planned Obsolescence and Modification of the Life-Cycle of the CAD/CAM/CAE Systems *European Journal of Sustainable Development* **7(3)** 217-224
[3] Goncharov P S, Eltsov M Yu, Korshikov S B, Laptev I V, Osiyuk V A 2010 NX for the designer-mechanical engineer *MQM Press* 503
[4] Torop D N, Terlikov V V 2011 Teamcenter. Beginning of work *DMK Press* 280
[5] Kong H, Gosselin C M 2007 Type Synthesis of Parallel Mechanisms *Springer* 272
[6] Merlet J-P, Gosselin C 2008 Springer Handbook of Robotics *Springer* 269-285
[7] Glazunov V A, Lastochkin A B, Terekhova A N, Bic V N 2007 On specific features of relative manipulators *Problems of mechanical engineering and machine reliability* **36(2)** 170-177.
[8] Rybak L A, Erzhukov V V, Chichvarin A V 2011 Effective methods for solving problems of kinematics and the dynamics of a robot machine of a parallel structure *Physical-Mathematical Literature* 148
[9] Khalapyan S Y, Rybak L A, Glushchenko A I, Mamaev Y A 2016 On neural network model development to solve parallel robots kinematics and control problems *International Journal*

IOP Conf. Series: Materials Science and Engineering **945** (2020) 012077 doi:10.1088/1757-899X/945/1/012077

of Pharmacy and Technology **8(4)** 25085-25095.

[10] Rybak L A, Mamaev Y A, Malyshev D I, Virabyan L G 2016 Software module of hexapod robot specified movement trajectory of output unit for 3d-printing of products *Bulletin of BSTU named after V G Shukhov* **8** 155-164

[11] Eltsov M Yu, Kozlov A A, Sedoykin A V, Shirokova L Yu 2010 Design in NX under Teamcenter *DMK Press* 783

Acknowledgments

This work was supported by the Russian Science Foundation, the agreement number № 16-19-00148, using the equipment of the Center for High Technologies of BSTU named after V G Shukhov.

IOP Conf. Series: Materials Science and Engineering **945** (2020) 012078 doi:10.1088/1757-899X/945/1/012078

Temperature modification of diatomite sludge of oil extraction production at obtaining a sorbent

I V Starostina[1*], Yu L Makridina[1], Zh A Sapronova[1] and I V Bomba[1]

[1]Belgorod State Technological University named after V.G. Shoukhov, 46, Kostukova Str., Belgorod, 308012, Russia

E-mail: fomina.katerina@mail.ru

Abstract. One of the most widely used methods of water treatment is adsorption. The use of agricultural waste is of great interest in obtaining economical and efficient sorption materials. In this paper, the features of temperature modification of diatomite sludge are studied. Diatomite sludge is a multi-tonnage waste of oil extraction production formed during the purification of sunflower oil with diatomite. The kinetics of thermal activation of diatomite sludge at temperatures of 450, 500, and 600 °C was studied. It was found that the temperature of 450 °C is optimal for thermal modification of waste. In this case, part of the organic compounds is preserved, as with a further increase in temperature, they completely disintegrate into gaseous products. IR spectra and energy dispersion analysis were analyzed, and the structure and physical and chemical properties of diatomite and diatomite sludge were studied before and after heat treatment. It was found that the water extract of the samples, both initial and thermomodified, has a pH close to neutral. With an increase in the firing temperature, a small suspension effect is observed, which means the presence of a charge in the DSH particles in the aqueous medium, which can intensify the convergence of molecules with a positive charge with the surface of the sorbent. The development of methods for the secondary use of agricultural raw material processing waste for wastewater treatment allows solving comprehensively the problems of reducing the anthropogenic load on natural ecosystems, is an urgent task of modern society for environmental protection, and has scientific and practical significance.

1. Introduction

Currently, waste that requires recycling or disposal is generated everywhere. Many scientists consider the term “waste” obsolete, as each waste can become a valuable secondary raw material in some technological process [1, 2].

In the process of processing agricultural raw materials, large amounts of organic and organomineral waste are generated. As the volume of food production is steadily increasing (due to the continued growth of the world’s population), waste is generated in large quantities every year [3, 4]. Most of it is not used, but is taken to landfills for storage. This leads to environmental pollution and alienation of valuable territories.

Global production of oils and fats is about 2.5-3 million tons/year, 75% of which is obtained from plants. During the separation of oils from oilseeds and subsequent cleaning of the resulting oils, a large amount of waste is generated, for which disposal is an urgent problem [5, 6].

IOP Conf. Series: Materials Science and Engineering **945** (2020) 012078 doi:10.1088/1757-899X/945/1/012078

An example of such waste is diatomite sludge (DS). Diatomite is a sedimentary rock formed by silica fragments of the shells of microscopic diatoms – diatoms and radillaries. The main part of the siliceous carapace (skeleton) is amorphous silica hydrates of various degrees of water content – varieties of opal of the $mSiO_2 \cdot nH_2O$ type, the crystal component is represented by quartz admixtures [7, 8]. Natural diatomite is subjected to thermochemical stabilization, treated with solutions of alkalis, calcined at a temperature of about 1000 °C, and then used as a filter material in the food industry, in particular, in brewing and oil extraction production [7, 9]. DS formed during the purification of vegetable oils is non-toxic, however, during its long-term storage, air and water bodies are polluted with aldehydes, ketones, oxide compounds and other substances - products of decomposition of organic components [10]. Taking into account the high specific surface area of sludge particles and the significant content of adsorbed vegetable oil, the risk of self-ignition is possible when the waste comes into contact with air oxygen [11].

Sunflower is a traditional oilseed crop for Russia and Eastern Europe. It is also grown in small quantities in India, the United States, and other countries [12, 13]. Table 1 shows data on the share of sunflower oil consumption in the world out of the total amount of vegetable oils, % [14, 15].

Table 1. Share of sunflower oil in global consumption of vegetable oils.

Oil	Consumption in the world, %	Oil	Consumption in the world, %
Palm	33.1	Cotton	4.3
Soybean	24.2	Palm kernel	3.6
Rapeseed	15.2	Coconut	2.9
Sunflower	8.5	Olive	2.7
Peanut	5.1	Corn	0.4

In small quantities DS can be used as a feed additive after oil purification, however, this product is very perishable [16], which makes it difficult to use it widely.

The main brand of diatomaceous earth used in the fat and oil industry and allowing achieving high filtration rates is "Celite 545 VO". Diatomite is used to extract waxes and sticky impurities from sunflower oil that degrade consumer qualities.

It is known that many organic and organomineral wastes are suitable raw materials for the production of sorption materials that can be used in water treatment [17-21]. This makes it possible to reduce significantly the cost of wastewater treatment and at the same time dispose of waste, which leads to a comprehensive reduction in the anthropogenic load on natural ecosystems.

One of the most frequently used methods of modifying such materials is carbonation, which allows forming a sorbing carbon layer and neutralizing simultaneously organic compounds that are subject to oxygen and bacterial decomposition.

To develop a sorption material based on DS, it is necessary to conduct a set of studies to assess the physical and chemical changes that occur during heat treatment.

2. Materials and methods

As an object of research, we used the spent DS formed during the purification of sunflower oil with diatomite of the Celite 545 VO brand in the conditions of EFKO LLC, Alekseevka, Belgorod region, Russia. DS is an oily paste-like mass with an organic content of up to 70%. The mineralogical composition is mainly represented by cristobalite, quartz is present in small quantities, and clay minerals in an amorphous state are noted. The bulk density is 450 kg/m^3, the true density is 2750 kg/m^3. The composition of substances adsorbed on the surface of diatomaceous earth during the filtration of vegetable oils also includes vegetable waxes, the content of which is 1.3-12 % [22].

Thermal modification of the DS was performed by heating the average sample without oxygen in an electric furnace of the SNOL 25/12 brand with holding at the maximum temperature for 1 hour.

Infrared spectra were obtained using the Vertex 70 instrument (USA), and then decoded using reference and scientific literature [23–25].

Energy dispersion analysis was performed using a scanning electron microscope TESCAN MIRA LMU (Czech Republic).

The pH values of water extracts were measured on the pH meter I-500 "Aquilon" (Russia).

3. Results and discussion

DS samples were fired at various temperatures, after which their structure and physical and chemical properties were studied.

In the course of research, it was found that at temperatures of 300-450°C, organic substances are not completely oxidized. The intermediates formed as a result of heat treatment settle on the surface of the diatomaceous earth particles in the form of a carbon layer. To determine the optimal conditions for the modification of the DS, it is necessary to obtain information about the transformations of the substances that make up its composition during heat treatment.

The results of studies of samples of materials in the IR spectrum are shown in figures 1,2.

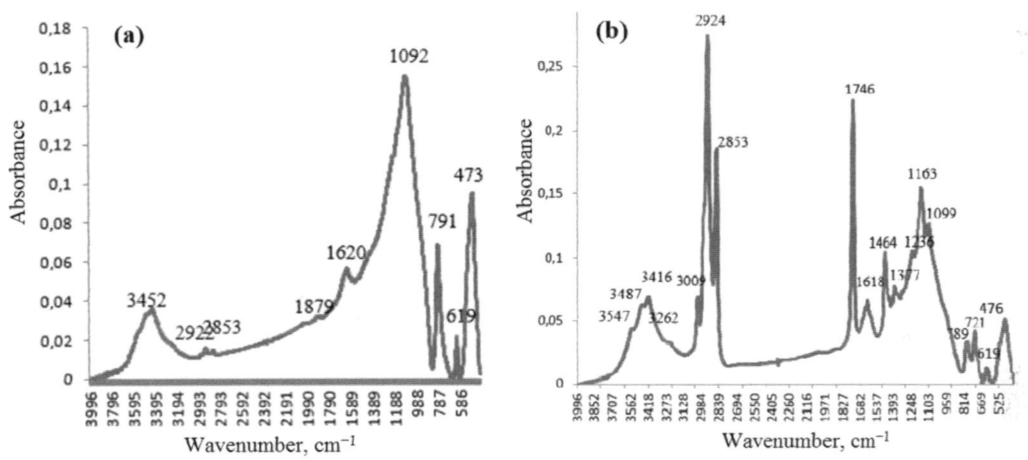

Figure 1. IR-spectrums: a – diatomite; b – DS.

In the region of 1300-400 cm^{-1}, minerals show bands of stretching and bending of Si-O bonds and bands of bending of O-H bonds. The shape and position of the bands depend very much on the location inside the layers.

If the majority of octahedral positions are occupied by divalent central atoms, the O-H bending bands are shifted towards smaller wave numbers in the range of 700-600 cm^{-1} [23].

The initial diatomite has expressed peaks of 1092 cm^{-1} and 473 cm^{-1}, which are related to Si-O bond fluctuations. Peaks of 3452 cm^{-1} show the presence of bound water; peaks of 791 cm^{-1} and 619 cm^{-1} can be attributed to the O-H of water molecules affected by hydrogen bonds [24]. In addition, the band at 791 cm^{-1} can be associated with valence vibrations of the ν_s (Al-O) clay impurity minerals that are part of the diatomite.

Crystallization water is often characterized by peaks in the range of 1670-1600 cm^{-1}, which is marked by a peak of 1620 cm^{-1} on the IR spectrum of diatomite [24].

The DS sample clearly shows a peak of 2924 cm^{-1}, which indicates the presence of asymmetrically attached CH_2 radicals, a peak of 2853 cm^{-1} can be attributed to a CH_2 bond with symmetric bonds, and a peak of 3009 cm^{-1} indicates $-CH_3$. A blurred peak in the range of 3600-3200 cm^{-1}, related to chemically bound water, still stands out. The intense peak of 1746 cm^{-1} is defined as C=O bond. Similar peaks can be noted in the IR spectra of linseed oil and vegetable wax (Table 2) [25].

Table 2. Characteristic peaks for vegetable wax and linseed oil.

Range, cm^{-1}	Bond between molecules	Type of oscillations
3600-3200	O -H	Band of valence vibrations
3000-2800	C-H	Band of valence vibrations
1780-1700	C=O	Band of valence vibrations
1480-1300	C-H	Bands of deformation vibrations
1300-900	C-O	Band of valence vibrations
750-700	C-H	Bands of rotational vibrations

The peak of 1464 cm^{-1} probably belongs to the deformation vibrations of C-H (1480-1300 cm^{-1}), similar to the IR spectrogram of linseed oil. The peaks of 1377 cm^{-1}, 1236 cm^{-1}, and 1163 cm^{-1} fall within the range of 1300-900 cm^{-1}, which corresponds to the presence of C-O bonds [23]. They can determine the peaks characteristic of pure diatomite: 1099 cm^{-1}, 476 cm^{-1}.

Further, samples that were heat-treated at 450 °C were examined (DS_{450}), when organic substances were not fully destroyed, and at 500 °C (DS_{500}), when most of the organic compounds decomposed.

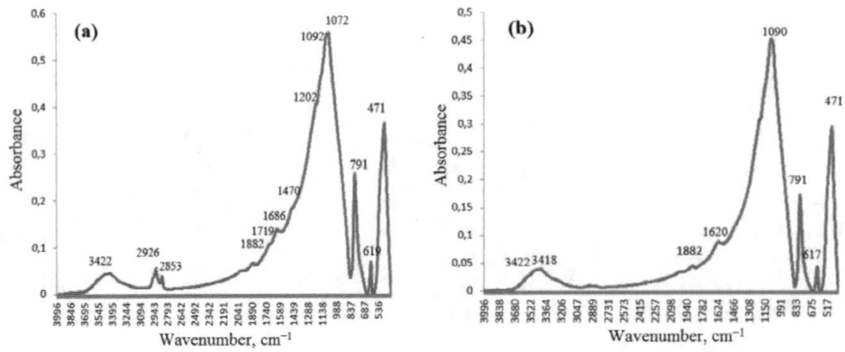

Figure 2. IR spectra of DS after heat treatment: a – DS_{450}; b – DS_{500}.

According to the results of the IR analysis shown in Figure 2, after heat treatment, the peaks related to vibrations in organic molecules become less expressed, and the peaks indicating the presence of silicon compounds (1092 cm^{-1},1090 cm^{-1}, 471 cm^{-1}) are more expressed. However, the peak of 1090-1092 cm^{-1} in both DS_{450} and DS_{500} is more intense than in the original diatomite. Presumably, it is enhanced by fluctuations in oxygen organic products of partial destruction of waxes and vegetable oil, and the presence of C-O-C (1270-1060 cm^{-1}) and C-O (1150-1000 cm^{-1}) bonds is possible.

There are no peaks characteristic of C=O bonds (1900-1550 cm^{-1}), which are among the first to be destroyed during heat treatment.

The intense bands at 2926 cm^{-1} and 2853 cm^{-1}, which corresponds to the region of 2850-3000 cm^{-1}, can be attributed to valence vibrations, and the medium intensity band at 1470 cm^{-1} can be attributed to deformation vibrations of methylene groups and characterize the presence of a Csp^3-H bond on the carbon surface of DS_{450}.

With an increase in the processing temperature to 500 °C, their intensity decreases to zero, which indicates the burnout of surface forms of carbon containing C-H fragments.

The valence fluctuations at 1719 cm^{-1} in DS_{450} correspond to the carbonyl group (C=O) of esters or lactone that are part of vegetable waxes and waxy substances, which explains the presence of hydrophobic properties of this material. The IR range of 3100-3700 cm^{-1} characterizes the valence vibrations of various types of hydroxyl groups, including in water molecules adsorbed on hydroxyl groups. Siloxane groups Si-O-Si are characterized by intense peaks of 1111-1000 cm^{-1} [24, 25].

Thus, it can be seen that the IR spectrograms of diatomite and DS_{500} are very similar to each other, the latter slightly differs in the intensity of the peaks, and their general shape and location are almost identical for the samples. Therefore, the temperature of 450 °C is the threshold for the heat treatment of DS, when a part of organic compounds is preserved, as with a further increase in temperature they completely break down into gaseous products.

The results of energy-dispersive studies are in good agreement with IR spectroscopy data (Figure 3). It can be noted that at a temperature of 450 °C, a significant amount of carbon is retained in the material. The ratio of oxygen and carbon changes already at a temperature of 500 °C – the processes of destruction of the bulk of organic compounds occur. The sample obtained at a temperature of 600 °C contains a small amount of carbon. The maximum peak of oxygen, which is obviously part of the inorganic oxides, is clearly recorded.

Previously, it was found that the source material has a significant hydrophobicity, as it contains a large amount of oil and wax residues. As the firing temperature increases, the hydrophobicity decreases markedly [26] due to the decomposition of organic substances. The hydrophobic properties of the sorption material are a positive factor in the recovery of hydrophobic pollutants, such as petroleum products. At the same time, such a sorption material will be ineffective when treating wastewater from heavy metal ions, dyes, and other ions and compounds with hydrophilic properties. The presence of a charge on the surface of the latter particles can be a positive factor when interacting with oppositely charged ions and polarized molecules.

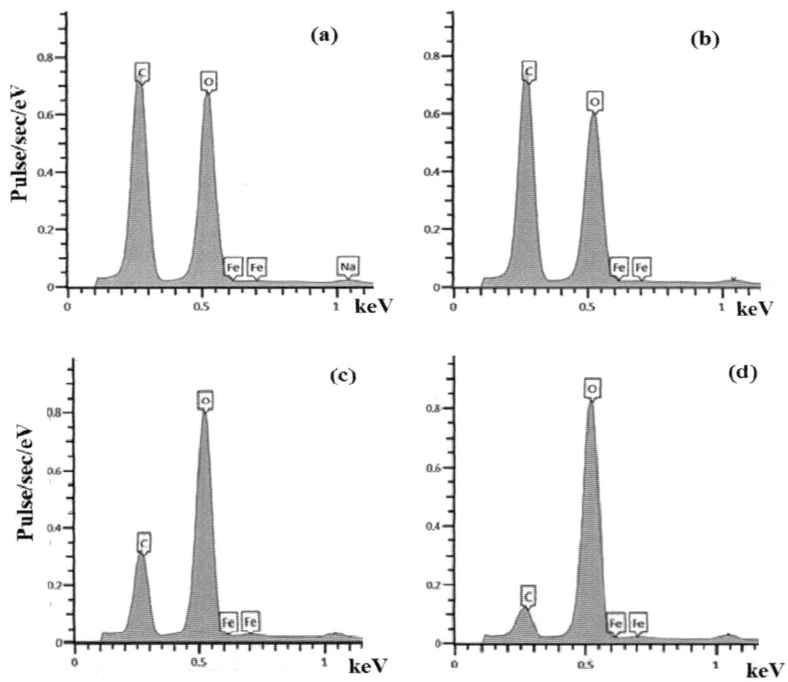

Figure 3. Energy dispersion analysis of DS: a – DS; b – DS_{450}; c – DS_{500}; d – DS_{600}.

To determine the presence of charge on the surface of the sorption material, the suspension effect was studied.

It was found that the water extract of samples, both initial and heat-treated, has, in general, close to neutral pH (Table 3). With an increase in the firing temperature, a small suspension effect is observed, which means the presence of a charge in the DS particles in the aqueous medium. As the suspension has a more acidic reaction than the filtrate, therefore, the particles have a small negative charge. Despite the fact that the charge of the particles is very small, its presence can facilitate the convergence of molecules with a positive charge with the surface of the sorbent. In particular, this applies to heavy metal ions.

IOP Conf. Series: Materials Science and Engineering **945** (2020) 012078 doi:10.1088/1757-899X/945/1/012078

Table 3. pH of suspension and filtrate of water extract DS.

Sample	pH of suspension	pH of filtrate
DS	6.44	6.45
DS_{400}	6.89	6.92
DS_{450}	6.90	6.93
DS_{500}	6.85	6.99
DS_{550}	6.81	7.3
DS_{600}	6.79	7.2
DS_{650}	6.77	7.3

The information obtained is of great interest, as the DS processed at different temperatures can serve as a sorption material for hydrophobic substances (firing temperature 400-450 °C) or positively charged ions, such as heavy metals or dyes (firing temperature 500-550 °C).

4. Summary

Studies were conducted to determine the optimal conditions for modifying the DS. The analysis of the IR spectra of diatomite and diatomite sludge of various degrees of modification allowed establishing that the temperature of 450 °C is optimal for the heat treatment of DS, when a part of organic compounds is preserved, as with a further increase in temperature they completely disintegrate into gaseous products. The results of energy-dispersive studies are in good agreement with IR spectroscopy data.

When the firing temperature increases, there are small changes in the pH of the water extract and a suspension effect occurs, due to the appearance of a negative charge around the sludge particles. Despite the fact that the charge of the particles is very small, its presence can facilitate the convergence of molecules with a positive charge with the surface of the sorbent. In particular, this applies to heavy metal ions. Thus, by processing at different temperatures, it is possible to obtain a sorption material that is more effective in extracting hydrophobic or hydrophilic pollutants.

5. References

[1] Blengini G A, Mathieux F, Mancini L et al 2019 Joint Research Centre Science for Policy report: Recovery of critical and other raw materials from mining waste and landfills: State of play on existing practices *Publications Office of the European Union* 125

[2] Delgado A L, Robla J I, Padilla I et al 2020 Zero-waste process for the transformation of a hazardous aluminum waste into a raw material to obtain zeolites *Journal of Cleaner Production* **255** 120178

[3] Zhanga Q, Zhanga D, Xu H 2020 Biochar filled high-density polyethylene composites with excellent properties: Towards maximizing the utilization of agricultural wastes *Industrial Crops & Products* **146** 112185

[4] Suffo M, De la Mata M, Molina S I 2020 A sugar-beet waste based thermoplastic agro-composite as substitute for raw materials *Journal of Cleaner Production* **257** 120382

[5] Benincasa M, Accorsini R F 2008 Pseudomonas aeruginosa LBI production as an integrated process using the wastes from sunflower-oil refining as a substrate *Bioresource Technology* **99** 3843–3849

[6] Sverguzova S V, Sapronova Zh A, Svyatchenko A V, Otiti T 2018 Adsorption of spindle oil by native and thermal modified chestnut tree waste *Construction Materials and Products* **1(1)** 4–11

[7] Kiryushina N Yu, Tarasova G I, Sverguzova S V 2010 Slag waste in water treatment *Bulletin of BSTU named after V G Shukhov* **4** 140–145

[8] Nikiforov E A, Ubaskina Y A, Ryabov G K, Fetyuhina E G 2011 Regarding diatomite special structure in the production of foamglassceramic thermal insulation products *Building materials and products* **1** 60–64

[9] Postolachi L, Rusu V, Lupascu T, Petuhov O 2018 Improvement of filtration properties of ghidirim diatomite (republic of Moldova) *Chemistry Journal Of Moldova* **13(1)** 117-121
[10] Zinoviev A A 1952 Chemistry of fats *Pishchepromizdat* 551
[11] Poletaeva M A, Susoeva N M 2014 Utilization of bleached clay in the production of vegetable oils *Polzunovsky Bulletin* **3** 249-251
[12] Ebrahimiana E, Seyyedia S M, Bybordi A, Damalas A C 2019 Seed yield and oil quality of sunflower, safflower, and sesame under different levels of irrigation water availability *Agricultural Water Management* **218** 149–157
[13] Shorstkii I, Khudyakov D 2020 Influence of pulsed electrical discharge, hydrostatic pressure and temperature on rheological properties of sunflower cake during oil pressing *Heliyon* **6** e03046
[14] Gunstone F D 2020 Vegetable Oils in Food Technology *Blackwell* 337
[15] O'Brien R D 2009 Fats and Oils Formulating and Processing for Applications *CRC Press* 744
[16] Kononenko S I, Chikov A E, Osepchuk D V 2009 Fat additive for broiler chickens from waste of the oil extraction industry *Problems of biology of productive animals* **3** 26-34
[17] Dai Y, Sun Q, Wang W et al 2018 Utilizations of agricultural waste as adsorbent for the removal of contaminants: A review *Chemosphere* **211** 235–253
[18] Sverguzova S V, Bomba I V, Voronina Yu S 2018 Purification of oily emulsions with leaf litter of cherry and rowan *Chemical Bulletin* **1(4)** 4–19
[19] Sapronova Z A, Svergusova S V, Fomina E V, Aksenov A K 2019 Physical and chemical features of aqueous media treatment process from components of cow milk with a milled waste of corn *IOP Conference Series: Earth and Environmental Science* **272** 032008
[20] Sapronova Z A,Sverguzova S V, Svyatchenko A V, Fomina E V, Voitovich E V 2019 Obtaining sorption material from the leaves of Aésculus hippocastanum L *Atlantis Highlights in Material Sciences and Technology* **1** 311–315
[21] Galimova R Z, Nguyen T K T, Shaikhiev I G, Sverguzova S V 2019 Sulfuric acid modified sawdust of acacia auricularis (acacia auriculiformis) for the purification of wastewater from copper ions (II) *Chemical Bulletin* **2(4)** 12–23
[22] Yarullin R N, Yarullin R R, Supyrev A V et al 2013 *Pat. 2488425 Russian Federation IPC B01D41/02* Method for regeneration of spent filter material **2011100770/05**
[23] Stuart B H 2004 Infrared spectroscopy: fundamentals and applications John *Wiley & Sons Ltd* 244
[24] Derrick M R, Stulik D, Landry J M 1999 Infrared Spectroscopy in Conservation Science *The Getty Conservation Institute Los Angeles* 248
[25] Larkin P 2011 Infrared and raman spectroscopy principles and spectral interpretation *Elsevier Inc* 230
[26] Starostina I V, Porozhnyuk E V, Stolyarov D V 2020 The processing of kieselguhr sludge with obtaining a new end product as reserve for reducing resources consumption of vegetable oil manufacture *IOP Conference Series: Materials Science and Engineering* 012075

Acknowledgments
This work was supported by a grant from the President of the Russian Federation for state support of young Russian scientists – candidates of sciences and doctors of sciences and leading scientific schools of the Russian Federation, application number MD-1249.2020.5. The work is realized in the framework of the Program of flagship university development on the base of the Belgorod State Technological University named after V.G. Shukhov, using equipment of High Technology Center at BSTU named after V G Shukhov.

IOP Conf. Series: Materials Science and Engineering **945** (2020) 012079 doi:10.1088/1757-899X/945/1/012079

Polymer composite material based on titanium hydride

N I Cherkashina[1*], E V Popova[2], E O Pilavidou[1] and E P Klopot[1]

[1]Belgorod State Technological University named after V.G. Shukhov, Belgorod, Russia
[2]Yu. A. Gagarin Research and Test Cosmonaut Training Center, Moscow, Russia

Email: cherkashina.ni@bstu.ru

Abstract. The article discusses the possibility of synthesizing highly filled polymer composites based on polystyrene and titanium hydride. The composites were prepared by solid-phase compaction at high specific pressure. For uniform distribution, titanium hydride was preliminarily milled in a jet-vortex mill. The particle size of titanium hydride after grinding is in the range from 0.1 to 107 μm. The modal particle diameter is 35.41 μm, and the specific surface area of the particles is 15279 cm^2 / cm^3. The titanium hydride content in the composites did not exceed 60 wt%. The article presents data on the density and microhardness of the obtained composites. It was found that the introduction of titanium hydride increases the physicomechanical characteristics of the composites. The Vickers microhardness of a pure polystyrene sample is 19.87 ± 0.20 HV, and a sample containing 60 wt% titanium hydride is 36.17 ± 0.78 HV (at a load of 200 g).

1. Introduction

Polymer materials are widely used in various fields of industry, including construction, automotive, nuclear, and others. [1-2] The use of polymers is due to their low cost, ease of processing and high physical and mechanical properties. The introduction of various fillers into polymers makes it possible to expand the field of application, to create functional materials with desired properties [3]. The introduction of radiation shielding fillers into cement and polymer matrices makes it possible to create composites that can be used in the nuclear and space industries [4-5]. It is known that the use of materials containing hydrogen significantly increases the protection against neutron radiation. Hydrogen is a very good neutron moderator and is the main constituent in most materials used for neutron shielding. The nucleus of a hydrogen atom - a proton has, with a good degree of accuracy, the same mass as a neutron, therefore, with each act of elastic scattering on a hydrogen nucleus, a neutron loses about half of its initial energy. A neutron barrier made of hydrogen has the best shielding properties per unit mass of shielding.

Of particular interest for radiation protection are metal hydrides, for example, titanium hydride [6-10]. In this work, the possibility of creating polymer composites based on titanium hydride and a thermoplastic polystyrene matrix is investigated. It is known that the main problem of creating effective radiation-protective composite materials is the uniform distribution of the filler in the matrix [11-15]. The work [16] studied the possibility of creating a hydrophobic radiation-protective filler for

IOP Conf. Series: Materials Science and Engineering **945** (2020) 012079 doi:10.1088/1757-899X/945/1/012079

introduction into a non-polar polymer matrix. In [17], data are presented on the modification of the surface of mineral fillers with organic particles for further introduction into a composite material. In this work, the possibility of preliminary grinding of the titanium hydride filler to improve the homogeneity of the components in the finished composite is considered.

2. Materials and methods

Amorphous impact-resistant polystyrene (UPS-803E) was used as a matrix. The structural structure of the used polystyrene is shown in Figure 1.

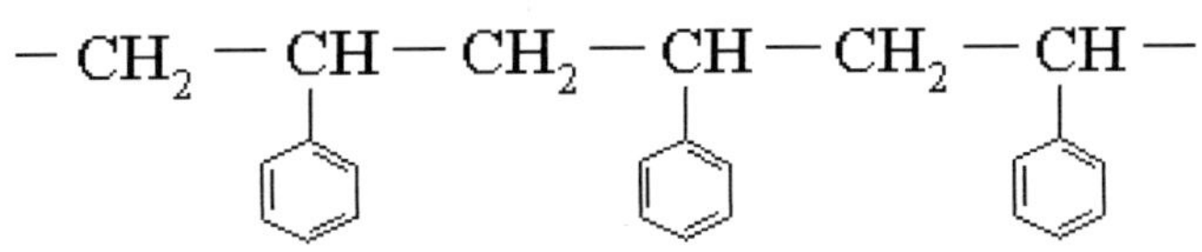

Figure 1. Structural structure of polystyrene.

The initial titanium hydride was spherical granules (shot) with a diameter of 1-4 mm. The true density of titanium hydride granules is 3800 kg/m^3, bulk density is 2526 kg/m^3, hydrogen content is 3.55 wt%. The elemental composition of titanium hydride is shown in Table 1.

Table 1. Elemental composition of titanium hydride.

Element	Ti	H	Al	Fe	O	Si	C	N
wt%	95.63	3.50	0.40	0.20	0.10	0.08	0.05	0.04

The grinding of titanium hydride shot to a powdery state was carried out in a jet-vortex mill for 15 minutes. The composites were synthesized by hot pressing at a polystyrene softening temperature of 180 ° C and a pressure of 80 MPa. To carry out pressing, the powdery components of polystyrene and titanium hydride were previously mixed and the mixture was placed in a mold.

The dispersion of the powders was monitored on a «Microsizer 201C» laser particle analyzer. The Vickers microhardness of the obtained samples was investigated on a NEXUS 4504 device at the same load of 200 g.

3. Results and discussion

Figure 2 shows the data on the granulometry of titanium hydride after grinding the shot in a jet-vortex mill for 15 minutes.

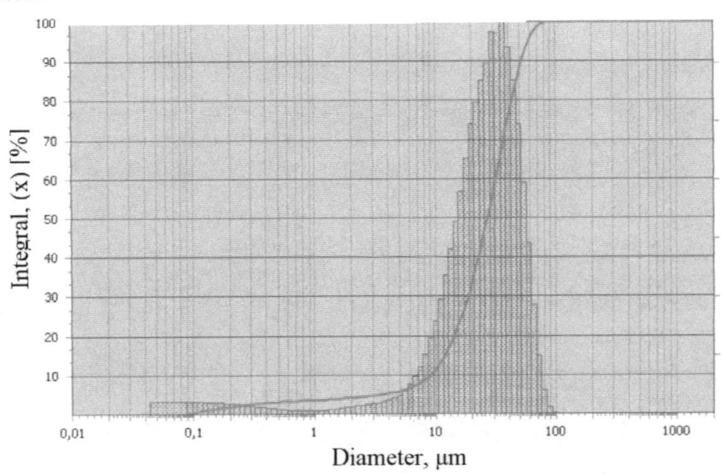

Figure 2. Fractional composition of titanium hydride powder.

IOP Conf. Series: Materials Science and Engineering **945** (2020) 012079 doi:10.1088/1757-899X/945/1/012079

Analysis of the data on the granulometry of the titanium hydride powder after grinding showed that the particle sizes are in the range from 0.1 to 107 μm (Fig. 2), with most of the particles in the range from 10 to 70 μm. The modal particle diameter is 35.41 μm, and the specific surface area of the particles is 15279 cm^2/cm^3. The use of titanium hydride powder with high dispersion is a necessary condition for obtaining homogeneous polymer composites. The use of preliminary grinding of titanium hydride shot will significantly improve the physical and mechanical characteristics of the final composites.

Ground titanium hydride was used to synthesize polymer composites. Composites with different filler contents up to 60 wt% were manufactured. The introduction of a filler of more than 60 wt% was not possible. With a small amount of polymer, the filler does not bind into a single composite, which leads to low physical and mechanical characteristics and destruction of the composite.

Table 2 presents data on the density of the obtained composites based on polystyrene, depending on the content of titanium hydride.

Table 2. Density of the developed composites.

Titanium hydride content, wt%	–	20	40	60
Density, g/cm^3	0.99	1.21	1.47	1.71

As expected, the addition of titanium hydride increases the density of the samples. A sample containing 60 wt% titanium hydride has the highest density of 1.71 g/cm^3, and a sample of polystyrene without filler has the lowest density.

One of the characteristics of the near-surface layers of the material is the Vickers microhardness (HV), which makes it possible to assess the nature of changes in the micromechanical properties and the degree of mechanical heterogeneity in composites. To assess the Vickers microhardness, a pyramidal diamond indenter was pressed into the surface of polystyrene composites under load for at least 15 seconds. The load in all measurements was the same - 200 g. Table 3 presents data on the microhardness by the Vickers method of polystyrene composites with different titanium hydride content. Figure 3 shows the obtained prints of a tetrahedral pyramid on the studied composites.

Table 3. Vickers microhardness (HV) of the developed composites.

Titanium hydride content, wt%	–	20	40	60
Vickers microhardness (HV)	19.87± 0.20	24.42± 0.56	29.26± 0.63	36.17± 0.78

Analysis of the data in Table 3 showed that the introduction of the proposed filler titanium hydride significantly increases the microhardness of composites, the higher the filler content, the higher the Vickers microhardness. An increase in the Vickers microhardness with an increase in the filler content is explained by the fact that the introduced filler titanium hydride has a higher hardness. The hardness of titanium compounds on the Mohs scale is 4-5, while the hardness on the Mohs scale for polymers, including polystyrene, is only 1 [18]. It is known that the introduction of inorganic particles significantly increases the mechanical characteristics of composites, including microhardness [19].

a b

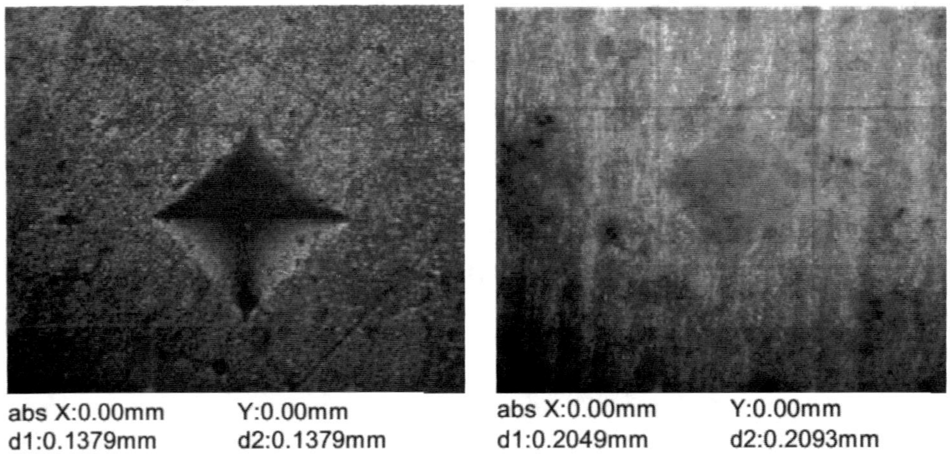

Figure 3. Image of the indentation of the indenter of a microhardness tester when measuring the microhardness at a load of 200 g: a - pure polystyrene, b - a composite with 60 wt% titanium hydride.

To calculate the hardness, the dimensions of the resulting print are measured. The size of the print is measured optically, for which the sample is unloaded and the indenter is removed. During these actions, elastic deformation is restored in the polymer material, as a result of which the size of the indentation differs from the true one achieved when the indenter was inserted. Perhaps, in this case, the true microhardness values are less than those calculated from the measurement results.

4. Summary

The work has established the possibility of synthesizing polymer composites based on thermoplastic polystyrene and ground titanium hydride. It has been established that grinding titanium hydride shot in a jet-vortex mill effectively grinds titanium hydride to a powdery state. The particle size of titanium hydride after grinding is in the range from 0.1 to 107 μm. The modal particle diameter is 35.41 μm, and the specific surface area of the particles is 15279 cm^2/cm^3.

Composite samples were obtained with different contents of titanium hydride filler: 20, 40, and 60 wt%. Samples containing more than 60 wt. % had low mechanical characteristics. After pressing, these samples did not form a single composite due to the lack of a binder, which led to low mechanical characteristics.

The introduction of titanium hydride increased the density of the samples. The sample containing 60 wt% titanium hydride has the highest density of 1.71 g/cm^3, and the sample of polystyrene without filler has the lowest density (0.99 g/cm^3).

The introduction of the proposed filler titanium hydride also significantly increases the microhardness of the composites, the higher the filler content, the higher the Vickers microhardness. The Vickers microhardness of a pure polystyrene specimen is 19.87 ± 0.20 HV, and a sample containing 60 wt% titanium hydride - 36.17 ± 0.78 HV (at a load of 200 g).

It was found in the work that the introduction of titanium hydride increases the physicomechanical characteristics of the composites. Further research will be aimed at studying the radiation-protective properties of the developed composites. The effect of neutron radiation on composites will be studied for the possibility of their use in biological protection of transport nuclear power plants.

5. References

[1] Pavlenko V I, Lipkanskij V M, Yastrebinskii R N 2004 Calculations of the Passage of Gamma-Quanta through a Polymer Radiation-Protective Composite *Journal of Engineering Physics and Thermophysics* **1(77)** 11-14

[2] Kesarwani S 2017 Polymer Composites in Aviation Sector. A Brief Review Article

International Journal of Engineering Research & Technology **6** 518-525

[3] Sorokin V V, Sharapov O N, Shunkin N M, Kiryushina N Yu 2019 New polymeric composites based on epoxy resin with techogenic wastes *Bulletin of BSTU named after V.G. Shukhov* **6** 8-13

[4] Yastrebinsky R N, Bondarenko G G, Pavlenko V I 2017 Radiation Hardening of Constructional Cement–Magnetite–Serpentinite Composite under Gamma Irradiation at Increased Dose *Inorganic Materials: Applied Research* **5(8)** 691-695

[5] Pavlenko V I, Bondarenko G G, Yastrebinsky R N 2017 Attenuation of Photon and Neutron Radiation Using Iron–Magnetite–Serpentinite Radiation-Protective Composite *Inorganic Materials: Applied Research* **2(8)** 275-278

[6] Mueller W M, Blackledge J P, Libowitz G G 1973 Titanium hydrides, in Metal Hydrides. Mueller W.M., New York - London: Academic Press. 336-383

[7] Mazitov A B, Oganov A R, Yanilkin A V 2018 Titanium-hydrogen interaction at high pressure *Journal of Applied Physics* **123** 235901

[8] Sakintuna B, Lamari-Darkrim F, Hirscher M 2007 Metal hydride materials for solid hydrogen storage: A review *International Journal of Hydrogen Energy* **32(9)** 1121–1140

[9] Yastrebinsky R N 2017 Decrease gripping gamma–radiation scale composite neutron and protective material on the basis of the modified hydride of the titan with various content of atoms of bor *Problems of Atomic Science and Technology* **4(110)** 103-106

[10] Pavlenko V I, Yastrebinskii R N, Voronov D V 2008 Investigation of heavy radiation–shielding concrete after activation by fast neutrons and gamma radiation *Journal of Engineering Physics and Thermophysics* **4(81)** 686–691

[11] Özkalayc F, Kaçal M R, Agar O, Polat H, Sharma A, Akman F 2020 Lead(II) chloride effects on nuclear shielding capabilities of polymer composites (Journal of Physics and Chemistry of Solids) **145** 109543

[12] Pavlenko V I, Bondarenko G G, Yastrebinsky R N 2016 Radiation resistance of structural radiation-protective composite material based on magnetite matrix *Inorganic Materials: Applied Research* **5(7)** 718-723

[13] Muthamma M V, Bubbly S G, Gudennavar S B 2020 Attenuation properties of epoxy-Ta_2O_5 and epoxy-Ta_2O_5-Bi_2O_3 composites at γ-ray energies 59.54 and 662 keV *Journal of Applied Polymer Science* **137(44)** 49366

[14] Pavlenko V I, Epifanovskii I S, Yastrebinskii R N, Kuprieva O V 2011 Thermoplastic constructional composite material for radiation protection *Inorganic Materials: Applied Research* **2(2)** 47-52

[15] Cherkashina N I, Pavlenko A V 2018 Influence of SiO_2 crystal structure on the thermal cycle of polymer composites *Construction Materials and Products* **1(4)** 21-29

[16] Yastrebinsky R N, Bondarenko G G, Pavlenko A V 2018 *Structural Features of Mineral Crystalline Phases and Defectiveness of Bismuth Organo-siliconate Crystals at High Temperatures* (Inorganic Materials: Applied Research) 5(9) 825-831.

[17] Yastrebinsky R N, Pavlenko V I, Matukhin P V, Cherkashina N I, Kuprieva O V 2013 Modifying the surface of iron-oxide minerals with organic and inorganic modifiers *Middle East Journal of Scientific Research* **18(10)** 1455-1462

[18] Encyclopedia of Materials: Science and Technology, Editors-in-Chief: K.H. Jürgen Buschow, Robert W. Cahn, ... Patrick Veyssière, 2001 Elsevier Ltd. 10388

[19] Zamfirova G 2009 *Total microhardness as a method for determination of the elastic properties of polymer materials 12-th International Conf." Mechanics and Tecnology of Composite Materials"* Varna, Bulgaria, 275-280

Acknowledgments

The work was carried out within the framework of the State Assignment of the Ministry of Education and Science of the Russian Federation, project No. 0625-2020-0011.

IOP Conf. Series: Materials Science and Engineering **945** (2020) 012080 doi:10.1088/1757-899X/945/1/012080

Heat-resistant structural composite material based on modified titanium hydride

R N Yastrebinsky[1*], A A Karnauhov[1], L V Denisova[1] and E O Pilavidou[1]

[1]Department of Life Safety, Belgorod State Technological University named after V.G. Shukhov, 46, Kostyukova str., Belgorod, 308012, Russia

E-mail: yrndo@mail.ru

Abstract. The paper presents the results of studies of technological parameters of production, physical, mechanical, thermophysical properties and processes of gas formation during thermal heating of a composite material based on a modified fraction of titanium hydride and a Portland cement binder. The introduction of a plasticizing additive into the composite makes it possible to improve the technological and strength characteristics of the protection material. It has been shown that the decrease in the physical and mechanical characteristics of the material does not exceed 10 - 20%, which indicates its thermal stability in terms of the main performance indicators. The introduction of a boron additive does not significantly affect the physical and mechanical properties of the composite. It was found that the main gas evolution occurs during the drying of composites and practically ends after its completion. Reheating the cooled block to the initial heating temperature (300-350 °C) practically does not cause gas or water evolution, if during the cooling period the material had no contact with water.

1. Introduction

To protect against reactor radiation, materials of various nuclear composition and density are used, according to which they are divided into light (with an increased content of hydrogen, boron, graphite), heavy (metals) and medium density (metal-hydrogen).

The most used light hydrogen-containing material is water, where the concentration of hydrogen nuclei is 6.7 x 10^{23} nuclei/cm^3. The relaxation length for fast neutrons of the fission spectrum near water is from 8.0 to 10.6 cm (for the equilibrium spectrum region of 10.0 cm). The main disadvantage of water is its radiolysis and non-structural nature (the need to use metal sheathing) [1-3].

Also effective light materials are polymers, in particular polyethylene, with a hydrogen nucleus concentration of $8 \cdot 10^{22}$ nuclei/cm^3. The relaxation length of fast neutrons (λ_b) for the equilibrium spectrum is 8.0 cm. The main disadvantage of polyethylene is its low operating temperature (up to 80 °C) and fire hazard [4-7].

To absorb thermal neutrons, boron carbide is widely used as an additive or filling due to the high absorption of the Maxwellian neutron spectrum and the insignificant formation of secondary gamma quanta. When the densities are comparable with graphite, λ_b for boron carbide is 10.2 cm [8-10].

Metals are mainly used as heavy protections, among which iron-containing ones (steel or cast iron) are the most relevant due to their high structural design and availability. The steel grade is determined by the operating condition of the protective structure and depends on the amount of carbon and

IOP Conf. Series: Materials Science and Engineering **945** (2020) 012080 doi:10.1088/1757-899X/945/1/012080

alloying elements. The use of titanium, tungsten, molybdenum and other heavy metal protections is mainly limited by their cost.

Metal-hydrogen materials are widely used due to their high efficiency with respect to both neutron and gamma radiation. Among the complex protections, iron-water and lead-water materials are used, represented, respectively, by steel or lead sheets in water. This type of shielding is used in VVER reactors located between the core and the steel reactor vessel. By a certain ratio of the water and metal components, one can achieve equality of relaxation lengths, both for neutron and gamma radiation. So for steel protection, the optimal concentration of the metal is 70-80 wt% (the relaxation length for neutrons and gamma quanta is 7.5 cm), and for lead protection 20 wt% (relaxation length 10 cm). The limitation of such protection is the radiolysis of water and the associated formation of hydrogen, the operating temperature and the need to organize heat removal [10].

A special type of metal-hydrogen protection is made up of metal hydrides, in particular, lithium, zirconium and titanium hydrides. Despite their higher cost, they are the most promising materials for heat-resistant structures of metal-hydrogen shields used for ship and space nuclear power. Thus, the relaxation lengths of the neutron dose rate for hydrogenated metal plates based on titanium and zirconium hydride for the equilibrium spectrum region are, respectively, 5.5 and 7.3 cm [11]. The disadvantage of hydrogenated titanium and zirconium metal matrices is the complexity of the production technology, insufficient thermal stability and low structural properties (brittleness).

One of the main characteristics of the radiation-protective properties of concretes is the content of chemically bound water, since physical mixing water (including hydrosilicate water) is removed from the material during its thermal drying. The operating temperature of reactor concrete is 280 °C. Thus, the content of chemically bound water in concrete should be determined by the hydrogen-containing aggregates used. At the same time, the achievement of the specified properties of the material is ensured not only by the correct choice of the type and amount of filler, but also by the observance of the technological parameters for the production and drying of the concrete composite [12].

One of the most effective fillers of reactor concrete is titanium hydride, which has better protective properties and a lower production temperature (titanium hydrogenation) than zirconium hydride. The problem of low thermal stability and brittleness of titanium hydride is partially solved by its use in the form of a shot. This technology allows titanium hydride to be used as a concrete filler, which provides structural protection properties. However, the possibility of cracking in hydrogenated titanium shot reduces its thermal stability to 250-300 °C due to hydrogen diffusion, which significantly limits the possibility of its use for radiation protection.

In this regard, the authors proposed the use of a modified fraction of titanium hydride obtained by chemical and electrochemical deposition of hydrogen traps on the surface of titanium hydride based on boron-containing systems or metallic titanium with a defect structure [13-17].

This paper presents the technological parameters of production, physical, mechanical, thermophysical properties and processes of gas evolution in a composite material based on a modified fraction of titanium hydride and a Portland cement binder.

2. Materials and methods

Modified titanium hydride shot, Portland cement binder, water, and plasticizing additive were used as the main components in the preparation of composites.

A titanium hydride shot with a hydrogen content of 3.5% by weight, modified by electrochemical deposition of titanium metal from a titanium oxalate solution, was used.

Portland cement grade CEM I 52.5N in accordance with GOST 31108-2003 with a Portland cement clinker content of at least 95% was used as a binder.

For mixing composite mixtures, drinking water is used that meets the requirements of GOST 2874-82.

Anionic plasticizer of the C-3 grade according to TU 5730-004-97474489-2007 was used as a plasticizing additive.

Used boron additive contains 0.78 wt.% Boron atoms.

IOP Conf. Series: Materials Science and Engineering **945** (2020) 012080 doi:10.1088/1757-899X/945/1/012080

The analysis of gases released during the thermal treatment of the composite was carried out by the chromatographic method on a 3700 series chromatograph.

3. Results and discussion

Taking into account the performed multi-group modeling to assess the effectiveness of the protective properties of materials based on titanium hydride with respect to neutron and gamma radiation, as well as studies of the thermal stability of titanium hydride fraction, optimal compositions and technology for obtaining a heat-resistant structural composite material for radiation protection have been developed.

Due to the thermal stability of the modified titanium hydride shot, composites based on it can be used in radiation shielding structures operating under prolonged thermal loads above 350 °C. The introduction of a plasticizing additive into the composite makes it possible to improve the technological and strength characteristics of the protection material. The use of a boron-containing additive can significantly reduce the yield of captured gamma radiation, as well as the activation of the material and surrounding structures.

Based on the design features of the radiation protection of a nuclear reactor (stacking the material in metal structures of various geometries), when developing the composition, priority was given to the maximum content of titanium hydride fraction as the main protection material, the density of the protective composite, the degree of compaction and mobility of the prepared composite mixture. The latter circumstance plays a decisive role in the filling of hard-to-reach volumes of metal structures of radiation protection and the possible formation of voids in it.

An important factor affecting the change in strength and deformability is the granulometry of the aggregate, which affects the workability of the composite concrete mixture. The filling of titanium hydride shot has increased fluidity, which makes it possible to better fill difficult-to-reach volumes. When using a cement bond, to obtain the highest density of composites, it is necessary to ensure the maximum packing of the shot particles.

If the particle sizes of polydisperse mixtures are limited to the values of the smallest and largest diameters, in principle two ways of obtaining the maximum packing density are possible. In the first case, two to four discrete fractions with a fixed particle size are mixed. The second way to achieve the maximum packing density is to introduce a polydisperse filler with a continuous size distribution in the range of the smallest and largest particle diameters.

Analysis of literary sources [3-5] reveals a number of features in the methods of stacking and compaction of bulk materials in the form of a shot.

First, to obtain a denser packing of granules, the ratio of the diameter (area) of the container to the diameter of the granules should be more than 15 units. Secondly, when compacting the single-fraction composition of spherical granules, the volume fraction of the packing does not exceed 70 - 74%.

When using two fractions of spherical granules in the package, the volume fraction increases and can be 83 - 86%. When using three fractions of spherical granules, the packing density can be increased up to 90%.

The optimization of the fractional composition of the titanium hydride shot, carried out on the basis of the considered theoretical provisions, made it possible to achieve a bulk density of up to 2.6 g/cm^3.

According to the results of the carried out multi-group modeling, technological and operational requirements, the optimal composition and technology for obtaining the radiation protection material have been developed.

The main characteristic of the workability of freshly prepared composite mixtures is the mobility coefficient determined by the truncated cone settlement method. It has been experimentally found that the necessary workability and mobility of the mixture is achieved when the value of the mobility coefficient is not less than 2.0 and 1.4, respectively, for fluid and high-density mixtures. For the developed composite composition, the mobility coefficient was 1.49 ± 0.08 (before the mixture was compacted).

Physicomechanical and thermophysical characteristics of the composite material are shown in table 1.

Table 1. Physicomechanical and thermophysical characteristics of the composite.

Indicator	Numerical value
Bulk density of the composite mixture, g/cm^3	3.32±0.03
Density (28 days of hardening), g/cm^3	3.30±0.03
Density (drying 110 ° C), g/cm^3	3.28±0.03
Density (drying 300 ° C), g/cm^3	3.20±0.03
Ultimate compressive strength after 28 days of hardening (28 days R_{com}), MPa	39.5±1.8
28 days R_{com} (drying 110 °C), MPa	42.7±1.8
28 days R_{com} (drying at 300 °C), MPa	36.4±1.8
28 days R_{com} (drying 350 °C), MPa	32.8±1.8
Mixture mobility coefficient	1.49±0.08
Thermal conductivity coefficient at 20 °C, W/(m·K)	3.27
Linear thermal expansion coefficient, m·deg^{-1}	$5.08 \cdot 10^{-6}$
Specific heat, kJ/(kg·K) at 20 °C	1.018
50 °C	1.024
100 °C	1.031
150 °C	0.894
200 °C	0.911
250 °C	0.918
300 °C	0.926

Sufficiently complete and reliable data have been obtained on the main physical and technical properties of the developed composite materials based on titanium hydride shot in the range of operating temperatures as applied to long-term operating conditions. Experimental statistical results of the obtained experimental-technological characteristics of the material are presented in table 2.

Table 2. Experimental and technological characteristics of radiation protection materials.

Indicator	Numerical value
The flammability group	is not flammable
Change in material mass (%) in oxygen atmosphere at a temperature of 300° C for: 500 h	-0.15
1000 h	-0.18
5000 h	-0.20
Change in material volume (%) in oxygen atmosphere at 300 °C for:	
500 h	+0.31
1000 h	+0.32
5000 h	+0.36
Change in material density (%) in an oxygen atmosphere at a temperature of 300° C for: 500 h	- 0.81
1000 h	-0.92
5000 h	-1.12
Change in material strength (%) in an oxygen atmosphere at a temperature of 300 °C for: 500 h	- 6.5
1000 h	-8.2
5000 h	-10.4

The results of the studies carried out showed a decrease in the mass of the samples within 1% during their heating, due to the loss of moisture collected by them from the air under standard conditions. This is confirmed by the DTA data, where exothermic peaks are noted in the temperature range 70-100, 150-200, 250-300 °C, corresponding to the stepwise dehydration of bound water from calcium hydroaluminate 2CaO • Al2O3 • $8H_2O$ (α-C2AH8), formed during cement hydration. This is accompanied by the formation of tricalcium aluminate 3CaO • Al_2O_3, which is capable of absorbing moisture from the ambient air. No other changes in the mass of the samples indicating the occurrence of thermooxidative reactions were found.

The change in the volume of the composite at an elevated temperature in the air atmosphere is of an average character, since for the same composition, both an increase and a decrease in the geometry of the samples within 0.35% are noted. This circumstance indicates the constancy of the geometric characteristics of the composite protection material.

The decrease in the density of the composite during prolonged heat treatment at 300-350 ° C is within 1% and corresponds to a decrease in weight due to the removal of moisture sorbed from the air.

The decrease in the strength of the composite with prolonged heating at a temperature of 300-350 °C was 14.8%. In this case, the residual compressive strength of the composite, even after holding for 1000 h, is not less than 30 MPa.

Thus, the decrease in the physical and mechanical characteristics of the material does not exceed 10 - 20%, which indicates its thermal stability in terms of the main performance indicators. The introduction of a boron additive does not significantly affect the physical and mechanical properties of the composite.

The results of computational and experimental studies of gas evolution from 1 ton of material, starting from the stage of mixing the mixture and until the composite warms up from 110 to 600 °C, are presented in table 3. The volumes of the evolved gas are reduced to normal conditions.

Table 3. Composition and volume of evolved gases during thermal heating of the composite material.

Temperature, ° C	Emitted gas volume, l/t				
	CO_2	CO	SO_2	Formaldehyde	H_2O ***
25	-	-	-	-	3.9
110	-	-	-	-	6.7
*300	6.27	0.08	0.10	0.014	20.7
**300	6.29	0.08	0.10	0.014	12.3
400-600	49.99	0.61	0.77	0.113	3.2
>600	563	6.95	8.68	1.277	-

* temperature rise from 110 to 300 °C and holding at 300 °C for 5 days.
** exposure at 300 °C for 20 days.
*** volume of condensed water.

From the data in Table 3, it follows that when drying 1 ton of material, based on the MPC standards of the working area, the volume of the room for heat treatment of 1 ton of material should be at least 69, 96 and 12 m^3 in terms of SO_2, formaldehyde and CO, respectively. In practice, the requirements of the MPC standards are met in the presence of supply and exhaust ventilation.

Thus, the main gas evolution occurs during the drying of composites and practically ends after its completion. Reheating the cooled block to the initial heating temperature (300-350 °C) practically does not cause gas or water evolution, if during the cooling period the material did not have contact with water.

4. Summary

1. The thermal stability of the modified titanium hydride shot allows the use of composites based on it in the structures of radiation protection operating under prolonged thermal loads above 350 ° C.
2. The introduction of a plasticizing additive into the composite makes it possible to improve the technological and strength characteristics of the protection material.

IOP Conf. Series: Materials Science and Engineering **945** (2020) 012080 doi:10.1088/1757-899X/945/1/012080

3. Under prolonged thermal loads, the decrease in the physicomechanical characteristics of the composite based on the modified fraction of titanium hydride does not exceed 10 - 20%, which indicates its thermal stability in terms of basic performance indicators.

4. The main gas evolution occurs during the drying of the composite material. Reheating the composite to the initial heating temperature does not cause gas evolution if the material did not come into contact with water during the cooling period.

5. References

[1] Sickafus K E 2007 Radiation-induced amorphization resistance and radiation tolerance in structurally related oxides *Nature materials* **6(3)** 217-23

[2] Fukai Y 1993 The Metal-Hydrogen System. Basic Bulk Properties *Springer Series in Materials Sciense* **21** 1237

[3] Kozhukhova N I, Strokova V V, Kozhukhova M I, Zhernovsky I V 2018 Structure formation in alkali activated aluminosilicate binding systems using natural raw materials with different crystallinity degree *Construction Materials and Products* **1(4)** 38–43

[4] Pavlenko V I, Cherkashina N I and Yastrebinsky R N 2019 Synthesis and radiation shielding properties of polyimide/Bi2O3 composites *Heliyon* **5** E01703

[5] Nasseri M M 2018 Comparison of HfB2 and ZrB2 behaviors for using in nuclear industry *Annals of Nuclear Energy* **114** 603-06

[6] Cherkashina N I and Pavlenko A V 2018 Modification of Optical Characteristics of a Polymer Composite Material under Irradiation *Tech. Phys.* **63** 571-75

[7] Pavlenko A V, Cherkashina N I and Yastrebinski R N 2016 Nanodisperse metalloorganosiloxane fillers of polymers *Nanotechnologies in Construction-a Scientific Internet-journal* **8(4)** 113-30

[8] Pavlenko A V, Cherkashina N I and Noskov A V 2017 Calculation of the frequency electronic transmission factors at the passage through the polymeric polyimide composite material filled by bismuth silicate *Problems of Atomic Science and Technology* **5** 21-26

[9] Cherkashina N I, Pavlenko V I and Yastrebinskii R N 2019 Phase Transitions and Electrophysical Properties of Tungsten(VI) Oxide in a 83-673 K Temperature Range *Russian Physics Journal* **62(5)** 870–75

[10] Pavlenko V I, Edamenko O D, Cherkashina N I, Kuprieva O V and Noskov A V 2015 Study of the attenuation coefficients of photon and neutron beams passing through titanium hydride *J. Surf. Invest-X-ray* **9** 546-49

[11] Pavlenko A V, Cherkashina N I and Yastrebinski R N 2016 Nanodisperse metalloorganosiloxane fillers of polymers *Nanotechnologies in Construction-a Scientific Internet-journal* **8(4)** 113-30

[12] Yastrebinsky R.N, Karnauhov A.A., Yastrebinskaya A.V. Improving the radiation-thermal stability of titanium hydride / Journal of Physics: Conference Series, 2020, pp.1-6.

[13] Cherkashina N I and Pavlenko A V 2018 Synthesis of Polymer Composite Based on Polyimide and Bi12SiO20 Sillenite *Polymer Plast. Tech. Eng.* **57** 1923-31

[14] Guseynov R M, Radzhabov R A, Magomedova U M 2019 The element with constant phase shift in galvanodynamic mode *Chemical Bulletin* **2(2)** 4-8

[15] Slyusar O A, Cherkashina N I and Yastrebinskaya A V 2015 Effect of Additives on Dispersed System Structure Formation *Refractories and Industrial Ceramics* **55(6)** 562-64

[16] Shestakov I Ya, Veretnova T A, Strekalova T A 2018 Investigation of water electrical activator with coaxial arrangement of electrodes *Chemical Bulletin* **4(1)** 12-18

[17] Choi E and Ejiri H 1996 Search for Time-Correlated Fast Neutrons from DD Fusion at Room Temperature *Japanese Journal of Applied Physics* A **5(35)** 2793-96

Acknowledgments
The work was carried out within the framework of the State Assignment of the Ministry of Education and Science of the Russian Federation, project No. 0625-2020-0011.

IOP Conf. Series: Materials Science and Engineering **945** (2020) 012081 doi:10.1088/1757-899X/945/1/012081

Use of glauconitic sand as a filler in the concrete mixes production

L N Botsman[1*], R S Lunev[1], M S Ageeva[1] and S V Nerovnaya[1]

[1]Department of Material science and material technology, Belgorod State Technological University named after V.G. Shukhov, Kostyukov St., 46, Belgorod, 308012, Russia

E-mail: lora80@list.ru

Abstract. In recent years, any production tries to minimize labor costs by automating processes. Because of the high rate of automation, machines have come to replace people. In addition to reducing the share of human labor and increasing machine labor, any manufacturer seeks to reduce the cost of raw materials, which contributes to the search for new solutions in the use of old materials. Russia's natural resource base is quite diverse, with special attention paid to the processing of technogenic waste and the creation of resource-saving technologies. Waste generation occurs at all stages of the movement of raw materials: from the moment of its extraction, when it is still a natural resource, to the end of operation of the product made from it. Waste dumps are a rich source of raw materials for the construction materials industry, as many of them are close to natural raw materials in their composition and properties. The complex physical and chemical composition and structure of waste from a number of industrial productions allow considering them as a real raw material base for the construction materials industry. The paper describes the principles of selecting the composition of concrete on glauconitic sand. Glauconitic sand is a waste product, which is why its use reduces the cost of concrete mixes and products made from them.

1. Introduction

Due to the active use of natural resources, the processing of huge volumes of rock mass, water and air is accompanied by a continuous increase in waste[1-2]. The sources of which include chemical, petrochemical, oil refining, energy, metallurgy, and other enterprises that allocate significant funds for their storage. One of the most extensive industries is the construction materials industry, where the share of raw materials reaches 50%. It allows using a large amount of waste, many of which are close to natural raw materials in their composition and properties [3-4]. At the same time, the accumulated scientific and practical experience in the use of industrial waste in Russia and abroad allows considering it as a valuable raw material for the production of construction materials [5-7].

In the Bryansk region, the agricultural and industrial enterprise LLC "PHOSPHATES" is developing the Polpinsky Deposit of natural phosphorite for the production of mineral fertilizers, such as phosphor flour. But a valuable mineral is not only phosphorite flour, which is used in agriculture as a fertilizer, but also the main waste from the extraction and processing of phosphorites – quartz-glauconitic sand. Quartz-glauconitic sand is removed during ore processing and stored in landfills. Currently, the accumulated sand is not used in practice.

2. Materials and methods

The most important component of this waste is glauconite. Glauconite is a mineral, an aqueous aluminum silicate of iron, silica and potassium oxide of non-permanent composition, belongs to the group of hydromica. The chemical composition of glauconites from various deposits varies widely. Features of the chemical composition and structure of the mineral have a significant impact on its properties [8]. Numerous studies have established that glauconite is characterized by a rich elemental composition. Unlike natural sands, which are monomineral in their composition and are mainly represented by quartz, glauconite sand contains illite, apatite, glauconite and small amounts of hematite. The chemical composition of sand was determined using x-ray phase analysis, which was performed on a general-purpose x-ray diffractometer (DRON-3M) using the powder method in the range of double angles 2θ 4÷56°. Diffraction patterns of samples were taken in the automated mode of the device using the PELDos program, which indicates the predominance of silicon oxide, which accounts for up to 90.1%.

Table 1 shows the main oxides that make up glauconitic sand.

Table 1. Chemical composition of glauconitic sand.

The content of oxides, %										
SiO_2	Al_2O_3	Fe_2O_3	FeO	TiO_2	CaO	MgO	K_2O+ Na_2O	P_2O_5	SO_3	F
83.2-90.1	1.1-1.3	1.9-2.2	0.41-0.46	0.13-0.15	1.5-2.0	0.3-0.5	1.1-1.2	1.2-1.4	0.2-0.4	0.11-0.13

The content of organic impurities does not exceed 1%, which is due to the hydraulic method of removing the host rock (glauconitic sand) during the extraction of apatite ores. The true density of glauconitic sand is 2620 kg/m^3. The bulk density of glauconitic sand is 1400 kg/m^3. The content of organic impurities does not exceed 1%. Voidness is 46.4%. Table 2 shows radiological studies of sand.

Table 2. Radiological studies of the sand of the CC "AIP-PHOSPHATES".

Determined indicators	Measure unit	Research results
Specific activity (Ara), ^{226}Ra	Bq/kg	132.8±18.8
Specific activity (ATh), ^{232}Th	Bq/kg	5.6±4.1
Specific activity (AK), ^{40}K	Bq/kg	118.8±47.5
Level of effective specific activity of natural radionuclides	Bq/kg	150.7±20.0

The grain composition of sand varies by Mk from 1.12 to 1.4 depending on the layers of production. The grain composition is shown in figure 1 as dispersion curves.

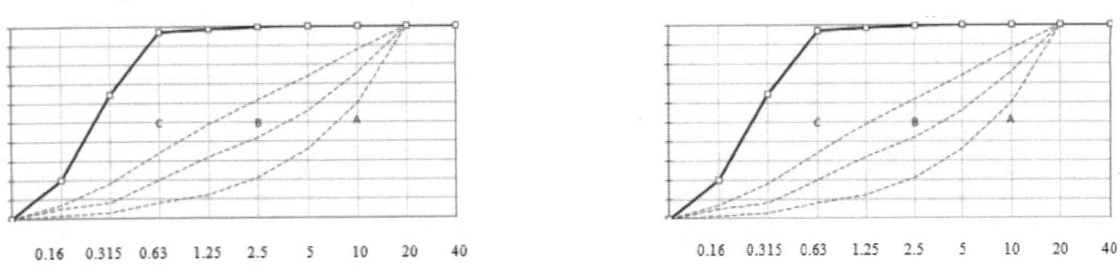

a b

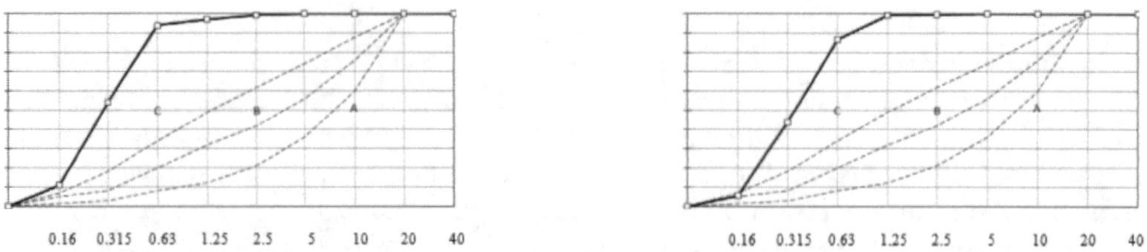

b d

Figure 1. Graphical representation of the screen analysis curve of glauconitic sand of different size modulus: a – Mk – 1.12; b – Mk – 1.2; c – Mk – 1.4; d – Mk – 1.65.

The microstructural features of sand were studied using a high-resolution scanning electron microscope (up to 1 nm) TESCAN MIRA 3 LMU, which includes an energy-dispersive spectrometer (EMF) X-MAX 50 Oxford Instruments NanoAnalysis for electron-probe microanalysis.

Analysis of the results showed that quartz grains have a rounded shape and this confirms its marine origin (Fig. 2).

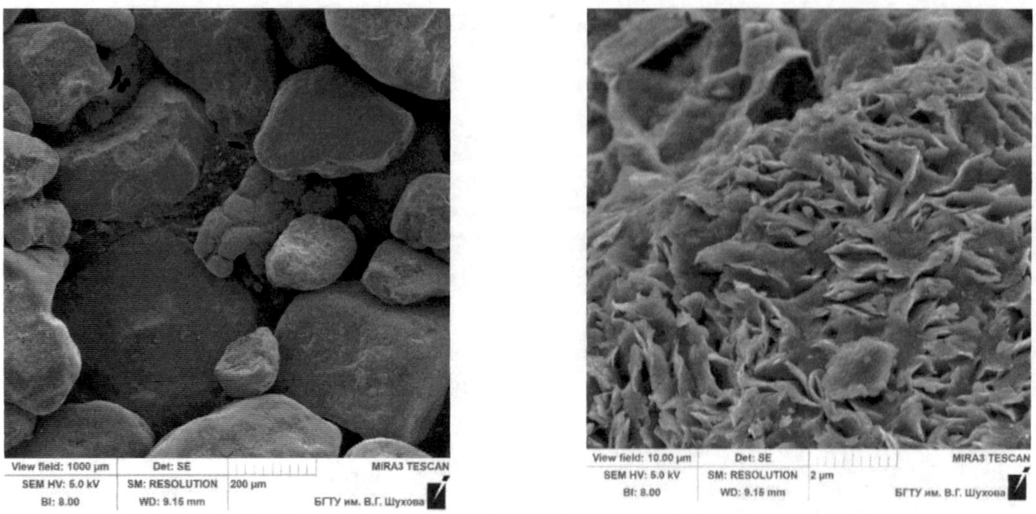

Figure 2. Surface morphology, size and nature of glauconite particles.

In addition to quartz grains, the sand consists of conglomerates of 100-800 microns in size, consisting of apatite-cemented quartz grains of 0.2-50 microns in size. There are also larger ones up to 200-300 microns.

At higher magnification, they can see the layered structure of this sand, as it is a product of biotite decomposition and belongs to ferruginous hydromica of variable composition. Its negatively charged lattice consists mainly of leafy aluminum silicate structures.

Thus, glauconitic sand from this quarry can be used in all types of construction, belongs to the 1st class of building materials, including for heavy and light concrete, for filling roads, without restrictions on the radiation factor [9-10]. All types of transport are allowed. Based on the results of radiological studies, it was found that the measured values of the content of natural radionuclides in the sand correspond to the standard values for 1 class building materials.

3. Results and discussion

According to the test results, it was found that the sand is classified according to GOST 8736-2014 "Sand for construction works. Technical conditions (as amended)" [11] as very small (Mk-1.12) or fine (Mk-1.65). It was also found that concretes on various types of cement with glauconitic sand have lower strength values compared to concrete on quartz sand. At the same time, the modulus of sand size was similar, and the mobility of the concrete mix with glauconitic sand was 15-25% less than on quartz sand. This is due to the ability of glauconite contained in sand to absorb a significant amount of water compared to quartz. The results of research have shown that the water demand of the concrete mixture for obtaining a given mobility is directly proportional to the amount of sand in the mixture [12]. Excessive water content contributes to the formation of a less dense structure and leads to a significant increase in porosity.

To obtain equal strength concrete in comparison with the traditional aggregate – quartz sand, it is necessary to increase the cement content, which leads to its overspending. Despite the overestimated cement consumption when using glauconitic sand in a concrete mix, the calculation of economic efficiency shows the feasibility of using this sand as a fine aggregate. According to the calculations (table 3) the cost of concrete on quartz sand is 4410 rubles/m^3.

Table 3. Cost calculation for B25 P4 using quartz sand.

Name of the material	**Material consumption per 1 m^3, kg**	**Price, ruble/kg**	**Involved cost, ruble/kg**
Cement	340	5.5	1870
Sand, Mkr=1.7	780	0.38	296.4
Rubble fractions 5-20 mm	1150	1.9	2185
LST	2.72	20	54.4
Water	185	0.02	3.7
Production cost of 1m^3			**4410**

When replacing quartz sand with glauconite, keeping all other parameters, we get the cost of 4152 rubles/m^3 (table 4).

Table 4. Cost calculation for B25 P2 using glauconitic sand.

Name of the material	**Material consumption per 1 m^3, kg**	**Price, ruble/kg**	**Involved cost, ruble/kg**
Cement	340	5.5	1870
Sand, Mkr=1.65	780	0.05	39
Rubble fractions 5-20 mm	1150	1.9	2185
LST	2.72	20	54.4
Water	185	0.02	3.7
Production cost of 1m^3			**4152**

It is worth noting that with glauconitic sand on a given amount of water, we got the mobility of P2 instead of P4. The net savings while maintaining all parameters is 257 rubles/m^3. However, due to the high water demand of sand, in order to maintain the water-cement ratio and obtain the required mobility of P4, it is necessary to adjust the cement consumption (table 5).

IOP Conf. Series: Materials Science and Engineering **945** (2020) 012081 doi:10.1088/1757-899X/945/1/012081

Table 5. Calculation of the cost of B25 P4 using glauconitic sand.

Name of the material	Material consumption per 1 m^3, kg	Price, ruble/kg	Involved cost, ruble/kg
Cement	360	5.5	1980
Sand, Mkr=1.65	760	0.05	38
Rubble fractions 5-20 mm	1150	1.9	2185
LST	2.88	20	57.6
Water	185	0.02	3.7
Production cost of 1m^3			**4264**

In this case, with increased cement consumption, the net savings are 145 rubles/m^3. The strength and mobility are similar to those of a concrete mix on quartz sand.

4. Summary

Thus, the use of technogenic glauconitic sand as a fine aggregate for ordinary concrete allows preserving all the physical, mechanical and operational characteristics of the mixture and finished products. At the same time, the use of this sand reduces the cost of concrete mix and solves the problem of storing technogenic waste.

5. References

[1] Kikava O Sh, Solomin I A Construction waste recycling *Moscow* 2000
[2] Kalgin A A, Fakhratov M A, Kikava O ASh, Baev V V Industrial waste production of building materials *Moscow* 2002
[3] Suleymanova L A, Erokhina I A, Suleymanov A G 2007 Resource-saving materials in construction *Proceedings of higher educational institutions. Construction* **7** 113–116
[4] Tolstoy A D, Lesovik B S, Milkina A S 2018 Features of the structure of new generation concrete using technogenic materials *Bulletin of the Siberian State Automobile and Road University* **15 (4)** 588-595
[5] Al-Bu-Ali U S, Lesovik R V, Kharkhardin A N, Tolstoy A D, Akhmed A A A, Alaskhanov A Kh, Aymenov Zh T 2020 Calculation and selection of grain composition of aggregate from high-density concrete scrap packaging *Bulletin of BSTU named after V G Shukhov* **6** 18-28
[6] Akhmed A A, Fedyuk R S, Liseytsev Yu L, Timokhin R A, Murali G 2020 Use of Iraqi concrete scrap as a filler and aggregate for heavy and light concrete *Construction Materials and Products* **3(3)** 28-39
[7] Onoprienko N N, Salnikova O N 2019 Improving the efficiency of construction solutions of domestic production *Bulletin of BSTU named after V G Shukhov* **8** 22-29
[8] Tolstoy A D 2020 Fine grained high strength concrete *Construction Materials and Products* **3 (1)** 39-43
[9] Bazhenov Yu M 1975 "Method for determining the composition of various types of concrete"
[10] Suleymanova L A, Lesovik R V, Glagolev E S, Sopin D M 2008 High-quality concretes on technogenic raw materials for critical products and structures *Bulletin of BSTU named after V G Shukhov* **4** 34–37
[11] GOST 8736-2014 "Sand for construction works. Technical specifications (as amended)"

IOP Conf. Series: Materials Science and Engineering **945** (2020) 012081 doi:10.1088/1757-899X/945/1/012081

[12] Fomina E V, Voytovich E V, Fomin A E, Lebedev M S, Kozhukhova N I 2015 Assessment of the radiation quality of OEMK slag for its use in construction composites *Bulletin of BSTU named after V G Shukhov* **6** 130–133.

Acknowledgments
The work is realized in the framework of the President Grant in Rissian Federation № NSh-2584.2020.8using the equipment of the Center for High Technologies of BSTU named after V.G. Shukhov.

IOP Conf. Series: Materials Science and Engineering **945** (2020) 012082 doi:10.1088/1757-899X/945/1/012082

Stabilization of metallurgical slug from arc steel-making furnaces

A N Bodyakov[1*], K V Meshkova[1] and G S Dukhovny[1]

[1]Belgorod State Technological University named after V.G. Shukhov, Kostyukov St., 46, Belgorod, 308012, Russia

E-mail: savaa72@mail.ru

Abstract. At the present time the deficit of stone construction materials for transport construction points to the need for improving the range of used materials. The experience of motorway construction in economically advanced countries shows that the deficit of materials and construction costs can be significantly reduced through extensive use of industrial wastes. One promising avenue of researches is the use of metallurgical production wastes. Main methods on the prevention of slag decomposition by its stabilization are considered. The principle of increasing the stability of stone slag materials by adding a stabilizer to the steel-making slag composition during slag removal from the arc steel-making furnace, ensuring slag stabilization and prevention of the silicate decomposition after cooling, is established. Processes of the slag structure formation during its stabilization are studied. Processes of crystallisation and change of structure of dicalcium silicate are shown. The decomposition of steel-making slag can be prevented by adding a chemical stabilizer into its composition. It is established that a stabilizing agent shall consist of metallurgical production wastes that allows to fully stabilize slag and keep high-temperature modifications of belite in it when cooled. The possibility of obtaining crushed slag with a stable structure achieved by a chemical stabilization of melted slag is shown.

1. Introduction

One of the primary reasons constraining the pace of motorway construction in the Russian Federation is a severe shortage of main road construction materials, including crushed stone. According to the general assessment [1, 2], steel and foundry industries in Russia collected in heaps more than 300 million tons of steel-making slags. As per the directive of European Commission and European Parliament on wastes and by-products, steel-making slag is classified as a nonwaste by-product [3]. Despite of a sufficient number of available slagheap recycling schemes, no solutions on the recycling of special so-called self-disintegrating slag of the iron and steel Industry have been found to date. Self-slaking slags from arc steel-making furnaces (ASF) cannot be recycled with the use of schemes implemented in the industry, as they quickly turn into fine powder during solidification and cooling processes.

The reason of metallurgical slag decomposition is the polymorphic transformation of dicalcium silicate (C2S, belite) which is a part of its composition. There are five polymorphic forms of C2S at room temperature [3]. Due to a big difference in the lattice structure, transformation of β-C2S into γ-C2S is accompanied by the increase in volume by approximately 12 % resulting in slag slaking.

IOP Conf. Series: Materials Science and Engineering **945** (2020) 012082 doi:10.1088/1757-899X/945/1/012082

To stabilize high-temperature modifications of belite and prevent its polymorphic transformation into a low-temperature modification, either quenching or addition of stabilizers is required. The general requirement to stabilizers is a similar size of ionic radiuses and charges of a stabilizing element and ions of dicalcium silicate[4 - 6]. While embedding into the structure of dicalcium silicate, they deform the belite lattice preventing the rotation of silicon–oxygen tetrahedrons and their transformation into low-temperature polymorphic modifications. [7, 8]

Therefore, the purpose of this work is to study the influence of stabilizer addition to the structure and properties of steel-making slags. New technical solutions on recycling of melted slugs allow to ecxlude removal of metallurgical slags to heaps and ensure their large-capacity use in various industries[9, 10].

2. Materials and methods

To study physical, chemical and processing behaviour, samples of ASF slag obtained with the use of dry recycling process and slag cooled by a hydraulic method with the use of current process were taken. The chemical composition of selected samples is given in Table 1. There are no significant differences in the composition of slag cooled hydraulically or by air. The main difference may be in the defect rate of particles of slag cooled hydraulically or by air.

Table 1. Chemical composition of slag and stabilizer.

Name	Content of oxides (%)											
	CaO	SiO_2	$F_{e2}O_3$	MgO	Al_2O_3	MnO	Cr_2O_3	P_2O_5	ZnO	SO_3	K_2O	Na_2O
Air-cooled slag	39.73	30.04	17.39	8.67	2.39	0.82	0.25	0.17	0.16	0.05	0.04	0.14
Hydraulically cooled slag	34.30	29.21	22.12	9.71	3.17	0.67	0.13	0.18	0.22	0.15	0.08	0.15
Stabilizer	10.87	6.51	54.64	4.35	0.48	2.33	0.35	–	3.85	1.16	6.19	8.05

Stabilizer chemical composition is given in Table 1. To stabilize slag, the weight of added portion of stabilizer granules was accepted as equal to 0–7 % of the slag weight in increments of 1 %. The portion was added into melted slag in the form of a paper envelope (to imitate a paper bag), affixed to the end of a 1.5-meter rod.

The chemical composition of used raw and synthesized materials was determined with the help of an x-ray fluorescence spectrometer of ARL 9900 WorkStation series with integrated diffraction system. The scanning electron microscopy (SEM) was carried out to study the structure and composition of synthesized stone materials. Studies were carried out using the high-resolution scanning election microscope TESCAN MIRA 3 LMU. The strength grade of crushed slag (crushability) was determined by the degree of destruction of grains in the water-saturated condition during the cylinder crushing test (GOST 32817-2014). Crushing strength grade was determined by weight loss of the crushed stone sample resulting from the impact of shock loads from steel balls and abrasion forces in the abrasion testing machine (GOST 32819-2014). Micro-Devaluation 1 abrasion grade was determined by weight loss of the crushed stone sample, resulting from abrasion of material grains in the presence of steel balls and water (GOST 32816-2014). Resistance of crushed slag grain structure to disintegration was determined by weight loss of the sample after storage in distilled water within 30 days and further three-time steam curing (GOST 32858-2014). The freeze resistance grade was evaluated by exposing slag samples to alternating cycles of freezing in the water-saturated condition and thawing in water (GOST 32863-2014). A stabilized part of slag was estimated visually. To estimate the percentage output of the stabilized fraction, sifting of unstabilized part through a sieve with a mesh width of 0.63 mm was carried out. The calculation of disintegrated fraction was evaluated by the weight of sieved substance.

IOP Conf. Series: Materials Science and Engineering **945** (2020) 012082 doi:10.1088/1757-899X/945/1/012082

3. Results and discussion

The crystal structure of slag in the initial state is uneven and zonal which is conditioned by a regular distribution of additives and inclusions (Fig. 1). Separate polycrystals of monticellite in the form of grains grow freely until contact with each other. Thickness of these crystals does not exceed 400 nm. Magnesioferrite is represented by isometric rounded crystals forming compact aggregates. Photomicrographs of slag clearly show merwinite crystals with a rounded shape, larnite crystals resemble merwinite crystals.

Fig. 1. Photomicrographs of initial slag after decomposition.

Addition of 1 % of stabilizer granulas into melted slag ensures stabilization of 46.2 % of slag (Table 1). Addition of granulas in the proportion from 2 to 4 % of the weight of recycled slag results in the stabilization of more than 90 % of slag. The maximum percentage of stable fraction (100 %) can be observed in case of addition of 5 % of a stabilizer. Increase in the weight of a stabilizer over 5 % is accompanied by a partial decomposition of cooled slag (3.6–5.1 %) due to decrease of solubility of the added stabilizer as a result of temperature reduction in the crucible.

Table 2. Results of selection of the stabilizer portion weight.

Stabilizer portion weight (% of slag weight)	0.0	1.0	2.0	3.0	4.0	5.0	6.0	7.0
Behaviour of metallurgical slag melted with the addition of a stabiliser	Slag fully decomposed	Slag partially decompose. There are unstabilized inclusions 53.8 % of weight 46.2 % stabilized	Slag didn't decompose. There are unstabilized inclusions 6.6 % of weight **93.7 % stabilized**	Slag didn't decompose. There are unstabilized inclusions 4.7 % of weight **95.6 % stabilized**	Slag didn't decompose. Separate pieces have cracks 1.5 % of weight **98.5 % stabilized**	Slag didn't decompose during cooling **100 % stabilized**	Slag partially decomposed during cooling (3.6 % of weight). A stabilizer didn't dissolve completely 96.4 % stabilized	Slag partially decomposed during cooling (5.1 % of weight) A stabilizer didn't dissolve completely 94.9 % stabilized

Fig. 2 and 3 show the slag microstructure stabilized with 2 % and 5 % of stabilizer granules respectively. Petrographic studies of stabilized slag samples have shown that it has a tight porous structure of rounded pores up to bubbles of different shape, perceptible to the eye. The slag microstructure resem-

IOP Conf. Series: Materials Science and Engineering **945** (2020) 012082 doi:10.1088/1757-899X/945/1/012082

bles a porphyraceous structure: respectively idiomorphic, bigger crystals of minerals are immersed into less decrystallised, cryptomerous mass of the same minerals. Individual specimens demonstrated a specific orientation of crystals, although variously oriented, deeply intertwined and alternating individuals and aggregates prevail (Fig. 2, 3).

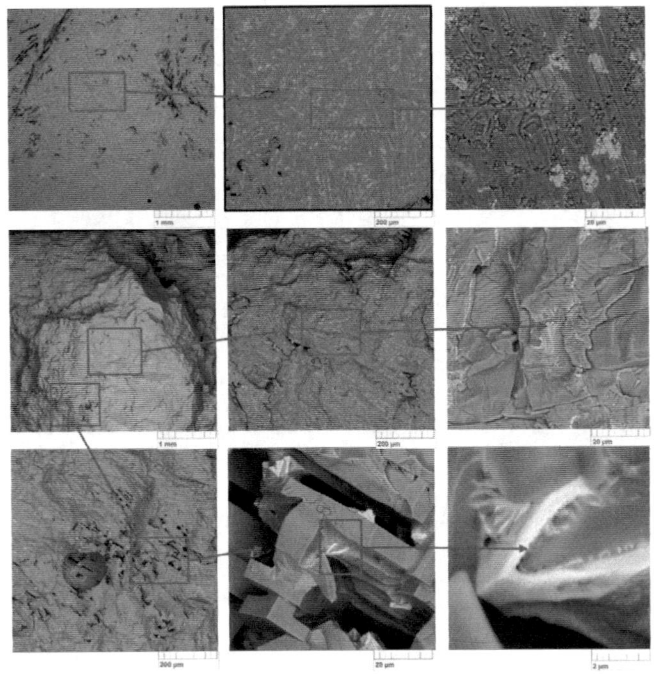

Fig. 2. Photomicrographs of slag specimens and cleavages stabilized with 2 % of an additive.

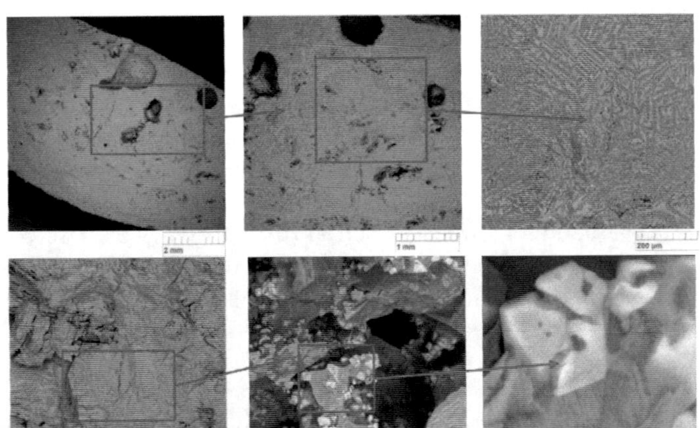

Fig. 3. Photomicrographs of slag specimens and cleavages stabilized with 5 % of an additive.

Two phases are clearly seen in the microstructure of stabilized samples: siliceous belite crystals (matrix) with inclusions of calcium ferrites (2CaO·Fe2O3). Calcium ferrites form a three-dimensional grid in the matrix, evenly distributing throughout the sample volume. The three-dimensional grid in slag with 2 % of a stabilizer is formed by thin films. Increase of a stabilizer content up to 5 % leads to the formation of disperse particles of calcium ferrite with bigger volume ratio and more even distribution in the matrix volume.

IOP Conf. Series: Materials Science and Engineering **945** (2020) 012082 doi:10.1088/1757-899X/945/1/012082

Physical and mechanical properties of the obtained stone material are given in Table 3. It is worth mentioning that increase in the amount of added stabilizer does not affect strength properties of ready products but rises the output percentage.

Table 3. Slag physical and mechanical properties.

Indicator name	Unstabilized slag	Stabilized slag
Strength grade	M300	M1000
Micro-Devaluation abrasion	Below the minimum permissible value	MD2
Abrasion grade in the abrasion testing machine	Below the minimum permissible value	I2
Freeze resistance grade	Below the minimum permissible value	F50
Strcuture stability	Below the minimum permissible value	US1

Physical and mechanical properties of broken slag not exposed to stabilization do not comply with minimum permissible requirements of the standard, such material cannot be used in the road construction. As per results of laboratory tests of samples, stabilized slag corresponds to strength grade in the water-saturated condition M1000, micro-Devaluation abrasion indicator MD2, abrasion grade in the abrasion testing machine I2, freeze resistance grade F50 and structure stability US1. Physical and mechanical properties of the obtained stone material allow to use it as base courses and surfacings in road structural layers.

4. Summary

Studies have shown that even distribution of a stabilizer in the volume of recycled liquid ASF slag ensures stabilization of high-temperature forms of dicalcium silicate (excluding silicate decomposition) during and after cooling. Addition of 5 % of a stabilizer to melted slag is the best solution. When adding a stabilizer in lower volume, an incomplete stabilization of liquid melt composition occurs, resulting in partial or full slag decomposition when cooled. Addition of a stabilizer in bigger volume results in incomplete dissolution and recovery of stabilizing components by melted slag. Slag stabilization is ensured by calcium ferrites formed when using a stabilizing substance close by its composition to recycled slag. The obtained stone material can be classified by its properties as crushed stone which can be used for construction and reconstruction of motorways as base courses and surfacings. Stabilization of melted slag will prevent formation of fine dust from γ-decomposition of dicalcium silicate which will lead to reduction of the negative impact of decomposition products on the environment.

5. References

[1] Smirnov L, Leontyev L, Sorokin Yu 2012 Processing and use of industrial waste from metallurgical production *Proceedings of the international Congress* 15-19

[2] Odenbah I 2020 Some new technologies in road construction. *Construction Materials and Products* **3(1)** 62–69.

[3] 2012 Position Paper on the Status of Ferrous Slag. Complying with the Waste Framework Directive 2008/98/CE URL: http://www.euroslag.org/status-of- slag/position-papers/(date of request 20.08.2020)

[4] Rizescu C, Bacinschi Z, Stoian V, Poinescu, A. 2010 Characterization of steel mill electric-arc furnace dust *Conferences(Tunisia)* 25

[5] Taranina T, Kabanova L, Korolev A, Zyrjanov F 2011 The results of jsc "chemk" stabilized low carbon ferro-chrome slag petrography research. *Mineralogy of technogenesis* **12** 62-71

[6] Beletskaya V, Rumyantseva E 2011 Prospects for the use of OEMK electric steelmaking slags *Bulletin of BSTU named after V G Shukhov* **3** 140-144
[7] Pontikes Y, Jones T, Geysen D, Blanpain B 2010 Options to Prevent Dicalcium Silicate-Driven Disintegration of Stainless Steel Slags *Archives Of Metallurgy And Materials* **55(4)** 1169–1172
[8] Sakamoto N 1996 Effect of cooling process to crystallization of stainless steel slag (development of control process of stainless steel slag dusting-1) *Current Advance in Materials and Processes* **9(4)** 803
[9] Pontikes Y, Kriskova L, Wang X 2011 Additions of industrial residues for hot stage engineering of stainless steel slags *2-nd International Slag Valorisation Symposium* 313–326
[10] Lopatin D, Chizhikova V, 2007 The new stability criterion of crystal-chemical stabilization of _β-C2S. *Advances in metallurgical processes and materials* 481–482

Acknowledgments
This work was realized in the framework of the Program of flagship university development on the base of the Belgorod State Technological University named after V G Shukhov, using equipment of High Technology Center at BSTU named after V G Shukhov.

Development of an algorithm for managing a multi-robot system for cargo transportation based on reinforcement learning in a virtual environment

L A Rybak[1*], L Behera[2], M A Averbukh[1] and A V Sapryka[1]

[1]Belgorod State Technological University named after V. G. Shukhov, 46, Kostyukova Str., Belgorod, 308012, Russia
[2]Indian Institute of Technology Kanpur, Kanpur, 208016, India

E-mail: rl_bgtu@intbel.ru

Abstract. This paper is devoted to the study of varieties of the Q-Learning algorithm – deep Q-networks and dueling Q-networks. These algorithms belong to the group of reinforcement learning algorithms. Neural network architectures are selected. The process of modeling the robot's operation in the problem of cargo delivery from a random point A to the green zone is described. The method of obtaining information about the environment by the robot using the Raycast method is described. A block diagram for controlling the robot movement has been developed, which consists of a block of positioning and state sensors, a neural network module, and a block for constructing a trajectory. The last two blocks together form a system for automatically controlling the agent's movement in the external environment. Modeling was performed in the Unity development environment. To work with ml agents, the special Unity ML-Agents tool is used. This tool is implemented using a modern DRL servo motor, which is based on the model of the optimization algorithm proximal policy optimization (PPO). A constructive simplification of the agent and environment to facilitate the reproduction of training scenes is implemented. An algorithm for training a robot in a random environment is presented. The optimal parameters of the algorithms under consideration are selected. Suggestions are made to improve the performance of algorithms for this problem.

1. Introduction

The problem of controlling robots in a previously unknown environment was solved at different times by different methods [1-7]. Most of the work in this area is devoted to path planning, where the environment was analyzed in order to find the most convenient path. The work of the foreign scientist S.V. Wilson [8] is considered one of the most significant works in this field and led to the birth of a whole class of problems devoted to animates (ANIMAT = ANIMAL + ROBOT), that is, robots trained using a reinforcement algorithm. The classic Wilson Animat works in a discrete world and is constantly trained in the same environment. The goal of the Animat is to learn how to reach the goal from any starting position in the minimum number of steps. There are works devoted to managing an autonomous agent in a continuous environment, where the agent learns to avoid obstacles, but does not achieve any goal [9] and a solution to the control problem is proposed, but only in a 2-dimensional environment [10]. All these works are characterized by the fact that they are constantly taught in the same environment, so it is only guaranteed that the robot can function effectively only in this environment and it is not known how effective its behavior will be if it changes slightly.

The paper will use an algorithm called time difference learning (TD, temporal difference), which belongs to the categories of learning algorithms without a model – without requiring information about the model, it can be applied to non-episodic tasks. Experiments will be presented with a robot that is trained in many typical environments, so it can function effectively in a completely unknown environment.

2. Materials and methods

As a platform for modeling, the Unity development environment was chosen with the special Unity ML–Agents tool connected – this is a new plugin in the Unity game engine that allows using Unity tools as the basis for creating an environment when training ml–agents. The robot has a task to reach the goal, avoiding collisions with obstacles. The environment is a three-dimensional space bounded around by walls. To facilitate the reproduction of scenes, all elements of the environment – agent, cargo, obstacles – are simplified and presented as geometric primitives. The agent (blue cube) appears in the center zone every time the scene is replayed. A cargo (a white parallelepiped) appears on the stage at a random location. The agent's task is to detect the cargo, approach it, and deliver it to the green zone in the fastest possible time (figure 1).

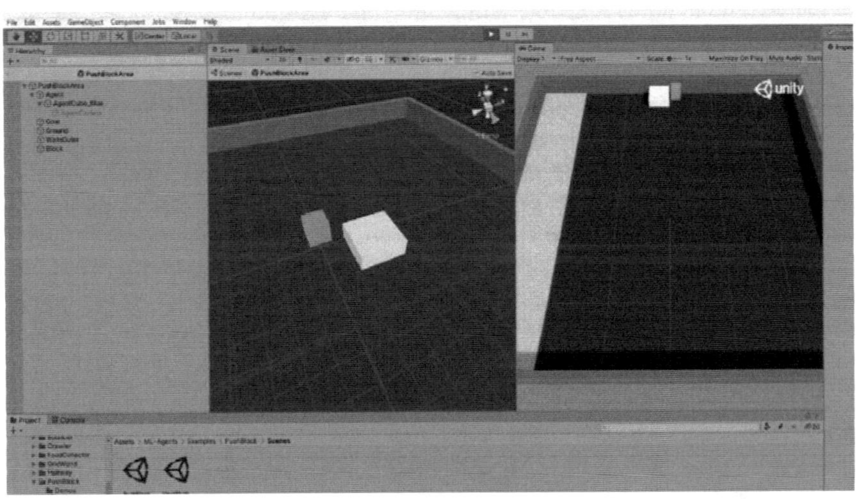

Figure 1. The agent (blue cube) and the load (white box) during training.

The robot receives information about the environment using a special Raycast method. A Raycast is a ray emitted from some object in some direction of some length (or infinite) to detect collisions with objects. After the ray is emitted, we get the object (or an array of objects, if we use Physics.RaycastAll) that it encountered, and then we can determine whether we hit the object we need. The operation diagram of the autonomous agent sensors is shown in figure 2 a, b.

IOP Conf. Series: Materials Science and Engineering **945** (2020) 012083 doi:10.1088/1757-899X/945/1/012083

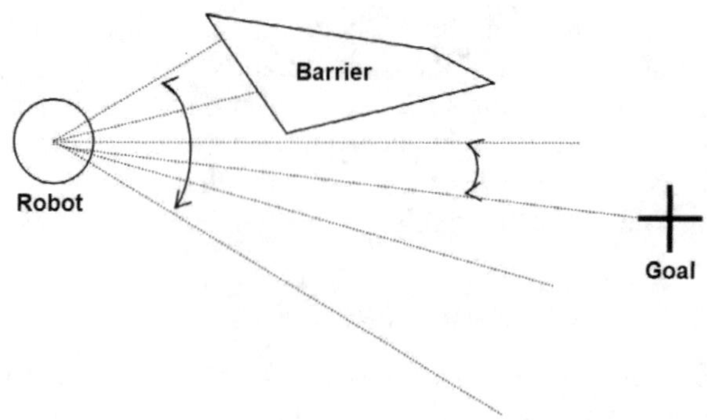

Figure 2a. Schematic representation of sensor operation.

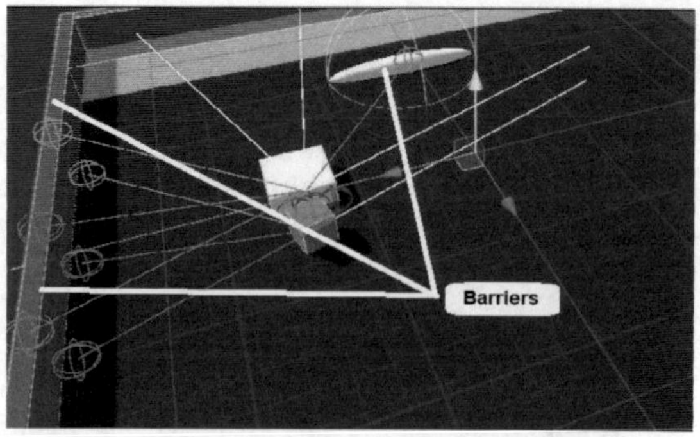

Figure 2b. Using Raycast rays to identify objects.

3. Results

3.1. Algorithm description

We will use the Q-learning algorithm. In Q-learning, an agent interacts with the environment iteratively by performing an action. The environment responds by informing the agent of the reward for this action and moving to the next state. This happens continuously until the environment is "resolved". Enhanced learning tries to learn the best sequence of actions. This is done by trying different combinations of actions, at first randomly, than using a policy based on what the model has learned from rewards up to this point. This happens until the environment reaches its final state [11, 12].

We also consider two modifications of this algorithm – Deep Q-network (DQN) and Duel Deep Q-network. As the standard Q-learning method is best suited for environments with a finite number of states and limited sets of actions, in which an exhaustive search for the optimal Q value was possible for all possible action/state pairs [10]. But most often they need to create an environment in which the number of states is very large, and in each state there are many actions available. Going through all the actions in each state would take too long. Another, more efficient approach is based on the approximation of a Q-function with some parameter θ (1).

$$Q(s,a;\theta) \approx Q^*(s,a) \tag{1}$$

They need to use a neural network with weights to approximate the value of Q for all possible actions in each state, it can be called a Q-network. However, the problem of network training and the problem of the type of objective function appear. The update rule for Q–learning looks like this:

$$Q(s,a) = Q(s,a) + \alpha(r + \gamma \max Q(s',a') - Q(s,a)) \tag{2}$$

where $r + y \max Q(s',a')$ – target value, and $Q(s,a)$ – predicted value. We strive to minimize this value by using the correct policy.

Similarly, in a Deep Q–network or DQN, they can define the loss function as the square of the difference between the target and predicted values. We also try to minimize losses by updating the weights θ.

$$E = (y_i - Q(s,a;\theta))^2 \tag{3}$$

where $y_i = r + \gamma \max_{a'} Q(s',a';\theta)$.

Gradient descent is used to update the weights and minimize losses. Therefore, in DQN neural networks are used as approximation functions for the Q-function, and gradient descent is used to minimize errors (figure 3).

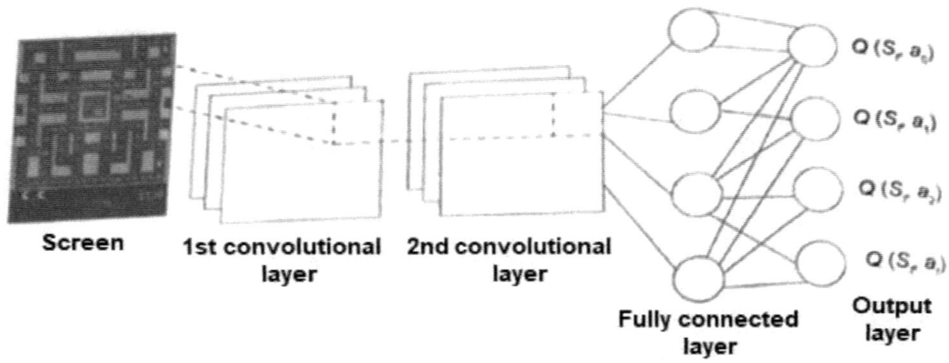

Figure 3. Deep q network architecture.

The Q function determines how good it is for the agent to perform action *a* in state *s*, and the value function determines how good it is for the agent to stay in state *s*. Let us introduce a new function called the advantage function. It can be defined as the difference between the value function and the average value of this state. The advantage function indicates how well the agent performs this action compared to other actions.

Thus, the value function sets the desirability of a state, and the advantage function sets the desirability of an action. When combining these functions, the result will tell us how desirable it is for the agent to perform action *a* in state *s*, that is, in fact, we get a Q-function, which can be defined as the sum of the value function and the advantage function:

$$Q(s,a) = V(s) + A(s) \tag{4}$$

Figure 4 shows the dueling network architecture DQN:

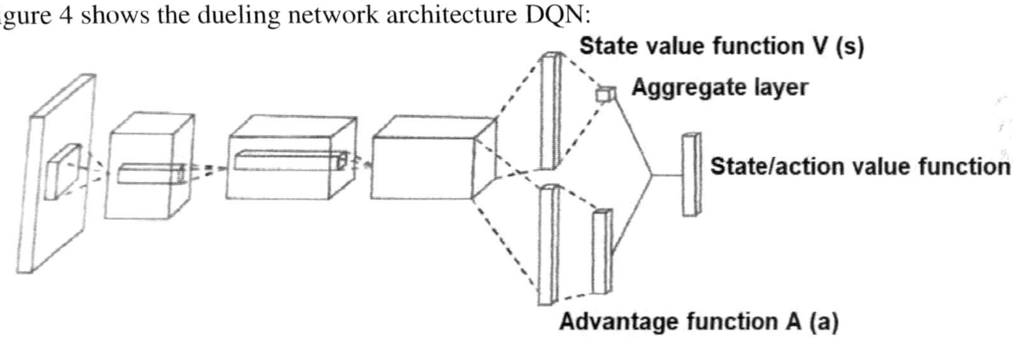

Figure 4. Dueling DQN architecture.

The dueling DQN architecture is almost identical to the DQN architecture, except that the fully connected layer is divided into two threads at the end. One thread calculates the value function, and the other calculates the advantage function. At the end, these two streams are combined by an aggregating layer, resulting in a Q-function [13].

3.2. Synthesis of the learning algorithm

The training process is divided into stages (tests). At each stage of training, the robot and the target are randomly placed at a new point in space. A new location for all obstacles is also randomly created. The reward that the robot received only at the end of the stage, in all other cases, the reward was zero. The stage ended when the robot reached the goal, when it encountered an obstacle, or because of a timeout. Different reward values are defined for all these three cases:

1) if the robot reaches the goal, it gets +1 to the reward;
2) when the robot collides with an obstacle, the reward value is calculated using the following formula:

IOP Conf. Series: Materials Science and Engineering **945** (2020) 012083 doi:10.1088/1757-899X/945/1/012083

$$r = 0.5 \cdot \exp(-2d_{goal} / l_{room}) \tag{5}$$

where d_{goal} – distance to the goal, and l_{room} – length of one side of the environment;

3) at inaction the robot receives a reward that is equal to the collision reward +0.3.

Four randomly generated convex quadrilaterals were used as obstacles in the learning process. At each stage of operation, the robot can choose one of five possible actions: move forward, move forward at an angle of 15 degrees to the left, move forward at an angle of 15 degrees to the right, turn 15 degrees to the left, and turn 15 degrees to the right. The robot motion control scheme is shown in figure 5.

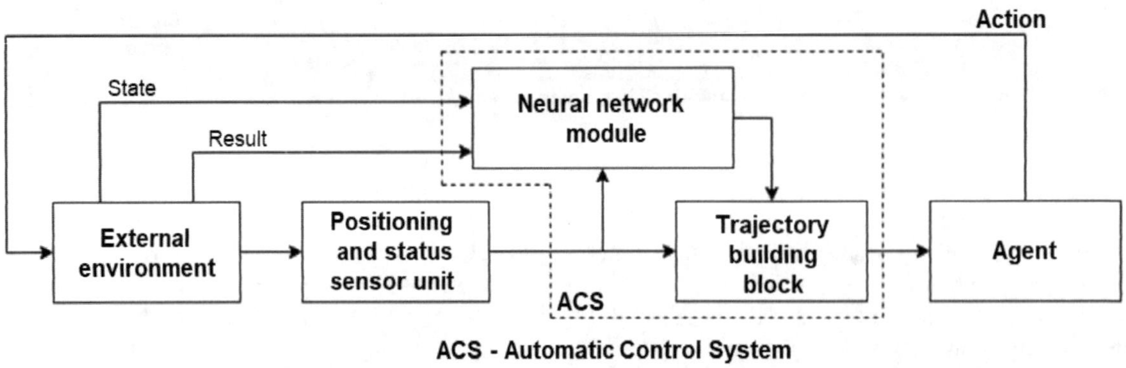

Figure 5. Scheme of the control system.

At each iteration of the robot appears in the generated environment, where it performs some actions based on the readings of the positioning and status sensor block. All data from this block is sent to the block of the neural network module and to the block of building the trajectory of movement, which gives commands to the robot to improve actions to achieve results. The robot agent performs actions in the external environment. In turn, the external environment transmits data about the agent's state and the result of its actions to the neural network module block.

3.3. Software implementation

The program uses several important hyperparameters that affect learning. The number of episodes is 2000, the discount factor is 0.99, the training series size is 64, and the number of hidden network layers is 256. Full characteristics of hyperparameters are shown in table 1.

Table 1. Network hyperparameters.

Hyperparameter	Value
Number of Episodes	2000
Number of Timesteps	1000
Print Checkpoint step every	4
Training Batch Size	64
Discount Rate / Gamma	0.99
Learning Rate / alpha	5,00E–04
Number of Hidden Layers	256
Fully Connected Layer 1 Units	64
Fully Connected Layer 2 Units	64
TAU	1,00E–03
Epsilon	0.1
Epsilon–Min	0.01
Epsilon–Decay	0.995

Figure 6 shows the working window of the simulator while training the agent to deliver cargo from a random point to the green zone without obstacles (purple parallelepipeds).

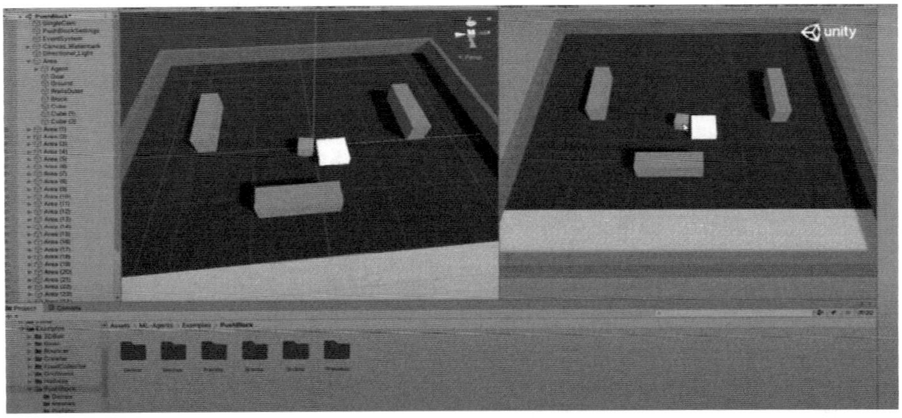

Figure 6. The process of training an agent in an environment with obstacles.

Figure 7 a, b shows the evaluation results obtained during model training. The two approaches had very different training behaviors. The Vanilla method showed a better start to the game score, with a faster increase in score over the first 250 episodes. However, after 500 episodes, both models converge on a similar pattern.

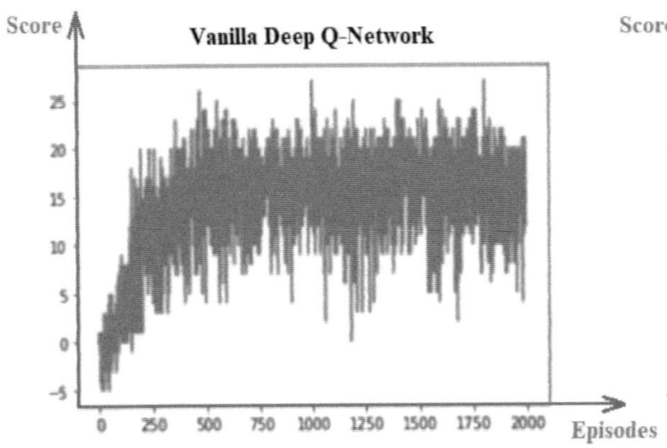

Figure 7a. Training quality indicators for a standard deep q-network.

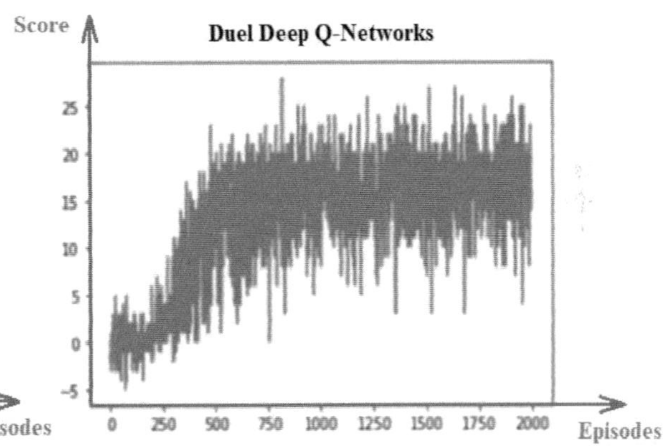

Figure 7b. Quality indicators of the dueling deep q-network.

Using the Vanilla method, a score of +15 was achieved in 260 episodes, while Dueling Networks achieved these score in 360 episodes.

4. Discussion

All the research discussed above is aimed at developing and testing a management system based on reinforcement learning. The analysis of reinforcement learning algorithms, namely the Q-learning algorithm and its modifications – Deep Q-learning and Duel Q-learning. Simulation of learning and experience reproduction by a robot in the Unity development environment with the ML-Agents module is performed. A block diagram of the developed system is constructed. The choice of hyperparameters of the studied neural network is made. The performance indicators of the standard deep Q-network and the dueling network are analyzed. Dueling networks were expected to improve on the standard method.

IOP Conf. Series: Materials Science and Engineering **945** (2020) 012083 doi:10.1088/1757-899X/945/1/012083

5. Summary

The resulting improvements can be explained by the fact that this model was not configured with any hyperparameter setting. However, regardless of the basic values of the hyperparameter, the agent is able to quickly and efficiently solve the environment.

6. References

[1] Rybak L A, Khalapyan S Y, Malyshev D I 2019 Optimization of the trajectory of movement of a parallel robot based on a modified ACO algorithm *IOP Conference Series: Materials Science and Engineering* **560(1)** 012036

[2] Khalapyan S Y, Rybak L A, Malyshev D I 2020 Two-Stage Method for Controlling the Movement of a Parallel Robot Based on a Planar Three-Revolute-Prismatic-Revolute Mechanism *Advances in Intelligent Systems and Computing* **1126** 368-379.

[3] Ballesteros-Escamilla M, Chairez I, Boltyanski V G, Poznyak A 2018 Realization of robust optimal control by dynamic neural-programming *IFAC-PapersOnLine* **51(13)** 468-473

[4] Rybak L A, Mamaev Y A, Virabyan L G 2016 Correction algorithms synthesis for the motion path of the hexapod robot output link based on the theory of artificial neural networks *Bulletin of BSTU named after V G Shukhov* **12** 142-151

[5] Reddy T K, Arora V, Behera L. 2020 HJB-Equation-Based Optimal Learning Scheme for Neural Networks with Applications in Brain-Computer Interface *IEEE Transactions on Emerging Topics in Computational Intelligence* **4(2)** 159-170

[6] Eremenko Y I, Khalapyan S Y, Anpilov A O 2020 Improving energy efficiency of iron ore concentrate dehydration through automation using neural technologies *Gornyi Zhurnal* **3** 62-66

[7] Ballesteros M, Chairez I, Poznyak A 2020 Robust optimal feedback control design for uncertain systems based on artificial neural network approximation of the Bellman's value function *Neurocomputing* **413** 134-144

[8] Wilson S W 1985 Knowlege growth in an artificial animal *Proceedings of the First International Conference on Genetic Algorithms and their Applications* 16-23.

[9] Prescott T J, Mayhew J E W 1992 Obstacles avoidance through reinforcement learning *Advances in Neural Information Processing system 4* 523-530.

[10] Kuzmin V 2003 Using neural networks in the q-learning algorithm *Transport and Telecommunication* **4(1)** 74-86.

[11] Testing the Q – learning algorithm on popular game platforms [Electronic resource]. URL: https://habr.com/ru/post/345656/ (date of access 05/09/2020).

[12] Yogi S C, Tripathi V K, Kamath A K, Behera L 2019 Q-learning Based Navigation of a Quadrotor using Non-singular Terminal Sliding Mode Control *IEEE International Conference on Robot and Human Interactive Communication* 8956365

[13] Bharat R, Bosag Zadeh R 2019 TensorFlow for Deep Learning *BHV-Petersburg* 256

Acknowledgments

The research was funded by RFBR within the research project No 18-57-45014, using equipment of High Technology Center at BSTU named after V G Shukhov.

IOP Conf. Series: Materials Science and Engineering **945** (2020) 012084 doi:10.1088/1757-899X/945/1/012084

TiO_2 coatings on the surface of buildings structural elements

V V Sirota[1*], S V Zaitsev[1], M V Limarenko[1] and D S Prochorenkov[1]

[1]Belgorod State Technological University named after V. G. Shukhov, Russia, Belgorod, Kostyukova 46

Email: zmas36@mail.ru

Abstract. The article presents the results of a study of the photocatalytic activity of coatings based on titanium dioxide on various structural elements of buildings. Fine-grained lightweight concrete, porcelain stoneware and window glass were used as a base. TiO2 coating was applied by reactive magnetron sputtering. The structure and phase composition of the coatings were investigated by scanning electron microscopy and X-ray phase analysis. The photocatalytic activity of the samples was evaluated according to ISO 27448: 2009 and ISO 10678: 2010. The photocatalytic efficiency of the obtained TiO2 coatings on building materials is shown. The data obtained allow us to make a preliminary assessment of the effectiveness of the application of this method of applying photocatalytic coatings to structural elements of buildings, which will be clarified in the course of future studies.

1. Introduction

Applied research in the field of structural and functional materials with photocatalytic activity has been developed in recent years in various industries, including the construction industry. Photocatalytic concretes and other structural materials can provide not only self-cleaning of the surface, but also contribute to reduce the concentration of pollutants in the air [1]. These building materials have been used in the building industry for several years. Many researchers have noted the effectiveness of the use of the photocatalytic effect in building [2, 3]. TiO_x layers are often included in the compositions of light-correcting coatings on window glasses, hardening coatings on glass containers and are the basis for self-cleaning coatings on various technical devices [4]. The magnetron sputtering method has a number of advantages: the ability to obtain dense coatings, the possibility of obtaining a coating with the same thickness over a large area of the product, the ability to vary the structure and phase composition of the coating without significantly changing the rate of deposition and heating of the substrate. In the overwhelming majority of applications of TiO_x coatings, an enormous role is played by their structural-phase condition, which depends on the deposition method and process parameters [5-8]. In connection with the above, the purpose of this research was to obtain building elements with photocatalytic activity and to assess their effectiveness.

2. Materials and methods

Float glass, porcelain stoneware and fine-grained lightweight concrete were used as objects of research. Substrates for TiO_2 coating were prepared from these materials. The size of the substrates is

50 × 50 × 5 mm. Before applying the coating, the substrates were decreased with alcohol and dried with nitrogen. To remove residual contaminants, the surface of the substrates was clean with argon ions in the chamber of vacuum unit for 10 min at a pressure of $8 \cdot 10^{-2}$ Pa and a voltage across the ion source of 2.2 kV. Coatings were deposited on the substrate installing UNICOAT-200 setting with a dual magnetrons system and pulsed power. As an ion source used two titanium plates (Ti -99,99 wt.%) by 154 cm^2 each. The working gas - Ar (99,999 vol%), the reaction gas - O_2 (99.999 vol%). The working pressure in the chamber is 0.17 ± 0.01 Pa. The volume fraction of the reactive gas (O_2) - 14%. The magnetrons were powered by bipolar pulses with a duration of 50 μs (18 kHz) in the mode of keeping the current at a level of 4 A for each magnetron. The distance from the substrates to the magnetron and the ion source was the same 70 mm. During coating, samples (substrates) were not additionally heated.

The phase composition of the coatings was studied on an ARL9900 Intellipower Workstation diffractometer (ThermoTechno) with Co-Kα radiation ($\lambda = 0.1541744$ nm). The phases were identified using the JCPDF database. The morphology of surfaces and chips of the coatings was investigated using a scanning electron microscope (TESCAN MIRA 3 LMU). SEM images were acquired at 5 kV acceleration voltages.

The intrinsic photocatalytic activity of the coatings was evaluated in accordance with ISO 10678:2010 by the relative decrease in the concentration of the indicator. Indicator - methylene blue.It decomposes in aqueous solution in the presence of a photocatalytically active surface and UV irradiation in the range of 320-400 nm. The samples were kept in a solution of indicator with a concentration of 2 μmol/l in the dark for 12 h for adsorption saturation of the surface. After exposure in the dark produced solution photometry at a wavelength of 664 nm and taking into account the optical thickness of the cuvette of 20.11 mm in a spectrophotometer SF-56. Then, the samples in solution were irradiated with UV radiation. During irradiation, the solutions were stirred every 20 min. Evaluation of photocatalytic activity was carried out every hour for a relative decrease in the optical density of the solution for 10 h exposure.

3. Results and discussion

All types of substrates were coated with TiO_2 under the same conditions for an hour. This makes it possible to compare the composition, structure, and photocatalytic properties of coatings formed on the surface of the materials under study. The synthesized phases were identified by comparing the X-ray diffraction patterns before and after the formation of the titanium dioxide layer for the same samples by the difference curve (figure 1). The difference curve is the diffractograms of the applied coating. We failed to find crystalline phases in titanium oxide films in concrete samples (figure 1 a). The peaks on the difference curve are exactly the same as those of the original material. This may be due to the growth of titanium oxide crystallites in the direction of orientation of mineral crystals on the concrete surface. On porcelain stoneware, in addition to X-ray amorphous phases, anatase was found by the main reflections at 29.5°, 43.9° and 63.0° (figure 1 b). Rutile was identified in the coating on float glass (figure 1 c), the formation of which may be associated with the features of the glass structure and its surface.

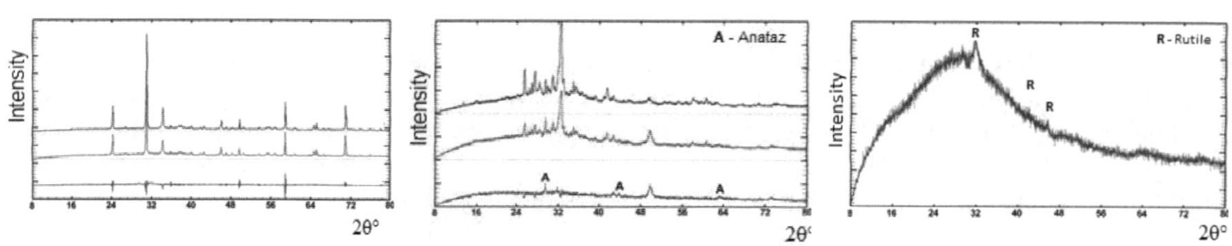

(a) (b) (c)

Figure 1. X-ray diffraction patterns of TiO_2 coatings on concrete (a), porcelain stoneware (b) and float glass (c).

IOP Conf. Series: Materials Science and Engineering **945** (2020) 012084 doi:10.1088/1757-899X/945/1/012084

The structure and texture of the films were evaluative from photographs obtained with a high-resolution scanning electron microscope TESCAN MIRA 3 LMU. The thickness of the synthesized coatings was estimated by SEM-images of transverse chips (figure 2).

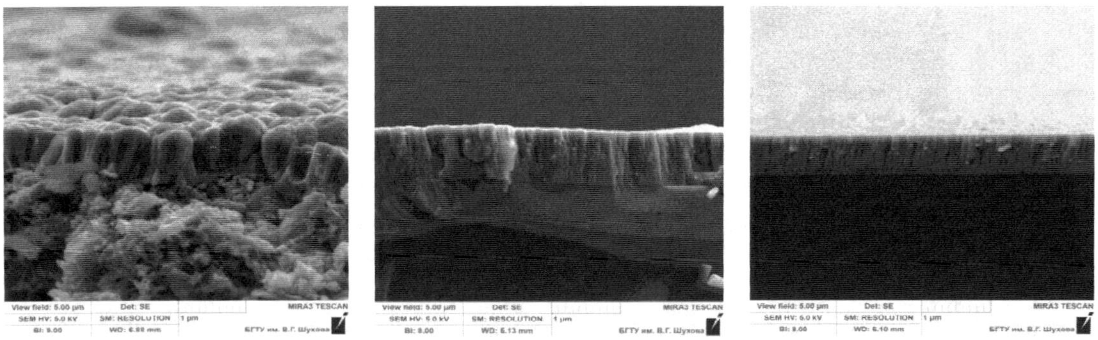

(a) (b) (c)

Figure 2. SEM micrographs of fractured surface of TiO_2 coating on concrete (a), porcelain stoneware (b) and on float glass (c).

Table 1 shows the average thicknesses of the synthesized titanium oxide coatings on the surface of the substrates.

Table 1. Average values of the thickness of titanium oxide coatings obtained by magnetron sputtering on the surface of the substrates.

	Fine-grained lightweight concrete	Porcelain stoneware	Float glass
Average thickness of the coating, nm	1000	900	700

On the surface of a concrete substrate, the structure of the coating is porous, which is explained by the peculiarities and composition of the substrate. Different forms of cement stone crystals and their mutual arrangement lead to the growth of titanium oxide crystallites in different directions, which leads to the formation of nano- and microporosity in the synthesized film. When using denser materials such as porcelain stoneware and glass as a substrate, the formed structure of the coating is completely different (Figures 2 b and 2 c). The flat and smooth surface of these materials contributes to the formation of a dense titanium oxide structure. The crystallite size is much larger here, especially in the case of porcelain stoneware (Figure 2b), which give single reflections on the diffractograms (Figure 1 b). From the SEM images of transverse chips (figures 2 b and 2 c), it can be seen that the coating consists of two layers - the amorphous lower (transitional) and the crystalline upper (main). The thickness of the transition layer on porcelain stoneware is approximately half of the entire coating, i.e. about 400-500 nm. A different picture on glass - here the transition layer makes up about a quarter of the entire coating, i.e. about 150 nm.

Table 2. Photocatalytic activity of TiO_2 coatings obtained by magnetron sputtering on the surface of the studied substrates by the degradation of methylene blue dye.

	Fine-grained lightweight concrete	Porcelain stoneware	Float glass
Photocatalytic activity of titanium oxide coatings EDMB, μmol /(м^2·h)	10.6	5.19	1.12

IOP Conf. Series: Materials Science and Engineering **945** (2020) 012084 doi:10.1088/1757-899X/945/1/012084

The photocatalytic activity values for the obtained materials, calculated in accordance with ISO 10678:2010 and shown in Table 2. The value of the photocatalytic activity for TiO_2 coatings on fine-grained concrete is overestimated. The reason lies in their porosity and high adsorption capacity. However, the value obtained on porcelain stoneware 5.19 μmol/(m^2·h) is consistent with the photocatalytic activity for Pilkington Aktiv (TM) glasses of 7 μmol/(m^2·h) [9].

In accordance with the Langmuir-Hinshelwood kinetic model, the rate constants of the photocatalytic reaction of methylene blue degradation on TiO_2 coatings are calculated. Kinetic constants for all samples are presented in Table 3.

Table 3. Rate constants of the photocatalytic reaction for TiO_2coatings obtained by magnetron sputtering on the surface of the studied substrates by the degradation of methylene blue dye.

	Fine-grained lightweight concrete	Porcelain stoneware	Float glass
k,·h^{-1}	0.1825	0.0196	0.0044

The obtained values of the rate constants of the photocatalytic reaction agree with the previously presented results. Photocatalytically active materials (porcelain stoneware, fine-grained concrete with titanium oxide coatings) have values comparable to the results obtained by other researchers [10].

It should be noted that the modified porcelain stoneware is characterized by a decrease in the contact angle of wetting by water drops after ultraviolet irradiation of the surface for 24 hours (figure 3). This is the effect of photoinduced hydrophilicity, which was noted by other researchers on the surface of photocatalytically active materials [11].

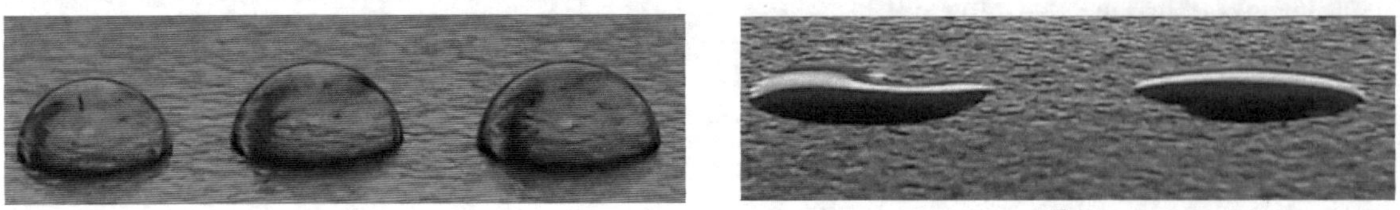

(a) (b)

Figure 3. Change in the contact angle of water wetting of the surface of a TiO_2 coating on porcelain stoneware as a result of exposure to ultraviolet radiation: without UV treatment (a), after UV treatment (b).

Such a large decrease in the contact angle from 74° to 20° is determined by the formation on the surface of the porcelain stoneware of anatase coating TiO_2. A similar effect, but to a much lesser extent, is observed on TiO_2 coatings on window glass (figure 4). Here, the decrease in the contact angle is 23° - from 74° to 51°.

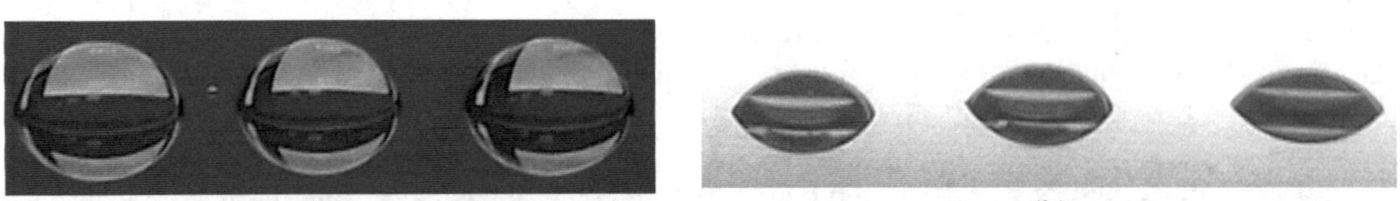

(a) (b)

Figure 4. Changes in the contact angle of water wetting of the surface of a TiO_2 coating on glass as a result of UV irradiation without UV treatment (a), after UV treatment (b).

4. Summary

The creation of photocatalytically active structural elements of buildings by the method of reactive magnetron sputtering is quite reasonable and technically feasible. The photocatalytic activity of the

TiO_2 coating samples obtained in this work turned out to be comparable with the results of other researchers and industrial samples. One may conclude that the obtained experimental samples is a promising material for use in construction.

5. References

[1] Janus M, Zajac K 2016. *Concretes with Photocatalytic Activity* (London: IntechOpen) p 141
[2] Witkowski H, Jackiewicz-Rek W, Chilmon K, Jarosławski J, Tryfon-Bojarska A. and Gasinski A. 2019 *Appl. Sci.* **9** 1735
[3] Yang L, Hakki A, Wang F, Macphee D E 2018 *Appl. Catal. B.* **222** 200.
[4] Diebold U. 2003 *Surf. Sci. Rep.* **43** 53.
[5] Nartsev V M, Atkarskaya A B, Zaĭtsev S V, Osipenko N V, Prokhorenkov D S, and Evtushenko E I 2016 *J. Opt. Technol.* **83(4)** 263.
[6] Mráz S, Schneider J M 2011 *J. Appl. Phys.* **109** id 023512 6 p.
[7] Mukherjee S K, Nebatti A, Mohtascham F 2014 *Thin Solid Films* **558** 443.
[8] Pansila P, Wititnun N, Chaiyakun S 2012 *Procedia Eng.* **32** 862.
[9] Navabpour P, Ostovarpour S, Tattershall C 014 *Coatings* **2**(4) 433.
[10] Bannier E, Darut G, Sánchez E, Denoirjean A, Bordes M C, Salvador M D, Rayón E, Ageorges H 2011 *Surf. Coat. Tech.* **206**(2-3) 378.
[11] Shibuya M, Miyauchi M 2009 *Adv. Mater.* **21** 1373.

Acknowledgments

The work was carried out within the framework of the State Assignment of the Ministry of Education and Science of the Russian Federation, project No. 0625-2020-0011. The study was carried out equipment of the Center of Advanced Technologies, Belgorod State Technological University named after V.G. Shoukhov.

IOP Conf. Series: Materials Science and Engineering **945** (2020) 012085 doi:10.1088/1757-899X/945/1/012085

Digital design of a cable-driven parallel robot

S I Anciferov[1*], E V Gaponenko[1], A V Karachevtseva[1] and L R Kholoshevskaya[1]

[1]Belgorod State Technological University named after V. G. Shukhov, 46, Kostyukova Str., Belgorod, 308012, Russia

E-mail: gaponenkobel@gmail.com

Abstract. The paper analyzes various computer-aided design systems. It is shown that the use of CAD / CAM / CAE-system NX allows to reduce the cost of the final product, increase the level of quality, and reduce the time to market. The article discusses the digital design of a cable-driven parallel robot (CDPR) for various applications using the NX CAD / CAM / CAE system. The mechanism is a rigidly fixed frame connected through tensioned cables to a platform. CDPRs have such advantages as a large working area, ease of assembly and disassembly, high mobility, large lifting capacity, and reconfiguration. The kinematic diagram of the device is presented. Based on the terms of reference and sketch documentation, a preliminary composition of the CDPR has been developed. During the design, the Top-Down method was used. The characteristics are analyzed, and the workspace of the mechanism is formed, taking into account the factors of limitation. Shown is a detailed design of the basic structure of a CDPR with an indication of the main operations. An electronic-digital model of the basic structure of a CDPR, which will later be used for kinematic and static calculations, as well as for making an experimental sample, is presented.

1. Introduction

Computer-aided design systems that solve a wide range of engineering problems have become widespread for completely new designs developing of equipment [1-3]. There are many digital design software on the market that have their features. The leading suppliers in the market of such software are Dassault Systems with their products with CAD / CAM / CAE-system CATIA and CAD / CAE-system SolidWorks, Siemens PLM Software with CAD / CAM / CAE-system NX and CAD / CAE-system Solid Edge, and Top Systems with CAD / CAM / CAE-system T-Flex.

But despite such a wide range of design software, one of the best can be considered the CAD / CAM / CAE-system NX, which provides an integrated solution for digital design, technological preparation of both additive and traditional manufacturing, and engineering analysis of mechanical systems, possessing powerful functionality for the development of competitive products.

NX CAD / CAM / CAE system allows to achieve your goal with maximum efficiency, reduce the cost of the final product, increase the level of quality and shorten its time to market. At the same time, the NX CAD / CAM / CAE system provides the user with a fully integrated manufacturing solution covering all stages, such as: part design, topological optimization of model geometry, model adaptation, analysis and verification of geometry for strength, production preparation, data and process management.

IOP Conf. Series: Materials Science and Engineering **945** (2020) 012085 doi:10.1088/1757-899X/945/1/012085

Consider the application of the NX CAD / CAM / CAE system for the digital design of a cable-driven parallel robot (CDPR).

2. Materials and methods

CDPRs are a special type of kinematic structure, consisting mainly of a working tool connected to a fixed base platform using cables [4-9]. Conventional CDPRs are composed of three parts, including a fixed platform, a mobile platform, and multiple cables that are used to connect the fixed platform and the mobile platform. The cable length can be changed through winches powered by motors mounted on a fixed platform.

One of the disadvantages of classic parallel robots is their limited workspace, mainly due to the limited travel of linear actuators and their accuracy. In CDPRs, this problem is solved by replacing the actuators with cables. In addition, CDPRs have such advantages as a large working area, easy assembly and disassembly, high mobility, large lifting capacity and easy reconfiguration. By controlling the length of the cables over a wide range, you can get access to a very large workspace from several tens of centimeters to several tens of meters or more. The use of cables instead of rigid links further reduces weight since the actuators do not change position and are fixed to a fixed base so that the only moving parts are the cables and the output link. As a result, a robot is obtained with higher speed and maneuverability and increased load capacity. The production costs of CDPRs are significantly lower than those of conventional manipulators. CDPRs are easy to install. Such a manipulator can be assembled using a number of inexpensive winches and cables. Also, since the motors do not need to be installed close to a moving platform, they are suitable for use in hazardous environments. Also, their lifting capacity is relatively high, even comparable to construction cranes.

The kinematic structure of the robot is shown in Fig. 1. The CDPR is designed to move loads in special conditions and contains four electric motors 1, 2, 3, 4, each of which includes a device for laying cables and a gearbox. Electric motors are driven into rotation by control device 5. Gearboxes of electric motors 1, 2, 3, 4 are connected to cables 6, 7, 8, 9 to move the load in the horizontal plane, respectively, through rollers 10, 11, 12, 13. Rollers 10, 11 , 12, 13, are located on the frame 14. The output link is a movable platform 15 connected to the cables 6, 7, 8, 9.

The CDPR works as follows. When the electric motors 1, 2, 3, 4 rotate, with devices for laying cables and gearboxes, the angular position of the shafts of which is set by the control device 5, the length of the cables 6, 7, 8, 9, supported by rollers 10, 11, 12, 13, changes, located on the frame 14, as a result of which the gripper moves to the desired point in space. In this case, it is assumed that the cables 6, 7, 8, 9 are capable of moving the gripper due to the output link, which is attached to the platform 15.

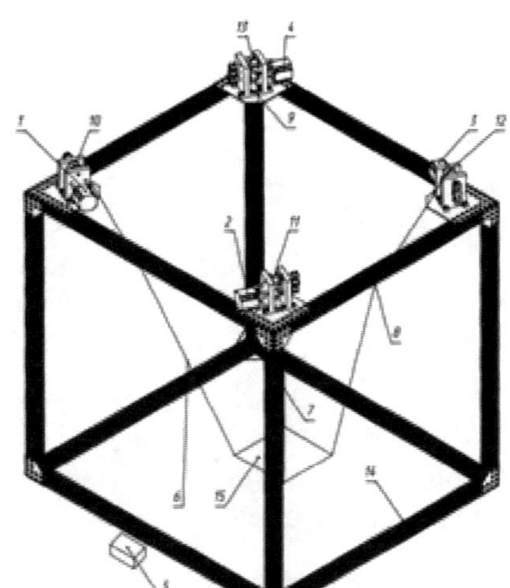

Figure 1. Kinematic diagram of a CDPR.

IOP Conf. Series: Materials Science and Engineering **945** (2020) 012085 doi:10.1088/1757-899X/945/1/012085

3. Results and discussion

The composition of the CDPR was developed on the basis of the technical task and sketch documentation for the design and construction work, as well as for the kinematic and static calculation.

The Top-Down method was used during the design [10]. The composition of the product was formed in the NX CAD / CAM / CAE system to facilitate the work. First, the geometry of the frame model is created. The frame consists of several elements: vertical post, lower profile, upper profile. Frame design starts with bottom profile.

A section of the profile is drawn using the "Sketch" command. We begin to draw one of the parts of this profile. The sketch is completely limited after mirroring the remaining parts (Fig. 2).

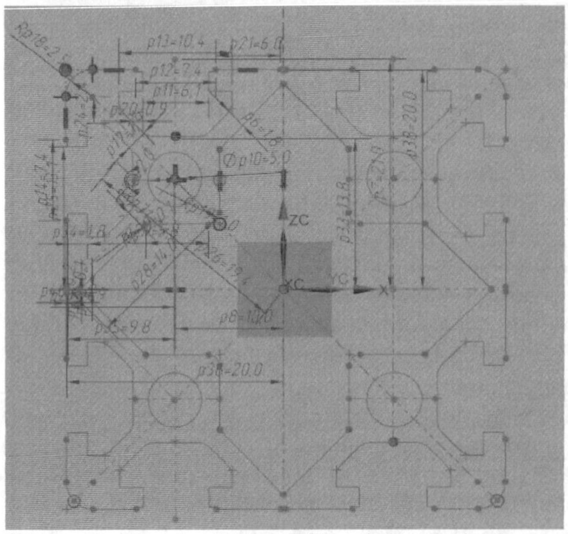

Figure 2. Sketch of the profile of the vertical post.

A solid is drawn using the Extrude command when the profile sketch is ready (Figure 3). The last action is to round off the edges using the "Rounding edges" command (Fig. 4).

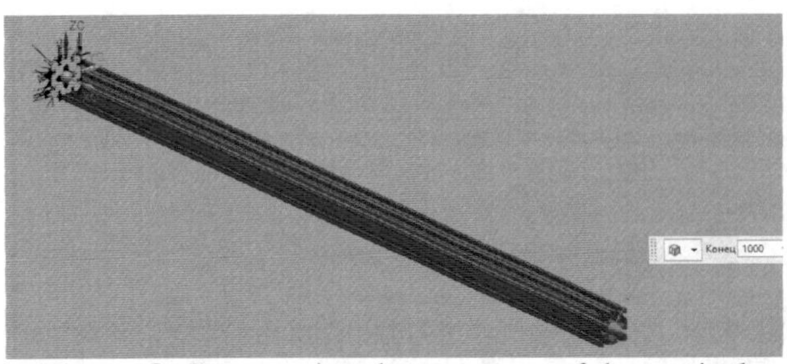

Figure 3. Constructing the geometry of the vertical rack.

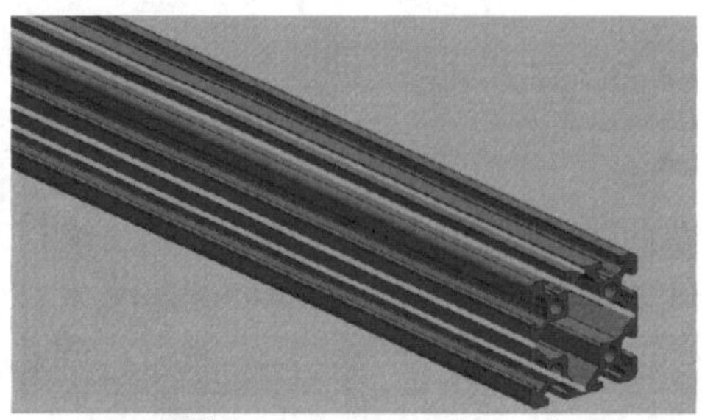

Figure 4. Rounding edges.

The remaining frame elements are completed after building the lower profile. A model of the frame assembly is assembled using the commands located on the "Assemblies" toolbar (Fig. 5). In product design, you mate major assembly components using the "Mates" command.

IOP Conf. Series: Materials Science and Engineering **945** (2020) 012085 doi:10.1088/1757-899X/945/1/012085

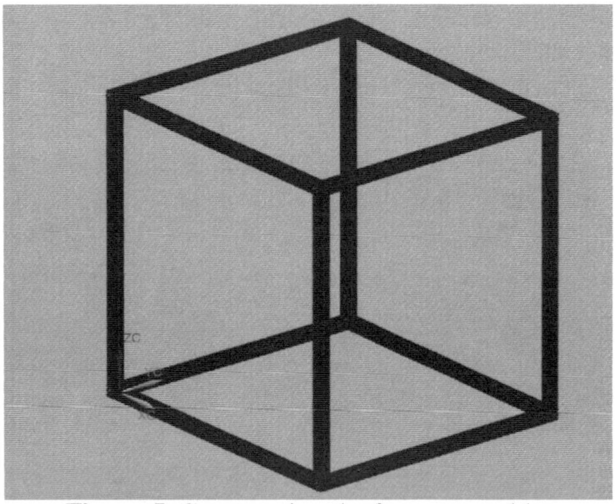

Figure 5. Constructing the frame assembly.

Then we build plates (Fig. 6) and attach them at the junction of the profiles that make up the frame. The next step is to build plates (Fig. 7) on which the motor, winch, pulleys, etc. will then be located.

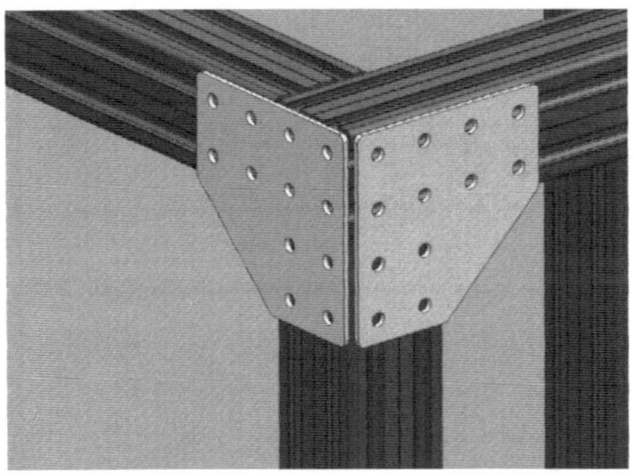

Figure 6. Creation of plates for profile joints.

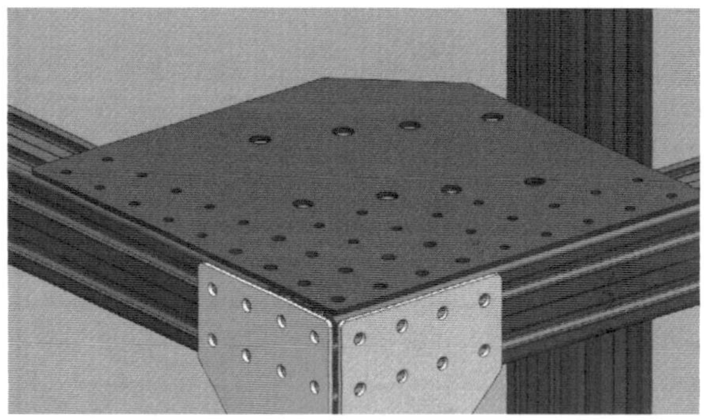

Figure 7. Construction of plates for mounting motors on them.

After all the components are built, fasteners are added to the frame assembly. We also model the corners (Fig. 8), in which bushings, couplings, pulleys, and coils will then be attached. After that, a motor of a standard size is built according to GOST (Fig. 9).

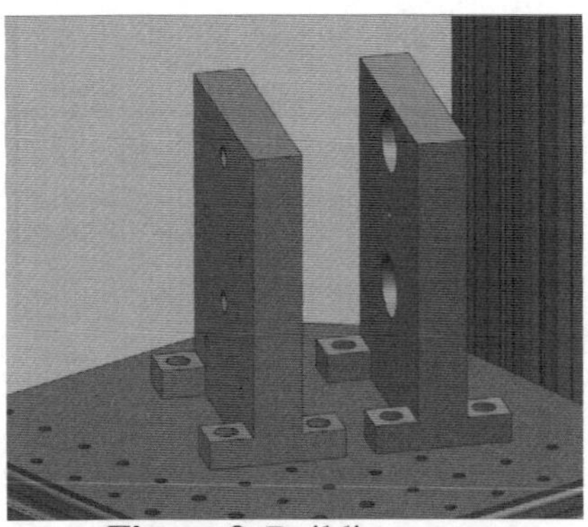

Figure 8. Building corners.

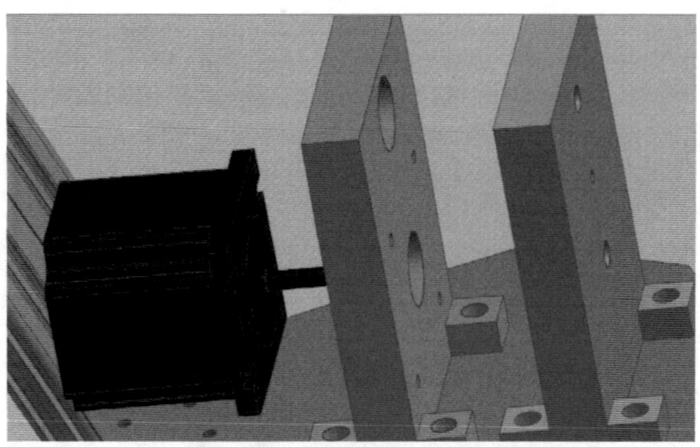

Figure 9. Building a motor.

IOP Conf. Series: Materials Science and Engineering **945** (2020) 012085 doi:10.1088/1757-899X/945/1/012085

The next step is to build bushings, couplings, bearings (the zz marking in the bearing means that it is closed on both sides by metal membranes).

Then we build shafts, pins and coils for the cable (1 layer of cable with a diameter of 1.2 mm is placed on the smallest coil, 1.6 m of the same cable should be placed on a coil of larger diameter) (Fig. 10-13).

Figure 10. Fastening shafts in bearings.

Figure 11. Fastening the studs.

Figure 12. Fastening cable spools.

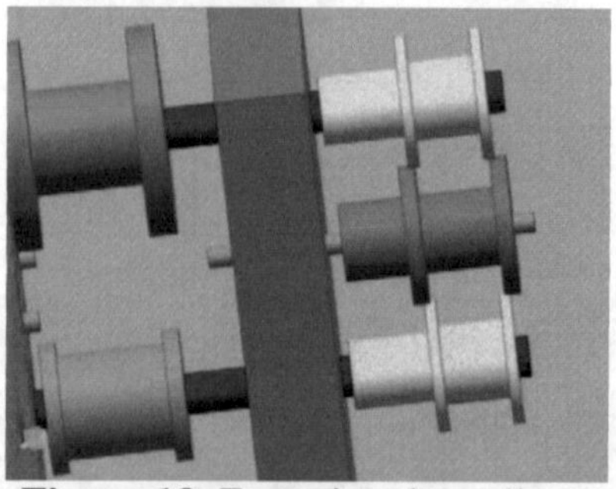

Figure 13. Fastening the pulleys.

After all the components are built, fasteners (bolts, nuts) are added to the frame assembly (Fig. 14).

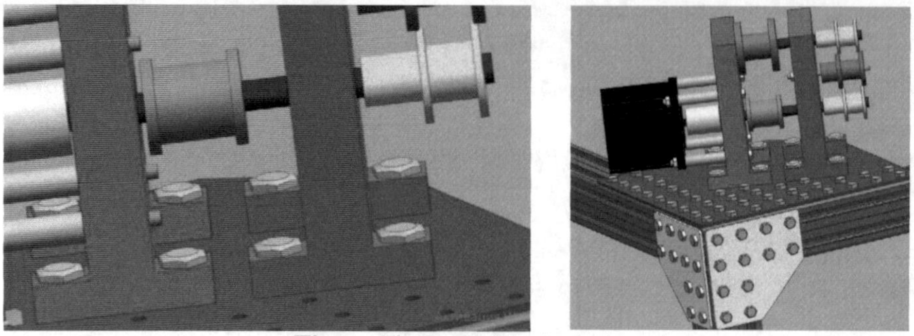

Figure14. Fastening the bolts.

Fig. 15 shows an electronic digital model of the basic structure of a CDPR for various applications. This model can be applied to various options for cables and grippers, depending on the task and environment of the robot.

Figure 15. An electronic digital model of the basic structure of a CDPR.

Commercial and standard products dominate the assembly of the CDPR, but some of the parts, such as bushings, reels, pulleys, and corners, will be made by one of the types of additive technologies used in various industries. The use of additive technologies in the manufacture of a CDPR will allow solving the problem of rapid prototyping of a complex mechanical system, which in turn can reduce the time spent on manufacturing and repairing a product and reduce the time for evaluating the physical model of the robot. Fast and cheap production of parts using additive technologies can significantly increase the competitiveness of new products on the market.

4. Summary

Thus, the CAD / CAM / CAE-system NX application allows you to design new products in the shortest possible time in a single environment. The resulting digital models of products are subsequently improved to create an optionally selectable range of products. This approach allows us to consider the wishes of all consumers of the market of robotic systems who need geometric shapes, colors, sizes of products, and materials from which they are made.

In the future, it is planned to carry out kinematic and static calculations based on the draft design and the constructed digital model, as well as to manufacture an experimental sample of a CDPR.

5. References

[1] Kyratsis P, Kakoulis K, Markopoulos A 2020 Advances in CAD/CAM/CAE Technologies *Machines* **8(1)** 13

[2] Milchev G, Miltchev R 2018 Development of Information Technologies, Planned Obsolescence and Modification of the Life-Cycle of the CAD/CAM/CAE Systems *European Journal of Sustainable Development* **7(3)** 217-224

[3] Goncharov P S, Eltsov M Yu, Korshikov S B, Laptev I V, Osiyuk V A 2010 NX for the designer-mechanical engineer *MQM Press* 503

[4] Pott A, Mütherich H, Kraus W, Schmidt V, Miermeister P, Verl A 2013 IPAnema: A family of Cable-Driven Parallel Robots for Industrial Applications *Springer* **12** 119-134

[5] Merlet J P 2010 A portable, modular parallel wire crane for rescue operations *Robotics and Automation* 2834-2839

[6] Gaponenko E V Structural analysis and classification of robotic systems with drive mechanisms based on cable elements *Bulletin of BSTU name after V.G.* **9** 126–136

[7] Pott A, Mütherich H, Kraus W, Schmidt V, Miermeister P, Verl A 2013 IPAnema: A family of Cable-Driven Parallel Robots for Industrial Applications *Springer* 119-134

[8] Zhang N, Shang W 2016 Dynamic trajectory planning of a 3-DOF under-constrained cable-driven parallel robot *Mechanism and Machine Theory* **98** 21-35

[9] Du J, Bao H, Cui C, Yang D 2012 Dynamic analysis of cable-driven parallel manipulators with time-varying cable lengths *Finite Elements in Analysis and Design* **48** 1392-1399
[10] Eltsov M Yu, Kozlov A A, Sedoykin A V, Shirokova L Yu 2010 Design in NX under Teamcenter *DMK Press* 783

Acknowledgments

The reported study was funded by RFBR according to the research project № 18-31-20060, using the equipment of the Center for High Technologies of BSTU named after V G Shukhov.

AUTHOR INDEX

Institute of Physics
Dirac House, Temple Back
Bristol BS1 6BE UK

ISSN: 1757-8981
ISBN 978-1-7138-4778-6

CURRAN ASSOCIATES INC.
proceedings.com